STUDENT'S SOLUTIONS MANUAL
SINGLE VARIABLE

MARK WOODARD
Furman University

CALCULUS
SECOND EDITION

William Briggs
University of Colorado at Denver

Lyle Cochran
Whitworth University

Bernard Gillett
University of Colorado at Boulder

with the assistance of
Eric Schulz
Walla Walla Community College

PEARSON

Boston Columbus Indianapolis New York San Francisco Upper Saddle River
Amsterdam Cape Town Dubai London Madrid Milan Munich Paris Montreal Toronto
Delhi Mexico City São Paulo Sydney Hong Kong Seoul Singapore Taipei Tokyo

Reproduced by Pearson from electronic files supplied by the author.

Copyright © 2015, 2011 Pearson Education, Inc.
Publishing as Pearson, 75 Arlington Street, Boston, MA 02116.

ISBN-13: 978-0-321-95495-4
ISBN-10: 0-321-95495-5

1 2 3 4 5 6 EBM 17 16 15 14

www.pearsonhighered.com

Contents

Chapter 1

Functions

1.1 Review of Functions

1.1.1 A function is a rule which assigns each domain element to a unique range element. The independent variable is associated with the domain, while the dependent variable is associated with the range.

1.1.3 The vertical line test is used to determine whether a given graph represents a function. (Specifically, it tests whether the variable associated with the vertical axis is a function of the variable associated with the horizontal axis.) If every vertical line which intersects the graph does so in exactly one point, then the given graph represents a function. If any vertical line $x = a$ intersects the curve in more than one point, then there is more than one range value for the domain value $x = a$, so the given curve does not represent a function.

1.1.5 Item i. is true while item ii. isn't necessarily true. In the definition of function, item i. is stipulated. However, item ii. need not be true – for example, the function $f(x) = x^2$ has two different domain values associated with the one range value 4, because $f(2) = f(-2) = 4$.

1.1.7 $f(g(2)) = f(-2) = f(2) = 2$. The fact that $f(-2) = f(2)$ follows from the fact that f is an even function.

$$g(f(-2)) = g(f(2)) = g(2) = -2.$$

1.1.9 When f is an even function, we have $f(-x) = f(x)$ for all x in the domain of f, which ensures that the graph of the function is symmetric about the y-axis.

1.1.11 Graph A does not represent a function, while graph B does. Note that graph A fails the vertical line test, while graph B passes it.

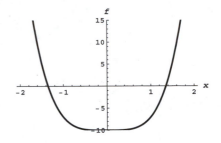

1.1.13   The domain of this function is the set of a real numbers. The range is $[-10, \infty)$.

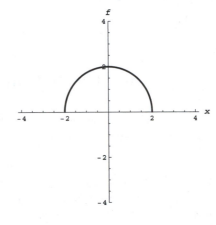

1.1.15   The domain of this function is $[-2, 2]$. The range is $[0, 2]$.

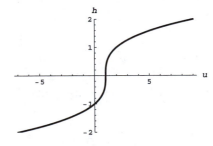

1.1.17   The domain and the range for this function are both the set of all real numbers.

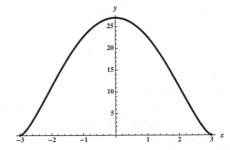

1.1.19   The domain of this function is $[-3, 3]$. The range is $[0, 27]$.

1.1.21 The independent variable t is elapsed time and the dependent variable d is distance above the ground. The domain in context is $[0, 8]$

1.1.23 The independent variable h is the height of the water in the tank and the dependent variable V is the volume of water in the tank. The domain in context is $[0, 50]$

1.1.25 $f(10) = 96$

1.1.27 $g(1/z) = (1/z)^3 = \frac{1}{z^3}$

1.1.29 $F(g(y)) = F(y^3) = \frac{1}{y^3-3}$

1.1.31 $g(f(u)) = g(u^2 - 4) = (u^2 - 4)^3$

1.1.33 $F(F(x)) = F\left(\frac{1}{x-3}\right) = \frac{1}{\frac{1}{x-3}-3} = \frac{1}{\frac{1}{x-3}-\frac{3(x-3)}{x-3}} = \frac{1}{\frac{10-3x}{x-3}} = \frac{x-3}{10-3x}$

1.1.35 $f(\sqrt{x+4}) = (\sqrt{x+4})^2 - 4 = x + 4 - 4 = x.$

1.1.37 $g(x) = x^3 - 5$ and $f(x) = x^{10}$. The domain of h is the set of all real numbers.

1.1.39 $g(x) = x^4 + 2$ and $f(x) = \sqrt{x}$. The domain of h is the set of all real numbers.

1.1.41 $(f \circ g)(x) = f(g(x)) = f(x^2 - 4) = |x^2 - 4|$. The domain of this function is the set of all real numbers.

1.1.43 $(f \circ G)(x) = f(G(x)) = f\left(\frac{1}{x-2}\right) = \left|\frac{1}{x-2}\right|$. The domain of this function is the set of all real numbers except for the number 2.

1.1.45 $(G \circ g \circ f)(x) = G(g(f(x))) = G(g(|x|)) = G(x^2 - 4) = \frac{1}{x^2-4-2} = \frac{1}{x^2-6}$. The domain of this function is the set of all real numbers except for the numbers $\pm\sqrt{6}$.

1.1.47 $(g \circ g)(x) = g(g(x)) = g(x^2 - 4) = (x^2 - 4)^2 - 4 = x^4 - 8x^2 + 16 - 4 = x^4 - 8x^2 + 12$. The domain is the set of all real numbers.

1.1.49 Because $(x^2 + 3) - 3 = x^2$, we may choose $f(x) = x - 3$.

1.1.51 Because $(x^2 + 3)^2 = x^4 + 6x^2 + 9$, we may choose $f(x) = x^2$.

1.1.53 Because $(x^2)^2 + 3 = x^4 + 3$, this expression results from squaring x^2 and adding 3 to it. Thus we may choose $f(x) = x^2$.

1.1.55

a. $(f \circ g)(2) = f(g(2)) = f(2) = 4.$

b. $g(f(2)) = g(4) = 1.$

c. $f(g(4)) = f(1) = 3.$

d. $g(f(5)) = g(6) = 3.$

e. $f(f(8)) = f(8) = 8.$

f. $g(f(g(5))) = g(f(2)) = g(4) = 1.$

1.1.57 $\frac{f(x+h)-f(x)}{h} = \frac{(x+h)^2-x^2}{h} = \frac{(x^2+2hx+h^2)-x^2}{h} = \frac{h(2x+h)}{h} = 2x + h.$

1.1.59 $\frac{f(x+h)-f(x)}{h} = \frac{\frac{2}{x+h}-\frac{2}{x}}{h} = \frac{\frac{2x-2(x+h)}{x(x+h)}}{h} = \frac{2x-2x-2h}{hx(x+h)} = -\frac{2h}{hx(x+h)} = -\frac{2}{x(x+h)}.$

1.1.61 $\frac{f(x+h)-f(x)}{h} = \frac{\frac{x+h}{x+h+1}-\frac{x}{x+1}}{h} = \frac{\frac{(x+h)(x+1)-x(x+h+1)}{(x+1)(x+h+1)}}{h} = \frac{x^2+x+hx+h-x^2-xh-x}{h(x+1)(x+h+1)} = \frac{h}{h(x+1)(x+h+1)} = \frac{1}{(x+1)(x+h+1)}$

1.1.63 $\frac{f(x)-f(a)}{x-a} = \frac{x^3-2x-(a^3-2a)}{x-a} = \frac{(x^3-a^3)-2(x-a)}{x-a} = \frac{(x-a)(x^2+ax+a^2)-2(x-a)}{x-a} = \frac{(x-a)(x^2+ax+a^2-2)}{x-a} = x^2 + ax + a^2 - 2.$

1.1.65 $\frac{f(x)-f(a)}{x-a} = \frac{\frac{-4}{x^2}-\frac{-4}{a^2}}{x-a} = \frac{\frac{-4a^2+4x^2}{a^2x^2}}{x-a} = \frac{4(x^2-a^2)}{(x-a)a^2x^2} = \frac{4(x-a)(x+a)}{(x-a)a^2x^2} = \frac{4(x+a)}{a^2x^2}.$

1.1.67

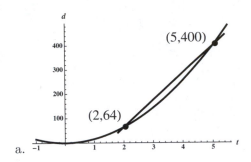

a.

b. The slope of the secant line is given by $\frac{400-64}{5-2} = \frac{336}{3} = 112$ feet per second. The object falls at an average rate of 112 feet per second over the interval $2 \le t \le 5$.

1.1.69

a.

b. The slope of the secant line is given by $\frac{1-4}{2-(1/2)} = -\frac{3}{3/2} = -2$ cubic cm per atmosphere. The volume decreases at an average rate of 2 cubic cm per atmosphere over the interval $0.5 \le p \le 2$.

1.1.71 This function is symmetric about the y-axis, because $f(-x) = (-x)^4 + 5(-x)^2 - 12 = x^4 + 5x^2 - 12 = f(x)$.

1.1.73 This function has none of the indicated symmetries. For example, note that $f(-2) = -26$, while $f(2) = 22$, so f is not symmetric about either the origin or about the y-axis, and is not symmetric about the x-axis because it is a function.

1.1.75 This curve (which is not a function) is symmetric about the x-axis, the y-axis, and the origin. Note that replacing either x by $-x$ or y by $-y$ (or both) yields the same equation. This is due to the fact that $(-x)^{2/3} = ((-x)^2)^{1/3} = (x^2)^{1/3} = x^{2/3}$, and a similar fact holds for the term involving y.

1.1.77 This function is symmetric about the origin. Note that $f(-x) = (-x)|(-x)| = -x|x| = -f(x)$.

1.1.79 Function A is symmetric about the y-axis, so is even. Function B is symmetric about the origin, so is odd. Function C is also symmetric about the y-axis, so is even.

1.1.81

a. True. A real number z corresponds to the domain element $z/2 + 19$, because $f(z/2 + 19) = 2(z/2 + 19) - 38 = z + 38 - 38 = z$.

b. False. The definition of function does not require that each range element comes from a unique domain element, rather that each domain element is paired with a unique range element.

c. True. $f(1/x) = \frac{1}{1/x} = x$, and $\frac{1}{f(x)} = \frac{1}{1/x} = x$.

d. False. For example, suppose that f is the straight line through the origin with slope 1, so that $f(x) = x$. Then $f(f(x)) = f(x) = x$, while $(f(x))^2 = x^2$.

e. False. For example, let $f(x) = x + 2$ and $g(x) = 2x - 1$. Then $f(g(x)) = f(2x - 1) = 2x - 1 + 2 = 2x + 1$, while $g(f(x)) = g(x + 2) = 2(x + 2) - 1 = 2x + 3$.

f. True. This is the definition of $f \circ g$.

g. True. If f is even, then $f(-z) = f(z)$ for all z, so this is true in particular for $z = ax$. So if $g(x) = cf(ax)$, then $g(-x) = cf(-ax) = cf(ax) = g(x)$, so g is even.

h. False. For example, $f(x) = x$ is an odd function, but $h(x) = x + 1$ isn't, because $h(2) = 3$, while $h(-2) = -1$ which isn't $-h(2)$.

i. True. If $f(-x) = -f(x) = f(x)$, then in particular $-f(x) = f(x)$, so $0 = 2f(x)$, so $f(x) = 0$ for all x.

1.1.83 We will make heavy use of the fact that $|x|$ is x if $x > 0$, and is $-x$ if $x < 0$. In the first quadrant where x and y are both positive, this equation becomes $x - y = 1$ which is a straight line with slope 1 and y-intercept -1. In the second quadrant where x is negative and y is positive, this equation becomes $-x - y = 1$, which is a straight line with slope -1 and y-intercept -1. In the third quadrant where both x and y are negative, we obtain the equation $-x - (-y) = 1$, or $y = x + 1$, and in the fourth quadrant, we obtain $x + y = 1$. Graphing these lines and restricting them to the appropriate quadrants yields the following curve:

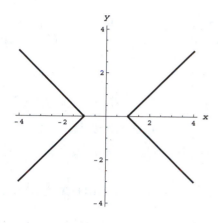

1.1.85 Because the composition of f with itself has first degree, f has first degree as well, so let $f(x) = ax + b$. Then $(f \circ f)(x) = f(ax + b) = a(ax + b) + b = a^2x + (ab + b)$. Equating coefficients, we see that $a^2 = 9$ and $ab + b = -8$. If $a = 3$, we get that $b = -2$, while if $a = -3$ we have $b = 4$. So the two possible answers are $f(x) = 3x - 2$ and $f(x) = -3x + 4$.

1.1.87 Let $f(x) = ax^2 + bx + c$. Then $(f \circ f)(x) = f(ax^2 + bx + c) = a(ax^2 + bx + c)^2 + b(ax^2 + bx + c) + c$. Expanding this expression yields $a^3x^4 + 2a^2bx^3 + 2a^2cx^2 + ab^2x^2 + 2abcx + ac^2 + abx^2 + b^2x + bc + c$, which simplifies to $a^3x^4 + 2a^2bx^3 + (2a^2c + ab^2 + ab)x^2 + (2abc + b^2)x + (ac^2 + bc + c)$. Equating coefficients yields $a^3 = 1$, so $a = 1$. Then $2a^2b = 0$, so $b = 0$. It then follows that $c = -6$, so the original function was $f(x) = x^2 - 6$.

1.1.89 $\frac{f(x+h)-f(x)}{h} = \frac{\sqrt{x+h}-\sqrt{x}}{h} = \frac{\sqrt{x+h}-\sqrt{x}}{h} \cdot \frac{\sqrt{x+h}+\sqrt{x}}{\sqrt{x+h}+\sqrt{x}} = \frac{(x+h)-x}{h(\sqrt{x+h}+\sqrt{x})} = \frac{1}{\sqrt{x+h}+\sqrt{x}}$.

$\frac{f(x)-f(a)}{x-a} = \frac{\sqrt{x}-\sqrt{a}}{x-a} = \frac{\sqrt{x}-\sqrt{a}}{x-a} \cdot \frac{\sqrt{x}+\sqrt{a}}{\sqrt{x}+\sqrt{a}} = \frac{x-a}{(x-a)(\sqrt{x}+\sqrt{a})} = \frac{1}{\sqrt{x}+\sqrt{a}}$.

1.1.91 $\frac{f(x+h)-f(x)}{h} = \frac{\frac{-3}{\sqrt{x+h}} - \frac{-3}{\sqrt{x}}}{h} = \frac{-3(\sqrt{x}-\sqrt{x+h})}{h\sqrt{x}\sqrt{x+h}} = \frac{-3(\sqrt{x}-\sqrt{x+h})}{h\sqrt{x}\sqrt{x+h}} \cdot \frac{\sqrt{x}+\sqrt{x+h}}{\sqrt{x}+\sqrt{x+h}} = \frac{-3(x-(x+h))}{h\sqrt{x}\sqrt{x+h}(\sqrt{x}+\sqrt{x+h})} = \frac{3}{\sqrt{x}\sqrt{x+h}(\sqrt{x}+\sqrt{x+h})}$.

$\frac{f(x)-f(a)}{x-a} = \frac{\frac{-3}{\sqrt{x}} - \frac{-3}{\sqrt{a}}}{x-a} = \frac{-3\left(\frac{\sqrt{a}-\sqrt{x}}{\sqrt{a}\sqrt{x}}\right)}{x-a} = \frac{(-3)(\sqrt{a}-\sqrt{x})}{(x-a)\sqrt{a}\sqrt{x}} \cdot \frac{\sqrt{a}+\sqrt{x}}{\sqrt{a}+\sqrt{x}} = \frac{(3)(x-a)}{(x-a)(\sqrt{a}\sqrt{x})(\sqrt{a}+\sqrt{x})} = \frac{3}{\sqrt{ax}(\sqrt{a}+\sqrt{x})}$.

1.1.93

a. The formula for the height of the rocket is valid from $t = 0$ until the rocket hits the ground, which is the positive solution to $-16t^2 + 96t + 80 = 0$, which the quadratic formula reveals is $t = 3 + \sqrt{14}$. Thus, the domain is $[0, 3 + \sqrt{14}]$.

b.

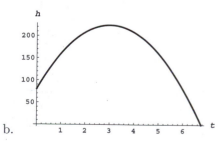

The maximum appears to occur at $t = 3$. The height at that time would be 224.

1.1.95 This would not necessarily have either kind of symmetry. For example, $f(x) = x^2$ is an even function and $g(x) = x^3$ is odd, but the sum of these two is neither even nor odd.

1.1.97 This would be an odd function, so it would be symmetric about the origin. Suppose f is even and g is odd. Then $\frac{f}{g}(-x) = \frac{f(-x)}{g(-x)} = \frac{f(x)}{-g(x)} = -\frac{f}{g}(x)$.

1.1.99 This would be an even function, so it would be symmetric about the y-axis. Suppose f is even and g is even. Then $f(g(-x)) = f(g(x))$, because $g(-x) = g(x)$.

1.1.101 This would be an even function, so it would be symmetric about the y-axis. Suppose f is even and g is odd. Then $g(f(-x)) = g(f(x))$, because $f(-x) = f(x)$.

1.1.103

  a. $f(g(-2)) = f(-g(2)) = f(-2) = 4$

  b. $g(f(-2)) = g(f(2)) = g(4) = 1$

  c. $f(g(-4)) = f(-g(4)) = f(-1) = 3$

  d. $g(f(5) - 8) = g(-2) = -g(2) = -2$

  e. $g(g(-7)) = g(-g(7)) = g(-4) = -1$

  f. $f(1 - f(8)) = f(-7) = 7$

1.2   Representing Functions

1.2.1 Functions can be defined and represented by a formula, through a graph, via a table, and by using words.

1.2.3 The domain of a rational function $\frac{p(x)}{q(x)}$ is the set of all real numbers for which $q(x) \neq 0$.

1.2.5

1.2.7 Compared to the graph of $f(x)$, the graph of $f(x+2)$ will be shifted 2 units to the left.

1.2.9 Compared to the graph of $f(x)$, the graph of $f(3x)$ will be scaled horizontally by a factor of 3.

1.2.11 The slope of the line shown is $m = \frac{-3-(-1)}{3-0} = -2/3$. The y-intercept is $b = -1$. Thus the function is given by $f(x) = (-2/3)x - 1$.

1.2.13

The slope is given by $\frac{5-3}{2-1} = 2$, so the equation of the line is $y - 3 = 2(x - 1)$, which can be written as $y = 2x - 2 + 3$, or $y = 2x + 1$.

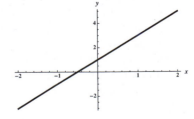

1.2.15 Using price as the independent variable p and the average number of units sold per day as the dependent variable d, we have the ordered pairs $(250, 12)$ and $(200, 15)$. The slope of the line determined by these points is $m = \frac{15-12}{200-250} = -\frac{3}{50}$. Thus the demand function has the form $d(p) = (-3/50)p + b$ for some constant b. Using the point $(200, 15)$, we find that $15 = (-3/50) \cdot 200 + b$, so $b = 27$. Thus the demand function is $d = (-3/50)p + 27$. While the domain of this linear function is the set of all real numbers, the formula is only likely to be valid for some subset of the interval $(0, 450)$, because outside of that interval either $p \leq 0$ or $d \leq 0$.

1.2.17 The slope is given by the rate of growth, which is 24. When $t = 0$ (years past 2015), the population is 500, so the point $(0, 500)$ satisfies our linear function. Thus the population is given by $p(t) = 24t + 500$. In 2030, we have $t = 15$, so the population will be approximately $p(15) = 360 + 500 = 860$.

1.2.19 For $x < 0$, the graph is a line with slope 1 and y-intercept 3, while for $x > 0$, it is a line with slope $-1/2$ and y-intercept 3. Note that both of these lines contain the point $(0, 3)$. The function shown can thus be written

$$f(x) = \begin{cases} x + 3 & \text{if } x < 0; \\ -\frac{1}{2}x + 3 & \text{if } x \geq 0. \end{cases}$$

1.2.21

The cost is given by

$$c(t) = \begin{cases} 0.05t & \text{for } 0 \leq t \leq 60 \\ 1.2 + 0.03t & \text{for } 60 < t \leq 120 \end{cases}.$$

1.2.23

1.2.25

1.2.27

1.2.29

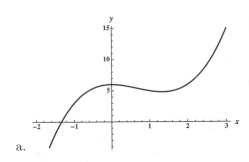

b. The function is a polynomial, so its domain is the set of all real numbers.

c. It has one peak near its y-intercept of $(0, 6)$ and one valley between $x = 1$ and $x = 2$. Its x-intercept is near $x = -4/3$.

a.

1.2.31

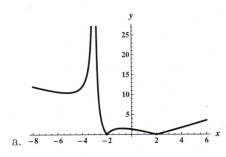

a.

b. The domain of the function is the set of all real numbers except -3.

c. There is a valley near $x = -5.2$ and a peak near $x = -0.8$. The x-intercepts are at -2 and 2, where the curve does not appear to be smooth. There is a vertical asymptote at $x = -3$. The function is never below the x-axis. The y-intercept is $(0, 4/3)$.

1.2.33

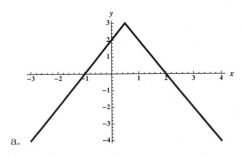

a.

b. The domain of the function is $(-\infty, \infty)$

c. The function has a maximum of 3 at $x = 1/2$, and a y-intercept of 2.

1.2.35 The slope of this line is constantly 2, so the slope function is $s(x) = 2$.

1.2.37 The slope function is given by $s(x) = \begin{cases} 1 & \text{if } x < 0; \\ -1/2 & \text{if } x > 0. \end{cases}$

1.2.39

a. Because the area under consideration is that of a rectangle with base 2 and height 6, $A(2) = 12$.

b. Because the area under consideration is that of a rectangle with base 6 and height 6, $A(6) = 36$.

c. Because the area under consideration is that of a rectangle with base x and height 6, $A(x) = 6x$.

1.2.41

a. Because the area under consideration is that of a trapezoid with base 2 and heights 8 and 4, we have $A(2) = 2 \cdot \frac{8+4}{2} = 12$.

b. Note that $A(3)$ represents the area of a trapezoid with base 3 and heights 8 and 2, so $A(3) = 3 \cdot \frac{8+2}{2} = 15$. So $A(6) = 15 + (A(6) - A(3))$, and $A(6) - A(3)$ represents the area of a triangle with base 3 and height 2. Thus $A(6) = 15 + 6 = 21$.

c. For x between 0 and 3, $A(x)$ represents the area of a trapezoid with base x, and heights 8 and $8 - 2x$. Thus the area is $x \cdot \frac{8+8-2x}{2} = 8x - x^2$. For $x > 3$, $A(x) = A(3) + A(x) - A(3) = 15 + 2(x-3) = 2x + 9$. Thus

$$A(x) = \begin{cases} 8x - x^2 & \text{if } 0 \le x \le 3; \\ 2x + 9 & \text{if } x > 3. \end{cases}$$

1.2.43 $f(x) = |x - 2| + 3$, because the graph of f is obtained from that of $|x|$ by shifting 2 units to the right and 3 units up.

$g(x) = -|x + 2| - 1$, because the graph of g is obtained from the graph of $|x|$ by shifting 2 units to the left, then reflecting about the x-axis, and then shifting 1 unit down.

1.2.45

a. Shift 3 units to the right.

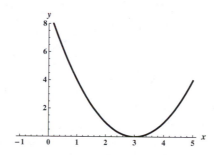

b. Horizontal scaling by a factor of 2, then shift 2 units to the right.

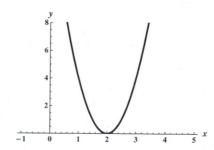

c. Shift to the right 2 units, vertical scaling by a factor of 3 and flip, shift up 4 units.

d. Horizontal scaling by a factor of $\frac{1}{3}$, horizontal shift right 2 units, vertical scaling by a factor of 6, vertical shift up 1 unit.

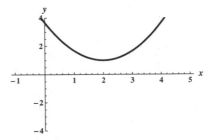

1.2.47   The graph is obtained by shifting the graph of x^2 two units to the right and one unit up.

1.2.49   This function is $-3f(x)$ where $f(x) = x^2$. Vertically scale the graph of f by a factor of 3 and then flip.

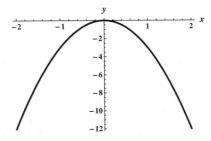

1.2.51   This function is $2f(x+3)$ where $f(x) = x^2$. Vertically scale the graph of f by a factor of 2 and then shift left 3 units.

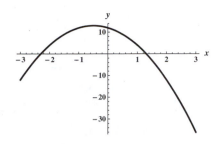

1.2.53 By completing the square, we have that $h(x) = -4(x^2 + x - 3) = -4\left(x^2 + x + \frac{1}{4} - \frac{1}{4} - 3\right) = -4(x + (1/2))^2 + 13$. So it is $-4f(x + (1/2)) + 13$ where $f(x) = x^2$. Vertically scale the graph of f by a factor of 4, then reflect about the x-axis, then shift left $1/2$ unit, and then up 13 units.

1.2.55

a. True. A polynomial $p(x)$ can be written as the ratio of polynomials $\frac{p(x)}{1}$, so it is a rational function. However, a rational function like $\frac{1}{x}$ is not a polynomial.

b. False. For example, if $f(x) = 2x$, then $(f \circ f)(x) = f(f(x)) = f(2x) = 4x$ is linear, not quadratic.

c. True. In fact, if f is degree m and g is degree n, then the degree of the composition of f and g is $m \cdot n$, regardless of the order they are composed.

d. False. The graph would be shifted two units to the left.

1.2.57 The points of intersection are found by solving $x^2 = -x^2 + 8x$. This yields the quadratic equation $2x^2 - 8x = 0$ or $(2x)(x - 4) = 0$. So the x-values of the points of intersection are 0 and 4. The actual points of intersection are $(0, 0)$ and $(4, 16)$.

1.2.59 $y = \sqrt{x} - 1$, because the y value is always 1 less than the square root of the x value.

1.2.61 The car moving north has gone $30t$ miles after t hours and the car moving east has gone $60t$ miles. Using the Pythagorean theorem, we have $s(t) = \sqrt{(30t)^2 + (60t)^2} = \sqrt{900t^2 + 3600t^2} = \sqrt{4500t^2} = 30\sqrt{5}t$ miles. The context domain could be $[0, 4]$.

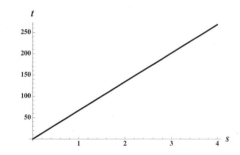

1.2.63 $y = \frac{3200}{x}$. Note that $\frac{x \text{ dollars per gallon}}{32 \text{ miles per gallon}} \cdot y \text{ miles}$ would represent the numbers of dollars, so this must be 100. So we have $\frac{xy}{32} = 100$, or $y = \frac{3200}{x}$. We certainly have $x > 0$, and a reasonable upper bound to imagine for x is \$5 (let's hope), so the context domain is $(0, 5]$.

1.2.65   **1.2.67**

1.2.69

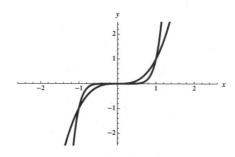

1.2.71

  a. The zeros of f are the points where the graph crosses the x-axis, so these are points A, D, F, and I.

  b. The only high point, or peak, of f occurs at point E, because it appears that the graph has larger and larger y values as x increases past point I and decreases past point A.

  c. The only low points, or valleys, of f are at points B and H, again assuming that the graph of f continues its apparent behavior for larger values of x.

  d. Past point H, the graph is rising, and is rising faster and faster as x increases. It is also rising between points B and E, but not as quickly as it is past point H. So the marked point at which it is rising most rapidly is I.

  e. Before point B, the graph is falling, and falls more and more rapidly as x becomes more and more negative. It is also falling between points E and H, but not as rapidly as it is before point B. So the marked point at which it is falling most rapidly is A.

1.2.73

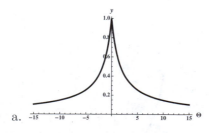   b. This appears to have a maximum when $\theta = 0$. Our vision is sharpest when we look straight ahead.

                                         c. For $|\theta| \le .19°$. We have an extremely narrow range where our eyesight is sharp.

1.2.75

  a. Using the points $(1986, 1875)$ and $(2000, 6471)$ we see that the slope is about 328.3. At $t = 0$, the value of p is 1875. Therefore a line which reasonably approximates the data is $p(t) = 328.3t + 1875$.

  b. Using this line, we have that $p(9) = 4830$.

1.2.77

a. Because you are paying \$350 per month, the amount paid after m months is $y = 350m + 1200$.

b. After 4 years (48 months) you have paid $350 \cdot 48 + 1200 = 18000$ dollars. If you then buy the car for \$10,000, you will have paid a total of \$28,000 for the car instead of \$25,000. So you should buy the car instead of leasing it.

1.2.79 The function makes sense for $0 \le h \le 2$.

1.2.81

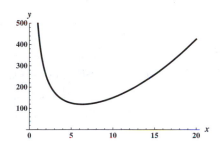

a. The volume of the box is $x^2 h$, but because the box has volume 125 cubic feet, we have that $x^2 h = 125$, so $h = \frac{125}{x^2}$. The surface area of the box is given by x^2 (the area of the base) plus $4 \cdot hx$, because each side has area hx. Thus $S = x^2 + 4hx = x^2 + \frac{4 \cdot 125 \cdot x}{x^2} = x^2 + \frac{500}{x}$.

b. By inspection, it looks like the value of x which minimizes the surface area is about 6.3.

1.2.83 Suppose that the parabola f crosses the x-axis at a and b, with $a < b$. Then a and b are roots of the polynomial, so $(x - a)$ and $(x - b)$ are factors. Thus the polynomial must be $f(x) = c(x - a)(x - b)$ for some non-zero real number c. So $f(x) = cx^2 - c(a + b)x + abc$. Because the vertex always occurs at the x value which is $\frac{-\text{coefficient on x}}{2 \cdot \text{coefficient on } x^2}$ we have that the vertex occurs at $\frac{c(a+b)}{2c} = \frac{a+b}{2}$, which is halfway between a and b.

1.2.85

a.

n	1	2	3	4	5
$n!$	1	2	6	24	120

b.

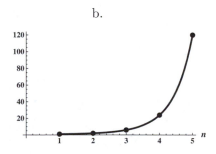

c. Using trial and error and a calculator yields that 10! is more than a million, but 9! isn't.

1.2.87

a.

n	1	2	3	4	5	6	7	8	9	10
$T(n)$	1	5	14	30	55	91	140	204	285	385

b. The domain of this function consists of the positive integers.

c. Using trial and error and a calculator yields that $T(n) > 1000$ for the first time for $n = 14$.

1.3   Trigonometric Functions

1.3.1 Let O be the length of the side opposite the angle x, let A be length of the side adjacent to the angle x, and let H be the length of the hypotenuse. Then $\sin x = \frac{O}{H}$, $\cos x = \frac{A}{H}$, $\tan x = \frac{O}{A}$, $\csc x = \frac{H}{O}$, $\sec x = \frac{H}{A}$, and $\cot x = \frac{A}{O}$.

1.3.3 The radian measure of an angle θ is the length of the arc s on the unit circle associated with θ.

1.3.5 $\sin^2 x + \cos^2 x = 1$, $1 + \cot^2 x = \csc^2 x$, and $\tan^2 x + 1 = \sec^2 x$.

1.3.7 The tangent function is undefined where $\cos x = 0$, which is at all real numbers of the form $\frac{\pi}{2} + k\pi$, k an integer. This is the set of odd multiples of $\pi/2$.

1.3.9

The point on the unit circle associated with $2\pi/3$ is $(-1/2, \sqrt{3}/2)$, so $\cos(2\pi/3) = -1/2$.

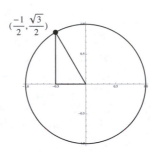

1.3.11

The point on the unit circle associated with $-3\pi/4$ is $(-\sqrt{2}/2, -\sqrt{2}/2)$, so $\tan(-3\pi/4) = 1$.

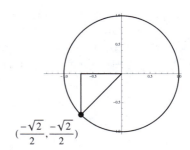

1.3.13

The point on the unit circle associated with $-13\pi/3$ is $(1/2, -\sqrt{3}/2)$, so $\cot(-13\pi/3) = -1/\sqrt{3} = -\sqrt{3}/3$.

1.3.15

The point on the unit circle associated with $-17\pi/3$ is $(1/2, \sqrt{3}/2)$, so $\cot(-17\pi/3) = 1/\sqrt{3} = \sqrt{3}/3$.

1.3.17 Because the point on the unit circle associated with $\theta = 0$ is the point $(1, 0)$, we have $\cos 0 = 1$.

1.3.19 Because $-\pi$ corresponds to a half circle clockwise revolution, the point on the unit circle associated with $-\pi$ is the point $(-1, 0)$. Thus $\cos(-\pi) = -1$.

1.3.21 Because $5\pi/2$ corresponds to one and a quarter counterclockwise revolutions, the point on the unit circle associated with $5\pi/2$ is the same as the point associated with $\pi/2$, which is $(0, 1)$. Thus $\sec 5\pi/2$ is undefined.

1.3.23 From our definitions of the trigonometric functions via a point $P(x, y)$ on a circle of radius $r = \sqrt{x^2 + y^2}$, we have $\sec\theta = \frac{r}{x} = \frac{1}{x/r} = \frac{1}{\cos\theta}$.

1.3.25 We have already established that $\sin^2\theta + \cos^2\theta = 1$. Dividing both sides by $\cos^2\theta$ gives $\tan^2\theta + 1 = \sec^2\theta$.

1.3.27

Using the triangle pictured, we see that

$$\sec(\pi/2 - \theta) = \frac{c}{a} = \csc\theta.$$

This also follows from the sum identity $\cos(a+b) = \cos a\cos b - \sin a\sin b$ as follows: $\sec(\pi/2 - \theta) = \frac{1}{\cos(\pi/2+(-\theta))} = \frac{1}{\cos(\pi/2)\cos(-\theta)-\sin(\pi/2)\sin(-\theta)} = \frac{1}{0-(-\sin(\theta))} = \csc(\theta)$.

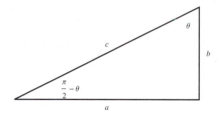

1.3.29 Using the fact that $\frac{\pi}{12} = \frac{\pi/6}{2}$ and the half-angle identity for cosine:

$$\cos^2(\pi/12) = \frac{1 + \cos(\pi/6)}{2} = \frac{1 + \sqrt{3}/2}{2} = \frac{2 + \sqrt{3}}{4}.$$

Thus, $\cos(\pi/12) = \sqrt{\frac{2+\sqrt{3}}{4}}$.

1.3.31 First note that $\tan x = 1$ when $\sin x = \cos x$. Using our knowledge of the values of the standard angles between 0 and 2π, we recognize that the sine function and the cosine function are equal at $\pi/4$. Then, because we recall that the period of the tangent function is π, we know that $\tan(\pi/4 + k\pi) = \tan(\pi/4) = 1$ for every integer value of k. Thus the solution set is $\{\pi/4 + k\pi, \text{where } k \text{ is an integer}\}$.

1.3.33 Given that $\sin^2\theta = \frac{1}{4}$, we have $|\sin\theta| = \frac{1}{2}$, so $\sin\theta = \frac{1}{2}$ or $\sin\theta = -\frac{1}{2}$. It follows that $\theta = \pi/6, 5\pi/6, 7\pi/6, 11\pi/6$.

1.3.35 The equation $\sqrt{2}\sin(x) - 1 = 0$ can be written as $\sin x = \frac{1}{\sqrt{2}} = \frac{\sqrt{2}}{2}$. Standard solutions to this equation occur at $x = \pi/4$ and $x = 3\pi/4$. Because the sine function has period 2π the set of all solutions can be written as:

$$\{\pi/4 + 2k\pi, \text{where } k \text{ is an integer}\} \cup \{3\pi/4 + 2l\pi, \text{where } l \text{ is an integer}\}.$$

1.3.37 As in the previous problem, let $u = 3x$. Then we are interested in the solutions to $\cos u = \sin u$, for $0 \le u < 6\pi$.

This would occur for $u = 3x = \pi/4, 5\pi/4, 9\pi/4, 13\pi/4, 17\pi/4$, and $21\pi/4$. Thus there are solutions for the original equation at

$$x = \pi/12, 5\pi/12, 3\pi/4, 13\pi/12, 17\pi/12, \text{ and } 7\pi/4.$$

1.3.39 If $\sin\theta\cos\theta = 0$, then either $\sin\theta = 0$ or $\cos\theta = 0$. This occurs for $\theta = 0, \pi/2, \pi, 3\pi/2$.

1.3.41

    a. False. For example, $\sin(\pi/2 + \pi/2) = \sin(\pi) = 0 \ne \sin(\pi/2) + \sin(\pi/2) = 1 + 1 = 2$.

    b. False. That equation has zero solutions, because the range of the cosine function is $[-1, 1]$.

    c. False. It has infinitely many solutions of the form $\pi/6 + 2k\pi$, where k is an integer (among others.)

    d. False. It has period $\frac{2\pi}{\pi/12} = 24$.

    e. True. The others have a range of either $[-1, 1]$ or $(-\infty, -1] \cup [1, \infty)$.

1.3.43 If $\cos\theta = 5/13$, then the Pythagorean identity gives $|\sin\theta| = 12/13$. But if $0 < \theta < \pi/2$, then the sine of θ is positive, so $\sin\theta = 12/13$. Thus $\tan\theta = 12/5$, $\cot\theta = 5/12$, $\sec\theta = 13/5$, and $\csc\theta = 13/12$.

1.3.45 If $\csc\theta = 13/12$, then $\sin\theta = 12/13$, and the Pythagorean identity gives $|\cos\theta| = 5/13$. But if $0 < \theta < \pi/2$, then the cosine of θ is positive, so $\cos\theta = 5/13$. Thus $\tan\theta = 12/5$, $\cot\theta = 5/12$, and $\sec\theta = 13/5$.

1.3.47 The amplitude is 3, and the period is $\frac{2\pi}{1/3} = 6\pi$.

1.3.49 The amplitude is 3.6, and the period is $\frac{2\pi}{\pi/24} = 48$.

1.3.51 Stretch the graph of $y = \cos x$ horizontally by a factor of 3 and vertically by a factor of 2, and reflect across the x-axis.

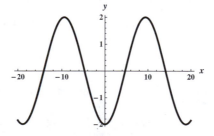

1.3.53 Stretch the graph of $y = \cos x$ horizontally by a factor of $\frac{24}{\pi}$, then stretch it vertically by a factor of 3.6 and then shift it up 2 units.

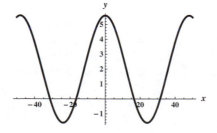

1.3.55 It is helpful to imagine first shifting the function horizontally so that the x intercept is where it should be, then stretching the function horizontally to obtain the correct period, and then stretching the function vertically to obtain the correct amplitude, and then shifting the whole graph up. Because the old x-intercept is at $x = 0$ and the new one should be at $x = 9$ (halfway between where the maximum and the minimum occur), we need to shift the function 9 units to the right. Then to get the right period, we need to multiply (before applying the sine function) by $\pi/12$ so that the new period is $\frac{2\pi}{\pi/12} = 24$. Finally, to get the right amplitude and to get the max and min at the right spots, we need to multiply on the outside by 3, and then shift the whole thing up 13 units. Thus, the desired function is:

$$f(x) = 3\sin((\pi/12)(x - 9)) + 13 = 3\sin((\pi/12)x - 3\pi/4) + 13.$$

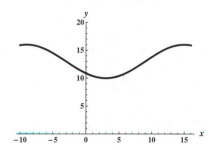

1.3.57 Let C be the circumference of the earth. Then the first rope has radius $r_1 = \frac{C}{2\pi}$. The circle generated by the longer rope has circumference $C + 38$, so its radius is $r_2 = \frac{C+38}{2\pi} = \frac{C}{2\pi} + \frac{38}{2\pi} \approx r_1 + 6$, so the radius of the bigger circle is about 6 feet more than the smaller circle.

1.3.59 We are seeking a function with amplitude 10 and period 1.5, and value 10 at time 0, so it should have the form $10\cos(kt)$, where $\frac{2\pi}{k} = 1.5$. Solving for k yields $k = \frac{4\pi}{3}$, so the desired function is $d(t) = 10\cos(4\pi t/3)$.

1.3.61 Let L be the line segment connecting the tops of the ladders and let M be the horizontal line segment between the walls h feet above the ground. Now note that the triangle formed by the ladders and L is equilateral, because the angle between the ladders is 60 degrees, and the other two angles must be equal and add to 120, so they are 60 degrees as well. Now we can see that the triangle formed by L, M and the right wall is similar to the triangle formed by the left ladder, the left wall, and the ground, because they are both right triangles with one angle of 75 degrees and one of 15 degrees. Thus $M = h$ is the distance between the walls.

1.3.63 To find $s(t)$ note that we are seeking a periodic function with period 365, and with amplitude 87.5 (which is half of the number of minutes between 7:25 and 4:30). We need to shift the function 4 days plus one fourth of 365, which is about 95 days so that the max and min occur at $t = 4$ days and at half a year later. Also, to get the right value for the maximum and minimum, we need to multiply by negative one and add 117.5 (which represents 30 minutes plus half the amplitude, because $s = 0$ corresponds to 4:00 AM.) Thus we have

$$s(t) = 117.5 - 87.5\sin\left(\frac{\pi}{182.5}(t - 95)\right).$$

A similar analysis leads to the formula

$$S(t) = 844.5 + 87.5\sin\left(\frac{\pi}{182.5}(t - 67)\right).$$

The graph pictured shows $D(t) = S(t) - s(t)$, the length of day function, which has its max at the summer solstice which is about the 172nd day of the year, and its min at the winter solstice.

1.3.65 The area of the entire circle is πr^2. The ratio $\frac{\theta}{2\pi}$ represents the proportion of the area swept out by a central angle θ. Thus the area of a sector of a circle is this same proportion of the entire area, so it is $\frac{\theta}{2\pi} \cdot \pi r^2 = \frac{r^2\theta}{2}$.

1.3.67 Note that $\sin A = \frac{h}{c}$ and $\sin C = \frac{h}{a}$, so $h = c\sin A = a\sin C$. Thus

$$\frac{\sin A}{a} = \frac{\sin C}{c}.$$

Now drop a perpendicular from the vertex A to the line determined by $\overline{BC}$, and let h_2 be the length of this perpendicular. Then $\sin C = \frac{h_2}{b}$ and $\sin B = \frac{h_2}{C}$, so $h_2 = b\sin C = c\sin B$. Thus

$$\frac{\sin C}{c} = \frac{\sin B}{b}.$$

Putting the two displayed equations together gives

$$\frac{\sin A}{a} = \frac{\sin B}{b} = \frac{\sin C}{c}.$$

Chapter One Review

1

  a. True. For example, $f(x) = x^2$ is such a function.

  b. False. For example, $\cos(\pi/2 + \pi/2) = \cos(\pi) = -1 \neq \cos(\pi/2) + \cos(\pi/2) = 0 + 0 = 0$.

  c. False. Consider $f(1 + 1) = f(2) = 2m + b \neq f(1) + f(1) = (m + b) + (m + b) = 2m + 2b$. (At least these aren't equal when $b \neq 0$.)

  d. True. $f(f(x)) = f(1 - x) = 1 - (1 - x) = x$.

  e. False. This set is the union of the disjoint intervals $(-\infty, -7)$ and $(1, \infty)$.

3

  a. This line has slope $\frac{2-(-3)}{4-2} = 5/2$. Therefore the equation of the line is $y - 2 = \frac{5}{2}(x - 4)$, so $y = \frac{5}{2}x - 8$.

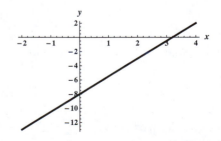

b. This line has the form $y = (3/4)x + b$, and because $(-4, 0)$ is on the line, $0 = (3/4)(-4) + b$, so $b = 3$. Thus the equation of the line is given by $y = (3/4)x + 3$.

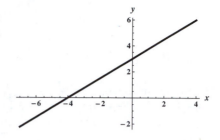

c. This line has slope $\frac{0-(-2)}{4-0} = \frac{1}{2}$, and the y-intercept is given to be -2, so the equation of this line is $y = (1/2)x - 2$.

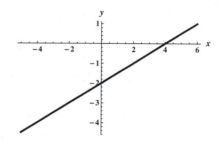

5

Because $|x| = \begin{cases} -x & \text{if } x < 0; \\ x & \text{if } x \geq 0, \end{cases}$

we have

$2(x - |x|) = \begin{cases} 2(x - (-x)) = 4x & \text{if } x < 0; \\ 2(x - x) = 0 & \text{if } x \geq 0. \end{cases}$

7

a. This is a straight line with slope 2/3 and y-intercept 10/3.

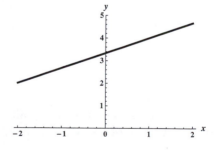

b. Completing the square gives $y = (x^2 + 2x + 1) - 4$, or $y = (x+1)^2 - 4$, so this is the standard parabola shifted one unit to the left and down 4 units.

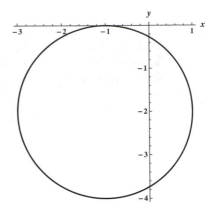

c. Completing the square, we have $x^2 + 2x + 1 + y^2 + 4y + 4 = -1 + 1 + 4$, so we have $(x+1)^2 + (y+2)^2 = 4$, a circle of radius 2 centered at $(-1, -2)$.

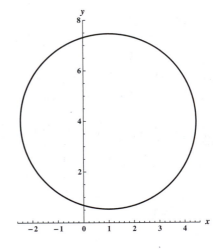

d. Completing the square, we have $x^2 - 2x + 1 + y^2 - 8y + 16 = -5 + 1 + 16$, or $(x-1)^2 + (y-4)^2 = 12$, which is a circle of radius $\sqrt{12}$ centered at $(1, 4)$.

9 The domain of $x^{1/7}$ is the set of all real numbers, as is its range. The domain of $x^{1/4}$ is the set of non-negative real numbers, as is its range.

11 We are looking for the line between the points $(0, 212)$ and $(6000, 200)$. The slope is $\frac{212-200}{0-6000} = -\frac{12}{6000} = -\frac{1}{500}$. Because the intercept is given, we deduce that the line is $B = f(a) = -\frac{1}{500}a + 212$.

13

a.

b.

c.

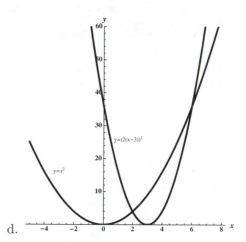

d.

15

a. $h(g(\pi/2)) = h(1) = 1$

b. $h(f(x)) = h(x^3) = x^{3/2}$.

c. $f(g(h(x))) = f(g(\sqrt{x})) = f(\sin(\sqrt{x})) = (\sin(\sqrt{x}))^3$.

d The domain of $g(f(x))$ is $\mathbb{R}$, because the domain of both functions is the set of all real numbers.

e. The range of $f(g(x))$ is $[-1, 1]$. This is because the range of g is $[-1, 1]$, and on the restricted domain $[-1, 1]$, the range of f is also $[-1, 1]$.

17 $\frac{f(x+h)-f(x)}{h} = \frac{(x+h)^2-2(x+h)-(x^2-2x)}{h} = \frac{x^2+2hx+h^2-2x-2h-x^2+2x}{h} = \frac{2hx+h^2-2h}{h} = 2x + h - 2.$

$\frac{f(x)-f(a)}{x-a} = \frac{x^2-2x-(a^2-2a)}{x-a} = \frac{(x^2-a^2)-2(x-a)}{x-a} = \frac{(x-a)(x+a)-2(x-a)}{x-a} = x + a - 2.$

19 $\frac{f(x+h)-f(x)}{h} = \frac{(x+h)^3+2-(x^3+2)}{h} = \frac{x^3+3x^2h+3xh^2+h^3+2-x^3-2}{h} = \frac{h(3x^2+3xh+h^2)}{h} = 3x^2 + 3xh + h^2.$

$\frac{f(x)-f(a)}{x-a} = \frac{x^3+2-(a^3+2)}{x-a} = \frac{x^3-a^3}{x-a} = \frac{(x-a)(x^2+ax+a^2)}{x-a} = x^2 + ax + a^2.$

21

a. Because $f(-x) = \cos -3x = \cos 3x = f(x)$, this is an even function, and is symmetric about the y-axis.

b. Because $f(-x) = 3(-x)^4 - 3(-x)^2 + 1 = 3x^4 - 3x^2 + 1 = f(x)$, this is an even function, and is symmetric about the y-axis.

c. Because replacing x by $-x$ and/or replacing y by $-y$ gives the same equation, this represents a curve which is symmetric about the y-axis and about the origin and about the x-axis.

23

a. A 135 degree angle measures $135 \cdot (\pi/180)$ radians, which is $3\pi/4$ radians.

b. A $4\pi/5$ radian angle measues $4\pi/5 \cdot (180/\pi)$ degrees, which is 144 degrees.

c. Because the length of the arc is the measure of the subtended angle (in radians) times the radius, this arc would be $4\pi/3 \cdot 10 = \frac{40\pi}{3}$ units long.

25

a. We need to scale the ordinary cosine function so that its period is 6, and then shift it 3 units to the right, and multiply it by 2. So the function we seek is $y = 2\cos((\pi/3)(t-3)) = -2\cos(\pi t/3)$.

b. We need to scale the ordinary cosine function so that its period is 24, and then shift it to the right 6 units. We then need to change the amplitude to be half the difference between the maximum and minimum, which would be 5. Then finally we need to shift the whole thing up by 15 units. The function we seek is thus $y = 15 + 5\cos((\pi/12)(t-6)) = 15 + 5\sin(\pi t/12)$.

27

a. $-\sin x$ is pictured in F.

b. $\cos 2x$ is pictured in E.

c. $\tan(x/2)$ is pictured in D.

d. $-\sec x$ is pictured in B.

e. $\cot 2x$ is pictured in C.

f. $\sin^2 x$ is pictured in A.

29 $\sin x = -\frac{1}{2}$ for $x = 7\pi/6$ and for $x = 11\pi/6$, so the intersection points are $(7\pi/6, -1/2)$ and $(11\pi/6, -1/2)$.

Chapter 2

Limits

2.1 The Idea of Limits

2.1.1 The average velocity of the object between time $t = a$ and $t = b$ is the change in position divided by the elapsed time: $v_{\text{av}} = \frac{s(b)-s(a)}{b-a}$.

2.1.3 The slope of the secant line between points $(a, f(a))$ and $(b, f(b))$ is the ratio of the differences $f(b) - f(a)$ and $b - a$. Thus $m_{\text{sec}} = \frac{f(b)-f(a)}{b-a}$.

2.1.5 Both problems involve the same mathematics, namely finding the limit as $t \to a$ of a quotient of differences of the form $\frac{g(t)-g(a)}{t-a}$ for some function g.

2.1.7 The average velocity is $\frac{s(3)-s(2)}{3-2} = 156 - 136 = 20$.

2.1.9

   a. Over $[1, 4]$, we have $v_{\text{av}} = \frac{s(4)-s(1)}{4-1} = \frac{256-112}{3} = 48$.

   b. Over $[1, 3]$, we have $v_{\text{av}} = \frac{s(3)-s(1)}{3-1} = \frac{240-112}{2} = 64$.

   c. Over $[1, 2]$, we have $v_{\text{av}} = \frac{s(2)-s(1)}{2-1} = \frac{192-112}{1} = 80$.

   d. Over $[1, 1+h]$, we have $v_{\text{av}} = \frac{s(1+h)-s(1)}{1+h-1} = \frac{-16(1+h)^2+128(1+h)-(112)}{h} = \frac{-16h^2-32h+128h}{h} = \frac{h(-16h+96)}{h} = 96 - 16h = 16(6 - h)$.

2.1.11

   a. $\frac{s(2)-s(0)}{2-0} = \frac{72-0}{2} = 36$.

   b. $\frac{s(1.5)-s(0)}{1.5-0} = \frac{66-0}{1.5} = 44$.

   c. $\frac{s(1)-s(0)}{1-0} = \frac{52-0}{1} = 52$.

   d. $\frac{s(0.5)-s(0)}{0.5-0} = \frac{30-0}{0.5} = 60$.

2.1.13

The slope of the secant line is given by $\frac{s(2)-s(0.5)}{2-0.5} = \frac{136-46}{1.5} = 60$. This represents the average velocity of the object over the time interval $[0.5, 2]$.

2.1.15

Time Interval	$[1, 2]$	$[1, 1.5]$	$[1, 1.1]$	$[1, 1.01]$	$[1, 1.001]$
Average Velocity	80	88	94.4	95.84	95.984

The instantaneous velocity appears to be 96 ft/s.

2.1.17 $\frac{s(1.01)-s(1)}{0.01} = 47.84$, while $\frac{s(1.001)-s(1)}{0.001} = 47.984$ and $\frac{s(1.0001)-s(1)}{0.0001} = 47.9984$. It appears that the instantaneous velocity at $t = 1$ is approximately 48.

2.1.19

Time Interval	$[2, 3]$	$[2.9, 3]$	$[2.99, 3]$	$[2.999, 3]$	$[2.9999, 3]$	$[2.99999, 3]$
Average Velocity	20	5.6	4.16	4.016	4.002	4.0002

The instantaneous velocity appears to be 4 ft/s.

2.1.21

Time Interval	$[3, 3.1]$	$[3, 3.01]$	$[3, 3.001]$	$[3, 3.0001]$
Average Velocity	-17.6	-16.16	-16.016	-16.002

The instantaneous velocity appears to be -16 ft/s.

2.1.23

Time Interval	$[0, 0.1]$	$[0, 0.01]$	$[0, 0.001]$	$[0, 0.0001]$
Average Velocity	79.468	79.995	80.000	80.0000

The instantaneous velocity appears to be 80 ft/s.

2.1.25

x Interval	$[2, 2.1]$	$[2, 2.01]$	$[2, 2.001]$	$[2, 2.0001]$
Slope of Secant Line	8.2	8.02	8.002	8.0002

The slope of the tangent line appears to be 8.

2.1.27

x Interval	$[-1, -0.9]$	$[-1, -0.99]$	$[-1, -0.999]$	$[-1, -0.9999]$
Slope of the Secant Line	0.524862	0.5025	0.50025	0.500025

The slope of the tangent line appears to be 0.5.

2.1.29

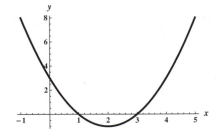

a. Note that the graph is a parabola with vertex $(2, -1)$.

b. At $(2, -1)$ the function has tangent line with slope 0.

c.

x Interval	$[2, 2.1]$	$[2, 2.01]$	$[2, 2.001]$	$[2, 2.0001]$
Slope of the Secant Line	.1	.01	.001	.0001

The slope of the tangent line at $(2, -1)$ appears to be 0.

2.1.31

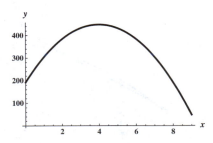

a. Note that the graph is a parabola with vertex $(4, 448)$.

b. At $(4, 448)$ the function has tangent line with slope 0, so $a = 4$.

c.

x Interval	$[4, 4.1]$	$[4, 4.01]$	$[4, 4.001]$	$[4, 4.0001]$
Slope of the Secant Line	-1.6	-0.16	-0.016	-0.0016

The slopes of the secant lines appear to be approaching zero.

d. On the interval $[0, 4)$ the instantaneous velocity of the projectile is positive.

e. On the interval $(4, 9]$ the instantaneous velocity of the projectile is negative.

2.1.33 For line AD, we have

$$m_{AD} = \frac{y_D - y_A}{x_D - x_A} = \frac{f(\pi) - f(\pi/2)}{\pi - (\pi/2)} = \frac{1}{\pi/2} \approx .63662.$$

For line AC, we have

$$m_{AC} = \frac{y_C - y_A}{x_C - x_A} = \frac{f(\pi/2 + 0.5) - f(\pi/2)}{(\pi/2 + 0.5) - (\pi/2)} = -\frac{\cos(\pi/2 + 0.5)}{0.5} \approx 0.958851.$$

For line AB, we have

$$m_{AB} = \frac{y_B - y_A}{x_B - x_A} = \frac{f(\pi/2 + 0.05) - f(\pi/2)}{(\pi/2 + 0.05) - (\pi/2)} = -\frac{\cos(\pi/2 + 0.05)}{0.05} \approx 0.999583.$$

Computing one more slope of a secant line:

$$m_{\text{sec}} = \frac{f(\pi/2 + 0.01) - f(\pi/2)}{(\pi/2 + 0.01) - (\pi/2)} = -\frac{\cos(\pi/2 + 0.01)}{0.01} \approx 0.999983.$$

Conjecture: The slope of the tangent line to the graph of f at $x = \pi/2$ is 1.

2.2   Definitions of Limits

2.2.1 Suppose the function f is defined for all x near a except possibly at a. If $f(x)$ is arbitrarily close to a number L whenever x is sufficiently close to (but not equal to) a, then we write $\lim_{x \to a} f(x) = L$.

2.2.3 Suppose the function f is defined for all x near a but greater than a. If $f(x)$ is arbitrarily close to L for x sufficiently close to (but strictly greater than) a, then we write $\lim_{x \to a^+} f(x) = L$.

2.2.5 It must be true that $L = M$.

2.2.7

a. $h(2) = 5$.

b. $\lim_{x \to 2} h(x) = 3$.

c. $h(4)$ does not exist.

d. $\lim_{x \to 4} f(x) = 1$.

e. $\lim_{x \to 5} h(x) = 2$.

2.2.9

a. $f(1) = -1$.

b. $\lim_{x \to 1} f(x) = 1$.

c. $f(0) = 2$.

d. $\lim_{x \to 0} f(x) = 2$.

2.2.11

a.

x	1.9	1.99	1.999	1.9999	2	2.0001	2.001	2.01	2.1
$f(x) = \frac{x^2-4}{x-2}$	3.9	3.99	3.999	3.9999	undefined	4.0001	4.001	4.01	4.1

b. $\lim\limits_{x \to 2} f(x) = 4$.

2.2.13

a.

t	8.9	8.99	8.999	9	9.001	9.01	9.1
$g(t) = \frac{t-9}{\sqrt{t}-3}$	5.98329	5.99833	5.99983	undefined	6.00017	6.00167	6.01662

b. $\lim\limits_{t \to 9} \dfrac{t-9}{\sqrt{t}-3} = 6$.

2.2.15

a.

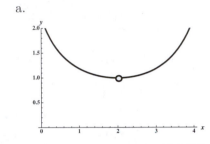

b.

x	1.8	1.9	1.99	2.01	2.1	2.2
$f(x)$	1.0067	1.00167	1.00002	1.00002	1.00167	1.0067

From both the graph and the table, the limit appears to be 1.

2.2.17

a.

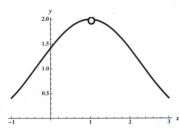

b.

x	0.9	0.99	0.999	1.001	1.01	1.1
$f(x)$	1.993342	1.999933	1.999999	1.999999	1.999933	1.993342

From both the graph and the table, the limit appears to be 2.

2.2.19

x	4.9	4.99	4.999	4.9999	5	5.0001	5.001	5.01	5.1
$f(x) = \frac{x^2-25}{x-5}$	9.9	9.99	9.999	9.9999	undefined	10.0001	10.001	10.01	10.1

$\lim\limits_{x \to 5^+} \dfrac{x^2-25}{x-5} = 10$, $\lim\limits_{x \to 5^-} \dfrac{x^2-25}{x-5} = 10$, and thus $\lim\limits_{x \to 5} \dfrac{x^2-25}{x-5} = 10$.

2.2.21

   a. $f(1) = 0$.

   b. $\lim_{x \to 1^-} f(x) = 1$.

   c. $\lim_{x \to 1^+} f(x) = 0$.

   d. $\lim_{x \to 1} f(x)$ does not exist, since the two one-sided limits aren't equal.

2.2.23

   a. $f(1) = 3$.

   b. $\lim_{x \to 1^-} f(x) = 2$.

   c. $\lim_{x \to 1^+} f(x) = 2$.

   d. $\lim_{x \to 1} f(x) = 2$.

   e. $f(3) = 2$.

   f. $\lim_{x \to 3^-} f(x) = 4$.

   g. $\lim_{x \to 3^+} f(x) = 1$.

   h. $\lim_{x \to 3} f(x)$ does not exist.

   i. $f(2) = 3$.

   j. $\lim_{x \to 2^-} f(x) = 3$.

   k. $\lim_{x \to 2^+} f(x) = 3$.

   l. $\lim_{x \to 2} f(x) = 3$.

2.2.25

a.

x	$\frac{2}{\pi}$	$\frac{2}{3\pi}$	$\frac{2}{5\pi}$	$\frac{2}{7\pi}$	$\frac{2}{9\pi}$	$\frac{2}{11\pi}$
$f(x) = \sin(1/x)$	1	−1	1	−1	1	−1

If $x_n = \frac{2}{(2n+1)\pi}$, then $f(x_n) = (-1)^n$ where n is a non-negative integer.

b. As $x \to 0$, $1/x \to \infty$. So the values of $f(x)$ oscillate dramatically between -1 and 1.

c. $\lim_{x \to 0} \sin(1/x)$ does not exist.

2.2.27

a. False. In fact $\lim_{x \to 3} \frac{x^2 - 9}{x - 3} = \lim_{x \to 3} (x + 3) = 6$.

b. False. For example, if $f(x) = \begin{cases} x^2 & \text{if } x \neq 0; \\ 5 & \text{if } x = 0 \end{cases}$ and if $a = 0$ then $f(a) = 5$ but $\lim_{x \to a} f(x) = 0$.

c. False. For example, the limit in part a of this problem exists, even though the corresponding function is undefined at $a = 3$.

d. False. It is true that the limit of $\sqrt{x}$ as x approaches zero from the right is zero, but because the domain of $\sqrt{x}$ does not include any numbers to the left of zero, the two-sided limit doesn't exist.

e. True. Note that $\lim_{x \to \pi/2} \cos x = 0$ and $\lim_{x \to \pi/2} \sin x = 1$, so $\lim_{x \to \pi/2} \frac{\cos x}{\sin x} = \frac{0}{1} = 0$.

2.2.29

2.2.31

2.2.33

h	0.01	0.001	0.0001	-0.01	-0.001	-0.0001
$\tan(3h)/h$	3.0009	3.000009	3.00000009	3.0009	3.000009	3.00000009

$$\lim_{h \to 0} \frac{\tan(3h)}{h} = 3$$

2.2.35

h	0.01	0.001	0.0001	-0.01	-0.001	-0.0001
$(1 - \cos(h))/h$	0.0049999	0.0005	0.00005	-0.0049999	-0.0005	-0.00005

$$\lim_{h \to 0} \frac{1 - \cos(h)}{h} = 0$$

2.2.37

a. $\lim_{x \to -1^-} \lfloor x \rfloor = -2$, $\lim_{x \to -1^+} \lfloor x \rfloor = -1$, $\lim_{x \to 2^-} \lfloor x \rfloor = 1$, $\lim_{x \to 2^+} \lfloor x \rfloor = 2$.

b. $\lim_{x \to 2.3^-} \lfloor x \rfloor = 2$, $\lim_{x \to 2.3^+} \lfloor x \rfloor = 2$, $\lim_{x \to 2.3} \lfloor x \rfloor = 2$.

c. In general, for an integer a, $\lim_{x \to a^-} \lfloor x \rfloor = a - 1$ and $\lim_{x \to a^+} \lfloor x \rfloor = a$.

d. In general, if a is not an integer, $\lim_{x \to a^-} \lfloor x \rfloor = \lim_{x \to a^+} \lfloor x \rfloor = \lfloor a \rfloor$.

e. $\lim_{x \to a} \lfloor x \rfloor$ exists and is equal to $\lfloor a \rfloor$ for non-integers a.

2.2.39 By zooming in closely, you should be able to convince yourself that the answer is 0.

2.2.41 By zooming in closely, you should be able to convince yourself that the answer is 16.

2.2.43

    a. Note that the function is piecewise constant.

    b. $\lim_{w \to 3.3} f(w) = .95$.

    c. $\lim_{w \to 1^+} f(w) = .61$ corresponds to the fact that for any piece of mail that weighs slightly over 1 ounce, the postage will cost 61 cents. $\lim_{w \to 1^-} f(w) = .44$ corresponds to the fact that for any piece of mail that weighs slightly less than 1 ounce, the postage will cost 44 cents.

    d. $\lim_{w \to 4} f(w)$ does not exist because the two corresponding one-side limits don't exist. (The limit from the left is .95, while the limit from the right is 1.12.)

2.2.45

a. Because of the symmetry about the y axis, we must have $\lim_{x \to -2^+} f(x) = 8$.

b. Because of the symmetry about the y axis, we must have $\lim_{x \to -2^-} f(x) = 5$.

2.2.47

a.

$$\lim_{x \to 0} \frac{\tan 2x}{\sin x} = 2.$$

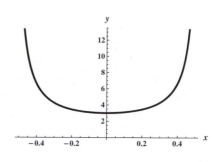

$$\lim_{x \to 0} \frac{\tan 3x}{\sin x} = 3.$$

$$\lim_{x \to 0} \frac{\tan 4x}{\sin x} = 4.$$

b. It appears that $\displaystyle\lim_{x \to 0} \frac{\tan(px)}{\sin x} = p.$

2.2.49

For $p = 8$ and $q = 2$, it appears that the limit is 4.

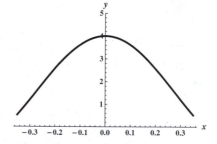

For $p = 12$ and $q = 3$, it appears that the limit is 4.

For $p = 4$ and $q = 16$, it appears that the limit is $1/4$.

For $p = 100$ and $q = 50$, it appears that the limit is 2.

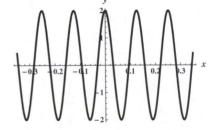

Conjecture: $\displaystyle\lim_{x \to 0} \frac{\sin px}{\sin qx} = \frac{p}{q}$.

2.3   Techniques for Computing Limits

2.3.1 If $f(x) = a_n x^n + a_{n-1} x^{n-1} + \cdots + a_1 x + a_0$, then $\displaystyle\lim_{x \to a} f(x) = \lim_{x \to a}(a_n x^n + a_{n-1} x^{n-1} + \cdots + a_1 x + a_0)$

$= a_n \left(\lim_{x \to a} x\right)^n + a_{n-1}\left(\lim_{x \to a} x\right)^{n-1} + \cdots + a_1 \lim_{x \to a} x + \lim_{x \to a} a_0$

$= a_n a^n + a_{n-1} a^{n-1} + \cdots + a_1 a + a_0 = f(a)$.

2.3.3 For a rational function $r(x)$, we have $\displaystyle\lim_{x \to a} r(x) = r(a)$ exactly for those numbers a which are in the domain of r.

2.3.5 Because $\frac{x^2 - 7x + 12}{x - 3} = \frac{(x-3)(x-4)}{x-3} = x - 4$ (for $x \neq 3$), we can see that the graphs of these two functions are the same except that one is undefined at $x = 3$ and the other is a straight line that is defined everywhere. Thus the function $\frac{x^2 - 7x + 12}{x - 3}$ is a straight line except that it has a "hole" at $(3, -1)$. The two functions have the same limit as $x \to 3$, namely $\displaystyle\lim_{x \to 3} \frac{x^2 - 7x + 12}{x - 3} = \lim_{x \to 3}(x - 4) = -1$.

2.3.7 If p and q are polynomials then $\displaystyle\lim_{x \to 0} \frac{p(x)}{q(x)} = \frac{\lim_{x \to 0} p(x)}{\lim_{x \to 0} q(x)} = \frac{p(0)}{q(0)}$. Because this quantity is given to be equal to 10, we have $\frac{p(0)}{2} = 10$, so $p(0) = 20$.

2.3.9 $\displaystyle\lim_{x \to 5} \sqrt{x^2 - 9} = \sqrt{\lim_{x \to 5}(x^2 - 9)} = \sqrt{16} = 4$.

2.3.11 $\displaystyle\lim_{x \to 4}(3x - 7) = 3 \lim_{x \to 4} x - 7 = 3 \cdot 4 - 7 = 5$.

2.3.13 $\displaystyle\lim_{x \to -9}(5x) = 5 \lim_{x \to -9} x = 5 \cdot -9 = -45$.

2.3.15 $\displaystyle\lim_{x \to 6} 4 = 4$.

2.3.17 $\lim\limits_{x\to1} 4f(x) = 4\lim\limits_{x\to1} f(x) = 4\cdot 8 = 32$. This follows from the Constant Multiple Law.

2.3.19 $\lim\limits_{x\to1}(f(x)-g(x)) = \lim\limits_{x\to1} f(x) - \lim\limits_{x\to1} g(x) = 8 - 3 = 5$. This follows from the Difference Law.

2.3.21 $\lim\limits_{x\to1}\dfrac{f(x)g(x)}{h(x)} = \dfrac{\lim\limits_{x\to1}(f(x)g(x))}{\lim\limits_{x\to1} h(x)} = \dfrac{\lim\limits_{x\to1} f(x) \cdot \lim\limits_{x\to1} g(x)}{\lim\limits_{x\to1} h(x)} = \dfrac{8\cdot3}{2} = 12$. This follows from the Quotient and Product Laws.

2.3.23 $\lim\limits_{x\to1}(h(x))^5 = \left(\lim\limits_{x\to1} h(x)\right)^5 = (2)^5 = 32$. This follows from the Power Law.

2.3.25 $\lim\limits_{x\to1}(2x^3 - 3x^2 + 4x + 5) = \lim\limits_{x\to1} 2x^3 - \lim\limits_{x\to1} 3x^2 + \lim\limits_{x\to1} 4x + \lim\limits_{x\to1} 5 = 2(\lim\limits_{x\to1} x)^3 - 3(\lim\limits_{x\to1} x)^2 + 4(\lim\limits_{x\to1} x) + 5 = 2(1)^3 - 3(1)^2 + 4\cdot1 + 5 = 8$.

2.3.27 $\lim\limits_{x\to1}\dfrac{5x^2+6x+1}{8x-4} = \dfrac{\lim\limits_{x\to1}(5x^2+6x+1)}{\lim\limits_{x\to1}(8x-4)} = \dfrac{5(\lim\limits_{x\to1} x)^2 + 6\lim\limits_{x\to1} x + \lim\limits_{x\to1} 1}{8\lim\limits_{x\to1} x - \lim\limits_{x\to1} 4} = \dfrac{5(1)^2+6\cdot1+1}{8\cdot1-4} = 3$.

2.3.29 $\lim\limits_{b\to2}\dfrac{3b}{\sqrt{4b+1}-1} = \dfrac{\lim\limits_{b\to2} 3b}{\lim\limits_{b\to2}(\sqrt{4b+1}-1)} = \dfrac{3\lim\limits_{b\to2} b}{\lim\limits_{b\to2}\sqrt{4b+1}-\lim\limits_{b\to2} 1} = \dfrac{3\cdot2}{\sqrt{\lim\limits_{b\to2}(4b+1)}-1} = \dfrac{6}{3-1} = 3$.

2.3.31 $\lim\limits_{x\to3}\dfrac{-5x}{\sqrt{4x-3}} = \dfrac{\lim\limits_{x\to3} -5x}{\lim\limits_{x\to3}\sqrt{4x-3}} = \dfrac{-5\lim\limits_{x\to3} x}{\sqrt{\lim\limits_{x\to3}(4x-3)}} = \dfrac{-5\cdot3}{\sqrt{4\lim\limits_{x\to3} x - \lim\limits_{x\to3} 3}} = \dfrac{-15}{\sqrt{4\cdot3-3}} = -5$.

2.3.33

a. $\lim\limits_{x\to-1^-} f(x) = \lim\limits_{x\to-1^-}(x^2+1) = (-1)^2 + 1 = 2$.

b. $\lim\limits_{x\to-1^+} f(x) = \lim\limits_{x\to-1^+}\sqrt{x+1} = \sqrt{-1+1} = 0$.

c. $\lim\limits_{x\to-1} f(x)$ does not exist.

2.3.35

a. $\lim\limits_{x\to2^+}\sqrt{x-2} = \sqrt{2-2} = 0$.

b. The domain of $f(x) = \sqrt{x-2}$ is $[2,\infty)$. Thus, any question about this function that involves numbers less than 2 doesn't make any sense, because those numbers aren't in the domain of f.

2.3.37 Using the definition of $|x|$ given, we have $\lim\limits_{x\to0^-}|x| = \lim\limits_{x\to0^-}(-x) = -0 = 0$. Also, $\lim\limits_{x\to0^+}|x| = \lim\limits_{x\to0^+} x = 0$. Because the two one-sided limits are both 0, we also have $\lim\limits_{x\to0}|x| = 0$.

2.3.39 $\lim\limits_{x\to1}\dfrac{x^2-1}{x-1} = \lim\limits_{x\to1}\dfrac{(x+1)(x-1)}{x-1} = \lim\limits_{x\to1}(x+1) = 2$.

2.3.41 $\lim\limits_{x\to4}\dfrac{x^2-16}{4-x} = \lim\limits_{x\to4}\dfrac{(x+4)(x-4)}{-(x-4)} = \lim\limits_{x\to4}[-(x+4)] = -8$.

2.3.43 $\lim\limits_{x\to b}\dfrac{(x-b)^{50} - x + b}{x-b} = \lim\limits_{x\to b}\dfrac{(x-b)^{50} - (x-b)}{x-b} = \lim\limits_{x\to b}\dfrac{(x-b)((x-b)^{49}-1)}{x-b} =$ $\lim\limits_{x\to b}[(x-b)^{49}-1] = -1$.

2.3.45 $\lim\limits_{x \to -1} \dfrac{(2x-1)^2 - 9}{x+1} = \lim\limits_{x \to -1} \dfrac{(2x-1-3)(2x-1+3)}{x+1} = \lim\limits_{x \to -1} \dfrac{2(x-2)2(x+1)}{x+1} = \lim\limits_{x \to -1} 4(x-2) = 4 \cdot (-3) = -12.$

2.3.47 $\lim\limits_{x \to 9} \dfrac{\sqrt{x}-3}{x-9} = \lim\limits_{x \to 9} \dfrac{(\sqrt{x}-3)(\sqrt{x}+3)}{(x-9)(\sqrt{x}+3)} = \lim\limits_{x \to 9} \dfrac{x-9}{(x-9)(\sqrt{x}+3)} = \lim\limits_{x \to 9} \dfrac{1}{\sqrt{x}+3} = \dfrac{1}{6}.$

2.3.49 $\lim\limits_{x \to a} \dfrac{x-a}{\sqrt{x}-\sqrt{a}} = \lim\limits_{x \to a} \dfrac{x-a}{\sqrt{x}-\sqrt{a}} \cdot \dfrac{\sqrt{x}+\sqrt{a}}{\sqrt{x}+\sqrt{a}} = \lim\limits_{x \to a} \dfrac{(x-a)(\sqrt{x}+\sqrt{a})}{x-a} = \lim\limits_{x \to a} (\sqrt{x}+\sqrt{a}) = 2\sqrt{a}.$

2.3.51 $\lim\limits_{h \to 0} \dfrac{\sqrt{16+h}-4}{h} = \lim\limits_{h \to 0} \dfrac{(\sqrt{16+h}-4)(\sqrt{16+h}+4)}{h(\sqrt{16+h}+4)} = \lim\limits_{h \to 0} \dfrac{(16+h)-16}{h(\sqrt{16+h}+4)} = \lim\limits_{h \to 0} \dfrac{h}{h(\sqrt{16+h}+4)}$
$= \lim\limits_{h \to 0} \dfrac{1}{(\sqrt{16+h}+4)} = \dfrac{1}{8}.$

2.3.53

a. The statement we are trying to prove can be stated in cases as follows: For $x > 0$, $-x \leq x \sin(1/x) \leq x$, and for $x < 0$, $x \leq x \sin(1/x) \leq -x$.

Now for all $x \neq 0$, note that $-1 \leq \sin(1/x) \leq 1$ (because the range of the sine function is $[-1,1]$). We will consider the two cases $x > 0$ and $x < 0$ separately, but in each case, we will multiply this inequality through by x, switching the inequalities for the $x < 0$ case.

For $x > 0$ we have $-x \leq x \sin(1/x) \leq x$, and for $x < 0$ we have $-x \geq x \sin(1/x) \geq x$, which are exactly the statements we are trying to prove.

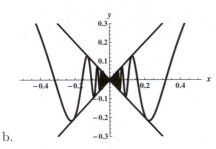

c. Because $\lim\limits_{x \to 0} -|x| = \lim\limits_{x \to 0} |x| = 0$, and because $-|x| \leq x \sin(1/x) \leq |x|$, the Squeeze Theorem assures us that $\lim\limits_{x \to 0} [x \sin(1/x)] = 0$ as well.

b.

2.3.55

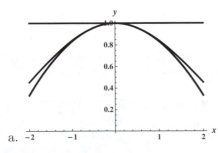

a.

b. Note that $\lim\limits_{x \to 0} \left[1 - \dfrac{x^2}{6} \right] = 1 = \lim\limits_{x \to 0} 1$. So because $1 - \dfrac{x^2}{6} \leq \dfrac{\sin x}{x} \leq 1$, the squeeze theorem assures us that $\lim\limits_{x \to 0} \dfrac{\sin x}{x} = 1$ as well.

2.3.57

a. False. For example, if $f(x) = \begin{cases} x & \text{if } x \neq 1; \\ 4 & \text{if } x = 1, \end{cases}$ then $\lim\limits_{x \to 1} f(x) = 1$ but $f(1) = 4$.

b. False. For example, if $f(x) = \begin{cases} x+1 & \text{if } x \leq 1; \\ x-6 & \text{if } x > 1, \end{cases}$ then $\lim\limits_{x \to 1^-} f(x) = 2$ but $\lim\limits_{x \to 1^+} f(x) = -5$.

c. False. For example, if $f(x) = \begin{cases} x & \text{if } x \neq 1; \\ 4 & \text{if } x = 1, \end{cases}$ and $g(x) = 1$, then f and g both have limit 1 as $\dot{x} \to 1$, but $f(1) = 4 \neq g(1)$.

d. False. For example $\lim\limits_{x \to 2} \dfrac{x^2 - 4}{x - 2}$ exists and is equal to 4.

e. False. For example, it would be possible for the domain of f to be $[1, \infty)$, so that the one-sided limit exists but the two-sided limit doesn't even make sense. This would be true, for example, if $f(x) = x - 1$.

2.3.59 $\lim\limits_{x \to 2} (5x - 6)^{3/2} = (5 \cdot 2 - 6)^{3/2} = 4^{3/2} = 2^3 = 8.$

2.3.61 $\lim\limits_{x \to 1} \dfrac{\sqrt{10x - 9} - 1}{x - 1} = \lim\limits_{x \to 1} \dfrac{(\sqrt{10x - 9} - 1)(\sqrt{10x - 9} + 1)}{(x - 1)(\sqrt{10x - 9} + 1)} = \lim\limits_{x \to 1} \dfrac{(10x - 9) - 1}{(x - 1)(\sqrt{10x - 9} + 1)} =$

$\lim\limits_{x \to 1} \dfrac{10(x - 1)}{(x - 1)(\sqrt{10x - 9} + 1)} = \lim\limits_{x \to 1} \dfrac{10}{(\sqrt{10x - 9} + 1)} = \dfrac{10}{2} = 5.$

2.3.63 $\lim\limits_{h \to 0} \dfrac{(5 + h)^2 - 25}{h} = \lim\limits_{h \to 0} \dfrac{25 + 10h + h^2 - 25}{h} = \lim\limits_{h \to 0} \dfrac{h(10 + h)}{h} = \lim\limits_{h \to 0} (10 + h) = 10.$

2.3.65 We have

$$\lim\limits_{w \to -k} \dfrac{w^2 + 5kw + 4k^2}{w^2 + kw} = \lim\limits_{w \to -k} \dfrac{(w + 4k)(w + k)}{(w)(w + k)} = \lim\limits_{w \to -k} \dfrac{w + 4k}{w} = \dfrac{-k + 4k}{-k} = -3.$$

If $k = 0$, we have $\lim\limits_{w \to -k} \dfrac{w^2 + 5kw + 4k^2}{w^2 + kw} = \lim\limits_{w \to 0} \dfrac{w^2}{w^2} = 1.$

2.3.67 In order for $\lim\limits_{x \to -1} g(x)$ to exist, we need the two one-sided limits to exist and be equal. We have $\lim\limits_{x \to -1^-} g(x) = \lim\limits_{x \to -1^-} (x^2 - 5x) = 6$, and $\lim\limits_{x \to -1^+} g(x) = \lim\limits_{x \to -1^+} (ax^3 - 7) = -a - 7$. So we need $-a - 7 = 6$, so we require that $a = -13$. Then $\lim\limits_{x \to -1} f(x) = 6$.

2.3.69 $\lim\limits_{x \to 1} \dfrac{x^6 - 1}{x - 1} = \lim\limits_{x \to 1} \dfrac{(x - 1)(x^5 + x^4 + x^3 + x^2 + x + 1)}{x - 1} = \lim\limits_{x \to 1} (x^5 + x^4 + x^3 + x^2 + x + 1) = 6.$

2.3.71 $\lim\limits_{x \to a} \dfrac{x^5 - a^5}{x - a} = \lim\limits_{x \to a} \dfrac{(x - a)(x^4 + ax^3 + a^2x^2 + a^3x + a^4)}{x - a} = \lim\limits_{x \to a} (x^4 + ax^3 + a^2x^2 + a^3x + a^4) = 5a^4.$

2.3.73 $\lim\limits_{x \to 1} \dfrac{\sqrt[3]{x} - 1}{x - 1} = \lim\limits_{x \to 1} \dfrac{\sqrt[3]{x} - 1}{(\sqrt[3]{x} - 1)(\sqrt[3]{x^2} + \sqrt[3]{x} + 1)} = \lim\limits_{x \to 1} \dfrac{1}{\sqrt[3]{x^2} + \sqrt[3]{x} + 1} = \dfrac{1}{3}.$

2.3.75 $\lim\limits_{x \to 1} \dfrac{x - 1}{\sqrt{x} - 1} = \lim\limits_{x \to 1} \dfrac{(x - 1)(\sqrt{x} + 1)}{(\sqrt{x} - 1)(\sqrt{x} + 1)} = \lim\limits_{x \to 1} \dfrac{(x - 1)(\sqrt{x} + 1)}{x - 1} = \lim\limits_{x \to 1} (\sqrt{x} + 1) = 2.$

2.3.77 $\lim\limits_{x \to 4} \dfrac{3(x-4)\sqrt{x+5}}{3 - \sqrt{x+5}} = \lim\limits_{x \to 4} \dfrac{3(x-4)(\sqrt{x+5})(3 + \sqrt{x+5})}{(3 - \sqrt{x+5})(3 + \sqrt{x+5})} = \lim\limits_{x \to 4} \dfrac{3(x-4)(\sqrt{x+5})(3 + \sqrt{x+5})}{9 - (x+5)} =$

$\lim\limits_{x \to 4} \dfrac{3(x-4)(\sqrt{x+5})(3 + \sqrt{x+5})}{-(x-4)} = \lim\limits_{x \to 4} [-3(\sqrt{x+5})(3 + \sqrt{x+5})] = (-3)(3)(3 + 3) = -54.$

2.3.79 Let $f(x) = x - 1$ and $g(x) = \dfrac{5}{x-1}$. Then $\lim\limits_{x \to 1} f(x) = 0$, $\lim\limits_{x \to 1} f(x)g(x) = \lim\limits_{x \to 1} \dfrac{5(x - 1)}{x - 1} = \lim\limits_{x \to 1} 5 = 5.$

2.3.81 Let $p(x) = x^2 + 2x - 8$. Then $\lim\limits_{x \to 2} \dfrac{p(x)}{x - 2} = \lim\limits_{x \to 2} \dfrac{(x - 2)(x + 4)}{x - 2} = \lim\limits_{x \to 2} (x + 4) = 6.$

The constants are unique. We know that 2 must be a root of p (otherwise the given limit couldn't exist), so it must have the form $p(x) = (x - 2)q(x)$, and q must be a degree 1 polynomial with leading coefficient 1 (otherwise p wouldn't have leading coefficient 1.) So we have $p(x) = (x - 2)(x + d)$, but because $\lim\limits_{x \to 2} \dfrac{p(x)}{x - 2} = \lim\limits_{x \to 2} (x + d) = 2 + d = 6$, we are forced to realize that $d = 4$. Therefore, we have deduced that the only possibility for p is $p(x) = (x - 2)(x + 4) = x^2 + 2x - 8$.

2.3.83 $\lim\limits_{S \to 0^+} r(S) = \lim\limits_{S \to 0^+} (1/2) \left(\sqrt{100 + \dfrac{2S}{\pi}} - 10 \right) = 0.$

The radius of the circular cylinder approaches zero as the surface area approaches zero.

2.3.85 $\lim\limits_{x \to 10} E(x) = \lim\limits_{x \to 10} \dfrac{4.35}{x\sqrt{x^2 + 0.01}} = \dfrac{4.35}{10\sqrt{100.01}} \approx .0435$ N/C.

2.3.87

a. As $x \to 0^+$, $(1 - x) \to 1^-$. So $\lim\limits_{x \to 0^+} g(x) = \lim\limits_{(1-x) \to 1^-} f(1 - x) = \lim\limits_{z \to 1^-} f(z) = 6.$ (Where $z = 1 - x$.)

b. As $x \to 0^-$, $(1 - x) \to 1^+$. So $\lim\limits_{x \to 0^-} g(x) = \lim\limits_{(1-x) \to 1^+} f(1 - x) = \lim\limits_{z \to 1^+} f(z) = 4.$ (Where $z = 1 - x$.)

2.3.89 $\lim\limits_{x \to a} p(x) = \lim\limits_{x \to a} (a_n x^n + a_{n-1} x^{n-1} + \cdots + a_1 x + a_0) = \lim\limits_{x \to a} (a_n x^n) + \lim\limits_{x \to a} (a_{n-1} x^{n-1}) + \cdots + \lim\limits_{x \to a} (a_1 x) +$
$\lim\limits_{x \to a} a_0 = a_n \lim\limits_{x \to a} x^n + a_{n-1} \lim\limits_{x \to a} x^{n-1} + \cdots + a_1 \lim\limits_{x \to a} x + a_0 = a_n (\lim\limits_{x \to a} x)^n + a_{n-1} (\lim\limits_{x \to a} x)^{n-1} + \cdots + a_1 (\lim\limits_{x \to a} x) + a_0 =$
$a_n a^n + a_{n-1} a^{n-1} + \cdots + a_1 a + a_0 = p(a).$

2.4   Infinite Limits

2.4.1

$\lim\limits_{x \to a^+} f(x) = -\infty$ means that when x is very close
to (but a little bigger than) a, the corresponding
values for $f(x)$ are negative numbers whose absolute value is very large.

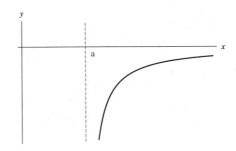

2.4.3 A vertical asymptote for a function f is a vertical line $x = a$ so that one or more of the following are true: $\lim\limits_{x \to a^-} f(x) = \pm\infty$, $\lim\limits_{x \to a^+} f(x) = \pm\infty$.

2.4.5 Because the numerator is approaching a non-zero constant while the denominator is approaching zero, the quotient of these numbers is getting big – at least the absolute value of the quotient is getting big. The quotient is actually always negative, because a number near 100 divided by a negative number is always negative. Thus $\lim\limits_{x \to 2} \dfrac{f(x)}{g(x)} = -\infty.$

2.4.7

x	$\dfrac{x+1}{(x-1)^2}$	x	$\dfrac{x+1}{(x-1)^2}$
1.1	210	.9	190
1.01	20,100	.99	19,900
1.001	2,001,000	.999	1,999,000
1.0001	200,010,000	.9999	199,990,000

From the data given, it appears that $\lim\limits_{x \to 1} f(x) = \infty.$

2.4.9

a. $\lim\limits_{x \to 1^-} f(x) = \infty.$

b. $\lim\limits_{x \to 1^+} f(x) = \infty.$

c. $\lim\limits_{x \to 1} f(x) = \infty.$

d. $\lim\limits_{x \to 2^-} f(x) = \infty.$

e. $\lim\limits_{x \to 2^+} f(x) = -\infty.$

f. $\lim\limits_{x \to 2} f(x)$ does not exist.

2.4.11

a. $\lim\limits_{x \to -2^-} h(x) = -\infty$.

b. $\lim\limits_{x \to -2^+} h(x) = -\infty$.

c. $\lim\limits_{x \to -2} h(x) = -\infty$.

d. $\lim\limits_{x \to 3^-} h(x) = \infty$.

e. $\lim\limits_{x \to 3^+} h(x) = -\infty$.

f. $\lim\limits_{x \to 3} h(x)$ does not exist.

2.4.13

a. $\lim\limits_{x \to 0^-} \dfrac{1}{x^2 - x} = \infty$.

b. $\lim\limits_{x \to 0^+} \dfrac{1}{x^2 - x} = -\infty$.

c. $\lim\limits_{x \to 1^-} \dfrac{1}{x^2 - x} = -\infty$.

d. $\lim\limits_{x \to 1^+} \dfrac{1}{x^2 - x} = \infty$.

2.4.15

2.4.17

a. $\lim\limits_{x \to 2^+} \dfrac{1}{x - 2} = \infty$.

b. $\lim\limits_{x \to 2^-} \dfrac{1}{x - 2} = -\infty$.

c. $\lim\limits_{x \to 2} \dfrac{1}{x - 2}$ does not exist.

2.4.19

a. $\lim\limits_{x \to 4^+} \dfrac{x - 5}{(x - 4)^2} = -\infty$.

b. $\lim\limits_{x \to 4^-} \dfrac{x - 5}{(x - 4)^2} = -\infty$.

c. $\lim\limits_{x \to 4} \dfrac{x - 5}{(x - 4)^2} = -\infty$.

2.4.21

a. $\lim\limits_{x \to 3^+} \dfrac{(x - 1)(x - 2)}{(x - 3)} = \infty$.

b. $\lim\limits_{x \to 3^-} \dfrac{(x - 1)(x - 2)}{(x - 3)} = -\infty$.

c. $\lim\limits_{x \to 3} \dfrac{(x - 1)(x - 2)}{(x - 3)}$ does not exist.

2.4.23

a. $\lim\limits_{x \to 2^+} \dfrac{x^2 - 4x + 3}{(x - 2)^2} = -\infty$.

b. $\lim\limits_{x \to 2^-} \dfrac{x^2 - 4x + 3}{(x - 2)^2} = -\infty$.

c. $\lim\limits_{x \to 2} \dfrac{x^2 - 4x + 3}{(x - 2)^2} = -\infty$.

2.4.25 $\lim\limits_{x \to 0} \dfrac{x^3 - 5x^2}{x^2} = \lim\limits_{x \to 0} \dfrac{x^2(x - 5)}{x^2} = \lim\limits_{x \to 0} (x - 5) = -5$.

2.4.27 $\lim\limits_{x \to 1^+} \dfrac{x^2 - 5x + 6}{x - 1} = \lim\limits_{x \to 1^+} \dfrac{(x - 2)(x - 3)}{x - 1} = \infty$. (Note that as $x \to 1^+$, the numerator is near 2, while the denominator is near zero, but is positive. So the quotient is positive and large.)

2.4.29

a. $\lim\limits_{x \to 5} \dfrac{x-5}{x^2-25} = \lim\limits_{x \to 5} \dfrac{1}{x+5} = \dfrac{1}{10}$, so there isn't a vertical asymptote at $x = 5$.

b. $\lim\limits_{x \to -5^-} \dfrac{x-5}{x^2-25} = \lim\limits_{x \to -5^-} \dfrac{1}{x+5} = -\infty$, so there is a vertical asymptote at $x = -5$.

c. $\lim\limits_{x \to -5^+} \dfrac{x-5}{x^2-25} = \lim\limits_{x \to -5^+} \dfrac{1}{x+5} = \infty$. This also implies that $x = -5$ is a vertical asymptote, as we already noted in part b.

2.4.31 $f(x) = \dfrac{x^2-9x+14}{x^2-5x+6} = \dfrac{(x-2)(x-7)}{(x-2)(x-3)}$. Note that $x = 3$ is a vertical asymptote, while $x = 2$ appears to be a candidate but isn't one. We have $\lim\limits_{x \to 3^+} f(x) = \lim\limits_{x \to 3^+} \dfrac{x-7}{x-3} = -\infty$ and $\lim\limits_{x \to 3^-} f(x) = \lim\limits_{x \to 3^-} \dfrac{x-7}{x-3} = \infty$, and thus $\lim\limits_{x \to 3} f(x)$ doesn't exist. Note that $\lim\limits_{x \to 2} f(x) = 5$.

2.4.33 $f(x) = \dfrac{x+1}{x^3-4x^2+4x} = \dfrac{x+1}{x(x-2)^2}$. There are vertical asymptotes at $x = 0$ and $x = 2$. We have $\lim\limits_{x \to 0^-} f(x) = \lim\limits_{x \to 0^-} \dfrac{x+1}{x(x-2)^2} = -\infty$, while $\lim\limits_{x \to 0^+} f(x) = \lim\limits_{x \to 0^+} \dfrac{x+1}{x(x-2)^2} = \infty$, and thus $\lim\limits_{x \to 0} f(x)$ doesn't exist.

Also we have $\lim\limits_{x \to 2^-} f(x) = \lim\limits_{x \to 2^-} \dfrac{x+1}{x(x-2)^2} = \infty$, while $\lim\limits_{x \to 2^+} f(x) = \lim\limits_{x \to 2^+} \dfrac{x+1}{x(x-2)^2} = \infty$, and thus $\lim\limits_{x \to 2} f(x) = \infty$ as well.

2.4.35 $\lim\limits_{\theta \to 0^+} \csc \theta = \lim\limits_{\theta \to 0^+} \dfrac{1}{\sin \theta} = \infty$.

2.4.37 $\lim\limits_{x \to 0^+} -10 \cot x = \lim\limits_{x \to 0^+} \dfrac{-10 \cos x}{\sin x} = -\infty$. (Note that as $x \to 0^+$, the numerator is near -10 and the denominator is near zero, but is positive. Thus the quotient is a negative number whose absolute value is large.)

2.4.39

a. $\lim\limits_{x \to (\pi/2)^+} \tan x = -\infty$.

b. $\lim\limits_{x \to (\pi/2)^-} \tan x = \infty$.

c. $\lim\limits_{x \to (-\pi/2)^+} \tan x = -\infty$.

d. $\lim\limits_{x \to (-\pi/2)^-} \tan x = \infty$.

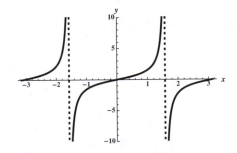

2.4.41

a. False. $\lim\limits_{x \to 1^-} f(x) = \lim\limits_{x \to 1^+} f(x) = \lim\limits_{x \to 1} f(x) = \lim\limits_{x \to 1} \dfrac{(x-1)(x-6)}{(x-1)(x+1)} = \dfrac{-5}{2}$.

b. True. For example, $\lim\limits_{x \to -1^+} f(x) = \lim\limits_{x \to -1^+} \dfrac{(x-1)(x-6)}{(x-1)(x+1)} = -\infty$.

c. False. For example $g(x) = \dfrac{1}{x-1}$ has $\lim\limits_{x \to 1^+} g(x) = \infty$, but $\lim\limits_{x \to 1^-} g(x) = -\infty$.

2.4.43 One example is $f(x) = \dfrac{1}{x-6}$.

2.4.45 $f(x) = \frac{x^2-3x+2}{x^{10}-x^9} = \frac{(x-2)(x-1)}{x^9(x-1)}$. f has a vertical asymptote at $x = 0$, because $\lim\limits_{x\to 0^+} f(x) = -\infty$ (and $\lim\limits_{x\to 0^-} f(x) = \infty$.) Note that $\lim\limits_{x\to 1} f(x) = -1$, so there isn't a vertical asymptote at $x = 1$.

2.4.47 $h(x) = \dfrac{\cos x}{(x+1)^3}$ has a vertical asymptote at $x = -1$, because $\lim\limits_{x\to -1^+} \dfrac{\cos x}{(x+1)^3} = \infty$, and because $\lim\limits_{x\to -1^-} h(x) = -\infty$.

2.4.49 $g(\theta) = \tan(\pi\theta/10) = \frac{\sin(\pi\theta/10)}{\cos(\pi\theta/10)}$ has a vertical asymptote at each $\theta = 10n + 5$ where n is an integer. This is due to the fact that $\cos(\pi\theta/10) = 0$ when $\pi\theta/10 = \pi/2 + n\pi$ where n is an integer, which is the same as $\{\theta : \theta = 10n + 5, n \text{ an integer}\}$. Note that at all of these numbers which make the denominator zero, the numerator isn't zero.

2.4.51 $f(x) = \frac{1}{\sqrt{x}\sec x} = \frac{\cos x}{\sqrt{x}}$ has a vertical asymptote at $x = 0$.

2.4.53

    a. Note that the numerator of the given expression factors as $(x-3)(x-4)$. So if $a = 3$ or if $a = 4$ the limit would be a finite number. In fact, $\lim\limits_{x\to 3} \dfrac{(x-3)(x-4)}{x-3} = -1$ and $\lim\limits_{x\to 4} \dfrac{(x-3)(x-4)}{x-4} = 1$.

    b. For any number other than 3 or 4, the limit would be either $\pm\infty$. Because $x - a$ is always positive as $x \to a^+$, the limit would be $+\infty$ exactly when the numerator is positive, which is for a in the set $(-\infty, 3) \cup (4, \infty)$.

    c. The limit would be $-\infty$ for a in the set $(3, 4)$.

2.4.55

    a. The slope of the secant line is $\frac{f(h)-f(0)}{h} = \frac{h^{2/3}}{h} = h^{-1/3}$.

    b. $\lim\limits_{h\to 0^+} \dfrac{1}{h^{1/3}} = \infty$, and $\lim\limits_{h\to 0^-} \dfrac{1}{h^{1/3}} = -\infty$. The tangent line is infinitely steep at the origin (i.e., it is a vertical line.)

2.5 Limits at Infinity

2.5.1

As $x < 0$ becomes large in absolute value, the corresponding values of f level off near 10.

2.5.3 If $f(x) \to 100,000$ as $x \to \infty$ and $g(x) \to \infty$ as $x \to \infty$, then the ratio $\frac{f(x)}{g(x)} \to 0$ as $x \to \infty$. (Because *eventually* the values of f are small compared to the values of g.)

2.5.5 $\lim\limits_{x\to\infty} (-2x^3) = -\infty$, and $\lim\limits_{x\to -\infty} (-2x^3) = \infty$.

2.5.7
$$\lim_{x\to\infty} \frac{1-x}{2x} = -\frac{1}{2}, \quad \lim_{x\to\infty} \frac{1-x}{x^2} = 0, \quad \lim_{x\to\infty} \frac{1-x^2}{2x} = -\infty$$

2.5.9 $\lim\limits_{x\to\infty}(3+10/x^2)=3+\lim\limits_{x\to\infty}(10/x^2)=3+0=3.$

2.5.11 $\lim\limits_{\theta\to\infty}\dfrac{\cos\theta}{\theta^2}=0.$ Note that $-1\le\cos\theta\le1$, so $-\dfrac{1}{\theta^2}\le\dfrac{\cos\theta}{\theta^2}\le\dfrac{1}{\theta^2}$. The result now follows from the squeeze theorem.

2.5.13 $\lim\limits_{x\to\infty}\dfrac{\cos x^5}{\sqrt{x}}=0.$ Note that $-1\le\cos x^5\le1$, so $\dfrac{-1}{\sqrt{x}}\le\dfrac{\cos x^5}{\sqrt{x}}\le\dfrac{1}{\sqrt{x}}$. Because $\lim\limits_{x\to\infty}\dfrac{1}{\sqrt{x}}=\lim\limits_{x\to\infty}\dfrac{-1}{\sqrt{x}}=$
0, we have $\lim\limits_{x\to\infty}\dfrac{\cos x^5}{\sqrt{x}}=0$ by the squeeze theorem.

2.5.15 $\lim\limits_{x\to\infty}x^{12}=\infty.$ Note that x^{12} is positive when $x>0$.

2.5.17 $\lim\limits_{x\to\infty}x^{-6}=\lim\limits_{x\to\infty}\dfrac{1}{x^6}=0.$

2.5.19 $\lim\limits_{x\to\infty}(3x^{12}-9x^7)=\infty.$

2.5.21 $\lim\limits_{x\to-\infty}(-3x^{16}+2)=-\infty.$

2.5.23 $\lim\limits_{x\to\infty}(-12x^{-5})=\lim\limits_{x\to\infty}-\dfrac{12}{x^5}=0.$

2.5.25 $\lim\limits_{x\to\infty}\dfrac{4x}{20x+1}=\lim\limits_{x\to\infty}\dfrac{4x}{20x+1}\cdot\dfrac{1/x}{1/x}=\lim\limits_{x\to\infty}\dfrac{4}{20+1/x}=\dfrac{4}{20}=\dfrac{1}{5}.$ Thus, the line $y=\frac{1}{5}$ is a horizontal asymptote.
$\lim\limits_{x\to-\infty}\dfrac{4x}{20x+1}=\lim\limits_{x\to-\infty}\dfrac{4x}{20x+1}\cdot\dfrac{1/x}{1/x}=\lim\limits_{x\to-\infty}\dfrac{4}{20+1/x}=\dfrac{4}{20}=\dfrac{1}{5}.$ This shows that the curve is also asymptotic to the asymptote in the negative direction.

2.5.27 $\lim\limits_{x\to\infty}\dfrac{(6x^2-9x+8)}{(3x^2+2)}\cdot\dfrac{1/x^2}{1/x^2}=\lim\limits_{x\to\infty}\dfrac{6-9/x+8/x^2}{3+2/x^2}=\dfrac{6-0+0}{3+0}=2.$ Similarly $\lim\limits_{x\to-\infty}f(x)=2.$ The line $y=2$ is a horizontal asymptote.

2.5.29 $\lim\limits_{x\to\infty}\dfrac{3x^3-7}{x^4+5x^2}=\lim\limits_{x\to\infty}\dfrac{3x^3-7}{x^4+5x^2}\cdot\dfrac{3/x^4}{1/x^4}=\lim\limits_{x\to\infty}\dfrac{1/x-(7/x^4)}{1+(5/x^2)}=\dfrac{0-0}{1+0}=0.$ Thus, the line $y=0$ (the x-axis) is a horizontal asymptote.
$\lim\limits_{x\to-\infty}\dfrac{3x^3-7}{x^4+5x^2}=\lim\limits_{x\to-\infty}\dfrac{3x^3-7}{x^4+5x^2}\cdot\dfrac{3/x^4}{1/x^4}=\lim\limits_{x\to-\infty}\dfrac{1/x-(7/x^4)}{1+(5/x^2)}=\dfrac{0-0}{1+0}=0.$ Thus, the curve is asymptotic to the x-axis in the negative direction as well.

2.5.31 $\lim\limits_{x\to\infty}\dfrac{(2x+1)}{(3x^4-2)}\cdot\dfrac{1/x^4}{1/x^4}=\lim\limits_{x\to\infty}\dfrac{2/x^3+1/x^4}{3-2/x^4}=\dfrac{0+0}{3-0}=0.$ Similarly $\lim\limits_{x\to-\infty}f(x)=0.$ The line $y=0$ is a horizontal asymptote.

2.5.33 $\lim\limits_{x\to\infty}\dfrac{(40x^5+x^2)}{(16x^4-2x)}\cdot\dfrac{1/x^4}{1/x^4}=\lim\limits_{x\to\infty}\dfrac{40x+1/x^2}{16-2/x^3}=\infty.$ Similarly $\lim\limits_{x\to-\infty}f(x)=-\infty.$ There are no horizontal asymptotes.

2.5.35

　　a. $f(x)=\dfrac{x^2-3}{x+6}=x-6+\dfrac{33}{x+6}.$ The oblique asymptote of f is $y=x-6.$

Because $\lim\limits_{x \to -6^+} f(x) = \infty$, there is a verti-

b. cal asymptote at $x = -6$. Note also that $\lim\limits_{x \to -6^-} f(x) = -\infty$.

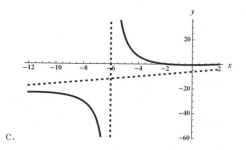

c.

2.5.37

a. $f(x) = \frac{x^2 - 2x + 5}{3x - 2} = (1/3)x - 4/9 + \frac{37}{9(3x - 2)}$. The oblique asymptote of f is $y = (1/3)x - 4/9$.

Because $\lim\limits_{x \to (2/3)^+} f(x) = \infty$, there is a verti-

b. cal asymptote at $x = 2/3$. Note also that $\lim\limits_{x \to (2/3)^-} f(x) = -\infty$.

c.

2.5.39

a. $f(x) = \frac{4x^3 + 4x^2 + 7x + 4}{1 + x^2} = 4x + 4 + \frac{3x}{1 + x^2}$. The oblique asymptote of f is $y = 4x + 4$.

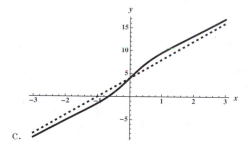

b. There are no vertical asymptotes.

c.

2.5.41 First note that $\sqrt{x^6} = x^3$ if $x > 0$, but $\sqrt{x^6} = -x^3$ if $x < 0$. We have $\lim\limits_{x \to \infty} \frac{4x^3 + 1}{(2x^3 + \sqrt{16x^6 + 1})} \cdot \frac{1/x^3}{1/x^3} =$

$\lim\limits_{x \to \infty} \frac{4 + 1/x^3}{2 + \sqrt{16 + 1/x^6}} = \frac{4 + 0}{2 + \sqrt{16 + 0}} = \frac{2}{3}.$

However, $\lim\limits_{x \to -\infty} \frac{4x^3 + 1}{(2x^3 + \sqrt{16x^6 + 1})} \cdot \frac{1/x^3}{1/x^3} = \lim\limits_{x \to -\infty} \frac{4 + 1/x^3}{2 - \sqrt{16 + 1/x^6}} = \frac{4 + 0}{2 - \sqrt{16 + 0}} = \frac{4}{-2} = -2.$

So $y = \frac{2}{3}$ is a horizontal asymptote (as $x \to \infty$) and $y = -2$ is a horizontal asymptote (as $x \to -\infty$).

2.5.43 First note that $\sqrt[3]{x^6} = x^2$ and $\sqrt{x^4} = x^2$ for all x (even when $x < 0$.) We have $\lim\limits_{x \to \infty} \frac{\sqrt[3]{x^6 + 8}}{(4x^2 + \sqrt{3x^4 + 1})} \cdot$

$\frac{1/x^2}{1/x^2} = \lim\limits_{x \to \infty} \frac{\sqrt[3]{1 + 8/x^6}}{4 + \sqrt{3 + 1/x^4}} = \frac{1}{4 + \sqrt{3 + 0}} = \frac{1}{4 + \sqrt{3}}.$

The calculation as $x \to -\infty$ is similar. So $y = \frac{1}{4 + \sqrt{3}}$ is a horizontal asymptote.

2.5.45

a. False. For example, the function $y = \frac{\sin x}{x}$ on the domain $[1, \infty)$ has a horizontal asymptote of $y = 0$, and it crosses the x-axis infinitely many times.

b. False. If f is a rational function, and if $\lim_{x \to \infty} f(x) = L \neq 0$, then the degree of the polynomial in the numerator must equal the degree of the polynomial in the denominator. In this case, both $\lim_{x \to \infty} f(x)$ and $\lim_{x \to -\infty} f(x) = \frac{a_n}{b_n}$ where a_n is the leading coefficient of the polynomial in the numerator and b_n is the leading coefficient of the polynomial in the denominator. In the case where $\lim_{x \to \infty} f(x) = 0$, then the degree of the numerator is strictly less than the degree of the denominator. This case holds for $\lim_{x \to -\infty} f(x) = 0$ as well.

c. True. There are only two directions which might lead to horizontal asymptotes: there could be one as $x \to \infty$ and there could be one as $x \to -\infty$, and those are the only possibilities.

2.5.47

a. $\lim_{x \to \infty} \frac{2x^3 + 10x^2 + 12x}{x^3 + 2x^2} \cdot \frac{(1/x^3)}{(1/x^3)} = \lim_{x \to \infty} \frac{2 + 10/x + 12/x^2}{1 + 2/x} = 2$. Similarly, $\lim_{x \to -\infty} f(x) = 2$. Thus, $y = 2$ is a horizontal asymptote.

b. Note that $f(x) = \frac{2x(x+2)(x+3)}{x^2(x+2)}$. So $\lim_{x \to 0^+} f(x) = \lim_{x \to 0^+} \frac{2(x+3)}{x} = \infty$, and similarly, $\lim_{x \to 0^-} f(x) = -\infty$. There is a vertical asymptote at $x = 0$. Note that there is no asymptote at $x = -2$ because $\lim_{x \to -2} f(x) = -1$.

2.5.49

a. We have $\lim_{x \to \infty} \frac{3x^4 + 3x^3 - 36x^2}{x^4 - 25x^2 + 144} \cdot \frac{(1/x^4)}{(1/x^4)} = \lim_{x \to \infty} \frac{3 + 3/x - 36/x^2}{1 - 25/x^2 + 144/x^4} = 3$. Similarly, $\lim_{x \to -\infty} f(x) = 3$. So $y = 3$ is a horizontal asymptote.

b. Note that $f(x) = \frac{3x^2(x+4)(x-3)}{(x+4)(x-4)(x+3)(x-3)}$. Thus, $\lim_{x \to -3^+} f(x) = -\infty$ and $\lim_{x \to -3^-} f(x) = \infty$. Also, $\lim_{x \to 4^-} f(x) = -\infty$ and $\lim_{x \to 4^+} f(x) = \infty$. Thus there are vertical asymptotes at $x = -3$ and $x = 4$.

2.5.51

a. $\lim_{x \to \infty} \frac{x^2 - 9}{x^2 - 3x} \cdot \frac{(1/x^2)}{(1/x^2)} = \lim_{x \to \infty} \frac{1 - 9/x^2}{1 - 3/x} = 1$. A similar result holds as $x \to -\infty$. So $y = 1$ is a horizontal asymptote.

b. Because $\lim_{x \to 0^+} f(x) = \lim_{x \to 0^+} \frac{x+3}{x} = \infty$ and $\lim_{x \to 0^-} f(x) = -\infty$, there is a vertical asymptote at $x = 0$.

2.5.53

a. First note that $f(x) = \frac{\sqrt{x^2+2x+6}-3}{x-1} \cdot \frac{\sqrt{x^2+2x+6}+3}{\sqrt{x^2+2x+6}+3} = \frac{x^2+2x+6-9}{(x-1)(\sqrt{x^2+2x+6}+3)} = \frac{(x-1)(x+3)}{(x-1)(\sqrt{x^2+2x+6}+3)}$.

Thus

$$\lim_{x \to \infty} f(x) = \lim_{x \to \infty} \frac{x+3}{\sqrt{x^2+2x+6}+3} \cdot \frac{1/x}{1/x} = \lim_{x \to \infty} \frac{1+3/x}{\sqrt{1+2/x+6/x^2}+3/x} = 1.$$

Using the fact that $\sqrt{x^2} = -x$ for $x < 0$, we have $\displaystyle\lim_{x \to -\infty} f(x) = -1$. Thus the lines $y = 1$ and $y = -1$ are horizontal asymptotes.

b. f has no vertical asymptotes.

2.5.55

a. Note that when $x > 1$, we have $|x| = x$ and $|x - 1| = x - 1$. Thus

$$f(x) = (\sqrt{x} - \sqrt{x-1}) \cdot \frac{\sqrt{x} + \sqrt{x-1}}{\sqrt{x} + \sqrt{x-1}} = \frac{1}{\sqrt{x} + \sqrt{x-1}}.$$

Thus $\displaystyle\lim_{x \to \infty} f(x) = 0$.

When $x < 0$, we have $|x| = -x$ and $|x - 1| = 1 - x$. Thus

$$f(x) = (\sqrt{-x} - \sqrt{1-x}) \cdot \frac{\sqrt{-x} + \sqrt{1-x}}{\sqrt{-x} + \sqrt{1-x}} = -\frac{1}{\sqrt{-x} + \sqrt{1-x}}.$$

Thus, $\displaystyle\lim_{x \to -\infty} f(x) = 0$. There is a horizontal asymptote at $y = 0$.

b. f has no vertical asymptotes.

2.5.57

One possible such graph is:

2.5.59 $\displaystyle\lim_{x \to 0^+} \frac{\cos x + 2\sqrt{x}}{\sqrt{x}} = \infty$. $\displaystyle\lim_{x \to \infty} \frac{\cos x + 2\sqrt{x}}{\sqrt{x}} = \lim_{x \to \infty} \left(2 + \frac{\cos x}{\sqrt{x}}\right) = 2$. There is a vertical asymptote at $x = 0$ and a horizontal asymptote at $y = 2$.

2.5.61 $\displaystyle\lim_{t \to \infty} p(t) = \lim_{t \to \infty} \frac{3500t}{t + 1} = 3500$. The steady state exists. The steady state value is 3500.

2.5.63 $\displaystyle\lim_{t \to \infty} a(t) = \lim_{t \to \infty} 2\left(\frac{t + \sin t}{t}\right) = \lim_{t \to \infty} 2\left(1 + \frac{\sin t}{t}\right) = 2$. The steady state exists. The steady state value is 2.

2.5.65 $\displaystyle\lim_{n \to \infty} f(n) = \lim_{n \to \infty} \frac{n - 1}{n} = \lim_{n \to \infty} [1 - (1/n)] = 1$.

2.5.67 $\displaystyle\lim_{n \to \infty} f(n) = \lim_{n \to \infty} \frac{n + 1}{n^2} = \lim_{n \to \infty} [1/n + 1/n^2] = 0$.

2.5.69

a. No. If $m = n$, there will be a horizontal asymptote, and if $m = n + 1$, there will be a slant asymptote.

b. Yes. For example, $f(x) = \frac{x^4}{\sqrt{x^6 + 1}}$ has a slant asymptote of $y = x$ as $x \to \infty$ and a slant asymptote of $y = -x$ as $x \to -\infty$.

2.6 Continuity

2.6.1

    a. $a(t)$ is a continuous function during the time period from when she jumps from the plane and when she touches down on the ground, because her position is changing continuously with time.

    b. $n(t)$ is not a continuous function of time. The function "jumps" at the times when a quarter must be added.

    c. $T(t)$ is a continuous function, because temperature varies continuously with time.

    d. $p(t)$ is not continuous – it jumps by whole numbers when a player scores a point.

2.6.3 A function f is continuous on an interval I if it is continuous at all points in the interior of I, and it must be continuous from the right at the left endpoint (if the left endpoint is included in I) and it must be continuous from the left at the right endpoint (if the right endpoint is included in I.)

2.6.5

    a. A function f is continuous from the left at $x = a$ if a is in the domain of f, and $\lim\limits_{x \to a^-} f(x) = f(a)$.

    b. A function f is continuous from the right at $x = a$ if a is in the domain of f, and $\lim\limits_{x \to a^+} f(x) = f(a)$.

2.6.7 The domain of $f(x) = \sqrt{1 - x^2}$ is the set of real numbers x so that $1 - x^2 \geq 0$, i.e. $[-1, 1]$. The function is continuous on its domain.

2.6.9 f is discontinuous at $x = 1$, at $x = 2$, and at $x = 3$. At $x = 1$, $f(1)$ does not exist (so the first condition is violated). At $x = 2$, $f(2)$ exists and $\lim\limits_{x \to 2} f(x)$ exists, but $\lim\limits_{x \to 2} f(x) \neq f(2)$ (so condition 3 is violated). At $x = 3$, $\lim\limits_{x \to 3} f(x)$ does not exist (so condition 2 is violated).

2.6.11 f is discontinuous at $x = 1$, at $x = 2$, and at $x = 3$. At $x = 1$, $\lim\limits_{x \to 1} f(x)$ does not exist, and $f(1)$ does not exist (so conditions 1 and 2 are violated). At $x = 2$, $\lim\limits_{x \to 2} f(x)$ does not exist (so condition 2 is violated). At $x = 3$, $f(3)$ does not exist (so condition 1 is violated).

2.6.13 The function is defined at 5, in fact $f(5) = \frac{50 + 15 + 1}{25 + 25} = \frac{66}{50} = \frac{33}{25}$. Also, $\lim\limits_{x \to 5} f(x) = \lim\limits_{x \to 5} \frac{2x^2 + 3x + 1}{x^2 + 5x} = \frac{33}{25} = f(5)$. The function is continuous at $a = 5$.

2.6.15 f is discontinuous at 1, because 1 is not in the domain of f.

2.6.17 f is discontinuous at 1, because $\lim\limits_{x \to 1} f(x) \neq f(1)$. In fact, $f(1) = 3$, but $\lim\limits_{x \to 1} f(x) = 2$.

2.6.19 f is discontinuous at 4, because 4 is not in the domain of f.

2.6.21 Because f is a polynomial, it is continuous on all of $\mathbb{R}$.

2.6.23 Because f is a rational function, it is continuous on its domain. Its domain is $(-\infty, -3) \cup (-3, 3) \cup (3, \infty)$.

2.6.25 Because f is a rational function, it is continuous on its domain. Its domain is $(-\infty, -2) \cup (-2, 2) \cup (2, \infty)$.

2.6.27 Because $f(x) = \left(x^8 - 3x^6 - 1\right)^{40}$ is a polynomial, it is continuous everywhere, including at 0. Thus $\lim\limits_{x \to 0} f(x) = f(0) = (-1)^{40} = 1$.

2.6.29 Because $f(x) = \left(\frac{x+5}{x+2}\right)^4$ is a rational function, it is continuous at all points in its domain, including at $x = 1$. Thus $\lim\limits_{x \to 1} f(x) = f(1) = 16$.

2.6.31 Because $x^3 - 2x^2 - 8x = x(x^2 - 2x - 8) = x(x - 4)(x + 2)$, we have (as long as $x \neq 4$)

$$\sqrt{\frac{x^3 - 2x^2 - 8x}{x - 4}} = \sqrt{x(x + 2)}.$$

Thus, $\lim\limits_{x \to 4} \sqrt{\frac{x^3 - 2x^2 - 8x}{x - 4}} = \lim\limits_{x \to 4} \sqrt{x(x + 2)} = \sqrt{24}$, using Theorem 2.11 and the fact that the square root is a continuous function.

2.6.33 Recall that $\lim\limits_{x \to 0} \frac{\sin x}{x} = 1$. Now noting that the function $f(x) = \cos 2x$ is continuous at 1, we have by Theorem 2.11 that $\lim\limits_{x \to 0} \cos \left(\frac{2 \sin x}{x}\right) = \cos \left(2 \left(\lim\limits_{x \to 0} \frac{\sin x}{x}\right)\right) = \cos(2 \cdot 1) = \cos 2$.

2.6.35 f is continuous on $[0, 1)$, on $(1, 2)$, on $(2, 3]$, and on $(3, 4]$.

2.6.37 f is continuous on $[0, 1)$, on $(1, 2)$, on $[2, 3)$, and on $(3, 5]$.

2.6.39

a. f is defined at 1. We have $f(1) = 1^2 + (3)(1) = 4$. To see whether or not $\lim\limits_{x \to 1} f(x)$ exists, we investigate the two one-sided limits. $\lim\limits_{x \to 1^-} f(x) = \lim\limits_{x \to 1^-} 2x = 2$, and $\lim\limits_{x \to 1^+} f(x) = \lim\limits_{x \to 1^+} (x^2 + 3x) = 4$, so $\lim\limits_{x \to 1} f(x)$ does not exist. Thus f is discontinuous at $x = 1$.

b. f is continuous from the right, because $\lim\limits_{x \to 1^+} f(x) = 4 = f(1)$.

c. f is continuous on $(-\infty, 1)$ and on $[1, \infty)$.

2.6.41 f is continuous on $(-\infty, -\sqrt{8}]$ and on $[\sqrt{8}, \infty)$.

2.6.43 Because f is the composition of two functions which are continuous everywhere, it is continuous everywhere.

2.6.45 Because f is the composition of two functions which are continuous everywhere, it is continuous everywhere.

2.6.47 $\lim\limits_{x \to 2} \sqrt{\frac{4x + 10}{2x - 2}} = \sqrt{\frac{18}{2}} = 3$.

2.6.49 $\lim\limits_{x \to 3} \sqrt{x^2 + 7} = \sqrt{9 + 7} = 4$.

2.6.51 $f(x) = \csc x$ isn't defined at $x = k\pi$ where k is an integer, so it isn't continuous at those points. So it is continuous on intervals of the form $(k\pi, (k+1)\pi)$ where k is an integer. $\lim\limits_{x \to \pi/4} \csc x = \sqrt{2}$. $\lim\limits_{x \to 2\pi^-} \csc x = -\infty$.

2.6.53 f isn't defined for any number of the form $\pi/2 + k\pi$ where k is an integer, so it isn't continuous there. It is continuous on intervals of the form $(\pi/2 + k\pi, \pi/2 + (k + 1)\pi)$, where k is an integer.
$$\lim\limits_{x \to \pi/2^-} f(x) = \infty. \quad \lim\limits_{x \to 4\pi/3} f(x) = \frac{1 - \sqrt{3}/2}{-1/2} = \sqrt{3} - 2.$$

2.6.55

a. Because A is a continuous function of r on $[0, .08]$, and because $A(0) = 5000$ and $A(.08) \approx 11098.2$, (and 7000 is an intermediate value between these two numbers) the Intermediate Value Theorem guarantees a value of r between 0 and .08 where $A(r) = 7000$.

b. Solving $5000(1 + (r/12))^{120} = 7000$ for r, we see that $(1 + (r/12))^{120} = 7/5$, so $1 + r/12 = \sqrt[120]{7/5}$, so $r = 12(\sqrt[120]{7/5} - 1) \approx 0.034$.

2.6.57

a. Note that $f(x) = 2x^3 + x - 2$ is continuous everywhere, so in particular it is continuous on $[-1, 1]$. Note that $f(-1) = -5 < 0$ and $f(1) = 1 > 0$. Because 0 is an intermediate value between $f(-1)$ and $f(1)$, the Intermediate Value Theorem guarantees a number c between -1 and 1 where $f(c) = 0$.

b. Using a graphing calculator and a computer algebra system, we see that the root of f is about 0.835.

c.

2.6.59

a. Note that $f(x) = x^3 - 5x^2 + 2x$ is continuous everywhere, so in particular it is continuous on $[-1, 5]$. Note that $f(-1) = -8 < -1$ and $f(5) = 10 > -1$. Because -1 is an intermediate value between $f(-1)$ and $f(5)$, the Intermediate Value Theorem guarantees a number c between -1 and 5 where $f(c) = -1$.

b. Using a graphing calculator and a computer algebra system, we see that there are actually three different values of c between -1 and 5 for which $f(c) = -1$. They are $c \approx -0.285$, $c \approx 0.778$, and $c \approx 4.507$.

c.

2.6.61

a. Note that $f(x) = \cos x - x$ is continuous on its domain, so in particular it is continuous on $[0, \pi/2]$. Note that $f(0) = 1 > 0$ and $f(\pi/2) = -\pi/2 < 0$. Because 0 is an intermediate value between $f(0)$ and $f(\pi/2)$, the Intermediate Value Theorem guarantees a number c between 0 and $\pi/2$ where $f(c) = 0$.

b. Using a graphing calculator and a computer algebra system, we see that the value of c guaranteed by the theorem is about 0.739085.

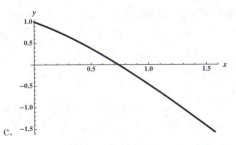

c.

2.6.63

a. True. If f is right continuous at a, then $f(a)$ exists and the limit from the right at a exists and is equal to $f(a)$. Because it is left continuous, the limit from the left exists — so we now know that the limit as $x \to a$ of $f(x)$ exists, because the two one-sided limits are both equal to $f(a)$.

b. True. If $\lim_{x \to a} f(x) = f(a)$, then $\lim_{x \to a^+} f(x) = f(a)$ and $\lim_{x \to a^-} f(x) = f(a)$.

c. False. The statement would be true if f were continuous. However, if f isn't continuous, then the statement doesn't hold. For example, suppose that $f(x) = \begin{cases} 0 & \text{if } 0 \le x < 1; \\ 1 & \text{if } 1 \le x \le 2, \end{cases}$ Note that $f(0) = 0$ and $f(2) = 1$, but there is no number c between 0 and 2 where $f(c) = 1/2$.

d. False. Consider $f(x) = x^2$ and $a = -1$ and $b = 1$. Then f is continuous on $[a, b]$, but $\frac{f(1) + f(-1)}{2} = 1$, and there is no c on (a, b) with $f(c) = 1$.

2.6.65 Because $f(x) = x^3 + 3x - 18$ is a polynomial, it is continuous on $(-\infty, \infty)$, and because the absolute value function is continuous everywhere, $|f(x)|$ is continuous everywhere.

2.6.67 Let $f(x) = \frac{1}{\sqrt{x}-4}$. Then f is continuous on $[0, 16) \cup (16, \infty)$. So $h(x) = |f(x)|$ is continuous on this set as well.

2.6.69 $\lim_{x \to \pi} \dfrac{\cos^2 x + 3\cos x + 2}{\cos x + 1} = \lim_{x \to \pi} \dfrac{(\cos x + 1)(\cos x + 2)}{\cos x + 1} = \lim_{x \to \pi} (\cos x + 2) = 1.$

2.6.71 $\lim_{x \to \pi/2} \dfrac{\sin x - 1}{\sqrt{\sin x} - 1} = \lim_{x \to \pi/2} (\sqrt{\sin x} + 1) = 2.$

2.6.73 $\lim_{x \to 0} \dfrac{\cos x - 1}{\sin^2 x} = \lim_{x \to 0} \dfrac{\cos x - 1}{1 - \cos^2 x} = \lim_{x \to 0} \dfrac{\cos x - 1}{(1 - \cos x)(1 + \cos x)} = \lim_{x \to 0} -\dfrac{1}{1 + \cos x} = -\dfrac{1}{2}.$

2.6.75 Because $-1 \le \sin t \le 1$ for all t, we have $\frac{-1}{t^2} \le \frac{\sin t}{t^2} \le \frac{1}{t^2}$, so by the Squeeze Theorem, $\lim_{t \to \infty} \dfrac{\sin t}{t^2} = 0.$

2.6.77

The graph shown isn't drawn correctly at the integers. At an integer a, the value of the function is 0, whereas the graph shown appears to take on all the values from 0 to 1.
Note that in the correct graph, $\lim_{x \to a^-} f(x) = 1$ and $\lim_{x \to a^+} f(x) = 0$ for every integer a.

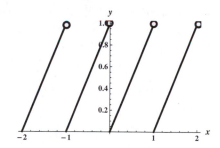

2.6.79 With slight modifications, we can use the examples from the previous two problems.

a. The function $y = x - \lfloor x \rfloor$ is defined at $x = 1$ but isn't continuous there.

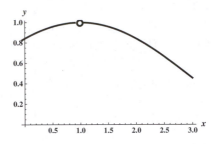

b. The function $y = \frac{\sin(x-1)}{x-1}$ has a limit at $x = 1$, but isn't defined there, so isn't continuous there.

2.6.81

a. In order for g to be continuous from the left at $x = 1$, we must have $\lim\limits_{x\to 1^-} g(x) = g(1) = a$. We have $\lim\limits_{x\to 1^-} g(x) = \lim\limits_{x\to 1^-} (x^2 + x) = 2$. So we must have $a = 2$.

b. In order for g to be continuous from the right at $x = 1$, we must have $\lim\limits_{x\to 1^+} g(x) = g(1) = a$. We have $\lim\limits_{x\to 1^+} g(x) = \lim\limits_{x\to 1^+} (3x + 5) = 8$. So we must have $a = 8$.

c. Because the limit from the left and the limit from the right at $x = 1$ don't agree, there is no value of a which will make the function continuous at $x = 1$.

2.6.83 Let $f(x) = 70x^3 - 87x^2 + 32x - 3$. Note that $f(0) < 0$, $f(.2) > 0$, $f(.55) < 0$, and $f(1) > 0$. Because the given polynomial is continuous everywhere, the Intermediate Value Theorem guarantees us a root on $(0, .2)$, at least one on $(.2, .55)$, and at least one on $(.55, 1)$. Because there can be at most 3 roots and there are at least 3 roots, there must be exactly 3 roots. The roots are $x_1 = 1/7$, $x_2 = 1/2$ and $x_3 = 3/5$.

2.6.85

a. Note that $A(.01) \approx 2615.55$ and $A(.1) \approx 3984.36$. By the Intermediate Value Theorem, there must be a number r_0 between .01 and .1 so that $A(r_0) = 3500$.

b. The desired value is $r_0 \approx 0.0728$ or 7.28%.

2.6.87 We can argue essentially like the previous problem, or we can imagine an identical twin to the original monk, who takes an identical version of the original monk's journey up the winding path while the monk is taking the return journey down. Because they must pass somewhere on the path, that point is the one we are looking for.

2.6.89 The discontinuity is not removable, because $\lim\limits_{x\to a} f(x)$ does not exist. The discontinuity pictured is a jump discontinuity.

2.6.91 Note that $\lim\limits_{x\to 2} \dfrac{x^2 - 7x + 10}{x - 2} = \lim\limits_{x\to 2} \dfrac{(x-2)(x-5)}{x-2} = \lim\limits_{x\to 2}(x - 5) = -3$. Because this limit exists, the discontinuity is removable.

2.6.93

a. Note that $-1 \le \sin(1/x) \le 1$ for all $x \ne 0$, so $-x \le x\sin(1/x) \le x$ (for $x > 0$. For $x < 0$ we would have $x \le x\sin(1/x) \le -x$.) Because both $x \to 0$ and $-x \to 0$ as $x \to 0$, the Squeeze Theorem tells us that $\lim\limits_{x\to 0} x\sin(1/x) = 0$ as well. Because this limit exists, the discontinuity is removable.

b. Note that as $x \to 0^+$, $1/x \to \infty$, and thus $\lim\limits_{x\to 0^+} \sin(1/x)$ does not exist. So the discontinuity is not removable.

2.6.95 Note that $h(x) = \frac{x^3 - 4x^2 + 4x}{x(x-1)} = \frac{x(x-2)^2}{x(x-1)}$. Thus $\lim\limits_{x\to 0} h(x) = -4$, and the discontinuity at $x = 0$ is removable. However, $\lim\limits_{x\to 1} h(x)$ does not exist, and the discontinuity at $x = 1$ is not removable (it is infinite.)

2.6.97

    a. Consider $g(x) = x + 1$ and $f(x) = \frac{|x-1|}{x-1}$. Note that both g and f are continuous at $x = 0$. However $f(g(x)) = f(x+1) = \frac{|x|}{x}$ is not continuous at 0.

    b. The previous theorem says that the composition of f and g is continuous at a if g is continuous at a and f is continuous at $g(a)$. It does not say that if g and f are both continuous at a that the composition is continuous at a.

2.6.99

    a. Using the hint, we have

$$\sin x = \sin(a + (x - a)) = \sin a \cos(x - a) + \sin(x - a) \cos a.$$

    Note that as $x \to a$, we have that $\cos(x - a) \to 1$ and $\sin(x - a) \to 0$.

    So,

$$\lim_{x \to a} \sin x = \lim_{x \to a} \sin(a + (x - a)) = \lim_{x \to a} (\sin a \cos(x - a) + \sin(x - a) \cos a) = (\sin a) \cdot 1 + 0 \cdot \cos a = \sin a.$$

    b. Using the hint, we have

$$\cos x = \cos(a + (x - a)) = \cos a \cos(x - a) - \sin a \sin(x - a).$$

    So,

$$\lim_{x \to a} \cos x = \lim_{x \to a} \cos(a + (x - a)) = \lim_{x \to a} ((\cos a) \cos(x - a) - (\sin a) \sin(x - a)) = (\cos a) \cdot 1 - (\sin a) \cdot 0 = \cos a.$$

2.7   Precise Definitions of Limits

2.7.1 Note that all the numbers in the interval $(1, 3)$ are within 1 unit of the number 2. So $|x - 2| < 1$ is true for all numbers in that interval. In fact, $\{x \colon 0 < |x - 2| < 1\}$ is exactly the set $(1, 3)$ with $x \neq 2$.

2.7.3

    $(3, 8)$ has center 5.5, so it is not symmetric about the number 5.
    $(1, 9)$ and $(4, 6)$ and $(4.5, 5.5)$ are symmetric about the number 5.

2.7.5 $\lim_{x \to a} f(x) = L$ if for any arbitrarily small positive number ϵ, there exists a number δ, so that $f(x)$ is within ϵ units of L for any number x within δ units of a (but not including a itself).

2.7.7 We are given that $|f(x) - 5| < .1$ for values of x in the interval $(0, 5)$, so we need to ensure that the set of x values we are allowing fall in this interval.

    Note that the number 0 is two units away from the number 2 and the number 5 is three units away from the number 2. In order to be sure that we are talking about numbers in the interval $(0, 5)$ when we write $|x - 2| < \delta$, we would need to have $\delta = 2$ (or a number less than 2). In fact, the set of numbers for which $|x - 2| < 2$ is the interval $(0, 4)$ which is a subset of $(0, 5)$.

    If we were to allow δ to be any number greater than 2, then the set of all x so that $|x - 2| < \delta$ would include numbers less than 0, and those numbers aren't on the interval $(0, 5)$.

2.7.9

    a. In order for f to be within 2 units of 5, it appears that we need x to be within 1 unit of 2. So $\delta = 1$.

    b. In order for f to be within 1 unit of 5, it appears that we would need x to be within 1/2 unit of 2. So $\delta = .5$.

2.7.11

    a. In order for f to be within 3 units of 6, it appears that we would need x to be within 2 units of 3. So $\delta = 2$.

    b. In order for f to be within 1 unit of 6, it appears that we would need x to be within 1/2 unit of 3. So $\delta = 1/2$.

2.7.13

    a. If $\epsilon = 1$, we need $|x^3 + 3 - 3| < 1$. So we need $|x| < \sqrt[3]{1} = 1$ in order for this to happen. Thus $\delta = 1$ will suffice, or any δ so that $0 < \delta \leq 1$.

    b. If $\epsilon = .5$, we need $|x^3 + 3 - 3| < .5$. So we need $|x| < \sqrt[3]{.5}$ in order for this to happen. Thus $\delta = \sqrt[3]{.5} \approx .079$ will suffice, or any δ so that $0 < \delta \leq 0.79$.

2.7.15

    a. For $\epsilon = 1$, the required value of δ would also be 1. A larger value of δ would work to the right of 2, but this is the largest one that would work to the left of 2. So we require $0 < \delta \leq 1$.

    b. For $\epsilon = 1/2$, the required value of δ would also be 1/2, so we require $0 < \delta \leq 1/2$.

    c. It appears that for a given value of ϵ with $0 < \epsilon < 2$, it would be wise to let δ satisfy $0 < \delta \leq \epsilon$.

2.7.17

    a. For $\epsilon = 2$, it appears that a value of $\delta = 1$ (or smaller) would work.

    b. For $\epsilon = 1$, it appears that a value of $\delta = 1/2$ (or smaller) would work.

    c. For an arbitrary ϵ, a value of $\delta = \epsilon/2$ or smaller appears to suffice.

2.7.19 For any $\epsilon > 0$, let $\delta = \epsilon/8$. Then if $0 < |x - 1| < \delta$, we would have $|x - 1| < \epsilon/8$. Then $|8x - 8| < \epsilon$, so $|(8x + 5) - 13| < \epsilon$. This last inequality has the form $|f(x) - L| < \epsilon$, which is what we were attempting to show. Thus, $\lim\limits_{x \to 1} (8x + 5) = 13$.

2.7.21 First note that if $x \neq 4$, $f(x) = \frac{x^2 - 16}{x - 4} = x + 4$.

    Now if $\epsilon > 0$ is given, let $\delta = \epsilon$. Now suppose $0 < |x - 4| < \delta$. Then $x \neq 4$, so the function $f(x)$ can be described by $x + 4$. Also, because $|x - 4| < \delta$, we have $|x - 4| < \epsilon$. Thus $|(x + 4) - 8| < \epsilon$. This last inequality has the form $|f(x) - L| < \epsilon$, which is what we were attempting to show. Thus, $\lim\limits_{x \to 4} \dfrac{x^2 - 16}{x - 4} = 8$.

2.7.23 Let $\epsilon > 0$ be given. Let $\delta = \sqrt{\epsilon}$. Then if $0 < |x - 0| < \delta$, we would have $|x| < \sqrt{\epsilon}$. But then $|x^2| < \epsilon$, which has the form $|f(x) - L| < \epsilon$. Thus, $\lim_{x \to 0} f(x) = 0$.

2.7.25 Let $\epsilon > 0$ be given.

Because $\lim_{x \to a} f(x) = L$, we know that there exists a $\delta_1 > 0$ so that $|f(x) - L| < \epsilon/2$ when $0 < |x - a| < \delta_1$. Also, because $\lim_{x \to a} g(x) = M$, there exists a $\delta_2 > 0$ so that $|g(x) - M| < \epsilon/2$ when $0 < |x - a| < \delta_2$.

Now let $\delta = \min(\delta_1, \delta_2)$.

Then if $0 < |x - a| < \delta$, we would have $|f(x) - g(x) - (L - M)| = |(f(x) - L) + (M - g(x))| \leq |f(x) - L| + |M - g(x)| = |f(x) - L| + |g(x) - M| \leq \epsilon/2 + \epsilon/2 = \epsilon$. Note that the key inequality in this sentence follows from the triangle inequality.

2.7.27

a. Let $\epsilon > 0$ be given. It won't end up mattering what δ is, so let $\delta = 1$. Note that the statement $|f(x) - L| < \epsilon$ amounts to $|c - c| < \epsilon$, which is true for any positive number ϵ, without any restrictions on x. So $\lim_{x \to a} c = c$.

b. Let $\epsilon > 0$ be given. Let $\delta = \epsilon$. Note that the statement $|f(x) - L| < \epsilon$ has the form $|x - a| < \epsilon$, which follows whenever $0 < |x - a| < \delta$ (because $\delta = \epsilon$). Thus $\lim_{x \to a} x = a$.

2.7.29 Let $N > 0$ be given. Let $\delta = 1/\sqrt{N}$. Then if $0 < |x - 4| < \delta$, we have $|x - 4| < 1/\sqrt{N}$. Taking the reciprocal of both sides, we have $\frac{1}{|x-4|} > \sqrt{N}$, and squaring both sides of this inequality yields $\frac{1}{(x-4)^2} > N$. Thus $\lim_{x \to 4} f(x) = \infty$.

2.7.31 Let $N > 1$ be given. Let $\delta = 1/\sqrt{N - 1}$. Suppose that $0 < |x - 0| < \delta$. Then $|x| < 1/\sqrt{N - 1}$, and taking the reciprocal of both sides, we see that $1/|x| > \sqrt{N - 1}$. Then squaring both sides yields $\frac{1}{x^2} > N - 1$, so $\frac{1}{x^2} + 1 > N$. Thus $\lim_{x \to 0} f(x) = \infty$.

2.7.33

a. False. In fact, if the statement is true for a specific value of δ_1, then it would be true for any value of $\delta < \delta_1$. This is because if $0 < |x - a| < \delta$, it would automatically follow that $0 < |x - a| < \delta_1$.

b. False. This statement is not equivalent to the definition – note that it says "for an arbitrary δ there exists an ϵ" rather than "for an arbitrary ϵ there exists a δ."

c. True. This is the definition of $\lim_{x \to a} f(x) = L$.

d. True. Both inequalities describe the set of x's which are within δ units of a.

2.7.35 Assume $|x - 3| < 1$, as indicated in the hint. Then $2 < x < 4$, so $\frac{1}{4} < \frac{1}{x} < \frac{1}{2}$, and thus $\left|\frac{1}{x}\right| < \frac{1}{2}$.

Also note that the expression $\left|\frac{1}{x} - \frac{1}{3}\right|$ can be written as $\left|\frac{x-3}{3x}\right|$.

Now let $\epsilon > 0$ be given. Let $\delta = \min(6\epsilon, 1)$. Now assume that $0 < |x - 3| < \delta$. Then

$$|f(x) - L| = \left|\frac{x - 3}{3x}\right| < \left|\frac{x - 3}{6}\right| < \frac{6\epsilon}{6} = \epsilon.$$

Thus we have established that $\left|\frac{1}{x} - \frac{1}{3}\right| < \epsilon$ whenever $0 < |x - 3| < \delta$.

2.7.37 Assume $|x - (1/10)| < (1/20)$, as indicated in the hint. Then $1/20 < x < 3/20$, so $\frac{20}{3} < \frac{1}{x} < \frac{20}{1}$, and thus $\left|\frac{1}{x}\right| < 20$.

Also note that the expression $\left|\frac{1}{x} - 10\right|$ can be written as $\left|\frac{10x-1}{x}\right|$.

Let $\epsilon > 0$ be given. Let $\delta = \min(\epsilon/200, 1/20)$. Now assume that $0 < |x - (1/10)| < \delta$. Then

$$|f(x) - L| = \left|\frac{10x - 1}{x}\right| < |(10x - 1) \cdot 20|$$

$$\leq |x - (1/10)| \cdot 200 < \frac{\epsilon}{200} \cdot 200 = \epsilon.$$

Thus we have established that $\left|\frac{1}{x} - 10\right| < \epsilon$ whenever $0 < |x - (1/10)| < \delta$.

2.7.39 Because we are approaching a from the right, we are only considering values of x which are close to, but a little larger than a. The numbers x to the right of a which are within δ units of a satisfy $0 < x - a < \delta$.

2.7.41

    a. Let $\epsilon > 0$ be given. let $\delta = \epsilon/2$. Suppose that $0 < x < \delta$. Then $0 < x < \epsilon/2$ and

$$|f(x) - L| = |2x - 4 - (-4)| = |2x| = 2|x|$$
$$= 2x < \epsilon.$$

    b. Let $\epsilon > 0$ be given. let $\delta = \epsilon/3$. Suppose that $0 < 0 - x < \delta$. Then $-\delta < x < 0$ and $-\epsilon/3 < x < 0$, so $\epsilon > -3x$. We have

$$|f(x) - L| = |3x - 4 - (-4)| = |3x| = 3|x|$$
$$= -3x < \epsilon.$$

    c. Let $\epsilon > 0$ be given. Let $\delta = \epsilon/3$. Because $\epsilon/3 < \epsilon/2$, we can argue that $|f(x) - L| < \epsilon$ whenever $0 < |x| < \delta$ exactly as in the previous two parts of this problem.

2.7.43 Let $\epsilon > 0$ be given, and let $\delta = \epsilon^2$. Suppose that $0 < x < \delta$, which means that $x < \epsilon^2$, so that $\sqrt{x} < \epsilon$. Then we have

$$|f(x) - L| = |\sqrt{x} - 0| = \sqrt{x} < \epsilon.$$

as desired.

2.7.45

    a. We say that $\lim_{x \to a^+} f(x) = \infty$ if for each positive number N, there exists $\delta > 0$ such that

$$f(x) > N \quad \text{whenever} \quad a < x < a + \delta.$$

    b. We say that $\lim_{x \to a^-} f(x) = -\infty$ if for each negative number N, there exists $\delta > 0$ such that

$$f(x) < N \quad \text{whenever} \quad a - \delta < x < a.$$

    c. We say that $\lim_{x \to a^-} f(x) = \infty$ if for each positive number N, there exists $\delta > 0$ such that

$$f(x) > N \quad \text{whenever} \quad a - \delta < x < a.$$

2.7.47 Let $N > 0$ be given. Let $\delta = 1/N$, and suppose that $1 - \delta < x < 1$. Then $\frac{N-1}{N} < x < 1$, so $\frac{1-N}{N} > -x > -1$, and therefore $1 + \frac{1-N}{N} > 1 - x > 0$, which can be written as $\frac{1}{N} > 1 - x > 0$. Taking reciprocals yields the inequality $N < \frac{1}{1-x}$, as desired.

2.7.49 Let $M < 0$ be given. Let $\delta = \sqrt[4]{-10/M}$. Suppose that $0 < |x + 2| < \delta$. Then $(x + 2)^4 < -10/M$, so $\frac{1}{(x+2)^4} > \frac{M}{-10}$, and $\frac{-10}{(x+2)^4} < M$, as desired.

2.7.51 Let $\epsilon > 0$ be given. Let $N = 1/\epsilon$. Suppose that $x > N$. Then $\frac{1}{x} < \epsilon$, and so $|f(x) - L| = |2 + \frac{1}{x} - 2| < \epsilon$.

2.7.53 Let $M > 0$ be given. Let $N = M - 1$. Suppose that $x > N$. Then $x > M - 1$, so $x + 1 > M$, and thus $\frac{x^2 + x}{x} > M$, as desired.

2.7.55 Let $\epsilon > 0$ be given. Let $N = \lfloor (1/\epsilon) \rfloor + 1$. By assumption, there exists an integer $M > 0$ so that $|f(x) - L| < 1/N$ whenever $|x - a| < 1/M$. Let $\delta = 1/M$.

    Now assume $0 < |x - a| < \delta$. Then $|x - a| < 1/M$, and thus $|f(x) - L| < 1/N$. But then

$$|f(x) - L| < \frac{1}{\lfloor (1/\epsilon) \rfloor + 1} < \epsilon,$$

as desired.

2.7.57 Let $f(x) = \frac{|x|}{x}$ and suppose $\lim\limits_{x \to 0} f(x)$ does exist and is equal to L. Let $\epsilon = 1/2$. There must be a value of δ so that when $0 < |x| < \delta$, $|f(x) - L| < 1/2$. Now consider the numbers $\delta/3$ and $-\delta/3$, both of which are within δ of 0. We have $f(\delta/3) = 1$ and $f(-\delta/3) = -1$. However, it is impossible for both $|1 - L| < 1/2$ and $|-1 - L| < 1/2$, because the former implies that $1/2 < L < 3/2$ and the latter implies that $-3/2 < L < -1/2$. Thus $\lim\limits_{x \to 0} f(x)$ does not exist.

2.7.59 Because f is continuous at a, we know that $\lim\limits_{x \to a} f(x)$ exists and is equal to $f(a) > 0$. Let $\epsilon = f(a)/3$. Then there is a number $\delta > 0$ so that $|f(x) - f(a)| < f(a)/3$ whenever $|x - a| < \delta$. Then whenever x lies in the interval $(a - \delta, a + \delta)$ we have $-f(a)/3 \le f(x) - f(a) \le f(a)/3$, so $2f(a)/3 \le f(x) \le 4f(a)/3$, so f is positive in this interval.

Chapter Two Review

1

a. False. Because $\lim\limits_{x \to 1} \frac{x - 1}{x^2 - 1} = \lim\limits_{x \to 1} \frac{1}{x + 1} = \frac{1}{2}$, f doesn't have a vertical asymptote at $x = 1$.

b. False. In general, these methods are too imprecise to produce accurate results.

c. False. For example, the function $f(x) = \begin{cases} 2x & \text{if } x < 0; \\ 1 & \text{if } x = 0; \\ 4x & \text{if } x > 0 \end{cases}$ has a limit of 0 as $x \to 0$, but $f(0) = 1$.

d. True. When we say that a limit exists, we are saying that there is a real number L that the function is approaching. If the limit of the function is ∞, it is still the case that there is no real number that the function is approaching. (There is no real number called "infinity.")

e. False. It could be the case that $\lim\limits_{x \to a^-} f(x) = 1$ and $\lim\limits_{x \to a^+} f(x) = 2$.

f. False.

g. False. For example, the function $f(x) = \begin{cases} 2 & \text{if } 0 < x < 1; \\ 3 & \text{if } 1 \le x < 2, \end{cases}$ is continuous on $(0, 1)$, and on $[1, 2)$, but isn't continuous on $(0, 2)$.

h. True. $\lim\limits_{x \to a} f(x) = f(a)$ if and only if f is continuous at a.

3 This function is discontinuous at $x = -1$, at $x = 1$, and at $x = 3$. At $x = -1$ it is discontinuous because $\lim\limits_{x \to -1} f(x)$ does not exist. At $x = 1$, it is discontinuous because $\lim\limits_{x \to 1} f(x) \ne f(1)$. At $x = 3$, it is discontinuous because $f(3)$ does not exist, and because $\lim\limits_{x \to 3} f(x)$ does not exist.

5

a.

x	$.9\pi/4$	$.99\pi/4$	$.999\pi/4$	$.9999\pi/4$
$f(x)$	1.4098	1.4142	1.4142	1.4142

x	$1.1\pi/4$	$1.01\pi/4$	$1.001\pi/4$	$1.0001\pi/4$
$f(x)$	1.4098	1.4142	1.4142	1.4142

The limit appears to be approximately 1.4142.

Copyright © 2015 Pearson Education, Inc.

b. $\lim\limits_{x \to \pi/4} \dfrac{\cos 2x}{\cos x - \sin x} = \lim\limits_{x \to \pi/4} \dfrac{\cos^2 x - \sin^2 x}{\cos x - \sin x} = \lim\limits_{x \to \pi/4} (\cos x + \sin x) = \sqrt{2}.$

7

There are infinitely many different correct functions which you could draw. One of them is:

9 $\lim\limits_{x \to 1} \sqrt{5x + 6} = \sqrt{11}.$

11 $\lim\limits_{x \to 1} \dfrac{x^3 - 7x^2 + 12x}{4 - x} = \dfrac{1 - 7 + 12}{4 - 1} = \dfrac{6}{3} = 2.$

13 $\lim\limits_{x \to 1} \dfrac{1 - x^2}{x^2 - 8x + 7} = \lim\limits_{x \to 1} \dfrac{(1 - x)(1 + x)}{(x - 7)(x - 1)} = \lim\limits_{x \to 1} \dfrac{-(x + 1)}{x - 7} = \dfrac{1}{3}.$

15

$$\lim_{x \to 3} \frac{1}{x - 3} \left(\frac{1}{\sqrt{x + 1}} - \frac{1}{2} \right) = \lim_{x \to 3} \frac{2 - \sqrt{x + 1}}{2(x - 3)\sqrt{x + 1}} \cdot \frac{(2 + \sqrt{x + 1})}{(2 + \sqrt{x + 1})}$$

$$= \lim_{x \to 3} \frac{4 - (x + 1)}{2(x - 3)(\sqrt{x + 1})(2 + \sqrt{x + 1})}$$

$$= \lim_{x \to 3} \frac{-(x - 3)}{2(x - 3)(\sqrt{x + 1})(2 + \sqrt{x + 1})}$$

$$= \lim_{x \to 3} -\frac{1}{2\sqrt{x + 1}(2 + \sqrt{x + 1})} = -\frac{1}{16}.$$

17 $\lim\limits_{x \to 3} \dfrac{x^4 - 81}{x - 3} = \lim\limits_{x \to 3} \dfrac{(x - 3)(x + 3)(x^2 + 9)}{x - 3} = \lim\limits_{x \to 3} (x + 3)(x^2 + 9) = 108.$

19 $\lim\limits_{x \to 81} \dfrac{\sqrt[4]{x} - 3}{x - 81} = \lim\limits_{x \to 81} \dfrac{\sqrt[4]{x} - 3}{(\sqrt{x} + 9)(\sqrt[4]{x} + 3)(\sqrt[4]{x} - 3)} = \lim\limits_{x \to 81} \dfrac{1}{(\sqrt{x} + 9)(\sqrt[4]{x} + 3)} = \dfrac{1}{108}.$

21 $\lim\limits_{x \to \pi/2} \dfrac{\frac{1}{\sqrt{\sin x}} - 1}{x + \pi/2} = \dfrac{0}{\pi} = 0.$

23

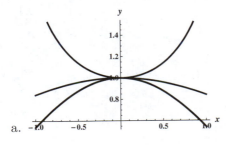

a.

b. Because $\lim\limits_{x \to 0} \cos x = \lim\limits_{x \to 0} \dfrac{1}{\cos x} = 1$, the squeeze theorem assures us that $\lim\limits_{x \to 0} \dfrac{\sin x}{x} = 1$ as well.

25 $\lim\limits_{x\to 5}\dfrac{x-7}{x(x-5)^2}=-\infty.$

27 $\lim\limits_{x\to 3^-}\dfrac{x-4}{x^2-3x}=\lim\limits_{x\to 3^-}\dfrac{x-4}{x(x-3)}=\infty.$

29 $\lim\limits_{x\to 0^-}\dfrac{2}{\tan x}=-\infty.$

31 $\lim\limits_{x\to\infty}\dfrac{2x-3}{4x+10}=\lim\limits_{x\to\infty}\dfrac{2-(3/x)}{4+(10/x)}=\dfrac{2}{4}=\dfrac{1}{2}.$

33 $\lim\limits_{x\to-\infty}(-3x^3+5)=\infty.$

35 $\lim\limits_{x\to\infty}\dfrac{\sqrt{25x^2+8}}{x+2}=\lim\limits_{x\to\infty}\dfrac{\sqrt{25+8/x^2}}{1+2/x}=5.$

37 $\lim\limits_{x\to\infty}\dfrac{4x^3+1}{1-x^3}=\lim\limits_{x\to\infty}\dfrac{4+(1/x^3)}{(1/x^3)-1}=\dfrac{4+0}{0-1}=-4.$ A similar result holds as $x\to-\infty$. Thus, $y=-4$ is a horizontal asymptote as $x\to\infty$ and as $x\to-\infty$.

39 $\lim\limits_{x\to\pm\infty}\dfrac{12x^2}{\sqrt{16x^4+7}}=\lim\limits_{x\to\pm\infty}\dfrac{12}{\sqrt{16+7/x^4}}=3.$

41 $\lim\limits_{x\to\pm\infty}\dfrac{x^2-x}{x^2-1}=\lim\limits_{x\to\pm\infty}\dfrac{x}{x+1}=\lim\limits_{x\to\pm\infty}\dfrac{1}{1+1/x}=1,$ so that $f(x)$ has a horizontal asymptote of $y=1$. There is a vertical asymptote where $x+1=0$, i.e. for $x=-1$.

43 $\lim\limits_{x\to\infty}\dfrac{3x^2+2x-1}{4x+1}=\lim\limits_{x\to\infty}\dfrac{3x^2+2x-1}{4x+1}\cdot\dfrac{1/x}{1/x}=\lim\limits_{x\to\infty}\dfrac{3x+2-1/x}{4+1/x}=\infty.$

$\lim\limits_{x\to-\infty}\dfrac{3x^2+2x-1}{4x+1}=\lim\limits_{x\to-\infty}\dfrac{3x^2+2x-1}{4x+1}\cdot\dfrac{1/x}{1/x}=\lim\limits_{x\to-\infty}\dfrac{3x+2-1/x}{4+1/x}=-\infty.$

By long division, we can write $f(x)$ as $f(x)=\dfrac{3x}{4}+\dfrac{5}{16}+\dfrac{-21/16}{4x+1}$, so the line $y=\dfrac{3x}{4}+\dfrac{5}{16}$ is the slant asymptote.

45 $\lim\limits_{x\to\infty}\dfrac{1+x-2x^2-x^3}{x^2+1}=\lim\limits_{x\to\infty}\dfrac{1+x-2x^2-x^3}{x^2+1}\cdot\dfrac{1/x^2}{1/x^2}=\lim\limits_{x\to\infty}\dfrac{1/x^2+1/x-2-x}{1+1/x^2}=-\infty.$

$\lim\limits_{x\to-\infty}\dfrac{1+x-2x^2-x^3}{x^2+1}=\lim\limits_{x\to-\infty}\dfrac{1+x-2x^2-x^3}{x^2+1}\cdot\dfrac{1/x^2}{1/x^2}=\lim\limits_{x\to-\infty}\dfrac{1/x^2+1/x-2-x}{1+1/x^2}=\infty.$

By long division, we can write $f(x)$ as $f(x)=-x-2+\dfrac{2x+3}{x^2+1}$, so the line $y=-x-2$ is the slant asymptote.

47 f is discontinuous at 5, because $f(5)$ does not exist, and also because $\lim\limits_{x\to 5}f(x)$ does not exist

49 h is not continuous at 3 because $\lim\limits_{x\to 3^-}h(x)$ does not exist, so $\lim\limits_{x\to 3}h(x)$ does not exist.

51 The domain of f is $(-\infty,-\sqrt{5}]$ and $[\sqrt{5},\infty)$, and f is continuous on that domain.

53 The domain of h is $(-\infty,-5)$, $(-5,0)$, $(0,5)$, $(5,\infty)$, and like all rational functions, it is continuous on its domain.

55 In order for g to be left continuous at 1, it is necessary that $\lim\limits_{x\to 1^-}g(x)=g(1)$, which means that $a=3$. In order for g to be right continuous at 1, it is necessary that $\lim\limits_{x\to 1^+}g(x)=g(1)$, which means that $a+b=3+b=3$, so $b=0$.

57

One such possible graph is pictured to the right.

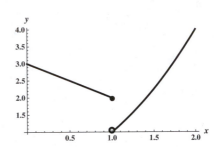

59

a. Any such rectangle with length x has width $y = 100/x$, so its perimeter is

$$P(x) = 2x + 2y = 2x + \frac{200}{x}.$$

b. P is continuous for $x > 0$. $P(10) = 40$ while $P(30) = 60 + \frac{200}{30} > 60$. Thus by the Intermediate Value Theorem, there must be some x with $10 \le x \le 30$ and $P(x) = 50$.

c. The rectangle with perimeter 50 can be determined by solving $P(x) = 50$.

$$2x + \frac{200}{x} = 50$$
$$2x^2 - 50x + 200 = 0$$
$$x = 5, 20$$

So a perimeter of 50 occurs for a 5×20 or a 20×5 rectangle.

d. If there were a rectangle with perimeter 30, it would be a solution to $P(x) = 30$, so

$$2x + \frac{200}{x} = 30, \qquad 2x^2 - 30x + 200 = 0$$

The discriminant of this quadratic is $900 - 1600 < 0$, so the quadratic has no (real) solution.

e. Plotting $P(x)$ for various viewing windows seems to show that the smallest possible perimeter is 40, for a 10×10 rectangle.

61 Let $\epsilon > 0$ be given. Let $\delta = \epsilon$. Now suppose that $0 < |x - 5| < \delta$.
Then

$$|f(x) - L| = \left| \frac{x^2 - 25}{x - 5} - 10 \right| = \left| \frac{(x - 5)(x + 5)}{x - 5} - 10 \right| = |x + 5 - 10|$$
$$= |x - 5| < \epsilon.$$

63 Let $N > 0$ be given. Let $\delta = 1/\sqrt[4]{N}$. Suppose that $0 < |x - 2| < \delta$. Then $|x - 2| < \frac{1}{\sqrt[4]{N}}$, so $\frac{1}{|x-2|} > \sqrt[4]{N}$, and $\frac{1}{(x-2)^4} > N$, as desired.

Chapter 3

Derivatives

3.1 Introducing the Derivative

3.1.1 The secant line through the points $(a, f(a))$ and $(x, f(x))$ for x near a, of the graph of f, is given by $m_{\text{sec}} = \frac{f(x) - f(a)}{x - a}$. As x approaches a, we obtain the limit $m_{\text{tan}} = \lim_{x \to a} \frac{f(x) - f(a)}{x - a} = \lim_{x \to a} m_{\text{sec}}$.

3.1.3 The average rate of change of f over $[a, x]$ is the slope of the secant line $m_{\text{sec}} = \frac{f(x) - f(a)}{x - a}$. As x approaches a, the length of the interval $x - a$ goes to zero, and in the limit we obtain the instantaneous rate of change of f at a given by $m_{\text{tan}} = \lim_{x \to a} \frac{f(x) - f(a)}{x - a}$.

3.1.5 $f'(a)$ is the value of the derivative of f at a. Also, $f'(a)$ is the slope of the tangent line to the graph of f at $(a, f(a))$. Furthermore, $f'(a)$ is the instantaneous rate of change of f at a.

3.1.7 $\frac{dy}{dx}$ is the limit of $\frac{\Delta y}{\Delta x}$ and is the rate of change of y with respect to x.

3.1.9

a. $m_{\text{tan}} = \lim_{x \to 3} \frac{x^2 - 5 - 4}{x - 3} = \lim_{x \to 3} \frac{x^2 - 9}{x - 3} =$

$\lim_{x \to 3} \frac{(x - 3)(x + 3)}{x - 3} = \lim_{x \to 3} (x + 3) = 6.$

b. Using the point-slope form of the equation of a line, we obtain $y - 4 = 6(x - 3)$, or $y = 6x - 14$.

c.

3.1.11

a. $m_{\text{tan}} = \lim_{x \to 1} \frac{-5x + 1 + 4}{x - 1} = \lim_{x \to 1} \frac{-5x + 5}{x - 1} =$

$\lim_{x \to 1} -5 \left(\frac{x - 1}{x - 1} \right) = -5.$

b. Using the point-slope form of the equation of a line, we get $y + 4 = -5(x - 1)$ which equals $y = -5x + 1$, the function itself.

c.

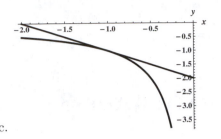

3.1.13   a. $m_{\tan} = \lim\limits_{x \to -1} \dfrac{\frac{1}{x} + 1}{x + 1} = \lim\limits_{x \to -1} \dfrac{\frac{1+x}{x}}{x + 1} = \lim\limits_{x \to -1} \dfrac{1}{x} = -1.$

b. $y - (-1) = -1(x + 1)$, or $y = -x - 2$.

c.

3.1.15

a. $m_{\tan} = \lim\limits_{h \to 0} \dfrac{2(0 + h) + 1 - 1}{h} = \lim\limits_{h \to 0} \dfrac{2h}{h} = 2.$

b. $y - 1 = 2x$, or $y = 2x + 1$.

3.1.17

a. $m_{\tan} = \lim\limits_{h \to 0} \dfrac{-7(-1 + h) - 7}{h} = \lim\limits_{h \to 0} -\dfrac{7h}{h} = \lim\limits_{h \to 0} -7 = -7.$

b. $y - 7 = -7(x + 1)$ or $y = -7x$.

3.1.19

a. $m_{\tan} = \lim\limits_{h \to 0} \dfrac{(2 + h)^2 - 4 - 0}{h} = \lim\limits_{h \to 0} \dfrac{4 + 4h + h^2 - 4}{h} = \lim\limits_{h \to 0}(4 + h) = 4.$

b. $y - 0 = 4(x - 2)$, or $y = 4x - 8$.

3.1.21

a. $m_{\tan} = \lim\limits_{h \to 0} \dfrac{(1 + h)^3 - 1}{h} = \lim\limits_{h \to 0} \dfrac{1 + 3h + 3h^2 + h^3 - 1}{h} = \lim\limits_{h \to 0} \dfrac{h(3 + 3h + h^2)}{h} = \lim\limits_{h \to 0}(3 + 3h + h^2) = 3.$

b. $y - 1 = 3(x - 1)$, or $y = 3x - 2$.

3.1.23

a. $m_{\tan} = \lim\limits_{h \to 0} \dfrac{\frac{1}{3 - 2(h - 1)} - \frac{1}{5}}{h} = \lim\limits_{h \to 0} \dfrac{\frac{5 - (3 - 2h + 2)}{15 - 10(h - 1)}}{h} = \lim\limits_{h \to 0} \dfrac{2}{15 - 10(h - 1)} = \dfrac{2}{25}.$

b. $y - \frac{1}{5} = \frac{2}{25}(x + 1)$, or $y = \frac{2}{25}x + \frac{7}{25}$.

3.1.25

a. $m_{\tan} = \lim\limits_{h \to 0} \dfrac{\sqrt{1 + h + 3} - 2}{h} = \lim\limits_{h \to 0} \dfrac{\sqrt{4 + h} - 2}{h} \cdot \dfrac{\sqrt{4 + h} + 2}{\sqrt{4 + h} + 2} = \lim\limits_{h \to 0} \dfrac{4 + h - 4}{h(\sqrt{4 + h} + 2)} =$
$\lim\limits_{h \to 0} \dfrac{1}{\sqrt{4 + h} + 2} = \dfrac{1}{4}.$

b. $y - 2 = \frac{1}{4}(x - 1)$, or $y = \frac{1}{4}x + \frac{7}{4}$.

3.1.27

a. $f'(-3) = \lim\limits_{h \to 0} \dfrac{8(-3 + h) + 24}{h} = \lim\limits_{h \to 0} \dfrac{8h}{h} = 8.$

b. $y - (-24) = 8(x + 3)$, or $y = 8x$.

3.1.29

a. $f'(-2) = \lim\limits_{h \to 0} \dfrac{4(-2+h)^2 + 2(-2+h) - 12}{h} = \lim\limits_{h \to 0} \dfrac{16 - 16h + 4h^2 - 4 + 2h - 12}{h} =$

$\lim\limits_{h \to 0} \dfrac{-14h + 4h^2}{h} = -14.$

b. $y - 12 = -14(x+2)$, or $y = -14x - 16.$

3.1.31

a. $f'\left(\dfrac{1}{4}\right) = \lim\limits_{h \to 0} \dfrac{\frac{1}{\sqrt{\frac{1}{4}+h}} - 2}{h} = \lim\limits_{h \to 0} \dfrac{1 - 2\sqrt{\frac{1}{4}+h}}{h\sqrt{\frac{1}{4}+h}} = \lim\limits_{h \to 0} \dfrac{\left(1 - 2\sqrt{\frac{1}{4}+h}\right)\left(1 + 2\sqrt{\frac{1}{4}+h}\right)}{h\sqrt{\frac{1}{4}+h}\left(1 + 2\sqrt{\frac{1}{4}+h}\right)} =$

$\lim\limits_{h \to 0} \dfrac{1 - 4\left(\frac{1}{4}+h\right)}{h\sqrt{\frac{1}{4}+h}\left(1 + 2\sqrt{\frac{1}{4}+h}\right)} = \lim\limits_{h \to 0} -\dfrac{4}{\sqrt{\frac{1}{4}+h}\left(1 + 2\sqrt{\frac{1}{4}+h}\right)} = -4.$

b. $y - 2 = -4\left(x - \frac{1}{4}\right)$, or $y = -4x + 3.$

3.1.33

a. $f'(4) = \lim\limits_{h \to 0} \dfrac{\sqrt{2(4+h)+1} - 3}{h} = \lim\limits_{h \to 0} \dfrac{\sqrt{9+2h} - 3}{h} \cdot \dfrac{\sqrt{9+2h} + 3}{\sqrt{9+2h} + 3} = \lim\limits_{h \to 0} \dfrac{9 + 2h - 9}{h(\sqrt{9+2h} + 3)} =$

$\lim\limits_{h \to 0} \dfrac{2}{\sqrt{9+2h} + 3} = \dfrac{1}{3}.$

b. $y - 3 = \frac{1}{3}(x - 4)$, or $y = \frac{1}{3}x + \frac{5}{3}.$

3.1.35

a. $f'(5) = \lim\limits_{h \to 0} \dfrac{\frac{1}{5+h+5} - \frac{1}{10}}{h} = \lim\limits_{h \to 0} \dfrac{10 - (10+h)}{10h(10+h)} = \lim\limits_{h \to 0} \dfrac{-1}{10(10+h)} = -\dfrac{1}{100}.$

b. $y - \frac{1}{10} = -\frac{1}{100}(x - 5)$, or $y = -\frac{1}{100}x + \frac{3}{20}.$

3.1.37

a.

$$f'(x) = \lim\limits_{h \to 0} \dfrac{3(x+h)^2 + 2(x+h) - 10 - (3x^2 + 2x - 10)}{h}$$

$$= \lim\limits_{h \to 0} \dfrac{3x^2 + 6xh + 3h^2 + 2x + 2h - 10 - 3x^2 - 2x + 10}{h}$$

$$= \lim\limits_{h \to 0} \dfrac{6xh + 2h + 3h^2}{h} = \lim\limits_{h \to 0} (6x + 2 + 3h) = 6x + 2.$$

b. We have $f'(1) = 8$, and the tangent line is given by $y + 5 = 8(x - 1)$, or $y = 8x - 13.$

c.

3.1.39

a. $f'(x) = \lim\limits_{h \to 0} \dfrac{5(x+h)^2 - 6(x+h) + 1 - (5x^2 - 6x + 1)}{h} =$

$\lim\limits_{h \to 0} \dfrac{5x^2 + 10xh + 5h^2 - 6x - 6h - 5x^2 + 6x}{h} = \lim\limits_{h \to 0} \dfrac{10xh + 5h^2 - 6h}{h} = \lim\limits_{h \to 0} (10x + 5h - 6) = 10x - 6.$

b. We have $f'(2) = 14$, so the tangent line is given by $y - 9 = 14(x - 2)$, or $y = 14x - 19$.

c.

3.1.41

a. $\dfrac{d}{dx}\left(ax^2 + bx + c\right) =$

$\lim\limits_{h \to 0} \dfrac{a(x+h)^2 + b(x+h) + c - \left(ax^2 + bx + c\right)}{h} = \lim\limits_{h \to 0} \dfrac{ax^2 + 2axh + ah^2 + bx + bh + c - ax^2 - bx - c}{h}$

$= \lim\limits_{h \to 0} \dfrac{2axh + ah^2 + bh}{h} = \lim\limits_{h \to 0} (2ax + ah + b) = 2ax + b.$

b. With $a = 4, b = -3, c = 10$ we have $\frac{d}{dx}(4x^2 - 3x + 10) = 2 \cdot 4 \cdot x + (-3) = 8x - 3.$

c. From part (b), $f'(1) = 8 \cdot 1 - 3 = 5.$

3.1.43 $m_{\text{tan}} = \lim\limits_{h \to 0} \dfrac{\frac{1}{1+h+1} - \frac{1}{2}}{h} = \lim\limits_{h \to 0} \dfrac{2 - (2+h)}{h(2+h)2} = \lim\limits_{h \to 0} \dfrac{-1}{(2+h)2} = -\dfrac{1}{4}.$

3.1.45 $m_{\text{tan}} = \lim\limits_{h \to 0} \dfrac{2\sqrt{25+h} - 1 - (2\sqrt{25} - 1)}{h} = \lim\limits_{h \to 0} \dfrac{2(\sqrt{25+h} - \sqrt{25})}{h} =$

$\lim\limits_{h \to 0} \dfrac{2(\sqrt{25+h} - \sqrt{25})(\sqrt{25+h} + \sqrt{25})}{h(\sqrt{25+h} + \sqrt{25})} = \lim\limits_{h \to 0} \dfrac{2(25 + h - 25)}{h(\sqrt{25+h} + \sqrt{25})} = \lim\limits_{h \to 0} \dfrac{2}{\sqrt{25+h} + \sqrt{25}} = \dfrac{1}{5}.$

3.1.47

a. True. Because the graph is a line, any secant line has the same graph as the function and thus the same slope.

b. False. For example, take $f(x) = x^2$, $P = (0,0)$ and $Q = (1,1)$. Then the secant line has slope $m_{\text{sec}} = \frac{1-0}{1-0} = 1$, but the the graph has a horizontal tangent at P so $m_{\text{tan}} = 0$ and $m_{\text{sec}} > m_{\text{tan}}$.

c. True. $m_{\text{sec}} = \frac{(x+h)^2 - x^2}{h} = \frac{2xh + h^2}{h} = 2x + h$, while $m_{\text{tan}} = \lim\limits_{h \to 0} (2x + h) = 2x$. Because we assume that $h > 0$, we have $m_{\text{sec}} = 2x + h > 2x = m_{\text{tan}}$.

3.1.49

a.

$$f'(x) = \lim\limits_{h \to 0} \dfrac{\sqrt{3(x+h) + 1} - \sqrt{3x+1}}{h}$$

$$= \lim\limits_{h \to 0} \dfrac{\sqrt{3(x+h) + 1} - \sqrt{3x+1}}{h} \cdot \dfrac{\sqrt{3x + 3h + 1} + \sqrt{3x+1}}{\sqrt{3x + 3h + 1} + \sqrt{3x+1}}$$

$$= \lim\limits_{h \to 0} \dfrac{3x + 3h + 1 - 3x - 1}{h(\sqrt{3x + 3h + 1} + \sqrt{3x+1})} = \lim\limits_{h \to 0} \dfrac{3}{\sqrt{3x + 3h + 1} + \sqrt{3x+1}} = \dfrac{3}{2\sqrt{3x+1}}.$$

b. We have $f'(8) = \frac{3}{10}$. Using the point-slope form, we get that the tangent line has equation $y - 5 = \frac{3}{10}(x - 8)$, which can be written as $y = \frac{3}{10}x + \frac{13}{5}$.

3.1.51

a. $f'(x) = \lim_{h \to 0} \dfrac{\frac{2}{3(x+h)+1} - \frac{2}{3x+1}}{h} = \lim_{h \to 0} \dfrac{6x + 2 - (6x + 6h + 2)}{h(3x+1)(3x+3h+1)} = \lim_{h \to 0} \dfrac{-6h}{h(3x+1)(3x+3h+1)} = -\dfrac{6}{(3x+1)^2}$.

b. We have $f'(-1) = -\frac{3}{2}$. Using the point-slope form, we get that the tangent line has equation $y + 1 = -\frac{3}{2}(x + 1)$, which can be written as $y = -\frac{3}{2}x - \frac{5}{2}$.

3.1.53

a. At C and D, the slope of the tangent line (and thus of the curve) is negative.

b. At A, B, and E, the slope of the curve is positive.

c. The graph is in its steepest ascent at A followed by B. At E it barely increases, at D it slightly decreases and at C it is decreasing the most, so the points in decreasing order of slope are A, B, E, D, C.

3.1.55

a. From the graph we approximate the derivative by the slope of a secant line: For example we see that $E(9) \approx 265\,\text{kWh}$ and $E(11) \approx 285\,\text{kWh}$, so the power after 10 hours is approximately the slope of the secant line through these points, so $P(10) \approx m_{sec} = \frac{E(11) - E(9)}{11 - 9} \approx \frac{285\,\text{kWh} - 265\,\text{kWh}}{2h} = 10\,\text{kW}$. Similarly, after 20 hours, using 19 hours and 21 hours, that $P(20) \approx m_{sec} = \frac{E(21) - E(19)}{21 - 19} \approx \frac{330\,\text{kWh} - 340\,\text{kWh}}{2h} = -5\text{kW}$.

b. The power is zero where the graph of $E(t)$ has a horizontal tangent line, which happens approximately at $t = 6$ hours and $t = 18$ hours.

c. The power has a maximum where the graph of $E(t)$ has the steepest increase, which is approximately at $t = 12$ hours.

3.1.57 Consider $a = 2$ and $f(x) = \frac{1}{x+1}$.

Then $f'(2) = \lim_{x \to 2} \dfrac{\frac{1}{x+1} - \frac{1}{3}}{x - 2}$ as desired.

We have $f'(2) = \lim_{x \to 2} \dfrac{\frac{1}{x+1} - \frac{1}{3}}{x - 2} = \lim_{x \to 2} \dfrac{3 - (x+1)}{(x-2)3(x+1)} = \lim_{x \to 2} \dfrac{-(x-2)}{(x-2)3(x+1)} = \lim_{x \to 2} \dfrac{-1}{3(x+1)} = -\dfrac{1}{9}$.

3.1.59 Consider $a = 2$ and $f(x) = x^4$.

Then $f'(2) = \lim_{h \to 0} \dfrac{(2+h)^4 - 16}{h}$ as desired.

We have

$$f'(2) = \lim_{h \to 0} \frac{(2+h)^4 - 16}{h} = \lim_{h \to 0} \frac{16 + 32h + 24h^2 + 8h^3 + h^4 - 16}{h}$$

$$= \lim_{h \to 0} \frac{h(32 + 24h + 8h^2 + h^3)}{h} = \lim_{h \to 0} (32 + 24h + 8h^2 + h^3) = 32.$$

3.1.61 It is not differentiable at $x = 2$. The denominator of f is zero when $x = 2$, so f is not defined at $x = 2$, and is therefore not differentiable there.

3.1.63 In order for f to be differentiable at $x = 1$, it would need to be continuous there. Thus, $\lim\limits_{x \to 1^-} f(x) =$ $\lim\limits_{x \to 1^-} 2x^2 = 2 = \lim\limits_{x \to 1^+} f(x) = \lim\limits_{x \to 1^+} ax - 2 = a - 2$, so the only possible value for a is 4. Now checking the differentiability at 1, we have (from the left)

$$\lim_{x \to 1^-} \frac{f(x) - f(1)}{x - 1} = \lim_{x \to 1^-} \frac{2x^2 - 2}{x - 1} = \lim_{x \to 1^-} 2(x + 1) = 4.$$

Also, from the right we have

$$\lim_{x \to 1^+} \frac{f(x) - f(1)}{x - 1} = \lim_{x \to 1^+} \frac{4x - 2 - 2}{x - 1} = \lim_{x \to 1^+} 4 = 4,$$

so f is differentiable at 1 for $a = 4$.

3.1.65

a. Note that the slope generated by the centered difference quotient is $\frac{f(4.5) - f(3.5)}{2(.5)} = \sqrt{4.5} - \sqrt{3.5} \approx$ 0.250492. The centered difference quotient line is very close to the tangent line, which closely approximates the function near the point of tangency.

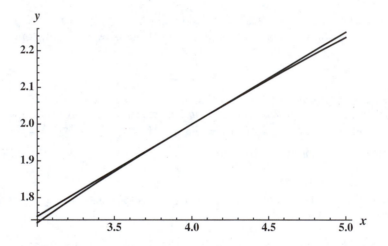

b.

h	Approximation	Error
0.1	0.25002	2.0×10^{-5}
0.01	$\approx .25000$	2.0×10^{-7}
0.001	$\approx .25000$	2.0×10^{-9}

c. The centered difference quotient is symmetric about zero, so using the corresponding negative values yields the same results.

d. The centered difference quotient appears to be more accurate than the approximation in the previous problem.

3.1.67

a. Forward:

$$\frac{\text{erf}(1.05) - \text{erf}(1)}{.05} = \frac{.862436 - .842701}{.05} = 0.3947.$$

Centered:

$$\frac{\text{erf}(1.05) - \text{erf}(.95)}{2(.05)} = \frac{.862436 - .820891}{0.1} = 0.41545.$$

b. Forward:
$$0.3947 - 0.4151075 \approx -0.0204075.$$

Centered:
$$0.41545 - 0.4151075 \approx 0.0003425.$$

3.2   Working with Derivatives

3.2.1 $f(x)$ refers to the value of the function at x, while $f'(x)$ refers to the slope of the graph. If the function is positive and increasing, such as $f(x) = x^2$ for $x = 2$, then $f(x) > 0$ and $f'(x) > 0$. But if the function is positive and decreasing, such as $f(x) = x^2$ for $x = -2$, then $f(x) > 0$ and $f'(x) < 0$.

3.2.3 Yes, differentiable functions are continuous by Theorem 3.1.

3.2.5

The function f is not differentiable at $x = -2, 0, 2$, so f' is not defined at those points. Elsewhere, the slope is constant.

3.2.7 (c) is the only line with negative slope, so it corresponds to derivative (A). Since (d) contains the points $(2,0)$ and $(0,-1)$, it has slope $\frac{1}{2}$, so it corresponds to derivative (B). Finally, lines (a) and (b) are parallel; since (b) contains the points $(0,1)$ and $(-1,0)$, it has slope 1, so that (a) has slope 1 as well. They both correspond to derivative (C).

3.2.9

a. The function has non-negative slope everywhere, and as there is a horizontal tangent at $x = 0$, so the derivative has to be zero at zero. The graph of the derivative has to be above the x-axis and touching it at $x = 0$, so (D) is the graph of the derivative.

b. The graph of this function has three horizontal tangent lines, at $x = -1, 0, 1$, and the matching graph of the derivative with three zeros is (C).

c. The function has negative slope on $(-1, 0)$, and positive slope on $(0, 1)$ and has a horizontal tangent at $x = 0$, so the derivative has to be negative on $(-1, 0)$, positive on $(0, 1)$ and zero at $x = 0$; the graph is (B).

d. The function has negative slope everywhere so the graph of the derivative has to be negative everywhere, which is graph (A).

3.2.11

The function always has non-negative slope, so the derivative is never below the x axis.
However, it does have slope zero at about $x = 2$.

3.2.13 Note that f is undefined at $x = -2$ and $x = 1$, but is differentiable elsewhere. It is decreasing, and increasingly rapidly, as x increases towards $x = -2$. It decreases, but increasingly slowly, and towards a zero slope, as x increases from 1. Finally, between $x = -2$ and $x = 1$, the function increases, but increasingly slowly, until $x = 0$ and then decreases, but increasingly rapidly, as x approaches 1. A graph of the derivative is

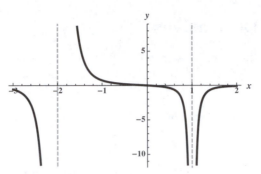

3.2.15

    a. The function f is not continuous at $x = 1$, because the graph has a jump there.

    b. The function f is not differentiable at $x = 1$ because it is not continuous at that point (Theorem 3.1 Alternate Version), and it is also not differentiable at $x = 2$ because the graph has a corner there.

    c.

3.2.17

    a. True. Differentiability implies continuity, by Theorem 3.1.

    b. True. Because the absolute value function is continuous, and $y = x + 1$ is continuous, and the composition of continuous functions is continuous, we know that this function is continuous. Note that it is not differentiable at $x = -1$ because the absolute value function is not differentiable at $x = 0$.

    c. False. In order for f to be differentiable on $[a, b]$, f would need to be defined at a and at b. Because the domain of f doesn't include these endpoints, this situation is not possible.

3.2.19

Because $f'(x) = x$ is negative for $x < 0$ and positive for $x > 0$, we have that the graph of f has to have negative slope on $(-\infty, 0)$ and positive slope on $(0, \infty)$ and has to have a horizontal tangent at $x = 0$. Because f' only gives us the slope of the tangent line and not the actual value of f, there are infinitely many graphs possible, they all have the same shape, but are shifted along the $y-$axis.

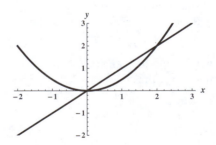

3.2.21 With $f(x) = 3x - 4$, we have

$$f'(x) = \lim_{h \to 0} \frac{f(x+h) - f(x)}{h} = \lim_{h \to 0} \frac{3(x+h) - 4 - (3x - 4)}{h} = \lim_{h \to 0} \frac{3h}{h} = \lim_{h \to 0} 3 = 3.$$

The slope of the tangent line at $(1, -1)$ is 3, so the slope of the normal line is $-\frac{1}{3}$. The equation of the normal line is thus $y - (-1) = -\frac{1}{3}(x - 1)$, or $y = -\frac{1}{3}x - \frac{2}{3}$.

3.2.23 With $f(x) = \frac{2}{x}$, we have $f'(x) = \lim\limits_{h \to 0} \dfrac{f(x+h) - f(x)}{h} = \lim\limits_{h \to 0} \dfrac{\frac{2}{x+h} - \frac{2}{x}}{h} = \lim\limits_{h \to 0} \dfrac{\frac{2x - (2(x+h))}{x(x+h)}}{h} =$

$\lim\limits_{h \to 0} \dfrac{-2h}{hx(x+h)} = \lim\limits_{h \to 0} \dfrac{-2}{x(x+h)} = -\dfrac{2}{x^2}$. At the point $(1, 2)$ the slope of the tangent line is -2, so the slope of the normal line is $\frac{1}{2}$. The equation of the normal line is $y - 2 = \frac{1}{2}(x - 1)$ or $y = \frac{x}{2} + \frac{3}{2}$.

3.2.25 With $f(x) = x^2 + 1$, we have $f'(x) = \lim\limits_{h \to 0} \dfrac{f(x+h) - f(x)}{h} = \lim\limits_{h \to 0} \dfrac{(x+h)^2 + 1 - (x^2 + 1)}{h} =$

$\lim\limits_{h \to 0} \dfrac{x^2 + 2hx + h^2 + 1 - x^2 - 1}{h} = \lim\limits_{h \to 0} \dfrac{2hx + h^2}{h} = \lim\limits_{h \to 0} \dfrac{h(2x + h)}{h} = 2x$. We are looking for points $(x, x^2 + 1)$ where the slope of the line between this point and $Q(3, 6)$ is equal to $2x$. So we seek solutions to the equation

$$\frac{6 - (x^2 + 1)}{3 - x} = 2x,$$

which can be written as $5 - x^2 = 2x(3 - x)$, or $x^2 - 6x + 5 = 0$. Factoring, we obtain $(x - 5)(x - 1) = 0$, so the solutions are $x = 5$ and $x = 1$. Note that at the point $(5, 26)$ on the curve the tangent line is $y - 26 = 10(x - 5)$ which does contain the point $Q(3, 6)$ and at the point $(1, 2)$ the equation of the tangent line is $y - 2 = 2(x - 1)$, which also contains the point $Q(3, 6)$.

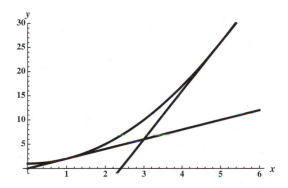

3.2.27 With $f(x) = \frac{1}{x}$, we have

$$f'(x) = \lim\limits_{h \to 0} \frac{f(x+h) - f(x)}{h} = \lim\limits_{h \to 0} \frac{\frac{1}{x+h} - \frac{1}{x}}{h} = \lim\limits_{h \to 0} \frac{\frac{x - (x+h)}{x(x+h)}}{h}$$

$$= \lim\limits_{h \to 0} \frac{-h}{hx(x+h)} = \lim\limits_{h \to 0} \frac{-1}{x(x+h)} = -\frac{1}{x^2}.$$

We are looking for points $(x, 1/x)$ where the slope of the line between this point and $Q(-2, 4)$ is equal to $\frac{-1}{x^2}$. So we seek solutions to the equation

$$\frac{1/x - 4}{x + 2} = -\frac{1}{x^2},$$

which can be written as $x - 4x^2 = -x - 2$, or $4x^2 - 2x - 2 = 0$, or $2x^2 - x - 1 = 0$. This factors as $(2x + 1)(x - 1) = 0$, so the solutions are $x = 1$ and $x = -1/2$. Note that at $x = 1$ the equation of the tangent line is $y - 1 = -1(x - 1)$ which does contain the point $Q(-2, 4)$, and at $x = -1/2$ the equation of the tangent line is $y + 2 = -4\left(x + \frac{1}{2}\right)$ or $y = -4x - 4$ which also contains the point $Q(-2, 4)$.

3.2.29

a. The tangent line for Q looks to be the steepest at $t = 0$.

b. All tangent lines for Q have positive slope, so Q' is positive for $t \geq 0$.

c. The tangent lines appear to be getting less steep as t increases, so Q' is decreasing.

d.

3.2.31

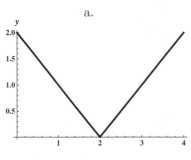

a.

b. $f'_+(2) = \lim\limits_{h \to 0^+} \dfrac{|2 + h - 2| - 0}{h} = \lim\limits_{h \to 0^+} \dfrac{h}{h} = 1$, because for $h > 0$, we have $|h| = h$. Similarly, $f'_-(2) = \lim\limits_{h \to 0^-} \dfrac{|2 + h - 2| - 0}{h} = \lim\limits_{h \to 0^-} -\dfrac{h}{h} = -1$, because for $h < 0$, we have $|h| = -h$.

c. Because f is defined at $a = 2$ and the graph of f does not jump, f is continuous at $a = 2$. Because the left-hand and right-hand derivatives are not equal, f is not differentiable at $a = 2$.

3.2.33

a.   The graph has a vertical tangent at $x = 2$.

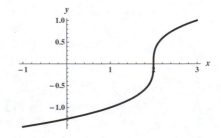

b. The graph has a vertical tangent at $x = -1$.

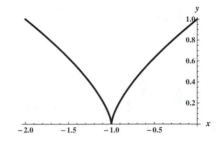

c. The graph has a vertical tangent at $x = 4$.

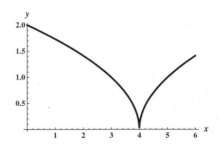

d. The graph has a vertical tangent at $x = 0$.

3.2.35 $f'(0) = \lim\limits_{h\to 0} \dfrac{h^{1/3}}{h} = \lim\limits_{h\to 0} \dfrac{1}{h^{2/3}} = +\infty$. Thus the graph of f has a vertical tangent at $x = 0$.

3.2.37

a.

b. For $x < 0$, we have $f'(x) = \lim\limits_{h\to 0} \dfrac{x+h-x}{h} = \lim\limits_{h\to 0} \dfrac{h}{h} = \lim\limits_{h\to 0} 1 = 1$.

c. For $x > 0$, we have $f'(x) = \lim\limits_{h\to 0} \dfrac{x+h+1-(x+1)}{h} = \lim\limits_{h\to 0} \dfrac{h}{h} = \lim\limits_{h\to 0} 1 = 1$.

d.

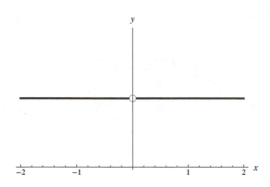

e. f is not differentiable at $x = 0$ as it is not continuous there. Also, if we were to compute the derivative of f from the right at 0 we would have

$$\lim_{h \to 0^+} \frac{f(0 + h) - f(0)}{h} = \lim_{h \to 0^+} \frac{h + 1 - 0}{h} = \lim_{h \to 0^+} \frac{h + 1}{h},$$

which does not exist.

3.3   Rules of Differentiation

3.3.1 Often the limit definition of f' is difficult to compute, especially for functions which are reasonably complicated. The rules for differentiation allow us to easily compute the derivatives of complex functions.

3.3.3 The slope at every point on the graph of the function is zero.

3.3.5 By the constant multiple rule, the derivative of the function cf where c is a constant and f is a function is cf'. That is, the derivative of a constant times a function is that same constant times the derivative of the function.

3.3.7 By the power rule, $y' = 5x^{5-1} = 5x^4$.

3.3.9 By the constant rule, $f'(x) = 0$.

3.3.11 By the power rule $h'(t) = 1t^{1-1} = t^0 = 1$.

3.3.13 By the constant multiple rule and the power rule, $f'(x) = 5 \cdot \frac{d}{dx} x^3 = 5 \cdot 3x^2 = 15x^2$.

3.3.15 By the constant multiple and power rules, $p'(x) = 8 \cdot \frac{d}{dx} x = 8 \cdot 1 = 8$.

3.3.17 By the constant multiple and power rules, $g'(t) = 100 \frac{d}{dt} t^2 = 100 \cdot 2t = 200t$.

3.3.19 $f'(x) = \frac{d}{dx}(3x^4 + 7x) = \frac{d}{dx}(3x^4) + \frac{d}{dx}(7x) = 12x^3 + 7$.

3.3.21 $f'(x) = \frac{d}{dx}(10x^4 - 32x + \frac{1}{2}) = \frac{d}{dx}(10x^4) - \frac{d}{dx}(32x) + \frac{d}{dx}\left(\frac{1}{2}\right) = 40x^3 - 32 - 0 = 40x^3 - 32$.

3.3.23 $g'(w) = \frac{d}{dw}(2w^3 + 3w^2 + 10w) = 2\frac{d}{dw}(w^3) + 3\frac{d}{dw}(w^2) + 10\frac{d}{dw}(w) = 2(3w^2) + 3(2w) + 10(1) = 6w^2 + 6w + 10$.

3.3.25 Expanding the product yields $f(x) = 6x^3 + 3x^2 + 4x + 2$. So

$$f'(x) = \frac{d}{dx}(6x^3 + 3x^2 + 4x + 2) = \frac{d}{dx}(6x^3) + \frac{d}{dx}(3x^2) + \frac{d}{dx}(4x) + \frac{d}{dx}(2)$$
$$= 18x^2 + 6x + 4.$$

3.3.27 Expanding the product yields $h(x) = x^4 + 2x^2 + 1$. So

$$h'(x) = \frac{d}{dx}(x^4 + 2x^2 + 1) = \frac{d}{dx}(x^4) + \frac{d}{dx}(2x^2) + \frac{d}{dx}(1)$$
$$= 4x^3 + 4x.$$

3.3.29 f simplifies as $f(w) = w^2 - 1$, so $f'(w) = 2w$ for $w \neq 0$.

3.3.31 g simplifies as $g(x) = \frac{(x-1)(x+1)}{x-1} = x + 1$. Thus $g'(x) = 1$ for $x \neq 1$.

3.3.33 y simplifies as $y = \frac{(\sqrt{x}-\sqrt{a})(\sqrt{x}+\sqrt{a})}{\sqrt{x}-\sqrt{a}} = \sqrt{x} + \sqrt{a}$. Thus $\frac{dy}{dx} = \frac{1}{2\sqrt{x}}$ for $x \neq a$.

3.3.35

a. $y' = -6x$, so the slope of the tangent line at $a = 1$ is -6. Thus, the tangent line at the point $(1, -1)$ is $y + 1 = -6(x - 1)$, or $y = -6x + 5$.

b.

3.3.37

a. $y' = \frac{1}{2\sqrt{x}}$, so the slope of the tangent line at $x = 4$ is $\frac{1}{4}$. Thus, the tangent line at the point $(4, 2)$ is $y - 2 = \frac{1}{4}(x - 4)$, or $y = \frac{1}{4}x + 1$.

b.

3.3.39

a. $f'(x) = 2x - 6$, so the slope is zero when $2x - 6 = 0$, which is at $x = 3$.

b. The slope is 2 when $2x - 6 = 2$ which is at $x = 4$.

3.3.41

a. The slope of the tangent line is given by $f'(x) = 6x^2 - 6x - 12$, and this quantity is zero when $x^2 - x - 2 = 0$, or $(x - 2)(x + 1) = 0$. The two solutions are thus $x = -1$ and $x = 2$, so the points on the graph are $(-1, 11)$ and $(2, -16)$.

b. The slope of the tangent line is 60 when $6x^2 - 6x - 12 = 60$, which is when $6x^2 - 6x - 72 = 0$. Simplifying this quadratic expression yields the equation $x^2 - x - 12 = 0$, which has solutions $x = -3$ and $x = 4$, so the points on the graph are $(-3, -41)$, and $(4, 36)$.

3.3.43

a. The slope of the tangent line is given by $\frac{2}{\sqrt{x}} - 1$. This is equal to zero when $\sqrt{x} = 2$, or $x = 4$. The point on the graph is $(4, 4)$.

b. The slope of the tangent line is $-\frac{1}{2}$ when $\frac{2}{\sqrt{x}} - 1 = -\frac{1}{2}$. Solving for x gives $x = 16$. The point on the graph is $(16, 0)$.

3.3.45 $f'(x) = 20x^3 + 30x^2 + 3$, $f''(x) = 60x^2 + 60x$, and $f^{(3)}(x) = 120x + 60$.

3.3.47 f simplifies as $f(x) = \frac{(x-8)(x+1)}{x+1} = x - 8$. So for $x \neq -1$, $f'(x) = 1$, $f''(x) = 0$, and $f^{(3)}(x) = 0$.

3.3.49

a. False. 10^5 is a constant, so the constant rule assures us that $\frac{d}{dx}(10^5) = 0$.

b. True. This follows because the slope is given by $f'(x) = 4 > 0$ for all x.

c. False. We have $\frac{d}{dx}(5x^3 + 2x + 5) = 15x^2 + 2$. Thus we have $\frac{d^2}{dx^2}(5x^3 + 2x + 5) = 30x$, and $\frac{d^3}{dx^3}(5x^3 + 2x + 5) = 30$. It is true that $\frac{d^n}{dx^n}(5x^3 + 2x + 5) = 0$ for $n \geq 4$.

3.3.51 First note that because the slope of $4x + 1$ is 4, it must be the case that $f'(2) = 4$. Also, at $x = 2$, we have $y = 4 \cdot 2 + 1 = 9$, so $f(2) = 9$. Because the line tangent to the graph of g at 2 has slope 3, we know that $g'(2) = 3$. The tangent line to g at $x = 2$ must be $y - (-2) = 3(x - 0)$, so the value of the tangent line at 2 (which must also be the value of $g(2)$) is 4. So $g(2) = 4$.

a. $y'(2) = f'(2) + g'(2) = 4 + 3 = 7$. The line contains the point $(2, f(2) + g(2)) = (2, 13)$. Thus, the equation of the tangent line is $y - 13 = 7(x - 2)$, or $y = 7x - 1$.

b. $y'(2) = f'(2) - 2g'(2) = 4 - 2 \cdot 3 = -2$. The line contains the point $(2, f(2) - 2g(2)) = (2, 1)$. Thus, the equation of the tangent line is $y - 1 = -2(x - 2)$, or $y = -2x + 5$.

c. $y'(2) = 4f'(2) = 4 \cdot 4 = 16$. The line contains the point $(2, 4f(2)) = (2, 36)$. Thus, the equation of the tangent line is $y - 36 = 16(x - 2)$, or $y = 16x + 4$.

3.3.53 For $f(x) = x^2 + bx + c$ we have $f'(x) = 2x + b$, so $f'(1) = 2 + b$. Because the slope of $4x + 2$ is 4, we require $2 + b = 4$, so $b = 2$. Also, because the value of $4x + 2$ at $x = 1$ is 6, we must have $f(1) = 1 + 2 + c = 6$, so $c = 3$. Thus the curve $f(x) = x^2 + 2x + 3$ has $y = 4x + 2$ as its tangent line at $x = 1$.

3.3.55 $G'(2) = 3f'(2) - g'(2) = 3(-3) - 1 = -10$.

3.3.57 $G'(5) = 3f'(5) - g'(5) = 3 \cdot 1 - (-1) = 4$.

3.3.59 $\frac{d}{dx}[1.5f(x)]_{x=2} = 1.5f'(2) = 1.5 \cdot 5 = 7.5$.

3.3.61

a. Let $f(x) = \sqrt{x}$ and $a = 9$. Then $\lim_{h \to 0} \frac{f(a+h) - f(a)}{h} = \lim_{h \to 0} \frac{\sqrt{9+h} - \sqrt{9}}{h} = f'(9)$.

b. Because $f'(x) = \frac{1}{2\sqrt{x}}$, we have $f'(9) = \frac{1}{6}$, so this is the value of the original limit.

3.3.63

a. Let $f(x) = x^{100}$ and $a = 1$. Then $\lim_{x \to 1} \frac{f(x) - f(1)}{x - 1} = f'(1)$.

b. Because $f'(x) = 100x^{99}$, we have $f'(1) = 100$, so this is the value of the original limit.

3.3.65

a. $d'(t) = 32t$ is the velocity of the stone after t seconds, measured in feet per second.

b. The stone travels $d(6) = 16 \cdot 6^2 = 576$ feet and strikes the ground with a velocity of $32 \cdot 6 = 192$ feet per second. Converting to miles per hour, we have $192 \cdot \frac{3600}{5280} \approx 130.9$ miles per hour.

3.3.67

a. $A'(t) = -\frac{1}{25}t + 2$ square miles per year.

b. First we must find when $A(t) = 38$. This occurs when $-\frac{t^2}{50} + 2t + 20 = 38$, which can be written as $t^2 - 100t + 900 = 0$. This factors as $(t - 90)(t - 10) = 0$, so the only solution on the given domain is $t = 10$. At this time, we have $A'(10) = -0.4 + 2 = 1.6$ square miles per year.

c. Note that $A'(20) = -0.8 + 2 = 1.2$ square miles per year. In order to maintain a density of 1000 people per square mile, we must multiply the density of people per square mile times the number of square miles per year in order to obtain the rate of people per year required to maintain that density. Thus the population growth rate must be $1000 \cdot 1.2 = 1200$ people per year.

3.3.69

$$\frac{d}{dx}x^n = \lim_{h \to 0} \frac{(x+h)^n - x^n}{h} = \lim_{h \to 0} \left(\frac{x^n + nx^{n-1}h + \frac{n(n-1)}{2}x^{n-2}h^2 + \cdots + nxh^{n-1} + h^n - x^n}{h} \right)$$

$$= \lim_{h \to 0} \left(nx^{n-1} + \frac{n(n-1)}{2}x^{n-2}h + \cdots + nxh^{n-2} + h^{n-1} \right) = nx^{n-1} + 0 + 0 + \cdots + 0 = nx^{n-1}$$

3.3.71

a. $\frac{d}{dx}(\sqrt{x}) = \frac{d}{dx}x^{1/2} = \frac{1}{2} \cdot x^{-1/2} = \frac{1}{2\sqrt{x}}$.

b.

$$\frac{d}{dx}x^{3/2} = \lim_{h \to 0} \frac{(x+h)^{3/2} - x^{3/2}}{h} = \lim_{h \to 0} \frac{((x+h)^{3/2} - x^{3/2})((x+h)^{3/2} + x^{3/2})}{h((x+h)^{3/2} + x^{3/2})}$$

$$= \lim_{h \to 0} \frac{(x+h)^3 - x^3}{h((x+h)^{3/2} + x^{3/2})} = \lim_{h \to 0} \frac{x^3 + 3x^2h + 3xh^2 + h^3 - x^3}{h((x+h)^{3/2} + x^{3/2})}$$

$$= \lim_{h \to 0} \frac{3x^2 + 3xh + h^2}{((x+h)^{3/2} + x^{3/2})} = \frac{3x^2 + 0 + 0}{x^{3/2} + x^{3/2}} = \frac{3x^2}{2x^{3/2}} = \frac{3}{2}x^{1/2}.$$

c.

$$\frac{d}{dx}x^{5/2} = \lim_{h \to 0} \frac{(x+h)^{5/2} - x^{5/2}}{h} = \lim_{h \to 0} \frac{((x+h)^{5/2} - x^{5/2})((x+h)^{5/2} + x^{5/2})}{h((x+h)^{5/2} + x^{5/2})}$$

$$= \lim_{h \to 0} \frac{(x+h)^5 - x^5}{h((x+h)^{5/2} + x^{5/2})} = \lim_{h \to 0} \frac{x^5 + 5x^4h + 10x^3h^2 + 10x^2h^3 + 5xh^4 + h^5 - x^5}{h((x+h)^{5/2} + x^{5/2})}$$

$$= \lim_{h \to 0} \frac{5x^4 + 10x^3h + 10x^2h^2 + 5xh^3 + h^4}{((x+h)^{5/2} + x^{5/2})}$$

$$= \frac{5x^4 + 0 + 0 + 0 + 0}{x^{5/2} + x^{5/2}} = \frac{5x^4}{2x^{5/2}} = \frac{5}{2}x^{3/2}.$$

d. It appears that $\frac{d}{dx}x^{n/2} = \frac{n}{2} \cdot x^{(n/2)-1}$.

3.4 The Product and Quotient Rules

3.4.1 The derivative of the product fg with respect to x is given by $f'(x)g(x) + f(x)g'(x)$.

3.4.3 $\dfrac{d}{dx}(x^n) = nx^{n-1}$ for all integers n.

3.4.5

$$\frac{d}{dx}(x^n \cdot x^{-n}) = x^{-n}\frac{d}{dx}x^n + x^n\frac{d}{dx}x^{-n} = nx^{-n}x^{n-1} - nx^nx^{-n-1} = nx^{-1} - nx^{-1} = 0.$$

3.4.7 $f'(x) = 12x^3(2x^2 - 1) + 3x^4 \cdot 4x = 24x^5 - 12x^3 + 12x^5 = 36x^5 - 12x^3$.

3.4.9 $f'(t) = 5t^4(\sqrt{t} + 1) + t^5\left(\frac{1}{2\sqrt{t}}\right) = \frac{11}{2}t^{9/2} + 5t^4$.

3.4.11 $h'(x) = (1)(x^3 + x^2 + x + 1) + (x - 1)(3x^2 + 2x + 1) = x^3 + x^2 + x + 1 + 3x^3 + 2x^2 + x - 3x^2 - 2x - 1 = 4x^3$.

3.4.13 $g'(w) = 3w^2(w^3 - 1) + (w^3 + 4) \cdot 3w^2 = 6w^5 + 9w^2 = 3w^2(2w^3 + 3)$.

3.4.15

    a. $f'(x) = 1(3x + 4) + (x - 1) \cdot 3 = 6x + 1$.

    b. $f'(x) = \frac{d}{dx}(3x^2 + x - 4) = 6x + 1$.

3.4.17

    a. $g'(y) = (12y^3 - 2y)(y^2 - 4) + (3y^4 - y^2) \cdot 2y = 18y^5 - 52y^3 + 8y$.

    b. $g'(y) = \frac{d}{dy}(3y^6 - 13y^4 + 4y^2) = 18y^5 - 52y^3 + 8y$.

3.4.19 $f'(x) = \frac{(x+1)\cdot 1 - x\cdot 1}{(x+1)^2} = \frac{1}{(x+1)^2}$.

3.4.21 $g'(w) = \frac{(\sqrt{w}+1) - (w-1)\left(\frac{1}{2\sqrt{w}}\right)}{(\sqrt{w}+1)^2} = \frac{2\sqrt{w}(\sqrt{w}+1) - (w-1)}{2\sqrt{w}(\sqrt{w}+1)^2} = \frac{w + 2\sqrt{w} + 1}{2\sqrt{w}(\sqrt{w}+1)^2} = \frac{(\sqrt{w}+1)^2}{2\sqrt{w}(\sqrt{w}+1)^2} = \frac{1}{2\sqrt{w}}$.

3.4.23 $h'(t) = \frac{(t^2+1) - t(2t)}{(t^2+1)^2} = \frac{1-t^2}{(t^2+1)^2}$.

3.4.25 $y' = \frac{d}{dt}\left(\frac{3t-1}{2t-2}\right) = \frac{(2t-2)\cdot 3 - (3t-1)\cdot 2}{(2t-2)^2} = -\frac{4}{(2t-2)^2} = -\frac{1}{(t-1)^2}$.

3.4.27 $g'(x) = \frac{(x^2-1)(4x^3) - (x^4+1)(2x)}{(x^2-1)^2} = \frac{2x^5 - 4x^3 - 2x}{(x^2-1)^2} = \frac{2x(x^4 - 2x^2 - 1)}{(x^2-1)^2}$.

3.4.29

    a. $f'(w) = \frac{w(3w^2-1) - (w^3-w)\cdot 1}{w^2} = \frac{2w^3}{w^2} = 2w$ for $w \neq 0$.

    b. For $w \neq 0$ this simplifies as $w^2 - 1$. $f'(w) = \frac{d}{dw}(w^2 - 1) = 2w$.

3.4.31

    a. $y' = \frac{(x-a)(2x) - (x^2-a^2)}{(x-a)^2} = \frac{2x^2 - 2ax - x^2 + a^2}{(x-a)^2} = \frac{x^2 - 2ax + a^2}{(x-a)^2} = \frac{(x-a)^2}{(x-a)^2} = 1$ for $x \neq a$.

    b. For $x \neq a$, this simplifies as $y = \frac{(x+a)(x-a)}{x-a} = x + a$. $y' = \frac{d}{dx}(x+a) = 1$.

3.4.33

    a. $y' = \frac{(x-1) - (x+5)}{(x-1)^2} = -\frac{6}{(x-1)^2}$.
       At $a = 3$ we have $y' = -\frac{6}{4} = -\frac{3}{2}$ and $y = 4$, so the
       equation of the tangent line is $y - 4 = -\frac{3}{2} \cdot (x - 3)$,
       or $y = -\frac{3}{2}x + \frac{17}{2}$.

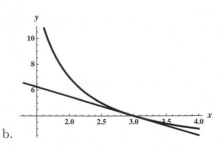

    b.

3.4.35

a. $y = x(2x^{-2} + 1) = 2x^{-1} + x$, so $y' = -2x^{-2} + 1$, and $y'(-1) = -2 + 1 = -1$. Also, $y(1) = -3$, so the equation of the tangent line is $y - (-3) = -1(x - (-1))$, or $y = -x - 4$.

b.

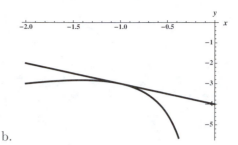

3.4.37 $f'(x) = (-9) \cdot 3 \cdot x^{-9-1} = -27x^{-10}$.

3.4.39 $g'(t) = \frac{d}{dt}(3t^2 + 6t^{-7}) = 6t - 42t^{-8}$.

3.4.41 $g'(t) = \frac{d}{dt}(1 + 3t^{-1} + t^{-2}) = -3t^{-2} - 2t^{-3}$.

3.4.43

a. $p'(t) = \frac{(t+2)200 - 200t}{(t+2)^2} = \frac{400}{(t+2)^2}$.

b. $p'(5) = \frac{400}{49} \approx 8.16$.

c. The value of p' is as large as possible when its denominator is as small as possible, which is when $t = 0$. The value of $p'(0)$ is 100.

d. $\lim_{t \to \infty} p(t) = \lim_{t \to \infty} \frac{200t}{(t+2)} = \lim_{t \to \infty} \frac{200}{1 + 2/t} = 200.$. This means that the population eventually approaches a steady state of 200.

e.

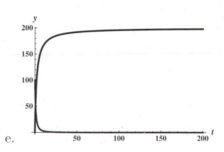

3.4.45

a. The slope is $f'(x) = \frac{(2x^2+1)(1-2x)-(x-x^2)(4x)}{(2x^2+1)^2} = \frac{2x^2-4x^3+1-2x-4x^2+4x^3}{(2x^2+1)^2} = \frac{1-2x-2x^2}{(2x^2+1)^2}$. This is zero when the numerator $1 - 2x - 2x^2$ is zero. Solving $-2x^2 - 2x + 1 = 0$ via the quadratic formula gives $\frac{-1 \pm \sqrt{3}}{2}$.

b. The graph of f has a horizontal tangent line at $x = \frac{-1+\sqrt{3}}{2}$ and at $x = \frac{-1-\sqrt{3}}{2}$.

3.4.47 Note that $g(x) = \frac{3-x}{2x}$. We have $g'(x) = \frac{2x(-1)-(3-x)2}{4x^2} = \frac{-x-3+x}{2x^2} = -\frac{3}{2x^2}$.

3.4.49 First note that h can be written as $h(x) = \frac{4}{(x+1)(1-x)} = \frac{4}{1-x^2}$. Then $h'(x) = \frac{(1-x^2)\cdot 0 - 4(-2x)}{(1-x^2)^2} = \frac{8x}{(1-x^2)^2}$.

3.4.51

a. False. In fact, because π^5 is a constant, its derivative is zero.

b. False. It is certainly a reasonable way to proceed, but one could also write the given quantity as $x + 3 + 2x^{-1}$, and then proceed using the sum rule and the power rule and the extended power rule.

c. False. $\frac{d}{dx}\left(\frac{1}{x^5}\right) = \frac{d}{dx}\left(x^{-5}\right) = -5x^{-6} = -\frac{5}{x^6}$.

3.4.53

$f'(x) = 2x(2 + x^{-3}) + x^2(-3x^{-4}) = 4x + 2x^{-2} - 3x^{-2} = 4x - x^{-2}$.
$f''(x) = 4 + 2x^{-3}$.
$f'''(x) = -6x^{-4}$.

3.4.55

$$f'(x) = \frac{d}{dx}\left(\frac{x^2 - 7x}{x+1}\right) = \frac{(x+1)(2x-7) - (x^2 - 7x)\cdot 1}{(x+1)^2} = \frac{x^2 + 2x - 7}{x^2 + 2x + 1} = \frac{x^2 + 2x - 7}{(x+1)^2}.$$

$$f''(x) = \frac{d}{dx}\left(\frac{x^2 + 2x - 7}{x^2 + 2x + 1}\right) = \frac{(x^2 + 2x + 1)(2x+2) - (x^2 + 2x - 7)(2x+2)}{(x+1)^4}$$

$$= \frac{(2x^2 + 4x + 2) - (2x^2 + 4x - 14)}{(x+1)^3} = \frac{16}{(x+1)^3}.$$

3.4.57 $f'(x) = \frac{d}{dx}\left(4x^2 - \frac{2x}{5x+1}\right) = 8x - \frac{(5x+1)2 - (2x)(5)}{(5x+1)^2} = 8x - \frac{2}{(5x+1)^2}.$

3.4.59

$$h'(r) = \frac{(r+1)(-1 - \frac{1}{2\sqrt{r}}) - (2 - r - \sqrt{r})\cdot 1}{(r+1)^2} = \frac{-r - \frac{\sqrt{r}}{2} - 1 - \frac{1}{2\sqrt{r}} - 2 + r + \sqrt{r}}{(r+1)^2}$$

$$= \frac{\frac{\sqrt{r}}{2} - \frac{1}{2\sqrt{r}} - 3}{(r+1)^2}\cdot\frac{2\sqrt{r}}{2\sqrt{r}} = \frac{r - 1 - 6\sqrt{r}}{2\sqrt{r}(r+1)^2}.$$

3.4.61 $h'(x) = (35x^6 + 5)(6x^3 + 3x^2 + 3) + (5x^7 + 5x)(18x^2 + 6x) = 15((7x^6 + 1)(2x^3 + x^2 + 1) + (x^7 + x)(6x^2 + 2x)) = 300x^9 + 135x^8 + 105x^6 + 120x^3 + 45x^2 + 15.$

3.4.63

a.

$y' = -\frac{54x}{(x^2+9)^2}$. At $x = 2$, $y' = -\frac{108}{169}$ and $y = \frac{27}{13}$.
Thus the tangent line is given by

$$y - \frac{27}{13} = -\frac{108}{169}(x - 2),$$

or $y = -\frac{108}{169}x + \frac{567}{169}.$

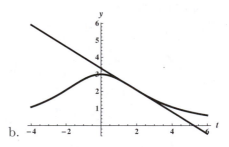

b.

3.4.65 $\frac{d}{dx}\left[\frac{f(x)}{g(x)}\right]\Big|_{x=2} = \frac{g(2)f'(2) - f(2)g'(2)}{(g(2))^2} = \frac{2\cdot 5 - 4\cdot 4}{4} = -\frac{3}{2}.$

3.4.67 $\frac{d}{dx}\left[\frac{f(x)}{x+2}\right]\Big|_{x=4} = \frac{(4+2)f'(4) - f(4)}{36} = \frac{6 - 2}{36} = \frac{1}{9}.$

3.4.69 $\frac{d}{dx}\left[\frac{f(x)g(x)}{x}\right]\Big|_{x=4} = \frac{4(f'(4)g(4) + f(4)g'(4)) - f(4)g(4)}{16} = \frac{4(1\cdot 3 + 2\cdot 1) - (2\cdot 3)}{16} = \frac{14}{16} = \frac{7}{8}.$

3.4.71

a. The instantaneous rate of change is $\frac{d}{dx}F(x) = -\frac{2kQq}{x^3}$ N/m $= -\frac{1.8\times 10^{10}Qq}{x^3}$ N/m.

b. $\left[\frac{d}{dx}F(x)\right]\Big|_{x=0.001} = -\frac{2(9\times 10^9)}{(0.001)^3} = -\frac{18\times 10^9}{10^{-9}} = -18\times 10^{18} = -1.8\times 10^{19}$ Newtons per meter.

c. Because the distance x appears in the denominator of $F'(x)$, the absolute value of the instantaneous rate of change decreases with the separation.

3.4.73

a. The tangent line at $x = a$ is $y - a^2 = 2a(x - a)$ and at $x = b$ is $y - b^2 = 2b(x - b)$.

These intersect when $a^2 + 2ax - 2a^2 = b^2 + 2bx - 2b^2$, or $(2a - 2b)x = a^2 - b^2$, which is met when $x = \frac{a+b}{2}$. So $c = \frac{a+b}{2}$.

b. The tangent line at $x = a$ is $y - \sqrt{a} = \frac{1}{2\sqrt{a}}(x - a)$ and at $x = b$ is $y - \sqrt{b} = \frac{1}{2\sqrt{b}}(x - b)$.

These intersect when $\sqrt{a} + \frac{1}{2\sqrt{a}}(x - a) = \sqrt{b} + \frac{1}{2\sqrt{b}}(x - b)$, or $\left(\frac{1}{2\sqrt{a}} - \frac{1}{2\sqrt{b}}\right)x = \frac{\sqrt{b}-\sqrt{a}}{2}$, which is met when $x = \sqrt{ab}$. So $c = \sqrt{ab}$.

c. The tangent line at $x = a$ is $y - \frac{1}{a} = -\frac{1}{a^2}(x - a)$ and at $x = b$ is $y - \frac{1}{b} = -\frac{1}{b^2}(x - b)$.

These intersect when $\frac{1}{a} + -\frac{1}{a^2}(x - a) = \frac{1}{b} - \frac{1}{b^2}(x - b)$, or $\left(\frac{2}{a} - \frac{x}{a^2}\right) = \left(\frac{2}{b} - \frac{x}{b^2}\right)$, which is met when $x\left(\frac{1}{b^2} - \frac{1}{a^2}\right) = \frac{2}{b} - \frac{2}{a}$, or $x \cdot \left(\frac{a^2-b^2}{a^2 b^2}\right) = \frac{2(a-b)}{ab}$. Thus we arrive at $x = \frac{2ab}{a+b}$. So $c = \frac{2ab}{a+b}$.

d. The tangent line at $x = a$ is $y - f(a) = f'(a)(x - a)$ and at $x = b$ is $y - f(b) = f'(b)(x - b)$.

These intersect when $f(a) + f'(a)(x - a) = f(b) + f'(b)(x - b)$, or $(f'(a) - f'(b))x = f(b) - f(a) - f'(b)b + f'(a)a$. Solving for x yields $x = \frac{f(b)-f(a)-f'(b)b+f'(a)a}{f'(a)-f'(b)}$ provided $f'(a) \neq f'(b)$ (which occurs when the tangent lines are parallel and don't intersect.)

3.4.75

$$\frac{d^2}{dx^2}(f(x)g(x)) = \frac{d}{dx}(f'(x)g(x) + f(x)g'(x)) = f''(x)g(x) + f'(x)g'(x) + f'(x)g'(x) + f(x)g''(x)$$
$$= f''(x)g(x) + 2f'(x)g'(x) + f(x)g''(x).$$

3.4.77

a.

$$\frac{d}{dx}[(f(x)g(x))h(x)] = \frac{d}{dx}[f(x)g(x)] \cdot h(x) + f(x)g(x) \cdot \frac{d}{dx}h(x)$$
$$= [f'(x)g(x) + f(x)g'(x)]h(x) + f(x)g(x)h'(x)$$
$$= f'(x)g(x)h(x) + f(x)g'(x)h(x) + f(x)g(x)h'(x).$$

b. $\frac{d}{dx}[\sqrt{x}(x - 1)(x + 3)] = \frac{(x-1)(x+3)}{2\sqrt{x}} + \sqrt{x}(x + 3) + \sqrt{x}(x - 1).$

3.5 Derivatives of Trigonometric Functions

3.5.1 A direct substitution would yield the quotient of zero with itself, which isn't defined

3.5.3 Because $\tan x = \frac{\sin x}{\cos x}$, and $\cot x = \frac{\cos x}{\sin x}$, we can use the quotient rule to compute these derivatives, because we know the derivatives of $\sin x$ and of $\cos x$.

3.5.5 $f'(x) = \cos x$ and $f'(\pi) = \cos \pi = -1$.

3.5.7 $\lim\limits_{x \to 0} \dfrac{\sin 3x}{x} = \lim\limits_{x \to 0} \dfrac{3 \sin 3x}{3x} = 3 \lim\limits_{x \to 0} \dfrac{\sin 3x}{3x} = 3 \cdot 1 = 3.$

3.5.9 $\lim\limits_{x \to 0} \dfrac{\sin 7x}{\sin 3x} = \lim\limits_{x \to 0} \dfrac{\frac{7 \sin 7x}{7x}}{\frac{3 \sin 3x}{3x}} = \dfrac{7}{3} \cdot \lim\limits_{x \to 0} \dfrac{\frac{\sin 7x}{7x}}{\frac{\sin 3x}{3x}} = \dfrac{7}{3} \cdot \dfrac{1}{1} = \dfrac{7}{3}.$

3.5.11 $\lim\limits_{x \to 0} \dfrac{\tan 5x}{x} = \lim\limits_{x \to 0} \dfrac{5 \sin 5x}{5x \cos 5x} = 5 \lim\limits_{x \to 0} \dfrac{\sin 5x}{5x} \cdot \lim\limits_{x \to 0} \dfrac{1}{\cos 5x} = 5 \cdot 1 \cdot 1 = 5.$

3.5.13 $\lim\limits_{x \to 0} \dfrac{\tan 7x}{\sin x} = \lim\limits_{x \to 0} \dfrac{\sin 7x}{\cos 7x \cdot \sin x} = \lim\limits_{x \to 0} \left(\dfrac{1}{\cos 7x} \cdot \dfrac{x}{\sin x} \cdot \dfrac{7 \sin 7x}{7x}\right) =$

$7 \cdot \lim\limits_{x \to 0} \dfrac{1}{\cos 7x} \cdot \lim\limits_{x \to 0} \dfrac{x}{\sin x} \cdot \lim\limits_{x \to 0} \dfrac{\sin 7x}{7x} = 7 \cdot 1 \cdot 1 \cdot 1 = 7.$

3.5.15 $\lim\limits_{x \to 2} \dfrac{\sin(x - 2)}{x^2 - 4} = \lim\limits_{x \to 2} \left(\dfrac{1}{x + 2} \cdot \dfrac{\sin(x - 2)}{x - 2}\right) = \lim\limits_{x \to 2} \dfrac{1}{x + 2} \cdot \lim\limits_{x \to 2} \dfrac{\sin(x - 2)}{x - 2} = \dfrac{1}{4} \cdot 1 = \dfrac{1}{4}.$

3.5.17 $y' = \cos x - \sin x.$

3.5.19 $y' = 12x^3 \sin x + 3x^4 \cos x = 3x^3(4\sin x + x\cos x).$

3.5.21 $y' = \sin x + x\cos x.$

3.5.23 $y' = \dfrac{(\sin x + 1)(-\sin x) - (\cos x)(\cos x)}{(1+\sin x)^2} = \dfrac{-1(\sin^2 x + \cos^2 x) - \sin x}{(1+\sin x)^2} = \dfrac{-1(1+\sin x)}{(1+\sin x)^2} = -\dfrac{1}{1+\sin x}.$

3.5.25 $y' = \cos x \cos x + \sin x \cdot (-\sin x) = \cos^2 x - \sin^2 x = \cos(2x).$

3.5.27 $y' = -\sin x \cos x + \cos x(-\sin x) = -2\sin x \cos x = -\sin(2x).$

3.5.29 $\dfrac{d}{dx}(\cot x) = \dfrac{d}{dx}\left(\dfrac{\cos x}{\sin x}\right) = \dfrac{\sin x(-\sin x) - \cos x(\cos x)}{\sin^2 x} = \dfrac{-(\sin^2 x + \cos^2 x)}{\sin^2 x} = -\dfrac{1}{\sin^2 x} = -\csc^2 x.$

3.5.31 $\dfrac{d}{dx}(\csc x) = \dfrac{d}{dx}\left(\dfrac{1}{\sin x}\right) = \dfrac{0 - \cos x}{\sin^2 x} = -\dfrac{1}{\sin x}\cdot\dfrac{\cos x}{\sin x} = -\csc x \cot x.$

3.5.33 $y' = \sec x \tan x - \csc x \cot x.$

3.5.35 $y' = \frac{\csc x}{2\sqrt{x}} + \sqrt{x}(-\csc x \cot x) = \frac{\csc x - 2x \csc x \cot x}{2\sqrt{x}} = \frac{\csc x(1 - 2x \cot x)}{2\sqrt{x}}.$

3.5.37

$$y' = \frac{(1+\csc x)(-\csc^2 x) - \cot x(-\csc x \cot x)}{(1+\csc x)^2} = \frac{-\csc^2 x - \csc^3 x + \csc x(\csc^2 x - 1)}{(1+\csc x)^2}$$

$$= \frac{-\csc x(1 + \csc x)}{(1+\csc x)^2} = -\frac{\csc x}{1 + \csc x}$$

3.5.39

$$y' = \frac{0 - (\sec z \tan z \csc z - \sec z \csc z \cot z)}{\sec^2 z \csc^2 z} = \frac{\sec z \csc z(\cot z - \tan z)}{\sec^2 z \csc^2 z}$$

$$= \frac{\cot z - \tan z}{\sec z \csc z} = \cos^2 z - \sin^2 z = \cos(2z).$$

3.5.41 $y' = \sin x + x\cos x$, so $y'' = \cos x + \cos x + -x\sin x = 2\cos x - x\sin x.$

3.5.43 $y' = \frac{x\cos x - \sin x}{x^2}$. Then $y'' = \frac{x^2(\cos x - x\sin x - \cos x) - 2x(x\cos x - \sin x)}{x^4} = \frac{2\sin x}{x^3} - \frac{2\cos x}{x^2} - \frac{\sin x}{x}.$

3.5.45 $y' = -\csc^2 x$ and $y'' = -((-\csc x \cot x)\csc x + \csc x(-\csc x \cot x)) = 2\cot x \csc^2 x.$

3.5.47

$$y' = \sec x \tan x \csc x - \sec x \csc x \cot x = \sec x \csc x(\tan x - \cot x) = \sec^2 x - \csc^2 x.$$

$$y'' = \sec x(\sec x \tan x) + (\sec x \tan x)\sec x - ((-\csc x \cot x)\csc x + \csc x(-\csc x \cot x))$$
$$= 2\sec^2 x \tan x + 2\csc^2 x \cot x.$$

3.5.49

a. False. $\frac{d}{dx}\sin^2 x = \sin x \cos x + \cos x \sin x = 2\sin x \cos x \neq \cos^2 x.$

b. False. $\frac{d^2}{dx^2}\sin x = \frac{d}{dx}\cos x = -\sin x \neq \sin x.$

c. True. $\frac{d^4}{dx^4}\cos x = \frac{d^3}{dx^3}(-\sin x) = \frac{d^2}{dx^2}(-\cos x) = \frac{d}{dx}\sin x = \cos x.$

d. True. In fact, $\pi/2$ isn't even in the domain of $\sec x$.

3.5.51 $\displaystyle\lim_{x\to 0}\frac{\sin ax}{\sin bx}=\lim_{x\to 0}\frac{a\sin ax}{ax}\cdot\frac{bx}{b\sin bx}=\frac{a}{b}\lim_{x\to 0}\frac{\sin ax}{ax}\cdot\lim_{x\to 0}\frac{bx}{\sin bx}=\frac{a}{b}\cdot 1\cdot 1=\frac{a}{b}.$

3.5.53 $\displaystyle\lim_{x\to 0}\frac{3\sec^5 x}{x^2+4}=\frac{3\sec^5(0)}{4}=\frac{3}{4}.$

3.5.55 $\displaystyle\lim_{x\to\pi/4}3\csc(2x)\cot(2x)=\lim_{x\to\pi/4}3\frac{1}{\sin 2x}\frac{\cos 2x}{\sin 2x}=3\frac{\cos\pi/2}{(\sin\pi/2)^2}=3\cdot\frac{0}{1}=0.$

3.5.57 $\frac{dy}{dx}=\cos x\sin x+x(-\sin x)\sin x+x\cos x\cos x=\sin x\cos x-x\sin^2 x+x\cos^2 x=\frac12\sin 2x+x\cos 2x.$

3.5.59 $\displaystyle\frac{dy}{dx}=\frac{(\sin x-\cos x)\cos x-\sin x(\cos x+\sin x)}{(\sin x-\cos x)^2}=\frac{-(\sin^2 x+\cos^2 x)}{(\sin x-\cos x)^2}=$

$\displaystyle\frac{-1}{\sin^2 x-2\sin x\cos x+\cos^2 x}=\frac{-1}{1-\sin 2x}=\frac{1}{\sin(2x)-1}=\frac{1}{2\sin x\cos x-1}.$

3.5.61 $\displaystyle\frac{dy}{dx}=\frac{(1+\cos x)\sin x-(1-\cos x)(-\sin x)}{(1+\cos x)^2}=\frac{2\sin x}{(1+\cos x)^2}.$

3.5.63

a. $y'=2\cos x$, so $y'(\pi/6)=\sqrt3$. $y(\pi/6)=2$. The tangent line is thus given by $y-2=\sqrt3(x-\pi/6)$, or $y=\sqrt3 x+2-\frac{\pi\sqrt3}{6}$.

b.

3.5.65

a. $y'=\dfrac{(1-\cos x)(-\sin x)-\cos x\sin x}{(1-\cos x)^2}=$
$-\dfrac{\sin x}{(1-\cos x)^2}$, so $y'(\pi/3)=-2\sqrt3$. $y(\pi/3)=1$. The tangent line is thus given by $y-1=-2\sqrt3(x-\pi/3)$, or $y=-2\sqrt3 x+\frac{2\sqrt3\pi}{3}+1$.

b.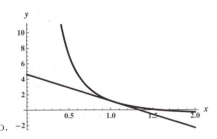

3.5.67 For a horizontal tangent line we need $f'(x)=1+2\sin x=0$, or $\sin x=-\frac12$. This occurs for $x=\frac{7\pi}{6}+2n\pi$ where n is any integer, or for $x=\frac{11\pi}{6}+2n\pi$ where n is any integer.

3.5.69

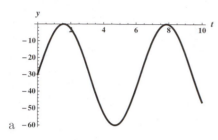

b. $v(t) = y'(t) = 30 \cos t$ cm per second.

a

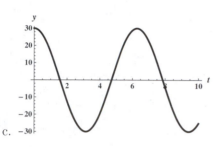

d. $v(t) = 30 \cos t = 0$ when $t = \frac{2n+1}{2} \cdot \pi$ where n is a non-negative integer. At those times, the position is given by $\begin{cases} 0 & \text{if } n \text{ is even} \\ -60 & \text{if } n \text{ is odd.} \end{cases}$

c.

e. The maximum velocity is 30 cm per second because $|\cos t| \leq 1$ for all t. We have $\cos t = 1$ for $t = 2n\pi$ for a positive integer n. At those times, $y(2n\pi) = -30$.

$a(t) = v'(t) = -30 \sin t.$

f.

3.5.71

a. $y'(t) = A \cos t$, $y''(t) = -A \sin t$, so $y''(t) + y(t) = -A \sin t + A \sin t = 0$ for all A and all t.

b. $y'(t) = -B \sin t$, $y''(t) = -B \cos t$, so $y''(t) + y(t) = -B \cos t + B \cos t = 0$ for all B and all t.

c. $y' = A \cos t - B \sin t$, $y'' = -A \sin t - B \cos t$, so $y''(t) + y(t) = -A \sin t - B \cos t + A \sin t + B \cos t = 0$ for all A, B, t.

3.5.73

$$\lim_{x \to 0} \frac{\cos x - 1}{x} = \lim_{x \to 0} \frac{(\cos x - 1)(\cos x + 1)}{x(\cos x + 1)} = \lim_{x \to 0} \frac{\cos^2 x - 1}{x(\cos x + 1)} = \lim_{x \to 0} -\frac{\sin^2 x}{x(\cos x + 1)}$$

$$= \lim_{x \to 0} \frac{\sin x}{x} \cdot \lim_{x \to 0} -\frac{\sin x}{(\cos x + 1)} = 1 \cdot \frac{0}{2} = 0.$$

3.5.75

$$\frac{d}{dx} \cos x = \lim_{h \to 0} \frac{\cos(x + h) - \cos x}{h} = \lim_{h \to 0} \frac{\cos x \cos h - \sin x \sin h - \cos x}{h}$$

$$= \cos x \left(\lim_{h \to 0} \frac{\cos h - 1}{h} \right) - \sin x \left(\lim_{h \to 0} \frac{\sin h}{h} \right) = \cos x \cdot 0 - \sin x \cdot 1 = -\sin x.$$

3.5.77 g is continuous at 0 if and only if $\lim\limits_{x \to 0} g(x) = g(0)$. Because $\lim\limits_{x \to 0} g(x) = \lim\limits_{x \to 0} \dfrac{1 - \cos x}{2x} = \dfrac{1}{2} \cdot 0 = 0$, we require $a = 0$ in order for g to be continuous.

3.5.79

a. $\dfrac{d}{dx} \sin^2 x = \sin x \cos x + \cos x \sin x = 2 \sin x \cos x$.

b. $\dfrac{d}{dx} \sin^3 x = \dfrac{d}{dx}(\sin^2 x)(\sin x) = (2 \sin x \cos x) \sin x + \sin^2 x \cdot \cos x = 3 \sin^2 x \cos x$.

c. $\dfrac{d}{dx} \sin^4 x = \dfrac{d}{dx}(\sin^3 x)(\sin x) = (3 \sin^2 x \cos x)(\sin x) + (\sin^3 x)(\cos x) = 4 \sin^3 x \cos x$.

d. We guess that $\dfrac{d}{dx} \sin^n x = n \sin^{n-1} x \cos x$.

   We have already seen that the claim is valid for $n = 2$. Suppose our guess is valid for a given positive integer n. Then

   $$\dfrac{d}{dx} \sin^{n+1} x = \dfrac{d}{dx}(\sin^n x)(\sin x) = (n \sin^{n-1} x \cos x)(\sin x) + \sin^n x \cos x = (n+1) \sin^n x \cos x.$$

   Thus by induction, the result holds for all n.

3.5.81

a. $f(x) = \sin x$, $a = \pi/6$.

b. $\lim\limits_{h \to 0} \dfrac{\sin(\pi/6 + h) - (1/2)}{h} = f'(\pi/6) = \cos(\pi/6) = \sqrt{3}/2$.

3.5.83

a. $f(x) = \cot x$, $a = \pi/4$.

b. $\lim\limits_{x \to \pi/4} \dfrac{\cot(x) - 1}{x - \pi/4} = f'(\pi/4) = -\csc^2(\pi/4) = -2$.

3.5.85 Because D is a difference quotient, and because $h = 0.01$ is small, D is a good approximation to f'. Therefore, the graph of D is nearly indistinguishable from the graph of $f'(x) = \cos x$.

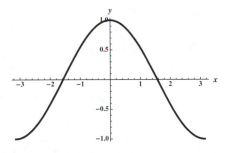

3.6 Derivatives as Rates of Change

3.6.1

The average rate of change over the interval $[a, a + \Delta x]$ is the slope of the line through $(a, f(a))$ and $(a + \Delta x, f(a + \Delta x))$, given by $m_{\text{avg}} = \dfrac{f(a+\Delta x) - f(a)}{\Delta x}$. The instantaneous rate of change is the limit of this expression as $\Delta x \to 0$, which is the slope of the tangent line at $(a, f(a))$.

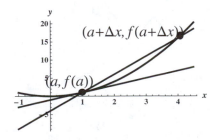

3.6.3 If $\frac{dy}{dx}$ is small, then small changes in x will result in relatively small changes in the value of y.

3.6.5 Acceleration is the instantaneous rate of change of the velocity; that is, if $s(t)$ is the position of an object at time t, then $s''(t) = \frac{d}{dt}(v(t)) = a(t)$ is the acceleration of the object at time t.

3.6.7 Each of the first 200 stoves cost on average \$70 to produce, while the 201st stove costs \$65 to produce.

3.6.9

a. $v_{\text{avg}} = \dfrac{f(0.75) - f(0)}{0.75} = \dfrac{30 - 0}{0.75} = 40$ mph.

b. $v_{\text{avg}} = \dfrac{f(0.75) - f(0.25)}{0.75 - 0.25} = \dfrac{30 - 10}{0.5} = 40$ mph.

   This is a pretty good estimate, since the graph is nearly linear over that time interval.

c. $v_{\text{avg}} = \dfrac{f(2.25) - f(1.75)}{2.25 - 1.75} = \dfrac{-14 - 16}{0.5} = -60$ mph.

   At 11 a.m. the velocity is $v(2) \approx -60$ mph. The car is moving south with a speed of approximately 60 mph.

d. From 9 a.m. until about 10:08 a.m., the car moves north, away from the station. Then it moves south, passing the station at approximately 11:02 a.m., and continues south until about 11:40 a.m. Then the car drives north until 12:00 noon stopping south of the station.

3.6.11

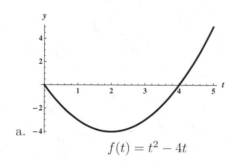

a.

$$f(t) = t^2 - 4t$$

b.

$$f'(t) = 2t - 4$$

b. $f'(t) = 0$ when $t = 2$ – that is when the object is stationary. For $0 \le t < 2$ we have $f'(t) < 0$ so the object is moving to the left. For $2 < t \le 5$ we have $f'(t) > 0$ so the object is moving to the right.

c. $f'(1) = -2$ ft/sec and $f''(t) = 2$ ft/sec^2, so in particular, $f''(1) = 2$ ft/sec^2.

d. $f'(t) = 0$ when $t = 2$ and $f''(2) = 2$ ft/sec^2.

e. On the interval $(2, 5]$ the velocity and acceleration are both positive, so the object's speed is increasing.

3.6.13

a.

$$f(t) = 2t^2 - 9t + 12$$

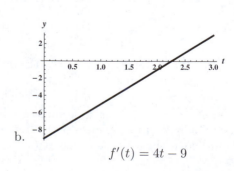

b.

$$f'(t) = 4t - 9$$

b. $f'(t) = 0$ when $t = 9/4$ – that is when the object is stationary. For $0 \le t < 9/4$ we have $f'(t) < 0$ so the object is moving to the left. For $9/4 < t \le 3$ we have $f'(t) > 0$ so the object is moving to the right.

c. $f'(1) = -5$ ft/sec and $f''(t) = 4$ ft/sec^2, so in particular, $f''(1) = 4$ ft/sec^2.

d. $f'(t) = 0$ when $t = 9/4$ and $f''(9/4) = 4$ ft/sec^2.

e. On the interval $(9/4, 3]$ both the velocity and acceleration are positive, so the object's speed is increasing.

3.6.15

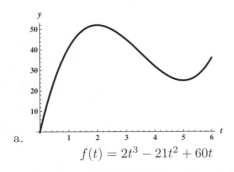

a. $f(t) = 2t^3 - 21t^2 + 60t$

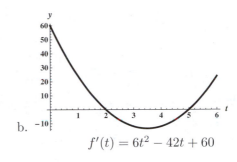

b. $f'(t) = 6t^2 - 42t + 60$

b. $f'(t) = 0$ when $6(t-2)(t-5) = 0$, which is at $t = 2$ and $t = 5$ – that is when the object is stationary. For $0 \le t < 2$ we have $f'(t) > 0$ so the object is moving to the right. For $2 < t < 5$ we have $f'(t) < 0$ so the object is moving to the left. For $5 < t \le 8$ we have $f'(t) > 0$, so the object is moving to the right again.

c. $f'(1) = 24$ ft/sec and $f''(t) = 12t - 42$, so $f''(1) = -30$ ft/sec^2.

d. $f'(t) = 0$ when $t = 2$ and $t = 5$. We have $f''(2) = -18$ ft/sec^2 and $f''(5) = 18$ ft/sec^2.

e. $f''(t) = 12t - 42$ is positive for $t > \frac{42}{12} = \frac{7}{2}$ and negative for $t < \frac{7}{2}$. So f' and f'' are both positive on $(5, 6]$ and are both negative on $(2, 3.5)$, so that is where the object is speeding up.

3.6.17

a. $v(t) = s'(t) = -32t + 64$ ft/sec.

b. The highest point is reached at the instant when the stone changes from moving upward (where $v > 0$) to moving downward (where $v < 0$), so it must occur when $v = 0$, which is at $t = 2$.

c. The height of the stone at its highest point is $s(2) = -16 \cdot 4 + 64 \cdot 2 + 32 = 96$ feet.

d. The stone strikes the ground when $s(t) = 0$ for $t > 0$. Using the quadratic formula we see that this occurs when $t = 2 + \sqrt{6} \approx 4.45$ seconds.

e. The velocity when the stone hits the ground is $v(2 + \sqrt{6}) = -32(2 + \sqrt{6}) + 64 = -32\sqrt{6} \approx -78.38$ feet per second.s

f. The acceleration due to gravity is always negative, so the object is speeding up when its velocity is negative; that is, on its downward journey during the interval $(2, 2 + \sqrt{6})$.

3.6.19

    a. The average growth rate from 1995 to 2005 is
$$\frac{p(10) - p(0)}{10 - 0} = \frac{8038 - 7055}{10} = 98.3 \text{ thousand}$$
people/year.

    b The instantaneous growth rate is $p'(t) = -0.54t + 101$. In 1997 we have $p'(2) = 99.92$ thousand people per year and in 2005 we have $p'(10) = 95.6$ thousand people per year.

    c.
The population was growing but the rate was slowing over this time interval.

3.6.21

    a. The average cost function is given by $\overline{C}(x) = \frac{C(x)}{x} = \frac{1000}{x} + .1$. The marginal cost function is given by $M(x) = C'(x) = .1$.

    b. At $a = 2000$ we have $\overline{C}(2000) = \frac{1000}{2000} + .1 = .6$, and $M(2000) = .1$.

    c. The average cost per item when producing 2000 items is \$0.60. The cost of producing the next item is \$0.10.

3.6.23

    a. The average cost function is given by $\overline{C}(x) = \frac{C(x)}{x} = \frac{100}{x} + 40 - 0.01x$. The marginal cost function is given by $M(x) = C'(x) = 40 - 0.02x$.

    b. At $a = 1000$ we have $\overline{C}(1000) = \frac{100}{1000} + 40 - (.01)(1000) = 30.1$, and $M(1000) = 20$.

    c. The average cost per item when producing 1000 items is \$30.10. The cost of producing the next item is \$20.00.

3.6.25

    a. $D(10) = 40 - 20 = 20$ DVDs per day.

    b. Demand is zero when $D(p) = 40 - 2p = 0$, which occurs for $p = 20$ dollars.

    c. The elasticity is $E(p) = \frac{dD}{dp}\frac{p}{D} = -2\left(\frac{p}{40-2p}\right) = \frac{p}{p-20}$.

    d. This quantity satisfies $-1 < E(p) < 0$ when $-1 < \frac{p}{p-20} < 0$ which occurs when $p < 20 - p$, or $p < 10$. So for prices in the interval $(0, 10)$ the demand is inelastic, while for prices in the interval $(10, 20)$ the demand is elastic.

    e. If the price goes up from 10 to 10.25, that is a $\frac{.25}{10} = .025 = 2.5\%$ increase in price.

    f. If the price goes up from 10 to 10.25, the demand goes from $D(10) = 40 - 20 = 20$ to $D(10.25) = 40 - 20.5 = 19.5$, which is a $\frac{.5}{20} = 2.5\%$ decrease.

3.6.27 $E(p) = \frac{dD}{dp}\frac{p}{D} = -\frac{a}{2}\frac{1}{p^{3/2}}\frac{p\sqrt{p}}{a} = -\frac{1}{2}$, so the demand is inelastic for all prices p.

3.6.29

    a. False. For example, when a ball is thrown up in the air near the surface of the earth, its acceleration is constant (due to gravity) but its velocity changes during its trip.

    b. True. If the rate of change of velocity is zero, then velocity must be constant.

   c. False. If the velocity is constant over an interval, then the average velocity is equal to the instantaneous velocity over the interval.

   d. True. For example, a ball dropped from a tower has negative acceleration and increasing speed as it falls toward the earth.

3.6.31 In each case, the stone reaches its maximum height when its velocity is zero.

On Mars, this occurs when $v(t) = s'(t) = 96 - 12t = 0$, or when $t = 8$ seconds. So the maximum height on Mars is $s(8) = 384$ feet.

On Earth, this occurs when $v(t) = s'(t) = 96 - 32t = 0$, or when $t = 3$ seconds. So the maximum height on Earth is $s(3) = 144$ feet.

The stone will travel $384 - 144 = 240$ feet higher on Mars.

3.6.33 The first stone reaches its maximum height when $f'(t) = -32t + 32 = 0$, so after 1 second, and its maximum height is therefore $f(1) = -16 + 32 + 48 = 64$ feet.

The second stone reaches its maximum height when $g'(t) = -32t + v_0 = 0$, so when $t = \frac{v_0}{32}$. Its height at that time is $g(v_0/32) = -16(v_0/32)^2 + (v_0^2/32) = \frac{v_0^2}{64}$. This is equal to 64 when $v_0 = 64$ feet per second.

3.6.35

   a. The velocity is zero at $t = 1, 2,$ and 3.

   b. The object is moving in the positive direction when the slope of s is positive, so from $t = 0$ to $t = 1$, and from $t = 2$ to $t = 3$. It is moving in the negative direction from $t = 1$ to $t = 2$, and for $t > 3$.

   c.

   d. The speed is increasing on $(0, 1/2)$ as the velocity is positive and the acceleration is positive there. On $(1/2, 1)$ the speed is decreasing as the velocity is positive but the acceleration is negative. On $(1, 3/2)$ the speed is increasing as the velocity is negative and the acceleration is negative, but on $(3/2, 2)$ the speed is decreasing as the velocity is negative but the acceleration is positive. On approximately $(2, 2.6)$ the speed is increasing as the velocity is positive and the acceleration is positive, but on about $(2.6, 3)$ the velocity is positive but the acceleration is negative, so the speed is decreasing. On $(3, \infty)$ both the velocity and the acceleration are negative, so the object is speeding up.

3.6.37

   a. $P(x) = xp(x) - C(x) = 100x + 0.02x^2 - 50x - 100 = 0.02x^2 + 50x - 100$.

   b. The average profit is $\overline{P}(x) = \frac{P(x)}{x} = 0.02x + 50 - \frac{100}{x}$. The marginal profit is $P'(x) = .04x + 50$.

   c. $\overline{P}(500) = 59.8$. $P'(500) = 70$.

   d. The average profit for the first 500 items sold is \$59.80, while the profit on the 501st item is \$70.00.

3.6.39

   a. $P(x) = xp(x) - C(x) = 100x + 0.04x^2 - 800$.

b. The average profit is $\overline{P}(x) = \frac{P(x)}{x} = .04x + 100 - \frac{800}{x}$. The marginal profit is $P'(x) = .08x + 100$.

c. $\overline{P}(1000) = 139.2$. $P'(1000) = 180$.

d. The average profit for the first 1000 items sold is \$139.20, while the profit on the 1001st item is \$180.00.

3.6.41

a. Because the graph represents the growth rate, the slowest rate (of about 1.1 million people per year) occurs at about $t = 30$, which corresponds to the year 1930.

b. The largest growth rate occurs at $t = 60$, so the year 1960 at the largest growth rate of about 2.9 million per year.

c. Because $p'(t) > 0$ for all t shown on the graph, $p(t)$ is never decreasing.

d. The population growth rate $p'(t)$ is increasing from about 1905 to 1915, from 1930 to 1960, and from 1980 to 1990.

3.6.43

a.

b. $v(t) = s'(t) = \frac{(t+1)100 - 100t \cdot 1}{(t+1)^2} = \frac{100}{(t+1)^2}$.

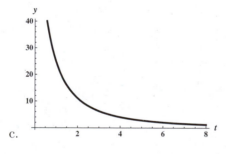

c.

The velocity of the marble is decreasing.

d. $s(t) = 80$ when $\frac{100t}{t+1} = 80$, or $100t = 80t + 80$, which occurs when $t = 4$ seconds.

e. $v(t) = 50$ when $\frac{100}{(t+1)^2} = 50$, or $(t+1)^2 = 2$. This occurs for $t = \sqrt{2} - 1 \approx 0.414$ seconds.

3.6.45

a. The average cost function is $\overline{C}(x) = \frac{C(x)}{25000} = 50 + \frac{5000}{x} + 0.00006x$. The marginal cost function is $C'(x) = -\frac{125000000}{x^2} + 1.5$.

The average cost decreases to about 50 per unit as the batch size increases, while the marginal cost is negative but increases.

b. $\overline{C}(5000) = 51.3$, $C'(5000) = -3.5$.

c. If the batch size is 5000, then the average cost of producing 25000 items is $51.30 per item. If the batch size is increased from 5000 to 5001, then the cost of producing 25000 items would decrease by about $3.50.

3.6.47

a.

$$R(p) = \frac{100p}{p^2+1}$$

b.

$$R'(p) = \frac{100(1-p^2)}{(p^2+1)^2}$$

c. $R'(p)$ is zero at $p=1$, and the maximum of $R(p)$ occurs at this same value of p, so that is the price to charge in order to maximize revenue. The revenue at this price is $50.00.

3.6.49

a. The mass oscillates about the equilibrium point.

b. $\frac{dx}{dt} = 10\cos t + 10\sin t$ is the velocity of the mass at time t.

c. $\frac{dx}{dt} = 0$ when $\sin t = -\cos t$, which occurs when $t = \frac{4n+3}{4} \cdot \pi$ where n is any positive integer.

d. The model is unrealistic as it ignores the effects of friction and gravity. In reality, the amplitude would decrease as the mass oscillates.

3.6.51

a. Juan starts out faster, but slows toward the end, while Jean starts slower but increases her speed toward the end.

b. Because both start and finish at the same time, they finish with the same average angular velocity.

c. It is a tie.

d. Jean's velocity is given by $\theta'(t) = \frac{\pi t}{4}$. At $t=2$, $\theta'(2) = \frac{\pi}{2}$ radians per minute. Her velocity is greatest at $t=4$.

e. Juan's velocity is given by $\phi'(t) = \pi - \frac{\pi t}{4}$. At $t=2$, $\phi'(2) = \frac{\pi}{2}$ radians per minute as well. His velocity is greatest at $t=0$.

3.6.53

a.

At the beginning the volume is 4,000,000 cubic meters.

b. The tank is empty when $V(t) = 100(200 - t)^2 = 0$, which occurs when $t = 200$.

c.

Because $V(t)$ can be written as $V(t) = 4,000,000 - 40,000t + 100t^2$, the flow rate is $V'(t) = -40,000 + 200t$ cubic meters per minute.

d. The flow rate is largest (in absolute value) when $t = 0$ and smallest when $t = 200$.

3.6.55

a. $T'(t) = 160 - 80x$, so $T'(1) = 80$, so the heat flux at 1 is -80. At $x = 3$ we have $T'(3) = -80$, so the heat flux at 3 is 80.

b. The heat flux $-T'(x)$ is negative for $0 \leq x < 2$ and positive for $2 < x \leq 4$.

c. At any point other than the midpoint of the rod, heat flows toward the closest end of the rod, and "out the end."

3.7   The Chain Rule

3.7.1 If $y = f(x)$ and $u = g(x)$ then $\dfrac{dy}{dx} = \dfrac{dy}{du} \cdot \dfrac{du}{dx}$. Alternatively, we have $\dfrac{d}{dx}(f(g(x))) = f'(g(x))g'(x)$.

3.7.3 The derivative of $f(g(x))$ equals f' evaluated at $g(x)$ multiplied by g' evaluated at x.

3.7.5 The inner function is $x^2 + 10$ and the outer function is u^{-5}, so with $y = f(u)$ and $u = g(x)$, we have $f(u) = u^{-5}$ and $g(x) = x^2 + 10$. Then $y = (x^2 + 10)^{-5}$.

3.7.7 With $u = 3x + 7$ and $y = u^{10}$ we have $\frac{dy}{dx} = \frac{dy}{dx} \cdot \frac{du}{dx} = 10u^9 \cdot 3 = 30(3x + 7)^9$.

3.7.9 With $u = \sin x$ and $y = u^5$ we have $\frac{dy}{dx} = \frac{dy}{du} \cdot \frac{du}{dx} = 5u^4 \cdot \cos x = 5 \sin^4 x \cos x$.

3.7.11 With $u = 2x^2 + 3$ and $y = \sqrt{u}$, we have $\frac{dy}{dx} = \frac{dy}{du}\frac{du}{dx} = 4x \cdot \frac{1}{2\sqrt{u}} = \frac{2x}{\sqrt{2x^2+3}}$.

3.7.13 With $u = x^2 + 1$ and $y = \sqrt{u}$ we have $\frac{dy}{dx} = \frac{dy}{du} \cdot \frac{du}{dx} = \frac{1}{2\sqrt{u}} \cdot (2x) = \frac{x}{\sqrt{x^2+1}}$.

3.7.15 With $u = 5x^2$ and $y = \tan u$ we have $\frac{dy}{dx} = \frac{dy}{du} \cdot \frac{du}{dx} = \sec^2 u \cdot (10x) = 10x \sec^2 5x^2$.

3.7.17 With $u = 5x + 1$ and $y = \sec u$, we have $\frac{dy}{dx} = \frac{dy}{du}\frac{du}{dx} = \sec u \tan u \cdot 5 = 5 \sec(5x + 1) \tan(5x + 1)$.

3.7.19 With $g(x) = 3x^2 + 7x$ and $f(u) = u^{10}$ we have $\frac{d}{dx}[f(g(x))] = f'(g(x))g'(x) = 10(3x^2 + 7x)^9(6x + 7)$.

3.7.21 With $g(x) = 10x + 1$ and $f(u) = \sqrt{u}$, we have $\frac{d}{dx}[f(g(x))] = f'(g(x))g'(x) = \frac{1}{2\sqrt{10x+1}} \cdot 10 = \frac{5}{\sqrt{10x+1}}$.

3.7.23 With $g(x) = 7x^3 + 1$ and $f(u) = 5u^{-3}$ we have $\frac{d}{dx}[f(g(x))] = f'(g(x))g'(x) = -15(7x^3+1)^{-4}(21x^2) = -315(7x^3 + 1)^{-4} \cdot x^2$.

3.7.25 With $g(x) = 3x + 1$ and $f(u) = \sec u$, we have $\frac{d}{dx}[f(g(x))] = f'(g(x))g'(x) = \sec(3x + 1)\tan(3x + 1) \cdot 3 = 3\sec(3x + 1)\tan(3x + 1)$.

3.7.27 With $g(w) = \sqrt{w}$ and $f(u) = \tan u$ we have $\frac{d}{dw}[f(g(w))] = f'(g(w))g'(w) = \sec^2\sqrt{w} \cdot \frac{1}{2\sqrt{w}} = \frac{\sec^2\sqrt{w}}{2\sqrt{w}}$.

3.7.29 With $g(x) = 4x^3 + 3x + 1$ and $f(u) = \sin u$ we have $\frac{d}{dx}[f(g(x))] = f'(g(x))g'(x) = \cos u \cdot (12x^2 + 3) = (12x^2 + 3) \cdot \cos(4x^3 + 3x + 1)$.

3.7.31 With $g(x) = 2\sqrt{x}$ and $f(u) = \sin u$, we have $\frac{d}{dx}[f(g(x))] = f'(g(x))g'(x) = \cos(2\sqrt{x}) \cdot \frac{1}{\sqrt{x}} = \frac{\cos(2\sqrt{x})}{\sqrt{x}}$.

3.7.33 With $g(x) = \sec x + \tan x$ and $f(u) = u^5$ we have $\frac{d}{dx}[f(g(x))] = f'(g(x))g'(x) = 5u^4 \cdot (\sec x \tan x + \sec^2 x) = 5(\sec x + \tan x)^4(\sec x \tan x + \sec^2 x) = 5\sec x(\sec x + \tan x)^5$.

3.7.35

a. $u = g(x) = \cos x$, $y = f(u) = u^3$. So $\frac{dy}{dx} = \frac{dy}{du} \cdot \frac{du}{dx} = 3\cos^2 x \cdot (-\sin x) = -3\cos^2 x \sin x$.

b. $u = g(x) = x^3$, $y = f(u) = \cos u$. So $\frac{dy}{dx} = \frac{dy}{du} \cdot \frac{du}{dx} = -\sin x^3 \cdot 3x^2 = -3x^2 \sin x^3$.

3.7.37

a. $h'(3) = f'(g(3))g'(3) = f'(1) \cdot 20 = 5 \cdot 20 = 100$.

b. $h'(2) = f'(g(2))g'(2) = f'(5) \cdot 10 = -10 \cdot 10 = -100$.

c. $p'(4) = g'(f(4))f'(4) = g'(1) \cdot (-8) = 2 \cdot (-8) = -16$.

d. $p'(2) = g'(f(2))f'(2) = g'(3) \cdot 2 = 20 \cdot 2 = 40$.

e. $h'(5) = f'(g(5))g'(5) = f'(2) \cdot 20 = 2 \cdot 20 = 40$.

3.7.39 Note that $a(70) = 13330$. $\frac{d}{dt} p(a(t))|_{t=70} = p'(a(70))a'(70) \approx \frac{738-765}{14330-13330} \cdot \frac{13440-13330}{80-70} = -.297$ hPa per minute.

3.7.41 Take $g(x) = 2x^6 - 3x^3 + 3$, and $n = 25$. Then $y' = n(g(x))^{n-1}g'(x) = 25(2x^6 - 3x^3 + 3)^{24}(12x^5 - 9x^2)$.

3.7.43 Take $g(x) = 1 + 2\tan x$, and $n = 15$. Then $y' = n(g(x))^{n-1}g'(x) = 15(1 + 2\tan x)^{14}(2\sec^2 x) = 30(1 + 2\tan x)^{14}\sec^2 x$.

3.7.45

$$\frac{d}{dx}\sqrt{1 + \cot^2 x} = \frac{1}{2\sqrt{1 + \cot^2 x}} \cdot \frac{d}{dx}(1 + \cot^2 x) = \frac{1}{2\sqrt{1 + \cot^2 x}} \cdot 2\cot x \cdot \frac{d}{dx}\cot x$$
$$= \frac{1}{2\sqrt{1 + \cot^2 x}} \cdot 2\cot x \cdot (-\csc^2 x) = -\frac{\cot x \csc^2 x}{\sqrt{1 + \cot^2 x}}.$$

3.7.47

$$\frac{d}{dx}\sin(\sin\sqrt{x}) = \cos(\sin\sqrt{x})\frac{d}{dx}\sin\sqrt{x}$$
$$= \cos(\sin\sqrt{x}) \cdot \cos(\sqrt{x}) \cdot \frac{1}{2\sqrt{x}}.$$

3.7.49

$$\frac{d}{dx}\sin^5(\cos 3x) = 5\sin^4(\cos 3x) \cdot \frac{d}{dx}(\sin(\cos 3x))$$

$$= 5\sin^4(\cos 3x) \cdot \cos(\cos 3x) \cdot \frac{d}{dx}\cos 3x$$

$$= 5\sin^4(\cos 3x) \cdot \cos(\cos 3x) \cdot (-\sin 3x) \cdot 3$$

$$= -15\sin^4(\cos 3x)\cos(\cos 3x)\sin 3x.$$

3.7.51

$$\frac{d}{dx}(\sin^2 x + 1)^4 = 4(\sin^2 x + 1)^3 \cdot \frac{d}{dx}(\sin^2 x + 1) = 4(\sin^2 x + 1)^3 \cdot 2\sin x \cdot \frac{d}{dx}\sin x$$

$$= 4(\sin^2 x + 1)^3 \cdot 2\sin x \cos x = 8(\sin^2 x + 1)^3 \sin x \cos x.$$

3.7.53 $\frac{d}{dx}\sqrt{x + \sqrt{x}} = \frac{1}{2\sqrt{x+\sqrt{x}}} \cdot \frac{d}{dx}(x + \sqrt{x}) = \frac{1}{2\sqrt{x+\sqrt{x}}} \cdot \left(1 + \frac{1}{2\sqrt{x}}\right).$

3.7.55 $\frac{d}{dx}f(g(x^2)) = f'(g(x^2)) \cdot \frac{d}{dx}(g(x^2)) = f'(g(x^2)) \cdot g'(x^2) \cdot 2x.$

3.7.57 $y' = 5\left(\frac{x}{x+1}\right)^4 \cdot \frac{(x+1)(1) - x(1)}{(x+1)^2} = \frac{5x^4}{(x+1)^6}.$

3.7.59 $y' = 1 \cdot (x^2 + 1)^3 + x \cdot 3(x^2 + 1)^2 \cdot 2x = (x^2 + 1)^2(x^2 + 1 + 6x^2) = (x^2 + 1)^2(7x^2 + 1).$

3.7.61 $\frac{dy}{d\theta} = 2\theta\sec 5\theta + \theta^2(5\sec 5\theta\tan 5\theta) = \theta\sec 5\theta(2 + 5\theta\tan 5\theta).$

3.7.63 $y' = 4((x+2)(x^2+1))^3 \cdot ((1)(x^2+1) + (x+2)(2x)) = 4((x+2)(x^2+1))^3(3x^2 + 4x + 1) = 4(x+2)^3(x^2+1)^3(3x+1)(x+1).$

3.7.65 $y' = \frac{1}{2}(x^4 + \cos 2x)^{-1/2}(4x^3 - 2\sin 2x) = \frac{2x^3 - \sin 2x}{\sqrt{x^4 + \cos 2x}}.$

3.7.67 $y' = 2(p+\pi)^1 \sin p^2 + (p+\pi)^2(\cos p^2)(2p) = (p+\pi)(2\sin p^2 + 2p^2\cos p^2 + 2p\pi\cos p^2) = 2(p+\pi)(\sin p^2 + p^2\cos p^2 + p\pi\cos p^2).$

3.7.69

a. True. The product rule alone will suffice.

b. True. This function is the composition of $x^2 + 1$ with x^{-12}.

c. True. The derivative of the composition $f(g(x))$ is the product of $f'(g(x))$ with $g'(x)$, so it is the product of two derivatives.

d. False. In fact, $\frac{d}{dx}P(Q(x)) = P'(Q(x))Q'(x)$.

3.7.71

$$\frac{d^2}{dx^2}\sin x^2 = \frac{d}{dx}(2x\cos x^2) = 2(\cos x^2 - 2x^2\sin x^2)$$

$$= 2\cos x^2 - 4x^2\sin x^2.$$

Note that in the middle of this calculation we used a result from the middle of the previous problem – namely the derivative of $x\cos x^2$.

3.7.73 $\frac{d^2}{dx^2}(x^2+1)^{-2} = \frac{d}{dx}\left(-2(x^2+1)^{-3} \cdot 2x\right) = -\frac{d}{dx}\left(\frac{4x}{(x^2+1)^3}\right) = -\frac{(x^2+1)^3 \cdot 4 - 4x \cdot 3(x^2+1)^2(2x)}{(x^2+1)^6} = -\frac{4(x^2+1) - 24x^2}{(x^2+1)^4} = -\frac{4 - 20x^2}{(x^2+1)^4} = \frac{4(5x^2-1)}{(x^2+1)^4}.$

3.7.75 $\frac{d}{dx}\sqrt{f(x)} = \frac{1}{2\sqrt{f(x)}} \cdot f'(x)$.

3.7.77

$y' = \frac{(x^3-6x-1)(2)(x^2-1)(2x)-(x^2-1)^2(3x^2-6)}{(x^3-6x-1)^2}$,

so $y'(3) = \frac{(27-18-1)(2)(8)(6)-(64)(21)}{64} = \frac{768-1344}{64} =$

$-\frac{576}{64} = -9$. The equation of the tangent line is

thus $y - 8 = -9(x - 3)$, or $y = -9x + 35$.

3.7.79

a. $g'(4) = 3$, $g(4) = 3 \cdot 4 - 5 = 7$. $f'(7) = -2$, $f(7) = -2 \cdot 7 + 23 = 9$. Thus, $h(4) = f(g(4)) = f(7) = 9$, and $h'(4) = f'(g(4))g'(4) = f'(7) \cdot 3 = -2 \cdot 3 = -6$.

b. The tangent line to h at $(4, 9)$ is given by $y - 9 = -6(x - 4)$, or $y = -6x + 33$.

3.7.81 $y' = 2\sec(2x)\tan(2x)$; $y(\pi/6) = \sec \pi/3 = 2$ and $y'(\pi/6) = 2\sec(\pi/3)\tan(\pi/3) = 4\sqrt{3}$, so the tangent line at $\pi/6$ is $y - 2 = 4\sqrt{3}\left(x - \frac{\pi}{6}\right)$, or $y = 4\sqrt{3}\left(x - \frac{\pi}{6}\right) + 2$.

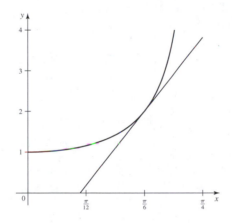

3.7.83 First, note that $g'(x) = \cos(\pi f(x)) \cdot \pi f'(x)$.

a. $g'(0) = \cos(\pi \cdot f(0)) \cdot \pi f'(0) = \cos(-3\pi) \cdot 3\pi = -3\pi$.

b. $g'(1) = \cos(\pi \cdot f(1)) \cdot \pi f'(1) = \cos(3\pi) \cdot 5\pi = -5\pi$.

3.7.85

a. $\frac{d^2y}{dt^2} = \frac{d}{dt}\left(-y_0\sqrt{\frac{k}{m}}\sin\left(t\sqrt{\frac{k}{m}}\right)\right) = -y_0 \cdot \frac{k}{m} \cdot \cos\left(t\sqrt{\frac{k}{m}}\right)$.

b. $-\frac{k}{m}y = -\frac{k}{m}\left(y_0\cos\left(t\sqrt{\frac{k}{m}}\right)\right) = \frac{d^2y}{dt^2}$.

3.7.87

a. Assuming a non leap year, March 1st corresponds to $t = 59$. We have $D(59) = 12 - 3\cos\left(\frac{2\pi(69)}{365}\right) \approx$ 10.88 hours.

b. $\frac{d}{dt}D(t) = 3 \cdot \frac{2\pi}{365}\sin\left(\frac{2\pi(t+10)}{365}\right)$ hours per day.

c. $D'(59) \approx 0.048$ hours per day ≈ 2 minutes and 52 seconds per day. This means that on March 1st, the days are getting longer by just shy of 3 minutes per day.

d.

e. The largest increase in the length of the days appears to be at about $t = 81$, and the largest decrease at about $t = 265$. These correspond to March 22nd and to September 22nd. The least rapid changes occur at about $t = 172$ and $t = 355$. These correspond to June 21st and December 21st.

3.7.89

a. $E'(t) = 400 + 200\cos\left(\frac{\pi t}{12}\right)$ MW.

b. Because the maximum value of $\cos\theta$ is 1, the maximum value of $E'(t)$ will be 600 MW, where $\cos\left(\frac{\pi t}{12}\right) = 1$, which is where $t = 0$, which corresponds to noon.

c. Because the minimum value of $\cos\theta$ is -1, the minimum value of $E'(t)$ will be 200 MW, where $\cos\left(\frac{\pi t}{12}\right) = -1$, which is where $\frac{\pi t}{12} = \pi$, or $t = 12$, which corresponds to midnight.

d.

3.7.91

a. $f'(x) = \frac{d}{dx}(\cos^2 x + \sin^2 x) = 2(\cos x)(-\sin x) + 2\sin x \cos x = 0.$

b. If $f(x)$ is a constant, then the output value must be the same at any input value, so we choose to evaluate f at a nice value like $x = 0$. We see that $f(0) = \cos^2 0 + \sin^2 0 = 1^2 + 0^2 = 1$, so we must have $\cos^2 x + \sin^2 x = 1$ for all x.

3.7.93 $\frac{d}{dx}\left(f(x)(g(x))^{-1}\right) = f'(x)(g(x))^{-1} + f(x)(-(g(x))^{-2}g'(x)) = \frac{f'(x)}{g(x)} - \frac{f(x)g'(x)}{(g(x))^2} = \frac{g(x)f'(x) - f(x)g'(x)}{(g(x))^2}.$

3.7.95

a. $h(x) = (x^2 - 3)^5$, $a = 2$.

b. $h'(x) = 5(x^2 - 3)^4(2x) = 10x(x^2 - 3)^4$, so the value of this limit is $h'(2) = 20$.

3.7.97

a. $h(x) = \sin x^2$, $a = \frac{\pi}{2}$.

b. $h'(x) = (\cos x^2)(2x)$, so the value of this limit is $h'\left(\frac{\pi}{2}\right) = \pi \cdot \cos\left(\frac{\pi^2}{4}\right) \approx -2.45.$

3.7.99 $\lim\limits_{x \to 5} \dfrac{f(x^2) - f(25)}{x - 5} = \dfrac{d}{dx}\left[f(x^2)\right]_{x=5} = 2 \cdot 5 \cdot f'(25) = 10f'(25).$

3.7.101

a. $\lim\limits_{v \to u} H(v) = \lim\limits_{v \to u}\left(\dfrac{f(v) - f(u)}{v - u} - f'(u)\right) = \lim\limits_{v \to u}\left(\dfrac{f(v) - f(u)}{v - u}\right) - f'(u) = f'(u) - f'(u) = 0.$

b. Suppose $u = v$. Then clearly both sides of the given expression are 0, so they are equal. Suppose $u \neq v$. Then $H(v) = \frac{f(v)-f(u)}{v-u} - f'(u)$, so $H(v) + f'(u) = \frac{f(v)-f(u)}{v-u}$, so the result holds by multiplying both sides of this equation by $v - u$.

c.

$$h'(a) = \lim_{x \to a} \frac{f(g(x)) - f(g(a))}{x-a} = \lim_{x \to a} \frac{H(g(x)) + f'(g(a))}{x-a} \cdot (g(x) - g(a))$$

$$= \lim_{x \to a} \left[(H(g(x)) + f'(g(a))) \cdot \frac{g(x)-g(a)}{x-a} \right].$$

d. $h'(a) = \lim_{x \to a} \left[(H(g(x)) + f'(g(a))) \cdot \frac{g(x)-g(a)}{x-a} \right] = (0 + f'(g(a))) \cdot g'(a) = f'(g(a))g'(a).$

3.8 Implicit Differentiation

3.8.1 Implicit differentiation gives a single unified derivative, whereas solving for y explicitly yields two different functions.

3.8.3 The result of implicit differentiation is often an expression involving both the dependent and independent variables, so one would need to know both in order to calculate the value of the derivative.

3.8.5

a. $4x^3 + 4y^3 \frac{dy}{dx} = 0$. Thus $4y^3 \frac{dy}{dx} = -4x^3$, so $\frac{dy}{dx} = -\frac{x^3}{y^3}$.

b. When $x = 1$ and $y = -1$, we have $\frac{dy}{dx} = \frac{-1}{-1} = 1$.

3.8.7

a. $2y\frac{dy}{dx} = 4$, so $\frac{dy}{dx} = \frac{2}{y}$.

b. $\left.\frac{dy}{dx}\right|_{(1,2)} = \frac{2}{2} = 1$.

3.8.9

a. $\frac{dy}{dx} \cos y = 20x^3$, so $\frac{dy}{dx} = \frac{20x^3}{\cos y}$.

b. $\left.\frac{dy}{dx}\right|_{(1,\pi)} = \frac{20}{\cos \pi} = -20$.

3.8.11

a. $-\frac{dy}{dx} \sin y = 1$, so $\frac{dy}{dx} = -\frac{1}{\sin y} = -\csc y$.

b. $\left.\frac{dy}{dx}\right|_{(0,\pi/2)} = -\csc(\pi/2) = -1$.

3.8.13 $(y + x\frac{dy}{dx})\cos(xy) = 1 + \frac{dy}{dx}$, so $y\cos(xy) + x\frac{dy}{dx}\cos(xy) = 1 + \frac{dy}{dx}$. If we rearrange terms in order to have the terms with a factor of $\frac{dy}{dx}$ all on the same side, we obtain $y\cos(xy) - 1 = \frac{dy}{dx} - x\frac{dy}{dx}\cos(xy)$. Factoring out the $\frac{dy}{dx}$ factor gives $y\cos(xy) - 1 = \frac{dy}{dx}(1 - x\cos(xy))$, so $\frac{dy}{dx} = \frac{y\cos(xy)-1}{1-x\cos(xy)}$.

3.8.15 $1 + \frac{dy}{dx} = -\sin y \cdot \frac{dy}{dx}$, so $\frac{dy}{dx} + (\sin y)\frac{dy}{dx} = -1$, and $\frac{dy}{dx} = -\frac{1}{1+\sin y}$.

3.8.17 $-2y\frac{dy}{dx}\sin y^2 + 1 = 2y\frac{dy}{dx}$, which we can write as $1 = \frac{dy}{dx}2y + 2y\frac{dy}{dx}\sin y^2$, or $1 = \frac{dy}{dx}(2y + 2y\sin y^2)$. Thus, $\frac{dy}{dx} = \frac{1}{2y+2y\sin y^2} = \frac{1}{2y(1+\sin y^2)}$.

3.8.19

$$3x^2 = \frac{(x-y)(1 + \frac{dy}{dx}) - (x+y)(1 - \frac{dy}{dx})}{(x-y)^2}$$

$$3x^2(x-y)^2 = x + x\frac{dy}{dx} - y - y\frac{dy}{dx} - x + x\frac{dy}{dx} - y + y\frac{dy}{dx}$$

$$3x^2(x-y)^2 + 2y = 2x\frac{dy}{dx}$$

$$\frac{dy}{dx} = \frac{3x^2(x-y)^2 + 2y}{2x}$$

3.8.21

$$18x^2 + 21\frac{dy}{dx}y^2 = 13(y + x\frac{dy}{dx})$$

$$21\frac{dy}{dx}y^2 - 13x\frac{dy}{dx} = 13y - 18x^2$$

$$\frac{dy}{dx} = \frac{13y - 18x^2}{21y^2 - 13x}.$$

3.8.23

$$\frac{4x^3 + 2y\frac{dy}{dx}}{2\sqrt{x^4 + y^2}} = 5 + 6y^2\frac{dy}{dx}$$

$$y\frac{dy}{dx} - 6\frac{dy}{dx}y^2\sqrt{x^4 + y^2} = 5\sqrt{x^4 + y^2} - 2x^3$$

$$\frac{dy}{dx} = \frac{5\sqrt{x^4 + y^2} - 2x^3}{y - 6y^2\sqrt{x^4 + y^2}}.$$

3.8.25

　　a. $2^2 + 2 \cdot 1 + 1^2 = 7$, so the point $(2, 1)$ does lie on the curve.

　　b. $2x + y + xy' + 2yy' = 0$, which can be written $(x + 2y)y' = -2x - y$. Solving for y' yields $y' = \frac{-2x-y}{x+2y}$. Thus, at the point $(2, 1)$ we have $y' = -\frac{5}{4}$. The equation of the tangent line is therefore $y - 1 = -\frac{5}{4}(x - 2)$, or $y = -\frac{5}{4}x + \frac{7}{2}$.

3.8.27

　　a. $\sin \pi + 5\frac{\pi^2}{5} = \pi^2$, so the point $(\pi^2/5, \pi)$ does lie on the curve.

　　b. $y'\cos y + 5 = 2yy'$, so $5 = y'(2y - \cos y)$, so $y' = \frac{5}{2y-\cos y}$. At the given point we have $y' = \frac{5}{2\pi+1}$. The equation of the tangent line is therefore $y - \pi = \frac{5}{1+2\pi}\left(x - \pi^2/5\right)$, or $y = \frac{5}{1+2\pi}x + \frac{\pi(1+\pi)}{1+2\pi}$.

3.8.29

　　a. $\cos(\pi/2 - \pi/4) + \sin(\pi/4) = (\sqrt{2}/2) + (\sqrt{2}/2) = \sqrt{2}$, so the point $(\pi/2, \pi/4)$ does lie on the curve.

　　b. $(1 - y')(-\sin(x - y)) + y'\cos y = 0$, which can be written as $y'(\cos y + \sin(x - y)) = \sin(x - y)$, so $y' = \frac{\sin(x-y)}{\cos y+\sin(x-y)}$. At the given point we have $y' = 1/2$. The equation of the tangent line is therefore $y - (\pi/4) = (1/2)(x - \pi/2)$, or $y = \frac{1}{2}x$.

3.8.31 $1 + 2yy' = 0$, so $y' = -\frac{1}{2y}$. Differentiating again, we obtain

$$y'' = -\frac{1}{2} \cdot \frac{-y'}{y^2} = \frac{y'}{2y^2} = \left(-\frac{1}{2y}\right) \cdot \frac{1}{2y^2} = -\frac{1}{4y^3}.$$

3.8.33 $1 + \frac{dy}{dx} = (\cos y)\frac{dy}{dx}$, so $1 = \frac{dy}{dx}(\cos y - 1)$, and thus $\frac{dy}{dx} = \frac{1}{\cos y-1}$. Thus

$$\frac{d^2y}{dx^2} = -\frac{1}{(\cos y - 1)^2} \cdot \left(-\sin y\frac{dy}{dx}\right) = \frac{\sin y}{(\cos y - 1)^2} \cdot \frac{1}{(\cos y - 1)} = \frac{\sin y}{(\cos y - 1)^3}.$$

3.8.35 $y'\cos y + 1 = y'$, so $1 = y'(1 - \cos y)$ and thus $y' = \frac{1}{1-\cos y} = (1 - \cos y)^{-1}$. Then

$$y'' = -1(1 - \cos y)^{-2}(y'\sin y) = -\frac{y'\sin y}{(1 - \cos y)^2} = -\frac{\sin y}{(1 - \cos y)^3}.$$

3.8.37 $\frac{dy}{dx} = \frac{5}{4}x^{\frac{5}{4}-1} = \frac{5}{4}x^{\frac{1}{4}}.$

3.8.39 $\dfrac{dy}{dx} = 5 \cdot \dfrac{2}{3}(5x+1)^{-\frac{1}{3}} = \dfrac{10}{3(5x+1)^{\frac{1}{3}}}.$

3.8.41 $\dfrac{dy}{dx} = \dfrac{1}{4}\left(\dfrac{2x}{4x-3}\right)^{-\frac{3}{4}} \cdot \dfrac{2(4x-3)-2x\cdot 4}{(4x-3)^2} = -\dfrac{3}{2}\left(\dfrac{4x-3}{2x}\right)^{\frac{3}{4}} \cdot \dfrac{1}{(4x-3)^2} = -\dfrac{3}{2^{7/4}x^{3/4}(4x-3)^{5/4}}.$

3.8.43 Note that $y = (1+x^{1/3})^{2/3}$, so $y' = (2/3)(1+x^{1/3})^{-1/3} \cdot (1/3)x^{-2/3} = \dfrac{2}{9x^{2/3}(1+x^{1/3})^{1/3}}.$

3.8.45 $\dfrac{1}{3}x^{-\frac{2}{3}} + \dfrac{4}{3}y^{\frac{1}{3}}y' = 0$, so at the given point we have $\dfrac{1}{3} + \dfrac{4}{3}y' = 0$, so $y' = -\dfrac{1}{4}.$

3.8.47 $y^{\frac{1}{3}} + \dfrac{1}{3}xy^{-\frac{2}{3}}y' + y' = 0$, so at the given point we have $2 + \dfrac{1}{3}\cdot 1 \cdot \dfrac{1}{4}y' + y' = 0$, so $\dfrac{13}{12}y' = -2$, so $y' = -\dfrac{24}{13}.$

3.8.49 $y + xy' + \dfrac{3}{2}x^{\frac{1}{2}}y^{-\frac{1}{2}} - \dfrac{1}{2}x^{\frac{3}{2}}y^{-\frac{3}{2}}y' = 0$, so at the given point we have $1 + y' + \dfrac{3}{2} - \dfrac{1}{2}y' = 0$, so $\dfrac{1}{2}y' = -\dfrac{5}{2}$, so $y' = -5.$

3.8.51

 a. False. For example, the equation $y\cos(xy) = x$, cannot be solved explicitly for y in terms of x.

 b. True. We have $2x + 2yy' = 0$, and the result follows by solving for y'.

 c. False. The equation $x = 1$ doesn't represent any sort of function – it is either just a number, or perhaps a vertical line, but it doesn't represent a differentiable function.

 d. False. $y + xy' = 0$, so $y' = -\dfrac{y}{x}$, $x \neq 0$.

3.8.53

 a. There are two points on the curve associated with $x = 1$. When $x = 1$, we have $1 + y^2 - y = 1$, so $y(y-1) = 0$. The two points are thus $(1,0)$ and $(1,1)$. Differentiating yields $1 + 2yy' - y' = 0$, so $y' = \dfrac{1}{1-2y}.$

 At $(1,0)$, we have $y' = 1$, so the tangent line is given by $y = x - 1$.

 At $(1,1)$, we have $y' = -1$, so the tangent line is given by $y - 1 = -1(x-1)$, or $y = -x + 2$.

 b.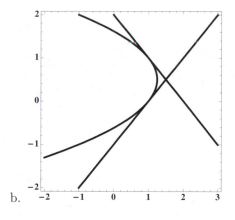

3.8.55

 a. $y(2x) + (x^2+4)y' = 0$, so $y' = -\dfrac{2xy}{x^2+4}.$

 b. At $y = 1$ we have $x^2 + 4 = 8$, so $x = \pm 2$. At the point $(2,1)$ we have $y' = -\dfrac{4}{8} = -\dfrac{1}{2}$. At the point $(-2,1)$ we have $y' = \dfrac{4}{8} = \dfrac{1}{2}$. Thus, the equations of the tangent lines are given by $y - 1 = -\dfrac{1}{2}(x-2)$ and $y - 1 = \dfrac{1}{2}(x+2)$, or $y = -\dfrac{1}{2}x + 2$ and $y = \dfrac{1}{2}x + 2$.

 c. $y = \dfrac{8}{x^2+4}$, so $y' = \dfrac{0-8\cdot 2x}{(x^2+4)^2} = -\dfrac{16x}{(x^2+4)^2}.$

 d. $y' = -\dfrac{16x}{(x^2+4)^2} = -\dfrac{2x}{x^2+4} \cdot \dfrac{8}{x^2+4} = -\dfrac{2x}{x^2+4} \cdot y = -\dfrac{2xy}{x^2+4}.$

3.8.57

a. From number 53, we have that $y' = \frac{1}{1-2y}$. A vertical tangent would occur at a point whose y value would make $1 - 2y$ equal to zero. So we are looking for where $2y = 1$ or $y = \frac{1}{2}$.

If $y = \frac{1}{2}$, then $x + \frac{1}{4} - \frac{1}{2} = 1$, so $x = \frac{5}{4}$, and there is a vertical tangent at $\left(\frac{5}{4}, \frac{1}{2}\right)$.

b. Because y' is never zero, there are no horizontal tangent lines.

3.8.59 Differentiating with respect to x gives $18x + 2y\frac{dy}{dx} - 36 + 6\frac{dy}{dx} = 0$, so

$$2y\frac{dy}{dx} + 6\frac{dy}{dx} = -18x + 36$$
$$(2y + 6)\frac{dy}{dx} = -18x + 36$$
$$\frac{dy}{dx} = \frac{-9x + 18}{y + 3}.$$

This is zero when $x = 2$. Using the original equation, we have $36 + y^2 - 72 + 6y + 36 = 0$, or $y^2 + 6y = 0$, or $y = 0$ and $y = -6$. Thus there are horizontal tangent lines at $(2, 0)$ and $(2, -6)$. So $y = 0$ and $y = -6$ are horizontal tangent lines.

This curve has vertical tangent lines when $y = -3$. Using the original equation, we have $9x^2 + 9 - 36x - 18 + 36 = 0$ or $9x^2 - 36x + 27 = 0$, or $x^2 - 4x + 3 = 0$. This factors as $(x - 3)(x - 1) = 0$, so the corresponding values of x are 3 and 1. Thus the vertical tangent lines occur at $(3, -3)$ and $(1, -3)$.

3.8.61

a. If we write $y^3 - 1 = xy - x$, we have $(y - 1)(y^2 + y + 1) = x(y - 1)$, so $y^2 + y + 1 = x$. Differentiating gives $2y\frac{dy}{dx} + \frac{dy}{dx} = 1$, so $\frac{dy}{dx} = \frac{1}{1+2y}$. Note also that $y = 1$ satisfies the equation; $y' = 0$ on this branch.

b.

$$y^3 - 1 = x(y - 1)$$
$$(y - 1)(y^2 + y + 1) = x(y - 1)$$
$$y^2 + y + 1 = x$$
$$y^2 + y + (1 - x) = 0,$$

so by the quadratic formula we have $y = \frac{-1 \pm \sqrt{4x-3}}{2}$. Note that this means that $\pm\sqrt{4x - 3} = 2y + 1$.

c.

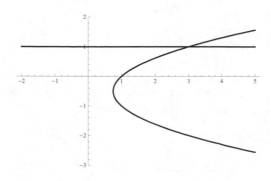

3.8.63

  a. $4x^3 = 4x - 4yy'$, so $y' = \frac{x - x^3}{y}$.

  b. $y = \pm\sqrt{x^2 - \frac{x^4}{2}}$.

  c.

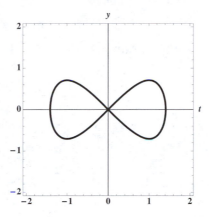

3.8.65

The slope of the normal line is the negative reciprocal of the slope of the tangent line. From 25:
$y' = -\frac{5}{4}$, so the slope of the normal line is $\frac{4}{5}$. At
the point $(2,1)$ we have the line $y - 1 = \frac{4}{5}(x - 2)$,
or $y = \frac{4}{5}x - \frac{3}{5}$.

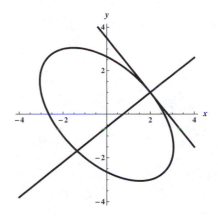

3.8.67

The slope of the normal line is the negative reciprocal of the slope of the tangent line. From 27:
$y' = \frac{5}{2\pi + 1}$, so the slope of the normal line is $\frac{-2\pi - 1}{5}$.
At the point $(\frac{\pi^2}{5}, \pi)$ we have the line $y - \pi = $
$\frac{-2\pi - 1}{5}(x - \frac{\pi^2}{5})$, or $y = \frac{-2\pi - 1}{5}x + \frac{2\pi^3 + \pi^2 + 25\pi}{25}$.

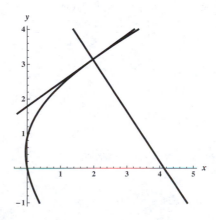

3.8.69

The slope of the normal line is the negative reciprocal of the slope of the tangent line. From 29: $y' = \frac{1}{2}$, so the slope of the normal line is -2. At the point $(\frac{\pi}{2}, \frac{\pi}{4})$ we have the line $y - \frac{\pi}{4} = -2(x - \frac{\pi}{2})$, or $y = -2x + \frac{5\pi}{4}$.

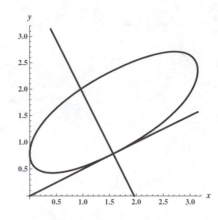

3.8.71

We have $9x^2 + 21y^2 y' = 10y'$, so at the point $(1,1)$ we have $9 + 21y' = 10y'$, so $y' = -\frac{9}{11}$.
Thus, the tangent line is given by $y - 1 = -\frac{9}{11}(x - 1)$, or $y = \frac{-9}{11}x + \frac{20}{11}$. The normal line is given by $y - 1 = \frac{11}{9}(x - 1)$, or $y = \frac{11}{9}x - \frac{2}{9}$.

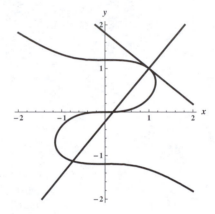

3.8.73

We have $2(x^2 + y^2 - 2x)(2x + 2yy' - 2) = 4x + 4yy'$, so at the point $(2,2)$ we have $2(4 + 4 - 4)(4 + 4y' - 2) = 8 + 8y'$, so $16 + 32y' = 8 + 8y'$, so $y' = -\frac{1}{3}$.
Thus, the tangent line is given by $y - 2 = -\frac{1}{3}(x - 2)$, or $y = \frac{-1}{3}x + \frac{8}{3}$. The normal line is given by $y - 2 = 3(x - 2)$, or $y = 3x - 4$.

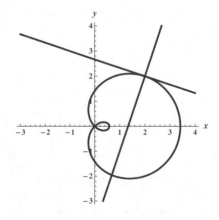

3.8.75

a. $1280 = 40L^{1/3}K^{2/3}$, so $0 = \frac{40}{3}L^{-2/3}K^{2/3} + \frac{80}{3}L^{1/3}K^{-1/3} \cdot \frac{dK}{dL}$. Multiplying both sides by $\frac{3}{40}L^{2/3}K^{1/3}$
yields
$$0 = K + 2L\frac{dK}{dL}, \text{ so } \frac{dK}{dL} = -\frac{1}{2}\frac{K}{L}.$$

b. With $L = 8$ and $K = 64$, $\frac{dK}{dL} = -\frac{64}{16} = -4$.

3.8.77 $V = \frac{\pi h^2(3r-h)}{3} = \frac{5\pi}{3}$. So

$$\frac{1}{3}[2\pi h(3r-h) + \pi h^2(3r'-1)] = 0,$$

$$6rh - 2h^2 + 3h^2 r' - h^2 = 0,$$

so $r' = 1 - \frac{2r}{h}$.

At $r = 2$ and $h = 1$, we have $r' = 1 - 4 = -3$.

3.8.79 Note for $y = mx$, $\frac{dy}{dx} = m = \frac{y}{x}$, and for $x^2 + y^2 = a^2$, $\frac{dy}{dx} = -\frac{x}{y}$. So for any point (x, y), we have $\frac{y}{x}$ and $-\frac{x}{y}$ are negative reciprocals.

3.8.81 For $xy = a$ we have $xy' + y = 0$, so $y' = -\frac{y}{x}$. For $x^2 - y^2 = b$, we have $2x - 2yy' = 0$, so $y' = \frac{x}{y}$. Let (c, d) be a point on both curves. Then the slope of the normal line to the first curve is $\frac{c}{d}$, but that is the slope of the tangent line to the second curve. Thus the two curves are orthogonal at any points of intersection.

3.8.83

$$(2x + 2yy')(x^2 + y^2 + x) + (x^2 + y^2)(2x + 2yy' + 1) = 8y^2 + 16xyy'$$
$$2yy'(x^2 + y^2 + x) + (x^2 + y^2)2yy' - 16xyy' = 8y^2 - 2x(x^2 + y^2 + x) - (x^2 + y^2)(2x + 1)$$
$$y' = \frac{8y^2 - 2x(x^2 + y^2 + x) - (x^2 + y^2)(2x + 1)}{2y(x^2 + y^2 + x) + 2y(x^2 + y^2) - 16xy}$$
$$= \frac{8y^2 - 2x^3 - 2xy^2 - 2x^2 - 2x^3 - x^2 - 2xy^2 - y^2}{2y(x^2 + y^2 + x + x^2 + y^2 - 8x)}$$
$$= \frac{7y^2 - 3x^2 - 4xy^2 - 4x^3}{2y(2x^2 + 2y^2 - 7x)}.$$

3.8.85 $\frac{y'}{2\sqrt{y}} + y + xy' = 0$, so $y' + 2x\sqrt{y}y' = -2y\sqrt{y}$, so $y' = \frac{-2y\sqrt{y}}{2x\sqrt{y}+1} = -\frac{2y^{3/2}}{2x\sqrt{y}+1}$.
Differentiating again we obtain

$$y'' = -\frac{(2x\sqrt{y}+1)(3\sqrt{y}y') - 2y^{3/2}\left(2\sqrt{y} + \frac{xy'}{\sqrt{y}}\right)}{(2x\sqrt{y}+1)^2} = \frac{-(2x\sqrt{y}+1)(3\sqrt{y}y') + 4y^2 + 2xyy'}{(2x\sqrt{y}+1)^2}$$

$$= \left(\frac{-(2x\sqrt{y}+1)(3\sqrt{y})\left(\frac{-2y\sqrt{y}}{2x\sqrt{y}+1}\right) + 4y^2 + 2xy\left(\frac{-2y\sqrt{y}}{1+2x\sqrt{y}}\right)}{(2x\sqrt{y}+1)^2}\right) \cdot \frac{2x\sqrt{y}+1}{2x\sqrt{y}+1}$$

$$= \frac{(2x\sqrt{y}+1)(6y^2) + 4y^2(1 + 2x\sqrt{y}) - 4xy^{5/2}}{(2x\sqrt{y}+1)^3} = \frac{10y^2 + 16xy^2\sqrt{y}}{(2x\sqrt{y}+1)^3}.$$

3.8.87 Differentiating with respect to x yields $2x(3y^2 - 2y^3) + x^2(6y - 6y^2)\frac{dy}{dx} = 0$. Solving for $\frac{dy}{dx}$ yields $\frac{dy}{dx} = \frac{-2x(3y^2-2y^3)}{6x^2(y-y^2)} = \frac{3y^2-2y^3}{3x(y^2-y)} = \frac{3y-2y^2}{3x(y-1)}$. The numerator is zero when $y = 0$ or when $y = \frac{3}{2}$. Using the original equation, there are no points where $y = 0$, and there are also no points where $y = \frac{3}{2}$ because we obtain the equation $x^2(\frac{27}{4} - \frac{27}{4}) = 4$, which has no solutions. So there are no horizontal tangent lines.

There could be vertical tangent lines where $x = 0$ or $y = 1$; in fact, when $y = 1$ the original equation becomes $x^2(3 - 2) = 4$, so $x = \pm 2$. There are vertical tangents at $(2, 1)$ and $(-2, 1)$. Letting $x = 0$ doesn't yield any points in the original equation.

3.8.89 Differentiating with respect to x yields $(1 - y^2) + x(-2yy') + 3y^2y' = 0$. Solving for y' yields $y' = \frac{y^2-1}{3y^2-2xy} = \frac{y^2-1}{y(3y-2x)}$. There could be a horizontal tangent line where $y = \pm 1$, but the original equation with $y = \pm 1$ yields $\pm 1 = 0$, so there are no horizontal tangent lines.

There could be a vertical tangent line for $y = 0$ or for $y = \frac{2}{3}x$. Using the original equation, letting $y = 0$ yields $x = 0$, and letting $y = \frac{2}{3}x$ yields $x\left(1 - \frac{4x^2}{9}\right) + \frac{8x^3}{27} = 0$. For $x \neq 0$, this yields $27 - 12x^2 + 8x^2 = 0$, or $x^2 = \frac{27}{4}$, so $x = \pm\frac{\sqrt{27}}{2} = \pm\frac{3\sqrt{3}}{2}$. The corresponding y values are $y = \pm\frac{2}{3}x = \pm\sqrt{3}$. So there are vertical tangent lines at $(0,0)$ and at $(3\sqrt{3}/2, \sqrt{3})$ and $(-3\sqrt{3}/2, -\sqrt{3})$.

3.9   Related Rates

3.9.1 The area of a circle of radius r is $A(r) = \pi r^2$. If the radius $r = r(t)$ changes with time, then the area of the circle is a function of r and r is a function of t, so ultimately A is a function of t. If the radius changes at rate $\frac{dr}{dt}$, then the area changes at rate $2\pi r \frac{dr}{dt}$.

3.9.3 Because area is width times height, if one increases, the other must decrease in order for the area to remain constant.

3.9.5

$A(x) = x^2$, $\frac{dx}{dt} = 2$ meters per second.

a. $\frac{dA}{dt} = 2x\frac{dx}{dt}$, so at $x = 10$ meters we have $\frac{dA}{dt} = 2 \cdot 10\text{m} \cdot 2\text{m/s} = 40\text{m}^2/\text{s}$.

b. At $x = 20$m we have $\frac{dA}{dt} = 2 \cdot 20\text{m} \cdot 2\text{m/s} = 80\text{m}^2/\text{s}$.

c.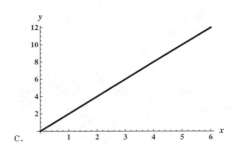

3.9.7

a. Let x be the length of a leg of an isosceles right triangle. Then $\frac{dx}{dt} = 2$ meters per second. The area is given by $A(x) = \frac{1}{2}x^2$. Thus, $\frac{dA}{dt} = \frac{dA}{dx}\frac{dx}{dt} = x \cdot 2 = 2x$ square meters per second. When $x = 2$, we have $\frac{dA}{dx} = 4$, so the area is increasing at 4 square meters per second.

b. When the hypotenuse is 1 meter long, the legs are $1/\sqrt{2}$ meters long. So $A'(1/\sqrt{2}) = 2 \cdot \frac{1}{\sqrt{2}} = \sqrt{2}$, so the area is increasing at $\sqrt{2}$ square meters per second.

c. If h is the length of the hypotenuse, then $x^2 + x^2 = h^2$, so $h = \sqrt{2}x$. So $\frac{dh}{dt} = \frac{dh}{dx}\frac{dx}{dt} = \sqrt{2}\frac{dx}{dt} = \sqrt{2} \cdot 2 = 2\sqrt{2}$ meters per second.

3.9.9

a. Let r be the radius of the circle and A the area, and note that we are given $\frac{dA}{dt} = 1$ square cm per second. Because $A = \pi r^2$, we have $\frac{dA}{dt} = \frac{dA}{dr}\frac{dr}{dt}$, so

$$1 = 2\pi r\frac{dr}{dt},$$

and thus

$$\frac{dr}{dt} = \frac{1}{2\pi r}.$$

When $r = 2$, we have $\frac{dr}{dt} = \frac{1}{4\pi}$ cm per second.

b. When $c = 2\pi r = 2$, we have $r = 1/\pi$. At this time, $\frac{dr}{dt} = \frac{1}{2\pi r} = \frac{1}{2\pi(1/\pi)} = \frac{1}{2}$ cm per second.

3.9.11 $A(x) = \pi x^2$, so $\frac{dA}{dt} = 2\pi x\frac{dx}{dt}$. At $x = 10$ ft and $\frac{dx}{dt} = -2$ ft/min we have $\frac{dA}{dt} = 2\pi \cdot 10 \cdot (-2) = -40\pi$ ft^2/min.

3.9.13 $V(r) = \frac{4}{3}\pi r^3$, so $\frac{dV}{dt} = 4\pi r^2\frac{dr}{dt} = 15$ in^3/min. At $r = 10$ inches we have $4\pi(10\,\text{in})^2\frac{dr}{dt} = 15$ in^3/min. Thus, $\frac{dr}{dt} = \frac{3}{80\pi}$ in/min ≈ 0.012 in/min.

3.9.15 $V(r) = \frac{4}{3}\pi r^3$, and $S(r) = 4\pi r^2$. $\frac{dV}{dt} = 4\pi r^2\frac{dr}{dt} = k \cdot 4\pi r^2$, so $\frac{dr}{dt} = k$, the constant of proportionality.

3.9.17 Using the results of the previous exercise, we are seeking the value of x where $\frac{dx}{dt} = \frac{dy}{dt} = 2x\frac{dx}{dt}$. This occurs for $x = 1/2$, so the desired points is $(1/2, 1/4)$.

3.9.19

By similar triangles, $\frac{2}{50} = \frac{h}{b}$, so $b = 25h$. Also, $A = \frac{1}{2}bh = 12.5h^2$, so the volume for $0 \le h \le 2$ is $V(h) = 12.5 \cdot h^2 \cdot 20 = 250h^2$. For $2 < h \le 3$, $V(h) = 250 \cdot 2^2 + 50 \cdot 20 \cdot (h - 2) = 1000h - 1000$. When $t = 250$ minutes, then $V = 250\,\text{min} \cdot 1\,\text{m}^3/\text{min} = 250\,\text{m}^3$. So $V(h) = 250h^2 = 250$, so $h = 1$ m. At that time $\frac{dV}{dt} = 500h\frac{dh}{dt} = 500 \cdot 1 \cdot \frac{dh}{dt} = 1\,\text{m}^3/\text{min}$. So $\frac{dh}{dt} = \frac{1}{500}$ m/min $= 0.002$ m/min $= 2$ mm/min.

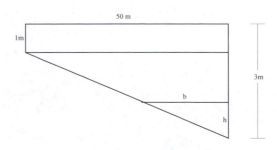

Fill time: The volume of the entire swimming pool is 2000 cubic meters, so at 1 cubic meter per minute, it will take 2000 minutes.

3.9.21

Let x be the distance the surface ship has traveled and D the depth of the submarine. We have $\frac{dx}{dt} = 10\,\mathrm{km/hr}$. Note that $\frac{D}{x} = \tan 20°$, so $D = x \cdot \tan 20° \approx 0.364x$. We have $\frac{dD}{dt} = 0.364\frac{dx}{dt} = 3.64\,\mathrm{km/hr}$. The depth of the submarine is increasing at a rate of 3.64 km/hr.

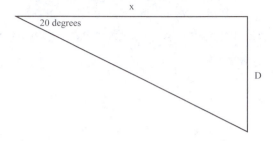

3.9.23

Let h be the vertical distance from the ground to the top of the ladder, and let x be the horizontal distance from the wall to the bottom of the ladder. By the Pythagorean Theorem, we have that $x^2 + h^2 = 169$. Thus, $2x\frac{dx}{dt} + 2h\frac{dh}{dt} = 0$, so $\frac{dh}{dt} = -\frac{x}{h}\frac{dx}{dt}$, and we are given that $\frac{dx}{dt} = 0.5$ feet per second. At $x = 5$ we have $h = \sqrt{169 - 25} = 12$ feet. Thus, $\frac{dh}{dt} = -\frac{5}{12} \cdot \frac{1}{2} = -\frac{5}{24}$ feet per second. So the top of the ladder slides down the wall at $\frac{5}{24}$ feet per second.

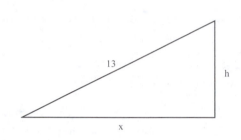

3.9.25

By similar triangles, $\frac{x+y}{20} = \frac{y}{5}$, so $x + y = 4y$, so $x = 3y$, and $\frac{dx}{dt} = 3\frac{dy}{dt}$. Because we are given that $\frac{dx}{dt} = -8$, we have $\frac{dy}{dt} = -\frac{8}{3}$ feet per second. The tip of her shadow is therefore moving at $-8 - \frac{8}{3} = -\frac{32}{3}$ feet per second.

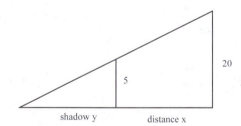

3.9.27 $V = \frac{1}{3}\pi r^2 h$ where $r = 3h$, so $V = 3\pi h^3$. We have that $\frac{dV}{dt} = 9\pi h^2 \frac{dh}{dt}$, and we given that $\frac{dh}{dt} = 2$ at the moment when $h = 12$, so at that time, $\frac{dV}{dt} = 9\pi \cdot 144\,\mathrm{cm^2} \cdot 2\,\mathrm{cm/sec} = 2592\pi\,\mathrm{cm^3/s}$. This is the rate at which the volume of the sandpile is increasing, so it must also be the rate at which the sand is leaving the bin, because there is no other sand involved.

3.9.29 Let h be the depth of the water in the tank at time t, and let r be the radius of the cone-shaped water at time t. By similar triangles, we have that $\frac{h}{r} = \frac{12}{6}$, so $h = 2r$. The volume of the water in the tank is given by $V = \frac{1}{3}\pi r^2 h = \frac{1}{3}\pi \frac{h^2}{4} \cdot h = \frac{\pi h^3}{12}$. Thus, $\frac{dV}{dt} = \frac{\pi h^2}{4}\frac{dh}{dt}$, and so when $h = 3$ we have $-2\,\mathrm{ft^3/s} = \frac{9\pi\,\mathrm{ft^2}}{4}\frac{dh}{dt}$, so $\frac{dh}{dt} = -\frac{8}{9\pi}\,\mathrm{ft/s}$. So the depth of the water is decreasing at a rate of $8/(9\pi)$ feet per second.

3.9.31 Let h be the depth of the water in the tank at time t, and let r be the radius of the cone-shaped water at time t. By similar triangles, we have that $\frac{h}{r} = \frac{12}{6}$, so $h = 2r$. The volume of the water in the tank is given by $V = \frac{1}{3}\pi r^2 h = \frac{1}{3}\pi \frac{h^2}{4} \cdot h = \frac{\pi h^3}{12}$. Thus, $\frac{dV}{dt} = \frac{\pi h^2}{4}\frac{dh}{dt}$. When $\frac{dh}{dt} = -1$, we have $\frac{dV}{dt} = -\frac{\pi h^2}{4}$. when $h = 6$, we have $\frac{dV}{dt} = -9\pi$, so the water is draining from the tank at 9π cubic feet per minute.

3.9.33 Let r be the radius of the exposed surface of the water of height h at time t. Consider the right triangle with legs of length r and $10 - h$ (from the center of the sphere measured down to where the water level is). The hypotenuse is given by the radius of the sphere, which is 10. By the Pythagorean theorem, we have

$$r^2 + (10 - h)^2 = 10^2,$$

which can be written as $r^2 + 100 - 20h + h^2 = 100$, so $r^2 + h^2 = 20h$. So $20\frac{dh}{dt} = 2h\frac{dh}{dt} + 2r\frac{dr}{dt}$. When $h = 5$, we have $10 \cdot \frac{1}{25\pi} = 5 \cdot \frac{1}{25\pi} + 5\sqrt{3}\frac{dr}{dt}$, so $\frac{dr}{dt} = \frac{\sqrt{3}}{75\pi}$. The surface area is given by $S = \pi r^2$, so $\frac{dS}{dt} = 2\pi r\frac{dr}{dt}$, so at this moment it is given by $\frac{dS}{dt} = 2\pi \cdot 5\sqrt{3} \cdot \frac{\sqrt{3}}{75\pi} = \frac{2}{5}$ square meters per minute.

3.9.35 Let x be the distance the motorcycle has traveled since the instant it went under the balloon, and let y be the height of the balloon above the ground t seconds after the motorcycle went under it. We have $x^2 + y^2 = D^2$ where D is the distance between the motorcycle and the balloon. Thus, $2x\frac{dx}{dt} + 2y\frac{dy}{dt} = 2D\frac{dD}{dt}$, and we are given that $\frac{dy}{dt} = 10$ feet per second, and $\frac{dx}{dt} = 40\,\text{mph} = \frac{176}{3}$ ft/s. After 10 seconds have passed, we have that $y = 150 + 100 = 250\,\text{ft}$, $x = \frac{1760}{3}$ ft and $D = \sqrt{250^2 + \left(\frac{1760}{3}\right)^2} \approx 638\,\text{ft}$. Thus, $\frac{dD}{dt} \approx \frac{1}{638}\left(\frac{1760}{3} \cdot \frac{176}{3} + 2500\right) \approx 57.86$ feet per second.

3.9.37

Let x be the distance between the fish and the fisherman's feet, and let D be the distance between the fish and the tip of the pole. Then $D^2 = x^2 + 144$, so $2D(dD/dt) = 2x(dx/dt)$. Note that $dD/dt = -1/3\,\text{ft/sec}$, so when $x = 20\,\text{ft}$, we have $dx/dt = \sqrt{400 + 144}/20 \cdot (-1/3) \approx -0.3887\,\text{ft/sec} \approx -4.66\,\text{in/sec}$. The fish is moving toward the fisherman at about 4.66 in/sec.

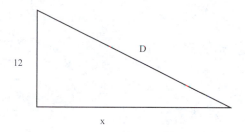

3.9.39

Let D be the length of the rope from the boat to the capstan, and let x be the horizontal distance from the boat to the dock. By the Pythagorean Theorem, $x^2 + 25 = D^2$, so $2x\frac{dx}{dt} = 2D\frac{dD}{dt}$, so $\frac{dx}{dt} = \frac{D}{x}\frac{dD}{dt}$. We are given that $\frac{dD}{dt} = -3$ feet per second, so when $x = 10$, we have $\frac{dx}{dt} = \frac{\sqrt{125}}{10} \cdot (-3) = -\frac{3\sqrt{5}}{2}$ feet per second. The boat is approaching the dock at $\frac{3\sqrt{5}}{2}$ feet per second.

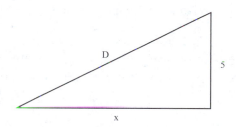

3.9.41 Let x be the distance the westbound airliner has traveled between noon and t hours after 1:00, and let y be the distance the northbound airliner has traveled t hours after 1:00, and let D be the distance between the planes. We have $D^2 = x^2 + y^2$, so $2D\frac{dD}{dt} = 2x\frac{dx}{dt} + 2y\frac{dy}{dt}$. We are given that $\frac{dx}{dt} = 500$ mph and $\frac{dy}{dt} = 550$ mph. At 2:30, we have that $x = 500 + 500 \cdot 1.5 = 1250$, and $y = 550 \cdot 1.5 = 825$ miles. $D = \sqrt{2243125} \approx 1497.7$ miles. Thus $\frac{dD}{dt} \approx \frac{1250 \cdot 500 + 825 \cdot 550}{1497.7} \approx 720.27$ miles per hour.

3.9.43 Let θ be the angle between the hands of the clock, and D the distance between the tips of the hands. By the law of cosines, $D^2 = 2.5^2 + 3^2 - 15\cos\theta$. So $2D\frac{dD}{dt} = 15\sin\theta\frac{d\theta}{dt}$. At 9:00 AM, we have $D^2 = 6.25 + 9$, so $D = \sqrt{15.25}$. Also, $\theta = \pi/2$ so $\sin\theta = 1$. Thus, $\frac{dD}{dt} = \frac{15}{2\sqrt{15.25}}\frac{d\theta}{dt}$. Now $\frac{d\theta}{dt} = \frac{d\theta_1}{dt} - \frac{d\theta_2}{dt}$ where $\frac{d\theta_1}{dt}$ is the angular change of the minute hand and $\frac{d\theta_2}{dt}$ is the angular change of the hour hand. We have $\frac{d\theta_1}{dt} = \frac{\pi}{30}$ radians per minute and $\frac{d\theta_2}{dt} = \frac{\pi}{360}$ radians per minute, so $\frac{d\theta}{dt} = \frac{11\pi}{360}$ radians per minute. Thus $\frac{dD}{dt} = \frac{15}{2\sqrt{15.25}} \cdot \frac{11\pi}{360} \approx .18436$ meters per minute, or about 11.06 meters per hour.

3.9.45

a. Let A be the point where the dragster started, let B be the point where camera 1 is located and let $C = y(t)$ be the position of the car at time t. Let θ be angle ABC. Note that $\tan\theta = \frac{y}{50}$, so $\sec^2\theta \cdot \frac{d\theta}{dt} = \frac{1}{50}\frac{dy}{dt}$. At time $t = 2$, we have that $\tan^2\theta = 4$, so $\sec^2\theta = \tan^2\theta + 1 = 5$. So $\frac{dy}{dt} = 5 \cdot 50 \cdot .75 = 187.5$ feet per second.

b. Let D be the point where camera 2 is located, and let ϕ be angle ADC. The $\phi = \tan^{-1}\left(\frac{y}{100}\right)$, so $\frac{d\phi}{dt} = \frac{1}{100\left(1+\left(\frac{y}{100}\right)^2\right)} \cdot \frac{dy}{dt}$. After 2 seconds, we know that $y = 100$ and $\frac{dy}{dt} = 187.5$. Thus $\frac{d\phi}{dt} = \frac{100}{20,000} \cdot 187.5 = .9375$ radians per second.

3.9.47 By the Law of Sines, $\frac{\sin\theta}{s} = \frac{\sin\left(\frac{3\pi}{4}-\theta\right)}{2}$, so $2\sin\theta = s\sin\left(\frac{3\pi}{4} - \theta\right) = s\left(\sin\left(\frac{3\pi}{4}\right)\cos\theta - \cos\left(\frac{3\pi}{4}\right)\sin\theta\right)$. We have

$$2\sin\theta = \frac{\sqrt{2}}{2}s(\sin\theta + \cos\theta)$$

$$2\tan\theta = \frac{\sqrt{2}}{2}s(\tan\theta + 1)$$

$$\tan\theta = \frac{(\sqrt{2}/2)\cdot s}{2-(\sqrt{2}/2)s} = \frac{\sqrt{2}s}{4-\sqrt{2}s}$$

$$\theta = \tan^{-1}\left(\frac{\sqrt{2}s}{4-\sqrt{2}s}\right).$$

Thus, $\frac{d\theta}{dt} = \frac{\sqrt{2}\cdot\frac{ds}{dt}}{4-2\sqrt{2}s+s^2}$. When $\frac{ds}{dt} = 15$ and $s = 7.5$ we arrive at $\frac{d\theta}{dt} = 0.54$ radians per hour.

3.9.49

Let h be the vertical distance between the point on the elevator shaft positioned directly opposite the observer and the point on the elevator shaft that the observer is observing. So $h > 0$ corresponds to $\theta > 0$ and $h < 0$ corresponds to $\theta < 0$. We have $\frac{h}{20} = \tan\theta$, so $\frac{1}{20}\frac{dh}{dt} = \sec^2\theta\frac{d\theta}{dt}$. We are given that $\frac{dh}{dt} = 5\,\text{m/s}$. At $h = -10$, we have $\tan\theta = -.5$, so $\sec^2\theta = 1 + \tan^2\theta = 1 + (.5)^2 = 1.25$. So $\frac{d\theta}{dt} = \frac{1}{20\cdot1.25}\cdot 5 = \frac{1}{5}$ radian per second. When $h = 20$, we have that $\tan\theta = 1$, so $\sec^2\theta = 1 + 1^2 = 2$, and thus $\frac{d\theta}{dt} = \frac{1}{20\cdot2}\cdot 5 = \frac{1}{8}$ radian per second.

3.9.51 Let x be the distance the eastbound boat has traveled at time t and let s be the distance the northeastbound boat has traveled. Note the diagram shown. By the Law of Sines, $\frac{\sin\left(\frac{\pi}{2}-\theta\right)}{s} = \frac{\sin\left(\frac{\pi}{4}+\theta\right)}{x}$. Thus,

$$x\left(\sin\left(\frac{\pi}{2}\right)\cos\theta - \cos\left(\frac{\pi}{2}\right)\sin\theta\right) =$$

$$s\left(\sin\left(\frac{\pi}{4}\right)\cos\theta + \cos\left(\frac{\pi}{4}\right)\sin\theta\right)$$

So

$$x\cos\theta = \frac{\sqrt{2}}{2}\cdot s\cdot\cos\theta + \frac{\sqrt{2}}{2}\cdot s\cdot\sin\theta,$$

$$x = \frac{\sqrt{2}}{2}\cdot s + \frac{\sqrt{2}}{2}\cdot s\cdot\tan\theta,$$

and thus $\tan\theta = \frac{x-\frac{\sqrt{2}}{2}s}{\frac{\sqrt{2}}{2}s} = \frac{\sqrt{2}x-s}{s}$, and therefore $\theta = \tan^{-1}\left(\frac{\sqrt{2}x-s}{s}\right)$.

We have

$$\frac{d\theta}{dt} = \frac{1}{1 + \left(\frac{\sqrt{2}x - s}{s}\right)^2} \cdot \frac{\left(\sqrt{2}\left(\frac{dx}{dt}\right) - \left(\frac{ds}{dt}\right)\right) \cdot s - \left(\sqrt{2}x - s\right) \cdot \frac{ds}{dt}}{s^2} = \frac{\sqrt{2}\left(s\frac{dx}{dt} - x\frac{ds}{dt}\right)}{s^2 + (\sqrt{2}x - s)^2}.$$

At time t, we have $s(t) = 15t$ and $x(t) = 12t$. Note that

$$s\frac{dx}{dt} - x\frac{ds}{dt} = 15t \cdot 12 - 12t \cdot 15 = 0.$$

Thus $\theta' = 0$ for every value of t, so that the angle is constant.

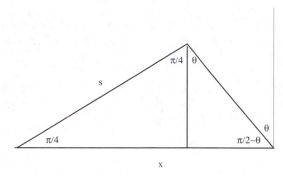

3.9.53 Let α be the angle between the line of sight to the bottom of the screen and the line of sight to the point 3 feet below where the floor and the wall meet. Note that $\cot \alpha = \frac{x}{3}$ and $\cot(\alpha + \theta) = \frac{x}{10}$, so $(-\csc^2 \alpha)\frac{d\alpha}{dt} = \frac{1}{3}\frac{dx}{dt}$ and $-\csc^2(\alpha + \theta)(\frac{d\alpha}{dt} + \frac{d\theta}{dt}) = \frac{1}{10}\frac{dx}{dt}$. At $x = 30$ feet, we have $\alpha \approx 0.0996687$ and then $\alpha + \theta \approx 0.321751$, so $\theta \approx 0.222082$. We obtain these results with a computer algebra system. When $\frac{dx}{dt} = 3$, we have $\frac{d\alpha}{dt} = -0.00990099$, and $\frac{d\alpha}{dt} + \frac{d\theta}{dt} \approx -0.0300001$, so $\frac{d\theta}{dt} \approx -0.0201$ radians per second.

3.9.55

a. The volume of the water in the tank (as a function of h — the depth of the water in the tank) is given by 5 times the area of the segment of water in a cross-sectional circle. For a tank of radius 1, the formula for such a segment is $\cos^{-1}(1 - h) - (1 - h)\sqrt{2h - h^2}$. Thus the volume of the water in the tank is given by $V = 5(\cos^{-1}(1 - h) - (1 - h)\sqrt{2h - h^2})$. We have

$$\frac{dV}{dt} = 5 \cdot \left(-\frac{1}{\sqrt{1 - (1-h)^2}} \cdot \left(-\frac{dh}{dt}\right) + \frac{dh}{dt}\sqrt{2h - h^2} - \frac{(1-h)^2}{\sqrt{2h - h^2}}\frac{dh}{dt}\right)$$

$$= 5\left(\sqrt{2h - h^2} + \frac{1 - (1-h)^2}{\sqrt{2h - h^2}}\right)\frac{dh}{dt}$$

$$= 5\left(\frac{2h - h^2 + 1 - 1 + 2h - h^2}{\sqrt{2h - h^2}}\right)\frac{dh}{dt}$$

$$= 5\left(\frac{2(2h - h^2)}{\sqrt{2h - h^2}}\right)\frac{dh}{dt}$$

$$= 10\sqrt{2h - h^2} \cdot \frac{dh}{dt}$$

When $h = .5$, we have $-\frac{3}{2} = \frac{dV}{dt} = 5\sqrt{3}\frac{dh}{dt}$, so $\frac{dh}{dt} = -\frac{\sqrt{3}}{10}$ meters per hr.

b. The surface area of the water is given by $S = 5 \cdot 2\sqrt{2h - h^2}$. So $\frac{dS}{dt} = 10 \cdot \frac{2 - 2h}{2\sqrt{2h - h^2}} \cdot \frac{dh}{dt}$, so at $h = .5$, we have $\frac{5}{\sqrt{3/4}} \cdot -\frac{\sqrt{3}}{10} = -1$ square meter per hr.

Chapter Three Review

1

    a. False. This function is not differentiable at $x = -\frac{1}{2}$. It is possible for a function to be continuous at a point and not differentiable at that point.

    b. False. For example, $f(x) = x^2 + 3$ and $g(x) = x^2 + 100$ have the same derivative, but aren't the same function.

    c. False. For example, $\frac{d}{dx}\left|\frac{1}{x^2}\right| = \frac{d}{dx}\left(\frac{1}{x^2}\right) = -\frac{2}{x^3} \neq \left|-\frac{2}{x^3}\right|$.

    d. False. For example, the function $f(x) = |x|$ has no derivative at 0, but there is no vertical tangent there.

    e. True. For example, a ball dropping from a high tower has acceleration due to gravity which is negative, but it is speeding up as it falls because the velocity (which is negative also) is in the same direction as the acceleration.

3

    a. $f'(1) = \lim_{h \to 0} \dfrac{5(1+h)^3 + (1+h) - 6}{h}$. Expanding yields $\lim_{h \to 0} \dfrac{5(1 + 3h^2 + 3h + h^3) + 1 + h - 6}{h} =$ $\lim_{h \to 0} \dfrac{5 - 15h^2 + 15h + 5h^3 + h - 5}{h}$. This can be written as $\lim_{h \to 0} \dfrac{5h^3 - 15h^2 + 16h}{h} = \lim_{h \to 0} (5h^2 - 15h + 16) = 16$.

    b. The tangent line at $(1, 6)$ is given by $y - 6 = 16(x - 1)$, or $y = 16x - 10$.

5

    a. $f'(0) = \lim_{h \to 0} \dfrac{\frac{1}{2\sqrt{3h+1}} - \frac{1}{2}}{h}$

    $= \lim_{h \to 0} \dfrac{1 - \sqrt{3h + 1}}{2\sqrt{3h + 1} \cdot h} =$ $\lim_{h \to 0} \dfrac{(1 - \sqrt{3h + 1})(1 + \sqrt{3h + 1})}{2\sqrt{3h + 1} \cdot h(1 + \sqrt{3h + 1})} =$ $\lim_{h \to 0} \dfrac{1 - (3h + 1)}{2\sqrt{3h + 1} \cdot h(1 + \sqrt{3h + 1})}$. Simplifying yields $\lim_{h \to 0} -\dfrac{3}{2\sqrt{3h + 1}(1 + \sqrt{3h + 1})}$, and this last limit can be seen to be $-\frac{3}{4}$.

    b. The tangent line at $\left(0, \frac{1}{2}\right)$ is given by $y - \frac{1}{2} = -\frac{3}{4}x$, or $y = -\frac{3}{4}x + \frac{1}{2}$.

7

    a. Average growth is $\frac{p(60) - p(50)}{10} = 2.7$ million people per year.

    b. The curve is pretty straight between $t = 50$ and $t = 60$, so the secant line between these two points is approximately as steep as the tangent line at a point in between.

    c. A reasonable estimate to the instantaneous grow rate at 1985 would be the slope of the secant line between $t = 80$ and $t = 90$. This is $\frac{p(90) - p(80)}{10} = 2.217$ million people per year.

9

   a. $v(15) \approx \frac{400-200}{5} = 40$ meters per second.

   b. Because the graph is a straight line for $t \geq 30$, $v(70) = \frac{D(90)-D(60)}{30} = \frac{1600-1400}{30} = \frac{20}{3}$ meters per second. The points at 60 and 90 were chosen because it is easier to detect the function values at those points using the given grid.

   c. The average velocity is $\frac{D(90)-D(20)}{70} \approx \frac{1600-550}{70} = 15$ meters per second.

   d.

e. The parachute was deployed.

11

$$g'(x) = \lim_{h \to 0} \frac{\sqrt{2(x+h)-3} - \sqrt{2x-3}}{h} = \lim_{h \to 0} \frac{\sqrt{2(x+h)-3} - \sqrt{2x-3}}{h} \cdot \frac{\sqrt{2(x+h)-3} + \sqrt{2x-3}}{\sqrt{2(x+h)-3} + \sqrt{2x-3}}$$

$$= \lim_{h \to 0} \frac{2(x+h)-3-(2x-3)}{h(\sqrt{2(x+h)-3} + \sqrt{2x-3})} = \lim_{h \to 0} \frac{2}{\sqrt{2(x+h)-3} + \sqrt{2x-3}} = \frac{2}{2\sqrt{2x-3}} = \frac{1}{\sqrt{2x-3}}.$$

13

15 $f'(x) = 2x^2 + 2\pi x + 7.$

17 $f'(t) = 10t \sin t + 5t^2 \cos t.$

19 $f'(\theta) = (4\sec^2(\theta^2 + 3\theta + 2)) \cdot (2\theta + 3) = (8\theta + 12)\sec^2(\theta^2 + 3\theta + 2).$

21 $f'(u) = \frac{(8u+1)(8u+1)-(4u^2+u)(8)}{(8u+1)^2} = \frac{64u^2+16u+1-32u^2-8u}{(8u+1)^2} = \frac{32u^2+8u+1}{(8u+1)^2}.$

23 $f'(\theta) = \sec^2(\sin \theta) \cdot \cos \theta.$

25 $f'(x) = 2(\sin x)\sqrt{3x-1} + 2x(\cos x)\sqrt{3x-1} + \frac{3x \sin x}{\sqrt{3x-1}}.$

27

$$\frac{d}{dx} \sin\left(\frac{x}{x+1}\right) = \cos\left(\frac{x}{x+1}\right) \cdot \frac{x+1-x}{(x+1)^2}$$

$$= \frac{\cos\left(\frac{x}{x+1}\right)}{(x+1)^2}.$$

29

$$\frac{d}{dw}\left(\frac{\tan^2 w}{\sec w + 1}\right) = \frac{(\sec w + 1)2(\tan w)(\sec^2 w) - \tan^2 w(\sec w \tan w)}{(\sec w + 1)^2}$$

$$= \frac{\sec w \tan w(2(\sec^2 w + \sec w) - \tan^2 w)}{(\sec w + 1)^2}$$

$$= \frac{\sec w \tan w(2(\sec^2 w + \sec w) - (\sec^2 w - 1))}{(\sec w + 1)^2}$$

$$= \frac{\sec w \tan w(\sec^2 w + 2\sec w + 1)}{(\sec w + 1)^2}$$

$$= \frac{\sec w \tan w(\sec w + 1)^2}{(\sec w + 1)^2} = \sec w \tan w.$$

31 $y' = \dfrac{(1 + \sin x)(-y' \sin y) - (\cos y)(\cos x)}{(1 + \sin x)^2}$ so that $y' = -\dfrac{\cos x \cos y}{(1 + \sin x)^2 + \sin y(1 + \sin x)}$.

33 $y'\sqrt{x^2 + y^2} + y \cdot \dfrac{x + yy'}{\sqrt{x^2+y^2}} = 0$, and thus $y'\left(\sqrt{x^2 + y^2} + \dfrac{y^2}{\sqrt{x^2+y^2}}\right) = -\dfrac{xy}{\sqrt{x^2+y^2}}$. This can be written as

$y'\left(\dfrac{x^2+2y^2}{\sqrt{x^2+y^2}}\right) = -\dfrac{xy}{\sqrt{x^2+y^2}}$, so $y' = -\dfrac{xy}{x^2+2y^2}$.

35 $y' = 9x^2 + \cos x$. At $x = 0$, $y' = 1$. So the tangent line is given by $y - 0 = 1(x - 0)$, or $y = x$.

37 $y' + \dfrac{y + xy'}{2\sqrt{xy}} = 0$. At the point $(1, 4)$, we have $y' + \dfrac{4 + y'}{4} = 0$, so $y' = -\dfrac{4}{5}$. The tangent line is given by $y - 4 = -\dfrac{4}{5}(x - 1)$, or $y = -\dfrac{4}{5}x + \dfrac{24}{5}$.

39 We are looking for values of x so that $y'(x) = 0$. We have $y' = \sqrt{6 - x} - \dfrac{x}{2\sqrt{6-x}}$, and this quantity is zero when $2(6 - x) - x = 0$, or $12 - 3x = 0$, so when $x = 4$. So at the point $(4, 4\sqrt{2})$ there is a horizontal tangent line. There is a vertical tangent line at $x = 6$, because $\lim\limits_{x \to 6^-} y'(x) = -\infty$.

41

$y' = \frac{1}{2}x^{-1/2}\cos\sqrt{x}$.

$y'' = -\frac{1}{4}x^{-3/2}\cos\sqrt{x} + -\frac{1}{4}x^{-1}\sin\sqrt{x}$.

$y''' = \frac{3}{8}x^{-5/2}\cos\sqrt{x} + \frac{1}{8}x^{-2}\sin\sqrt{x} + \frac{1}{4}x^{-2}\sin\sqrt{x} - \frac{1}{8}x^{-3/2}\cos\sqrt{x} = \frac{3}{8}x^{-5/2}\cos\sqrt{x} + \frac{3}{8}x^{-2}\sin\sqrt{x} - \frac{1}{8}x^{-3/2}\cos\sqrt{x}$.

43 $\frac{d}{dx}[x^2 f(x)] = 2xf(x) + x^2 f'(x)$.

45 $\dfrac{d}{dx}\left(\dfrac{xf(x)}{g(x)}\right) = \dfrac{(f(x) + xf'(x))g(x) - xf(x)g'(x)}{g(x)^2}$.

47

a. $\frac{d}{dx}\left[f(x) + 2g(x)\right]_{x=3} = f'(3) + 2g'(3) = 9 + 2 \cdot 9 = 27$.

b. $\frac{d}{dx}\left[\dfrac{xf(x)}{g(x)}\right]_{x=1} = \dfrac{g(1)(1 \cdot f'(1) + f(1)) - 1 \cdot f(1) \cdot g'(1)}{(g(1))^2} = \dfrac{9 \cdot [7+3] - 15}{81} = \dfrac{25}{27}$.

c. $\frac{d}{dx} f(g(x^2))\big|_{x=3} = f'(g(9)) \cdot g'(9) \cdot 2 \cdot 3 = f'(1) \cdot 7 \cdot 6 = 7 \cdot 42 = 294$.

d. $\frac{d}{dx}(f(x))^3\big|_{x=5} = 3f(5)^2 f'(5) = 3(9)^2 \cdot 5 = 1215$.

e. $(g^{-1})'(7) = \dfrac{1}{g'(3)} = \dfrac{1}{9}$.

49 Let $a = 5$ and $f(x) = \tan(\pi\sqrt{3x - 11})$. Note that $f'(5) = \dfrac{3\pi}{2}\dfrac{\sec^2(2\pi)}{2} = \dfrac{3\pi}{4}$.

So $\displaystyle\lim_{x \to 5} \frac{f(x) - f(5)}{x - 5} = \lim_{x \to 5} \frac{\tan(\pi\sqrt{3x - 11}) - 0}{x - 5} = f'(5) = \frac{3\pi}{4}$.

51

a. $\dfrac{d}{dx}\left(\dfrac{x^3}{f(x)}\right)\Big|_{x=2} = \left(\dfrac{f(x)(3x^2) - x^3 f'(x)}{(f(x))^2}\right)\Big|_{x=2} = \dfrac{5 \cdot 12 - 8 \cdot 3}{5^2} = \dfrac{36}{25}$.

b. $\dfrac{d}{dx}\left(\sqrt{f(x)}\right)\Big|_{x=1} = \left(\dfrac{f'(x)}{2\sqrt{f(x)}}\right)\Big|_{x=1} = \dfrac{1}{2\sqrt{3}}$.

c. $\dfrac{d}{dx}\left(\cos(\pi f(x))\right)\Big|_{x=3} = \left(-\sin(\pi f(x))(\pi f'(x))\right)\Big|_{x=3} = -4\pi\sin(7\pi) = 0$.

53

a. The average cost is $\dfrac{C(3000)}{3000} = \dfrac{1025000}{3000} \approx \$341.67$. The marginal cost is $C'(3000) = -0.04(3000) + 400 = \$280$.

b. The average cost of producing 3000 lawnmowers is \$341.67 per mower. The cost of producing the 3001st lawnmower is approximately \$280.

55

a. The average growth rate is $\dfrac{p(50) - p(0)}{50} = \dfrac{407500 - 80000}{50} = 6550$ people per year.

b. The growth rate in 1990 is $p'(40) = -5.1(40^2) + 144 \cdot 40 + 7200 = 4800$ people per year.

57

Let x be the distance the eastbound boat has traveled, and y the distance the southbound boat has traveled. By the Pythagorean Theorem, $D^2 = x^2 + y^2$, so $2D\frac{dD}{dt} = 2x\frac{dx}{dt} + 2y\frac{dy}{dt}$, so $\frac{dD}{dt} = \frac{x \cdot x' + y \cdot y'}{D}$. We are given that $x' = 40$, $y' = 30$, and at $t = .5$ hours, we have $x = 20$, $y = 15$, and $D = 25$. Thus, $\frac{dD}{dt} = \frac{20 \cdot 40 + 30 \cdot 15}{25} = 50$ mph.

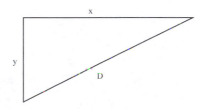

59 Let h be the elevation of the balloon, and s the length of the rope. We have $h = s\sin(65°)$, so $h' = s'\sin(65°) = -5 \cdot \sin(65°) \approx -4.53$ feet per second.

61

Let x be the distance the jet has flown since it went over the spectator. Let θ be the angle of elevation between the ground and the line from the spectator to the jet. Note that θ is also the angle pictured, and that $\cot\theta = \frac{x}{500}$. Thus, $\theta = \cot^{-1}\left(\frac{x}{500}\right)$. We are given that $x' = 450\,\text{mph} = 660\,\text{ft/sec}$.

$\theta' = -\dfrac{x'}{500 \cdot \left(1 + \left(\frac{x}{500}\right)^2\right)} = -\dfrac{500x'}{250,000 + x^2}$. After 2 seconds, $x = 1320$ feet, so at this time $\theta' = -\dfrac{500 \cdot 660}{250,000 + (1320)^2} \approx -0.166$ radians per second.

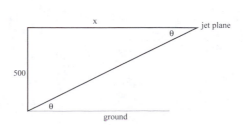

Chapter 4

Applications of the Derivative

4.1 Maxima and Minima

4.1.1 A number $M = f(c)$ where $c \in [a, b]$ with the property that $f(x) \leq M$ for all $x \in [a, b]$ is an absolute maximum for f on $[a, b]$, and a number $m = f(d)$ where $d \in [a, b]$ with the property that $f(x) \geq m$ for all $x \in [a, b]$ is an absolute minimum for f on $[a, b]$.

4.1.3 The function must be a continuous function defined on a closed interval.

4.1.5 The function shown has no absolute minimum on $[0, 3]$ because $\lim\limits_{x \to 0^+} f(x) = -\infty$. It has an absolute maximum near $x = 1$ and a local minimum near $x = 2.5$.

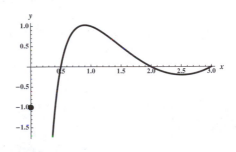

4.1.7 Note the existence of a horizontal tangent line at $x = 0$ where the maximum occurs.

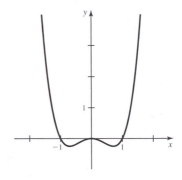

4.1.9 First find all the critical points by seeking all points x in the domain of f so that $f'(x) = 0$ or $f'(x)$ doesn't exist. Now compare the y-values of all of these points, together with the y-values of the endpoints. The largest y-value from among these is the maximum, and the smallest is the minimum.

4.1.11 $y = h(x)$ has an absolute maximum at $x = b$ and an absolute minimum at $x = c_2$.

4.1.13 $y = g(x)$ has no absolute maximum, but has an absolute minimum at $x = a$.

4.1.15 $y = f(x)$ has an absolute maximum at $x = b$ and an absolute minimum at $x = a$. It has local maxima at $x = p$ and $x = r$, and local minima at $x = q$ and $x = s$.

4.1.17 $y = g(x)$ has an absolute minimum at $x = b$ and an absolute maximum at $x = p$. It has local maxima at $x = p$ and $x = r$. It has a local minimum at $x = q$.

4.1.19 Note the horizontal tangent lines at 1 and 2, and the minimum at 0 and the maximum at 4.

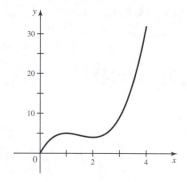

4.1.21 Note the horizontal tangent line at $x = 2$, and the "corners" at $x = 1$ and $x = 3$. Also note the absolute maximum at $x = 3$ and the absolute minimum at $x = 4$.

4.1.23

a. $f'(x) = 6x - 4$, which is zero when $x = 2/3$.

b. At $x = 2/3$ there is a local minimum.

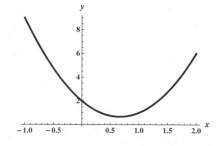

4.1.25

a. $f'(x) = x^2 - 9$, which is zero for $x = \pm 3$.

b. There is a local maximum at $x = -3$ and a local minimum at $x = 3$.

4.1.27

a. $f'(x) = 9x^2 + 3x - 2 = (3x+2)(3x-1)$, which is zero for $x = -2/3$ and $x = 1/3$.

b. There is a local maximum at $x = -2/3$ and a local minimum at $x = 1/3$.

4.1.29

a. $f'(x) = \frac{(x^2+1)\cdot 2(x+1) - (x+1)^2(2x)}{(x^2+1)^2}$. Factoring the numerator yields $\frac{(x+1)(2x^2+2 - (x+1)(2x))}{(x^2+1)^2}$, which is equal to $\frac{(x+1)(2x^2+2-2x^2-2x)}{(x+1)^2} = \frac{2(x+1)(1-x)}{(x^2+1)^2}$. This quantity is zero for $x = -1$ and for $x = 1$.

b. At $x = -1$ there is a local minimum (which is also an absolute minimum) and at $x = 1$ there is a local maximum (which is also an absolute maximum.)

4.1.31

a. $f'(x) = \frac{1}{\sqrt{x}} - 1$, which is zero when $x = 1$.

b. There is a local (and absolute) maximum at $x = 1$.

4.1.33

a. $f'(x) = 2x\sqrt{x+1} + x^2 \cdot \frac{1}{2\sqrt{x+1}} = \frac{4x(x+1)}{2\sqrt{x+1}} + \frac{x^2}{2\sqrt{x+1}} = \frac{5x^2+4x}{2\sqrt{x+1}}$. This is zero when $5x^2 + 4x = x(5x+4)$ is zero, which occurs for $x = 0$ and $x = -4/5$. The critical points are $(0,0)$ and $(-4/5, 16/(25\sqrt{5}))$.

b. There is a local maximum at $x = -4/5$ and a local minimum at $x = 0$.

4.1.35

a. $f'(x) = 2x$, which is zero for $x = 0$.

b. We have that $f(-2) = -6$, $f(0) = -10$, and $f(3) = -1$, so the maximum value of f on this interval is -1 and the minimum is -10.

c.

4.1.37

a. $f'(x) = -2\cos x \sin x$, which is zero for $x = 0$, $x = \pi/2$, and $x = \pi$. Because there are endpoints at $x = 0$ and $x = \pi$, only $(\pi/2, 0)$ is a critical point.

b. We have that $f(0) = 1$, $f(\pi/2) = 0$, and $f(\pi) = 1$, so the maximum value of f on this interval is 1 and the minimum is 0.

c.

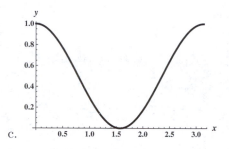

4.1.39

a. $f'(x) = 3\cos 3x$, which is zero when

$$3x = \ldots, -\pi/2, \pi/2, 3\pi/2, \ldots,$$

so when $x = \ldots, -\pi/6, \pi/6, \pi/2, \ldots$. The only such values on the given interval are $x = -\pi/6$ and $x = \pi/6$.

b. We have

$$f(-\pi/4) = -\sqrt{2}/2 \approx -0.707,$$

$f(-\pi/6) = -1$, $f(\pi/6) = 1$, and $f(\pi/3) = 0$, so the absolute maximum of f is 1 and the absolute minimum is -1.

c.

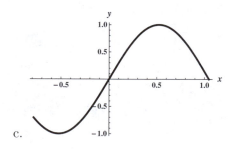

4.1.41

a. $f'(x) = 12x^2 - 42x + 36 = 6(2x^2 - 7x + 6) = 6(2x-3)(x-2)$. This is zero for $x = 2$ and $x = 3/2$.

b. $f(1) = 19$, $f(3/2) = 20.25$, $f(2) = 20$, and $f(3) = 27$. The absolute maximum is 27 at $x = 3$ and the absolute minimum is 19 at $x = 1$.

c.

4.1.43

a. $f'(x) = 6x^2 - 30x + 24 = 6(x^2 - 5x + 4) = 6(x - 4)(x - 1)$. This is zero at $x = 4$ and $x = 1$.

b. $f(1) = 11$ and $f(4) = -16$. At the endpoints we have $f(0) = 0$ and $f(5) = -5$. The absolute maximum is 11 and the absolute minimum is -16.

c.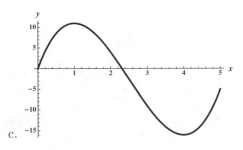

4.1.45

a. $f'(x) = 4x^2 + 10x - 6 = 2(2x^2 + 5x - 3) = 2(x + 3)(2x - 1)$. This is zero when $x = -3$ and when $x = 1/2$.

b. $f(-3) = 27$ and $f(1/2) = -19/12$. At the endpoints we have $f(-4) = 56/3 \approx 18.7$, and $f(1) = 1/3$. The absolute maximum is 27 and the absolute minimum is $-19/12$.

c.

4.1.47 The stone will reach its maximum height when its velocity is zero, which occurs at the only critical point for this inverted parabola. We have that $v(t) = s'(t) = -32t + 64$, which is zero when $t = 2$. The height at this time is $s(2) = 256$, the maximum height.

4.1.49

a. Note that $P(n) = 50n - .5n^2 - 100$, so $P'(n) = 50 - n$, which is zero when $n = 50$. It is clear that this is a maximum, since the graph of P is an inverted parabola.

b. Given a domain of $[0, 45]$, since the only critical point is not in the domain, the maximum must occur at an endpoint. Because $P(0) = -100$, and $P(45) = \$1137.50$, he should take 45 people on the tour.

4.1.51

a. False. The derivative $f'(x) = \frac{1}{2\sqrt{x}}$ is never zero, and the function has no critical points.

b. False. For example, the function $f(x) = \begin{cases} \sin x & \text{if } -5 \le x \le 0, \\ -8 & \text{if } 0 < x \le 5 \end{cases}$ is not continuous on $[-5, 5]$, but has an absolute maximum of 1.

c. False. For example, the function $f(x) = (x - 2)^3$ satisfies $f'(2) = 0$, but it has neither a maximum nor a minimum at $x = 2$.

d. True. This follows from the theorems in this section.

4.1.53

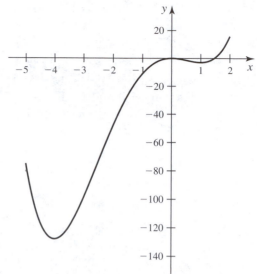

a. $f(x) = x^4 + 4x^3 - 8x^2$, so $f'(x) = 4x^3 + 12x^2 - 16x = 4x(x^2+3x-4) = 4x(x+4)(x-1)$. Thus the critical points of $f(x)$ in $[-5, 2]$ are $x = -4, 0, 1$.

b. $f(-5) = -75$, $f(-4) = -128$, $f(0) = 0$, $f(1) = -3$, $f(2) = 16$, so the absolute minimum of $f(x)$ on $[-5, 2]$ is -128 at $x = -4$; its absolute maximum is 16 at $x = 2$.

c.

4.1.55

a. $f'(x) = \sec x \tan x$ which is zero when $\tan x = 0$ (since $\sec x$ is never zero.) So we are looking for where $\frac{\sin x}{\cos x} = 0$, which is when $\sin x = 0$, which is at $x = 0$

b. $f(-\pi/4) = \sqrt{2} = f(\pi/4)$ and $f(0) = 1$. So the absolute maximum for f is $\sqrt{2}$ and the absolute minimum is 1.

c.

4.1.57

a. $f'(x) = \dfrac{\sqrt{x-4}\cdot 1 - x\cdot\frac{1}{2\sqrt{x-4}}\cdot 1}{x-4} = \dfrac{\sqrt{x-4}\cdot 1 - x\cdot\frac{1}{2\sqrt{x-4}}\cdot 1}{x-4}$. $\dfrac{2\sqrt{x-4}}{2\sqrt{x-4}} = \dfrac{2x-8-x}{2(x-4)^{3/2}} = \dfrac{x-8}{2(x-4)^{3/2}}$. This expression is 0 when $x = 8$. So $(8, 4)$ is the only critical point.

b. $f(6) = 3\sqrt{2} = f(12)$, and $f(8) = 4$. Note that $3\sqrt{2} \approx 4.24 > 4$. So the absolute maximum of f on this interval is $3\sqrt{2}$, and the absolute minimum is 4.

c.

4.1.59 $f'(x) = \sqrt{x-a} + \dfrac{x}{2\sqrt{x-a}} = \dfrac{2x-2a+x}{2\sqrt{x-a}} = \dfrac{3x-2a}{2\sqrt{x-a}}$. This expression is zero when $x = \frac{2a}{3}$; however, that number is not in the domain of f if $a > 0$. However, if $a < 0$, then $\frac{2a}{3}$ is in the domain, and thus gives a critical point.

4.1.61 $f'(x) = x^4 - a^4$, which is zero when $x^4 = a^4$, or $|x| = a$. So there are critical points at $x = a$ and at $x = -a$.

4.1.63

a. $f'(\theta) = 2\cos\theta - \sin\theta$, which is zero when $\tan\theta = 2$. So one critical point occurs at $\theta = \tan^{-1}(2) \approx 1.107$. And since the tangent function is periodic with period π, there are will also be solutions at this number plus or minus integer multiples of π. On the given interval, these are located at approximately $1.107 - 2\pi \approx -5.176$, at $1.107 - \pi \approx -2.034$, and at $1.107 + \pi \approx 4.249$.

b. From the graph, it appears that there is a local minimum at about $\theta = -2.034$ and at $\theta = 4.249$, and there is a local maximum at about $\theta = -5.176$, and at about $\theta = 1.107$.

c. From the graph, it appears that the local minimum at about $\theta = -2.034$ is also an absolute minimum, as is the one at $\theta = 4.249$. The local maximum at about $\theta = -5.176$, and at about $\theta = 1.107$ are also absolute maxima. The value of the absolute maximum appears to be about 2.24 and the value of the absolute minimum appears to be about -2.24.

4.1.65

a. $f'(x) = (x-3)^{5/3} + (x+2) \cdot \frac{5}{3}(x-3)^{2/3} = \frac{(x-3)^{2/3}}{3}(3x - 9 + 5x + 10) = \frac{(x-3)^{2/3}}{3}(8x + 1)$. This is zero when $x = 3$ and when $x = -\frac{1}{8}$. There are critical points at $x = -\frac{1}{8}$ and at $x = 3$.

b. From the graph, it appears that there is a local minimum of about -12.52 at $x = -\frac{1}{8}$.

c. The local minimum mentioned above is also an absolute minimum. The absolute maximum occurs at the left endpoint $x = -4$, where the value of f is about 51.23.

4.1.67

a. $h'(x) = \frac{(x^2+2x-3)(-1)-(5-x)(2x+2)}{(x^2+2x-3)^2} = \frac{x^2-10x-7}{(x^2+2x-3)^2}$. By the quadratic formula, the numerator is zero (making the quotient zero) when $x = \frac{10\pm\sqrt{100-4(-7)}}{2} = 5 \pm \frac{1}{2}\sqrt{128} = 5 \pm 4\sqrt{2}$. Note that $5 + 4\sqrt{2}$ isn't in the domain, so the only critical point is at $x = 5 - 4\sqrt{2}$.

b. From the graph, it appears that the one critical point mentioned above yields a local maximum.

c. The function has no absolute maximum and no absolute minimum on the given interval.

4.1.69

Note that

$$g(x) = \begin{cases} 3 - x + 2x + 2 = x + 5 & \text{if } -2 \le x \le -1, \\ 3 - x - 2x - 2 = 1 - 3x & \text{if } -1 \le x \le 3. \end{cases}$$

There is an absolute maximum of 4 and an absolute minimum of -8. The absolute maximum occurs at $x = -1$, and the absolute minimum occurs at a $x = 3$.

4.1.71

a. Because distance is rate times time, the time will be distance over rate. The swim distance is given by $\sqrt{2500 + x^2}$ meters, so the time for swimming is $\frac{\sqrt{2500 + x^2}}{2}$. For running, the distance is $50 - x$, so the time is $\frac{50 - x}{4}$. Thus we have $T(x) = \frac{\sqrt{2500 + x^2}}{2} + \frac{50 - x}{4}$.

b. $T'(x) = \frac{1}{2} \cdot \frac{1}{2} \left(x^2 + 2500\right)^{-1/2} \cdot 2x - \frac{1}{4} = \frac{x}{2\sqrt{x^2 + 2500}} - \frac{1}{4}$. This expression is zero when $\frac{x^2}{x^2 + 2500} = \frac{1}{4}$, so when $4x^2 = x^2 + 2500$, which occurs when $x^2 = \frac{2500}{3}$. So $x = \sqrt{\frac{2500}{3}} \approx 28.868$.

d.

c. $T(0) = 37.5$, $T(28.868) \approx 34.151$, and $T(50) = 25\sqrt{2} \approx 35.355$. The absolute minimum occurs at the only critical point. The minimal crossing time is approximately 34.151 seconds.

4.1.73

a. Note that since there is a local extreme value at 2 for f and since f is differentiable everywhere, we must have $f'(2) = 0$.

$g(2) = 2f(2) + 1 = 1$.

$h(2) = 2f(2) + 2 + 1 = 3$.

$g'(2) = 2 \cdot f'(2) + f(2) = 0$.

$h'(2) = 2f'(2) + f(2) + 1 = 1$.

b. h doesn't, since its derivative isn't zero at $x = 2$. However g might: for example, if $f(x) = (x - 2)^2$ then $g(x) = x(x - 2)^2 + 1$ has a local minimum at $x = 2$.

4.1.75

a. Because of the symmetry about the y-axis for an even function, a minimum at $x = c$ will correspond to a minimum at $x = -c$ as well.

b. Because of the symmetry about the origin, a minimum at $x = c$ will correspond to a maximum at $x = -c$. It is helpful to think about the symmetry about the origin as being the result of flipping about the y-axis and then flipping about the x-axis.

4.1.77

a. If $f(c)$ is a local maximum, then when x is near c but not equal to c, $f(c) \ge f(x)$, so $f(x) - f(c) \le 0$.

b. When x is near to c but a little bigger than c, $x - c > 0$. So in this case, $\frac{f(x)-f(c)}{x-c} \leq 0$, since the numerator is negative (or 0) and the denominator is positive.

Thus, $\lim\limits_{x \to c^+} \dfrac{f(x) - f(c)}{x - c} = f'(c) \leq 0$

c. When x is near to c but a little smaller than c, $x - c < 0$. So in this case, $\frac{f(x)-f(c)}{x-c} \geq 0$, since the numerator is negative (or 0) and the denominator is negative, making the quotient positive (or 0).

Thus, $\lim\limits_{x \to c^-} \dfrac{f(x) - f(c)}{x - c} = f'(c) \geq 0$.

d. From the above, we have that $f'(c) \leq 0$ and $f'(c) \geq 0$, so $f'(c) = 0$.

4.2 What Derivatives Tell Us

4.2.1 If f' is positive on an interval, f is increasing on that interval. If f' is negative on an interval, f is decreasing on that interval.

4.2.3

One such example is $f(x) = x^3$ at $x = 0$.

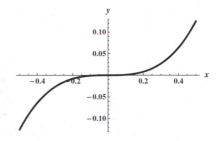

4.2.5 Under these circumstances the tangent lines lie below the graph of f.

4.2.7 An inflection point is a point on the graph of a function where the concavity changes. Thus, if $(c, f(c))$ is an inflection point, either $f''(x) < 0$ for x a little less than c and $f''(x) > 0$ for x a little bigger than c, or vice versa.

4.2.9

Yes, for example, consider $f(x) = 100 - x^2$ on the interval $(-8, 0)$. It is above the x axis, increasing, and concave down on that interval.

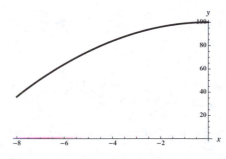

4.2.11

Such a function would be decreasing until $x = 2$, then increasing until $x = 5$, and then decreasing again after that.

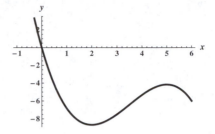

4.2.13

Such a function has extrema (minima) at 0 and 4, where the y value is zero. The function should never go below the x axis.

4.2.15

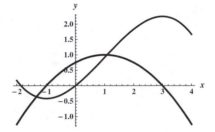

4.2.17

$f'(x) = -2x$, which is zero exactly when $x = 0$. On $(-\infty, 0)$ we note that $f' > 0$, so that f is increasing on this interval. On $(0, \infty)$, we note that $f' < 0$, so f is decreasing on this interval.

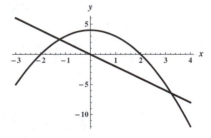

4.2.19

$f'(x) = 2(x-1)$, which is zero exactly when $x = 1$. On $(-\infty, 1)$ we note that $f' < 0$, so that f is decreasing on this interval. On $(1, \infty)$, we note that $f' > 0$, so f is increasing on this interval.

4.2.21

$f'(x) = 1 - 2x$, which is 0 when $x = 1/2$. On $(-\infty, 1/2)$ $f' > 0$ so f is increasing on this interval, while on $(1/2, \infty)$ $f' < 0$, so f is deceasing on this interval.

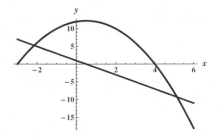

4.2.23

$f'(x) = -x^3 + 3x^2 - 2x = -x(x^2 - 3x + 2) = -x(x-1)(x-2)$. This is zero when $x = 0$, $x = 1$, and $x = 2$. Note that $f'(-1) > 0$, and $f'(1.5) > 0$, while $f'(.5) < 0$, and $f'(3) < 0$. So f is increasing on $(-\infty, 0)$ and on $(1, 2)$, while it is decreasing on $(0, 1)$ and on $(2, \infty)$.

4.2.25 $f'(x) = -9\sin 3x$, which is 0 for $3x = -3\pi, -2\pi, -\pi, 0, \pi, 2\pi,$ and 3π, which corresponds to $x = -\pi, -2\pi/3, -\pi/3, 0, \pi/3, 2\pi/3$ and π. Note that $f'(-5\pi/6) = 9 > 0$, $f'(-\pi/2) = -9 < 0$, $f'(-\pi/6) = 9 > 0$, $f'(\pi/6) = -9 < 0$, $f'(\pi/2) = 9 > 0$, and $f'(5\pi/6) = -9 < 0$. Thus f is increasing on $(-\pi, -2\pi/3)$, on $(-\pi/3, 0)$, and on $(\pi/3, 2\pi/3)$, while f is decreasing on $(-2\pi/3, -\pi/3)$, on $(0, \pi/3)$, and on $(2\pi/3, \pi)$.

4.2.27 $f'(x) = (2/3)x^{-1/3}(x^2 - 4) + x^{2/3} \cdot 2x = \frac{2(x^2-4)}{3x^{1/3}} + \frac{2x^{5/3}}{1} \cdot \frac{3x^{1/3}}{3x^{1/3}} = \frac{8x^2-8}{3x^{1/3}} = \frac{8(x+1)(x-1)}{3x^{1/3}}$. This is zero for $x = \pm 1$ and is undefined for $x = 0$. Note that $f'(-8) = -84 < 0$, $f'(-1/8) = \frac{21}{4} > 0$, $f'(1/8) = \frac{-21}{4} < 0$, and $f''(8) = 84 > 0$. Thus f is decreasing on $(-\infty, -1)$ and on $(0, 1)$, while f is increasing on $(-1, 0)$ and on $(1, \infty)$.

4.2.29 $f'(x) = -60x^4 + 300x^3 - 240x^2 = -60x^2(x^2 - 5x + 4) = -60x^2(x-4)(x-1)$. This is 0 for $x = 0$, $x = 1$, and $x = 4$. Note that $f'(-1) = -600 < 0$, $f'(1/2) = -26.25 < 0$, $f'(2) = 480 > 0$, and $f'(5) = -6000 < 0$. Thus f is increasing on $(1, 4)$ and is decreasing on $(-\infty, 1)$ and on $(4, \infty)$.

4.2.31 $f'(x) = -8x^3 + 2x = -2x(4x^2 - 1) = -2x(2x+1)(2x-1)$. This is zero for $x = 0$ and $x = \pm 1/2$. Note that $f'(-1) > 0$, $f'(-1/4) < 0$, $f'(1/4) > 0$, and $f'(1) < 0$, so f is increasing on $(-\infty, -1/2)$ and on $(0, 1/2)$, while it is decreasing on $(-1/2, 0)$ and on $(1/2, \infty)$.

4.2.33 $f'(x) = \cos x - (\cos x - x\sin x) = x\sin x$. This is zero on the given interval at $x = \pi$. Note that $f' > 0$ on $(0, \pi)$ (so f is increasing there) and $f' < 0$ on $(\pi, 2\pi)$ (so f is decreasing there).

4.2.35

a. $f'(x) = 2x$, so $x = 0$ is the only critical point.

b. Note that $f' < 0$ for $x < 0$ and $f' > 0$ for $x > 0$, so f has a local minimum of $f(0) = 3$ at $x = 0$.

c. Note that $f(-3) = 12$, $f(0) = 3$ and $f(2) = 7$, so the absolute maximum is 12 and the absolute minimum is 3.

4.2.37

  a. $f'(x) = x \cdot \frac{1}{2}(4 - x^2)^{-1/2} \cdot (-2x) + \sqrt{4 - x^2} \cdot 1 = \frac{4 - 2x^2}{\sqrt{4 - x^2}}$, which exists everywhere on $(-2, 2)$ and is zero only for $x = \pm\sqrt{2}$, so those are the only critical points.

  b. Note that $f'(-1.5) < 0$, $f'(0) > 0$ and $f'(1.5) < 0$, so f has a local minimum of $f(-\sqrt{2}) = -2$ and a local maximum of $f(\sqrt{2}) = 2$.

  c. Note that $f(-2) = 0 = f(2)$. So the absolute maximum is 2 at $x = \sqrt{2}$ and the absolute minimum is -2 at $x = -\sqrt{2}$.

4.2.39

  a. $f'(x) = -3x^2 + 9$, which is zero when $9 = 3x^2$, or $x^2 = 3$. So the critical points are at $x = \pm\sqrt{3}$.

  b. Note that $f'(-2) < 0$, $f'(0) > 0$, and $f'(2) < 0$, so there is a local minimum of $f(-\sqrt{3}) = -6\sqrt{3}$ and a local maximum of $f(\sqrt{3}) = 6\sqrt{3}$.

  c. There is an absolute maximum of 28 at $x = -4$ and an absolute minimum of $-6\sqrt{3}$ at $x = -\sqrt{3}$.

4.2.41

  a. $f'(x) = x^{2/3} + (x - 5) \cdot \frac{2}{3}x^{-1/3} = \frac{5x - 10}{3x^{1/3}}$, which is undefined at $x = 0$ and is 0 at $x = 2$. So these are the two critical points.

  b. Note that $f'(-1) > 0$ and $f'(1) < 0$, and $f'(3) > 0$ so f has a local maximum at $x = 0$ of $f(0) = 0$ and a local minimum at $x = 2$ of $-3\sqrt[3]{4} \approx -4.762$.

  c. Note that $f(-5) = -10\sqrt[3]{25} \approx -29.24$, $f(0) = 0$, and $f(5) = 0$, so the absolute maximum of f on $[-5, 5]$ is 0 and the absolute minimum is $-10\sqrt[3]{25}$.

4.2.43 $f'(x) = -6x + 2$, so the only critical point is at $x = 1/3$. The domain of f is $(-\infty, \infty)$, and f is everywhere continuous. Thus $1/3$ is the only critical point in the domain. Because $f'(x) > 0$ for $x < 1/3$ and $f'(x) < 0$ for $x > 1/3$, there is a local maximum at $x = 1/3$. By Theorem 4.5, there is also an absolute maximum there, with $f(1/3) = -14/3$.

4.2.45 Note that A is continuous on $(0, \infty)$.

    $A'(r) = -\frac{24}{r^2} + 4\pi r = \frac{4\pi r^3 - 24}{r^2}$, which is 0 for $r = \sqrt[3]{6/\pi}$, so there is only one critical point on the stated interval.

    Note that $A' < 0$ on $(0, \sqrt[3]{6/\pi})$ and $A' > 0$ on $(\sqrt[3]{6/\pi}, \infty)$, so there is a local minimum of $A(\sqrt[3]{6/\pi}) = 36\sqrt[3]{\pi/6}$.

    The local minimum mentioned above is an absolute minimum. There is no absolute maximum, because A is unbounded as $r \to \infty$.

4.2.47

The function sketched should be increasing and concave up everywhere.

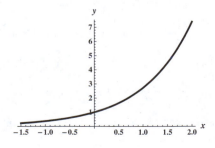

4.2.49

The function sketched should be decreasing every-where, concave down for $x < 0$, and concave up for $x > 0$.

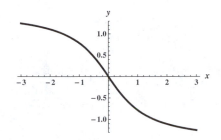

4.2.51 $f'(x) = 4x^3 - 6x^2$, so $f''(x) = 12x^2 - 12x = 12x(x-1)$. Note that f'' is zero when $x = 0$ and $x = 1$, so these are potential inflection points. Also note that $f''(-1) > 0$, $f''(.5) < 0$, and $f''(2) > 0$, so f is concave up on $(-\infty, 0)$ and on $(1, \infty)$, and is concave down on $(0, 1)$. There are inflection points at $(0, 1)$ and $(1, 0)$.

4.2.53 $f'(x) = 20x^3 - 60x^2$, and $f''(x) = 60x^2 - 120x = 60x(x-2)$. This is 0 for $x = 0$ and for $x = 2$. Note that $f''(-1) > 0$, $f''(1) < 0$, and $f''(3) > 0$. So f is concave up on $(-\infty, 0)$, concave down on $(0, 2)$, and concave up on $(2, \infty)$. There are inflection points at $x = 0$ and $x = 2$.

4.2.55 $f'(x) = \frac{(t+3)-(t-2)}{(t+3)^2} = \frac{5}{(t+3)^2}$, $f''(x) = \frac{-10}{(t+3)^3}$. Note that $f''(x) > 0$ for $x < -3$ and $f''(x) < 0$ for $x > -3$ so that f is concave up on $(-\infty, -3)$ and concave down on $(-3, \infty)$. Because neither $f'(x)$ nor $f''(x)$ is ever zero, there are no critical points and no inflection points.

4.2.57 $f'(\theta) = \sin\theta + \theta\cos\theta - 2\sin\theta = \theta\cos\theta - \sin\theta$, and $f''(\theta) = \cos\theta + -\theta\sin\theta - \cos\theta = -\theta\sin\theta$. On the given interval, this is zero for $\theta = \pi$. Note that $f''(\theta) < 0$ for $0 < \theta < \pi$ and $f''(\theta) > 0$ for $\pi < \theta < 2\pi$, so f is concave down on $(0, \pi)$ and is concave up on $(\pi, 2\pi)$, and there is a point of inflection at $\theta = \pi$.

4.2.59 $f'(x) = 10(x+1)^{3/2}(3x-4) + 12(x+1)^{5/2} = 2(x+1)^{3/2}(15x-20+6x+6) = 2(x+1)^{3/2}(21x-14) = 14(x+1)^{3/2}(3x-2)$. Then $f''(x) = 21(x+1)^{1/2}(3x-2) + 14(x+1)^{3/2}(3) = 7(x+1)^{1/2}(9x-6+6x+6) = 105x\sqrt{x+1}$. This is zero on the given interval for $x = 0$. Note that $f'' < 0$ on $(-1, 0)$ (so f is concave down on that interval) and $f'' > 0$ on $(0, \infty)$ (so f is concave up on that interval). There is a point of inflection at $x = 0$.

4.2.61 $g'(t) = 15t^4 - 120t^3 + 240t^2$, and $g''(t) = 60t^3 - 360t^2 + 480t = 60t(t-2)(t-4)$. Note that g'' is 0 for $t = 0, 2$, and 4. Note also that $g'' < 0$ on $(-\infty, 0)$ and on $(2, 4)$, so g is concave down on those intervals, while $g'' > 0$ on $(0, 2)$ and on $(4, \infty)$, so g is concave up there. There are inflection points at $t = 0, 2$, and 4.

4.2.63 $f'(x) = 3x^2 - 6x = 3x(x-2)$. This is zero when $x = 0$ and when $x = 2$, and these are the critical points. $f''(x) = 6x - 6$. Note that $f''(0) < 0$ and $f''(2) > 0$. Thus by the Second Derivative Test, there is a local maximum at $x = 0$ and a local minimum at $x = 2$.

4.2.65 $f'(x) = -2x$, so $x = 0$ is a critical point. $f''(x) = -2$, so $f''(0) = -2$ and the critical point yields a local maximum.

4.2.67 $f'(x) = 6x^2 - 6x = 6x(x-1)$, so $x = 0$ and $x = 1$ are critical points. $f''(x) = 12x - 6$, so $f''(1) = 6 > 0$, so the critical point at $x = 1$ yields a local minimum. Also, $f''(0) = -6 < 0$, so the critical point at 0 yields a local maximum.

4.2.69 $f'(x) = 12x^3 - 12x^2 = 12x^2(x-1)$, so there are critical points at $x = 0$ and $x = 1$. $f''(x) = 36x^2 - 24x = 12x(3x-2)$. Note that $f''(0) = 0$, so the second derivative test is inconclusive at $x = 0$, while $f''(1) = 12 > 0$, so there is a local minimum at $x = 1$. Using the First Derivative Test, it can be determined that there is neither a maximum nor a minimum at $x = 0$.

4.2.71 $f'(x) = 3x^2 - \sin x$ and $f''(x) = 6x - \cos x$. By inspection, we see that here is a critical number at $x = 0$. Using a CAS, we also find the critical number near $x = .327$. Note that $f''(0) = -1 < 0$, so there is a local maximum at $x = 0$, and note also that $f''(.327) \approx 1.02 > 0$, so there is local minimum at that point.

4.2.73

a. True. $f'(x) > 0$ implies that f is increasing, and $f''(x) < 0$ implies that f' is decreasing. So f is increasing, but at a decreasing rate.

b. False. In fact, if $f'(c)$ exists and isn't zero, then there isn't any kind of local extrema at $x = c$.

c. True. In fact, if two functions differ by a constant, then all of their derivatives are the same.

d. False. For example, consider $f(x) = x$ and $g(x) = x - 10$. Both are increasing, but $f(x)g(x) = x^2 - 10x$ is decreasing on $(-\infty, 5)$.

e. False. A continuous function with two local maxima must have a local minimum in between.

4.2.75

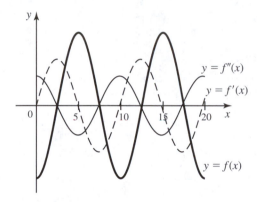

4.2.77 The graphs match as follows: (a) – (f) – (g); (b) – (e) –(i); (c) – (d) –(h). Note that (a) is always increasing, so its derivative must be always positive, and (f) switches from decreasing to increasing at 0, so its derivative must be negative for $x < 0$ and positive for $x > 0$.

Note that (b) has three extrema where there are horizontal tangent lines, so its derivative must cross the x-axis three times, and (e) has two extrema, so its derivative must cross the x-axis two times.

4.2.79

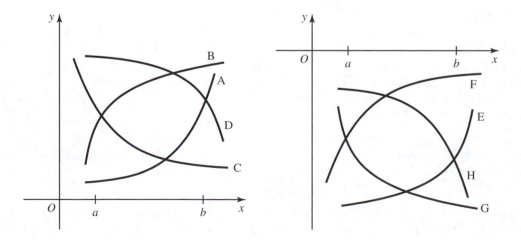

4.2.81

The graph sketched must have a flat tangent line at $x = -3/2$, $x = 0$, and $x = 1$, and must contain the points $(-2, 0)$, $(0, 0)$, and $(1, 0)$. The example to the right is only one possible such graph.

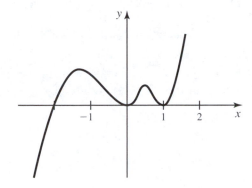

4.2.83

The graph sketched must be concave up on $(-\infty, -2)$ and on $(1, 3)$, and concave down on $(-2, 1)$ and on $(3, \infty)$. The example to the right is only one possible such graph.

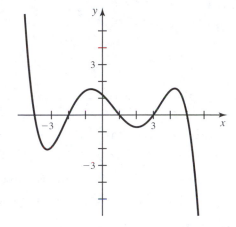

4.2.85

a. f is increasing on $(-2, 2)$. It is decreasing on $(-3, -2)$.

b. There are critical points of f at $x = -2$ and at $x = 0$. There is a local minimum at $x = -2$ and no extremum at $x = 0$.

c. There are inflection points of f at $x = -1$ and at $x = 0$.

d. f is concave up on $(-3, -1)$ and on $(0, 2)$, while it is concave down on $(-1, 0)$.

e.

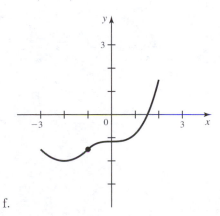

f.

4.2.87 $f'(x) = x^3 - 5x^2 - 8x + 48 = (x-4)^2(x+3)$. (This can be obtained by using trial-and-error to determine that $x = 4$ is a root, and then using long division of polynomials to see that $f'(x) = (x-4)(x^2 - x - 12)$.) Note that $f''(x) = 3x^2 - 10x - 8$, so $f''(-3) = 49 > 0$ and $f''(4) = 0$. So there is a local minimum at $x = -3$, but the test is inconclusive for $x = 4$. The first derivative test shows that there is neither a maximum nor a minimum at $x = 4$.

4.2.89 $f'(x) = 3x^2 + 4x + 4$, which is never 0. (Note that the discriminant $4^2 - 4 \cdot 3 \cdot 4 < 0$, so this quadratic has no real roots.) So there are no critical points.

4.2.91

a. $E = \frac{dD}{dp} \cdot \frac{p}{D} = -10 \frac{p}{500 - 10p} = \frac{p}{p - 50}$.

b. $E = \frac{12}{12 - 50} \cdot .045 = -1.42\%$.

c. If $D(p) = a - bp$, then $E(p) = -b \cdot \frac{p}{a - bp} = \frac{bp}{bp - a}$. So $E'(p) = \frac{(bp-a)b - bpb)}{(bp-a)^2} = -\frac{ab}{(bp-a)^2}$, which is less than 0 for $a, b > 0$ and $p \neq a/b$.

d. If $D(p) = \frac{a}{p^b}$, then $E(p) = -\frac{ab}{p^{b+1}} \cdot \frac{p}{a/p^b} = -b$.

4.2.93

a. $\lim\limits_{t \to \infty} \frac{300t^2}{t^2 + 30} \cdot \frac{1/t^2}{1/t^2} = \lim\limits_{t \to \infty} \frac{300}{1 + (30/t^2)} = 300$.

b. Note that $P'(t) = \frac{(t^2+30)(600t) - 300t^2(2t)}{(t^2+30)^2} = \frac{18000t}{(t^2+30)^2}$. We want to maximize this, so we compute its derivative $P''(t) = \frac{(t^2+30)^2 \cdot 18000 - 18000t \cdot 2(t^2+30) \cdot 2t}{(t^2+30)^4} = \frac{54000(10-t^2)}{(t^2+30)^3}$. This is 0 for $t = \sqrt{10}$, and an analysis of $P''(t)$ reveals that $P''(t) > 0$ for $t < \sqrt{10}$ and $P''(t) < 0$ for $t > \sqrt{10}$ so there is a local maximum for $P'(t)$ at $t = \sqrt{10}$.

c. Following the outline from the previous problem, we see that $P'(t) = \frac{2bKt}{(t^2+b)^2}$, and $P''(t) = \frac{2bK(b-3t^2)}{(t^2+b)^3}$. $P''(t)$ is 0 for $t = \sqrt{b/3}$, and the first derivative test reveals that this is a local maximum.

4.2.95

a. $f'(x) = 3x^2 + 2ax + b$, and $f''(x) = 6x + 2a$, which is 0 only for $x = -\frac{a}{3}$. Note that the sign of $f''(x)$ is different for $x < -a/3$ and $x > -a/3$, so this does represent an inflection point.

b.

$$f(x^*) - f(x^* + x)$$
$$= f(-a/3) - f(-a/3 + x) = (-a/3)^3 + a(-a/3)^2 + b(-a/3) + c$$
$$- ((-a/3 + x)^3 + a(-a/3 + x)^2 + b(-a/3 + x) + c)$$
$$= (-a/3)^3 + a(-a/3)^2 + b(-a/3) + c$$
$$- (-a/3)^3 - 3(-a/3)^2 x - 3(-a/3)x^2 - x^3 - a(-a/3)^2 - 2a(-a/3)x - ax^2 - b(-a/3) - bx - c$$
$$= -x^3 + \left(\frac{a^2}{3} - b \right) x.$$

Also,

$$f(x^* - x) - f(x^*) = f(-a/3 - x) - f(-a/3)$$
$$= (-a/3)^3 + 3(-a/3)^2(-x) + 3(-a/3)(-x)^2 + (-x)^3$$
$$+ a(-a/3)^2 + 2a(-a/3)(-x) + a(-x)^2$$
$$+ b(-a/3) + b(-x) + c - (-a/3)^3 - a(-a/3)^2 - b(-a/3) - c$$
$$= -x^3 + \left(\frac{a^2}{3} - b \right) x.$$

Thus the two expressions are the same for all x.

4.2.97

a. $f(-x) = \frac{1}{((-x)^2)^n+1} = \frac{1}{x^{2n}+1} = f(x)$, so f is even.

b. Note that $f(\pm 1) = \frac{1}{((-1)^2)^n+1} = \frac{1}{2}$, for all n.

c. $f'(x) = -(x^{2n}+1)^{-2}(2nx^{2n-1}) = -2nx^{2n-1}(x^{2n}+1)^{-2}$. So $f''(x) = -2nx^{2n-1}\cdot(-2(x^{2n}+1)^{-3})\cdot$ $2nx^{2n-1} + (x^{2n}+1)^{-2}\cdot(-2n(2n-1)x^{2n-2}) = \frac{-2nx^{2n-2}((-2n-1)x^{2n}+2n-1)}{(x^{2n}+1)^3}$. This is 0 when $x = 0$ for $n \geq 2$, and when $x = \pm\sqrt[2n]{\frac{2n-1}{2n+1}}$ for all positive n. An analysis of the sign of f'' shows that there is no sign change at $x = 0$, and f has inflection points at $x = \pm\sqrt[2n]{\frac{2n-1}{2n+1}}$.

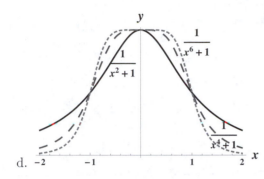

e. As n increases, the inflection points move further away from the y-axis, and closer to the point $(1, 1/2)$, although this movement is slight. The graphs become steeper and more "box-like."

d.

4.2.99 $f'(x) = 4x^3 + 3ax^2 + 2bx + c$, and $f''(x) = 12x^2 + 6ax + 2b = 2(6x^2 + 3ax + b)$. Note that $f''(x) = 0$ exactly when $x = \frac{-3a\pm\sqrt{9a^2-24b}}{12}$. This represents no real solutions when $9a^2 - 24b < 0$, which occurs when $b > 3a^2/8$. When $b = 3a^2/8$, there is one root, but in this case the sign of f'' doesn't change at the double root $x = -a/4$, so there are no inflection points for f. In the case $b < 3a^2/8$, there are two roots of f'', both of which yield inflection points of f, as can be seen by the change in sign of f'' at its two roots.

4.3   Graphing Functions

4.3.1 Because the intervals of increase and decrease and the intervals of concavity must be subsets of the domain, it is helpful to know what the domain is at the outset.

4.3.3 No. Polynomials are continuous everywhere, so they have no vertical asymptotes. Also, polynomials in x always tend to $\pm\infty$ as $x \to \pm\infty$.

4.3.5 The maximum and minimum must occur at either an endpoint or a critical point. So to find the absolute maximum and minimum, it suffices to find all the critical points, and then compare the values of the function at those points and at the endpoints. The largest such value is the maximum and the smallest is the minimum.

4.3.7

The function sketched should be decreasing and concave down for $x < 3$ and decreasing and concave up for $x > 3$.

4.3.9 The domain of f is $(-\infty, \infty)$, and there is no symmetry. The y intercept is $f(0) = 0$, and the x-intercepts are 0 and 3 because $f(x) = x(x-3)^2$. $f'(x) = 3x^2 - 12x + 9 = 3(x^2 - 4x + 3) = 3(x-3)(x-1)$. This is zero when $x = 1$ and $x = 3$. Note that $f'(0) > 0$, $f'(2) < 0$ and $f'(4) > 0$, so f is increasing on

$(-\infty, 1)$ and on $(3, \infty)$. It is decreasing on $(1, 3)$. Note that $f''(x) = 6x - 12$ which is zero at $x = 2$. Because $f''(1) < 0$ and $f''(3) > 0$, we conclude that f is concave down on $(-\infty, 2)$ and concave up on $(2, \infty)$. There is an inflection point at $(2, 2)$, a local maximum at $(1, 4)$ and a local minimum at $(3, 0)$.

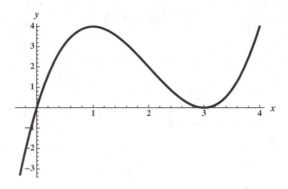

4.3.11

The domain of f is $(-\infty, \infty)$, and there is even symmetry, because $f(-x) = f(x)$. $f'(x) = 4x^3 - 12x = 4x(x^2 - 3)$. This is 0 when $x = \pm\sqrt{3}$ and when $x = 0$. $f''(x) = 12x^2 - 12 = 12(x^2 - 1)$, which is 0 when $x = \pm 1$. Note that $f'(-2) < 0$, $f'(-1) > 0$, $f'(1) < 0$, and $f'(2) > 0$. So f is decreasing on $(-\infty, -\sqrt{3})$ and on $(0, \sqrt{3})$. It is increasing on $(-\sqrt{3}, 0)$ and on $(\sqrt{3}, \infty)$. There is a local maximum of 0 at $x = 0$ and local minima of -9 at $x = \pm\sqrt{3}$. Note also that $f''(x) > 0$ for $x < -1$ and for $x > 1$ and $f''(x) < 0$ for $-1 < x < 1$, so there are inflection points at $x = \pm 1$. Also, f is concave down on $(-1, 1)$ and concave up on $(-\infty, -1)$ and on $(1, \infty)$. There is a y-intercept at $f(0) = 0$ and x-intercepts where $f(x) = x^4 - 6x^2 = x^2(x^2 - 6) = 0$, which is at $x = \pm\sqrt{6}$ and $x = 0$.

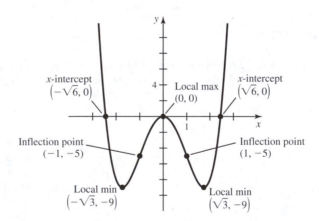

4.3.13 The domain of f is $(-\infty, \infty)$, and there is no symmetry. The y- intercept is $f(0) = -216$. The x-intercepts are 6 and -6.

$f'(x) = (x + 6)^2 + (x - 6)2(x + 6) = (x + 6)(x + 6 + 2x - 12) = (x + 6)(3x - 6) = 3(x + 6)(x - 2)$. The critical numbers are -6 and 2. Note that $f'(-7) > 0$, $f'(-2) < 0$, and $f'(3) > 0$, so f is increasing on $(-\infty, -6)$ and on $(2, \infty)$. It is decreasing on $(-6, 2)$. There is a local maximum of 0 at -6 and a local minimum of -256 at $x = 2$.

$f''(x) = 3(x - 2) + 3(x + 6) = 3(x - 2 + x + 6) = 3(2x + 4) = 6(x + 2)$, which is zero for $x = -2$. Note that $f''(x) < 0$ for $x < -2$ and $f''(x) > 0$ for $x > -2$, so f is concave down on $(-\infty, -2)$ and concave up on $(-2, \infty)$. The point $(-2, -128)$ is an inflection point.

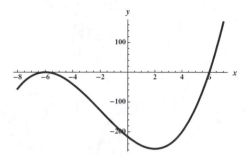

4.3.15

The domain of f is $(-\infty, 2) \cup (2, \infty)$, and there is no symmetry. Note that $\lim_{x \to 2^+} f(x) = \infty$ and $\lim_{x \to 2^-} f(x) = -\infty$, so there is a vertical asymptote at $x = 2$. There isn't a horizontal asymptote, because $\lim_{x \to \pm\infty} f(x) = \pm\infty$.

$f'(x) = \frac{(x-2) \cdot 2x - x^2}{(x-2)^2} = \frac{x(x-4)}{(x-2)^2}$. This is 0 when $x = 4$ and when $x = 0$.

$f''(x) = \frac{(x-2)^2(2x-4) - (x^2-4x) \cdot 2 \cdot (x-2)}{(x-2)^4} = \frac{8}{(x-2)^3}$. This is never 0.

Note that $f'(-1) > 0$, $f'(1) < 0$, $f'(3) < 0$ and $f'(5) > 0$. So f is decreasing on $(0, 2)$ and on $(2, 4)$. It is increasing on $(-\infty, 0)$ and on $(4, \infty)$. There is a local maximum of 0 at $x = 0$ and a local minimum of 8 at $x = 4$.

Note that $f''(x) > 0$ for $x > 2$ and $f''(x) < 0$ for $x < 2$, So f is concave up on $(2, \infty)$ and concave down on $(-\infty, 4)$. There are no inflection points, because the only change in concavity occurs at a vertical asymptote. The only intercept is $(0, 0)$.

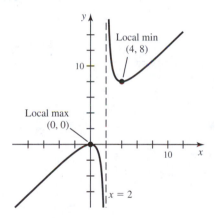

4.3.17

The domain of f is $(-\infty, -1) \cup (-1, 1) \cup (1, \infty)$, and there is odd symmetry, because $f(-x) = \frac{3(-x)}{(-x)^2 - 1} = -\frac{3x}{x^2 - 1} = -f(x)$. The only intercept is $(0, 0)$ which is both the y- and x-intercept.

Note that $\lim_{x \to -1^+} f(x) = \infty$ and $\lim_{x \to -1^-} f(x) = -\infty$, so there is a vertical asymptote at $x = -1$. Also, $\lim_{x \to 1^+} f(x) = \infty$ and $\lim_{x \to 1^-} f(x) = -\infty$, so there is a vertical asymptote at $x = 1$

Note that $\lim_{x \to \pm\infty} \frac{3x}{x^2 - 1} \cdot \frac{1/x^2}{1/x^2} = \lim_{x \to \pm\infty} \frac{3/x}{1 - (1/x^2)} = 0$, so $y = 0$ is a horizontal asymptote.

$f'(x) = \frac{(x^2-1) \cdot 3 - 3x \cdot 2x}{(x^2-1)^2} = \frac{-3x^2 - 3}{(x^2-1)^2} = -\frac{3(x^2+1)}{(x^2-1)^2}$. This is never 0, and is in fact negative wherever it is defined. Thus, f is decreasing on $(-\infty, -1)$, on $(-1, 1)$, and on $(1, \infty)$. There are no extrema.

$f''(x) = \frac{(x^2-1)^2(-6x) + 3(x^2+1) \cdot 2 \cdot (x^2-1) \cdot 2x}{(x^2-1)^4} = \frac{-6x^3 + 6x + 12x^3 + 12x}{(x^2-1)^3} = \frac{6x(x^2+3)}{(x^2-1)^3}$. This is 0 for $x = 0$. The point $(0, 0)$ is an point of inflection, because it is an interior point on the domain, and the second derivative changes from positive to negative there. The other concavity changes take place at the asymptotes. Note that f is concave down on $(-\infty, -1)$ and on $(0, 1)$, and concave up on $(-1, 0)$ and on $(1, \infty)$.

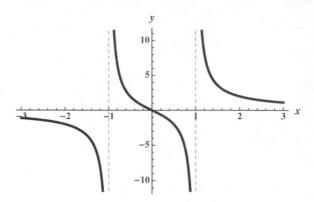

4.3.19

The domain of f is $(-\infty, -1/2) \cup (-1/2, \infty)$, and there is no symmetry.

Because $\displaystyle \lim_{x \to \pm\infty} \frac{x^2 + 12}{2x + 1} \cdot \frac{1/x}{1/x} = \lim_{x \to \pm\infty} \frac{x + (12/x)}{2 + (1/x)} = \pm\infty$, there is no horizontal asymptote. However,

there is an oblique asymptote of $y = \frac{x}{2} - \frac{1}{4}$, because we can write f as $f(x) = \frac{x}{2} - \frac{1}{4} + \frac{49/4}{2x+1}$ by long division.

Also, because $\lim_{x \to (-1/2)^-} f(x) = -\infty$ and $\lim_{x \to (-1/2)^+} f(x) = \infty$, there is a vertical asymptote at $x = -1/2$.

$f'(x) = \frac{(2x+1) \cdot 2x - (x^2+12) \cdot 2}{(2x+1)^2} = \frac{2x^2 + 2x - 24}{(2x+1)^2} = \frac{2(x+4)(x-3)}{(2x+1)^2}$. This is 0 for $x = -4$ and $x = 3$. $f''(x) = \frac{(2x+1)^2(4x+2) - (2x^2+2x-24) \cdot 2(2x+1) \cdot 2}{(2x+1)^4} = \frac{98}{(2x+1)^3}$, which is never 0.

Note that $f'(x) > 0$ on $(-\infty, -4)$ and on $(3, \infty)$. So f is increasing on $(-\infty, -4)$ and on $(3, \infty)$. Also, $f'(x) < 0$ on $(-4, -1/2)$ and on $(-1/2, 3)$. So f is decreasing on those intervals. There is a local maximum of -4 at $x = -4$ and a local minimum of 3 at $x = 3$.

Note also that $f''(x) < 0$ for $x < -1/2$, and $f''(x) > 0$ for $x > -1/2$, so f is concave down on $(-\infty, -1/2)$ and is concave up on $(-1/2, \infty)$. There are no inflection points because the only change in concavity occurs at the vertical asymptote. There are no x-intercepts because $x^2 + 12 > 0$ for all x, and the y-intercept is $f(0) = 12$.

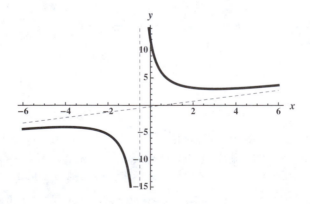

4.3.21

Note that the function has x-intercepts at $x = 0$ and $x = 3$. Because both $\sqrt{x}$ and $x - 3$ increase without bound as $x \to \infty$, there is no horizontal asymptote.

$f'(x) = \frac{x-3}{2\sqrt{x}} + \sqrt{x} = \frac{3x-3}{2\sqrt{x}}$, which is zero for $x = 1$. Note that $f'(x) < 0$ on $(0, 1)$ and $f'(x) > 0$ on $(1, \infty)$, so f is decreasing on $(0, 1)$ and increasing on $(1, \infty)$, and there is a local (and absolute) minimum of $f(1) = -2$ at $x = 1$.

$f''(x) = \frac{6\sqrt{x} - \frac{3x-3}{\sqrt{x}}}{4x} = \frac{6x - (3x-3)}{4x^{3/2}} = \frac{3x+3}{4x^{3/2}}$. This is never zero on the given domain, and is in fact always positive, so f is everywhere concave up on its domain.

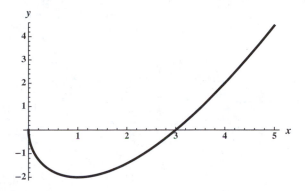

4.3.23

The domain of f is given to be $[-2\pi, 2\pi]$, and there is no symmetry, and no vertical asymptotes. There are no horizontal asymptotes to consider on this restricted domain.

$f'(x) = 1 - 2\sin x$. This is 0 when $\sin x = 1/2$, which occurs on the given interval for $x = -11\pi/6$, $-7\pi/6$, $\pi/6$, and $5\pi/6$. $f''(x) = -2\cos x$, which is 0 for $x = -3\pi/2, -\pi/2, \pi/2$, and $3\pi/2$.

Note that $f'(x) > 0$ on $(-2\pi, -11\pi/6)$, and on $(-7\pi/6, \pi/6)$, and on $(5\pi/6, 2\pi)$. So f is increasing on those intervals, while $f'(x) < 0$ on $(-11\pi/6, -7\pi/6))$ and on $(\pi/6, 5\pi/6)$, so f is decreasing there. f has local maxima at $x = -11\pi/6$ and at $x = \pi/6$ and local minima at $x = -7\pi/6$ and at $x = 5\pi/6$. Note also that $f''(x) < 0$ on $(-2\pi, -3\pi/2)$ and on $(-\pi/2, \pi/2)$ and on $(3\pi/2, 2\pi)$, so f is concave down on those intervals, while $f''(x) > 0$ on $(-3\pi/2, -\pi/2)$ and on $(\pi/2, 3\pi/2)$, so f is concave up there and there are inflection points at $x = \pm 3\pi/2$ and $x = \pm\pi/2$. The y-intercept is $f(0) = 2$ and the x-intercept is at approximately -1.030.

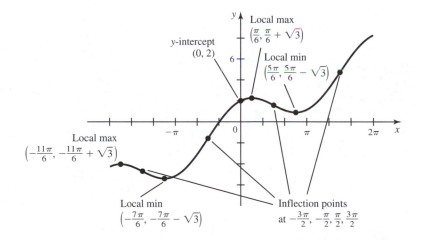

4.3.25

The domain of f is $(-\infty, \infty)$. There are no asymptotes. There are x-intercepts at $(0,0)$ and $(\pm 3\sqrt{3}, 0)$. f does have odd symmetry, because $f(-x) = -x - 3((-x)^{1/3}) = -(x - 3x^{1/3}) = -f(x)$.

$f'(x) = 1 - \frac{1}{\sqrt[3]{x^2}} = \frac{\sqrt[3]{x^2}-1}{\sqrt[3]{x^2}}$. This is undefined at $x = 0$, and is equal to zero at ± 1. Note that $f'(-2) > 0$, $f'(-1/2) < 0$, $f'(1/2) < 0$, $f'(2) > 0$. Thus, f is increasing on $(-\infty, -1)$ and on $(1, \infty)$. Because f is continuous at 0 (even though f' doesn't exist there), we can combine the intervals $(-1, 0)$ and $(0, 1)$ and state that f is decreasing on $(-1, 1)$. There is a local maximum at $(-1, 2)$ and a local minimum at $(1, -2)$.

$f''(x) = \frac{2}{3\sqrt[3]{x^5}}$, which is never zero, but is undefined at $x = 0$. Note that $f''(x) < 0$ for $x < 0$ and $f''(x) > 0$ for $x > 0$, so f is concave down on $(-\infty, 0)$ and is concave up on $(0, \infty)$. There is an inflection point at $(0, 0)$.

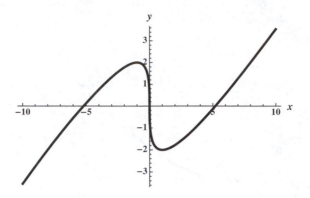

4.3.27

The domain of f is given to be $[0, 2\pi]$, so questions about symmetry and horizontal asymptotes aren't relevant. There are no vertical asymptotes.

$f'(x) = \cos x - 1$. This is never 0 on $(0, 2\pi)$. $f''(x) = -\sin x$, which is 0 on the given interval only for $x = \pi$. Note that $f'(x) < 0$ on $(0, 2\pi)$, so f is decreasing on the given interval and there are no relative extrema. Note also that $f''(x) < 0$ on $(0, \pi)$ and $f''(x) > 0$ on $(\pi, 2\pi)$, so f is concave down on $(0, \pi)$ and is concave up on $(\pi, 2\pi)$, and there is an inflection point at $x = \pi$. The only intercept is the origin $(0, 0)$.

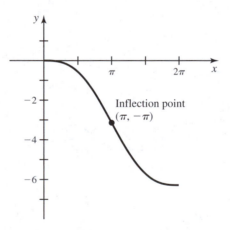

Inflection point
$(\pi, -\pi)$

4.3.29

The domain of f is given to be $(0, \infty)$.

There is no symmetry or asymptotes. (Note that $\lim\limits_{x \to \infty} \dfrac{1+x}{\sqrt{x}} = \infty$.)

$f'(x) = \dfrac{\sqrt{x} - (1+x)\frac{1}{2\sqrt{x}}}{x} \cdot \dfrac{2\sqrt{x}}{2\sqrt{x}} = \dfrac{2x-1-x}{2x^{3/2}} = \dfrac{x-1}{2x^{3/2}}$. This is zero for $x = 1$, and $f'(x) < 0$ on the interval $(0, 1)$ while $f'(x) > 0$ on $(1, \infty)$. Thus f is decreasing on $(0, 1)$ and is increasing on $(1, \infty)$, and there is a local (and absolute) minimum of 2 at $x = 1$.

$f''(x) = \dfrac{2x^{3/2} - (x-1)(3\sqrt{x})}{4x^3} \cdot \dfrac{\sqrt{x}}{\sqrt{x}} = \dfrac{2x^2 - 3x^2 + 3x}{4x^{7/2}} = \dfrac{-x+3}{4x^{5/2}}$. This is zero for $x = 3$, and an analysis of the sign of f'' reveals that $f'' > 0$ on $(0, 3)$, while $f'' < 0$ on $(3, \infty)$. The graph of f is therefore concave up on $(0, 3)$ and concave down on $(3, \infty)$, and there is an inflection point at $(3, 4/\sqrt{3})$.

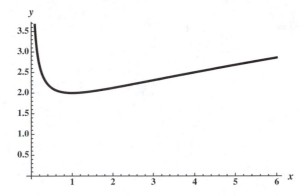

4.3.31

The domain of f is given to be $(-3\pi/2, 3\pi/2)$, but note that the function is not defined at $x = \pm\pi/2$, so the domain is actually $(-3\pi/2, -\pi/2) \cup (-\pi/2, \pi/2) \cup (\pi/2, 3\pi/2)$. Note that $f(-x) = -x + \tan(-x) = -x - \tan x = -(x + \tan x) = -f(x)$, so f has odd symmetry. f has vertical asymptotes at $x = \pm - 3\pi/2$ and at $x = \pm\pi/2$, because the tangent function increases or decreases without bound as x approaches these values.

$f'(x) = 1 + \sec^2 x$ which is always greater than 0. Thus f is increasing on each interval on which it is defined, and it has no extrema. $f''(x) = 2\sec x \cdot \sec x \tan x = 2\sec^2 x \tan x$. This is 0 at $x = \pm\pi$ and $x = 0$.

Note that $f''(x)$ is positive on $(-\pi, -\pi/2)$ and on $(0, \pi/2)$ and on $(\pi, 3\pi/2)$, so f is concave up on these intervals. Also, $f''(x)$ is negative on $(-3\pi/2, -\pi)$ and on $(-\pi/2, 0)$, and on $(\pi/2, \pi)$, so f is concave down on these intervals. There are points of inflection at $x = \pm\pi$ and at $x = 0$. The other changes in concavity occur at the vertical asymptotes.

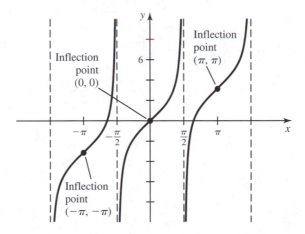

4.3.33

The domain of f is $(-\infty, \infty)$, and there is no symmetry. $f'(x) = x^2 - 4x - 5 = (x - 5)(x + 1)$. This is 0 when $x = -1, 5$. $f''(x) = 2x - 4$, which is 0 when $x = 2$. Note that $f'(-2) > 0$, $f'(0) < 0$, and $f'(6) > 0$. So f is increasing on $(-\infty, -1)$ and on $(5, \infty)$. It is decreasing on $(-1, 5)$. There is a local maximum of $14/3$ at $x = -1$ and a local minimum of $-94/3$ at $x = 5$. Note also that $f''(x) < 0$ for $x < 2$ and $f''(x) > 0$ for $x > 2$, so there is an inflection point at $(2, -40/3)$, and f is concave down on $(-\infty, 2)$ and concave up on $(2, \infty)$. The y intercept is $f(0) = 2$.

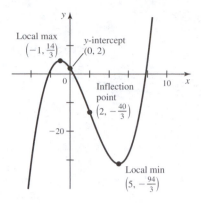

4.3.35

The domain of f is $(-\infty, \infty)$, and there is no symmetry. $f'(x) = 12x^3 + 12x^2 - 24x = 12x(x+2)(x-1)$. This is 0 when $x = -2$, when $x = 1$, and when $x = 0$. $f''(x) = 36x^2 + 24x - 24 = 12(3x^2 + 2x - 2)$, which is 0 when $x = \frac{-1 \pm \sqrt{7}}{3}$. These values are at approximately -1.215 and 0.549. Note that $f'(-3) < 0$, $f'(-1) > 0$, $f'(.5) < 0$, and $f'(2) > 0$. So f is decreasing on $(-\infty, -2)$ and on $(0, 1)$. It is increasing on $(-2, 0)$ and on $(1, \infty)$. There is a local maximum of 0 at $x = 0$ and a local minimum of -32 at $x = -2$ and a local minimum of -5 at $x = 1$. Let $r_1 < r_2$ be the two roots of $f''(x)$ mentioned above. Note that $f''(x) > 0$ for $x < r_1$ and for $x > r_2$ and $f''(x) < 0$ for $r_1 < x < r_2$, so there are inflection points at $x = r_1$ and at $x = r_2$. Also, f is concave down on (r_1, r_2) and concave up on $(-\infty, r_1)$ and on (r_2, ∞). There is a y-intercept at $f(0) = 0$ and x-intercepts where $f(x) = 3x^4 + 4x^3 - 12x^2 = x^2(3x^2 + 4x - 12) = 0$, which is at $x = \frac{-2 \pm 2\sqrt{10}}{3}$ and $x = 0$.

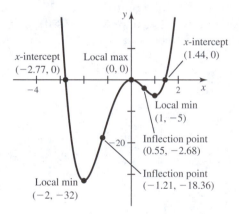

4.3.37

The domain of f is $(-\infty, -1) \cup (-1, 1) \cup (1, \infty)$, and there is no symmetry. Note that $\lim\limits_{x \to -1^+} f(x) = \infty$ and $\lim\limits_{x \to -1^-} f(x) = -\infty$, so there is a vertical asymptote at $x = -1$. Also, $\lim\limits_{x \to 1^+} f(x) = -\infty$ and $\lim\limits_{x \to 1^-} f(x) = \infty$, so there is a vertical asymptote at $x = 1$

Note that $\lim\limits_{x \to \pm\infty} \dfrac{3x - 5}{x^2 - 1} \cdot \dfrac{1/x^2}{1/x^2} = \lim\limits_{x \to \pm\infty} \dfrac{3/x}{1 - (1/x^2)} = 0$, so $y = 0$ is a horizontal asymptote.

$f'(x) = \dfrac{(x^2-1) \cdot 3 - (3x-5) \cdot 2x}{(x^2-1)^2} = \dfrac{-3x^2 + 10x - 3}{(x^2-1)^2} = \dfrac{(-3x+1)(x-3)}{(x^2-1)^2}$. This is 0 when $x = 3$ and when $x = 1/3$.

$f''(x) = \dfrac{(x^2-1)^2(-6x+10) - (-3x^2+10x-3)(2)(x^2-1) \cdot 2x}{(x^2-1)^4} = \dfrac{2(3x^3 - 15x^2 + 9x - 5)}{(x^2-1)^3}$. This is 0 for $x \approx 4.405$. Let r be this root of $f''(x)$.

Note that $f'(-2) < 0$, $f'(-1/2) < 0$, $f'(1/2) > 0$, $f'(2) > 0$ and $f'(4) < 0$. So f is decreasing on $(-\infty, -1)$, on $(-1, 1/3)$ and on $(3, \infty)$. It is increasing on $(1/3, 1)$ and on $(1, 3)$. There is a local maximum of $1/2$ at $x = 3$ and a local minimum of $9/2$ at $x = 1/3$.

Note that $f''(x) < 0$ for $x < -1$ and $f''(x) < 0$ for $1 < x < r$, while $f''(x) > 0$ for $-1 < x < 1$, and for $x > r$. Thus f is concave up on $(-1, 1)$ and on (r, ∞) and concave down on $(-\infty, -1)$ and on $(1, r)$. There is an inflection point at r. There is a y-intercept at $f(0) = 5$ and an x-intercept at $(5/3, 0)$.

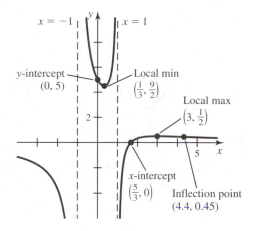

4.3.39

a. False. Maxima and minima can also occur at points where $f'(x)$ doesn't exist. Also, it is possible to have a zero of f' which doesn't correspond to an extreme point.

b. False. Inflection points can also occur at points where $f''(x)$ doesn't exist, and a zero of f'' might not correspond to an inflection point.

c. False. For example, $f(x) = \frac{(x^2-9)(x^2-16)}{(x+3)(x-4)}$ doesn't have a vertical asymptote at $x = -3$ or $x = 4$.

d. True. The limit of a rational function as $x \to \infty$ is a finite number when the degree of the denominator is greater than or equal to that of the numerator. If they both have the same degree, the limit is the ratio of the leading coefficients, and this is also true of the limit as $x \to -\infty$. In the case where the denominator has greater degree than the numerator, the limit is 0 as $x \to -\infty$ and as $x \to \infty$.

4.3.41

$f'(x)$ is 0 on the interior of the given interval at $x = \pm 3\pi/2$, $x = \pm\pi$, $x = \pm\pi/2$, and at $x = 0$.
$f'(x) > 0$ on $(-2\pi, -3\pi/2)$, $(-\pi, -\pi/2)$, $(0, \pi/2)$, and on $(\pi, 3\pi/2)$, so f is increasing on those intervals.
$f'(x) < 0$ on $(-3\pi/2, -\pi)$, $(-\pi/2, 0)$, $(\pi/2, \pi)$, and on $(3\pi/2, 2\pi)$, so f is decreasing on those intervals. There are local maxima at $x = \pm 3\pi/2$ and $x = \pm\pi/2$, and local minima at $x = 0$ and at $x = \pm\pi$. An example of such a function is sketched.

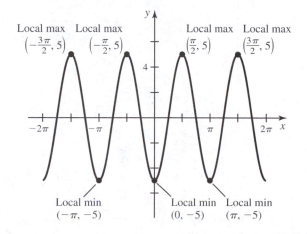

4.3.43

$f'(x)$ is 0 at $x = 0$, $x = -2$, and $x = 1$. $f'(x) > 0$ on $(-\infty, -2)$ and on $(1, \infty)$, so f is increasing on those intervals. $f'(x) < 0$ on $(-2, 0)$ and on $(0, 1)$, so f is decreasing on those intervals. There is a local maximum at $x = -2$ and a local minimum at $x = 1$. There isn't an extremum at $x = 0$.

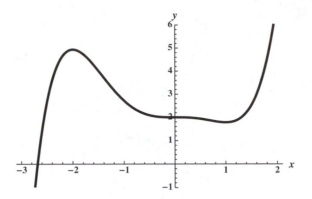

4.3.45

$f'(x)$ is 0 at $x = 1$ and $x = 3$.

$f'(x) > 0$ on $(0, 1)$ and on $(3, 4)$, so f is increasing on those intervals. $f'(x) < 0$ on $(1, 3)$, so f is decreasing on that interval. There is a local maximum at $x = 1$ and a local minimum at $x = 3$.

$f''(x)$ changes sign at $x = 2$ from negative to positive, so $x = 2$ is an inflection point where the concavity of f changes from down to up. An example of such a function is sketched.

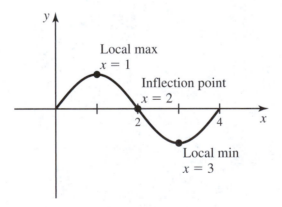

4.3.47

The domain of f is $(-\infty, \infty)$ and there is no symmetry. There are no asymptotes because f is a polynomial.

$f'(x) = 3x^2 - 12x - 135 = 3(x - 9)(x + 5)$, which is 0 for $x = 9$ and $x = -5$. $f'(x) > 0$ on $(-\infty, -5)$ and on $(9, \infty)$, so f is increasing on those intervals. $f'(x) < 0$ on $(-5, 9)$, so f is decreasing on that interval. There is a local maximum at $x = -5$ and a local minimum at $x = 9$.

$f''(x) = 6x - 12$, which is 0 for $x = 2$. $f''(x) > 0$ on $(2, \infty)$, so f is concave up on that interval. $f''(x) < 0$ on $(-\infty, 2)$, so f is concave down on that interval. There is a point of inflection at $x = 2$. The y-intercept is 0 and the x-intercepts are -9 and 15.

4.3.49

4.3.51

4.3.53

The domain of f is $(-\infty, -2) \cup (2, \infty)$ and there is no symmetry. There is a vertical asymptote at $x = 2$ because $\lim_{x \to 2^+} -\frac{x\sqrt{x^2-4}}{x-2} = -\infty$. There are no horizontal asymptotes.

$f'(x) = \frac{(x-2)(-x^2(x^2-4)^{-1/2}+(x^2-4)^{1/2}(-1))-((-x)(x^2-4)^{1/2})}{(x-2)^2}$. This can be written as $\frac{-x^2+2x+4}{(x-2)\sqrt{x^2-4}}$, and this quantity is 0 on the given domain only for $x = 1 + \sqrt{5} \approx 3.236$.

$f'(x) > 0$ on $(-\infty, -2)$ and on $(2, 1+\sqrt{5})$, so f is increasing on those intervals. $f'(x) < 0$ on $(1+\sqrt{5}, \infty)$, so f is decreasing on that interval. There is a local maximum at $x = 1 + \sqrt{5}$.

$f''(x)$ simplifies to be $-\frac{4(x+4)}{(x-2)^2(x+2)\sqrt{x^2-4}}$, which is 0 at $x = -4$

$f''(x) > 0$ on $(-4, -2)$, so f is concave up on that interval. $f''(x) < 0$ on $(-\infty, -4)$ and on $(1+\sqrt{5}, \infty)$, so f is concave down on those intervals. There is a point of inflection at $x = -4$.

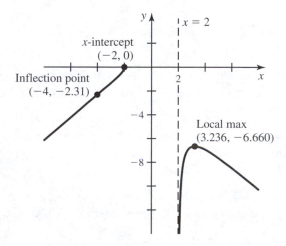

4.3.55

The domain of f is $(-\infty, \infty)$ and there is no symmetry. There are no asymptotes because f is a polynomial.

$f'(x) = 12x^3 - 132x^2 + 120x = 12x(x - 10)(x - 1)$, which is 0 for $x = 0$, $x = 1$, and $x = 10$.

$f'(x) > 0$ on $(0, 1)$ and on $(10, \infty)$, so f is increasing on those intervals. $f'(x) < 0$ on $(-\infty, 0)$ and on $(1, 10)$, so f is decreasing on those intervals. There is a local maximum at $x = 1$, and local minima at $x = 0$ and at $x = 10$.

$f''(x) = 36x^2 - 264x + 120 = 12(3x^2 - 22x + 10)$. This is 0 at approximately $x = .487$ and $x = 6.846$. Let these two roots be r_1 and r_2 with $r_1 < r_2$. $f''(x) > 0$ on $(-\infty, r_1)$ and on (r_2, ∞), so f is concave up on those intervals. $f''(x) < 0$ on (r_1, r_2), so f is concave down on that interval. There are points of inflection at $x = r_1$ and $x = r_2$.

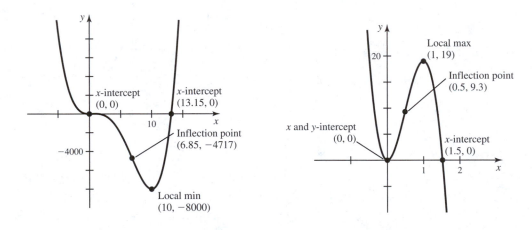

4.3.57

The domain of f is $(-\infty, \infty)$ and there is no symmetry. There are no asymptotes because f is a polynomial.

$f'(x) = 60x^5 - 180x^4 - 300x^3 + 900x^2 + 240x - 720 = 60(x + 2)(x + 1)(x - 1)(x - 2)(x - 3)$, which is 0 for $x = -2$, $x = -1$, $x = 1$, $x = 2$, and $x = 3$.

$f'(x) > 0$ on $(-2, -1)$, $(1, 2)$ and on $(3, \infty)$, so f is increasing on those intervals. $f'(x) < 0$ on $(-\infty, -2)$, $(-1, 2)$ and on $(2, 3)$, so f is decreasing on those intervals. There are local minima at $x = -2$, $x = 1$, and $x = 3$, and local maxima at $x = -1$ and $x = 2$.

$f''(x) = 300x^4 - 720x^3 - 900x^2 + 1800x + 240 = 60(5x^4 - 12x^3 - 15x^2 + 30x + 4)$. Using a computer algebra system, we find that this has 4 real roots, which we will call $r_1 < r_2 < r_3 < r_4$. They values of these roots are approximately -1.605, -0.125, 1.502, and 2.629.

$f''(x) > 0$ on $(-\infty, r_1)$, and on (r_2, r_3) and (r_4, ∞), so f is concave up on those intervals. $f''(x) < 0$ on (r_1, r_2) and on (r_3, r_4), so f is concave down on those intervals. There are points of inflection at each r_i for i from 1 to 4.

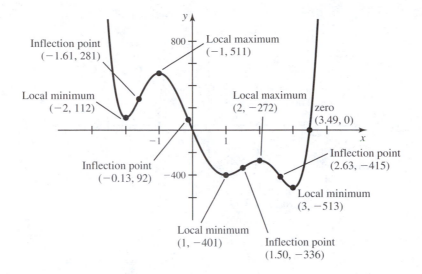

4.3.59

The domain of f is $(-\infty, \infty)$ and there is odd symmetry, because $f(-x) = -f(x)$. There are no vertical asymptotes, but $y = 0$ is a horizontal asymptote, because

$$\lim_{x \to \infty} \frac{x\sqrt{|x^2 - 1|}}{x^4 + 1} \cdot \frac{1/x^4}{1/x^4} = \lim_{x \to \infty} \frac{\sqrt{(1/x^4) - (1/x^6)}}{1 + (1/x^4)} = 0.$$

Note that $f(x) = \begin{cases} \frac{x\sqrt{x^2-1}}{x^4+1} \text{if } |x| \geq 1; \\ \frac{x\sqrt{1-x^2}}{x^4+1} \text{if } |x| < 1. \end{cases}$

Differentiating each part of the above and simplifying yields

$$f'(x) = \begin{cases} \frac{-2x^6 + 3x^4 + 2x^2 - 1}{(x^4 + 1)^2 \sqrt{x^2 - 1}} \text{if } |x| > 1; \\ \frac{2x^6 - 3x^4 - 2x^2 + 1}{(x^4+1)^2 \sqrt{1-x^2}} \text{if } |x| < 1. \end{cases}$$

The roots of this expression (on the respective domains) are approximately ± 1.374 and ± 0.596. Also, this derivative doesn't exist at $x = \pm 1$. Let the roots of f' be $\pm r_1$ and $\pm r_2$ where $0 < r_1 < r_2$. An analysis of the sign of f' shows that f is increasing on $(-r_2, -1)$ on $(-r_1, r_1)$ and on $(1, r_2)$, while f is decreasing on $(-\infty, -r_2)$, $(-1, -r_1)$, $(r_1, 1)$ and on (r_2, ∞), so there are local maxima at $x = -1$, $x = r_1$, and $x = r_2$, and local minima at $x = -r_1$, $x = -r_2$, and $x = 1$.

An analysis without computer of $f''(x)$ is not for the fainthearted. In the case $|x| > 1$ the second derivative is given by $\dfrac{x\left(6x^{10} - 19x^8 - 12x^6 + 42x^4 - 18x^2 - 3\right)}{(x^2 - 1)^{3/2}\left(x^4 + 1\right)^3}$, and for the case $|x| < 1$ we have $\dfrac{x\left(6x^{10} - 19x^8 - 12x^6 + 42x^4 - 18x^2 - 3\right)}{(1 - x^2)^{3/2}\left(x^4 + 1\right)^3}$. There is a root of approximately ± 1.790, and 0 is a root as well. Let the non-zero roots be $\pm r_3$ where $r_3 > 0$. An analysis of the sign of f'' reveals that f is concave down on $(-\infty, -r_3)$, $(0, 1)$ and on $(1, r_3)$, while it is concave up on $(-r_3, -1)$, $(-1, 0)$ and on (r_3, ∞). There are inflection points at $\pm r_3$ and at 0. The x-intercepts are ± 1 and 0.

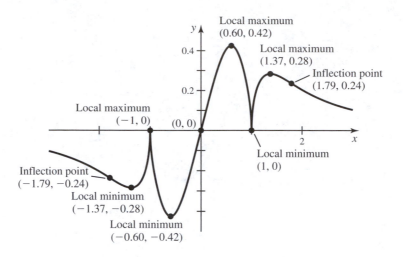

4.3.61

a. f has even symmetry, so we will analyze the function on $(0, 2\pi]$, and use the symmetry to graph the function over $[-2\pi, 0)$.

$f'(x) = \frac{x^2(-3\cos^2 x(-\sin x))-(1-\cos^3 x)\cdot 2x}{x^4} = \frac{3x\sin x \cos^2 x - 2(1-\cos^3 x)}{x^3}$. This has no roots on $(0, 2\pi)$, and in fact is always negative, so f is decreasing on $(0, 2\pi)$.

$f''(x)$ when simplified is given by $\frac{3\left((x^2-2)\cos^3(x)-2x^2\sin^2(x)\cos(x)-4x\sin(x)\cos^2(x)+2\right)}{x^4}$. The roots of f'' on $(0, 2\pi)$ are $r_1 \approx .89$, $r_2 \approx 2.47$, $r_3 \approx 3.48$, $r_4 \approx 4.76$, and $r_5 \approx 5.5$. An analysis of the sign of f'' reveals that there is a change in concavity at each of these roots, starting with concavity downward on $(-r_1, r_1)$. So each of $\pm r_i$ is an inflection point.

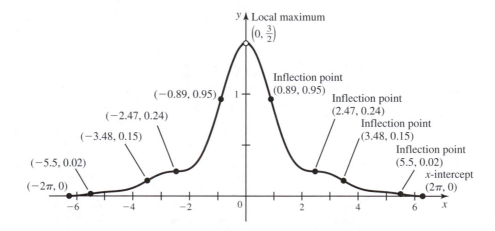

b. f has even symmetry, so we will analyze the function on $(0, 2\pi]$, and use the symmetry to graph the function over $[-2\pi, 0)$.

$$f'(x) = \frac{x^2(-5\cos^4 x(-\sin x)) - (1-\cos^5 x)(2x)}{x^4} = \frac{5x\cos^4 x \sin x - 2(1-\cos^5 x)}{x^3}.$$

This has roots on $(0, 2\pi)$ of $r_1 \approx 2.41$ and $r_2 \approx 2.83$. An analysis of the sign of f' shows that f is decreasing on $(0, r_1)$, increasing on (r_1, r_2), decreasing on $(r_2, 2\pi)$, so there is a local minimum at r_1 and a local maximum at r_2.

$f''(x)$ when simplified is given by

$$\frac{(5x^2-6)\cos^5(x) - 20x^2\sin^2(x)\cos^3(x) - 20x\sin(x)\cos^4(x) + 6}{x^4}.$$

The roots of f'' on $(0, 2\pi)$ are $r_3 \approx .63$, $r_4 \approx 2.62$, $r_5 \approx 3.45$, $r_6 \approx 4.96$, and $r_7 \approx 5.74$. An analysis of the sign of f'' reveals that there is a change in concavity at each of these roots, starting with concavity downward on $(-r_3, r_3)$. So each of $\pm r_i$ is an inflection point for $i = 3, 4, 5, 6$ and 7.

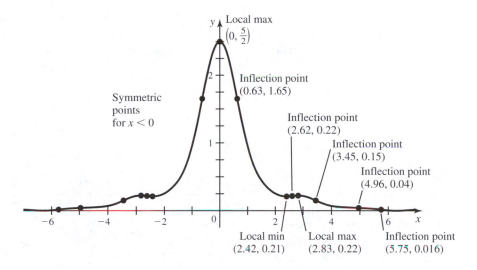

4.3.63

(a)   (b)   (c)

a.

(d)   (e)   (f)

b. The water is being poured in at a constant rate, so the depth is always increasing, so $y = h(t)$ is an increasing function.

c.

(a) No concavity

(b) Always concave down.

(c) Always concave up.

(d) Concave down for for the first half and concave up for the second half.

(e) At the beginning, in the middle, and at the end, there is no concavity. In the lower middle it is concave down and in the upper middle it will be concave up.

(f) This is concave down for the first half, and concave up for the second half.

d. (a) $h'(t)$ is constant, so there is no local max/min.   (b) $h'(t)$ is maximal at $t = 0$.   (c) $h'(t)$ is maximal at $t = 10$.

(d) $h'(t)$ is maximal at $t = 0$ and $t = 10$.   (e) $h'(t)$ is maximal on the first and last straight parts of $h(t)$.   (f) $h'(t)$ is maximal at $t = 0$ and $t = 10$.

4.3.65

If $f''(x) > 0$ on $(-\infty, 0)$ and on $(0, \infty)$, then $f'(x)$ is increasing on both of those intervals. But if there is a local max at 0, the function f must be switching from increasing to decreasing there. This means that f' must be switching from positive to negative. But if f' is switching from positive to negative, but increasing, there must be a cusp at $x = 0$, so $f'(0)$ does not exist.

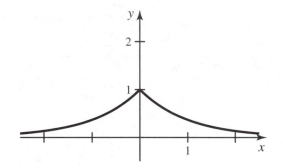

4.3.67

The equation is valid on only for $|x| \leq 1$ and $|y| \leq 1$. Using implicit differentiation, we have $(2/3)x^{-1/3} + (2/3)y^{-1/3}y' = 0$, so $y' = \frac{-y^{1/3}}{x^{1/3}}$. This is 0 for $y = 0$ (in which case $x = \pm 1$) and doesn't exist for $x = 0$ (in which case $x = \pm 1$.) In the first quadrant the curve is decreasing, in the 2nd it is increasing, in the 3rd it is decreasing, and in the 4th it is increasing. Differentiating y' yields $y'' = \frac{x^{1/3}(-1/3)y^{-2/3}y' + y^{1/3}(1/3)(x^{-2/3})}{x^{2/3}} = \frac{y^{2/3}+x^{2/3}}{3x^{4/3}y^{1/3}}$, which is positive when y is positive and negative when y is negative, so the curve is concave up in the first and 2nd quadrants, and concave down in the 3rd and 4th.

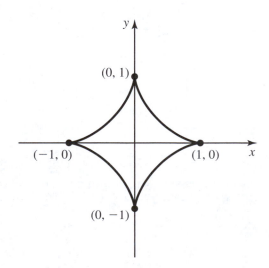

4.3.69

First note that the expression is symmetric when x and y are switched, so the curve should be symmetric about the line $y = x$. Also, if $y = x$, then $2x^3 = 3x^2$, so either $x = 0$ or $x = 3/2$, so this is where the curve intersects the line $y = x$.

Differentiating implicitly yields $3x^2 + 3y^2 y' = 3xy' + 3y$, so $y' = \frac{y - x^2}{y^2 - x}$. This is 0 when $y = x^2$, but this occurs on the curve when $x^3 + x^6 = 3x^3$, which yields $x = 0$ (and $y = 0$), or $x^3 = 2$, so $x = \sqrt[3]{2} \approx 1.260$. Note also that the derivative doesn't exist when $x = y^2$, which again yields $(0, 0)$ and $y^6 + y^3 = 3y^3$, or $y = \sqrt[3]{2}$. So there should be a flat tangent line at approximately $(1.260, 1.587)$ and a vertical tangent line at about $(1.587, 1.260)$.

Differentiating again and solving for y'' yields $y''(x) = \frac{2xy(x^3 - 3xy + y^3 + 1)}{(x - y^2)^3} = \frac{2xy}{(x - y^2)^3}$. In the first quadrant, when $x > y^2$, the curve is concave up, when $x < y^2$, the curve is concave down. In both the 2nd and 4th quadrants, the curve is concave up.

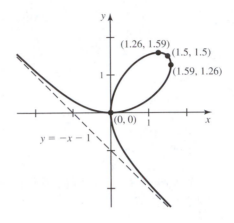

4.3.71

Note that the curve is symmetric about both the x-axis and the y-axis, so we can just consider the first quadrant, and obtain the rest by reflection.

Differentiating implicitly and solving for y' yields $y'(x) = \frac{2x^3 - 5x}{2y(y^2 - 2)}$. The numerator is negative on $(0, \sqrt{5/2})$ and positive on $(\sqrt{5/2}, \infty)$, while the denominator is negative for $0 < y < \sqrt{2}$ and positive for $y > \sqrt{2}$. Thus the relation is increasing in the rectangle $(0, \sqrt{5/2}) \times (0, \sqrt{2})$ and in the region $(\sqrt{5/2}, \infty) \times (\sqrt{2}, \infty)$, while it is decreasing in the other regions in the first quadrant. There are vertical tangent lines when $y = \sqrt{2}$. When $y = 2$ and $x = 0$ there is a horizontal tangent line.

Note that when $x = 0$, we have $y = 0$ or $y = \pm 2$, while if $y = 0$, we have $x = 0$ or $x = \pm \sqrt{5}$. Also, if $y = \sqrt{2}$ then $x = 1$ or $x = 2$. So some sample points to plot are $(0, 0)$, $(\pm\sqrt{5}, 0)$, $(0, \pm 2)$, $(\pm 1, \pm\sqrt{2})$, and $(\pm\sqrt{5}, \pm\sqrt{2})$. Also, when $1 < x < 2$, there are no corresponding y values on the curve.

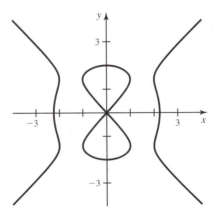

4.3.73

Note that the curve requires $-1 \le x < 1$.
Note also that the curve is symmetric about both the
x-axis and the y-axis, so we can just consider the first
quadrant, and obtain the rest by reflection.
Differentiating implicitly yields $4x^3 - 2x + 2yy' = 0$,
so $y' = \frac{x-2x^3}{y}$. This is 0 in the first quadrant for
$x = \sqrt{2}/2$. Note also that there is a vertical tangent
line at the point $(1,0)$. The derivative is positive on
$(0, \sqrt{2}/2)$ and negative on $(\sqrt{2}/2, 1)$, so in the first
quadrant the curve is increasing on that first interval
and decreasing on the second.
Differentiating again and solving for y'' (and rewrit-
ing) yields $y''(x) = \frac{x^4(2x^2-3)}{y}$, which is negative in the
first quadrant for $0 < x < 1$, so this curve is concave
down in the first quadrant.
The rest of the curve can be found by reflection.

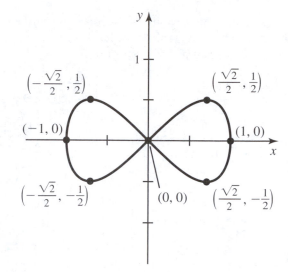

4.3.75

As n increases, the curves retain their symmetry, but
move "outward." That is, the curves enclose a greater
area. It appears that the figures approach the 2×2
square centered at the origin with sides parallel to the
coordinate axes.

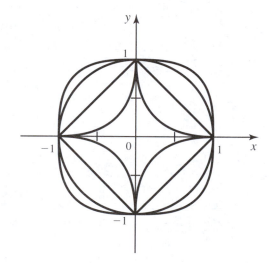

4.3.77

The domain of f is given to be $[-2\pi, 2\pi]$. There are no vertical asymptotes. Note that $f(-x) = \frac{(-x)(\sin(-x))}{((-x)^2+1)} = \frac{x \sin x}{x^2+1} = f(x)$, f has even symmetry. Questions about horizontal asymptotes aren't relevant because the given domain is an interval with finite length.

$f'(x) = \frac{(x^2+1)(x\cos x+\sin x)-x\sin x\cdot(2x)}{(x^2+1)^2}$, which can be simplified to $\frac{x(x^2+1)\cos x+(1-x^2)\sin x}{(x^2+1)^2}$, and with the aid of a computer algebra system, the roots of this expression can be found to be approximately ± 4.514 and ± 1.356, as well as $x = 0$. We will call the non-zero roots $\pm r_1$ and $\pm r_2$ where $0 < r_1 < r_2$. Note that $f'(x) < 0$ on $(-2\pi, -r_2)$ and on $(-r_1, 0)$ and (r_1, r_2), so f is decreasing there, while $f'(x) > 0$ on $(-r_2, -r_1)$, on $(0, r_1)$, and on $(r_2, 2\pi)$, so f is increasing on these intervals. There are local maxima at $x = \pm r_1$ and local minima at $x = 0$ and at $x = \pm r_2$.

$f''(x)$ has numerator $(x^2+1)^2((x^3+x)(-\sin x)+\cos x(3x^2+1)+(1-x^2)(\cos x)+\sin x(-2x)) - ((x^3+x)\cos x+(1-x^2)\sin x)(4x)(x^2+1)$ and denominator $(x^2+1)^4$. This simplifies to

$$f''(x) = \frac{(-x^5 - 7x)\sin x + (-2x^4 + 2)\cos x}{(x^2+1)^3},$$

which is 0 at approximately ± 5.961 and ± 2.561 and ± 0.494. We will call these 6 roots $\pm r_3$, $\pm r_4$ and $\pm r_5$ where $0 < r_3 < r_4 < r_5$. Note that $f''(x) < 0$ on $(-2\pi, -r_5)$ and on $(-r_4, -r_3)$, and on (r_3, r_4), and on $(r_5, 2\pi)$, so f is concave down on these intervals, while $f''(x) > 0$ on $(-r_5, -r_4)$, and on $(-r_3, r_3)$, and on (r_4, r_5), so f is concave up on these intervals. There are points of inflection at each of $\pm r_3$, $\pm r_4$, and $\pm r_5$.

There is an x-intercept at $(0, 0)$, which is also the y-intercept, as well as x-intercepts at $\pm 2\pi$.

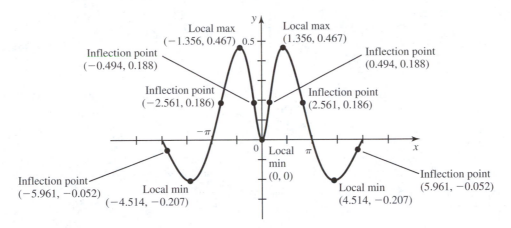

4.4 Optimization Problems

4.4.1 ...objective...constraints....

4.4.3 The constraint is $x + y = 10$, so we can express $y = 10 - x$ or $x = 10 - y$. Therefore the objective function can be expressed $Q = x^2(10 - x)$ or $Q = (10 - y)^2 y$.

4.4.5 Let x and y be the dimensions of the rectangle. The perimeter is $2x + 2y$, so the constraint is $2x + 2y = 10$, which gives $y = 5 - x$. The objective function to be maximized is the area of the rectangle, $A = xy$. Thus we have $A = xy = x(5 - x) = 5x - x^2$. We have $x, y \geq 0$, which also implies $x \leq 5$ (otherwise $y < 0$). Therefore we need to maximize $A(x) = 5x - x^2$ for $0 \leq x \leq 5$. The critical points of the objective function satisfy $A'(x) = 5 - 2x = 0$, which has the solution $x = 5/2$. To find the absolute maximum of A, we check the endpoints of $[0, 5]$ and the critical point $x = 5/2$. Because $A(0) = A(5) = 0$ and $A(5/2) = 25/4$, the absolute maximum occurs when $x = y = 5/2$, so width = length = $5/2$ m.

4.4.7 Let x and y be the dimensions of the rectangle. The area is $xy = 100$, so the constraint is $y = 100/x$. The objective function to be minimized is the perimeter of the rectangle, $P = 2x + 2y$. Using $y = 100/x$, we have $P = 2x + 2y = 2x + \frac{200}{x}$. Because $xy = 100 > 0$ we must have $x > 0$, so we need to minimize $P(x) = 2x + 200/x$ on the interval $(0, \infty)$. The critical points of the objective function satisfy $P'(x) = 2 - \frac{200}{x^2} = 0$, which has the solution $x = 10$. By the First (or Second) Derivative Test, this critical point corresponds to a local minimum, and by Theorem 4.5, this solitary local minimum is also the absolute minimum on the interval $(0, \infty)$. Therefore the dimensions of the rectangle with minimum perimeter are $x = 10$ and $y = \frac{100}{10} = 10$, so width = length = 10.

4.4.9 Let x and y be the two non-negative numbers. The constraint is $x + y = 23$, which gives $y = 23 - x$. The objective function to be maximized is the product of the numbers, $P = xy$. Using $y = 23 - x$, we have $P = xy = x(23 - x) = 23x - x^2$. Now x must be at least 0, and cannot exceed 23 (otherwise $y < 0$). Therefore we need to maximize $P(x) = 23x - x^2$ for $0 \leq x \leq 23$. The critical points of the objective function satisfy $P'(x) = 23 - 2x = 0$, which has the solution $x = 23/2$. To find the absolute maximum of P, we check the endpoints of $[0, 23]$ and the critical point $x = 23/2$. Because $P(0) = P(23) = 0$ and $P(23/2) = (23/2)^2$, the absolute maximum occurs when $x = y = 23/2$.

4.4.11 Let x and y be the two positive numbers. The constraint is $xy = 50$, which gives $y = 50/x$. The objective function to be minimized is the sum of the numbers, $S = x + y$. Using $y = 50/x$, we have $S = x + y = x + \frac{50}{x}$. Now x can be any positive number, so we need to maximize $S(x) = x + 50/x$ on the interval $(0, \infty)$. The critical points of the objective function satisfy $S'(x) = 1 - \frac{50}{x^2} = 0$, which has the solution $x = \sqrt{50} = 5\sqrt{2}$. By the First (or Second) Derivative Test, this critical point corresponds to a local minimum, and by Theorem 4.5, this solitary local minimum is also the absolute minimum on the interval $(0, \infty)$. Therefore the numbers with minimum sum are $x = 5\sqrt{2}$ and $y = \frac{50}{5\sqrt{2}} = \frac{10}{\sqrt{2}} = 5\sqrt{2}$, so $x = y = 5\sqrt{2}$.

4.4.13 We seek to minimize $S = 2x + y$ subject to the constraint $y = 12/x$. Substituting gives $S = 2x + 12/x$, so $S'(x) = 2 - 12/x^2$. This is zero when $x^2 = 6$, or $x = \sqrt{6}$. Note that for $0 < x < \sqrt{6}$ we have $S'(x) < 0$, and for $x > \sqrt{6}$ we have $S'(x) > 0$, so we have a minimum at $x = \sqrt{6}$. Note that when $x = \sqrt{6}$, we have $y = 12/x = 12/\sqrt{6} = 2\sqrt{6}$.

4.4.15 Let x be the length of the sides of the base of the box and y be the height of the box. The volume is $x \cdot x \cdot y = 100$, so the constraint is $x^2 y = 100$, which gives $y = 100/x^2$. The objective function to be minimized is the surface area S of the box, which consists of $2x^2$ (for the top and base) $+ 4xy$ (for the 4 sides); therefore $S = 2x^2 + 4xy$. Using $y = 100x^2$, we have $S = 2x^2 + 4xy = 2x^2 + 4x \cdot \frac{100}{x^2} = 2x^2 + \frac{400}{x}$. The base side length can be any $x > 0$, so we need to maximize $S(x) = 2x^2 + 400/x$ on the interval $(0, \infty)$. The critical points of the objective function satisfy $S'(x) = 4x - \frac{400}{x^2} = 0$; clearing denominators gives $4x^3 = 400$ so $x = \sqrt[3]{100}$. By the First (or Second) Derivative Test, this critical point corresponds to a local minimum, and by Theorem 4.5, this solitary local minimum is also the absolute minimum on the interval $(0, \infty)$. Therefore the dimensions of the box with minimum surface area are $x = \sqrt[3]{100}$ and $y = 100/\sqrt[3]{100}^2 = \sqrt[3]{100}$, so length = width = height = $\sqrt[3]{100}$ m.

4.4.17 Let x be the length of the sides of the base of the box and y be the height of the box. The volume of the box is $x \cdot x \cdot y = x^2 y$, so the constraint is $x^2 y = 16$, which gives $y = 16/x^2$. Let c be the cost per square foot of the material used to make the sides. Then the cost to make the base is $2cx^2$, the cost to make the 4 sides is $4cxy$, and the cost to make the top is $\frac{1}{2}cx^2$. The objective function to be minimized is the total cost, which is $C = 2cx^2 + 4cxy + \frac{1}{2}cx^2 = \frac{5}{2}cx^2 + 4cx \cdot \frac{16}{x^2} = c\left(\frac{5x^2}{2} + \frac{64}{x}\right)$. The base side length can be any $x > 0$, so we need to maximize $C(x) = c(5x^2/2 + 64/x)$ on the interval $(0, \infty)$. The critical points of the objective function satisfy $5x - \frac{64}{x^2} = 0$, which gives $x^3 = 64/5$ or $x = 4/\sqrt[3]{5}$. By the First (or Second) Derivative Test, this critical point corresponds to a local minimum, and by Theorem 4.5, this solitary local minimum is also the absolute minimum on the interval $(0, \infty)$. Therefore the box with minimum cost has base $4/\sqrt[3]{5}$ ft by $4/\sqrt[3]{5}$ ft and height $y = 16/(4/\sqrt[3]{5})^2 = 5^{2/3}$ ft.

4.4.19 The distance between $(x, 3x)$ and $(50, 0)$ is $d(x) = \sqrt{(3x - 0)^2 + (x - 50)^2}$. Instead of working with the distance, we can instead work with the square of the distance, because these two functions have minima which occur at the same place. So consider

$$(d(x))^2 = D(x) = (3x)^2 + (x - 50)^2 = 9x^2 + x^2 - 100x + 2500 = 10(x^2 - 10x + 250).$$

$\frac{dD}{dx} = 10(2x - 10)$, which is zero for $x = 5$. Because $\frac{d^2D}{dx^2} = 20 > 0$, we see that the critical point at $x = 5$ is a minimum. So the minimum of D (and d) occurs at $x = 5$. The value of d at the point $(5, 15)$, is $d(5) = \sqrt{15^2 + (-45)^2} = 15\sqrt{10} \approx 47.4$.

4.4.21

a. Let x be the distance from the point on the shoreline nearest to the boat to the point where the woman lands on shore; then the remaining distance she must travel on shore is $6 - x$. By the Pythagorean theorem, the distance the woman must row is $\sqrt{x^2 + 16}$. So the time for the rowing leg is $\frac{\text{distance}}{\text{rate}} = \frac{\sqrt{x^2+16}}{2}$ and the time for the walking leg is $\frac{\text{distance}}{\text{rate}} = \frac{6-x}{3}$. The total travel time for the trip is the objective function $T(x) = \frac{\sqrt{x^2+16}}{2} + \frac{6-x}{3}$. We wish to minimize this function for $0 \le x \le 6$. The critical points of the objective function satisfy $T'(x) = \frac{x}{2\sqrt{x^2+16}} - \frac{1}{3} = 0$, which when simplified gives $5x^2 = 64$, so $x = 8/\sqrt{5}$ is the only critical point in $(0, 6)$. From the First Derivative Test we see that T has a local minimum at this point, so $x = 8/\sqrt{5}$ must give the minimum value of T on $[0, 6]$.

b. Let $v > 0$ be the woman's rowing speed. Then the total travel time is now given by $T(x) = \frac{\sqrt{x^2+16}}{v} + \frac{6-x}{3}$. The derivative of the objective function is $T'(x) = \frac{x}{v\sqrt{x^2+16}} - \frac{1}{3}$. If we try to solve the equation $T'(x) = 0$ as in part (a) above, we see that there is at most one solution $x > 0$. Therefore there can be at most one critical point of T in the interval $(0, 6)$. Observe also that $T'(0) = -1/3 < 0$ so the absolute minimum of T on $[0, 6]$ cannot occur at $x = 0$. So one of two things must happen: there is a unique critical point for T in $(0, 6)$ which is the absolute minimum for T on $[0, 6]$, and then $T'(6) > 0$; or, T is decreasing on $[0, 6]$, and then $T'(6) \leq 0$ (the quickest way to the restaurant is to row directly in this case). The condition $T'(6) \leq 0$ is equivalent to $\frac{6}{\sqrt{6^2+16}} \leq \frac{v}{3}$ which gives $v \geq 9/\sqrt{13}$ mi/hr.

4.4.23 Let L be the ladder length and x be the distance between the foot of the ladder and the fence. The Pythagorean theorem gives the relationship $L^2 = (x+5)^2 + b^2$, where b is the height of the top of the ladder. We see that $b/(x+5) = 8/x$ by similar triangles, which gives $b = 8(x+5)/x$. Substituting in the expression for L^2 above gives $L^2 = (x+5)^2 + 64\frac{(x+5)^2}{x^2} = (x+5)^2\left(1 + \frac{64}{x^2}\right)$. It suffices to minimize L^2 instead of L. However in this case x and b must satisfy $x, b \leq 20$. Solving $20 = 8(x+5)/x$ for x gives $x = 10/3$, so the condition $b \leq 20$ corresponds to $x \geq 10/3$, and we see that we must minimize L^2 for $10/3 \leq x \leq 20$. We have $\frac{d}{dx}L^2 = (x+5)^2\left(-\frac{128}{x^3}\right) + 2(x+5)\left(1 + \frac{64}{x^2}\right) = \frac{2(x+5)(x^3-320)}{x^3}$. Because $x > 0$, the only critical point is $x = \sqrt[3]{320} \approx 6.840$. By the First Derivative Test, this critical point corresponds to a local minimum, and by Theorem 4.5, this solitary local minimum is also the absolute minimum on the interval $[10/3, 20]$. Substituting $x \approx 6.840$ in the expression for L^2 we find the length of the shortest ladder $L \approx 18.220$ ft.

4.4.25 Let the coordinates of the base of the rectangle be $(x, 0)$ and $(-x, 0)$ where $0 \leq x \leq 5$. Then the width of the rectangle is $2x$ and the height is $\sqrt{25 - x^2}$, so the area A is given by $A(x) = 2x\sqrt{25 - x^2}$. The critical points of this function satisfy $A'(x) = 2\sqrt{25 - x^2} + \frac{2x \cdot (-x)}{\sqrt{25-x^2}} = \frac{2(25-2x^2)}{\sqrt{25-x^2}} = 0$, which has unique solution $x = 5/\sqrt{2}$ in $(0, 5)$. We have $A(0) = A(5) = 0$, so the rectangle of maximum area has width $2x = 10/\sqrt{2}$ cm, height $y = \sqrt{25 - (25/2)} = 5/\sqrt{2}$ cm.

4.4.27 If we remove a sector of angle θ from a circle of radius 20, the remaining circumference is $2\pi \cdot 20 - \theta \cdot 20 = 20(2\pi - \theta)$, so the base of the cone formed has radius $r = \frac{20(2\pi-\theta)}{2\pi} = \frac{10(2\pi-\theta)}{\pi}$. As θ varies from 0 to 2π, the radius ranges from 0 to 20, but all possible cones formed have side length 20. The height h of the cone is given by the Pythagorean theorem: $h^2 + r^2 = 20^2$, so $h = \sqrt{400 - r^2}$. The volume of the cone given by $V = \frac{\pi}{3}r^2h = \frac{\pi}{3}r^2\sqrt{400 - r^2}$.

Thus

$$V'(r) = \frac{2\pi}{3}r\sqrt{400 - r^2} + \frac{\pi}{3}r^2(400 - r^2)^{-1/2} \cdot \frac{1}{2} \cdot (-2r)$$
$$= \frac{\pi}{3} \cdot \frac{2r(400 - r^2) - r^3}{\sqrt{400 - r^2}}$$
$$= \frac{\pi}{3} \cdot \frac{800r - 3r^3}{\sqrt{400 - r^2}}.$$

The only positive critical point occurs where $r = \sqrt{\frac{800}{3}} = 20\sqrt{\frac{2}{3}}$. An application of the First Derivative Test shows that this is a maximum. So

$$h = \sqrt{400 - \left(20\sqrt{\frac{2}{3}}\right)^2} = \sqrt{400 - 400 \cdot \frac{2}{3}} = 20\sqrt{\frac{1}{3}}.$$

4.4.29 Let x and y be the dimensions of the flower garden; the area of the flower garden is 30, so we have the constraint $xy = 30$ which gives $y = 30/x$. The dimensions of the garden and borders are $x + 4$ and $y + 2$, so the objective function to be minimized for $x > 0$ is $A = (x+4)(y+2) = (x+4)\left(\frac{30}{x} + 2\right) = 2x + \frac{120}{x} + 38$. The critical points of $A(x)$ satisfy $A'(x) = 2 - \frac{120}{x^2} = 0$, which has unique solution $x = \sqrt{60} = 2\sqrt{15}$. By the First (or Second) Derivative test, this critical point gives a local minimum, which by Theorem 4.5 must be the absolute minimum of A over $(0, \infty)$. The corresponding value of y is $30/2\sqrt{15} = \sqrt{15}$, so the dimensions are $\sqrt{15}$ by $2\sqrt{15}$ m.

4.4.31 The radius r and height h of the barrel satisfy the constraint $r^2 + h^2 = d^2$, which we can rewrite as $r^2 = d^2 - h^2$. The volume of the barrel is given by $V = \pi r^2 h = \pi(d^2 - h^2)h = \pi(d^2 h - h^3)$. The height h must satisfy $0 \le h \le d$, so we need to maximize $V(h)$ on the interval $[0, d]$. The critical points of V satisfy $V'(h) = \pi(d^2 - 3h^2) = 0$. The only critical point in $(0, d)$ is $h = d/\sqrt{3}$, which gives the maximum volume because at the endpoints $V(0) = V(d) = 0$. The corresponding r value satisfies $r^2 = d^2 - d^2/3 = 2d^2/3$, so $r = \sqrt{2}d/\sqrt{3}$ and we see that the ratio r/h that maximizes the volume is $\sqrt{2}$.

4.4.33 Let h be the height of the cylindrical tower and r the radius of the dome. The cylinder has volume $\pi r^2 h$, and the hemispherical dome has volume $2\pi r^3/3$ (half the volume of a sphere of radius r). The total volume is 750, so we have the constraint $\pi r^2 h + \frac{2\pi r^3}{3} = 750$ which gives $h = \frac{750}{\pi r^2} - \frac{2r}{3}$. We must have $h \ge 0$, which is equivalent to $r \le \sqrt[3]{1125/\pi}$. The objective function to be maximized is the cost of the metal to make the silo, which is proportional to the surface area of the cylinder ($= 2\pi rh$) plus 1.5 times the surface area of the hemisphere ($= 2\pi r^2$). So we can take as objective function $C = 2\pi rh + 1.5 \cdot 2\pi r^2 = 2\pi r\left(\frac{750}{\pi r^2} - \frac{2r}{3}\right) + 3\pi r^2 = \frac{1500}{r} + \frac{5}{3}\pi r^2$. The critical points of $C(r)$ satisfy $C'(r) = -\frac{1500}{r^2} + \frac{10}{3}\pi r = 0$, which gives $\pi r^3 = 450$ and hence $r = \sqrt[3]{450/\pi}$. The corresponding value of h is $h = \frac{750}{\pi r^2} - \frac{2r}{3} = \frac{750r}{\pi r^3} - \frac{2r}{3} = \left(\frac{750}{450} - \frac{2}{3}\right)r = r$. By the First (or Second) Derivative Test, this critical point corresponds to a local minimum, and by Theorem 4.5, this solitary local minimum is also the absolute minimum on the interval $[0, \sqrt[3]{1125/\pi}]$. Therefore the dimensions that minimize the cost are $r = h = \sqrt[3]{450/\pi}$ m.

4.4.35 Let x be the distance between the point and the weaker light source; then $12 - x$ is the distance to the stronger light source. The intensity is proportional to $I(x) = \frac{1}{x^2} + \frac{2}{(12-x)^2}$, so we can take this as our objective function to be minimized for $0 < x < 12$. The critical points of $I(x)$ satisfy $I'(x) = -\frac{2}{x^3} + \frac{4}{(12-x)^3} = 0$ which gives $\left(\frac{12-x}{x}\right)^3 = 2$, or $\frac{12-x}{x} = \sqrt[3]{2}$, or $x = \frac{12}{\sqrt[3]{2}+1} \approx 5.310$. By the First (or Second) Derivative Test, this critical point corresponds to a local minimum, and by Theorem 4.5, this solitary local minimum is also the absolute minimum on the interval $(0, 12)$. Therefore the intensity is weakest at the point $12/(\sqrt[3]{2}+1) \approx 5.310$ m from the weaker source.

4.4.37 Let x be the distance from the point on shore nearest the island to the point where the underwater cable meets the shore, and let y be the be the length of the underwater cable. By the Pythagorean theorem, $y = \sqrt{x^2 + 3.5^2}$. The objective function to be minimized is the cost given by $C(x) = 2400\sqrt{x^2 + 3.5^2} + 1200 \cdot (8-x) = 2400\sqrt{x^2 + 3.5^2} - 1200x + 9600$. We wish to minimize this function for $0 \le x \le 8$. The critical points of $C(x)$ satisfy $C'(x) = \frac{2400x}{\sqrt{x^2+3.5^2}} - 1200 = 1200\left(\frac{2x}{\sqrt{x^2+3.5^2}} - 1\right) = 0$, which we solve to obtain $x = 7\sqrt{3}/6$. By the First Derivative Test, this critical point corresponds to a local minimum, and by Theorem 4.5, this solitary local minimum is also the absolute minimum on the interval $[0, 8]$. Therefore the optimal point on shore has distance $x = 7\sqrt{3}/6$ mi from the point on shore nearest the island, in the direction of the power station.

4.4.39

a. Using the Pythagorean theorem, we find that the height of this triangle is 2. Let x be the distance from the point P to the base of the triangle; then the distance from P to the top vertex is $2 - x$ and the distance to each of the base vertices is $\sqrt{x^2 + 4}$, again by the Pythagorean theorem. Therefore the sum of the distances to the three vertices is given by $S(x) = 2\sqrt{x^2 + 4} + 2 - x$. We wish to minimize this function for $0 \le x \le 2$. The critical points of $S(x)$ satisfy $S'(x) = \frac{2x}{\sqrt{x^2+4}} - 1 = 0$, which has unique solution $x = 2/\sqrt{3}$ in $(0, 2)$. By the First Derivative Test, this critical point corresponds to a local minimum, and by Theorem 4.5, this solitary local minimum is also the absolute minimum on the interval $[0, 2]$. Therefore the optimal location for P is $2/\sqrt{3}$ units above the base.

b. In this case the objective function to be minimized is $S(x) = 2\sqrt{x^2 + 4} + h - x$ where $0 \le x \le h$. Exactly as above, we find that the only critical point $x > 0$ is $x = 2/\sqrt{3}$. This will give the absolute minimum on $[0, h]$ as long as $h \ge 2/\sqrt{3}$. When $h < 2/\sqrt{3}$, $S(x)$ is decreasing on $[0, h]$ and the minimum occurs at the endpoint $x = h$.

4.4.41 The critical points of the function $a(\theta)$ satisfy

$$a'(\theta) = \omega^2 r\left(-\sin\theta - \frac{2r\sin 2\theta}{L}\right) = -\omega^2 r\sin\theta\left(1 + \frac{4r\cos\theta}{L}\right) = 0,$$

using the identity $\sin 2\theta = 2\sin\theta\cos\theta$. There are two cases to consider separately: (a) $0 < L < 4r$ and (b) $L \geq 4r$. In case (a) the critical points in $[0, 2\pi]$ are $\theta = 0, \pi, 2\pi$ and also $\theta = \cos^{-1}(-L/(4r))$ and $2\pi - \cos^{-1}(-L/(4r))$. Comparing the values of $a(\theta)$ at these points shows that the maximum acceleration occurs at $\theta = 0$ and 2π and the minimum occurs at $\theta = \cos^{-1}(-L/(4r))$ and $2\pi - \cos^{-1}(-L/(4r))$. (There is a local maximum at $\theta = \pi$.) In case (b) the only critical points are $\theta = 0, \pi$ and 2π, and comparing the values of $a(\theta)$ at these points shows that the maximum acceleration occurs at $\theta = 0$ and 2π as in case (a), whereas the minimum occurs at $\theta = \pi$ in this case.

4.4.43

a. Let r and h be the radius and height of the can. The volume of the can is $V = \pi r^2 h$, which gives the constraint $\pi r^2 h = 354$ or $h = 354/(\pi r^2)$. The objective function to be minimized is the surface area, which consists of $2\pi r^2$ (for the top and bottom of the can) and $2\pi rh$ (for the side of the can). Therefore the objective function to be minimized is $A = 2\pi r^2 + 2\pi rh = 2\pi\left(r^2 + r\left(\frac{354}{\pi r^2}\right)\right) = 2\pi\left(r^2 + \frac{354}{\pi r}\right)$. We need to minimize $A(r)$ for $r > 0$. The critical points of $A(r)$ satisfy $A'(r) = 2\pi\left(2r - \frac{354}{\pi r^2}\right) = 0$, which gives $r = \sqrt[3]{(177/\pi)} \approx 3.834$ cm. The corresponding value of h is $h = \frac{354}{\pi r^2} = \frac{354r}{\pi r^3} = 2r \cdot \frac{177}{\pi r^3} = 2r$, so $h = 2\sqrt[3]{(177/\pi)} \approx 7.667$ cm. By the First (or Second) Derivative Test, this critical point corresponds to a local minimum, and by Theorem 4.5, this solitary local minimum is also the absolute minimum on the interval $(0, \infty)$.

b. We modify the objective function in part (a) above to account for the fact that the top and bottom of the can have double thickness: $A = 4\pi r^2 + 2\pi rh = 2\pi\left(2r^2 + r\left(\frac{354}{\pi r^2}\right)\right) = 4\pi\left(r^2 + \frac{177}{\pi r}\right)$. We need to minimize $A(r)$ for $r > 0$. The critical points of $A(r)$ satisfy $A'(r) = 4\pi\left(2r - \frac{177}{\pi r^2}\right) = 0$, which gives $r = \sqrt[3]{(177/2\pi)} \approx 3.043$ cm. The corresponding value of h is $h = \frac{354}{\pi r^2} = \frac{354r}{\pi r^3} = 4r \cdot \frac{177}{2\pi r^3} = 4r$, so $h = 4\sqrt[3]{(177/2\pi)} \approx 12.171$ cm. These dimensions are closer to those of a real soda can.

4.4.45 Let r and h be the radius and height of the cone; then we have the constraint $r^2 + h^2 = 3^2 = 9$, which gives $r^2 = 9 - h^2$. The objective function to be maximized is the volume of the cone, given by $V = \frac{\pi}{3}r^2 h = \frac{\pi}{3}(9 - h^2)h = \frac{\pi}{3}(9h - h^3)$. Because $r, h \geq 0$ we must have $0 \leq h \leq 3$. Therefore we need to maximize $V(h)$ over $[0, 3]$. The critical points of $V(h)$ satisfy $V'(h) = \frac{\pi}{3}(9 - 3h^2) = \pi(3 - h^2) = 0$, so $h = \sqrt{3}$ is the only critical point in $[0, 3]$. Because $V(0) = V(3) = 0$, the cone of maximum volume has height $h = \sqrt{3}$ and radius $r = \sqrt{6}$.

4.4.47 Let the radius of the Ferris wheel have length r, and let α be the angle the specific seat on the Ferris wheel makes with the center of the wheel (see the figure in the text). This point has coordinates $(r\cos\alpha, r + r\sin\alpha)$ so the distance from the seat to the base of the wheel is

$$d = \sqrt{r^2\cos^2\alpha + r^2(1 + \sin\alpha)^2} = \sqrt{2}r\sqrt{1 + \sin\alpha}.$$

Therefore the observer's angle satisfies $\tan\theta = \frac{r\sqrt{2}}{20}\sqrt{1 + \sin\alpha}$. Think of θ and α as functions of time t and differentiate: $\sec^2\theta\frac{d\theta}{dt} = \frac{r\sqrt{2}}{20} \cdot \frac{\cos\alpha}{2\sqrt{1+\sin\alpha}}\frac{d\alpha}{dt} = \frac{\pi r\sqrt{2}}{40} \cdot \frac{\cos\alpha}{\sqrt{1+\sin\alpha}}$. Therefore

$$\frac{d\theta}{dt} = \frac{\pi r\sqrt{2}}{40}\frac{\cos^2\theta\cos\alpha}{\sqrt{1 + \sin\alpha}}.$$

Observe that $\left|\frac{d\theta}{dt}\right| = \frac{\pi r\sqrt{2}}{40}\frac{\cos^2\theta|\cos\alpha|}{\sqrt{1+\sin\alpha}}\frac{\sqrt{1-\sin\alpha}}{\sqrt{1-\sin\alpha}}$, which can be written as $\frac{\pi r\sqrt{2}}{40}\cos^2\theta\sqrt{1 - \sin\alpha}$. When the seat on the Ferris wheel is at its lowest point we have $\theta = 0$ and $\alpha = -\pi/2$, which gives $\cos^2\theta = 1$ and $\sqrt{1 - \sin\alpha} = \sqrt{2}$. At any other point on the wheel we have $\cos^2\theta \leq 1$ and $\sqrt{1 - \sin\alpha} < \sqrt{2}$, so θ is changing most rapidly when the seat is at its lowest point.

4.4.49 Let r and h be the radius and height of the cylinder. The distance d from the centroid of the cylinder (the midpoint of the cylinder's axis of rotation) to any point on the top or bottom edge satisfies $d^2 = r^2 + \left(\frac{h}{2}\right)^2$ so the constraint is $r^2 + (h/2)^2 = R^2$. The volume of the cylinder is given by $V = \pi r^2 h = \pi\left(R^2 - \left(\frac{h}{2}\right)^2\right)h = \pi\left(R^2 h - \frac{h^3}{4}\right)$. Because $r, h \geq 0$ we must have $0 \leq h \leq 2R$. We wish to maximize $V(h)$ on this interval. The critical points of $V(h)$ satisfy $V'(h) = \pi\left(R^2 - \frac{3h^2}{4}\right) = 0$ which gives $h = 2R/\sqrt{3}$, and from the constraint we obtain $r = \sqrt{2}R/\sqrt{3}$. The volume $V(h) = 0$ at the endpoints $h = 0$ and $h = 2R$, so the maximum volume must occur at this critical point.

4.4.51

a. Let r and h be the radius and height of the inscribed cylinder. The region that lies above the cylinder inside the cone is a cone with radius r and height $H - h$; by similar triangles we have $\frac{H-h}{r} = \frac{H}{R}$ so $h = \frac{H}{R}(R - r)$. The volume of the cylinder is $V = \pi r^2 h = \frac{\pi H}{R}\left(Rr^2 - r^3\right)$, which we must maximize over $0 \leq r \leq R$. The critical points of $V(r)$ satisfy $V'(r) = \frac{\pi H}{R}\left(2Rr - 3r^2\right) = 0$, which has unique solution $r = 2R/3$ in $(0, R)$. Because $V(r) = 0$ at the endpoints $r = 0$ and $r = R$, the cylinder with maximum volume has radius $r = 2R/3$, height $h = H/3$ and volume $V = \pi r^2 h = \frac{4\pi}{27}R^2 H = \frac{4}{9}\cdot\frac{\pi}{3}R^2 H$; i.e. $4/9$ the volume of the cone.

b. The lateral surface area of the cylinder is $A = 2\pi r h = 2\pi r \cdot \frac{H}{R}(R - r) = \frac{2\pi H}{R}r(R - r)$. This function takes its maximum over $0 \leq r \leq R$ at $r = R/2$, so the cylinder with maximum lateral surface area has dimensions $r = R/2$ and $h = H/2$.

4.4.53 Let R and H be the radius and height of the larger cone and let r and h be the radius and height of the smaller inscribed cone. The region that lies above the smaller cone inside the larger cone is a cone with radius r and height $H - h$; by similar triangles we have $\frac{H-h}{r} = \frac{H}{R}$ so $h = \frac{H}{R}(R - r)$. The volume of the smaller cone is $V = \frac{\pi}{3}r^2 h = \frac{\pi H}{3R}\left(Rr^2 - r^3\right)$, which we must maximize over $0 \leq r \leq R$. The critical points of $V(r)$ satisfy $V'(r) = \frac{\pi H}{3R}\left(2Rr - 3r^2\right) = 0$ which has unique solution $r = 2R/3$ in $(0, R)$. Because $V(r) = 0$ at the endpoints $r = 0$ and $r = R$, the smaller cone with maximum volume has radius $r = 2R/3$ and height $h = H/3$, so the optimal ratio of the heights is 3:1.

4.4.55

Following the hint, place two points P and Q above the midpoint of the base of the square, at distances x and y to the sides (see figure), where $0 \leq x, y \leq 1/2$. Then join the bottom vertices of the square to P, the upper vertices to Q and join P to Q. This road system has total length $L = 2\sqrt{x^2 + \frac{1}{4}} + 2\sqrt{y^2 + \frac{1}{4}} + (1 - x - y) = 1 + \left(\sqrt{4x^2 + 1} - x\right) + \left(\sqrt{4y^2 + 1} - y\right)$. We can minimize the contributions from x and y separately; the critical points of the function $f(x) = \sqrt{4x^2 + 1} - x$ satisfy $f'(x) = \frac{4x}{\sqrt{4x^2+1}} - 1 = 0$ which gives $\sqrt{4x^2 + 1} = 4x$, so $12x^2 = 1$ and $x = 1/(2\sqrt{3})$. By the First (or Second) Derivative Test, this critical point corresponds to a local minimum, and by Theorem 4.5, this solitary local minimum is also absolute minimum on the interval $[0, 1/2]$. The minimum value of $f(x)$ on this interval is $f(1/(2\sqrt{3})) = \sqrt{3}/2$, so the shortest road system has length $L = 1 + 2\cdot\frac{\sqrt{3}}{2} = 1 + \sqrt{3} \approx 2.732$.

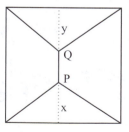

4.4.57 Let x be the distance between the point on the track nearest your initial position to the point where you catch the train. If you just catch the back of the train, then the train will have travelled $x + 1/3$ miles, which will require time $T = \frac{\text{distance}}{\text{rate}} = \frac{x + \frac{1}{3}}{20}$. The distance you must run is $\sqrt{x^2 + 1/(16)}$, so your running speed must be $v = \frac{\text{distance}}{\text{time}} = \frac{20\sqrt{x^2 + \frac{1}{16}}}{x + \frac{1}{3}}$. We wish to minimize this function for $x \geq 0$. The derivative of $v(x)$ can be written $v'(x) = \left(\frac{x}{x^2 + \frac{1}{16}} - \frac{1}{x + \frac{1}{3}} \right) v(x)$, so the critical points of $v(x)$ satisfy $\frac{x}{x^2 + \frac{1}{16}} = \frac{1}{x + \frac{1}{3}}$ so $x\left(x + \frac{1}{3}\right) = x^2 + \frac{1}{16}$ which gives $x = 3/16$ mi. By the First (or Second) Derivative Test, this critical point corresponds to a local minimum, and by Theorem 4.5, this solitary local minimum is also the absolute minimum on the interval $[0, \infty)$. The minimum running speed is $v\left(\frac{3}{16}\right) = \frac{20\sqrt{\left(\frac{3}{16}\right)^2 + \frac{1}{16}}}{\frac{3}{16} + \frac{1}{3}} = \frac{60\sqrt{9 + 16}}{9 + 16} = 12$ mph.

4.4.59

a. A point on the line $y = 3x + 4$ has the form $(x, 3x + 4)$, which has distance L to the origin given by $L^2 = x^2 + (3x + 4)^2 = 10x^2 + 24x + 16$. Because L is positive, it suffices to minimize L^2. The quadratic function $10x^2 + 24x + 16$ takes its minimum at $x = -24/20 = -6/5$, and the corresponding value of $y = 2/5$. Therefore the point closest to the origin on this line is $(-6/5, 2/5)$.

b. A point on the parabola $y = 1 - x^2$ has the form $(x, 1 - x^2)$, which has distance L to the point $(1,1)$ given by $L^2 = (x-1)^2 + \left(1 - (1 - x^2)\right)^2 = x^4 + x^2 - 2x + 1$. Because L is positive, it suffices to minimize L^2. The critical points of L^2 satisfy $\frac{dL^2}{dx} = 4x^3 + 2x - 2 = 2(2x^3 + x - 1) = 0$ This cubic equation has a unique root $x \approx 0.590$, so the point closest to $(1,1)$ on this parabola is approximately $(0.590, 0.652)$.

c. A point on the curve $y = \sqrt{x}$ has the form $(x, \sqrt{x})$, which has distance L to the point $(p,0)$ given by $L^2 = (x - p)^2 + \sqrt{x}^2 = x^2 + (1 - 2p)x + p^2$. Because L is positive, it suffices to minimize L^2 for $x \geq 0$. This quadratic function takes its minimum at $x = -(1 - 2p)/2 = p - 1/2$, so in case (i) the minimum occurs at the point $(p - 1/2, \sqrt{p - 1/2})$ and in case (ii) there are no critical points for $x > 0$, the function L^2 is increasing on $[0, \infty)$ so the minimum occurs at $(0, 0)$.

4.4.61

a. We find $g(0) = 0$, $g(40) = 30$ and $g(60) = 25$ miles per gallon. The value at $v = 0$ is reasonable because when a car first starts moving it needs a lot of power from its engine, so the gas mileage is very low. The decline from 30 to 25 mi/gal as v increases from 40 mi/hr to 60 mi/hr reflects the fact that gas mileage tends to decrease at speeds over 55 mi/hr.

b. The quadratic function $g(v) = (85v - v^2)/60$ takes its maximum value at $v = 85/2 = 42.5$ mi/hr.

c. At speed v the amount of gas needed to drive one mile is $1/g(v)$ and the time it takes is $1/v$. Hence the cost of gas for one mile is $p/g(v)$ and the cost for the driver is w/v, and so the cost for L miles is $C(v) = Lp/g(v) + Lw/v$.

d. We have $C(v) = 400\left(\frac{4}{g(v)} + \frac{20}{v} \right) = 1600\left(\frac{1}{g(v)} + \frac{5}{v} \right)$. The critical points of $C(v)$ satisfy $\frac{g'(v)}{g(v)^2} + \frac{5}{v^2} = 0$, which simplifies to $v^2 g'(v) + 5g(v)^2 = 0$. Substituting the formula for $g(v)$ above and using $g'(v) = (85 - 2v)/60$, we can factor out v^2 and reduce to the quadratic equation $v^2 - 194v + 8245 = 0$, which has roots $v \approx 62.883, 131.117$. The First (or Second) Derivative Test shows that $C(v)$ has a local minimum at $v \approx 62.9$, which is the unique critical point for $0 \leq v \leq 131$. Therefore the cost is minimized at this value of v.

e. Because L is a constant factor in the cost function $C(v)$, changing L will not change the critical points of $C(v)$.

f. The critical points of $C(v)$ now satisfy the equation $\frac{4.2g'(v)}{g(v)^2} + \frac{20}{v^2} = 0$, which simplifies to $4.2v^2 g'(v) + 20g(v)^2 = 0$. As above, substituting the formula for $g(v)$ above and using $g'(v) = (85 - 2v)/60$, we can factor out v^2 and reduce to the quadratic equation $v^2 - 195.2v + 8296 = 0$, which has roots $v \approx 62.532, 132.668$. As in part (d), the minimum cost occurs for $v \approx 62.532$, slightly less than the speed in part (d).

g. The critical points of $C(v)$ now satisfy the equation $\frac{4g'(v)}{g(v)^2} + \frac{15}{v^2} = 0$, which simplifies to $4v^2 g'(v) + 15g(v)^2 = 0$. As above, substituting the formula for $g(v)$ above and using $g'(v) = (85 - 2v)/60$, we can factor out v^2 and reduce to the quadratic equation $v^2 - 202v + 8585 = 0$, which has roots $v \approx 60.800, 141.200$. As in part (d), the minimum cost occurs for $v \approx 60.8$, less than the speed in part (d).

4.4.63

a. Let x, $d-x$ be the distances from the point where the rope meets the ground to the poles of height m, n respectively. Then the rope has length $L(x) = \sqrt{x^2 + m^2} + \sqrt{(d-x)^2 + n^2}$. We wish to minimize this function for $0 \le x \le d$. The critical points of $L(x)$ satisfy $L'(x) = \frac{x}{\sqrt{x^2+m^2}} - \frac{d-x}{\sqrt{(d-x)^2+n^2}} = 0$, which is equivalent to $\frac{x}{\sqrt{x^2+m^2}} = \frac{d-x}{\sqrt{(d-x)^2+n^2}}$, or in terms of the angles θ_1 and θ_2 in the figure, $\sec\theta_1 = \sec\theta_2$ and therefore $\theta_1 = \theta_2$. Observe that $L'(0) < 0$ and $L'(d) > 0$, so the minimum value of $L(x)$ must occur at some $x \in (0, d)$. There must be exactly one critical point, because as x ranges from 0 to d, θ_1 decreases and θ_2 increases, and so $\theta_1 = \theta_2$ can occur for at most one value of x.

b. Because the speed of light is constant, travel time is minimized when distance is minimized, which we saw in part (a) occurs when $\theta_1 = \theta_2$.

4.4.65 Let the angle of the cuts with the horizontal be ϕ_1 and ϕ_2, where $\phi_1 + \phi_2 = \theta$. The volume of the notch is proportional to $\tan\phi_1 + \tan\phi_2 = \tan\phi_1 + \tan(\theta - \phi_1)$, so it suffices to minimize the objective function $V(\phi_1) = \tan\phi_1 + \tan(\theta - \phi_1)$ for $0 \le \phi_1 \le \theta$. The critical points of $V(\phi_1)$ satisfy $\sec^2\phi_1 - \sec^2(\theta - \phi_1) = 0$, which is equivalent to the condition $\cos\phi_1 = \cos(\theta - \phi_1)$. This is satisfied if $\phi_1 = \theta - \phi_1$ which gives $\phi_1 = \theta/2$. There are no other solutions in $(0, \theta)$, because $\cos\phi_1$ is decreasing and $\cos(\theta - \phi_1)$ is increasing on $(0, \theta)$ and therefore can intersect at most once. So the only critical point occurs when $\phi_1 = \phi_2 = \theta/2$, and the First Derivative Test shows that this critical point is a local minimum; by Theorem 4.5, this must be the absolute minimum on $[0, \theta]$.

4.4.67 Let x and y be the lengths of the sides of the pen, with y the side parallel to the barn. The diagonal has length $\sqrt{x^2 + y^2}$, by the Pythagorean theorem. Therefore the constraint is $2x + y + \sqrt{x^2 + y^2} = 200$, which we rewrite as $2x + y = 200 - \sqrt{x^2 + y^2}$. Square both sides to obtain $4x^2 + 4xy + y^2 = 40{,}000 - 400\sqrt{x^2 + y^2} + x^2 + y^2$, which simplifies to $3x^2 + 4xy = 40{,}000 - 400\sqrt{x^2 + y^2}$. Now substitute $\sqrt{x^2 + y^2} = 200 - 2x - y$ in this equation and simplify to obtain $(3x - 200)(x - 200) = 4(100 - x)y$ so $y = \frac{(3x-200)(x-200)}{4(100-x)}$. The objective function to be maximized is the area of the pen, $A = xy$. Using the expression above for y in terms of x, we have $A = xy = \frac{x(3x-200)(x-200)}{4(100-x)} = -\frac{1}{4} \cdot \frac{x(3x-200)(x-200)}{(x-100)}$. The length x must be at least 0, and because the diagonal is at least as long as x, we must have $3x \le 200$; so x cannot exceed $200/3$. Therefore we need to maximize the function $A(x)$ defined above for $0 \le x \le 200/3$. We have $A'(x) = -\frac{1}{4} \cdot \left(\frac{(3x-200)(x-200)}{(x-100)} + \frac{x\cdot 3(x-200)}{(x-100)} + \frac{x(3x-200)}{(x-100)} - \frac{x(3x-200)(x-200)}{(x-100)^2} \right) = \left(\frac{1}{x} + \frac{3}{3x-200} + \frac{1}{x-200} - \frac{1}{x-100} \right) A(x)$. Because $A(x) > 0$ for $0 < x < 200/3$, the critical points of the objective function satisfy $\frac{1}{x} + \frac{3}{3x-200} + \frac{1}{x-200} = \frac{1}{x-100}$ which when simplified gives the equation $6x^3 - 1700x^2 + 160{,}000x - 4{,}000{,}000 = 0$. Using a numerical solver, we find that this equation has exactly one solution in the interval $(0, 200/3)$, which is $x \approx 38.81$. To find the absolute maximum of A, we check the endpoints of $[0, 200/3]$ and the critical point $x \approx 38.814$. We have $A(0) = A(200/3) = 0$, so the absolute maximum occurs when $x \approx 38.814$ m; using the formula for y in terms of x above gives $y \approx 55.030$ m.

4.5 Linear Approximation and Differentials

4.5.1

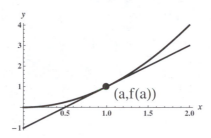

4.5.3 If f is differentiable at the point, then near that point, f is approximately linear, so the function nearly coincides with the tangent line at that point.

4.5.5 The relationship is given by $dy = f'(x)dx$, which is the linear approximation of the change Δy in $y = f(x)$ corresponding to a change dx in x.

4.5.7 The approximate average speed is $L(-1) = 60 - (-1) = 61$ miles per hour. The exact speed is $\frac{3600}{59}$ miles per hour which is about 61.02 miles per hour.

4.5.9 Let $T(x) = \frac{60D}{60+x}$. Then $T'(x) = -\frac{60D}{(60+x)^2}$, so $T'(0) = -\frac{D}{60}$. The linear approximation is given by $L(x) = T(0) - \frac{D}{60}(x - 0)$, or $L(x) = D\left(1 - \frac{x}{60}\right)$.

4.5.11 With $D = 80$, we have $T(-3) \approx L(-3) = 80(1 - \frac{-3}{60}) = 80 + 4 = 84$ minutes. The exact time required is $T(-3) = \frac{60 \cdot 80}{60 - 3} \approx 84.211$ minutes.

4.5.13

a. Note that $f(a) = f(2) = 8$ and $f'(a) = -2a = -4$, so the linear approximation has equation

$$y = L(x) = f(a) + f'(a)(x-a) = 8 + (-4)(x-2) = -4x + 16.$$

c. We have $f(2.1) \approx L(2.1) = 7.6$.

d. The percentage error is $100 \cdot \frac{|7.6 - 7.59|}{7.59} \approx 0.13\%$.

b.

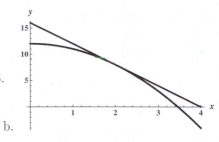

4.5.15

a. $f(0) = 1$ and $f'(x) = -1/(1+x)^2$ so that $f'(0) = -1$. Thus the linear approximation has the equation $y = L(x) = f(0) + f'(0)(x - 0) = 1 - x$.

c. $f(-0.1) \approx L(-0.1) = 1 - (-0.1) = 1.1$.

d. Noting that $f(-0.1) = 1/.9 = 10/9$, the percentage error is

$$100 \cdot \frac{|f(-0.1) - L(-0.1)|}{f(-0.1)} = 100 \cdot \frac{10/9 - 11/10}{10/9} = 1\%.$$

b.

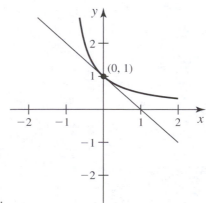

4.5.17

  a. Note that $f(a) = f(0) = \cos 0 = 1$ and $f'(a) = -\sin a = 0$, so the linear approximation has equation

  $$y = L(x) = f(a) + f'(a)(x-a) = 1.$$

  c. We have $f(-0.01) \approx L(-0.01) = 1.$

  d. The percentage error is $100 \cdot \frac{|1-\cos(-0.01)|}{\cos(-0.01)} \approx 0.005\%.$

  b.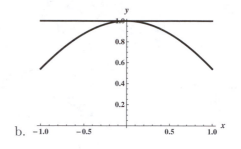

4.5.19

  a. Note that $f(a) = 8^{-1/3} = 1/2$ and $f'(a) = (-1/3)(8)^{-4/3} = -1/48$, so the linear approximation has equation

  $$y = L(x) = f(a) + f'(a)(x-a) = \frac{1}{2} + \frac{-1}{48}x.$$

  c. We have $f(-0.1) \approx L(-0.1) \approx .50208333$

  d. The percentage error is $100 \cdot \frac{|7.9^{-1/3} - .50208333|}{(7.9)^{-1/3}} \approx 0.003\%.$

  b.

4.5.21 Let $f(x) = 1/x$, $a = 200$. Then $f(a) = 0.005$ and $f'(a) = -1/a^2 = -0.000025$, so the linear approximation to f near $a = 200$ is $L(x) = f(a) + f'(a)(x-a) = 0.005 - 0.000025(x-200)$. Therefore $\frac{1}{203} = f(203) \approx L(203) = .004925.$

4.5.23 Let $f(x) = \sqrt{x}$, $a = 144$. Then $f(a) = 12$ and $f'(a) = 1/(2\sqrt{a}) = 1/24$, so the linear approximation to f near $a = 144$ is $L(x) = f(a) + f'(a)(x-a) = 12 + \frac{1}{24}(x-144)$. Therefore $\sqrt{146} = f(146) \approx L(146) = \frac{145}{12}.$

4.5.25 Let $f(x) = \frac{1}{x}$, $a = 1$. Then $f(a) = 1$ and $f'(a) = -\frac{1}{a^2} = -1$, so the linear approximation to f near $a = 1$ is $L(x) = f(a) + f'(a)(x-a) = 1 - (x-1) = -x+2$. Therefore $\frac{1}{1.05} = f(1.05) \approx L(1.05) = 2 - 1.05 = 0.95.$

4.5.27 Let $f(x) = \sin(x + \pi/4)$ with $a = 0$. Then $f'(x) = \cos(x + \pi/4)$, so that $f'(0) = f(0) = \sqrt{2}/2$. Then $f(x) \approx L(x) = f(0) + f'(0)(x-0) = (x+1)\sqrt{2}/2$. Thus $\sin\left(\frac{\pi}{4} + 0.1\right) = f(0.1) \approx L(0.1) = \frac{1.1\sqrt{2}}{2} \approx 0.7778.$

4.5.29 Let $f(x) = 1/\sqrt[3]{x}$, $a = 512$. Then $f(a) = 1/8$ and $f'(a) = -1/(3a^{4/3}) = -1/12{,}288$, so the linear approximation to f near $a = 512$ is $L(x) = f(a) + f'(a)(x-a) = \frac{1}{8} - \frac{1}{12{,}288}(x-512)$. Therefore $\frac{1}{\sqrt[3]{510}} = f(510) \approx L(510) = \frac{769}{6144} \approx 0.1252.$

4.5.31

  a. With $f(x) = \frac{2}{x}$ and $a = 1$, we have $f(a) = 2$ and $f'(a) = -\frac{2}{a^2} = -2$. Thus the linear approximation to $f(x)$ at $x = 1$ is $L(x) = f(1) + f'(1)(x-1) = 2 + -2(x-1) = -2x+4.$

  b. A plot of f with L in gray:

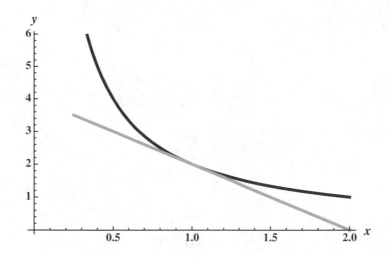

c. The linear approximation in part (b) appears to be an underestimate everywhere, because it lies below the graph of f.

d. Because $f'(x) = -\frac{2}{x^2}$, we have $f''(x) = \frac{4}{x^3}$, so that $f''(1) > 0$, and f is concave up at $x = 1$. This is consistent with L being an underestimate near $x = 1$.

4.5.33

a. With $f(x) = \frac{1}{\sqrt{x}} = x^{-1/2}$ and $a = 1$ we have $f(a) = 1$ and $f'(a) = -\frac{1}{2} \cdot a^{-3/2} = -\frac{1}{2a^{3/2}}$, so $f'(1) = -\frac{1}{2}$. Thus the linear approximation to f at $a = 1$ is $L(x) = 1 - \frac{1}{2}(x - 1) = -\frac{1}{2}x + \frac{3}{2}$.

b. A plot of f with L in gray:

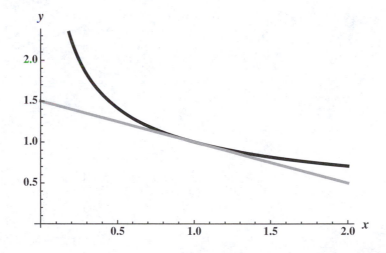

c. The linear approximation in part (b) appears to be an underestimate everywhere, because it lies below the graph of f.

d. $f''(a) = \frac{3}{4} \cdot a^{-5/2}$ which is positive for $a > 0$, so f is concave up on the interval $(0, \infty)$. This is consistent with L being an underestimate for f.

4.5.35 Note that $V'(r) = 4\pi r^2$, so $\Delta V \approx V'(a)\Delta r = 4\pi a^2 \Delta r$. Substituting $a = 5$ and $\Delta r = 0.1$ gives $\Delta V \approx 4\pi \cdot 25 \cdot 0.1 = 10\pi \approx 31.416$ ft^3.

4.5.37 Note that $S'(r) = \pi\sqrt{r^2 + h^2} + \pi r \cdot \frac{r}{\sqrt{r^2 + h^2}} = \pi \frac{2r^2 + h^2}{\sqrt{r^2 + h^2}}$, so $\Delta S \approx S'(a)\Delta r = \pi \frac{2a^2 + h^2}{\sqrt{a^2 + h^2}}\Delta r$. Substituting $h = 6$, $a = 10$ and $\Delta r = -0.1$ gives $\Delta S \approx \pi \frac{236}{\sqrt{136}}(-0.1) = \frac{-59\pi}{5\sqrt{34}} \approx -6.358$ m^2.

4.5.39 We have $f'(x) = 2$, so $dy = 2\,dx$.

4.5.41 We have $f'(x) = -3/x^4$, so $dy = -\frac{3}{x^4}dx$.

4.5.43 We have $f'(x) = a\sin x$, so $dy = a\sin x\,dx$.

4.5.45 We have $f'(x) = 9x^2 - 4$, so $dy = (9x^2 - 4)\,dx$.

4.5.47

a. True. Note that $f(0) = 0$ and $f'(0) = 0$, so the linear approximation at 0 is in fact $L(x) = 0$.

b. False. The function $f(x) = |x|$ is not differentiable at $x = 0$, so there is no good linear approximation at 0.

c. True. For linear functions, the linear approximation at any point and the function are equal.

4.5.49 We have $L(x) = f(4)+f'(4)(x-4) = 3+2(x-4)$. So $f(3.85) \approx L(3.85) = 3+2(3.85-4) = 3-.3 = 2.7$.

4.5.51   Note that $f(a) = f(0) = 1$ and $f'(a) = -1/(1+a)^2 = -1$, so the linear approximation has equation $y = L(x) = f(a) + f'(a)(x - a) = 1 - x$.

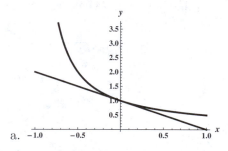

b. The linear approximation to $1/1.1$ is $\frac{1}{1.1} \approx$ $L(0.1) = 0.9$.

c. The percentage error is $100 \cdot \frac{|0.9 - \frac{1}{1.1}|}{\frac{1}{1.1}} = 1\%$.

4.5.53

a. $f(0) = 4$. $f'(x) = \dfrac{1}{3}(x + 64)^{-2/3}$, so $f'(0) = \frac{1}{48}$. Thus $L(x) = f(0) + f'(0)(x - 0) = 4 + \frac{x}{48}$.

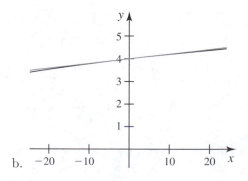

c. $\sqrt[3]{62.5} = f(-1.5) \approx L(-1.5) = 4 - \frac{1}{32} = 3.96875$.

d. $f(-1.5) = \sqrt[3]{62.5} \approx 3.9685$, so the percent error is

$$100 \cdot \frac{|3.96875 - 3.9685|}{3.9685} \approx 0.006\%.$$

4.5.55 $E(x) = |L(x) - s(x)| = \left|60 - x - \frac{3600}{60-x}\right|$. A graph is shown below, for x from -7.26 to 8.26.

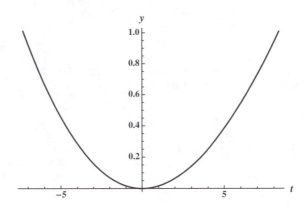

$E(x) \leq 1$ when $-7.26 \leq x \leq 8.26$, which corresponds to driving times for 1 mi from about 53 s to 68 s. Therefore, $L(x)$ gives approximations to $s(x)$ that are within 1 mi/hr of the true value when you drive 1 mile in t seconds, where $53 < t < 68$.

4.5.57 Note that $f(a) = f(8) = 2$ and $f'(a) = (1/3)a^{-2/3} = 1/12$, so the linear approximation has equation $y = L(x) = f(a) + f'(a)(x-a) = 2 + \frac{1}{12}(x-8) = \frac{x}{12} + \frac{4}{3}$.

x	Linear approx	Exact value	Percent error
8.1	$2.008\overline{3}$	2.00829885	1.717×10^{-3}
8.01	$2.0008\overline{3}$	2.000832986	1.734×10^{-5}
8.001	$2.00008\overline{3}$	2.00008333	1.736×10^{-7}
8.0001	$2.0000083\overline{3}$	2.00000833	1.736×10^{-9}
7.9999	$1.9999991\overline{6}$	1.999991667	1.736×10^{-9}
7.999	$1.999991\overline{6}$	1.999916663	1.736×10^{-7}
7.99	$1.99991\overline{6}$	1.999166319	1.738×10^{-5}
7.9	$1.9991\overline{6}$	1.991631701	1.756×10^{-3}

The percentage errors become extremely small as x approaches 8 In fact, each time we decrease Δx by a factor of 10, the percentage error decreases by a factor of 100.

4.5.59

a. The linear approximation near $x = 1$ is more accurate for f because the rate at which f' is changing at 1 is smaller than the rate at which g' is changing at 1. The graph of f bends away from the linear function more slowly than the graph of g.

b. The larger the value of $|f''(a)|$, the greater the deviation of the curve $y = f(x)$ from the tangent line at points near $x = a$.

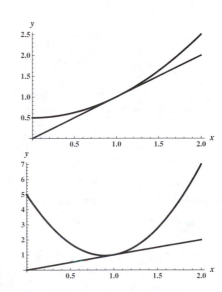

4.6   Mean Value Theorem

4.6.1   If f is a continuous function on the closed interval $[a, b]$ and is differentiable on (a, b) and the slope of the secant line that joins $(a, f(a))$ and $(b, f(b))$ is zero, then there is at least one value c in (a, b) at which the slope of the line tangent to f at $(c, f(c))$ is also zero.

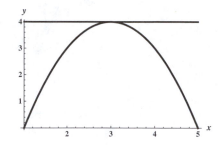

4.6.3   The function $f(x) = |x|$ is not differentiable at 0.

4.6.5   We seek a function over an interval for which it isn't true that there is a tangent line parallel to the secant line between the endpoints.

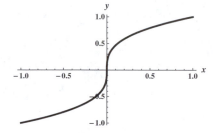

4.6.7   The function f is differentiable on $[0, 1]$ and $f(0) = f(1) = 0$, so Rolle's theorem applies. We wish to find a point x in $(0, 1)$ such that $f'(x) = 0$; we have $f'(x) = (x - 1)^2 + 2x(x - 1) = (x - 1)(3x - 1)$, so $x = 1/3$ satisfies the conclusion of Rolle's theorem.

4.6.9   The function f is differentiable on $[\pi/8, 3\pi/8]$ and $f(\pi/8) = f(3\pi/8) = 0$, so Rolle's theorem applies. We wish to find a point x in $(\pi/8, 3\pi/8)$ such that $f'(x) = 0$; we have $f'(x) = -4\sin 4x$, so $x = \pi/4$ satisfies the conclusion of Rolle's theorem.

4.6.11   The function f is not differentiable at $x = 0$, so Rolle's theorem does not apply.

4.6.13 g is continuous on $[-1, 3]$ and differentiable on $(-1, 3)$, and $g(-1) = 0 = g(3)$, so Rolle's theorem does apply. $g'(x) = 3x^2 - 2x - 5 = (x + 1)(3x - 5)$. This is zero for $x = -1$ (which is not on $(-1, 3)$) and for $x = 5/3$ (which is on $(-1, 3)$.) So $x = 5/3$ satisfies the conclusion of Rolle's theorem.

4.6.15   The average rate of change of the temperature from 3.2 km to 6.1 km is $\frac{-10.3 - 8.0}{6.1 - 3.2} \approx -6.3°/\text{km}$. Based on this, we cannot conclude that the lapse rate exceeds the critical value of $7°/$ km.

4.6.17

a. The function f is differentiable on $[-1, 2]$ so the Mean Value Theorem applies.

b. The average rate of change of f on $[-1, 2]$ is $\frac{f(2) - f(-1)}{2 - (-1)} = \frac{3 - 6}{3} = -1$. We wish to find a point c in $(-1, 2)$ such that $f'(c) = -1$, or equivalently $-2c = -1$ which gives $c = 1/2$.

c.

4.6.19

a. $\sqrt{x}$ is continuous on $[1,4]$ and differentiable on $(1,4)$, so the Mean Value Theorem applies.

b. The average rate of change of $f(x)$ on $[1,4]$ is

$$\frac{f(4) - f(1)}{4 - 1} = \frac{1}{3}$$

and $f'(x) = \frac{1}{2\sqrt{x}}$. The required c is therefore such that $2\sqrt{c} = 3$, so $c = \frac{9}{4}$.

c.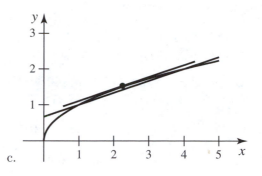

4.6.21

a. $x^{-1/3}$ is continuous on $[1/8, 8]$ and is differentiable on $(1/8, 8)$. Thus the Mean Value Theorem applies.

b. The average rate of change of $f(x)$ on $[1/8, 8]$ is

$$\frac{f(8) - f(1/8)}{8 - 1/8} = \frac{1/2 - 2}{63/8} = -\frac{4}{21}$$

and $f'(x) = -1/(3x^{4/3})$, so we want c such that $-1/(3c^{4/3}) = -4/21$, or $21 = 12c^{4/3}$. Thus $c = \left(\frac{7}{4}\right)^{3/4} \approx 1.52$.

c.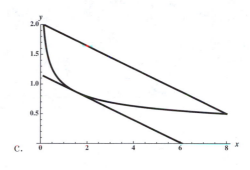

4.6.23

a. The Mean Value Theorem does not apply because the function f is not differentiable at $x = 0$.

b. Even though the Mean Value Theorem doesn't apply, it still happens to be the case that there are numbers c between -8 and 8 where the tangent line has slope $\frac{f(8) - f(-8)}{8 - (-8)} = \frac{1}{2}$. This occurs where $\frac{2}{3}c^{-2/3} = 1/2$, which gives $c = \pm\frac{8}{9} \cdot \sqrt{3}$.

c.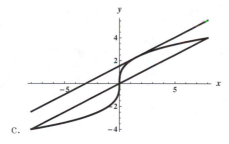

4.6.25

a. False. The function f is not differentiable at $x = 0$.

b. True. If $f(x) - g(x) = c$ is constant, then $f'(x) - g'(x) = 0$.

c. False. If $f'(x) = 0$ then we can conclude that $f(x) = c$ for some constant.

4.6.27 The functions $h(x)$ and $p(x)$ have the same derivative as $f(x)$ because they differ from $f(x)$ by a constant.

4.6.29 The secant line between the endpoints has slope $\frac{f(4) - f(-4)}{4 - (-4)} = \frac{4-1}{8} = \frac{3}{8}$.

The slope of the tangent line to the graph appears to have this value at approximately -2.5 and at about 2.6. These are eyeballed estimates, so your personal estimate may differ.

4.6.31 Because $f(1) \approx 2$ and $f(3) \approx 2$, the average rate of change of f on $[1,3]$ is $\frac{f(3)-f(1)}{3-1} = 0$. However, the tangent to f between $x = 1$ and $x = 2$ is the graph of f itself, which has slope 2, and the tangent to f between $x = 2$ and $x = 3$ is also the graph of f, which has slope 1. So there is no point in $(1,3)$ where the tangent line has slope 0. This does not contradict the Mean Value Theorem because f is not continuous everywhere on $[1, 3]$, nor differentiable everywhere on $(1, 3)$ both of these hypotheses fail at x = 2.

4.6.33 The average speed of the car over the 28 minute period ($= 28/60$ hr) is $\frac{30-0}{28/60} \approx 64.3$mi/hr, so the officer can conclude by the Mean Value Theorem that at some point the car exceeded the speed limit.

4.6.35 The runner's average speed is $6.2/(32/60) \approx 11.6$ mi /hr. By the Mean Value Theorem, the runner's speed was 11.6 mi/hr at least once. By the intermediate value theorem, all speeds between 0 and 11.6 mi/hr were reached. Because the initial and final speed was 0 mi/hr, the speed of 11 mi/hr was reached at least twice.

4.6.37 Observe that

$$\frac{f(b) - f(a)}{b - a} = \frac{A(b^2 - a^2) + B(b - a)}{b - a} = A(a + b) + B$$

and $f'(c) = 2Ac + B$, so the point c that satisfies the conclusion of the Mean Value Theorem is $c = (a+b)/2$.

4.6.39 Note that $f'(x) = 2\tan x \sec^2 x$ and $g'(x) = 2\sec x \sec x \tan x = 2\tan x \sec^2 x$, so $f'(x) = g'(x)$. This implies that $f - g$ is a constant, which also follows from the trigonometric identity $\sec^2 x = \tan^2 x + 1$.

4.6.41 Bolt's average speed during the race was $\frac{100}{9.58}$ m/s $= \frac{100}{9.58} \cdot \frac{3600}{1000}$ km/hr ≈ 37.58 km/hr, so by the Mean Value Theorem he must have exceeded 37 km/hr during the race.

4.6.43

a. If $g(x) = x$ then $g'(x) = 1$ and hence $\frac{f(b)-f(a)}{g(b)-g(a)} = \frac{f(b)-f(a)}{b-a} = \frac{f'(c)}{g'(c)} = f'(c)$.

b. We have $\frac{f(b)-f(a)}{g(b)-g(a)} = \frac{0-(-1)}{6-2} = \frac{1}{4}$; $\frac{f'(c)}{g'(c)} = \frac{2c}{4} = \frac{c}{2}$; so $c = 1/2$.

4.7   L'Hôpital's Rule

4.7.1 If $\lim_{x \to a} f(x) = 0$ and $\lim_{x \to a} g(x) = 0$, then we say $\lim_{x \to a} f(x)/g(x)$ is of indeterminate form 0/0.

4.7.3 Take the limit of the quotient of the derivatives of the numerator and denominator.

4.7.5 If $\lim_{x \to a} f(x)g(x)$ has the form $0 \cdot \infty$, then $\lim_{x \to a} \frac{f(x)}{1/g(x)}$ has the indeterminate form 0/0 or ∞/∞.

4.7.7 Consider $\lim_{x \to 0} \frac{\sin x}{x} = 1$.

4.7.9 It cannot be applied because it doesn't have one of the desired forms. In particular, $\lim_{x \to \infty} \sin x$ does not exist.

4.7.11 This is a $0 \cdot \infty$ form.

4.7.13 L'Hôpital's rule gives $\lim_{x \to 2} \frac{x^2 - 2x}{8 - 6x + x^2} = \lim_{x \to 2} \frac{2x - 2}{-6 + 2x} = \frac{2}{-2} = -1$.

4.7.15 L'Hôpital's rule gives $\lim_{x \to 1} \frac{x - 1}{\sqrt{x} - 1} = \lim_{x \to 1} \frac{1}{\frac{1}{2\sqrt{x}}} = \lim_{x \to 1} 2\sqrt{x} = 2$.

4.7.17 L'Hôpital's rule gives $\lim_{x \to 1} \frac{4\sin \pi x}{\cos \pi x + x} = \lim_{x \to 1} \frac{4\pi \cos \pi x}{-\pi \sin \pi x + 1} = -4\pi$.

4.7.19 L'Hôpital's rule gives $\displaystyle\lim_{x\to 0}\frac{3\sin 4x}{5x} = \lim_{x\to 0}\frac{12\cos 4x}{5} = \frac{12}{5}$.

4.7.21 L'Hôpital's rule gives $\displaystyle\lim_{u\to\pi/4}\frac{\tan u - \cot u}{u - \pi/4} = \lim_{u\to\pi/4}\frac{\sec^2 u + \csc^2 u}{1} = 2 + 2 = 4$.

4.7.23 Apply l'Hôpital's rule twice: $\displaystyle\lim_{x\to 0}\frac{1-\cos 3x}{8x^2} = \lim_{x\to 0}\frac{3\sin 3x}{16x} = \lim_{x\to 0}\frac{9\cos 3x}{16} = \frac{9}{16}$.

4.7.25 Apply l'Hôpital's rule twice:

$$\lim_{x\to\pi}\frac{\cos x + 1}{(x-\pi)^2} = \lim_{x\to\pi}\frac{-\sin x}{2(x-\pi)} = \lim_{x\to\pi}\frac{-\cos x}{2} = \frac{1}{2}.$$

4.7.27 Apply l'Hôpital's rule three times:

$$\lim_{x\to 0}\frac{\tan x - x}{x^3} = \lim_{x\to 0}\frac{\sec^2 x - 1}{3x^2} = \lim_{x\to 0}\frac{2\sec^2 x \tan x}{6x} = \lim_{x\to 0}\frac{4\sec^2 x\tan^2 x + 2\sec^4 x}{6} = \frac{1}{3}.$$

4.7.29 Apply l'Hôpital's rule twice:

$$\lim_{x\to -1}\frac{x^3 - x^2 - 5x - 3}{x^4 + 2x^3 - x^2 - 4x - 2} = \lim_{x\to -1}\frac{3x^2 - 2x - 5}{4x^3 + 6x^2 - 2x - 4} = \lim_{x\to -1}\frac{6x - 2}{12x^2 + 12x - 2} = 4.$$

4.7.31 L'Hôpital's rule gives $\displaystyle\lim_{v\to 3}\frac{v - 1 - \sqrt{v^2 - 5}}{v - 3} = \lim_{v\to 3}\frac{1 - \dfrac{v}{\sqrt{v^2 - 5}}}{1} = -\frac{1}{2}$.

4.7.33 Apply l'Hôpital's rule twice:

$$\lim_{x\to 2}\frac{x^2 - 4x + 4}{\sin^2 \pi x} = \lim_{x\to 2}\frac{2x - 4}{2\pi(\sin \pi x)(\cos \pi x)} = \lim_{x\to 2}\frac{2x - 4}{\pi\sin 2\pi x} = \lim_{x\to 2}\frac{2}{2\pi^2\cos 2\pi x} = \frac{2}{2\pi^2} = \frac{1}{\pi^2}.$$

4.7.35 Apply l'Hôpital's rule three times:

$$\lim_{x\to\infty}\frac{3x^4 - x^2}{6x^4 + 12} = \lim_{x\to\infty}\frac{12x^3 - 2x}{24x^3} = \lim_{x\to\infty}\frac{36x^2 - 2}{72x^2} = \lim_{x\to\infty}\frac{72x}{144x} = \frac{1}{2}.$$

4.7.37 Apply l'Hôpital's rule three times:

$$\lim_{x\to\pi/2^-}\frac{\tan x}{3/(2x-\pi)} = \lim_{x\to\pi/2^-}\frac{\sec^2 x}{-6/(2x-\pi)^2} = -(1/6)\lim_{x\to\pi/2^-}\frac{(2x-\pi)^2}{\cos^2 x} = (-1/6)\lim_{x\to\pi/2^-}\frac{4(2x-\pi)}{2\cos x(-\sin x)} =$$

$$(-1/6)\lim_{x\to\pi/2^-}\frac{8x - 4\pi}{-\sin(2x)} = (-1/6)\lim_{x\to\pi/2^-}\frac{8}{-2\cos(2x)} = -\frac{2}{3}.$$

4.7.39 Applying l'Hôpital's rule gives:

$$\lim_{x\to\pi/2^-}\frac{1 + 2\sec x}{1 + \tan x} = \lim_{x\to\pi/2^-}\frac{2\sec x \tan x}{\sec^2 x} = \lim_{x\to\pi/2^-}\frac{2\tan x}{\sec x} = \lim_{x\to\pi/2^-} 2\sin x = 2.$$

4.7.41 Observe that $\displaystyle\lim_{x\to 0} x\csc x = \lim_{x\to 0}\frac{x}{\sin x} = \lim_{x\to 0}\frac{1}{\cos x} = 1$, by l'Hôpital's rule.

4.7.43 Observe that the given limit can be written $\displaystyle\lim_{x\to 0}\frac{\sin 7x}{\sin 6x} = \lim_{x\to 0}\frac{7\cos 7x}{6\cos 6x} = \frac{7}{6}$.

4.7.45 Observe that $\displaystyle\lim_{x\to(\pi/2)^-}\left(\frac{\pi}{2} - x\right)\sec x = \lim_{x\to(\pi/2)^-}\frac{\pi/2 - x}{\cos x} = \lim_{x\to(\pi/2)^-}\frac{-1}{-\sin x} = 1$ by l'Hôpital's rule.

4.7.47 Observe that $\lim\limits_{x\to 0}\left(\cot x - \dfrac{1}{x}\right) = \lim\limits_{x\to 0}\left(\dfrac{\cos x}{\sin x} - \dfrac{1}{x}\right) = \lim\limits_{x\to 0}\dfrac{x\cos x - \sin x}{x\sin x}$. Apply l'Hôpital's rule twice:

$$\lim_{x\to 0}\frac{x\cos x - \sin x}{x\sin x} = \lim_{x\to 0}\frac{\cos x - x\sin x - \cos x}{\sin x + x\cos x}$$

$$= -\lim_{x\to 0}\frac{x\sin x}{\sin x + x\cos x} = -\lim_{x\to 0}\frac{\sin x + x\cos x}{\cos x + \cos x - x\sin x} = -\frac{0}{2} = 0.$$

4.7.49 Observe that $\lim\limits_{\theta\to(\pi/2)^-}(\tan\theta - \sec\theta) = \lim\limits_{\theta\to(\pi/2)^-}\left(\dfrac{\sin\theta}{\cos\theta} - \dfrac{1}{\cos\theta}\right) = \lim\limits_{\theta\to(\pi/2)^-}\dfrac{\sin\theta - 1}{\cos\theta}$. By l'Hôpital's rule $\lim\limits_{\theta\to(\pi/2)^-}\dfrac{\sin\theta - 1}{\cos\theta} = \lim\limits_{\theta\to(\pi/2)^-}\dfrac{\cos\theta}{-\sin\theta} = \dfrac{0}{-1} = 0.$

4.7.51

a. False; $\lim\limits_{x\to 2}x^2 - 1 = 3$, so l'Hôpital's rule does not apply. In fact, $\lim\limits_{x\to 2}\dfrac{x - 2}{x^2 - 1} = \dfrac{0}{3} = 0.$

b. False; l'Hôpital's rule does not say $\lim\limits_{x\to a}f(x)g(x) = \lim\limits_{x\to a}f'(x)\lim\limits_{x\to a}g'(x)$. In fact, $\lim\limits_{x\to 0}x\sin x = 0\cdot 0 = 0.$

c. False. $\lim_{x\to 2}\frac{x^3 - 2x^2 + x - 2}{x^2 - 1} = \frac{0}{3} = 0$, and $\lim_{x\to 2}\frac{3x^2 - 4x + 1}{2x} = \frac{5}{4}$, so the limits are not equal.

d. True. We use L'Hôpital's rule to evaluate the left-hand limit since it has the form $0/0$.

$$\lim_{x\to 2}\frac{x^3 - 2x^2 + x - 2}{x^2 - 4} = \lim_{x\to 2}\frac{3x^2 - 4x + 1}{2x} = \frac{5}{4},$$

and this is exactly the limit on the right-hand side.

4.7.53 Observe that $\lim\limits_{x\to\infty}\dfrac{2x^3 - x^2 + 1}{5x^3 + 2x} = \dfrac{2}{5}$ by Theorem 2.7. We can also use l'Hôpital's rule:

$$\lim_{x\to\infty}\frac{2x^3 - x^2 + 1}{5x^3 + 2x} = \lim_{x\to\infty}\frac{6x^2 - 2x}{15x^2 + 2} = \lim_{x\to\infty}\frac{12x - 2}{30x} = \lim_{x\to\infty}\frac{12}{30} = \frac{2}{5}.$$

4.7.55 By l'Hôpital's rule, $\lim\limits_{x\to 6}\dfrac{(5x+2)^{1/5} - 2}{x^{-1} - 6^{-1}} = \lim\limits_{x\to 6}\dfrac{(5x+2)^{-4/5}}{-x^{-2}} = -\dfrac{9}{4}.$

4.7.57 Observe that $(\sqrt{x-2} - \sqrt{x-4})\cdot\dfrac{\sqrt{x-2}+\sqrt{x-4}}{\sqrt{x-2}+\sqrt{x-4}} = \dfrac{x-2-(x-4)}{\sqrt{x-2}+\sqrt{x-4}} = \dfrac{2}{\sqrt{x-2}+\sqrt{x-4}}$, so $\lim\limits_{x\to\infty}\sqrt{x-2} - \sqrt{x-4} = \lim\limits_{x\to\infty}\dfrac{2}{\sqrt{x-2}+\sqrt{x-4}} = 0.$

4.7.59 Make the substitution $t = 1/x$; then $\lim\limits_{x\to\infty}x^3\left(\dfrac{1}{x} - \sin\dfrac{1}{x}\right) = \lim\limits_{t\to 0+}\dfrac{t - \sin t}{t^3} = \lim\limits_{t\to 0+}\dfrac{1 - \cos t}{3t^2} = \lim\limits_{t\to 0+}\dfrac{\sin t}{6t} = \lim\limits_{t\to 0+}\dfrac{\cos t}{6} = \dfrac{1}{6}$, using l'Hôpital's rule.

4.7.61 Observe that $\lim\limits_{x\to 1+}\left(\dfrac{1}{x-1} - \dfrac{1}{\sqrt{x-1}}\right) = \lim\limits_{x\to 1+}\dfrac{1 - \sqrt{x-1}}{x-1} = \infty.$

4.7.63 Use the identity $1 + 2 + \cdots + n = \dfrac{n(n+1)}{2}$; then

$$\lim_{n\to\infty}\frac{1 + 2 + \cdots + n}{n^2} = \lim_{n\to\infty}\frac{n(n+1)}{2n^2} = \lim_{n\to\infty}\frac{n+1}{2n} = \frac{1}{2}.$$

4.7.65 Let $z = \frac{1}{n}$. Then as $n \to \infty$ we have $z \to 0^+$. Then

$$\lim_{n \to \infty} n \cot(1/n) - n^2 = \lim_{z \to 0^+} \frac{\cot z}{z} - \frac{1}{z^2}$$

$$= \lim_{z \to 0^+} \frac{z \cos z - \sin z}{z^2 \sin z}$$

$$= \lim_{z \to 0^+} \frac{\cos z - z \sin z - \cos z}{2z \sin z + z^2 \cos z}$$

$$= \lim_{z \to 0^+} \frac{-\sin z}{2 \sin z + z \cos z}$$

$$= \lim_{z \to 0^+} \frac{-\cos z}{2 \cos z + \cos z - z \sin z}$$

$$= -\frac{1}{2 + 1 - 0} = -\frac{1}{3},$$

where the third and fifth equalities follow from l'Hôpital's rule.

4.7.67 L'Hôpital's rule gives $\displaystyle\lim_{x \to \infty} \frac{\sqrt{ax+b}}{\sqrt{cx+d}} = \lim_{x \to \infty} \frac{a}{\sqrt{ax+b}} \cdot \frac{\sqrt{cx+d}}{c} = \frac{a}{c} \lim_{x \to \infty} \frac{\sqrt{cx+d}}{\sqrt{ax+b}}$, which is the same form as the original limit, so l'Hôpital's rule fails in this case. We can evaluate this limit as follows: first observe that $\displaystyle\lim_{x \to \infty} \frac{ax+b}{cx+d} = \frac{a}{c}$ by l'Hôpital's rule; therefore $\displaystyle\lim_{x \to \infty} \frac{\sqrt{ax+b}}{\sqrt{cx+d}} = \lim_{x \to \infty} \sqrt{\frac{ax+b}{cx+d}} = \sqrt{\frac{a}{c}}$.

4.7.69 The triangle ABP has base $1 - \cos\theta$ and height $\sin\theta$, so its area is $f(\theta) = \frac{1}{2}\sin\theta(1 - \cos\theta)$. The sector OBP has area $\theta/2$, and the triangle OBP has base 1 and height $\sin\theta$; therefore $g(\theta) = \frac{1}{2}(\theta - \sin\theta)$. We have $\displaystyle\lim_{\theta \to 0} \frac{g(\theta)}{f(\theta)} = \lim_{\theta \to 0} \frac{\theta - \sin\theta}{\sin\theta(1 - \cos\theta)} = \lim_{\theta \to 0} \frac{\theta - \sin\theta}{\sin\theta - (1/2)\sin 2\theta}$. Three applications of l'Hôpital's rule gives $\displaystyle\lim_{\theta \to 0} \frac{g(\theta)}{f(\theta)} = \lim_{\theta \to 0} \frac{1 - \cos\theta}{\cos\theta - \cos 2\theta} = \lim_{\theta \to 0} \frac{\sin\theta}{2\sin 2\theta - \sin\theta} = \lim_{\theta \to 0} \frac{\cos\theta}{4\cos 2\theta - \cos\theta} = \frac{1}{3}$.

4.8 Newton's Method

4.8.1 Newton's method generates a sequence of x-intercepts of lines tangent to the graph of f to approximate the roots of f.

4.8.3 Generally, if two successive Newton approximations agree in their first p digits, then those approximations have p digits of accuracy. The method is terminated when the desired accuracy is reached.

4.8.5 Because $f'(x_n) = 2x_n$, we have

$$x_{n+1} = x_n - \frac{x_n^2 - 6}{2x_n} = \frac{2x_n^2 - x_n^2 + 6}{2x_n} = \frac{x_n^2 + 6}{2x_n}.$$

$x_1 = 2.5$ and $x_2 = 2.45$.

4.8.7 Because $f'(x_n) = -\sec^2 x_n$, we have

$$x_{n+1} = x_n - \frac{2 - \tan x_n}{\sec^2 x_n} = x_n - \left(\cos^2 x_n (2 - \tan x_n)\right).$$

$x_1 = 1.129204$, $x_2 = 1.108129$.

n	x_n
0	4.0
1	3.25
2	3.163462
3	3.162278
4	3.162278
5	3.162278
6	3.162278
7	3.162278
8	3.162278
9	3.162278
10	3.162278

Because $f'(x_n) = 2x_n$, we have

4.8.9 $x_{n+1} = x_n - \frac{x_n^2 - 10}{2x_n} = \frac{2x_n^2 - (x_n^2 - 10)}{2x_n} = \frac{x_n^2 + 10}{2x_n}$.

n	x_n
0	1.5
1	0.101436
2	0.501114
3	0.510961
4	0.510973
5	0.510973
6	0.510973
7	0.510973
8	0.510973
9	0.510973
10	0.510973

Because $f'(x_n) = \cos(x_n) + 1$, we have

4.8.11 $x_{n+1} = x_n - \frac{\sin x_n + x_n - 1}{\cos x_n + 1}$.

n	x_n
0	1.5
1	1.44389
2	1.36198
3	1.26818
4	1.19618
5	1.16857
6	1.16559
7	1.16556
8	1.16556
9	1.16556
10	1.16556

Because $f'(x_n) = \sec^2(x_n) - 2$, we have

4.8.13 $x_{n+1} = x_n - \frac{\tan x_n - 2x_n}{\sec^2(x_n) - 2}$.

4.8.15 A preliminary sketch of the two curves seems to indicate that they intersect once near $x = 2$.

Let $f(x) = \sin x - x/2$. Then $f'(x_n) = \cos x_n - 1/2$. The Newton's method formula becomes

$$x_{n+1} = x_n - \frac{\sin x_n - x_n/2}{\cos x_n - 1/2}.$$

If we use an initial estimate of $x_0 = 2$, we obtain $x_1 = 1.901$, $x_2 = 1.89551$, $x_3 = 1.89549$ and $x_4 = 1.89549$, so the point of intersection appears to be at approximately $x = 1.89549$.

4.8.17 A preliminary sketch of the two curves seems to indicate that they intersect three times, once between -2.5 and -2, once between 0 and $1/2$, and once between 1.5 and 2.

Let $f(x) = 4 - x^2 - 1/x$. Then $f'(x_n) = -2x_n + (1/x_n^2)$. The Newton's method formula becomes

$$x_{n+1} = x_n - \frac{4 - x_n^2 - (1/x_n)}{-2x_n + (1/x_n^2)}.$$

If we use an initial estimate of $x_0 = -2.25$, we obtain $x_1 = -2.11843$, $x_2 = -2.11491$, $x_3 = -2.11491$, so there appears to be a point of intersection near $x = -2.115$.

If we use an initial estimate of $x_0 = .25$, we obtain $x_1 = .254032$, $x_2 = .254102$, so there appears to be a point of intersection near $x = .254$.

If we use an initial estimate of $x_0 = 1.75$, we obtain $x_1 = 1.86535$, $x_2 = 1.86081$, so there appears to be another point of intersection near $x = 1.86$.

4.8.19 A preliminary sketch of the two curves seems to indicate that they intersect twice, once just to the right of 0, and once between 2 and 2.5.

Let $f(x) = 4\sqrt{x} - (x^2 + 1)$. Then $f'(x_n) = 2/\sqrt{x_n} - 2x_n$. The Newton's method formula becomes

$$x_{n+1} = x_n - \frac{4\sqrt{x_n} - (x_n^2 + 1)}{2/\sqrt{x_n} - 2x_n}.$$

If we use an initial estimate of $x_0 = .1$, we obtain $x_1 = .0583788$, $x_2 = .0629053$, $x_3 = .0629971$, so there appears to be a point of intersection near $x = .06299$.

If we use an initial estimate of $x_0 = 2.25$, we obtain $x_1 = 2.23026$, $x_2 = 2.23012$, $x_3 = 2.23012$, so there appears to be a point of intersection near $x = 2.23012$.

4.8.21 $f'(x) = \frac{-x \sin x - \cos x}{x^2}$, which is zero when $x \sin x + \cos x = 0$. Note that $f'(1) < 0$ and $f'(\pi) > 0$, so there must be a local minimum on the interval $(1, \pi)$. Let $g(x) = x \sin x + \cos x$. Then $g'(x_n) = \sin x_n + x_n \cos x_n - \sin x_n = x_n \cos x_n$, and the Newton's method formula becomes

$$x_{n+1} = x_n - \frac{x_n \sin x_n + \cos x_n}{x_n \cos x_n}.$$

If we use an initial estimate of $x_0 = 2.5$, we obtain $x_1 = 2.84702$, $x_2 = 2.79918$, $x_3 = 2.79839$, $x_4 = 2.79839$, so the smallest local minimum of f on $(0, \infty)$ occurs at approximately 2.79839.

4.8.23 $f'(x) = 9x^4 - 30x^3 + 7x^2 + 60x$ and $f''(x) = 36x^3 - 90x^2 + 14x + 60 = 2(18x^3 - 45x^2 + 7x + 30)$. We are seeking roots of $f''(x)$. If we apply Newton's method to f'' we obtain the recursion

$$x_{n+1} = x_n - \frac{36x_n^3 - 90x_n^2 + 14x_n + 60}{108x_n^2 - 180x_n + 14}.$$

Starting with an initial estimate of $x_0 = 1$, we obtain $x_1 = 1.34483$, $x_2 = 1.45527$, $x_3 = 1.49284$, $x_4 = 1.49974$, and $x_5 \approx 1.5$. We check directly that 1.5 is a root of f'', so $2x - 3$ is a factor of f'', and using long division, we see that $f''(x) = 2(2x - 3)(9x^2 - 9x - 10) = 2(2x - 3)(3x + 2)(3x - 5)$. So the potential inflection points of f are located at $x = -2/3$, $x = 3/2$, and $x = 5/3$. A check of the sign of f'' on the various intervals confirms that these are all the locations of inflection points.

4.8.25 The recursion for $f(x)$ is $x_{n+1} = x_n - \frac{(x_n-1)^2}{2x_n-2} = \frac{2x_n^2 - 2x_n - (x_n^2 - 2x_n + 1)}{2x_n - 2} = \frac{x_n^2 - 1}{2(x_n - 1)} = \frac{x_n + 1}{2}$.

The recursion for $g(x)$ is $y_{n+1} = y_n - \frac{y_n^2 - 1}{2y_n} = \frac{2y_n^2 - (y_n^2 - 1)}{2y_n} = \frac{y_n^2 + 1}{2y_n}$. The comparison below shows that Newton's method converges much faster for $g(x) = x^2 - 1$. This is because it is steeper near the root $x = 1$ – the value of $g'(1) = 2$, while $f'(1) = 0$. The flatness of f near 1 causes slow convergence.

n	x_n
0	2
1	1.5
2	1.25
3	1.125
4	1.0625
5	1.03125
6	1.01563
7	1.00781
8	1.00391
9	1.00195
10	1.00098

n	y_n
0	2
1	1.25
2	1.025
3	1.0003
4	≈ 1

4.8.27

a. True.

b. False. The quadratic formula gives exact values.

c. False. It sometime fails depending on factors such as the shape of the curve and the closeness of the initial estimate.

4.8.29 Let $g(x) = x^3/10 + 1 - x$. Fixed points of f are roots of g. The Newton's method recursion for g is given by

$$x_{n+1} = x_n - \frac{x_n^3/10 + 1 - x_n}{3x_n^2/10 - 1} = x_n - \frac{x_n^3 + 10 - 10x_n}{3x_n^2 - 10} = \frac{2(x_n^3 - 5)}{3x_n^2 - 10}.$$

A preliminary sketch of g indicates that there are three roots, near -3.5, 1, and 2.5.

n	x_n
0	-3.5
1	-3.57944
2	-3.57709
3	-3.57709

n	x_n
0	1
1	1.14286
2	1.1534
3	1.15347
4	1.15347

n	x_n
0	2.5
1	2.42857
2	2.42365
3	2.42362
4	2.42362

The fixed points of f are approximately -3.57709, 1.15347, and 2.42362.

4.8.31 Let $g(x) = 2x\cos x - x$. Fixed points of f are roots of g. Clearly $x = 0$ is a root of g. The Newton's method recursion for g is given by

$$x_{n+1} = x_n - \frac{2x_n \cos x_n - x_n}{2\cos x_n - 2x \sin x_n - 1}.$$

A preliminary sketch of g indicates that there is only one nonzero root on $[0, 2]$, near $x = 1$. We have:

n	x_n
0	1
1	1.0503
2	1.04721
3	1.0472
4	1.0472

The fixed points are 0 and approximately 1.0472.

4.8.33 A preliminary sketch of f indicates that there are two roots, near $x = -0.4$. and $x = 1.3$
The Newton's method recursion for f is given by

$$x_{n+1} = x_n - \frac{\cos(2x_n) - x_n^2 + 2x_n}{-2\sin(2x_n) - 2x_n + 2}.$$

We have:

n	x_n
0	-0.4
1	-0.337825
2	-0.335412
3	-0.335408
4	-0.335408

n	x_n
0	1.3
1	1.33256
2	1.33306
3	1.33306

The roots are approximately -0.335408 and 1.33306.

4.8.35 A preliminary sketch of f indicates that there are three roots, near $x = 1.2$. $x = 2.8$ and $x = 5.8$
The Newton's method recursion for f is given by

$$x_{n+1} = x_n - \frac{2 - \sin x_n - \sqrt{x_n}}{-\cos x_n - \frac{1}{2\sqrt{x_n}}} = x_n - \frac{4\sqrt{x_n} - 2\sqrt{x_n}\sin x_n - 2x_n}{-2\sqrt{x_n}\cos x_n - 1}.$$

We have:

n	x_n
0	1.2
1	1.16643
2	1.16717
3	1.16717

n	x_n
0	2.8
1	2.81291
2	2.81286
3	2.81286

n	x_n
0	5.8
1	5.85149
2	5.85098
3	5.85098

The roots are approximately 1.16717, 2.81286, and 5.85098.

4.8.37 A preliminary sketch of f indicates that there are three roots, near $x = -0.2$, near $x = 0.7$, and near $x = 1.2$.
The Newton's method recursion for f is given by

$$x_{n+1} = x_n - \frac{\cos 3x_n + \sin 5x_n + x_n}{-3\sin 3x_n + 5\cos 5x_n + 1}.$$

We have:

n	x_n
0	-0.2
1	-0.159941
2	-0.161642
3	-0.161645
4	-0.161645

n	x_n
0	0.7
1	0.675186
2	0.675756
3	0.675756

n	x_n
0	1.2
1	1.19666
2	1.19665
3	1.19665

The roots are approximately -0.161645, 0.675756, and 1.19665.

4.8.39 Because the residuals become small quickly, the convergence of x_n is quite slow. This is related to the extreme flatness of the graph of x^{10} between 0 and $1/2$.

n	x_n	Error	Residual
0	.5	.5	.000976563
1	.45	.45	.000340506
2	.405	.405	.000118727
3	.3645	.3645	.0000413976
4	.32805	.32805	.0000144345
5	.295245	.295245	5.03298×10^{-6}
6	.265721	.265721	1.75489×10^{-6}
7	.239148	.239148	6.11893×10^{-7}
8	.215234	.215234	2.13354×10^{-7}
9	.19371	.19371	7.43919×10^{-8}
10	.174339	.174339	2.59389×10^{-8}

4.8.41

a. The Newton's method formula would be:

$$x_{n+1} = x_n - \frac{1/x_n - a}{-1/x_n^2} = x_n + x_n - ax_n^2 = 2x_n - ax_n^2 = (2 - ax_n)x_n.$$

b. The approximation to $1/7$ is $.14285714$.

n	x_n
0	.1
1	.13
2	.1417
3	.14284777
4	.14285714
5	.14285714

4.8.43

Let $f(r) = m(r) - 600$. We are seeking a root of f. Using an initial guess of $r_0 = 0.05$ and the recursion

$$r_{n+1} = r_n - \frac{m(r_n) - 600}{m'(r_n)},$$

we obtain

n	x_n
0	0.05
1	0.0603375
2	0.0600701
3	0.0600699

The desired interest rate is about 6%.

4.8.45 Let $f(\lambda) = \tan(\pi\lambda) - \lambda$. We are looking for the first three positive roots of f. A preliminary sketch indicates that they are located near 1.4, 2.4, and 3.4. The Newton's method recursion is given by

$$x_{n+1} = x_n - \frac{\tan(\pi x_n) - x_n}{\pi \sec^2(\pi x_n) - 1}.$$

We obtain the following results:

n	x_n
0	1.4
1	1.34741
2	1.30555
3	1.29121
4	1.29012
5	1.29011
6	1.29011

n	x_n
0	2.4
1	2.37876
2	2.37331
3	2.37305
4	2.37305

n	x_n
0	3.4
1	3.4101
2	3.40919
3	3.40918
4	3.40918

The first three positive eigenvalues are approximately 1.29011, 2.37305, and 3.40918.

4.8.47 This problem can be solved (approximately) by setting up a computer or calculator program to run Newton's method, and then experimenting with different starting values. If this is done, it can be seen that any initial estimate between -4 and the local maximum at approximately -1.53 converges to the root at -2. Initial values between approximately -1.52 and -1.486 converge to the root at 3, while starting values between -1.485 and -1.475 converge to the root at -2. From -1.474 to approximately $.841$, starting values lead to convergence to the root at -1, while from $.842$ to $.846$ they lead to convergence to the root at -2. From about $.847$ to $.862$ they lead to convergence to the root at 3, while from $.863$ to the local minimum at 1.528 they lead to convergence to the root at -2. From about 1.528 to 4, the convergence is to the root at 3. Thus the approximate basis of convergence for -2 is $[-4, -1.53] \cup [-1.485, -1.475] \cup [.842, .846]$. For -1 the approximate basis of convergence is $[-1.474, .841]$, and for 3, it is $[-1.52, -1.486] \cup [.847, .862] \cup [1.53, 4]$.

4.9   Antiderivatives

4.9.1 Derivative, antiderivative.

4.9.3 $x + C$, where C is any constant.

4.9.5 $\dfrac{x^{p+1}}{p+1} + C$, where C is any real number and $p \neq -1$.

4.9.7 $\sqrt{x} + C$, where C is any constant.

4.9.9 Observe that $F(-1) = 4 + C = 4$, so $C = 0$.

4.9.11 The antiderivatives of $5x^4$ are $x^5 + C$. Check: $\frac{d}{dx}(x^5 + C) = 5x^4$.

4.9.13 The antiderivatives of $\sin 2x$ are $-(1/2)\cos 2x + C$. Check: $\frac{d}{dx}(-\frac{1}{2}\cos 2x + C) = \sin 2x$.

4.9.15 The antiderivatives of $3\sec^2 x$ are $3\tan x + C$. Check: $\frac{d}{dx}(3\tan x + C) = 3\sec^2 x$.

4.9.17 The antiderivatives of $-2/y^3 = -2y^{-3}$ are $y^{-2} + C$. Check: $\frac{d}{dy}(y^{-2} + C) = -2y^{-3}$.

4.9.19 The antiderivatives of $\frac{7}{2}x^{5/2}$ are $x^{7/2} + C$. Check: $\frac{d}{dx}(x^{7/2} + C) = \frac{7}{2}x^{5/2}$.

4.9.21 $\int (3x^5 - 5x^9)\,dx = 3\cdot\frac{x^6}{6} - 5\cdot\frac{x^{10}}{10} + C = \frac{1}{2}x^6 - \frac{1}{2}x^{10} + C$. Check: $\frac{d}{dx}(\frac{1}{2}x^6 - \frac{1}{2}x^{10} + C) = 3x^5 - 5x^9$.

4.9.23 $\int\left(4\sqrt{x} - \frac{4}{\sqrt{x}}\right)dx = \int(4x^{1/2} - 4x^{-1/2})\,dx = 4\cdot\frac{x^{3/2}}{3/2} - 4\cdot\frac{x^{1/2}}{1/2} + C = \frac{8}{3}x^{3/2} - 8x^{1/2} + C$. Check: $\frac{d}{dx}(\frac{8}{3}x^{3/2} - 8x^{1/2} + C) = 4\sqrt{x} - \frac{4}{\sqrt{x}}$.

4.9.25 Note that $\frac{d}{ds}(5s+3)^3 = 15(5s+3)^2$; therefore $\int (5s+3)^2\,ds = \frac{(5s+3)^3}{15} + C$.

4.9.27 $\int(3x^{1/3} + 4x^{-1/3} + 6)\,dx = 3\cdot\frac{3}{4}x^{4/3} + 4\cdot\frac{3}{2}x^{2/3} + 6x + C = \frac{9}{4}x^{4/3} + 6x^{2/3} + 6x + C$. Check: $\frac{d}{dx}(\frac{9}{4}x^{4/3} + 6x^{2/3} + 6x + C) = 3x^{1/3} + 4x^{-1/3} + 6$.

4.9.29 $\int (3x+1)(4-x)\,dx = \int(12x - 3x^2 + 4 - x)\,dx = \int(-3x^2 + 11x + 4)\,dx = -x^3 + \frac{11}{2}x^2 + 4x + C$. Check: $\frac{d}{dx}\left(\frac{9}{2}x^{4/3} + C\right) = 6\sqrt[3]{x}$.

4.9.31 $\int (3x^{-4} + 2 - 3x^{-2})\,dx = -x^{-3} + 2x + 3x^{-1} + C$. Check: $\frac{d}{dx}\left(-x^{-3} + 2x + 3x^{-1} + C\right) = 3x^{-4} + 2 - 3x^{-2}$.

4.9.33 $\int \frac{4x^4 - 6x^2}{x}\,dx = \int\left(\frac{4x^4}{x} - \frac{6x^2}{x}\right)dx = \int(4x^3 - 6x)\,dx = x^4 - 3x^2 + C$. Check: $\frac{d}{dx}\left(x^4 - 3x^2 + C\right) = 4x^3 - 6x$.

4.9.35 Using Table 4.9 (formulas 1 and 2), $\int(\sin 2y + \cos 3y)\,dy = -\frac{1}{2}\cos 2y + \frac{1}{3}\sin 3y + C$. Check: $\frac{d}{dy}(-\frac{1}{2}\cos 2y + \frac{1}{3}\sin 3y + C) = \sin 2y + \cos 3y$.

4.9.37 Using Table 4.9 (formula 3), $\int(\sec^2 x - 1)\,dx = \tan x - x + C$. Check: $\frac{d}{dx}(\tan x - x + C) = \sec^2 x - 1$.

4.9.39 Using Table 4.9 (formulas 3 and 5), $\int(\sec^2\theta + \sec\theta\tan\theta)\,d\theta = \tan\theta + \sec\theta + C$. Check: $\frac{d}{d\theta}(\tan\theta + \sec\theta + C) = \sec^2\theta + \sec\theta\tan\theta$.

4.9.41 $\int(3t^2 + \sec^2 2t)\,dt = t^3 + (1/2)\tan 2t + C$.

4.9.43 $\int \sec 4\theta\tan 4\theta\,d\theta = \frac{1}{4}\sec 4\theta + C$. Check: $\frac{d}{d\theta}(\frac{1}{4}\sec 4\theta + C) = \frac{1}{4}(4\sec 4\theta\tan 4\theta) = \sec 4\theta\tan 4\theta$.

4.9.45 We have $F(x) = \int(x^5 - 2x^{-2} + 1)\,dx = \frac{x^6}{6} + 2x^{-1} + x + C$; substituting $F(1) = 0$ gives $\frac{1}{6} + 2 + 1 + C = 0$, so $C = -\frac{19}{6}$, and thus $F(x) = \frac{x^6}{6} + \frac{2}{x} + x - \frac{19}{6}$.

4.9.47 We have $F(v) = \int \sec v\tan v\,dv = \sec v + C$; substituting $F(0) = 2$ gives $\sec 0 + C = 1 + C = 2$, so $C = 1$, and thus $F(v) = \sec v + 1$.

4.9.49 We have $F(x) = \int(8x^3 - 2x^{-2})\,dx = 2x^4 + 2x^{-1} + C$; substituting $F(1) = 5$ gives $2 + 2 + C = 5$, so $C = 1$, and thus $F(x) = 2x^4 + 2x^{-1} + 1$.

4.9.51 We have $f(x) = \int(2x - 3)\,dx = x^2 - 3x + C$; substituting $f(0) = 4$ gives $C = 4$, so $f(x) = x^2 - 3x + 4$.

4.9.53 We have $g(x) = \int 7x\left(x^6 - \frac{1}{7}\right)dx = \int(7x^7 - x)\,dx = \frac{7}{8}x^8 - \frac{x^2}{2} + C$; substituting $g(1) = 2$ gives $\frac{7}{8} - \frac{1}{2} + C = 2$, so $C = \frac{13}{8}$, and thus $g(x) = \frac{7}{8}x^8 - \frac{x^2}{2} + \frac{13}{8}$.

4.9.55 We have $f(u) = \int 4(\cos u - \sin 2u)\,du = 4(\sin u + \frac{1}{2}\cos 2u) + C = 4\sin u + 2\cos 2u + C$; substituting $f(\pi/6) = 0$ gives $4\sin\frac{\pi}{6} + 2\cos\frac{\pi}{3} + C = 2 + 1 + C = 0$, so $C = -3$, and thus $f(u) = 4\sin u + 2\cos 2u - 3$.

4.9.57 We have $f(x) = \int (2x - 5)\, dx = x^2 - 5x + C$; substituting $f(0) = 4$ gives $C = 4$, so $f(x) = x^2 - 5x + 4$.

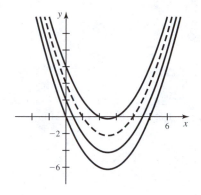

4.9.59 We have $f(x) = \int (3x + \sin \pi x)\, dx = \frac{3x^2}{2} - \frac{\cos \pi x}{\pi} + C$; substituting $f(2) = 3$ gives $6 - \frac{\cos 2\pi}{\pi} + C = 6 - \frac{1}{\pi} + C = 3$, so $C = \frac{1}{\pi} - 3$, and thus $f(x) = \frac{3x^2}{2} - \frac{\cos \pi x}{\pi} + \frac{1-3\pi}{\pi}$.

4.9.61 We have $f(t) = \int \frac{1}{t^2}\, dt = -\frac{1}{t} + C$. $f(1) = 4$ gives $4 = -1 + C$ so $C = 5$. Thus $f(t) = 5 - \frac{1}{t}$.

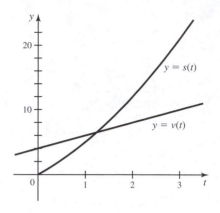

4.9.63   We have $s(t) = \int (2t + 4)\,dt = t^2 + 4t + C$; substituting $s(0) = 0$ gives $C = 0$, so $s(t) = t^2 + 4t$.

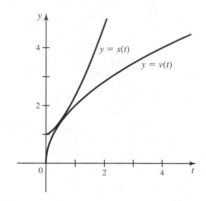

4.9.65   We have $s(t) = \int 2\sqrt{t}\,dt = 2 \cdot \frac{2}{3}t^{3/2} + C = \frac{4}{3}t^{3/2} + C$; substituting $s(0) = 1$ gives $C = 1$, so $s(t) = \frac{4}{3}t^{3/2} + 1$.

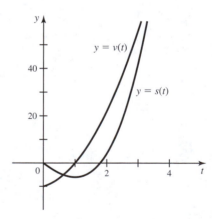

4.9.67   We have $s(t) = \int (6t^2 + 4t - 10)\,dt = 2t^3 + 2t^2 - 10t + C$; substituting $s(0) = 0$ gives $C = 0$, so $s(t) = 2t^3 + 2t^2 - 10t$.

4.9.69 $v(t) = \int a(t)\,dt = \int -32\,dt = -32t + C_1$. Because $v(0) = 20$, we have $0 + C_1 = 20$, so $v(t) = -32t + 20$.
  $s(t) = \int v(t)\,dt = \int (-32t + 20)\,dt = -16t^2 + 20t + C_2$. Because $s(0) = 0$, we have $C_2 = 0$, and thus $s(t) = -16t^2 + 20t$.

4.9.71 $v(t) = \int a(t)\,dt = \int .2t\,dt = .1t^2 + C_1$. Because $v(0) = 0$, we have $C_1 = 0$.
  $s(t) = \int v(t)\,dt = \int .1t^2\,dt = \frac{1}{30}t^3 + C_2$. Because $s(0) = 1$, we have $C_2 = 1$, and thus $s(t) = \frac{1}{30}t^3 + 1$.

4.9.73 Runner A has position function $s(y) = \int \sin t\, dt = -\cos t + C$; the initial condition $s(0) = 0$ gives $C = 1$, so $s(t) = 1 - \cos t$. Runner B has position function $S(t) = \int \cos t\, dt = \sin t + C$; the initial condition $S(0) = 0$ gives $C = 0$, so $S(t) = \sin t$. The smallest $t > 0$ where $s(t) = S(t)$ is $t = \pi/2$ s.

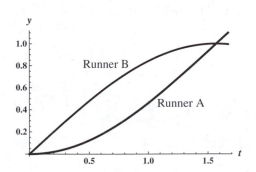

4.9.75

a. We have $v(t) = -9.8t + v_0$ and $v_0 = 30$, so $v(t) = -9.8t + 30$.

b. The height of the softball above ground is given by $s(t) = \int (-9.8t + 30)\, dt = -4.9t^2 + 30t + s_0 = -4.9t^2 + 30t$.

c. The ball reaches its maximum height when $v(t) = -9.8t + 30 = 0$, which gives $t = 30/9.8 \approx 3.06$ s; the maximum height is $s(30/9.8) \approx 45.92$ m.

d. The ball strikes the ground when $s(t) = 0$ (and $t > 0$), which gives $t(30 - 4.9t) = 0$, so $t = 30/4.9 \approx 6.12$ s.

4.9.77

a. We have $v(t) = -9.8t + v_0$ and $v_0 = 10$, so $v(t) = -9.8t + 10$.

b. The height of the payload above ground is given by $s(t) = \int (-9.8t + 10)\, dt = -4.9t^2 + 10t + s_0 = -4.9t^2 + 10t + 400$.

c. The payload reaches its maximum height when $v(t) = -9.8t + 10 = 0$, which gives $t = 10/9.8 \approx 1.02$ s; the maximum height is $s(10/9.8) \approx 405.10$ m.

d. The payload strikes the ground when $s(t) = 0$ (and $t > 0$), which gives $-4.9t^2 + 10t + 400 = 0$, so $t \approx 10.11$ s.

4.9.79

a. True. This is because $F'(x) = G'(x)$.

b. False. f is the derivative of F.

c. True. $\int f(x)\, dx$ is the most general antiderivative of $f(x)$, which is $F(x) + C$.

d. False. A function cannot have more than one derivative.

e. False. One can only conclude that $F(x)$ and $G(x)$ differ by a constant.

4.9.81 $\int \frac{\sqrt{2x} + \sqrt[3]{8x}}{x}\, dx = \int \sqrt{2}x^{-1/2} + 2x^{-2/3}\, dx = 2\sqrt{2x} + 6\sqrt[3]{x} + C$. Check: $\frac{d}{dx}(2\sqrt{2x} + 6\sqrt[3]{x} + C) = \frac{d}{dx}(2\sqrt{2}x^{1/2} + 6x^{1/3} + C) = \sqrt{2}x^{-1/2} + 2x^{-2/3} = \frac{\sqrt{2x} + \sqrt[3]{8x}}{x}$.

4.9.83 $\int (\csc^2 \theta + 2\theta^2 - 3\theta)\, d\theta = -\cot \theta + \frac{2}{3}\theta^3 - \frac{3}{2}\theta^2 + C$. Check: $\frac{d}{d\theta}(-\cot \theta + \frac{2}{3}\theta^3 - \frac{3}{2}\theta^2 + C) = \csc^2 \theta + 2\theta^2 - 3\theta$.

4.9.85 $\int \frac{1 + \sqrt{x}}{x^2}\, dx = \int (x^{-2} + x^{-3/2})\, dx = -x^{-1} - 2x^{-1/2} + C$. Check: $\frac{d}{dx}(-x^{-1} - 2x^{-1/2} + C) = x^{-2} + x^{-3/2} = \frac{1 + \sqrt{x}}{x^2}$.

4.9.87 $\int \sqrt{x}(2x^6 - 4\sqrt[3]{x})\, dx = \int (2x^{13/2} - 4x^{5/6})\, dx = 2 \cdot \frac{2}{15}x^{15/2} - 4 \cdot \frac{6}{11}x^{11/6} + C = \frac{4}{15}x^{15/2} - \frac{24}{11}x^{11/6} + C.$
Check: $\frac{d}{dx}\left(\frac{4}{15}x^{15/2} - \frac{24}{11}x^{11/6} + C\right) = 2x^{13/2} - 4x^{5/6} = \sqrt{x}(2x^6 - 4\sqrt[3]{x}).$

4.9.89 We have $F'(x) = \int \cos x\, dx = \sin x + C;$ $F'(0) = 3$ so $C = 3.$ Then $F(x) = \int (\sin x + 3)\, dx = -\cos x + 3x + C;$ $F(\pi) = 4$ gives $1 + 3\pi + C = 4$ so $C = 3 - 3\pi$ and $F(x) = -\cos x + 3x + 3 - 3\pi.$

4.9.91 We have $F''(x) = \int (672x^5 + 24x)\, dx = 672 \cdot \frac{x^6}{6} + 24 \cdot \frac{x^2}{2} + C;$ $F''(0) = 0$ so $C = 0.$ Next $F'(x) = \int (112x^6 + 12x^2)\, dx = 112 \cdot \frac{x^7}{7} + 12 \cdot \frac{x^3}{3} + C;$ $F'(0) = 2$ so $C = 2.$ Finally $F(x) = \int (16x^7 + 4x^3 + 2)\, dx = 16 \cdot \frac{x^8}{8} + 4 \cdot \frac{x^4}{4} + 2x + C;$ $F(0) = 1$ so $C = 1$ and $F(x) = 2x^8 + x^4 + 2x + 1.$

4.9.93

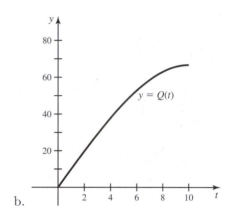

a. We have $Q(t) = \int 0.1(100 - t^2)\, dt = 0.1\left(100t - \frac{t^3}{3}\right) + C;$ $Q(0) = 0,$ so $C = 0$ and $Q(t) = 10t - t^3/30$ gal.

b.

c. $Q(10) = 200/3 \approx 67$ gal.

4.9.95 $\int \sin^2 x\, dx = \frac{1}{2}\int (1 - \cos 2x)\, dx = \frac{1}{2}\left(x - \frac{\sin 2x}{2}\right) + C = \frac{x}{2} - \frac{\sin 2x}{4} + C;$ $\int \cos^2 x\, dx = \frac{1}{2}\int (1 + \cos 2x)\, dx = \frac{1}{2}\left(x + \frac{\sin 2x}{2}\right) + C = \frac{x}{2} + \frac{\sin 2x}{4} + C.$

4.9.97 Check that $\frac{d}{dx}(\sqrt{x^2 + 1}) = \frac{2x}{2\sqrt{x^2+1}} = \frac{x}{\sqrt{x^2+1}}.$

4.9.99 Check that $\frac{d}{dx}\left(-\frac{1}{2(x^2-1)}\right) = -\frac{1}{2}\frac{d}{dx}(x^2 - 1)^{-1} = -\frac{1}{2}(-1)(x^2 - 1)^{-2}(2x) = \frac{x}{(x^2-1)^2}.$

Chapter Four Review

1

a. False. The point $(c, f(c))$ is a critical point for f, but is not necessarily a local maximum or minimum. Example: $f(x) = x^3$ at $c = 0.$

b. False. The fact that $f''(c) = 0$ does not necessarily imply that f changes concavity at c. Example: $f(x) = x^4$ at $c = 0.$

c. True. Both are antiderivatives of $2x.$

d. True. The function has a maximum on the closed interval determined by the two local minima, and the only way the maximum can occur at the endpoints is if the function is constant, in which case every point is a local max and min.

e. True. The slope of the linearization is given by $f'(0) = \cos(0) = 1,$ and the line has y-intercept $(0, f(0)) = (0, 0).$

f. False. For example, $\lim_{x \to \infty} x^2 = \infty$ and $\lim_{x \to \infty} x = \infty,$ but $\lim_{x \to \infty}(x^2 - x) = \infty.$

3   **5**

                    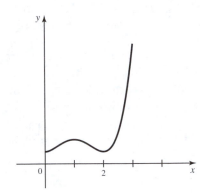

7 The critical points satisfy $f'(x) = 6x^2 - 6x - 36 = 6(x-3)(x+2) = 0$, so the critical points are $x = 3, -2$. Because $\lim\limits_{x \to \infty} f(x) = \infty$ and $\lim\limits_{x \to -\infty} f(x) = -\infty$, this function has no absolute max or min on $(-\infty, \infty)$.

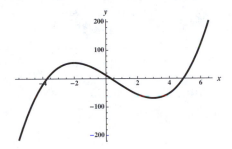

9 $f'(x) = \frac{(x+1)(2x) - (x^2+8)(1)}{(x+1)^2} = \frac{x^2+2x-8}{(x+1)^2}$. The critical points are the points where $f'(x) = 0$ or $f'(x)$ is undefined; this happens when $x^2 + 2x - 8 = (x+4)(x-2) = 0$ or when $(x+1)^2 = 0$, so the critical points are at $x = -4, 2$ and at $x = -1$. On the given interval, $[-5, 5]$, the function is unbounded since it has a vertical asymptote at $x = -1$. Because $\lim_{x \to -1^-} f(x) = -\infty$ and $\lim_{x \to -1^+} f(x) = \infty$, it has no absolute minimum or maximum.

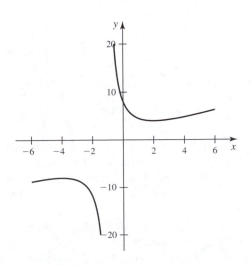

11

All points in the interval $[-3, 2]$ are critical points. The absolute max occurs at $(4, 9)$; there are no local maxima. All points $(x, 5)$ for x in the interval $[-3, 2]$ are absolute and local minima.

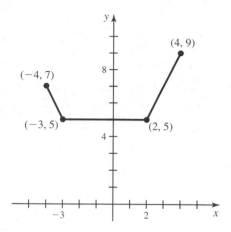

13 The derivatives of f are $f'(x) = 2x^3 - 6x + 4$, $\quad f''(x) = 6x^2 - 6$. Observe that $f'(x) = 2(x-1)^2(x+2)$, so we have critical points $x = 1, -2$. Solving $f''(x) = 6(x^2 - 1) = 0$ gives possible inflection points at $x = \pm 1$. Testing the sign of $f'(x)$ shows that f is decreasing on the interval $(-\infty, -2)$ and is increasing on $(-2, \infty)$. The First Derivative Test shows that a local minimum occurs at $x = -2$, and that the critical point $x = 1$ is neither a local max or min.

Testing the sign of $f''(x)$ shows that f is concave down on the interval $(-1, 1)$ and is concave up on the intervals $(-\infty, -1)$ and $(1, \infty)$. Therefore inflection points occur at $x = \pm 1$. Using a numerical solver, we see that the graph has x-intercepts at $x \approx -2.917, -0.215$. We also observe that $\lim\limits_{x \to \pm\infty} f(x) = \infty$, so f has no absolute maximum and an absolute minimum at $x = -2$.

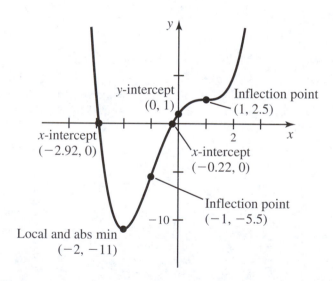

15 The derivatives of f are $f'(x) = -4\pi \sin[\pi(x-1)]$, $\quad f''(x) = -4\pi^2 \cos[\pi(x-1)]$. Solving $f'(x) = 0$ gives critical point $x = 1$, and solving $f''(x) = 0$ gives possible inflection points at $x = 1/2, 3/2$. Testing the sign of $f'(x)$ shows that f is decreasing on the interval $(1, 2)$ and increasing on $(0, 1)$. The First Derivative Test shows that a local maximum occurs at $x = 1$. By Theorem 4.5, this solitary local maximum must be the absolute maximum for f on the interval $[0, 2]$.

Testing the sign of $f''(x)$ shows that f is concave down on the interval $(1/2, 3/2)$ and concave up on the intervals $(0, 1/2)$ and $(3/2, 2)$. Therefore inflection points occur at $x = 1/2, 3/2$. These points are also the x-intercepts of the graph. We also observe that $f(0) = f(2) = -4$, so f takes its absolute minimum at these points.

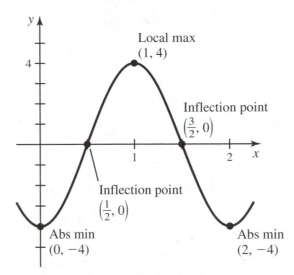

17 Note that the domain of f is the interval $(0, \infty)$. The derivatives of f are $f'(x) = \frac{1}{3}x^{-2/3} - \frac{1}{2}x^{-1/2}$, and $f''(x) = -\frac{2}{9}x^{-5/3} + \frac{1}{4}x^{-3/2}$. Solving $f'(x) = 0$ gives critical point $x = (2/3)^6$, and solving $f''(x) = 0$ gives a possible inflection point at $x = (8/9)^6$. Testing the sign of $f'(x)$ shows that f is increasing on the interval $(0, (2/3)^6)$ and decreasing on the interval $((2/3)^6, \infty)$. The First Derivative Test shows that a local maximum occurs at $x = (2/3)^6$ and By Theorem 4.5 this solitary local maximum must be the absolute maximum of f on the interval $(0, \infty)$.

Testing the sign of $f''(x)$ shows that f is concave down on the interval $(0, (8/9)^6)$ and concave up on the interval $((8/9)^6, \infty)$. Therefore an inflection point occurs at $x = (8/9)^6$. Using a numerical solver, we find that the graph has x-intercept at $x \approx 23.767$. Because $\lim_{x \to \infty} f(x) = -\infty$, f has no absolute minimum on the interval $(0, \infty)$.

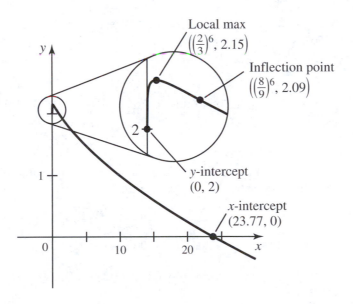

19 The derivatives of f are $f'(x) = \frac{2}{3}x^{-1/3} + \frac{1}{3}(x + 2)^{-2/3}$, $f''(x) = -\frac{2}{9}\left(x^{-4/3} + (x + 2)^{-5/3}\right)$. Solving $f'(x) = 0$ gives critical points $x \approx -2.57, -1.559$; we also have critical points at $x = -2, 0$ because $f'(x)$ is undefined at these points. Solving $f''(x) = 0$ numerically gives a possible inflection point at $x \approx -6.434$. We also have possible inflection points at $x = -2, 0$ because $f''(x)$ is undefined at these points. Testing the sign of $f'(x)$ shows that f is decreasing on the intervals $(-\infty, -2.566)$ and $(-1.559, 0)$ and increasing on $(-2.566, -1.559)$ and $(0, \infty)$. The First Derivative Test shows that local mins occur at $x \approx -2.566$ and $x = 0$ and a local max occurs at $x \approx -1.559$.

Testing the sign of $f''(x)$ shows that f is concave down on the intervals $(-\infty, -6.434)$, $(-2, 0)$ and $(0, \infty)$ and concave up on the interval $(-6.434, -2)$. Therefore inflection points occur at $x \approx -6.434$ and $x = -2$. Because $\lim\limits_{x \to \pm\infty} f(x) = \infty$, f has no absolute maximum. The absolute minimum occurs at $x \approx -2.566$.

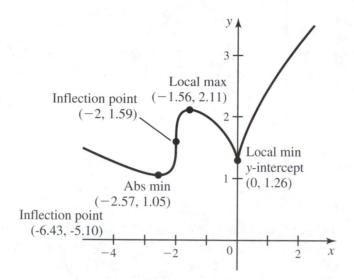

21 The objective function to be maximized is the volume of the cone, given by $V = \pi r^2 h/3$. By the Pythagorean theorem, r and h satisfy the constraint $h^2 + r^2 = 16$, which gives $r^2 = 16 - h^2$. Therefore $V(h) = \frac{\pi}{3}h(16 - h^2) = \frac{\pi}{3}(16h - h^3)$. We must maximize this function for $0 \le h \le 4$. The critical points of $V(h)$ satisfy $V'(h) = \frac{\pi}{3}(16 - 3h^2) = 0$, which has unique solution $h = 4/\sqrt{3} = 4\sqrt{3}/3$ in $(0, 4)$. Because $V(0) = V(4) = 0$, $h = 4\sqrt{3}/3$ gives the maximum value of $V(h)$ on $[0, 4]$. The corresponding value of r satisfies $r^2 = 16 - \frac{16}{3} = \frac{32}{3}$, so $r = \frac{4\sqrt{2}}{\sqrt{3}} = \frac{4\sqrt{6}}{3}$.

23 We have that $xy = 98$, and we want to maximize $p = (y-2)(x-1) = (y-2)(98/y-1) = 98 - y - 196/y + 2$. Note that $p'(y) = -1 + 196/y^2$, which is zero when $y^2 = 196$, so $y = \sqrt{196} = 14$. Also note that $p'(13) > 0$ and $p'(15) < 0$, so there is a local (in fact, absolute) maximum at $y = 14$. The value of x when $y = 14$ is $x = 98/14 = 7$.

25 The area of the triangle is $\frac{1}{2}pq$, and the constraint is $\sqrt{p^2 + q^2} = 10$, or $p^2 + q^2 = 100$. So we can write the area of the triangle as $A(p) = (1/2)p\sqrt{100 - p^2}$. We have $A'(p) = (1/2)\sqrt{100 - p^2} + \frac{1}{2}p \cdot \frac{-p}{\sqrt{100-p^2}} = \frac{100-p^2-p^2}{2\sqrt{100-p^2}} = \frac{100-2p^2}{2\sqrt{100-p^2}}$. This is zero for $p = \sqrt{50} = 5\sqrt{2}$. An application of the First Derivative Test shows that there is a local (in fact, absolute) maximum at this value of p. The value of q for this value of p is $\sqrt{100 - 50} = 5\sqrt{2}$ as well. So the area of the triangle is maximized when $p = q = 5\sqrt{2}$.

27

a. $f'(x) = \frac{2}{3}x^{-1/3}$, so $f'(27) = \frac{2}{3\sqrt[3]{27}} = \frac{2}{9}$. Thus $L(x) - 9 = \frac{2}{9}(x - 27)$, or $L(x) = 9 + \frac{2}{9}(x - 27) = \frac{2}{9}x + 3$.

b. $f(29) \approx L(29) = 9 + \frac{4}{9} \approx 9.44$. This is an overestimate, because $f''(27) < 0$. Note that the calculator value of $f(29)$ is about 9.43913.

29 Let $f(x) = 1/x^2$ and let $a = 4$. Then $f'(x) = \frac{-2}{x^3}$ so $f'(4) = \frac{-2}{64} = \frac{-1}{32}$. The linearization is $L(x) = \frac{1}{16} + (-1/32)(x - 4)$. Then $f(4.2) = 1/(4.2)^2 \approx L(4.2) = \frac{1}{16} - \frac{1}{32} \cdot \frac{2}{10} = .05625$.

31 $\Delta h \approx h'(a)\Delta t = -32 \cdot 5 \cdot 0.7 = -112$ feet.

33 It is possible to note that 1 is a root by inspection. Then by long division by $x - 1$, we have $f(x) = (x - 1)(3x^2 - x - 1)$. We apply Newton's method to the function $g(x) = 3x^2 - x - 1$.

The Newton's method recursion is given by $x_{n+1} = x_n - \frac{3x_n^2 - x_n - 1}{6x_n - 1}$. Applying this recursion to the initial estimates of $-.5$ and $.8$ yields:

n	x_n
0	$-.5$
1	$-.4375$
2	$-.434267$
3	$-.434259$
4	$-.434259$

n	x_n
0	$.8$
1	$.768421$
2	$.767592$
3	$.767592$
4	$.767592$

The roots are thus 1, and approximately -.434259 and .767592.

35 First note that $f'(x) = 10x^4 - 18x^2 - 4$ and $f''(x) = 40x^3 - 36x = 4x(10x^2 - 9)$. This is clearly 0 when $x = 0$, and when $10x^2 - 9 = 0$. Applying Newton's method to the function $g(x) = 10x^2 - 9$ with initial estimates of -1 and 1 yields:

n	x_n
0	-1
1	$-.95$
2	$-.948684$
3	$-.948683$
4	$-.948683$
5	$-.948683$

n	x_n
0	1
1	$.95$
2	$.948684$
3	$.948683$
4	$.948683$
5	$.948683$

Checking the signs of $f''(x)$ on the appropriate intervals leads to the conclusion that these are all the locations of inflection points of f. So the inflection points of f are located at 0 and approximately $\pm.948683$.

37 L'Hôpital's rule gives $\lim\limits_{t \to 0} \dfrac{1 - \cos 6t}{2t} = \lim\limits_{t \to 0} \dfrac{6 \sin 6t}{2} = 0$.

39 Observe that $\lim\limits_{\theta \to 0} \dfrac{3 \sin^2 2\theta}{\theta^2} = 3 \left(\lim\limits_{\theta \to 0} \dfrac{\sin 2\theta}{\theta} \right)^2$; l'Hôpital's rule gives $\lim\limits_{\theta \to 0} \dfrac{\sin 2\theta}{\theta} = \lim\limits_{\theta \to 0} \dfrac{2 \cos 2\theta}{1} = 2$, so $\lim\limits_{\theta \to 0} \dfrac{3 \sin^2 2\theta}{\theta^2} = 3 \cdot 2^2 = 12$.

41 Observe that $2\theta \cot 3\theta = 2 \cos 3\theta \cdot \dfrac{\theta}{\sin 3\theta}$; $\lim\limits_{\theta \to 0} \dfrac{\theta}{\sin 3\theta} = \lim\limits_{\theta \to 0} \dfrac{1}{3 \cos 3\theta} = \dfrac{1}{3}$ by l'Hôpital's rule, so

$$\lim_{\theta \to 0} 2\theta \cot 3\theta = 2 \cdot 1 \cdot \frac{1}{3} = \frac{2}{3}.$$

43 Apply l'Hôpital's rule twice:

$$\lim_{x \to 1} \frac{x^4 - x^3 - 3x^2 + 5x - 2}{x^3 + x^2 - 5x + 3} = \lim_{x \to 1} \frac{4x^3 - 3x^2 - 6x + 5}{3x^2 + 2x - 5} = \lim_{x \to 1} \frac{12x^2 - 6x - 6}{6x + 2} = 0.$$

45 $\int (2x + 1)^2 \, dx = \int (4x^2 + 4x + 1) \, dx = \frac{4}{3}x^3 + 2x^2 + x + C$.

47 $\int \left(\frac{1}{x^2} - \frac{2}{x^{5/2}} \right) dx = \int (x^{-2} - 2x^{-5/2}) \, dx = -x^{-1} - 2 \cdot \left(-\frac{2}{3} \right) x^{-3/2} = -\frac{1}{x} + \frac{4}{3}x^{-3/2} + C$.

49 Using Table 4.9 (formula 1), $\int (1 + \cos 3\theta) \, d\theta = \theta + \frac{\sin 3\theta}{3} + C$.

51 Using Table 4.9 (formula 5), $\int \sec 2x \tan 2x \, dx = \frac{1}{2} \sec 2x + C$.

53 $\int \left(\sqrt[4]{x^3} + \sqrt{x^5} \right) dx = \int (x^{3/4} + x^{5/2}) \, dx = \frac{4}{7} x^{7/4} + \frac{2}{7} x^{7/2} + C.$

55 We have $f(t) = \int (\sin t + 2t) \, dt = -\cos t + t^2 + C;$ $f(0) = -1 + C = 5,$ so $C = 6$ and $f(t) = -\cos t + t^2 + 6.$

57 Using the identity $\sin^2 x = (1 - \cos 2x)/2$ and Table 4.9 (formula 1), we see that $h(x) = \frac{1}{2} \int (1 - \cos 2x) \, dx = \frac{1}{2} \left(x - \frac{\sin 2x}{2} \right) + C = \frac{x}{2} - \frac{\sin 2x}{4} + C.$ We have $h(1) = \frac{1}{2} - \frac{\sin 2}{4} + C = 1,$ so $C = \frac{1}{2} + \frac{\sin 2}{4}$ and so $h(x) = \frac{x}{2} - \frac{\sin 2x}{4} + \frac{1}{2} + \frac{\sin 2}{4}.$

59 The velocity of the rocket is given by $v(t) = -9.8t + v_0 = -9.8t + 120,$ and the position function of the rocket is $s(t) = \int (-9.8t + 120) \, dt = -4.9t^2 + 120t + s_0 = -4.9t^2 + 120t + 125.$ The rocket reaches its maximum height when $v(t) = 0,$ which occurs at $t = 120/9.8 \approx 12.24$ s; the maximum height is $s(120/9.8) \approx 859.69$ m. The rocket hits the ground when $s(t) = 0;$ solving this quadratic equation gives $t \approx 25.49$ s.

61 Observe that $\lim\limits_{x \to \infty} \dfrac{2x^5 - x + 1}{5x^6 + x} = 0$ by Theorem 2.7. We can also use l'Hôpital's rule:

$$\lim_{x \to \infty} \frac{2x^5 - x + 1}{5x^6 + x} = \lim_{x \to \infty} \frac{10x^4 - 1}{30x^5 + 1} = \lim_{x \to \infty} \frac{40x^3}{150x^4} = \lim_{x \to \infty} \frac{4}{15x} = 0.$$

63 We have $\lim\limits_{x \to 0} \dfrac{\sin 3x}{5x} = \dfrac{3}{5} \lim\limits_{x \to 0} \dfrac{\sin 3x}{3x} = \dfrac{3}{5} \cdot 1 = \dfrac{3}{5}.$ We can also apply l'Hôpital's rule: $\lim\limits_{x \to 0} \dfrac{\sin 3x}{5x} = \lim\limits_{x \to 0} \dfrac{3 \cos 3x}{5} = \dfrac{3}{5}.$

Chapter 5

Integration

5.1 Approximating Areas under Curves

5.1.1

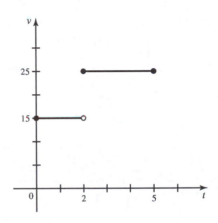

In the first 2 seconds, the object moves $15 \cdot 2 = 30$ meters. In the next three seconds, the object moves $25 \cdot 3 = 75$ meters, so the total displacement is $75 + 30 = 105$ meters.

5.1.3 Subdivide the interval from 0 to $\pi/2$ into subintervals. On each subinterval, pick a sample point (like the left endpoint, or the right endpoint, or the midpoint, for example) and call the first sample point x_1 and the second sample point x_2 and so on. For each sample point x_i, calculate the area of the rectangle which lies over the subinterval and has height $f(x_i) = \cos x_i$. Do this for each subinterval, and add the areas of the corresponding rectangles together. This will give an approximation to the area under the curve, with generally a better approximation occurring as n increases – where n is the number of subintervals used.

5.1.5 Because the interval $[1, 3]$ has length 2, if we subdivide it into 4 subintervals, each will have length $\Delta x = \frac{2}{4} = \frac{1}{2}$. The grid points will be $x_0 = 1$, $x_1 = 1 + \Delta x = 1.5$, $x_2 = 1 + 2\Delta x = 2$, $x_3 = 1 + 3\Delta x = 2.5$, and $x_4 = 1 + 4\Delta x = 3$.

If we use the left-hand side of each subinterval, we will use 1, 1.5, 2, and 2.5.

If we use the right-hand side of each subinterval, we will use 1.5, 2, 2.5, and 3.

If we use the midpoint of each subinerval, we will use 1.25, 1.75, 2.25, and 2.75.

5.1.7 It is an underestimate. If we use the right-hand side of each subinterval to determine the height of the rectangles, the height of each rectangle will be the minimum of f over the subinterval, so the sum of the areas of the rectangles will be less than the corresponding area under the curve.

5.1.9

a. On the first subinterval, the midpoint is .5, and $v(.5) = 1.75$. On the 2nd subinterval, the midpoint is 1.5 and $v(1.5) = 7.75$. Continuing in this manner, we obtain the estimate to the displacement of

$$v(.5) \cdot 1 + v(1.5) \cdot 1 + v(2.5) \cdot 1 + v(3.5) \cdot 1 = 1.75 + 7.75 + 19.75 + 37.75 = 67.$$

b. This time the midpoints are at .25, .75, 1.25 .... Each subinterval has length $\frac{1}{2}$. Thus, the estimate is given by

$$v(.25) \cdot .5 + v(.75) \cdot .5 + v(1.25) \cdot .5 + v(1.75) \cdot .5 + v(2.25) \cdot .5 + v(2.75) \cdot .5 + v(3.25) \cdot .5 + v(3.75) \cdot .5$$
$$= .5(1.1875 + 2.6875 + 5.6875 + 10.1875 + 16.1875 + 23.6875 + 32.6875 + 43.1875)$$
$$= .5(135.5) = 67.75.$$

5.1.11

The left-hand grid points are 0 and 4. The length of each subinterval is $8/2 = 4$. So the left Riemann sum is given by $v(0) \cdot 4 + v(4) \cdot 4 = 4 \cdot (1 + 9) = 40$.

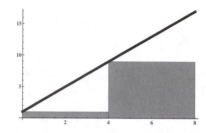

5.1.13

The left-hand grid points are 0, 2, 4, and 6. The length of each subinterval is 2. So the left Riemann sum is given by $v(0) \cdot 2 + v(2) \cdot 2 + v(4) \cdot 2 + v(6) \cdot 2 = 2 \cdot \left(\frac{1}{1} + \frac{1}{5} + \frac{1}{9} + \frac{1}{13}\right) \approx 2.776$.

5.1.15

The left-hand grid points are 0, 3, 6, 9, and 12. The length of each subinterval is 3. So the left Riemann sum is given by $\displaystyle\sum_{k=1}^{5} v(3(k-1)) \cdot 3 = 12 + 24 + 12\sqrt{7} + 12\sqrt{10} + 12\sqrt{13} \approx 148.963$.

5.1.17 The left Riemann sum is given by $f(1) + f(2) + f(3) + f(4) + f(5) = 2 + 3 + 4 + 5 + 6 = 20$.

The right Riemann sum is given by $f(2) + f(3) + f(4) + f(5) + f(6) = 3 + 4 + 5 + 6 + 7 = 25$.

5.1.19

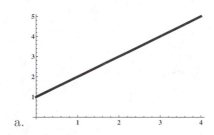

b. We have $\Delta x = \frac{4-0}{4} = 1$. The grid points are $x_0 = 0$, $x_1 = 1$, $x_2 = 2$, $x_3 = 3$, and $x_4 = 4$.

a.

c.

d. The left Riemann sum is $1 \cdot 1 + 2 \cdot 1 + 3 \cdot 1 + 4 \cdot 1 = 10$, which is an underestimate of the area under the curve. The right Riemann sum is $2 \cdot 1 + 3 \cdot 1 + 4 \cdot 1 + 5 \cdot 1 = 14$ which is an overestimate of the area under the curve.

5.1.21

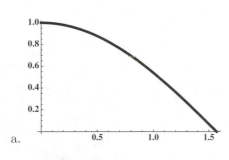

b. We have $\Delta x = \frac{\pi/2 - 0}{4} = \frac{\pi}{8}$. The grid points are $x_0 = 0$, $x_1 = \frac{\pi}{8}$, $x_2 = \frac{\pi}{4}$, $x_3 = \frac{3\pi}{8}$, and $x_4 = \frac{\pi}{2}$.

a.

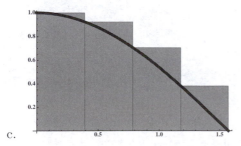

c.

d. The left Riemann sum is $1 \cdot \frac{\pi}{8} + \cos(\pi/8) \cdot \frac{\pi}{8} + \cos(\pi/4) \cdot \frac{\pi}{8} + \cos(3\pi/8) \cdot \frac{\pi}{8} \approx 1.185$, which is an overestimate of the area under the curve. The right Riemann sum is $\cos(\pi/8) \cdot \frac{\pi}{8} + \cos(\pi/4) \cdot \frac{\pi}{8} + \cos(3\pi/8) \cdot \frac{\pi}{8} + 0 \cdot \frac{\pi}{8} \approx 0.791$ which is an underestimate of the area under the curve.

5.1.23

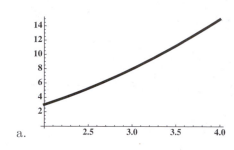

b. We have $\Delta x = \frac{4-2}{4} = \frac{1}{2}$. The grid points are $x_0 = 2$, $x_1 = 2.5$, $x_2 = 3$, $x_3 = 3.5$, and $x_4 = 4$.

a.

c.

d. The left Riemann sum is $(3 + 5.25 + 8 + 11.25) \cdot 0.5 = 13.75$, which is an underestimate of the area under the curve. The right Riemann sum is $(5.25 + 8 + 11.25 + 15) \cdot 0.5 = 19.75$ which is an overestimate of the area under the curve.

5.1.25

b. We have $\Delta x = \frac{3-0}{6} = \frac{1}{2}$. The grid points are $x_0 = 0$, $x_1 = \frac{1}{2}$, $x_2 = 1$, $x_3 = \frac{3}{2}$, $x_4 = 2$, $x_5 = \frac{5}{2}$, and $x_6 = 3$.

a.

c.

d. The left Riemann sum is $\frac{1}{2}\left(0 + \sqrt{1/2} + \sqrt{1} + \sqrt{3/2} + \sqrt{2} + \sqrt{5/2}\right) \approx 2.9636$, which is an underestimate of the area under the curve. The right Riemann sum is $\frac{1}{2}\left(\sqrt{1/2} + \sqrt{1} + \sqrt{3/2} + \sqrt{2} + \sqrt{5/2} + \sqrt{3}\right) \approx 3.82963$ which is an overestimate of the area under the curve.

5.1.27 We have $\Delta x = 2$, so the midpoints are 1, 3, 5, 7, and 9. So the midpoint Riemann sum is $2(f(1) + f(3) + f(5) + f(7) + f(9)) = 670$.

5.1.29

a.

c.

b. We have $\Delta x = \frac{4-0}{4} = 1$. The gridpoints are $x_0 = 0$, $x_1 = 1$, $x_2 = 2$, $x_3 = 3$, and $x_4 = 4$, so the midpoints are .5, 1.5, 2.5, and 3.5.

d. The midpoint Riemann sum is $1(2+4+6+8) = 20$.

5.1.31

a.

c.

b. We have $\Delta x = \frac{3-1}{4} = \frac{1}{2}$. So the midpoints are 1.25, 1.75, 2.25, and 2.75.

d. The midpoint Riemann sum is $.5(\sqrt{1.25} + \sqrt{1.75} + \sqrt{2.25} + \sqrt{2.75}) \approx 2.800$.

5.1.33

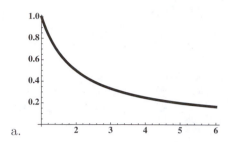

a.

b. We have $\Delta x = \frac{6-1}{5} = 1$. So the midpoints are 1.5, 2.5, 3.5, 4.5, and 5.5.

c.

d. The midpoint Riemann sum is $2/3 + 2/5 + 2/7 + 2/9 + 2/11 \approx 1.756$.

5.1.35 Note that $\Delta x = \frac{2-0}{4} = .5$. So the left Riemann sum is given by $.5(5 + 3 + 2 + 1) = 5.5$ and the right Riemann sum is given by $.5(3 + 2 + 1 + 1) = 3.5$.

5.1.37

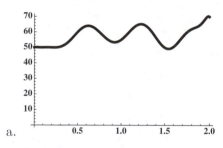

a.

b. With $n = 2$, we have $\Delta x = \frac{2-0}{2} = 1$, so the midpoints are .5 and 1.5. So the midpoint Riemann sum is $60 + 50 = 110$. For $n = 4$, we have $\Delta x = \frac{2-0}{4} = \frac{1}{2}$, so the midpoints are .25, .75, 1.25, and 1.75. The midpoint Riemann sum in this case is $0.5(50 + 60 + 65 + 60) = \frac{235}{2} = 117.5$.

5.1.39

a. $\displaystyle\sum_{k=1}^{5} k$.

b. $\displaystyle\sum_{k=1}^{6} (k + 3)$.

c. $\displaystyle\sum_{k=1}^{4} k^2$.

d. $\displaystyle\sum_{k=1}^{4} \frac{1}{k}$.

5.1.41

a. $\displaystyle\sum_{k=1}^{10} k = 1 + 2 + 3 + \ldots + 10 = 55$.

b. $\displaystyle\sum_{k=1}^{6} (2k + 1) = 3 + 5 + 7 + 9 + 11 + 13 = 48$.

c. $\displaystyle\sum_{k=1}^{4} k^2 = 1 + 4 + 9 + 16 = 30$.

d. $\displaystyle\sum_{n=1}^{5} (1 + n^2) = 2 + 5 + 10 + 17 + 26 = 60$.

e. $\displaystyle\sum_{m=1}^{3} \frac{2m + 2}{3} = \frac{4}{3} + \frac{6}{3} + \frac{8}{3} = 6$.

f. $\displaystyle\sum_{j=1}^{3} (3j - 4) = -1 + 2 + 5 = 6$.

g. $\displaystyle\sum_{p=1}^{5} (2p + p^2) = 3 + 8 + 15 + 24 + 35 = 85$.

h. $\displaystyle\sum_{n=0}^{4} \sin\frac{n\pi}{2} = 0 + 1 + 0 + (-1) + 0 = 0$.

5.1.43 Note that $\Delta x = \frac{1}{10}$, and $x_k = a + k\Delta x = \frac{k}{10}$.

a. The left Riemann sum is given by $\displaystyle\sum_{k=1}^{40} \sqrt{(k-1)/10} \cdot (1/10) \approx 5.227$.

The right Riemann sum is given by $\displaystyle\sum_{k=1}^{40} \sqrt{k/10} \cdot (1/10) \approx 5.427$.

The midpoint Riemann sum is given by $\displaystyle\sum_{k=1}^{40} \sqrt{\frac{2k-1}{20}} \cdot (1/10) \approx 5.335$.

b. It appears that the actual area is about $5 + 1/3$.

5.1.45 Note that $\Delta x = \frac{1}{15}$, and $x_k = 2 + k\Delta x = 2 + (k/15)$. So $f(x_k) = (2 + (k/15))^2 - 1$.

a. The left Riemann sum is given by $\displaystyle\sum_{k=1}^{74}[(2 + ((k-1)/15))^2 - 1] \cdot (1/15) \approx 105.170$.

The right Riemann sum is given by $\displaystyle\sum_{k=1}^{75}[(2 + (k/15))^2 - 1] \cdot (1/15) \approx 108.170$.

The midpoint Riemann sum is given by $\displaystyle\sum_{k=1}^{75}[((61/30) + ((k-1)/15))^2 - 1] \cdot (1/15) \approx 106.665$.

b. It appears that the actual area is about $106 + 2/3$.

5.1.47 The right Riemann sum is given by $A_n = \displaystyle\sum_{k=1}^{n} f(x_k)\Delta x = \sum_{k=1}^{n} \left(4 - \left(-2 + \frac{4k}{n}\right)^2\right) \cdot \frac{4}{n}$.

n	A_n
10	10.560
30	10.655
60	10.664
80	10.665

It appears that A_n is approaching $10\frac{2}{3}$.

5.1.49 The right Riemann sum is given by $A_n = \displaystyle\sum_{k=1}^{n} f(x_k)\Delta x = \sum_{k=1}^{n} \left(2 - 2\sin\left(\frac{-\pi}{2} + \frac{\pi k}{n}\right)\right) \cdot \frac{\pi}{n}$.

n	A_n
10	5.6549
30	6.0737
60	6.1785
80	6.2046

It appears that A_n is approaching 2π.

5.1.51

a. True. Because the curve is a straight line, the region under the curve and over each subinterval is a trapezoid. The formula for the area of such a trapezoid over $[x_i, x_{i+1}]$ is $\frac{f(x_i)+f(x_{i+1})}{2} \cdot \Delta x = \frac{2x_i + 5 + 2x_{i+1} + 5}{2} \cdot \Delta x = (x_i + x_{i+1} + 5)\Delta x$ and the area given by using the midpoint formula is $f\left(\frac{x_i + x_{i+1}}{2}\right)\Delta x = (x_i + x_{i+1} + 5)\Delta x$. So the area under the curve is exactly given by the midpoint Riemann sum. Note that this holds for any straight line.

b. False. The left Riemann sum will underestimate the area under an increasing function.

c. True. The value of f at the midpoint will always be between the value of f at the endpoints, if f is monotonic increasing or monotonic decreasing.

5.1.53 $\displaystyle\sum_{k=1}^{50}\left(\frac{2k}{25}+1\right)\cdot\frac{2}{25}=12.16.$

5.1.55 $\displaystyle\sum_{k=1}^{32}\left(3+\frac{1}{8}+\frac{k-1}{4}\right)^3\cdot\frac{1}{4}\approx 3639.125.$

5.1.57 This is the right Riemann sum for f on the interval $[1,5]$ for $n=4$.

5.1.59 This is the midpoint Riemann sum for f on the interval $[2,6]$ for $n=4$.

5.1.61 For all of the calculations below, we have $\Delta x=\frac{1}{2}$, and grid points $x_0=0$, $x_1=.5$, $x_2=1$, $x_3=2.5$, and $x_4=2$.

a.

The left Riemann sum is given by $\frac{1}{2}\left(f(0)+f(.5)+f(1)+f(1.5)\right)$ which is equal to $\frac{1}{2}(2+2.25+3+4.25)=5.75$.

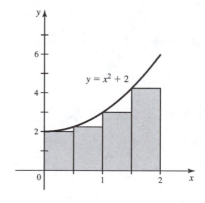

b.

The midpoint Riemann sum is given by $\frac{1}{2}\left(f(.25)+f(.75)+f(1.25)+f(1.75)\right)$ which is equal to $\frac{1}{2}(2.0625+2.5625+3.5625+5.0625)=6.625$.

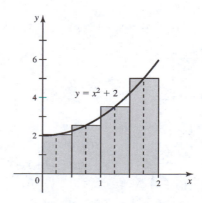

c.

The right Riemann sum is given by $\frac{1}{2}(f(.5) + f(1) + f(1.5) + f(2))$ which is equal to $\frac{1}{2}(2.25 + 3 + 4.25 + 6) = 7.75$.

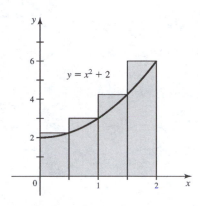

5.1.63 We have $\Delta x = \frac{7-1}{6} = 1$. The left Riemann sum is given by $1(10 + 9 + 7 + 5 + 2 + 1) = 34$ and the right Riemann sum is given by $1(9 + 7 + 5 + 2 + 1 + 0) = 24$.

5.1.65

a. During the first second, the velocity steadily increases from 0 to 20, then it remains constant until $t = 3$. From $t = 3$ until $t = 5$ it steadily decreases, and then remains constant until $t = 6$.

b. Between $t = 0$ and $t = 2$ the area under the curve is $\frac{1}{2} \cdot 1 \cdot 20 + 1 \cdot 20 = 30$.

c. Between $t = 2$ and $t = 5$ the displacement is the sum of the area of a rectangle with area 20 and a trapezoid with area 30, so the displacement is 50 meters.

d. Between $t = 0$ and $t = 5$ the displacement is 80. Between $t = 5$ and any time $t \geq 5$ the displacement is $10(t - 5)$ so the displacement between $t = 0$ and $t \geq 5$ is $80 + 10(t - 5)$.

5.1.67

a. Between 0 and 5, the area under the curve is given by the area of a square of area 4 and the area of a trapezoid of area 10.5, so the total area is 14.5.

b. Between 5 and 10, the area under the curve is given by the area of a trapezoid of area 5.5 and the area of a rectangle of area $4 \cdot 6 = 24$, so the total area is 29.5.

c. The mass of the entire rod would be the total area under the curve from 0 to 10, which would be $14.5 + 29.5 = 44$ grams.

d. At $x = \frac{19}{3}$ there is a mass of 22 on each side. Note that from 0 to 6 the mass is 20 grams, so the center of mass is a little greater than 6.

5.1.69 If $0 \leq t \leq 2$, the displacement is $30t$. If $2 \leq t \leq 2.5$, the displacement is $60 + 50(t - 2)$. If $2.5 \leq t \leq 3$, the displacement is $85 + 44(t - 2.5)$.

Thus, $d(t) = \begin{cases} 30t & \text{if } 0 \leq t \leq 2, \\ 50t - 40 & \text{if } 2 \leq t \leq 2.5 \\ 44t - 25 & \text{if } 2.5 \leq t \leq 3. \end{cases}$

5.1.71 Using the left Riemann sum

$$\sum_{k=1}^{n} \left| \left(-1 + \frac{2k-2}{n} \right) \left(\left(-1 + \frac{2k-2}{n} \right)^2 - 1 \right) \right| \cdot \frac{2}{n},$$

we have

n	16	32	64
A_n	.492188	.498047	.499512

It appears that the areas are approaching .5.

5.1.73 Using the left Riemann sum

$$\sum_{k=1}^{n} \left| 1 - \left(-1 + \frac{3k-3}{n} \right)^3 \right| \cdot \frac{3}{n},$$

we have

n	16	32	64
A_n	4.33054	4.52814	4.63592

It appears that the areas are approaching 4.75.

5.1.75 The midpoint Riemann sum gives

$$\sum_{k=1}^{n} f\left(a + \frac{b-a}{2n} + \frac{(k-1)(b-a)}{n} \right) \cdot \frac{b-a}{n} = \sum_{k=1}^{n} \left(m\left(a + \frac{b-a}{2n} + \frac{(k-1)(b-a)}{n} \right) + c \right) \cdot \frac{b-a}{n} =$$

$$m \cdot a \cdot n \cdot \frac{b-a}{n} + \frac{m(b-a)^2 n}{2n^2} + \frac{(n-1)n}{2} \cdot \frac{m(b-a)^2}{n^2} + \frac{cn(b-a)}{n} =$$

$$m \cdot a \cdot (b-a) + \frac{m(b-a)^2}{2n} + \frac{m(b-a)^2}{2} - \frac{m(b-a)^2}{2n} + c(b-a) =$$

$$m \cdot a \cdot (b-a) + \frac{m(b-a)^2}{2} + c(b-a) = (b-a) \cdot \left(\frac{m(a+b)}{2} + c \right).$$

This proves that the midpoint Riemann sum is independent of n. Because the region in question is a trapezoid, we know that the exact area is given by the width of the subinterval times the average value at the endpoints, which is $(b-a)\left(\frac{f(a)+f(b)}{2} \right) = (b-a)\left(\frac{ma+c+mb+c}{2} \right) = (b-a)\left(\frac{m(a+b)}{2} + c \right)$.

5.1.77 For a function that is concave up and increasing, each rectangle of the right Riemann sum will have its top edge above the curve, since the value of the function at the right edge of the rectangle will be larger than at any other point in the rectangle. Thus this will be an overestimate. For a function that is concave up and decreasing, however, each rectangle will lie wholly below the curve, since the value of the function at the right edge will be smaller than at any other point in the rectangle. Thus this will be an underestimate. For a function that is concave down and increasing, each rectangle of the right Riemann sum will have its top edge above the curve, since the value of the function at the right edge of the rectangle will be larger than at any other point in the rectangle. Thus this will be an overestimate. Finally, for a function that is concave down and decreasing, however, each rectangle will lie wholly below the curve, since the value of the function at the right edge will be smaller than at any other point in the rectangle. Thus this will be an underestimate. Graphs of each of the four situations are below:

Concave up and increasing

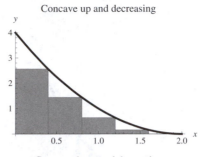

Concave up and decreasing

Concave down and increasing

Concave down and decreasing

So the answer is

	Increasing on $[a, b]$	Decreasing on $[a, b]$
Concave up on $[a, b]$	Overestimate	Underestimate
Concave down on $[a, b]$	Overestimate	Underestimate

5.2   Definite Integrals

5.2.1 The net area is the difference between the area above the x-axis and below the curve, and below the x-axis and above the curve.

5.2.3 When the function is strictly above the x-axis, the net area is equal to the area. The net area differs from the area when the function dips below the x-axis so that the area below the x-axis and above the curve is nonzero.

5.2.5 Because each of the functions $\sin x$ and $\cos x$ have the same amount of area above the x-axis as below between 0 and 2π, these both have value 0.

5.2.7 Because a region "from $x = a$ to $x = a$" has no width, its area is zero. This is akin to asking for the area of a one-dimensional object.

5.2.9 This integral represents the area under $y = x$ between $x = 0$ and $x = a$, which is a right triangle. The length of the base of the triangle is a and the height is a, so the area is $\frac{1}{2} \cdot a^2$, so $\int_0^a x \, dx = \frac{a^2}{2}$.

5.2.11

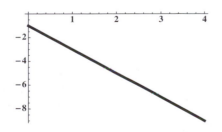

The left Riemann sum is $f(0) \cdot 1 + f(1) \cdot 1 + f(2) \cdot 1 + f(3) \cdot 1 = -1 - 3 - 5 - 7 = -16$.
The right Riemann sum is $f(1) \cdot 1 + f(2) \cdot 1 + f(3) \cdot 1 + f(4) \cdot 1 = -3 - 5 - 7 - 9 = -24$.
The midpoint Riemann sum is $f(.5) \cdot 1 + f(1.5) \cdot 1 + f(2.5) \cdot 1 + f(3.5) \cdot 1 = -2 - 4 - 6 - 8 = -20$.

5.2.13

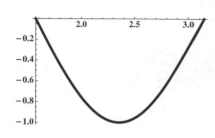

We have $\Delta x = \frac{\pi/2}{4} = \frac{\pi}{8}$. The left Riemann sum is $f(\pi/2) \cdot \frac{\pi}{8} + f(5\pi/8) \cdot \frac{\pi}{8} + f(6\pi/8) \cdot \frac{\pi}{8} + f(7\pi/8) \cdot \frac{\pi}{8} = \left(0 - \sqrt{2}/2 - 1 - \sqrt{2}/2\right) \cdot \frac{\pi}{8} = \frac{\pi}{8} \cdot (-1 - \sqrt{2}) \approx -.948$.

The right Riemann sum is $f(5\pi/8) \cdot \frac{\pi}{8} + f(6\pi/8) \cdot \frac{\pi}{8} + f(7\pi/8) \cdot \frac{\pi}{8} + f(\pi) \cdot \frac{\pi}{8} = \left(-\sqrt{2}/2 - 1 - \sqrt{2}/2 - 0\right) \cdot \frac{\pi}{8} = \frac{\pi}{8} \cdot (-1 - \sqrt{2}) \approx -.948$.

The midpoint Riemann sum is $f(9\pi/16) \cdot \frac{\pi}{8} + f(11\pi/16) \cdot \frac{\pi}{8} + f(13\pi/16) \cdot \frac{\pi}{8} + f(15\pi/16) \cdot \frac{\pi}{8} \approx \frac{\pi}{8} \cdot (-0.382683 - .92388 - .92388 - 0.382683) \approx -1.026$.

5.2.15

a.

b. The left Riemann sum $\sum_{k=1}^{4} f(x_{k-1}) \cdot 1 = 4$.

The right Riemann sum $\sum_{k=1}^{4} f(x_k) \cdot 1 = -4$.

The midpoint Riemann sum $\sum_{k=1}^{4} f(x_k^*) \cdot 1 = 0$.

c. The rectangles whose height is $f(x_k)$ contribute positively when $x_k < 2$ and negatively when $x_k > 2$.

5.2.17

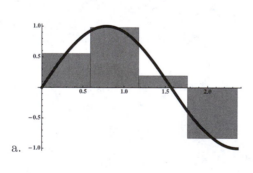

a.

b. The left Riemann sum $\sum_{k=1}^{4} f(x_{k-1}) \cdot \frac{3\pi}{16} \approx .7353$.

The right Riemann sum $\sum_{k=1}^{4} f(x_k) \cdot \frac{3\pi}{16} \approx 0.146$.

The midpoint Riemann sum $\sum_{k=1}^{4} f(x_k^*) \cdot \frac{3\pi}{16} \approx 0.530$.

c. The rectangles whose height is $f(x_k)$ contribute positively when $x_k < \frac{\pi}{2}$ and negatively when $x_k > \frac{\pi}{2}$.

5.2.19 This is $\int_0^2 (x^2 + 1)\, dx$.

5.2.21 This is $\int_1^2 x \cos x\, dx$.

5.2.23

The region in question is a triangle with base 4 and height 8, so the area is $\frac{1}{2} \cdot 8 \cdot 4 = 16$, and this is therefore the value of the definite integral as well.

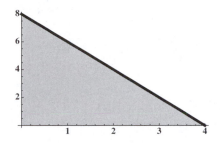

5.2.25

The region consists of two triangles, both below the axis. One has base 1 and height 1, the other has base 2 and height 2, so the net area is $-\frac{1}{2} \cdot 1 \cdot 1 - \frac{1}{2} \cdot 2 \cdot 2 = -2.5$.

5.2.27

The region consists of a quarter circle of radius 4, situated above the axis. So the net area is $\frac{\pi \cdot 4^2}{4} = 4\pi$.

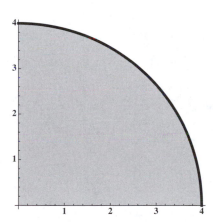

5.2.29

The region consists of a rectangle of area 10 above the axis, and a trapezoid of area 16 above the axis, so the net area is $10 + 16 = 26$.

5.2.31 $\displaystyle\int_0^a f(x)\,dx = 16.$

5.2.33 $\displaystyle\int_a^c f(x)\,dx = 11 - 5 = 6.$

5.2.35 $\displaystyle\int_0^\pi x \sin x \, dx = A(R_1) + A(R_2) = 1 + \pi - 1 = \pi.$

5.2.37 $\displaystyle\int_0^{2\pi} x \sin x \, dx = A(R_1) + A(R_2) - A(R_3) - A(R_4) = 1 + \pi - 1 - \pi - 1 - 2\pi + 1 = -2\pi.$

5.2.39

a. $\displaystyle\int_4^0 3x(4-x)\,dx = -\int_0^4 3x(4-x)\,dx = -32.$

b. $\displaystyle\int_0^4 x(x-4)\,dx = \frac{-1}{3}\int_0^4 3x(4-x)\,dx = \frac{-1}{3}\cdot 32 = \frac{-32}{3}.$

c. $\displaystyle\int_4^0 6x(4-x)\,dx = -2 \cdot \int_0^4 3x(4-x)\,dx = -2 \cdot 32 = -64.$

d. $\displaystyle\int_0^8 3x(4-x)\,dx = \int_0^4 3x(4-x)\,dx + \int_4^8 3x(4-x)\,dx = 32 + \int_4^8 3x(4-x)\,dx.$ It is not possible to
evaluate the given integral from the information given.

5.2.41

a. $\displaystyle\int_0^3 5f(x)\,dx = 5\int_0^3 f(x)\,dx = 5 \cdot 2 = 10.$

b. $\displaystyle\int_3^6 (-3g(x))\,dx = -3\int_3^6 g(x)\,dx = -3 \cdot 1 = -3.$

c. $\displaystyle\int_3^6 (3f(x) - g(x))\,dx = 3\int_3^6 f(x)\,dx - \int_3^6 g(x)\,dx = 3(-5) - 1 = -16.$

d. $\displaystyle\int_6^3 [f(x) + 2g(x)]\,dx = -\left[\int_3^6 f(x)\,dx + 2\int_3^6 g(x)\,dx\right] = -[-5 + 2 \cdot 1] = 3.$

5.2.43

a. $\displaystyle\int_0^1 (4x - 2x^3)\,dx = -2\int_0^1 (x^3 - 2x)\,dx = -2 \cdot \frac{-3}{4} = \frac{3}{2}.$

b. $\displaystyle\int_1^0 (2x - x^3)\,dx = \int_0^1 (x^3 - 2x)\,dx = \frac{-3}{4}.$

5.2.45

$$\int_0^2 (2x+1)\,dx = \lim_{n\to\infty}\sum_{k=1}^n f(x_k)\Delta x = \lim_{n\to\infty}\sum_{k=1}^n \left[2\left(\frac{2k}{n}\right) + 1\right]\frac{2}{n}$$

$$= \lim_{n\to\infty}\left[\frac{8}{n^2}\sum_{k=1}^n k + \frac{2}{n}\sum_{k=1}^n 1\right]$$

$$= \lim_{n\to\infty}\left[\frac{8}{n^2}\cdot\frac{n(n+1)}{2} + \frac{2}{n}\cdot n\right]$$

$$= \lim_{n\to\infty}\left[\frac{4(n+1)}{n} + 2\right] = 4 + 2 = 6.$$

5.2.47

$$\int_3^7 (4x + 6)\, dx = \lim_{n\to\infty} \sum_{k=1}^n f(x_k)\Delta x = \lim_{n\to\infty} \sum_{k=1}^n \left[4\left(3 + \frac{4k}{n}\right) + 6\right]\frac{4}{n}$$

$$= \lim_{n\to\infty}\left[\frac{64}{n^2}\sum_{k=1}^n k + \frac{72}{n}\sum_{k=1}^n 1\right]$$

$$= \lim_{n\to\infty}\left[\frac{64}{n^2}\cdot\frac{n(n+1)}{2} + \frac{72}{n}\cdot n\right]$$

$$= \lim_{n\to\infty}\left[\frac{32(n+1)}{n} + 72\right] = 104.$$

5.2.49

$$\int_1^4 (x^2 - 1)\, dx = \lim_{n\to\infty} \sum_{k=1}^n f(x_k)\Delta x = \lim_{n\to\infty} \sum_{k=1}^n \left[\left(1 + \frac{3k}{n}\right)^2 - 1\right]\frac{3}{n}$$

$$= \lim_{n\to\infty}\left[\frac{18}{n^2}\sum_{k=1}^n k + \frac{27}{n^3}\sum_{k=1}^n k^2\right]$$

$$= \lim_{n\to\infty}\left[\frac{18}{n^2}\cdot\frac{n(n+1)}{2} + \frac{27}{n^3}\cdot\frac{n(n+1)(2n+1)}{6}\right]$$

$$= \lim_{n\to\infty}\left[\frac{9(n+1)}{n} + \frac{18n^2 + 27n + 9}{2n^2}\right] = 9 + 9 = 18.$$

5.2.51

a. True. See problem 76 in the previous section for a proof.

b. True. See problem 77 in the previous section for a proof.

c. True. Because both of those function are periodic with period $\frac{2\pi}{a}$, and both have the same amount of area above the axis as below for one period, the net area of each between 0 and $\frac{2\pi}{a}$ is zero.

d. False. For example $\int_0^{2\pi} \sin x\, dx = 0 = \int_{2\pi}^0 \sin x\, dx$, but $\sin x$ is not a constant function.

e. False. Because x is not a constant, it can not be factored outside of the integral. For example $\int_0^1 x\cdot 1\, dx \neq x\int_0^1 1\, dx.$

5.2.53

a.

b. $\Delta x = \frac{1}{2}$, so the grid points are at 3, 3.5, 4, 4.5, 5, 5.5, and 6.

c. The left Riemann sum is $.5(-5-6-7-8-9-10) = -22.5$. The right Riemann sum is $.5(-6-7-8-9-10-11) = -25.5$.

d. The left Riemann sum overestimates the true value, while the right Riemann sum underestimates it.

5.2.55

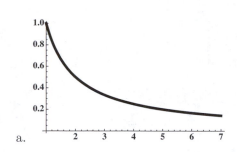

a.

b. $\Delta x = 1$, so the grid points are at 1, 2, 3, 4, 5, 6, and 7.

c. The left Riemann sum is approximately $1 + 0.5 + 0.333333 + 0.25 + -.2 + 0.166666 = 2.45$. The right Riemann sum is approximately $.5 + 0.333333 + 0.25 + 0.2 + 0.166666 + 0.142857 \approx 1.593$.

d. The left Riemann sum overestimates the true value, while the right Riemann sum underestimates it.

5.2.57

a. $\displaystyle\sum_{k=1}^{20}\left(\left(\frac{k-1}{20}\right)^2+1\right)\cdot\frac{1}{20}\approx 1.309.$ $\displaystyle\sum_{k=1}^{20}\left(\left(\frac{k}{20}\right)^2+1\right)\cdot\frac{1}{20}\approx 1.359.$

$\displaystyle\sum_{k=1}^{50}\left(\left(\frac{k-1}{50}\right)^2+1\right)\cdot\frac{1}{50}\approx 1.323.$ $\displaystyle\sum_{k=1}^{50}\left(\left(\frac{k}{50}\right)^2+1\right)\cdot\frac{1}{50}\approx 1.343.$

$\displaystyle\sum_{k=1}^{100}\left(\left(\frac{k-1}{100}\right)^2+1\right)\cdot\frac{1}{100}\approx 1.328.$ $\displaystyle\sum_{k=1}^{100}\left(\left(\frac{k}{100}\right)^2+1\right)\cdot\frac{1}{100}\approx 1.338.$

b. It appears that the integral's value is about $\frac{4}{3}$.

5.2.59

a.

$$\frac{3}{20}\sum_{k=1}^{20}\frac{1}{2\left(1+\frac{3}{20}(k-1)\right)}\approx 0.722 \qquad \frac{3}{20}\sum_{k=1}^{20}\frac{1}{2\left(1+\frac{3}{20}k\right)}\approx 0.666$$

$$\frac{3}{50}\sum_{k=1}^{50}\frac{1}{2\left(1+\frac{3}{50}(k-1)\right)}\approx 0.705 \qquad \frac{3}{50}\sum_{k=1}^{50}\frac{1}{2\left(1+\frac{3}{50}k\right)}\approx 0.682$$

$$\frac{3}{100}\sum_{k=1}^{100}\frac{1}{2\left(1+\frac{3}{100}(k-1)\right)}\approx 0.699 \qquad \frac{3}{100}\sum_{k=1}^{100}\frac{1}{2\left(1+\frac{3}{100}k\right)}\approx 0.688$$

b. It appears that the integral's value is about 0.69.

5.2.61

a. $\displaystyle\sum_{k=1}^{n}2\sqrt{1+\frac{3}{2n}+\frac{3(k-1)}{n}}\cdot\frac{3}{n}.$

b.

n	20	50	100
Midpoint Sum	9.3338	9.33341	9.33335

It appears that the integral's value is about $\frac{28}{3}$.

5.2.63

a. $\displaystyle\sum_{k=1}^{n}\left(4\left(\frac{2}{n}+\frac{4(k-1)}{n}\right)-\left(\frac{2}{n}+\frac{4(k-1)}{n}\right)^2\right)\cdot\frac{4}{n}.$

b.

n	20	50	100
Midpoint Sum	10.68	10.6688	10.6672

It appears that the integral's value is about $\frac{32}{3}$.

5.2.65

a. $\displaystyle\int_1^4 3f(x)\,dx = 3\int_1^4 f(x)\,dx = 3\cdot\left(\int_1^6 f(x)\,dx - \int_4^6 f(x)\,dx\right) = 3\cdot(10-5) = 15.$

b. $\displaystyle\int_1^6 (f(x)-g(x))\,dx = \int_1^6 f(x)\,dx - \int_1^6 g(x)\,dx = 10-5 = 5.$

c. $\displaystyle\int_1^4 (f(x)-g(x))\,dx = \int_1^4 f(x)\,dx - \int_1^4 g(x)\,dx = \left(\int_1^6 f(x)\,dx - \int_4^6 f(x)\,dx\right) - 2 = (10-5)-2 = 3.$

d. $\displaystyle\int_4^6 (g(x)-f(x))\,dx = \int_4^6 g(x)\,dx - \int_4^6 f(x)\,dx = \left(\int_1^6 g(x)\,dx - \int_1^4 g(x)\,dx\right) - 5 = (5-2)-5 = -2.$

e. $\displaystyle\int_4^6 8g(x)\,dx = 8\left(\int_1^6 g(x)\,dx - \int_1^4 g(x)\,dx\right) = 8(5-2) = 24.$

f. $\displaystyle\int_4^1 2f(x)\,dx = -2\int_1^4 f(x)\,dx = -2\cdot\left(\int_1^6 f(x)\,dx - \int_4^6 f(x)\,dx\right) = -2(10-5) = -10.$

5.2.67

The region above the axis is a triangle with base 2 and height $f(-2) = 6$, and the region below the axis is a triangle with base 2 and height $-f(2) = 6$, so the net area is 0, and the area is 12.

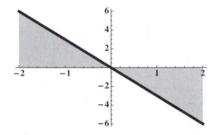

5.2.69

The region above the axis is a triangle with base 2 and height $f(0) = 1$, while the region below the axis consists of two triangles each with base 1 and height 1, so the net area is 0, and the area is 2.

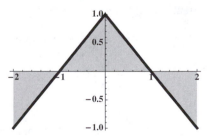

5.2.71

The region in question consists of two triangles above the axis, one with base 1 and height 2, and one with base 4 and height 8, so the net area is $\frac{1}{2}\cdot 1\cdot 2 + \frac{1}{2}\cdot 4\cdot 8 = 17.$

5.2.73

The region in question is a semicircle above the axis with radius 5, so the area is $\frac{1}{2}\pi \cdot 5^2 = \frac{25\pi}{2}$.

5.2.75 $\displaystyle\int_0^{10} f(x)\,dx = \int_0^5 2\,dx + \int_5^{10} 3\,dx = 10 + 15 = 25.$

5.2.77 $\displaystyle\int_1^5 x\lfloor x\rfloor\,dx = \int_1^2 x\lfloor x\rfloor\,dx + \int_2^3 x\lfloor x\rfloor\,dx + \int_3^4 x\lfloor x\rfloor\,dx + \int_4^5 x\lfloor x\rfloor\,dx = \int_1^2 x\,dx + \int_2^3 2x\,dx + \int_3^4 3x\,dx +$

$\displaystyle\int_4^5 4x\,dx.$ Each of these integrals represents the area of a trapezoid with base 1. The value of the integral is $\frac{1+2}{2} + \frac{4+6}{2} + \frac{9+12}{2} + \frac{16+20}{2} = 35.$

5.2.79

$$\int_a^b cf(x)\,dx = \lim_{\Delta\to 0}\sum_{k=1}^n cf(x_k^*)\Delta x_k$$

$$= \lim_{\Delta\to 0} c\sum_{k=1}^n f(x_k^*)\Delta x_k$$

$$= c\lim_{\Delta\to 0}\sum_{k=1}^n f(x_k^*)\Delta x_k = c\int_a^b f(x)\,dx.$$

5.2.81 Let n be a positive integer. Let $\Delta x = \frac{1}{n}$. Note that each grid point $\frac{k}{n}$ for $0 \le k \le n$ where i is an integer is a rational number. So $f(x_k) = 1$ for each grid point. So the right Riemann sum is $\displaystyle\sum_{k=1}^n f(x_k)\frac{1}{n} = \frac{1}{n}\sum_{k=1}^n 1 = \frac{1}{n}\cdot n = 1.$ The left Riemann sum calculation is similar, as is the midpoint Riemann sum calculation (because the grid midpoints are also rational numbers – they are the average of two rational numbers and hence are rational as well).

5.2.83

a. Note that for all values of $k = 1, 2, \ldots, n$, we have $x_{k-1}x_{k-1} \le x_{k-1}x_k$, so $\sqrt{x_{k-1}x_{k-1}} \le \sqrt{x_{k-1}x_k}$, and thus $x_{k-1} \le \sqrt{x_{k-1}x_k}$. Similarly, $x_{k-1}x_k \le x_kx_k$, so $\sqrt{x_{k-1}x_k} \le \sqrt{x_kx_k} = x_k$, so $\sqrt{x_{k-1}x_k} \le x_k$. Thus $x_{k-1} \le \sqrt{x_{k-1}x_k} \le x_k$ for all $k = 1, 2, \ldots, n$.

b. $\frac{1}{x_{k-1}} - \frac{1}{x_k} = \frac{x_k}{x_{k-1}x_k} - \frac{x_{k-1}}{x_{k-1}x_k} = \frac{x_k - x_{k-1}}{x_{k-1}x_k} = \frac{\Delta x_k}{x_{k-1}x_k}$, for all $k = 1, 2, \ldots, n$.

c. The Riemann sum is $\displaystyle\sum_{k=1}^n \frac{\Delta x_k}{\overline{x_k}^2}$. Using $\overline{x_k} = \sqrt{x_{k-1}x_k}$, we have

$$\sum_{k=1}^n \frac{\Delta x_k}{x_{k-1}x_k} = \sum_{k=1}^n\left(\frac{1}{x_{k-1}} - \frac{1}{x_k}\right),$$

where the last equality follows from part (b). Now note that the sum telescopes (that is, has many canceling terms).

$$\sum_{k=1}^{n}\left(\frac{1}{x_{k-1}}-\frac{1}{x_k}\right)=\left(\frac{1}{x_0}-\frac{1}{x_1}\right)+\left(\frac{1}{x_1}-\frac{1}{x_2}\right)+\cdots+\left(\frac{1}{x_{n-2}}-\frac{1}{x_{n-1}}\right)+\left(\frac{1}{x_{n-1}}-\frac{1}{x_n}\right)$$

$$=\frac{1}{x_0}+\left(-\frac{1}{x_1}+\frac{1}{x_1}\right)+\left(-\frac{1}{x_2}+\frac{1}{x_2}\right)+\cdots+\left(-\frac{1}{x_{n-1}}+\frac{1}{x_{n-1}}\right)-\frac{1}{x_n}$$

$$=\frac{1}{x_0}-\frac{1}{x_n}$$

$$=\frac{1}{a}-\frac{1}{b}.$$

d. The integral is the limit of the Riemann sum as $n \to \infty$. Thus we have

$$\int_a^b \frac{dx}{x^2}=\lim_{n\to\infty}\sum_{k=1}^{n}\frac{\Delta x_k}{\overline{x}_k{}^2}=\lim_{n\to\infty}\left(\frac{1}{a}-\frac{1}{b}\right)=\frac{1}{a}-\frac{1}{b}.$$

5.3 Fundamental Theorem of Calculus

5.3.1 A is also an antiderivative of f.

5.3.3 The fundamental theorem says that $\int_a^b f(x)\,dx = F(b) - F(a)$ where F is any antiderivative of f. So to evaluate $\int_a^b f(x)\,dx$, one could find an antiderivative $F(x)$, and then evaluate this at a and b and then subtract, obtaining $F(b) - F(a)$.

5.3.5

$A(x) = \int_0^x (3-t)\,dt$ represents the area between 0 and x and below this curve. As x increases (but remains less than 3), the trapezoidal region's area increases, so the area function increases until x is 3.

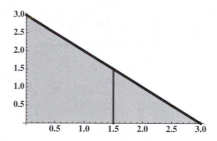

5.3.7 $\dfrac{d}{dx}\displaystyle\int_a^x f(t)\,dt = f(x)$, and $\displaystyle\int f'(x)\,dx = f(x) + C$.

5.3.9 $\dfrac{d}{dx}\displaystyle\int_a^x f(t)\,dt = f(x)$, and $\dfrac{d}{dx}\displaystyle\int_a^b f(t)\,dt = 0$. The latter is the derivative of a constant, the former follows from the Fundamental Theorem.

5.3.11

a. $A(-2) = \int_{-2}^{-2} f(t)\,dt = 0$.

b. $F(8) = \int_4^8 f(t)\,dt = -9$.

c. $A(4) = \int_{-2}^4 f(t)\,dt = 8 + 17 = 25$.

d. $F(4) = \int_4^4 f(t)\,dt = 0$.

e. $A(8) = \int_{-2}^8 f(t)\,dt = 25 - 9 = 16$.

5.3.13

a. $A(x) = \int_0^x f(t)\,dt = \int_0^x 5\,dt = 5x.$

b. $A'(x) = 5 = f(x).$

5.3.15

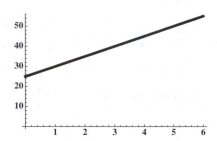

a. $A(x) = \int_{-5}^x f(t)\,dt = \int_{-5}^x 5\,dt = 5(x+5).$

b. $A'(x) = 5 = f(x).$

5.3.17

a. $A(2) = \int_0^2 t\,dt = 2$. $A(4) = \int_0^4 t\,dt = 8$. Because the region whose area is $A(x) = \int_0^x t\,dt$ is a triangle with base x and height x, its value is $\frac{1}{2}x^2$.

b. $F(4) = \int_2^4 t\,dt = 6$. $F(6) = \int_2^6 t\,dt = 16$. Because the region whose area is $A(x) = \int_2^x t\,dt$ is a trapezoid with base $x - 2$ and $h_1 = 2$ and $h_2 = x$, its value is $(x-2)\frac{2+x}{2} = \frac{x^2-4}{2} = \frac{x^2}{2} - 2.$

c. We have $A(x) - F(x) = \frac{x^2}{2} - \left(\frac{x^2}{2} - 2\right) = 2$, a constant.

5.3.19

a. The region is a triangle with base $x + 5$ and height $x + 5$, so its area is $A(x) = \frac{1}{2}(x+5)^2.$

b. $A'(x) = x + 5 = f(x).$

5.3.21

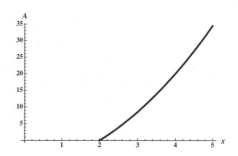

a. The region is a trapezoid with base $x - 2$ and heights $h_1 = f(2) = 7$ and $h_2 = f(x) = 3x + 1$, so its area is $A(x) = (x - 2) \cdot \frac{7 + 3x + 1}{2} = (x - 2) \cdot (\frac{3}{2}x + 4) = \frac{3}{2}x^2 + x - 8$.

b. $A'(x) = 3x + 1 = f(x)$.

5.3.23 $\int_0^1 (x^2 - 2x + 3)\, dx = \left(\frac{x^3}{3} - x^2 + 3x \right) \Big|_0^1 = \frac{1}{3} - 1 + 3 - (0 - 0 + 0) = \frac{7}{3}$. It does appear that the area is between 2 and 3.

5.3.25

$\int_{-2}^3 (x^2 - x - 6)\, dx = \left(\frac{x^3}{3} - \frac{x^2}{2} - 6x \right) \Big|_{-2}^3 = -\frac{125}{6}$.

5.3.27

$\int_0^5 (x^2 - 9)\, dx = \left(\frac{x^3}{3} - 9x \right) \Big|_0^5 = \frac{125}{3} - 45 - (0 - 0) = -\frac{10}{3}$.

5.3.29 $\int_0^2 4x^3\, dx = x^4 \Big|_0^2 = 16 - 0 = 16$.

5.3.31 $\int_0^1 (x + \sqrt{x})\, dx = \left(\frac{x^2}{2} + \frac{2x^{3/2}}{3} \right) \Big|_0^1 = \frac{1}{2} + \frac{2}{3} - (0 + 0) = \frac{7}{6}$.

5.3.33 $\int_1^9 \frac{2}{\sqrt{x}}\, dx = \int_1^9 2x^{-1/2}\, dx = 4x^{1/2} \Big|_1^9 = 12 - 4 = 8$.

5.3.35 $\int_{-2}^2 (x^2 - 4)\, dx = \left(\frac{x^3}{3} - 4x \right) \Big|_{-2}^2 = \frac{8}{3} - 8 - \left(-\frac{8}{3} + 8 \right) = \frac{16}{3} - 16 = -\frac{32}{3}$.

5.3.37 $\int_{1/2}^1 (x^{-3} - 8)\, dx = \left(\frac{x^{-2}}{-2} - 8x \right) \Big|_{1/2}^1 = -\frac{1}{2} - 8 - (-2 - 4) = -\frac{5}{2}$.

5.3.39 $\displaystyle\int_0^{\pi/4} \sec^2\theta\, d\theta = \tan\theta \,\Big|_0^{\pi/4} = 1 - 0 = 1.$

5.3.41 $\displaystyle\int_{-2}^{-1} x^{-3}\, dx = \frac{x^{-2}}{-2}\,\Big|_{-2}^{-1} = -\frac{1}{2x^2}\,\Big|_{-2}^{-1} = -\frac{1}{2} - \left(-\frac{1}{8}\right) = -\frac{3}{8}.$

5.3.43 $\displaystyle\int_1^4 (1-x)(x-4)\, dx = \int_1^4 (-x^2 + 5x - 4)\, dx = \left(-\frac{x^3}{3} + \frac{5x^2}{2} - 4x\right)\Big|_1^4 = \frac{9}{2}.$

5.3.45 $\displaystyle\int_1^2 \frac{3}{w^2}\, dw = \left(-\frac{3}{w}\right)\Big|_1^2 = \left(-\frac{3}{2}\right) - \left(-\frac{3}{1}\right) = \frac{3}{2}.$

5.3.47 $\displaystyle\int_0^{\pi/8} \cos 2x\, dx = \left(\frac{\sin 2x}{2}\right)\Big|_0^{\pi/8} = \frac{\sqrt{2}/2 - 0}{2} = \frac{\sqrt{2}}{4}.$

5.3.49

The area (and net area) of this region is given by $\displaystyle\int_1^4 \sqrt{x}\, dx = \frac{2}{3}x^{3/2}\,\Big|_1^4 = \frac{16}{3} - \frac{2}{3} = \frac{14}{3}.$

5.3.51

The net area of this region is given by $\displaystyle\int_{-2}^2 (x^4 - 16)\, dx = \left(\frac{x^5}{5} - 16x\right)\Big|_{-2}^2 = \frac{32}{5} - 32 - \left(-\frac{32}{5} + 32\right) = \frac{64}{5} - 64 = -\frac{256}{5}.$ Thus the area is $\dfrac{256}{5}.$

5.3.53 Because this region is below the axis, the area of it is given by $-\int_2^4 (x^2 - 25)\, dx = -\left(\frac{x^3}{3} - 25x\right)\Big|_2^4 = -\left(\frac{64}{3} - 100 - \left(\frac{8}{3} - 50\right)\right) = 50 - \frac{56}{3} = \frac{94}{3}.$

5.3.55 Because this region is below the axis, the area of it is given by

$$-\int_{-2}^{-1} x^{-5}\, dx = -\left(-\frac{1}{4}x^{-4}\right)\Big|_{-2}^{-1} = \frac{15}{64}.$$

5.3.57 The area is given by

$$-\int_{-\pi/4}^0 \sin x\, dx + \int_0^{3\pi/4} \sin x\, dx = \left(\cos x\,\Big|_{-\pi/4}^0\right) + \left(-\cos x\,\Big|_0^{3\pi/4}\right) = \left(1 - \frac{\sqrt{2}}{2}\right) + \left(1 + \frac{\sqrt{2}}{2}\right) = 2.$$

5.3.59 By a direct application of the Fundamental Theorem, this is $x^2 + x + 1$.

5.3.61 By the Fundamental Theorem and the chain rule, this is $\frac{1}{x^6} \cdot 3x^2 = \frac{3}{x^4}$.

5.3.63 This is $-\dfrac{d}{dx}\displaystyle\int_1^x \sqrt{t^4+1}\,dt = -\sqrt{x^4+1}.$

5.3.65 This can be written as

$$\frac{d}{dx}\left(\int_{-x}^0 \sqrt{1+t^2}\,dt + \int_0^x \sqrt{1+t^2}\,dt\right) = \frac{d}{dx}\left(-\int_0^{-x}\sqrt{1+t^2}\,dt + \int_0^x \sqrt{1+t^2}\,dt\right)$$

$$= -\sqrt{1+(-x)^2}(-1) + \sqrt{1+x^2} = 2\sqrt{1+x^2}.$$

5.3.67

(a) matches with (C) – its area function is increasing linearly.

(b) matches with (B) – its area function increases then decreases.

(c) matches with (D) – its area function is always increasing on $[0,b]$, although not linearly.

(d) matches with (A) – its area function decreases at first and then eventually increases.

5.3.69

a. It appears that $A(x) = 0$ for $x = 0$ and at about $x = 3$.

b. A has a local minimum at about $x = 1.5$ where the area function changes from decreasing to increasing, and a local max at around $x = 8.5$ where the area function changes from increasing to decreasing.

c.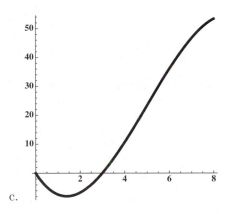

5.3.71

a. It appears that $A(x) = 0$ for $x = 0$ and $x = 10$.

b. A has a local maximum at $x = 5$ where the area function changes from increasing to decreasing.

c.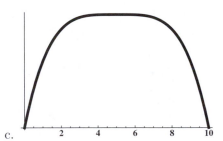

5.3.73 $A(2) = -\frac{1}{4}\pi \cdot 2^2 = -\pi.$ $A(5) = -\pi + \frac{1}{2}\cdot 3 \cdot 3 = \frac{9}{2} - \pi.$ $A(8) = \frac{9}{2} - \pi + \frac{1}{2}\cdot 3 \cdot 3 = 9 - \pi.$
$A(12) = 9 - \pi - \frac{1}{2}\cdot 4 \cdot 2 = 5 - \pi.$

5.3.75

a. $A(x) = \int_0^x \cos t\, dt = \sin x - \sin 0 = \sin x$

c. $A(\pi/2) = \sin \pi/2 = 1$ and $A(\pi) = \sin \pi = 0$. This reflects the net signed area under the curve $\cos x$ from 0 to $\pi/2$ and π respectively.

b.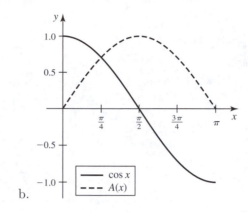

5.3.77

a. $A(x) = \int_0^x \cos \pi t\, dt = \dfrac{1}{\pi} \sin \pi t \Big|_0^x = \dfrac{\sin \pi x}{\pi}$.

c. $A(1/2) = \frac{1}{\pi}$. $A(1) = \frac{\sin \pi}{\pi} = 0$. The area bounded between the x-axis and the curve on $[0, 1/2]$ is equal to the area bounded between the x-axis and the curve on $[1/2, 1]$.

b.

5.3.79

a.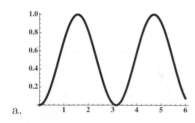

b. $g'(x) = \sin^2 x$.

c.

Note that g' is always positive, so g is always increasing. There are inflection points where g' changes from increasing to decreasing, and vice versa.

5.3.81

b. $g'(x) = \sin(\pi x^2)$.

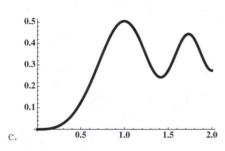

Note that g is increasing where $g' > 0$ and g is decreasing when $g' < 0$. Also, where g' is increasing, g is concave up and where g' is decreasing, g is concave down.

5.3.83

a. True. The net area under the curve increases as x increases, as long as f is above the axis.

b. True. The net area decreases as x increases, as long as f is below the axis.

c. False. These do not have the same derivative, so they are not antiderivatives of the same function.

d. True, because the two functions differ by a constant, and thus have the same derivative.

5.3.85

$$\int_1^4 \frac{x-2}{\sqrt{x}}\,dx = \int_1^4 \left(\frac{x}{\sqrt{x}} - \frac{2}{\sqrt{x}}\right) dx = \int_1^4 \left(x^{1/2} - 2x^{-1/2}\right) dx$$

$$= \left(\frac{2}{3}x^{3/2} - 4x^{1/2}\right)\Big|_1^4 = \frac{16}{3} - 8 - \left(\frac{2}{3} - 4\right) = \frac{14}{3} - \frac{12}{3} = \frac{2}{3}.$$

5.3.87 $\displaystyle\int_0^{\pi/3} \sec x \tan x\,dx = \sec x \Big|_0^{\pi/3} = 2 - 1 = 1.$

5.3.89 $\displaystyle\int_1^8 \sqrt[3]{y}\,dy = \frac{3}{4}y^{4/3}\Big|_1^8 = 12 - \frac{3}{4} = \frac{45}{4}.$

5.3.91

We can use geometry – there is a triangle with base 4 and height 2 and a triangle with base 2 and height 2, so the total area is $\frac{1}{2}\cdot 4\cdot 2 + \frac{1}{2}\cdot 2\cdot 2 = 6$.

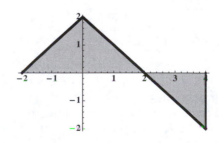

5.3.93

Because the region is below the axis on $[1, \sqrt{2}]$ and above on $[\sqrt{2}, 4]$ we need to compute $\displaystyle\int_{\sqrt{2}}^{4} (x^4 - 4)\, dx - \int_{1}^{\sqrt{2}} (x^4 - 4)\, dx =$

$\left(\dfrac{x^5}{5} - 4x\right)\bigg|_{\sqrt{2}}^{4} - \left(\dfrac{x^5}{5} - 4x\right)\bigg|_{1}^{\sqrt{2}} = \dfrac{1024}{5} -$

$16 - (4\sqrt{2}/5 - 4\sqrt{2}) - (4\sqrt{2}/5 - 4\sqrt{2}) + \dfrac{1}{5} - 4 =$

$185 + \dfrac{32\sqrt{2}}{5}$

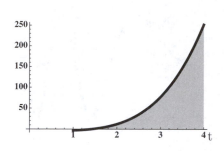

5.3.95 $\displaystyle\int_{3}^{8} f'(t)\, dt = f(8) - f(3).$

5.3.97 $\dfrac{d}{dx} \displaystyle\int_{0}^{\cos x} (t^4 + 6)\, dt = -(\cos^4 x + 6)\sin x.$

5.3.99 $\dfrac{d}{dt} \left(\displaystyle\int_{1}^{t} \dfrac{3}{x}\, dx - \int_{t^2}^{1} \dfrac{3}{x}\, dx \right) = \dfrac{d}{dt} \int_{1}^{t} \dfrac{3}{x}\, dx + \dfrac{d}{dt} \int_{1}^{t^2} \dfrac{3}{x}\, dx = \dfrac{3}{t} + \dfrac{6t}{t^2} = \dfrac{9}{t}.$

5.3.101

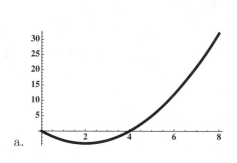

a.

b. We seek b so that $\displaystyle\int_{0}^{b} (x^2 - 4x) = 0$ for $b > 0$. We have $\left(\dfrac{x^3}{3} - 2x^2\right)\bigg|_{0}^{b} = \dfrac{b^3}{3} - 2b^2 = 0$, which occurs for $\dfrac{b}{3} = 2$, or $b = 6$.

c. We seek b so that $\displaystyle\int_{0}^{b} (x^2 - ax) = 0$ for $b > 0$. We have $\left(\dfrac{x^3}{3} - \dfrac{ax^2}{2}\right)\bigg|_{0}^{b} = \dfrac{b^3}{3} - \dfrac{ab^2}{2} = 0$, which occurs for $\dfrac{b}{3} = \dfrac{a}{2}$, or $b = \dfrac{3a}{2}$.

5.3.103 Because $\dfrac{d}{db} \displaystyle\int_{-1}^{b} x^2 (3 - x)\, dx = b^2 (3 - b)$ we see that this function of b has critical points at $b = 0$ and $b = 3$. Note also that the integrand is positive on $[0, 3]$, but is negative on $[3, \infty)$. So the maximum for this area function occurs at $b = 3$.

5.3.105 Differentiating both sides of the given equation yields $f(x) = -2\sin x + 3$.

5.3.107

Using a computer or calculator, we obtain:

x	500	1000	1500	2000
$S(x)$	1.5726	1.57023	1.57087	1.57098

This appears to be approaching $\dfrac{\pi}{2}$.
Note that between 0 and π, the area is approximately half the area of a rectangle with height 1 and base π, and then from π on there is approximately as much area above the axis as below.

5.3.109 By the Fundamental Theorem, $S'(x) = \sin x^2$, so $S''(x) = 2x \cos x^2$, so

$$S'(x)^2 + \left(\frac{S''(x)}{2x}\right)^2 = \sin^2 x^2 + \cos^2 x^2 = 1.$$

5.3.111

a. By definition of Reimann sums, $\int_a^b f'(x)\,dx$ is approximated by $\sum_{k=1}^n f'(x_{k-1})\Delta x$. But $f'(x_{k-1}) = \lim_{h\to 0} \frac{f(x_{k-1}+h) - f(x_{k-1})}{h}$. If $h = \Delta x$, then we have $f'(x_{k-1}) \approx \frac{f(x_{k-1}+\Delta x) - f(x_{k-1})}{\Delta x} = \frac{f(x_k) - f(x_{k-1})}{\Delta x}$, so that

$$\int_a^b f'(x)\,dx \approx \sum_{k=1}^n \frac{f(x_k) - f(x_{k-1})}{\Delta x} \cdot \Delta x.$$

b. Canceling the Δx factors we obtain

$$\int_a^b f'(x)\,dx \approx \sum_{k=1}^n \frac{f(x_k) - f(x_{k-1})}{\Delta x} \cdot \Delta x$$

$$= \sum_{k=1}^n (f(x_k) - f(x_{k-1}))$$

$$= (f(x_1) - f(x_0)) + (f(x_2) - f(x_1)) + \cdots + (f(x_{n-1}) - f(x_{n-2})) + (f(x_n) - f(x_{n-1}))$$

$$= f(x_n) - f(x_0) = f(b) - f(a).$$

c. The analogy between the two situations is that both (a) the sum of difference quotients and (b) integral of a derivative are equal to the difference in function values at the endpoints.

5.4 Working with Integrals

5.4.1 If f is odd, it is symmetric about the origin, which guarantees that between $-a$ and a, there is as much area above the axis and under f as there is below the axis and above f, so the net area must be 0.

5.4.3 $f(x) = x^{12}$ is an even function, because $f(-x) = (-x)^{12} = x^{12} = f(x)$. $g(x) = \sin x^2$ is also even, because $g(-x) = \sin((-x)^2) = \sin x^2 = g(x)$.

5.4.5 The average value of a continuous function on a closed interval $[a, b]$ will always be between the maximum and the minimum value of f on that interval. Because the function is continuous, the Intermediate Value Theorem assures us that the function will take on each value between the maximum and the minimum somewhere on the interval.

5.4.7 Because x^9 is an odd function, $\int_{-2}^{2} x^9\,dx = 0$.

5.4.9 $\int_{-2}^{2} (3x^8 - 2)\,dx = 2\int_{0}^{2} (3x^8 - 2)\,dx = 2\left(\frac{x^9}{3} - 2x\right)\Big|_0^2 = \left(\frac{1024}{3}\right) - 8 = \frac{1000}{3}$.

5.4.11 Note that the first two terms of the integrand form an odd function, and the last two terms form an even function. $\int_{-2}^{2} (x^9 - 3x^5 + 2x^2 - 10)\,dx = 2\int_{0}^{2} (2x^2 - 10)\,dx = 2\left(\frac{2x^3}{3} - 10x\right)\Big|_0^2 = \frac{32}{3} - 40 = -\frac{88}{3}$.

5.4.13 $\int_{-10}^{10} \frac{x}{\sqrt{200 - x^2}}\,dx = 0$ because the integrand is an odd function.

5.4.15 Because the integrand is an odd function and the interval is symmetric about 0, this integral's value is 0.

5.4.17

Because the integrand is an odd function and the interval is symmetric about 0, this integral's value is 0.

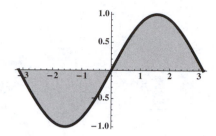

5.4.19

Because of the symmetry of the cosine function, the net area is zero between 0 and π.

5.4.21

The average value is $\dfrac{1}{1-(-1)}\displaystyle\int_{-1}^{1} x^3\,dx =$ $\dfrac{1}{2}\left(x^4/4\right)\Big|_{-1}^{1} = 0.$

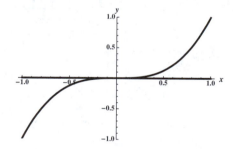

5.4.23

The average value is $\dfrac{1}{4-1}\displaystyle\int_{1}^{4} \dfrac{1}{\sqrt{x}}\,dx =$ $\dfrac{1}{3}\left(2\sqrt{x}\right)\Big|_{1}^{4} = \dfrac{4-2}{3} = \dfrac{2}{3}.$

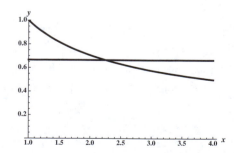

5.4.25

The average value is $\dfrac{1}{\pi} \displaystyle\int_{-\pi/2}^{\pi/2} \cos x \, dx =$

$\dfrac{1}{\pi} \left(\sin x \right) \Big|_{-\pi/2}^{\pi/2} = \dfrac{1}{\pi} \cdot (1 - -1) = \dfrac{2}{\pi} \approx 0.6366.$

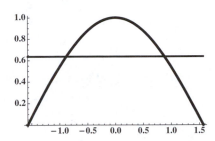

5.4.27

The average value is $\dfrac{1}{1} \displaystyle\int_{0}^{1} x^n \, dx =$

$\left(\dfrac{x^{n+1}}{n+1} \right) \Big|_{0}^{1} = \dfrac{1}{n+1}$. The picture shown is for the case $n = 3$.

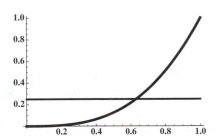

5.4.29 The average distance to the axis is given by $\dfrac{1}{20} \displaystyle\int_{0}^{20} 30x(20 - x) \, dx$. This is equal to $\dfrac{1}{20} \displaystyle\int_{0}^{20} (600x -$

$30x^2) \, dx = \dfrac{1}{20} \left(300x^2 - 10x^3 \right) \Big|_{0}^{20} = 2000.$

5.4.31 The average height is $\dfrac{1}{\pi} \displaystyle\int_{0}^{\pi} 10 \sin x \, dx = \dfrac{1}{\pi} \left(-10 \cos x \right) \Big|_{0}^{\pi} = \dfrac{1}{\pi} (10 - -10) = \dfrac{20}{\pi}.$

5.4.33 The average value is $\dfrac{1}{4} \displaystyle\int_{0}^{4} (8 - 2x) \, dx = \dfrac{1}{4} \left(8x - x^2 \right) \Big|_{0}^{4} = 4$. The function has a value of 4 when $8 - 2x = 4$, which occurs when $x = 2$.

5.4.35 The average value is $\dfrac{1}{a} \displaystyle\int_{0}^{a} \left(1 - \dfrac{x^2}{a^2} \right) dx = \dfrac{1}{a} \left(x - \dfrac{x^3}{3a^2} \right) \Big|_{0}^{a} = \dfrac{2}{3}$. The function attains this value when $\dfrac{2}{3} = 1 - \dfrac{x^2}{a^2}$, which is when $x^2 = \dfrac{a^2}{3}$, which on the given interval occurs for $x = \sqrt{3}a/3$.

5.4.37 The average value is $\dfrac{1}{2} \displaystyle\int_{-1}^{1} (1 - |x|) \, dx = \dfrac{1}{2} \displaystyle\int_{-1}^{0} (1 + x) \, dx + \dfrac{1}{2} \displaystyle\int_{0}^{1} (1 - x) \, dx = \dfrac{1}{2} \left(x + \dfrac{x^2}{2} \right) \Big|_{-1}^{0} +$

$\dfrac{1}{2} \left(x - \dfrac{x^2}{2} \right) \Big|_{0}^{1} = \dfrac{1}{4} + \dfrac{1}{4} = \dfrac{1}{2}$. The function attains this value twice, once on $[-1, 0]$ when $1 + x = \dfrac{1}{2}$ which occurs when $x = -\dfrac{1}{2}$, and once on $[0, 1]$ when $1 - x = \dfrac{1}{2}$ which occurs when $x = \dfrac{1}{2}$.

5.4.39

a. True. Because of the symmetry, the net area between 0 and 4 will be twice the net area between 0 and 2.

b. True. This follows because the symmetry implies that the net area from a to $a + 2$ is the opposite of the net area from $a - 2$ to a.

c. True. If $f(x) = cx + d$ on $[a, b]$ the value at the midpoint is $c \cdot \dfrac{a+b}{2} + d$, and the average value is

$\dfrac{1}{b-a} \displaystyle\int_{a}^{b} (cx + d) \, dx = \dfrac{1}{b-a} \left(\dfrac{cx^2}{2} + dx \right) \Big|_{a}^{b} = \dfrac{1}{b-a} \left(\dfrac{cb^2}{2} + db - \left(\dfrac{ca^2}{2} + da \right) \right) = \dfrac{c}{2} \cdot (a + b) + d.$

d. False, for example, when $a = 1$, we have that the maximum value of $x - x^2$ on $[0, 1]$ occurs at $\frac{1}{2}$ and is equal to $\frac{1}{4}$, but the average value is $\int_0^1 (x - x^2)\, dx = \left(\frac{x^2}{2} - \frac{x^2}{3} \right) \Big|_0^1 = \frac{1}{2} - \frac{1}{3} = \frac{1}{6}$.

5.4.41 $\sec^2 x$ is even, so the value of this integral is $2 \int_0^{\pi/4} \sec^2 x\, dx = 2\, (\tan x) \Big|_0^{\pi/4} = 2 \cdot (1 - 0) = 2$.

5.4.43 The integrand is an odd function, so the value of this integral is zero.

5.4.45 The average height of the arch is given by

$$\frac{1}{630} \int_{-315}^{315} \left(630 - \frac{630}{315^2} x^2 \right) dx = \frac{630}{630} \left(x - \frac{x^3}{3 \cdot 315^2} \right) \Big|_{-315}^{315} = (315 - 105 - (-315 + 105)) = 420 \text{ ft}.$$

5.4.47

a. $d^2 = x^2 + y^2 = x^2 + b^2(1 - (x^2/a^2))$. The average value of d^2 is $\dfrac{1}{2a} \displaystyle\int_{-a}^{a} \left(b^2 + \left(1 - \frac{b^2}{a^2} \right) x^2 \right) dx =$

$\dfrac{1}{2a} \left(b^2 x + \dfrac{(1 - (b^2/a^2)) x^3}{3} \right) \Big|_{-a}^{a} = \dfrac{1}{2a} \left(b^2 a + \dfrac{a^3}{3} - \dfrac{b^2 a}{3} - \left(-b^2 a - \dfrac{a^3}{3} + \dfrac{b^2 a}{3} \right) \right) = \dfrac{2b^2}{3} + \dfrac{a^2}{3}$.

b. If $a = b = R$, the above becomes $\frac{2R^2}{3} + \frac{R^2}{3} = R^2$.

c. $D^2 = (x - \sqrt{a^2 - b^2})^2 + y^2 = x^2 - 2x\sqrt{a^2 - b^2} + y^2 + a^2 - b^2 = \left(1 - \frac{b^2}{a^2} \right) x^2 - 2\sqrt{a^2 - b^2}\, x + a^2$. So the

average value of D^2 is $\dfrac{1}{2a} \displaystyle\int_{-a}^{a} D^2\, dx = \dfrac{1}{2a} \displaystyle\int_{-a}^{a} \left[\left(1 - \frac{b^2}{a^2} \right) x^2 + a^2 \right] dx - \dfrac{1}{a} \displaystyle\int_{-a}^{a} x\sqrt{a^2 - b^2}\, dx =$

$\dfrac{1}{a} \displaystyle\int_0^a \left[\left(1 - \frac{b^2}{a^2} \right) x^2 + a^2 \right] dx + 0 = \dfrac{1}{3}(a^2 - b^2) + a^2 = \dfrac{4a^2 - b^2}{3}$.

5.4.49

a. Because $\displaystyle\int_{-8}^{8} f(x)\, dx = 18 = 2 \displaystyle\int_0^8 f(x)\, dx$, we have $\displaystyle\int_0^8 f(x)\, dx = \dfrac{18}{2} = 9$.

b. Because $x f(x)$ is an odd function when $f(x)$ is even, we have $\displaystyle\int_{-8}^{8} x f(x)\, dx = 0$.

5.4.51 $f(g(-x)) = f(g(x))$, so $f(g(x))$ is an even function, and $\displaystyle\int_{-a}^{a} f(g(x))\, dx = 2 \displaystyle\int_0^a f(g(x))\, dx$.

5.4.53 $p(g(-x)) = p(g(x))$, so $p(g(x))$ is an even function, and $\displaystyle\int_{-a}^{a} p(g(x))\, dx = 2 \displaystyle\int_0^a p(g(x))\, dx$.

5.4.55

a. The average value is $\displaystyle\int_0^1 (ax - ax^2)\, dx = \left(\dfrac{ax^2}{2} - \dfrac{ax^3}{3} \right) \Big|_0^1 = \dfrac{a}{2} - \dfrac{a}{3} = \dfrac{a}{6}$.

b. The function is equal to its average value when $\frac{a}{6} = ax - ax^2$ which occurs when $6x - 6x^2 = 1$, so when $6x^2 - 6x + 1 = 0$. On the given interval, this occurs for $x = \frac{6 \pm \sqrt{12}}{12} = \frac{3 \pm \sqrt{3}}{6}$.

5.4.57

a. The area of the triangle is $\frac{1}{2} \cdot 2a \cdot a^2 = a^3$. The area under the parabola is $\int_{-a}^{a} (a^2 - x^2)\, dx =$

$\left(a^2 x - \dfrac{x^3}{3} \right) \Big|_{-a}^{a} = a^3 - \dfrac{a^3}{3} - \left(-a^3 + \dfrac{a^3}{3} \right) = 2a^3 - \dfrac{2a^3}{3} = \dfrac{4a^3}{3}$, as desired. The diagram shown is for $a = 2$.

b. The area of the rectangle described is $2a \cdot a^2 = 2a^3$, and $\frac{2}{3}$ of this is $\frac{4a^3}{3}$, which is the area under the parabola derived above.

5.4.59 $\int_{0}^{c} x(x-c)^2\, dx = \int_{0}^{c} (x^3 - 2cx^2 + c^2 x)\, dx = \left(\dfrac{x^4}{4} - 2c\dfrac{x^3}{3} + \dfrac{c^2 x^2}{2} \right) \Big|_{0}^{c} = \dfrac{c^4}{4} - \dfrac{2c^4}{3} + \dfrac{c^4}{2} = \dfrac{c^4}{12}$. This is one when $c = \sqrt[4]{12}$.

5.4.61

a. The left Riemann sum is given by $\dfrac{\pi}{2n} \displaystyle\sum_{k=0}^{n-1} \sin((k\pi)/(2n))$.

b.

$$\lim_{\theta \to 0} \theta \left(\frac{\cos\theta + \sin\theta - 1}{2(1 - \cos\theta)} \right) \left(\frac{1 + \cos\theta}{1 + \cos\theta} \right) = \lim_{\theta \to 0} \frac{\theta}{2} \left(\frac{(1 + \cos\theta)(\cos\theta + \sin\theta - 1)}{\sin^2\theta} \right)$$

$$= \left(\frac{1}{2} \lim_{\theta \to 0} \frac{\theta}{\sin\theta} \cdot \frac{1 + \cos\theta}{1} \right) \left(\lim_{\theta \to 0} \frac{\cos\theta - 1}{\sin\theta} + \lim_{\theta \to 0} \frac{\sin\theta}{\sin\theta} \right)$$

$$= \frac{1}{2} \cdot 1 \cdot 2 \left(\lim_{\theta \to 0} \frac{\frac{\cos -1}{\theta}}{\frac{\sin\theta}{\theta}} + 1 \right) = 1(0 + 1) = 1.$$

c. Using the previous result, the left Riemann sum is given by $\dfrac{\pi}{2n} \left(\dfrac{\cos(\pi/(2n)) + \sin(\pi/(2n)) - 1}{2(1 - \cos(\pi/(2n)))} \right)$. Let $\theta = \dfrac{\pi}{2n}$. Then as $n \to \infty$, $\theta \to 0$, and the limit of the left Riemann sum as $n \to \infty$ is 1.

5.4.63 Suppose f is even, so that $f(-x) = f(x)$. Then $f^n(x) = f^n(-x)$, so that f^n is an even function, no matter what the parity of n is.

Suppose g is an odd function, so that $g(-x) = -g(x)$. Then $g^n(-x) = (-1)^n g^n(x)$, so g^n is even when n is even, and is odd when n is odd.

Summarizing, we have:

	f is even	f is odd
n is even	f^n is even	f^n is even
n is odd	f^n is even	f^n is odd

5.4.65

a. Because of the symmetry, $\int_{c-a}^{c} (f(x) - d)\, dx = \int_{c}^{c+a} (d - f(x))\, dx$. Thus, $\int_{c-a}^{c+a} f(x)\, dx = \int_{c-a}^{c} (f(x) - d + d)\, dx + \int_{c}^{c+a} f(x)\, dx = \int_{c-a}^{c} d\, dx + \int_{c-a}^{c} (f(x) - d)\, dx + \int_{c}^{c+a} f(x)\, dx = ad + \int_{c}^{c+a} (d - f(x))\, dx + \int_{c}^{c+a} f(x)\, dx = ad + ad + 0 = 2ad$.

The curve is symmetric about $(\pi/4, 1/2)$. To see this, we will show that if $\sin^2(\pi/4 - x) = (1/2) - r$, then $\sin^2(\pi/4 + x) = (1/2) + r$. Using a double angle identity, we have $\sin^2(\pi/4 - x) = \frac{1}{2} - \frac{1}{2}\cos\left(\frac{\pi}{2} - 2x\right) = \frac{1}{2} - \frac{1}{2}\sin 2x$. On the other hand, $\sin^2(\pi/4 + x) = \frac{1}{2} - \frac{1}{2}\cos\left(\frac{\pi}{2} + 2x\right) = \frac{1}{2} + \frac{1}{2}\sin(2x)$.

c. Using the idea from part a), the area under f over this interval must be equal to the area of the rectangle over the interval $[0, \pi/2]$ with height $1/2$. Thus the area is $\frac{\pi}{2} \cdot \frac{1}{2} = \frac{\pi}{4}$.

5.4.67

a. Note that h is continuous and differentiable on $[a, b]$, and that $h(a) = (a - b)\int_a^a f(t)\,dt + (a - a)\int_a^b g(t)\,dt = 0 + 0 = 0$. Also, $h(b) = (b - b)\int_a^b f(t)\,dt + (b - a)\int_b^b g(t)\,dt = 0 + 0 = 0$. So by Rolle's theorem, there exists c between a and b so that $h'(c) = 0$. By the Product and Sum Rules, and the Fundamental Theorem of Calculus, we have $h'(x) = \int_a^x f(t)\,dt + (x-b)f(x) + \int_x^b g(t)\,dt - (x-a)g(x)$. So at the promised number c we have $h'(c) = \int_a^c f(t)\,dt + (c - b)f(c) + \int_c^b g(t)\,dt - (c - a)g(c) = 0$, so

$$\int_a^c f(t)\,dt + \int_c^b g(t)\,dt = (b - c)f(c) + (c - a)g(c).$$

b. Given f continuous on $[a, b]$, let g be the constant zero function, that is, $g(x) = 0$ for all x. Applying the result of part (a), we have that there exists a c between a and b with

$$\int_a^c f(t)\,dt + 0 = f(c)(b - c) + 0,$$

so

$$\int_a^c f(t)\,dt = f(c)(b - c),$$

as desired.

c. There exists a rectangle with base from c to b and height $f(c)$ so that the area of the rectangle is equal to the value of the integral of f from a to c.

d. Given a function f continuous on $[a,b]$, let $g = f$. Then by part (a) there exists c between a and b so that

$$\int_a^c f(t)\,dt + \int_c^b f(t)\,dt = f(c)(b-c) + f(c)(c-a),$$

so

$$\int_a^b f(t)\,dt = f(c)(b-c+c-a) = f(c)(b-a),$$

so

$$\frac{1}{b-a}\int_a^b f(t)\,dt = f(c).$$

5.5 Substitution Rule

5.5.1 It is based on the Chain Rule for differentiation.

5.5.3 Typically u is substituted for the inner function, so $u = g(x)$.

5.5.5 The new integral is $\displaystyle\int_{g(a)}^{g(b)} f(u)\,du$.

5.5.7 Using the identity $\cos^2 x = \frac{1}{2} + \frac{\cos 2x}{2}$, we have $\displaystyle\int \cos^2 x\,dx = \int \left(\frac{1}{2} + \frac{\cos 2x}{2}\right) dx = \frac{x}{2} + \frac{\sin 2x}{4} + C$.

5.5.9 $\displaystyle\int (x+1)^{12}\,dx = \frac{(x+1)^{13}}{13} + C$, because $\frac{d}{dx}\left(\frac{(x+1)^{13}}{13} + C\right) = (x+1)^{12}$.

5.5.11 $\displaystyle\int \sqrt{2x+1}\,dx = \frac{(2x+1)^{3/2}}{3} + C$, because $\frac{d}{dx}\left(\frac{(2x+1)^{3/2}}{3} + C\right) = \frac{3}{2}\cdot\frac{1}{3}\cdot(2x+1)^{1/2}\cdot 2 = \sqrt{2x+1}$.

5.5.13 Because $u = x^2 + 1$, $du = 2x\,dx$. Substituting yields $\displaystyle\int u^4\,du = \frac{u^5}{5} + C = \frac{(x^2+1)^5}{5} + C$.

5.5.15 Because $u = \sin x$, $du = \cos x\,dx$. Substituting yields $\displaystyle\int u^3\,du = \frac{u^4}{4} + C = \frac{\sin^4(x)}{4} + C$.

5.5.17 Let $u = x^2 - 1$. Then $du = 2x\,dx$. Substituting yields $\displaystyle\int u^{99}\,du = \frac{u^{100}}{100} + C = \frac{(x^2-1)^{100}}{100} + C$.

5.5.19 Let $u = 1 - 4x^3$. Then $du = -12x^2\,dx$, so $-\frac{1}{6}du = 2x^2\,dx$. Substituting yields $-\frac{1}{6}\displaystyle\int \frac{1}{\sqrt{u}}\,du = -\frac{1}{3}\cdot\sqrt{u} + C = -\frac{1}{3}\cdot\sqrt{1-4x^3} + C$.

5.5.21 Let $u = x^2 + x$. Then $du = 2x + 1\,dx$. Substituting yields $\displaystyle\int u^{10}\,du = \frac{u^{11}}{11} + C = \frac{(x^2+x)^{11}}{11} + C$.

5.5.23 Let $u = x^4 + 16$. Then $du = 4x^3\,dx$, so $\frac{1}{4}du = x^3\,dx$. Substituting yields $\frac{1}{4}\displaystyle\int u^6\,du = \frac{1}{4}\cdot\frac{u^7}{7} + C = \frac{(x^4+16)^7}{28} + C$.

5.5.25 Let $u = 4 - 9x^2$. Then $du = -18x\,dx$ so that $\displaystyle\int \frac{x}{\sqrt{4-9x^2}}\,dx = -\frac{1}{18}\int u^{-1/2}\,du = -\frac{\sqrt{u}}{9} + C = -\frac{\sqrt{4-9x^2}}{9} + C$.

5.5.27 Let $u = x^6 - 3x^2$. Then $du = (6x^5 - 6x)\,dx$, so $\frac{1}{6}du = (x^5 - x)\,dx$. Substituting yields $\frac{1}{6}\int u^4\,du =$ $\frac{1}{6}\cdot\frac{u^5}{5} + C = \frac{(x^6 - 3x^2)^5}{30} + C$.

5.5.29 Let $u = t^4$. Then $du = 4t^3\,dt$. Substituting yields $\int t^3\sin t^4\,dt = \frac{1}{4}\int \sin u\,du = -\frac{1}{4}\cos u + C =$ $-\frac{1}{4}\cos t^4 + C$.

5.5.31 Let $u = \sec w + 3$. Then $du = \sec w\tan w\,dw$. Substituting yields $\int(\sec w + 3)^9\sec w\tan w\,dw =$ $\int u^9\,du = \frac{u^{10}}{10} + C = \frac{(\sec w + 3)^{10}}{10} + C$.

5.5.33 Let $u = x - 4$, so that $u + 4 = x$. Then $du = dx$. Substituting yields $\int \frac{u+4}{\sqrt{u}}\,du =$ $\int\left(\frac{u}{\sqrt{u}} + \frac{4}{\sqrt{u}}\right)du = \int u^{1/2} + 4u^{-1/2}\,du = \frac{2}{3}u^{3/2} + 8u^{1/2} + C = \frac{2}{3}\cdot(x-4)^{3/2} + 8\sqrt{x-4} + C$.

5.5.35 Let $u = x + 4$, so that $u - 4 = x$. Then $du = dx$. Substituting yields

$$\int \frac{u-4}{\sqrt[3]{u}}\,du = \int\left(u^{2/3} - 4u^{-1/3}\right)du = \frac{3}{5}u^{5/3} + -6u^{2/3} + C$$
$$= \frac{3}{5}(x+4)^{5/3} - 6(x+4)^{2/3} + C.$$

5.5.37 Let $u = 2x + 1$. Then $du = 2dx$ and $x = \frac{u-1}{2}$. Substituting yields $\frac{1}{2}\int \frac{u-1}{2}\cdot\sqrt[3]{u}\,du = \frac{1}{4}\int(u^{4/3} -$ $u^{1/3})\,du = \frac{1}{4}\left(\frac{3}{7}u^{7/3} - \frac{3}{4}u^{4/3}\right) + C = \frac{3(2x+1)^{7/3}}{28} - \frac{3(2x+1)^{4/3}}{16} + C$.

5.5.39 Let $u = 4 - x^2$. Then $du = -2x\,dx$. Also, when $x = 0$ we have $u = 4$ and when $x = 1$ we have $u = 3$. Substituting yields $-\int_4^3 u\,du = \int_3^4 u\,du = \left(\frac{u^2}{2}\right)\Big|_3^4 = 8 - 4.5 = 3.5$.

5.5.41 Let $u = \sin\theta$. Then $du = \cos\theta\,d\theta$. Also, when $\theta = 0$ we have $u = 0$ and when $\theta = \pi/2$ we have $u = 1$. Substituting yields $\int_0^1 u^2\,du = \left(\frac{u^3}{3}\right)\Big|_0^1 = \frac{1}{3}$.

5.5.43 Let $u = 2y$. Then $du = 2\,dy$. When $y = -\pi/12$ we have $u = -\pi/6$, and when $y = \pi/8$ we have $u = \pi/4$. Thus

$$\int_{-\pi/12}^{\pi/8}\sec^2 2y\,dy = \frac{1}{2}\int_{-\pi/6}^{\pi/4}\sec^2 u\,du = \frac{1}{2}(\tan\pi/4 - \tan(-\pi/6)) = \frac{\sqrt{3}}{6} + \frac{1}{2}.$$

5.5.45 Let $u = \sin x$. Then $du = \cos x\,dx$. Also, when $x = \pi/4$ we have $u = \sqrt{2}/2$ and when $x = \pi/2$ we have $u = 1$. Substituting yields $\int_{\sqrt{2}/2}^1 \frac{1}{u^2}\,du = \left(-\frac{1}{u}\right)\Big|_{\sqrt{2}/2}^1 = \left(-1 - \left(-\frac{2}{\sqrt{2}}\right)\right) = \sqrt{2} - 1$.

5.5.47 Let $u = 2x - 3$. Then $du = 2\,dx$ and $x = (u+3)/2$. When $x = 2$, we have $u = 1$, and when $x = 6$ we have $u = 9$. Thus

$$\int_2^6 \frac{x}{\sqrt{2x-3}}\,dx = \frac{1}{4}\int_1^9 \frac{u+3}{\sqrt{u}}\,du = \frac{1}{4}\left(\frac{2}{3}u^{3/2} + 6\sqrt{u}\right)\Big|_1^9 = \frac{22}{3}.$$

5.5.49 Let $u = w^2 + 1$. Then $du = 2w \, dw$. Also, when $w = 0$ we have $u = 1$ and when $w = \sqrt{3}$ we have $u = 4$. Substituting yields $\displaystyle\int_0^{\sqrt{3}} \frac{w}{\sqrt{w^2 + 1}} \, dw = \frac{1}{2} \int_1^4 u^{-1/2} \, du = (\sqrt{u})\big|_1^4 = 2 - 1 = 1$.

5.5.51 Let $u = 2p + 1$. Then $du = 2 \, dp$. Also when $p = 0$ we have $u = 1$ and when $p = 1$ we have $u = 3$. Substituting yields $\displaystyle\int_0^1 (2p + 1)^3 \, dp = \frac{1}{2} \int_1^3 u^3 \, du = \frac{1}{8} (u^4)\big|_1^3 = \frac{81 - 1}{8} = 10$.

5.5.53 $\displaystyle\int_{-\pi}^{\pi} \cos^2 x \, dx = 2 \int_0^{\pi} \frac{1 + \cos 2x}{2} \, dx = \left(x + \frac{\sin 2x}{2} \right) \Big|_0^{\pi} = \pi$.

5.5.55 $\displaystyle\int \sin^2\left(\theta + \frac{\pi}{6}\right) d\theta = \frac{1}{2} \int \left(1 - \cos\left(2\theta + \frac{\pi}{3}\right)\right) d\theta = \frac{\theta}{2} - \frac{\sin\left(2\theta + \frac{\pi}{3}\right)}{4} + C$.

5.5.57 $\displaystyle\int_{-\pi/4}^{\pi/4} \sin^2 2\theta \, d\theta = 2 \int_0^{\pi/4} \sin^2 2\theta \, d\theta = 2 \int_0^{\pi/4} \frac{1 - \cos 4\theta}{2} \, d\theta = \left(\theta - \frac{\sin 4\theta}{4}\right)\Big|_0^{\pi/4} = \frac{\pi}{4}$.

5.5.59 Let $u = \sin^2 y + 2$ so that $du = 2 \sin y \cos y \, dy = \sin 2y \, dy$. Substituting yields $\displaystyle\int_2^{9/4} \frac{1}{u^2} \, du = \left(-\frac{1}{u}\right)\Big|_2^{9/4} = -\frac{4}{9} + \frac{1}{2} = \frac{1}{18}$.

5.5.61

a. True. This follows by substituting $u = f(x)$ to obtain the integral $\displaystyle\int u \, du = \frac{u^2}{2} + C = \frac{f(x)^2}{2} + C$.

b. True. Again, this follows from substituting $u = f(x)$ to obtain the integral $\displaystyle\int u^n \, du = \frac{u^{n+1}}{n + 1} + C = \frac{(f(x))^{n+1}}{n + 1} + C$ where $n \neq -1$.

c. False. If this were true, then $\sin 2x$ and $2 \sin x$ would have to differ by a constant, which they do not. In fact, $\sin 2x = 2 \sin x \cos x$.

d. False. The derivative of the right hand side is $(x^2 + 1)^9 \cdot 2x$ which is not the integrand on the left hand side.

e. False. If we let $u = f'(x)$, then $du = f''(x) \, dx$. Substituting yields $\displaystyle\int_{f'(a)}^{f'(b)} u \, du = \left(\frac{u^2}{2}\right)\Big|_{f'(a)}^{f'(b)} = \frac{(f'(b))^2}{2} - \frac{(f'(a))^2}{2}$.

5.5.63 Let $u = 10x$. Then $du = 10 \, dw$. Substituting yields $\dfrac{1}{10} \displaystyle\int \sec^2 u \, du = \frac{1}{10} \tan u + C = \frac{1}{10} \tan 10x + C$.

5.5.65 Let $u = \cot x$. Then $du = -\csc^2 x \, dx$. Substituting yields $-\displaystyle\int u^{-3} \, du = \frac{1}{2u^2} + C = \frac{1}{2 \cot^2 x} + C$.

5.5.67 Note that $\sin x \sec^8 x = \dfrac{\sin x}{\cos^8 x}$. Let $u = \cos x$, so that $du = -\sin x \, dx$. Substituting yields $-\displaystyle\int u^{-8} \, du = \frac{1}{7u^7} + C = \frac{1}{7 \cos^7 x} + C = \frac{\sec^7 x}{7} + C$.

5.5.69 Let $u = 1 - x^2$. Then $du = -2x \, dx$. Also note that when $x = 0$ we have $u = 1$, and when $x = 1$ we have $u = 0$. Substituting yields $-\dfrac{1}{2} \displaystyle\int_1^0 \sqrt{u} \, du = \frac{1}{2} \int_0^1 \sqrt{u} \, du = \left(\frac{u^{3/2}}{3}\right)\Big|_0^1 = \frac{1}{3}$.

5.5.71 Let $u = x^2 - 1$, so that $du = 2x\,dx$. Also note that when $x = 2$ we have $u = 3$, and when $x = 3$ we have $u = 8$. Substituting yields $\dfrac{1}{2}\displaystyle\int_3^8 u^{-1/3}\,du = \dfrac{1}{2}\left(\dfrac{3u^{2/3}}{2}\right)\Big|_3^8 = \dfrac{3}{4}\left(4 - \sqrt[3]{9}\right)$.

5.5.73 Let $u = 16 - x^4$. Then $du = -4x^3\,dx$. Also note that when $x = 0$ we have $u = 16$, and when $x = 2$ we have $u = 0$. Substituting yields $\dfrac{1}{4}\displaystyle\int_0^{16}\sqrt{u}\,du = \dfrac{1}{4}\left(\dfrac{2u^{3/2}}{3}\right)\Big|_0^{16} = \dfrac{32}{3}$.

5.5.75 Let $u = 3x + 1$ so that $du = 3\,dx$. Note that $9x^2 + 6x + 1 = (3x+1)^2 = u^2$, and also that when $x = 1$, $u = 4$ and when $x = 2$, $u = 7$. Substituting yields $\dfrac{4}{3}\displaystyle\int_4^7 \dfrac{1}{u^2}\,du = \dfrac{4}{3}\left(-\dfrac{1}{u}\right)\Big|_4^7 = \dfrac{4}{3}\left(-\dfrac{1}{7} - \left(-\dfrac{1}{4}\right)\right) = \dfrac{4}{3}\cdot\dfrac{3}{28} = \dfrac{1}{7}$.

5.5.77 $A(x) = \displaystyle\int_0^{\sqrt{\pi}} x\sin x^2\,dx$. Let $u = x^2$, so that $du = 2x\,dx$. Also, when $x = 0$ we have $u = 0$ and when $x = \sqrt{\pi}$ we have $u = \pi$. Substituting yields $\dfrac{1}{2}\displaystyle\int_0^{\pi}\sin u\,du = \dfrac{1}{2}\left(-\cos u\right)\Big|_0^{\pi} = 1$.

5.5.79 $A(x) = \displaystyle\int_2^6 (x-4)^4\,dx = \dfrac{(x-4)^5}{5}\Big|_2^6 = \dfrac{2^5}{5} - \left(-\dfrac{(2)^5}{5}\right) = \dfrac{64}{5}$.

5.5.81 $A(a) = \displaystyle\int_0^a \left(\dfrac{1}{a} - \dfrac{x^2}{a^3}\right)dx = \left(\dfrac{x}{a} - \dfrac{x^3}{3a^3}\right)\Big|_0^a = 1 - \dfrac{1}{3} = \dfrac{2}{3}$. This is a constant function.

5.5.83

a. Let $u = \sin px$, so that $du = p\cos px\,dx$. Note that when $x = 0$, $u = 0$, and when $x = \pi/2p$, $u = 1$. Substituting yields $\dfrac{1}{p}\displaystyle\int_0^1 f(u)\,du = \dfrac{\pi}{p}$.

b. Let $u = \sin x$ so that $du = \cos x\,dx$. Note that when $x = -\pi/2$, $u = -1$ and when $x = \pi/2$, $u = 1$. Substituting yields $\displaystyle\int_{-1}^1 f(u)\,du = 0$, because f is an odd function. Alternatively, we could note that when f is odd, $\cos x\cdot f(\sin x)$ is also odd, because $\sin x$ is odd and $\cos x$ is even. Thus the given integral must be zero because it is the definite integral of an odd function over a symmetric interval about 0.

5.5.85

a. $\displaystyle\int_0^4 \dfrac{200}{(t+1)^2}\,dt = \left(-\dfrac{200}{t+1}\right)\Big|_0^4 = -40 + 200 = 160$.

b. $\displaystyle\int_0^6 \dfrac{200}{(t+1)^3}\,dt = \left(-\dfrac{200}{2(t+1)^2}\right)\Big|_0^6 = -\dfrac{100}{49} + 100 = \dfrac{4800}{49}$.

c. $\Delta P = \displaystyle\int_0^T \dfrac{200}{(t+1)^r}\,dt$. This decreases as r increases, because $\dfrac{200}{(t+1)^r} > \dfrac{200}{(t+1)^{r+1}}$.

d. Suppose $\displaystyle\int_0^{10} \dfrac{200}{(t+1)^r}\,dt = 350$. Then $\left(\dfrac{200(t+1)^{-r+1}}{1-r}\right)\Big|_0^{10} = 350$, so $11^{1-r} - 1 = \dfrac{350(1-r)}{200}$, and thus $\dfrac{11}{11^r} = \dfrac{7-7r}{4} + \dfrac{4}{4} = \dfrac{11-7r}{4}$, and $11^r = \dfrac{44}{11-7r}$. Using trial and error to find r, we arrive at $r \approx 1.278$.

e. $\displaystyle\int_0^T \dfrac{200}{(t+1)^3}\,dt = \left(-\dfrac{200}{2(t+1)^2}\right)\Big|_0^T = -\dfrac{100}{(T+1)^2} + 100$. As $T \to \infty$, this expression $\to 100$, so in the long run, the bacteria approaches a finite limit.

5.5.87 $\dfrac{1}{\pi/k - 0} \displaystyle\int_0^{\pi/k} \sin kx\, dx = \dfrac{k}{\pi} \cdot \left(-\dfrac{\cos kx}{k} \right) \Big|_0^{\pi/k} = \dfrac{1}{\pi}(1 - (-1)) = \dfrac{2}{\pi}.$

5.5.89 The area on the left is given by $\displaystyle\int_4^9 \dfrac{(\sqrt{x}-1)^2}{2\sqrt{x}}\, dx.$ If we let $u = \sqrt{x} - 1$ so that $du = \dfrac{1}{2\sqrt{x}}\, dx,$ we obtain the equivalent integral $\displaystyle\int_1^2 u^2\, du$ which represents the area on the right.

5.5.91 Let $u = f(x)$, so that $du = f'(x)\, dx$. Substituting yields

$$\int_4^5 (5u^3 + 7u^2 + u)\, du = \left(\dfrac{5u^4}{4} + \dfrac{7u^3}{3} + \dfrac{u^2}{2} \right) \Big|_4^5 = \dfrac{7297}{12}.$$

5.5.93 Let $u = f^{(p)}(x)$ so that $du = f^{(p+1)}(x)\, dx$. Substituting yields

$$\int u^n\, du = \dfrac{u^{n+1}}{n+1} + C = \dfrac{1}{n+1} \left(f^{(p)}(x) \right)^{n+1} + C.$$

5.5.95 If we let $u = \sqrt{x+a}$, then $u^2 = x + a$ and $2u\, du = dx$. Substituting yields $\displaystyle\int_{\sqrt{a}}^{\sqrt{1+a}} (u^2 - a) \cdot u \cdot 2u\, du =$

$\displaystyle\int_{\sqrt{a}}^{\sqrt{1+a}} (2u^4 - 2au^2)\, du = \left(\dfrac{2u^5}{5} - \dfrac{2au^3}{3} \right) \Big|_{\sqrt{a}}^{\sqrt{1+a}} = \dfrac{2(\sqrt{1+a})^5}{5} - \dfrac{2a(\sqrt{1+a})^3}{3} - \dfrac{2a^{5/2}}{5} + \dfrac{2a^{5/2}}{3}.$

If we let $u = x + a$, then $u - a = x$ and $du = dx$. Substituting yields $\displaystyle\int_a^{a+1} (u - a)\sqrt{u}\, du =$

$\left(\dfrac{2u^{5/2}}{5} - \dfrac{2au^{3/2}}{3} \right) \Big|_a^{a+1} = \dfrac{2(a+1)^{5/2}}{5} - \dfrac{2a(a+1)^{3/2}}{3} - \dfrac{2a^{5/2}}{5} + \dfrac{2a^{5/2}}{3}.$

Note that the two results are the same.

5.5.97 If we let $u = \cos\theta$, then $du = -\sin\theta\, d\theta$. Substituting yields $\displaystyle\int -u^{-4}\, du = \dfrac{1}{3u^3} + C = \dfrac{1}{3\cos^3\theta} + C = \dfrac{\sec^3\theta}{3} + C.$

If we let $u = \sec\theta$, then $du = \sec\theta\tan\theta\, d\theta$. Substituting yields $\displaystyle\int u^2\, du = \dfrac{u^3}{3} + C = \dfrac{\sec^3\theta}{3} + C.$ Note that the two results are the same.

5.5.99

a. Because $\sin 2x = 2\sin x \cos x$, we can write $(\sin x \cos x)^2 = \left(\dfrac{\sin 2x}{2} \right)^2 = \dfrac{\sin^2 2x}{4}$. Then we have $I = \dfrac{1}{4}\displaystyle\int \sin^2 2x\, dx = \dfrac{1}{4}\left(\dfrac{x}{2} - \dfrac{\sin 4x}{8} \right) + C = \dfrac{x}{8} - \dfrac{\sin 4x}{32} + C.$ Note that we used the result of the previous problem during this derivation.

b. $I = \dfrac{1}{4}\displaystyle\int (1 - \cos 2x)(1 + \cos 2x)\, dx = \dfrac{1}{4}\displaystyle\int (1 - \cos^2 2x)\, dx = \dfrac{1}{4}\displaystyle\int \sin^2 2x\, dx = \dfrac{1}{4}\left(\dfrac{x}{2} - \dfrac{\sin 4x}{8} \right) + C = \dfrac{x}{8} - \dfrac{\sin 4x}{32} + C.$

c. The results are consistent. The work involved is similar in each method.

5.5.101

a. Let $u = cx$. Note that $du = c \cdot dx$. Substitution yields $\int_a^b f(cx)\, dx = \dfrac{1}{c} \int_{ac}^{bc} f(u)\, du$.

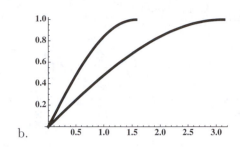

b.

5.5.103 Let $u = \sqrt{x+1}$ so that $u^2 = x+1$. Then $2u\, du = dx$. Substituting yields $\int 2 \cdot \dfrac{u\, du}{\sqrt{1+u}}$. Now let $v = \sqrt{1+u}$ so that $v^2 = 1+u$ and $2v\, dv = du$. Now a substitution yields $4 \int \dfrac{(v^2-1)v}{v}\, dv = 4 \int (v^2 - 1)\, dv = \dfrac{4v^3}{3} - 4v + C = \dfrac{4}{3}(1+u)^{3/2} - 4\sqrt{1+u} + C = \dfrac{4}{3}\left(1 + \sqrt{x+1}\right)^{3/2} - 4\sqrt{1 + \sqrt{x+1}} + C$.

5.5.105 Let $u = 1 - \sqrt{x}$. Then $x = (1-u)^2$ and $dx = 2(1-u)(-du) = -2(1-u)\cdot du$. Then we have

$$\int_0^1 \sqrt{x - x\sqrt{x}}\, dx = \int_1^0 (1-u)\sqrt{u}\cdot(-2)(1-u)\, du$$

$$= 2\int_0^1 (1-u)^2 \sqrt{u}\, du = 2\int_0^1 \left(1 - 2u + u^2\right)\sqrt{u}\, du$$

$$= 2\int_0^1 \left(u^{1/2} - 2u^{3/2} + u^{5/2}\right) du$$

$$= 2\left(\frac{2u^{3/2}}{3} - \frac{4u^{5/2}}{5} + \frac{2u^{7/2}}{7}\right)\Bigg|_0^1$$

$$= 2\left(\frac{2}{3} - \frac{4}{5} + \frac{2}{7}\right) = \frac{32}{105}.$$

5.5.107 Let $u = \cos\theta$, so that $du = -\sin\theta\, d\theta$. This substitution yields $\int_0^1 \dfrac{u}{\sqrt{u^2 + 16}}\, du$. Now let $v = u^2 + 16$, so that $dv = 2u\, du$. Now a substitution yields $\dfrac{1}{2}\int_{16}^{17} v^{-1/2}\, dv = \sqrt{v}\,\Big|_{16}^{17} = \sqrt{17} - 4$.

Chapter Five Review

1

a. True. The antiderivative of a linear function is a quadratic function.

b. False. $A'(x) = f(x)$, not $F(x)$.

c. True, Note that f is an antiderivative of f', so this follows from the Fundamental Theorem.

d. True. Because $|f(x)| \ge 0$ for all x, this integral must be positive, unless f is constantly 0.

e. False. For example, the average value of $\sin x$ on $[0, 2\pi]$ is zero.

f. True. This is equal to $2\int_a^b f(x)\, dx - 3\int_a^b g(x)\, dx = 2\int_a^b f(x)\, dx + 3\int_b^a g(x)\, dx$.

g. True. The derivative of the right hand side is $f'(g(x))g'(x)$ by the Chain Rule.

3

a. This region can be divided up into a 4×2 rectangle and a right triangle with base and height equal to 1. Thus, the integral is equal to $8 + \frac{1}{2} = 8.5$.

b. $\int_6^4 f(x)\,dx = -\int_4^6 f(x)\,dx$. The region whose area is $\int_4^6 f(x)\,dx$ consists of a 1×3 rectangle, together with a right triangle with base 1 and height 3, so $\int_4^6 f(x)\,dx = 3 + \frac{3}{2} = 4.5$, and $\int_6^4 f(x)\,dx = -4.5$.

c. $\int_5^7 f(x)\,dx = \int_5^6 f(x)\,dx + \int_6^7 f(x)\,dx$. The region lying over $[5, 6]$ is a right triangle with height 3 and base 1, so its area is $\frac{3}{2}$. The region lying under $[6, 7]$ has the same area, but is below the x-axis, so $\int_6^7 f(x)\,dx = \frac{-3}{2}$. So $\int_5^7 f(x)\,dx = \frac{3}{2} + \frac{-3}{2} = 0$.

d. Note that $\int_4^5 f(x)\,dx = 3$, because the area represented is that of a 1×3 rectangle. Now by the work above, $\int_0^7 f(x)\,dx = \int_0^4 f(x)\,dx + \int_4^5 f(x)\,dx + \int_5^7 f(x)\,dx = 8.5 + 3 + 0 = 11.5$.

5 $\int_0^4 \sqrt{8x - x^2}\,dx = \int_0^4 \sqrt{16 - (x^2 - 8x + 16)}\,dx = \int_0^4 \sqrt{16 - (x - 4)^2}\,dx$. This represents one quarter of the area inside the circle centered at $(4, 0)$ with radius 4, so its value is $\frac{1}{4} \cdot 16\pi = 4\pi$.

7

a. The right Riemann sum is $4 \cdot 1 + 7 \cdot 1 + 10 \cdot 1 = 21$.

b. The right Riemann sum is $\displaystyle\sum_{k=1}^n \left(3\left(1 + \frac{3k}{n}\right) - 2 \right) \cdot \frac{3}{n}$.

c. The sum evaluates as $\displaystyle\sum_{k=1}^n \frac{3}{n} + \sum_{k=1}^n \frac{27k}{n^2} = 3 + \frac{27}{n^2} \cdot \frac{n(n+1)}{2}$. As $n \to \infty$, the limit of this expression is $3 + 13.5 = 16.5$.

d. The area consists of a trapezoid with base 3 and heights 1 and 10, so the value is $(3)\left(\frac{1+10}{2}\right) = \frac{33}{2} = 16.5$. The Fundamental Theorem assures us that $\int_1^4 (3x - 2)\,dx = \left(\frac{3x^2}{2} - 2x\right)\Big|_1^4 = (24 - 8) - (3/2 - 2) = 16 + 1/2 = 16.5$.

9 Let $\Delta x = \frac{2-0}{n} = \frac{2}{n}$. Let $x_k = 0 + k\Delta x = \frac{2k}{n}$. Then $f(x_k) = \frac{4k^2}{n^2} - 4$. Thus,

$$\lim_{n \to \infty} \sum_{k=1}^n f(x_k)\Delta x = \lim_{n \to \infty} \sum_{k=1}^n \left(\frac{4k^2}{n^2} - 4\right) \cdot \frac{2}{n} = \lim_{n \to \infty} \left(\frac{8}{n^3} \sum_{k=1}^n k^2 - \frac{8}{n} \sum_{k=1}^n 1\right) =$$

$$\lim_{n \to \infty} \left(\frac{8}{n^3} \cdot \frac{n(n+1)(2n+1)}{6} - 8\right) = \frac{8}{3} - 8 = \frac{-16}{3}.$$

11 Let $\Delta x = \frac{4-0}{n} = \frac{4}{n}$. Let $x_k = 0 + k\Delta x = 0 + \frac{4k}{n}$. Then $f(x_k) = \frac{64k^3}{n^3} - \frac{4k}{n}$. Thus,

$$\lim_{n\to\infty} \sum_{k=1}^{n} f(x_k)\Delta x = \lim_{n\to\infty} \sum_{k=1}^{n} \left(\frac{64k^3}{n^3} - \frac{4k}{n}\right) \cdot \frac{4}{n} = \lim_{n\to\infty} \left(\frac{256}{n^4} \sum_{k=1}^{n} k^3 - \frac{16}{n^2} \sum_{k=1}^{n} k\right) =$$

$$\lim_{n\to\infty} \left(\frac{256}{n^4} \cdot \frac{n^2(n+1)^2}{4} - \frac{16}{n^2} \cdot \frac{n(n+1)}{2}\right) = 64 - 8 = 56.$$

13 This sum is equal to $\int_0^4 (x^5 + 1)\,dx = \left(\frac{x^6}{6} + x\right)\bigg|_0^4 = \frac{4^6}{6} + 4 = \frac{2060}{3}$.

15 $\int_{-2}^2 (3x^4 - 2x + 1)\,dx = \left(\frac{3x^5}{5} - x^2 + x\right)\bigg|_{-2}^2 = \frac{96}{5} - 4 + 2 - \left(-\frac{96}{5} - 4 - 2\right) = \frac{192}{5} + 4 = \frac{212}{5}$.

17 $\int_0^2 (x+1)^3\,dx = \left(\frac{(x+1)^4}{4}\right)\bigg|_0^2 = \frac{81}{4} - \frac{1}{4} = 20$.

19 $\int (9x^8 - 7x^6)\,dx = x^9 - x^7 + C$.

21 $\int_0^1 (x + \sqrt{x})\,dx = \left(\frac{x^2}{2} + \frac{2x^{3/2}}{3}\right)\bigg|_0^1 = \frac{1}{2} + \frac{2}{3} = \frac{7}{6}$.

23 Let $u = 4 - x^2$. Then $du = -2x\,dx$. Note that when $x = 0$, $u = 4$, and when $x = 1$, $u = 3$. We have

$$\int_0^1 \frac{6x}{(4-x^2)^{3/2}}\,dx = -3\int_0^1 \frac{-2x\,dx}{(4-x^2)^{3/2}} = 3\int_3^4 u^{-3/2}\,du = -6\left(u^{-1/2}\right)\bigg|_3^4 = -3 + \frac{6}{\sqrt{3}} = 2\sqrt{3} - 3.$$

25 Let $u = 25 - x^2$, and note that $du = -2x\,dx$. Substituting yields $-\frac{1}{2}\int_{25}^{16} u^{-1/2}\,du = -\sqrt{u}\,\bigg|_{25}^{16} = 5 - 4 = 1$.

27 $\int_0^\pi \sin^2 5\theta\,d\theta = \int_0^\pi \frac{1 - \cos 10\theta}{2}\,d\theta = \left(\frac{\theta}{2} - \frac{\sin 10\theta}{20}\right)\bigg|_0^\pi = \frac{\pi}{2}$.

29 Let $u = x^3 + 3x^2 - 6x$. Then $du = (3x^2 + 6x - 6)\,dx$. We have $\int \frac{x^2 + 2x - 2}{(x^3 + 3x^2 - 6x)^2}\,dx$

$= \frac{1}{3}\int \frac{(3x^2 + 6x - 6)\,dx}{(x^3 + 3x^2 - 6x)^2} = \frac{1}{3}\int \frac{du}{u^2} = -\frac{1}{3u} + C = -\frac{1}{3(x^3 + 3x^2 - 6x)} + C.$

31 The area is given by $\int_{-4}^4 (16 - x^2)\,dx = \left(16x - x^3/3\right)\bigg|_{-4}^4 = 64 - 64/3 - (-64 + 64/3) = 256/3$.

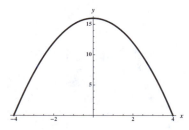

33 The area is given by $2\int_0^{2\pi} \sin(x/4)\,dx = 2\left(-4\cos(x/4)\right)\bigg|_0^{2\pi} = -8(0 - 1) = 8$.

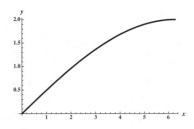

35

i. $\displaystyle\int_{-1}^{1} (x^4 - x^2)\, dx = \left(x^5/5 - x^3/3\right)\Big|_{-1}^{1} = (1/5 - 1/3) - (-1/5 - (-1/3)) = -4/15.$

ii. Because the region lies completely below the x-axis, the area bounded by the curve and the x-axis is
$$-\int_{-1}^{1} (x^4 - x^2)\, dx = 4/15.$$

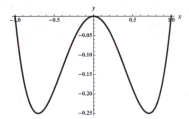

37

a. $\displaystyle\int_{-4}^{4} f(x)\, dx = 2\int_{0}^{4} f(x)\, dx = 2\cdot 10 = 20.$

b. $\displaystyle\int_{-4}^{4} 3g(x)\, dx = 3\cdot 0 = 0.$

c. $\displaystyle\int_{-4}^{4} 4f(x) - 3g(x)\, dx = 2\cdot 4\cdot\int_{0}^{4} f(x)\, dx - 3\cdot 0 = 8\cdot 10 - 0 = 80.$

d. Let $u = 4x^2$, so that $du = 8x\, dx$. Substituting yields $\displaystyle\int_{0}^{4} f(u)\, du = 10.$

e. Because f is an even function, $3xf(x)$ is an odd function. Thus, $\displaystyle\int_{-2}^{2} 3xf(x)\, dx = 0.$

39 $\displaystyle\int_{1}^{4} 3f(x)\, dx = 3\int_{1}^{4} f(x)\, dx = 3\cdot 6 = 18.$

41 $\displaystyle\int_{1}^{4} (3f(x) - 2g(x))\, dx = 3\int_{1}^{4} f(x)\, dx - 2\int_{1}^{4} g(x)\, dx = 3\cdot 6 - 2\cdot 4 = 18 - 8 = 10.$

43 There is not enough information to compute this integral.

45 The displacement is $\displaystyle\int_{0}^{2} 5\sin \pi t\, dt = \left(\frac{-5}{\pi}\cos \pi t\right)\Big|_{0}^{2} = 0.$
The distance traveled is
$$\int_{0}^{2} 5|\sin \pi t|\, dt = 5\int_{0}^{1} \sin \pi t\, dt + 5\int_{1}^{2} (-\sin \pi t)\, dt = \left(\frac{-5}{\pi}\cos \pi t\right)\Big|_{0}^{1} + \left(\frac{5}{\pi}\cos \pi t\right)\Big|_{1}^{2} = \frac{10}{\pi} + \frac{10}{\pi} = \frac{20}{\pi}.$$

47

   a. The average value is 2.5. This is because for a straight line, the average value occurs at the midpoint of the interval, which is at the point $(3.5, 2.5)$, so $c = 3.5$.

   b. The average value is 3 over the interval $[2, 4]$ and 3 over the interval $[4, 6]$, so is 3 over the interval $[2, 6]$. The function takes on this value at $c = 3$ and $c = 5$.

49 Let $u = 2x$, so that $du = 2\, dx$. We have $\dfrac{1}{2}\displaystyle\int_2^4 f'(u)\, du = \dfrac{1}{2} \cdot (f(4) - f(2)) = \dfrac{f(4)}{2} - 2$. Because we are given that this quantity is 10, we have $f(4) = 24$.

51

By the Fundamental Theorem, $f'(x) = \frac{1}{x}$, which is always positive for $x > 1$. Thus f is always increasing. Also, $f(1) = \displaystyle\int_1^1 \frac{1}{t}\, dt = 0$. Also, $f''(x) = \frac{-1}{x^2}$ which is always negative, so f is always concave down.

53

   a.

   b. The area is $2 \cdot 30 + 0 + 15 \cdot 1 = 75$.

   c. The area represents the distance that the diver ascends.

55

   a. $F(-2) = \displaystyle\int_{-1}^{-2} f(t)\, dt = \int_{-1}^{-2} t\, dt = \left.\frac{t^2}{2}\right|_{-1}^{-2} = \frac{3}{2}$.

$F(2) = \displaystyle\int_{-1}^{2} f(t)\, dt = \int_{-1}^{0} f(t)\, dt + \int_{0}^{2} f(t)\, dt = \int_{-1}^{0} t\, dt + \int_{0}^{2} \frac{t^2}{2}\, dt = \left.\frac{t^2}{2}\right|_{-1}^{0} + \left.\frac{t^3}{6}\right|_{0}^{2} = \left(0 - \frac{1}{2}\right) + \left(\frac{8}{6} - 0\right) = \frac{5}{6}$.

   b. By the Fundamental Theorem of Calculus, $F'(x) = f(x)$, so for $-2 \le x < 0$ we have $F''(x) = x$.

   c. By the Fundamental Theorem of Calculus, $F'(x) = f(x)$, so for $0 \le x < 2$ we have $F''(x) = \frac{x^2}{2}$.

   d. $F'(-1) = -1$ and $F'(1) = \frac{1}{2}$. These represent the rate of change of F at the given points. Because the graph in the exercise is the derivative of F, this is just the value of f at the given points.

e. Note that $F''(x) = \begin{cases} 1 & \text{if } -2 \le x < 0, \\ x & \text{if } 0 \le x \le 2. \end{cases}$ Thus we have $F''(-1) = 1$ and $F''(1) = 1$.

f. The difference $F(x) - G(x) = \displaystyle\int_{-1}^{-2} f(t)\,dt = \frac{3}{2}$, as noted in part (a).

57 By L'hôpital's rule, we have

$$\lim_{x \to 0} \frac{\int_0^x 2\cos t^2\,dt}{x} = \lim_{x \to 0} \frac{2\cos x^2}{1} = 2.$$

59

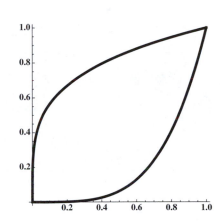

Because x^n and $\sqrt[n]{x}$ are inverse functions of each other, they are symmetric in the square $[0,1] \times [0,1]$ about the line $y = x$. Together, the two regions completely fill up the 1×1 square, so these two areas add to one.

61 Let $u = x^2$. Then $du = 2x\,dx$ so that

$$\int \sec^8(\tan x^2)\sin(\tan x^2)x\sec^2 x^2\,dx = \frac{1}{2}\int \sec^8(\tan u)\sin(\tan u)\sec^2 u\,du.$$

Now substitute $v = \tan u$. Then $dv = (1 + \tan^2 u)\,du = (1 + v^2)\,du$, so that (using the fact that $\sec^2 u = 1 + \tan^2 u$)

$$\frac{1}{2}\int \sec^8(\tan u)\sin(\tan u)\sec^2 u\,du = \frac{1}{2}\int \sec^8 v\sin v(1 + v^2)\frac{1}{1 + v^2}\,dv = \frac{1}{2}\int \frac{\sin v}{\cos^8 v}\,dv$$

$$= \frac{1}{14}\sec^7 v + C = \frac{1}{14}\sec^7(\tan u) + C = \frac{1}{14}\sec^7(\tan x^2) + C.$$

63 Let $u = \frac{1}{x}$. Then $du = -\frac{1}{x^2}\,dx$. Substituting yields $-\displaystyle\int \sin u\,du = \cos u + C = \cos\left(\frac{1}{x}\right) + C$.

65 Let $u = \sqrt{x} + 1$. Then $\sqrt{x} = u - 1$ and $du = \frac{1}{2\sqrt{x}}\,dx$, which can be written as $dx = 2\sqrt{x}\,du = (2u - 2)\,du$. Also when $x = 0$ we have $u = 1$, and when $x = 1$ we have $u = 2$. Substituting yields
$$\int_1^2 \frac{(u-1)(2u-2)}{u^4}\,du = 2\int_1^2 \frac{u^2 - 2u + 1}{u^4}\,du = 2\int_1^2 \left(u^{-2} - 2u^{-3} + u^{-4}\right)\,du = 2\left(-\frac{1}{u} + \frac{1}{u^2} - \frac{1}{3u^3}\right)\Big|_1^2 =$$
$$2\left(-\frac{1}{2} + \frac{1}{4} - \frac{1}{24} - \left(-1 + 1 - \frac{1}{3}\right)\right) = \frac{1}{12}.$$

67 This follows by differentiating each side of the equation. $\dfrac{d}{dx}\left(u(x) + 2\displaystyle\int_0^x u(t)\,dt\right) = u'(x) + 2u(x)$, and $\dfrac{d}{dx}10 = 0$. The reverse is not true, because if $u(x) + 2\displaystyle\int_0^x u(t)\,dt = C$ for any constant C, then it would satisfy the second equation, even if $C \ne 10$.

69   Note that $f'(x) = (x-1)^{15}(x-2)^9$, and that the zeros of f' are at $x = 1$ and $x = 2$.

   a. f' is positive and thus f is increasing on $(-\infty, 1)$ and on $(2, \infty)$, while f' is negative and f is decreasing on $(1, 2)$.

   b. $f''(x) = 15(x-1)^{14}(x-2)^9 + (x-1)^{15} \cdot 9(x-2)^8 = 3(x-1)^{14}(x-2)^8(8x-13)$.

     f is concave up on $\left(\frac{13}{8}, \infty\right)$ and concave down on $\left(-\infty, \frac{13}{8}\right)$.

   c. f has a local maximum at $x = 1$ and a local minimum at $x = 2$.

   d. f has an inflection point at $x = \frac{13}{8}$.

Chapter 6

Applications of Integration

6.1 Velocity and Net Change

6.1.1 The position of an object is the coordinate of the object on the line at a given time, often denoted $s(t)$. The displacement over an interval $[a, b]$ is $s(b) - s(a)$, the difference of the object's ending position and beginning position. It can be written as $\int_a^b v(t)\, dt$ where $v(t)$ is the object's velocity at time t. The distance traveled by the object is $\int_a^b |v(t)|\, dt$, the sum of the distance traveled along the line to the right and the distance traveled along the line to the left over the given time interval.

6.1.3 The displacement is given by $\int_a^b v(t)\, dt$, because this quantity is equal to $s(b) - s(a)$.

6.1.5 The value of Q at time t will be given by $Q(t) = Q(0) + \int_0^t Q'(x)\, dx$.

6.1.7

    a. The velocity is positive for $0 \le t < 1$ and for $3 < t < 5$, so the object is moving in the positive direction on those intervals.

    b. The displacement is $\displaystyle\int_0^3 v(t)\, dt = 12 - 16 = -4$.

    c. The distance traveled is $\displaystyle\int_1^5 |v(t)|\, dt = 16 + 10 = 26$.

    d. The displacement is $\displaystyle\int_0^5 v(t)\, dt = 12 - 16 + 10 = 6$.

    e. After five hours the object's position is 6 miles from the original position in the positive direction.

6.1.9

    a. The motion is positive for $0 \le t < 3$ and negative for $3 < t \le 6$.

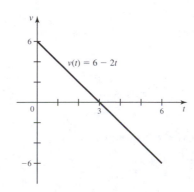

b. The displacement is $\int_0^6 (6 - 2t)\, dt = \left(6t - t^2\right)\big|_0^6 = 0$ m.

c. The distance traveled is $\int_0^3 (6 - 2t)\, dt + \int_3^6 (2t - 6)\, dt = 9 + 9 = 18$ m.

6.1.11

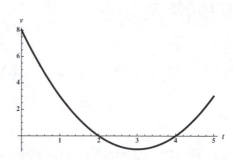

a. The motion is positive for $0 \le t < 2$ and $4 < t < 5$, and negative for $2 < t < 4$.

b. The displacement is $\displaystyle\int_0^5 (t^2 - 6t + 8)\, dt = \left(t^3/3 - 3t^2 + 8t\right)\big|_0^5 = 125/3 - 75 + 40 = \dfrac{20}{3}$ m.

c. The distance traveled from time 0 to 2 is $\displaystyle\int_0^2 (t^2 - 6t + 8)\, dt = \left(t^3/3 - 3t^2 + 8t\right)\big|_0^2 = 8/3 - 12 + 16 = \dfrac{20}{3}$.

   From time 2 to 4 is $-\displaystyle\int_2^4 (t^2 - 6t + 8)\, dt = -\left(t^3/3 - 3t^2 + 8t\right)\big|_2^4 = -(64/3 - 48 + 32) + (8/3 - 12 + 16) = \dfrac{4}{3}$

   m. From time 4 to 5 is $\displaystyle\int_4^5 (t^2 - 6t + 8)\, dt = \left(t^3/3 - 3t^2 + 8t\right)\big|_4^5 = (125/3 - 75 + 40) - (64/3 - 48 + 32) = \dfrac{4}{3}$

   m. Thus the total distance traveled is $\dfrac{20}{3} + \dfrac{4}{3} + \dfrac{4}{3} = \dfrac{28}{3}$ m.

6.1.13

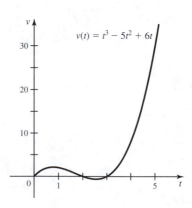

$v(t) = t^3 - 5t^2 + 6t$

a. The motion is positive for $0 < t < 2$ and $3 < t \le 5$, and negative for $2 < t < 3$.

b. The displacement is $\int_0^5 (t^3 - 5t^2 + 6t)\, dt = \left(\frac{t^4}{4} - \frac{5t^3}{3} + 3t^2\right)\Big|_0^5 = \frac{275}{12}$ m.

c. The distance traveled is $\int_0^2 v(t)\, dt - \int_2^3 v(t)\, dt + \int_3^5 v(t)\, dt = \left(\frac{t^4}{4} - \frac{5t^3}{3} + 3t^2\right)\Big|_0^2 - \left(\frac{t^4}{4} - \frac{5t^3}{3} + 3t^2\right)\Big|_2^3 + \left(\frac{t^4}{4} - \frac{5t^3}{3} + 3t^2\right)\Big|_3^5 = \frac{8}{3} + \frac{5}{12} + \frac{62}{3} = \frac{95}{4} = 23.75$ m.

6.1.15

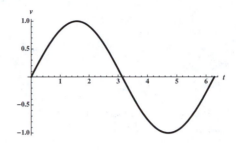

a. The motion is positive for $0 < t < \pi$, and negative for $\pi < t < 2\pi$.

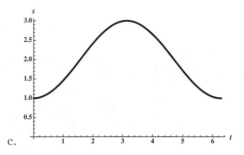

b. $s(t) = \int \sin t \, dt = -\cos t + C$, and because $s(0) = 1$, we must have $C = 2$. Thus, $s(t) = 2 - \cos t$. Also, $s(t) = s(0) + \int_0^t \sin x \, dx = 1 + (-\cos x)\big|_0^t = 2 - \cos t$.

c.

6.1.17

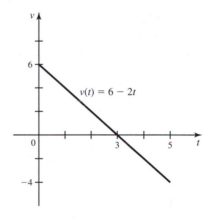

a. The motion is positive for $0 \leq t < 3$, and negative for $3 < t \leq 5$.

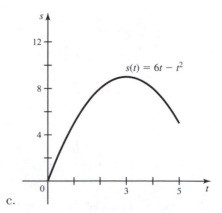

b. $s(t) = \int (6 - 2t) \, dt = 6t - t^2 + C$, and because $s(0) = 0$, we must have $C = 0$. Thus, $s(t) = 6t - t^2$. Also, $s(t) = s(0) + \int_0^t (6 - 2x) \, dx = (6x - x^2)\big|_0^t = 6t - t^2$.

c.

6.1.19

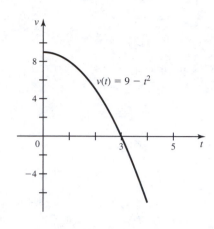

a.   The motion is positive for $0 \leq t < 3$, and negative for $3 < t \leq 4$.

b. $s(t) = \int (9 - t^2)\, dt = 9t - t^3/3 + C$, and because $s(0) = -2$, we must have $C = -2$. Thus, $s(t) = 9t - \frac{t^3}{3} - 2$. Also, $s(t) = s(0) + \int_0^t (9 - x^2)\, dx = -2 + \left(9x - x^3/3\right)\big|_0^t = 9t - \frac{t^3}{3} - 2$.

c.

6.1.21

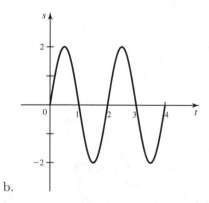

a. $s(t) = s(0) + \int_0^t 2\pi \cos \pi x\, dx = 2 \sin \pi x\big|_0^t = 2 \sin \pi t$.

b.

c. The mass reaches its lowest point at $t = 1.5$, $t = 3.5$ and $t = 5.5$.

d. The mass reaches its highest point at $t = .5$, $t = 2.5$, and $t = 4.5$.

6.1.23

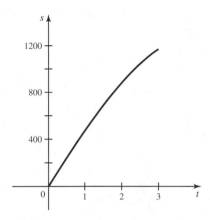

a. $s(t) = s(0) + \int_0^t 30(16 - x^2)\,dx = \left(480x - 10x^3\right)\big|_0^t = 480t - 10t^3 = 10t(48 - t^2).$

b. Because the velocity is positive, this is given by $s(2) - s(0) = 960 - 80 = 880$ miles.

c. The velocity is 400 when $480 - 30t^2 = 400$, or $t = \sqrt{8/3}$. At this point the plane has traveled $s(\sqrt{8/3}) = 480\sqrt{8/3} - 10\sqrt{8/3}^3 \approx 740.290$ miles.

6.1.25

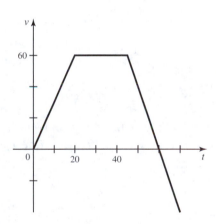

a. The velocity has a maximum of 60 for $20 \leq t \leq 45$. The velocity is 0 at $t = 0$ and at $t = 60$.

b. $\int_0^{20} 3t\,dt + \int_{20}^{30} 60\,dt = 1200$ m.

c. $1200 + \int_{30}^{45} 60\,dt + \int_{45}^{60}(240 - 4t)\,dt = 1200 + 900 + \left(240t - 2t^2\right)\big|_{45}^{60} = 2550$ m.

d. At time $t = 60$ the automobile is at position 2550. In the following 15 seconds, it moves $\int_{60}^{75}(240 - 4t)\,dt = 450$ feet in the opposite direction, so it is at position $2550 - 450 = 2100$.

6.1.27 $v(t) = \int a(t)\,dt = \int(-32)\,dt = -32t + C$, and because $v(0) = 70$, we have $C = 70$, so $v(t) = -32t + 70$. $s(t) = \int v(t)\,dt = \int(70 - 32t)\,dt = 70t - 16t^2 + D$, and because $s(0) = 10$, we must have $D = 10$. Thus $s(t) = -16t^2 + 70t + 10$.

6.1.29 $v(t) = \int a(t)\,dt = \int(-9.8)\,dt = -9.8t + C$, and because $v(0) = 20$, we have $C = 20$, so $v(t) = -9.8t + 20$. $s(t) = \int v(t)\,dt = \int(20 - 9.8t)\,dt = 20t - 4.9t^2 + D$, and because $s(0) = 0$, we must have $D = 0$. Thus $s(t) = 20t - 4.9t^2$.

6.1.31 $v(t) = \int a(t)\, dt = \int (-0.01t)\, dt = -.005t^2 + C$, and because $v(0) = 10$, we have $C = 10$, so $v(t) = 10 - .005t^2$.

$s(t) = \int v(t)\, dt = \int (10 - .005t^2)\, dt = 10t - \frac{1}{600}t^3 + D$, and because $s(0) = 0$, we must have $D = 0$. Thus $s(t) = 10t - \frac{1}{600}t^3$.

6.1.33

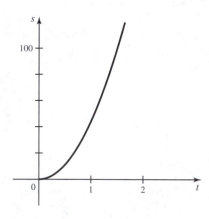

a.
The velocity is given by $\int 88\, dt = 88t + C$, and $C = 0$ because $v(0) = 0$, so $v(t) = 88t$ ft/s. The position is given by $\int 88t\, dt = 44t^2 + D$, but $D = 0$ because $s(0) = 0$, so $s(t) = 44t^2$ ft.

b. The car travels $s(4) = 44 \cdot 16 = 704$ feet.

c. Because a quarter mile is 1320 feet, we need $44t^2 = 1320$, so $t = \sqrt{30} \approx 5.477$ seconds.

d. We need $44t^2 = 300$, so $t \approx 2.611$ seconds.

e. It reaches that speed when $88t = 178$, or $t = 89/44$ seconds. At that time the racer has traveled $s(89/44) = 44(89/44)^2 = 89^2/44 \approx 180.023$ feet.

6.1.35 $v(t) = \int a(t)\, dt = \int -\frac{1280}{(1+8t)^3}\, dt = \frac{80}{(1+8t)^2} + C$, and $C = 0$, because $v(0) = 80$.

$s(t) = \int v(t)\, dt = \int \frac{80}{(1+8t)^2}\, dt = -\frac{10}{1+8t} + D$, but we can take $D = 0$ because the initial position is unspecified. Then in the first .2 seconds the train travels $s(.2) - s(0) = -\frac{10}{2.6} - (-10) = 10 - \frac{50}{13} = \frac{80}{13} \approx 6.154$ miles. Between time 0.2 and 0.4 the train travels $s(.4) - s(.2) = -\frac{10}{4.2} - \left(-\frac{10}{2.6}\right) = \frac{50}{13} - \frac{50}{21} = \frac{400}{273} \approx 1.465$ miles.

6.1.37

a. In the first 35 days the number of barrels produced is $\int_0^{30} 800\, dt + \int_{30}^{35} (2600 - 60t)\, dt = 24000 + 3250 = 27250$.

b. In the first 50 days the number of barrels produced is $27250 + \int_{35}^{40}(2600 - 60t)\, dt + \int_{40}^{50} 200\, dt = 27250 + 1750 + 2000 = 31000$.

c. A constant 200 barrels per day times 20 days yields 4000 barrels.

6.1.39

a. $P(20) = 250 + \int_0^{20}(30 + 30\sqrt{t})\, dt = 250 + \left(30t + 20t^{3/2}\right)\big|_0^{20} = 250 + 600 + 800\sqrt{5} = 850 + 800\sqrt{5} \approx 2639$ people.

b. $P(t) = 250 + \int_0^t (30 + 30\sqrt{x})\, dx = 250 + \left(30x + 20x^{3/2}\right)\big|_0^t = 250 + 30t + 20t^{3/2}$ people.

6.1.41

a. After 14 days the population is $N(14) = 1500 + \int_0^{14} 200(t+2)^{-1/2}\, dt = 1500 + \left(400\sqrt{t+2}\right)\big|_0^{14} = 3100 - 400\sqrt{2}$. After 34 days the population is $N(34) = 1500 + \int_0^{34} 200(t+2)^{-1/2}\, dt = 1500 + \left(400\sqrt{t+2}\right)\big|_0^{34} = 3900 - 400\sqrt{2} \approx 3334$ cells.

b. $N'(t) = 200(t + 2)^{-1/2}$ gives $N(t) = 400\sqrt{t + 2} + C$; the initial condition gives $1500 = 400\sqrt{2} + C$ so that $C = 1500 - 400\sqrt{2}$ and $N(t) = 400\sqrt{t + 2} + 1500 - 400\sqrt{2}$.

6.1.43

a. The additional cost is $\int_{100}^{150} (2000 - .5x)\,dx = \left(2000x - \frac{x^2}{4}\right)\Big|_{100}^{150} = 96875$ dollars.

b. The additional cost is $\int_{500}^{550} (2000 - .5x)\,dx = \left(2000x - \frac{x^2}{4}\right)\Big|_{500}^{550} = 86875$ dollars.

6.1.45

a. The additional cost is $\int_{100}^{150} (300 + 10x - .01x^2)\,dx = \left(300x + 5x^2 - \frac{x^3}{300}\right)\Big|_{100}^{150} = 69583.33$ dollars.

b. The additional cost is $\int_{500}^{550} (300 + 10x - .01x^2)\,dx = \left(300x + 5x^2 - \frac{x^3}{300}\right)\Big|_{500}^{550} = 139583.33$ dollars.

6.1.47

a. False. This would only be the case if the motion was all in the same direction. If the object changes direction at all, then the distance traveled is greater than the displacement.

b. True. This is because $v(t) = |v(t)|$ in this case.

c. True. This is because $R(t) > 0$ for $0 < t < 10$, but $R(t) < 0$ for $t > 10$.

d. True. The cost of increasing production from A to B is given by $\int_A^B C'(t)\,dt$, which is geometrically the area under the curve $y = C'(x)$ from A to B. If C' is positive and decreasing, there is more area under the curve from A to $2A$ than from $2A$ to $3A$.

6.1.49

a. The displacement is the net area, which is $\frac{1}{2} \cdot 2 \cdot 2 - \frac{1}{2} \cdot \frac{4}{3} \cdot 1 + \frac{1}{2} \cdot \frac{5}{3} \cdot 2 = 3$.

b. The distance traveled is $\frac{1}{2} \cdot 2 \cdot 2 + \frac{1}{2} \cdot \frac{4}{3} \cdot 1 + \frac{1}{2} \cdot \frac{5}{3} \cdot 2 = \frac{13}{3}$

c. $s(5) = s(0) + \int_0^5 v(t)\,dt = 0 + 3 = 3$.

d. $s(t) = \begin{cases} \int_0^t (-x + 2)\,dx & \text{if } 0 \le t \le 3, \\ \frac{3}{2} + \int_3^t (3x - 10)\,dx & \text{if } 3 < t \le 4, \\ 2 + \int_4^t (-2x + 10)\,dx & \text{if } 4 < t \le 5 \end{cases} = \begin{cases} -\frac{t^2}{2} + 2t & \text{if } 0 \le t \le 3, \\ \frac{3t^2}{2} - 10t + 18 & \text{if } 3 < t \le 4, \\ -t^2 + 10t - 22 & \text{if } 4 < t \le 5. \end{cases}$

6.1.51 The distance traveled is $\int_0^4 (1 - (t^2/16))\,dt = \left(t - (t^3/48)\right)\Big|_0^4 = \frac{8}{3}$. So the same distance could have been traveled over the given time period at a constant velocity of $\frac{8/3}{4} = \frac{2}{3}$.

6.1.53 The distance traveled is $\int_0^5 t\sqrt{25 - t^2}\,dt = \frac{1}{2} \int_0^{25} \sqrt{u}\,du = \left(\frac{1}{3}u^{3/2}\right)\Big|_0^{25} = \frac{125}{3}$. So the same distance could have been traveled over the given time period at a constant velocity of $\frac{125/3}{5} = \frac{25}{3}$.

6.1.55

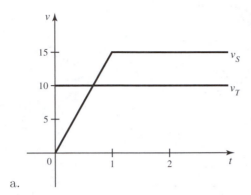

b. After 1 hour, Theo has ridden $1 \cdot 10 = 10$ miles, and Sasha has ridden $\frac{1}{2} \cdot 15 = 7.5$ miles, so Theo has ridden farther.

a.

c. After 2 hours, Theo has ridden $2 \cdot 10 = 20$ miles, and Sasha has ridden $7.5 + 15 \cdot 1 = 22.5$ miles, so Sasha has ridden farther.

d. The times when they arrive at the various mile markers are in the following table:

	10	15	20
Theo	1	3/2	2
Sasha	7/6	3/2	11/6

Note that Theo hits the 10 mile marker first, then they are tied as they hit the 15 mile marker, and Sasha hits the 20 mile marker first. The area under v_S is the same as the area under v_T for $t = 1.5$, for $t < 1.5$ the area under v_T is greater, and for $t > 1.5$, the area under v_S is greater.

e. Theo will then hit the 20 mile mark in $18.8/10 = 1.88$ hours. Sasha hits the 20 mile mark at $t = 11/6 \approx 1.833$ hours, so Sasha will win.

f. A head start of .2 hours is equivalent for Theo of $10 \cdot .2 = 2$ miles. It will take him $18/10 = 1.8$ hours to ride the other 18 miles, while it still takes Sasha about 1.83 hours to cover 20 miles, so Theo will win.

6.1.57

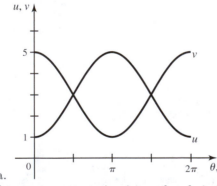

b. The average of Abe's speed function is $\frac{1}{2\pi} \int_0^{2\pi} (3 - 2\cos\varphi) \, d\varphi = \frac{1}{2\pi} (3\theta - 2\sin\varphi)|_0^{2\pi} = 3$ mph, and The average of Bess' speed function is $\frac{1}{2\pi} \int_0^{2\pi} (3 + 2\cos\theta) \, d\theta = \frac{1}{2\pi} (3\theta + 2\sin\theta)|_0^{2\pi} = 3$ mph. They have the same average speed.

a.

Abe starts out running into a headwind, Bess starts out running with a tailwind.

c. The track is 1/10 mile in radius. We have $u = \frac{ds}{dt}$ where $s = \frac{1}{10}\varphi$. Thus, $u = \frac{ds}{dt} = \frac{1}{10}\frac{d\varphi}{dt} = 3 - 2\cos\varphi$, so $dt = \frac{d\varphi}{10(3-2\cos\varphi)}$. The time T for one lap is then

$$T = \int_0^T dt = \int_0^{2\pi} \frac{d\varphi}{10(3 - 2\cos\varphi)} = \frac{\pi}{5\sqrt{5}} = \frac{\pi\sqrt{5}}{25}.$$

You may need to use a computer algebra system to compute the integral, or wait until you have studied chapter 7. The calculation for Bess' velocity function produces the same time. They tie the race. Both have average speed $\frac{2\pi}{10T} = \sqrt{5}$.

6.1.59

a. $\int_0^2 20(1 + \cos(\pi t/12))\, dt = \left(20t + \frac{240}{\pi}\sin(\pi t/12)\right)\big|_0^2 = 40 + \frac{240}{\pi}\cdot\frac{1}{2} = 40 + \frac{120}{\pi} \approx 78.2\,\mathrm{m}^3.$

b.

$Q(t) = \int_0^t 20(1 + \cos(\pi x/12))\, dx$, which is equal to $\left(20x + \frac{240}{\pi}\sin(\pi x/12)\right)\big|_0^t = 20t + \frac{240}{\pi}\cdot \sin(\pi t/12)\,\mathrm{m}^3.$

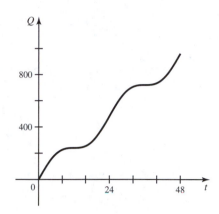

c. The reservoir is full when $20T + \frac{240}{\pi}\sin(\pi T/12) = 2500$, which occurs for $T \approx 122.6$ hours.

6.1.61

a. Note that $\displaystyle\int -\frac{\pi}{2}\sin\frac{\pi t}{2}\, dt = \cos\frac{\pi t}{2} + C.$ The value of this function at $t = 0$ is $1 + C = 6$, so we require $C = 5$, and we have

$$V(t) = 5 + \cos\frac{\pi t}{2}.$$

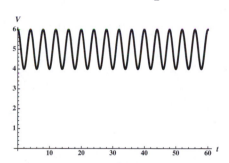

b. Because $\sin\frac{\pi t}{2}$ is periodic with period 4, the breathing cycle repeats every 4 seconds, so there are $\frac{60}{4} = 15$ breaths per minute.

c. The lungs are full at $t = 0$, at which time $V(0) = 6$ L, so this is the capacity of the lungs. The tidal volume is the difference between this amount and the amount in the lungs after each exhalation, which is the minimum value of $V(t)$. This minimum occurs at the first positive zero of $V'(t)$, which is at t=2. Because $V(2) = 4$, the tidal volume is $6 - 4 = 2$L.

6.1.63

a. $E = \int_0^{24}(300 - 200\sin(\pi t/12))\, dt = \left(300t + \frac{2400}{\pi}\cos(\pi t/12)\right)\big|_0^{24} = 7200$ MWh. This is equivalent to $7.2 \times 10^6 \cdot 3.6 \times 10^6 = 2.592 \times 10^{13}$ Joules.

b. For one day, $\frac{7.2\times10^6\,\mathrm{KWh}}{450\,\mathrm{Kwh/kg}} = 16,000$ kg coal needed.

   For one year, $16000\,\mathrm{kg} \times 365 = 5,840,000$ kg coal needed.

c. For one day, $\frac{7.2\times10^6\,\mathrm{KWh}}{1.6\times10^4\,\mathrm{Kwh/g}} = 450$ g U-235 need.

   For one year, $450 \times 365 = 164,250$ g needed.

d. $\dfrac{7.2\times10^6\,\mathrm{KWh/day}}{(200\,\mathrm{KW/turbine})\cdot(24\,\mathrm{hours/day})} = 1500$ turbines.

6.1.65 Using the Fundamental Theorem, we have

$$\int_a^b f'(x)\,dx = f(b) - f(a) = g(b) - g(a) = \int_a^b g'(x)\,dx.$$

6.1.67 If $f(x)$ is the elevation of trail one at position x and $g(x)$ is the elevation of trail two at position x, then because we are given that $f(a) = g(a)$ and $f(b) = g(b)$, we must have that

$$\int_a^b f'(x)\,dx = \int_a^b g'(x)\,dx,$$

so the two trails have the same net change in elevation.

6.2 Regions Between Curves

6.2.1 If f and g intersect at $x = a$ and $x = b$ with $a < b$ and if $f(x) \geq g(x)$ on $[a, b]$, then the area between these curves is given by $\int_a^b (f(x) - g(x))\,dx$.

6.2.3

6.2.5 The curves intersect when $x = x^2 - 2$, or $(x + 1)(x - 2) = 0$, so at $x = -1$ and $x = 2$. The area is
$\int_{-1}^2 (x - (x^2 - 2)) = \left. \left(2x + x^2/2 - x^3/3\right) \right|_{-1}^2 = 4.5.$

6.2.7 The two curves intersect when $3 - x = x^2 + 1$ or $x^2 + x - 2 = 0$, so at $x = 1$ and $x = -2$. The intersection point given in the picture is clearly the one at $x = 1$. Thus the area is

$$\int_0^1 (3 - x) - (x^2 + 1)\,dx = \left. \left(-\frac{1}{3}x^3 - \frac{1}{2}x^2 + 2x\right) \right|_0^1 = \frac{7}{6}.$$

The nonvertical lines intersect when $2x + 2 = 3x + 3$, or $x = -1$. The vertical line $x = 4$ intersects both nonvertical lines when $x = 4$. The area is given by $\int_{-1}^{4}(3x + 3 - (2x + 2))\,dx = \int_{-1}^{4}(x + 1)\,dx = (x^2/2 + x)\Big|_{-1}^{4} = 8 + 4 - ((1/2) - 1) = 12.5$.

6.2.9

These curves intersect where $2x^2 = x^2 + 4$, so at $x = \pm 2$. Additionally, $x^2 + 4 \geq 2x^2$ on $[-2, 2]$, so the area between the curves is

6.2.11

$$\int_{-2}^{2}(x^2 + 4) - 2x^2\,dx = \left(-\frac{1}{3}x^3 + 4x\right)\Big|_{-2}^{2} = \frac{32}{3}.$$

The curves intersect when $2x^2 - 4 = 2x$, or

$$2x^2 - 2x - 4 = 2(x + 1)(x - 2) = 0.$$

6.2.13

So the intersections occur when $x = -1$ and $x = 2$. The area is $\int_{-1}^{2}(2x - 2x^2 + 4)\,dx = \left(x^2 - 2x^3/3 + 4x\right)\Big|_{-1}^{2} = (4 - 16/3 + 8 - (1 + 2/3 - 4)) = 9$.

6.2.15 The curves intersect at $\pi/4$, so the area is given by $\int_{0}^{\pi/4}\sin x\,dx + \int_{\pi/4}^{\pi/2}\cos x\,dx = (-\cos x)\Big|_{0}^{\pi/4} +$

$(\sin x)\Big|_{\pi/4}^{\pi/2} = (1 - \sqrt{2}/2) + (1 - \sqrt{2}/2) = 2 - \sqrt{2}.$

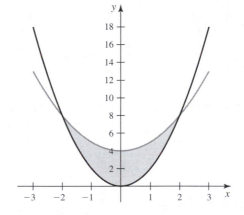

This is the region bounded by $x = 1$ and $x = 2$ above $y = 1/x^2$ and below $y = x$. The area is

6.2.17

$$\int_1^2 \left(x - \frac{1}{x^2} \right) dx = \left(\frac{1}{2}x^2 + \frac{1}{x} \right) \Big|_1^2 = 1.$$

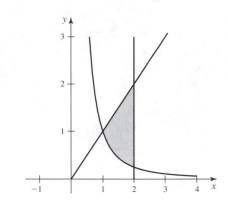

6.2.19 A plot of the region is

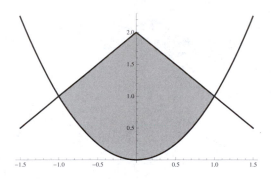

The curves intersect at $x = \pm 1$, and the area is twice the area from 0 to 1, so it is

$$2 \int_0^1 (2 - x - x^2)\, dx = 2 \left(2x - \frac{1}{2}x^2 - \frac{1}{3}x^2 \right) \Big|_0^1 = \frac{7}{3}.$$

6.2.21

$\int_2^6 (x/2 - |x - 3|)\, dx = \int_2^3 (x/2 - (3 - x))\, dx + \int_3^6 (x/2 - (x - 3))\, dx = \int_2^3 (3x/2 - 3)\, dx + \int_3^6 (3 - x/2)\, dx = (3x^2/4 - 3x) \Big|_2^3 + (3x - x^2/4) \Big|_3^6 = 27/4 - 9 - (3 - 6) + (18 - 9 - (9 - 9/4)) = 3/4 + 9/4 = 3.$

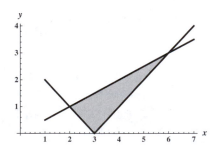

6.2.23 The two curves meet at $x = 0$; at that point $y = 1$. So the integration is from $y = 0$ to $y = 1$. Solving for x gives the curves $x = 2y^2 - 2$ and $x = 1 - y^2$. Thus the area of the region is

$$\int_0^1 ((1 - y^2) - (2y^2 - 2))\, dy = \int_0^1 (-3y^2 + 3)\, dy = (-y^3 + 3y) \Big|_0^1 = 2.$$

6.2.25 The curves $x = y^2 - 3y + 12$ and $x = -2y^2 - 6y + 30$ intersect where $y^2 - 3y + 12 = -2y^2 - 6y + 30$, or $3y^2 + 3y - 18 = 3(y+3)(y-2) = 0$. So the points of intersection are $y = -3$ and $y = 2$. Thus the area is

$$\int_{-3}^{2} ((-2y^2 - 6y + 30) - (y^2 - 3y + 12))\, dy = \int_{-3}^{2} (-3y^2 - 3y + 18)\, dy$$

$$= \left(-y^3 - \frac{3}{2}y^2 + 18y\right)\Big|_{-3}^{2}$$

$$= -8 - 6 + 36 - 27 + \frac{27}{2} + 54 = \frac{125}{2}.$$

6.2.27

  a. The area is given by $\displaystyle\int_{-\sqrt{2}}^{-1} (-(x^2 - 2))\, dx + \int_{-1}^{0} (-x)\, dx$.

  b. The area can also be written as $\displaystyle\int_{-1}^{0} (y - (-\sqrt{y+2}))\, dy$.

6.2.29

  a. The area is given by $\int_{-3}^{-2} (\sqrt{x+3} - (-\sqrt{x+3}))\, dx + \int_{-2}^{6} (\sqrt{x+3} - x/2)\, dx$.

  b. The area can also be written $\int_{-1}^{3} (2y - (y^2 - 3))\, dy$.

6.2.31 The region is plotted below.

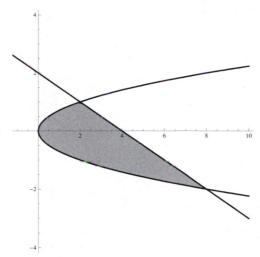

Note that the two curves intersect where $y = 2 - \frac{x}{2} = 2 - y^2$, so for $y^2 + y - 2 = (y+2)(y-1) = 0$. So they intersect at the points $(2, 1)$ and $(8, -2)$.

  a. Integrating with respect to x, we must solve the second equation for y to obtain $y = \pm\sqrt{\frac{x}{2}}$. Then the integral must be split into two integrals, and the area is

$$\int_{0}^{2} \left(\sqrt{\frac{x}{2}} - \left(-\sqrt{\frac{x}{2}}\right)\right)\, dx + \int_{2}^{8} \left(2 - \frac{x}{2} - \left(-\sqrt{\frac{x}{2}}\right)\right)\, dx$$

$$= \frac{2}{\sqrt{2}} \int_{0}^{2} x^{1/2}\, dx + \int_{2}^{8} \left(2 - \frac{x}{2} + \frac{1}{\sqrt{2}} x^{1/2}\right)\, dx$$

$$= \frac{2}{\sqrt{2}} \left(\frac{2}{3} x^{3/2}\right)\Big|_{0}^{2} + \left(2x - \frac{x^2}{4} + \frac{2}{3\sqrt{2}} x^{3/2}\right)\Big|_{2}^{8}$$

$$= \sqrt{2} \cdot \frac{2}{3} \cdot 2\sqrt{2} + 16 - 16 + \frac{2}{3\sqrt{2}} \cdot 16\sqrt{2} - 4 + 1 - \frac{2}{3\sqrt{2}} \cdot 2\sqrt{2}$$

$$= \frac{3}{2} + \frac{32}{3} - 3 - \frac{4}{3} = 9.$$

b. With respect to y, we have

$$\int_{-2}^{1} (4 - 2y - 2y^2)\, dy = \left(4y - y^2 - \frac{2}{3}y^3\right)\Bigg|_{-2}^{1} = \left(4 - 1 - \frac{2}{3}\right) - \left(-8 - 4 + \frac{16}{3}\right) = 9.$$

6.2.33

The area is given by $\int_0^8 (4 - x^{2/3})\, dx =$
$\left(4x - 3x^{5/3}/5\right)\Big|_0^8 = 32 - 96/5 = 64/5.$

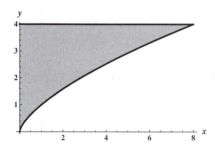

6.2.35

The curves intersect where $x = 4y = y^3$, so at $y = -2, 0, 2$. The area is twice the area from $y = 0$ to $y = 2$, so is

$$2\int_0^2 (4y - y^3)\, dy = \left(4y^2 - \frac{1}{2}y^4\right)\Bigg|_0^2 = 8.$$

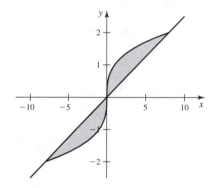

6.2.37

The area is given by $\int_0^{\sqrt{3}/2} (2x\sqrt{1 - x^2} - x)\, dx = \left(-\frac{2}{3}(1 - x^2)^{3/2} - \frac{x^2}{2}\right)\Bigg|_0^{\sqrt{3}/2} = \frac{-2}{24} - \frac{3}{8} + \frac{2}{3} = \frac{5}{24}.$

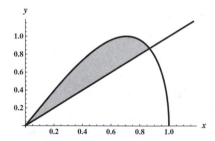

6.2.39

a. False. This can be done either with respect to x or with respect to y. For the latter, the relevant integral is $\int_0^1 (y - y^2)\, dy$.

b. False. On the interval $(0, \pi/4)$ the cosine function is greater, but on $(\pi/4, \pi/2)$ the sine function is greater. The area is $\int_0^{\pi/4} (\cos x - \sin x)\, dx + \int_{\pi/4}^{\pi/2} (\sin x - \cos x)\, dx$.

c. True. They both represent the area of the region in the first quadrant under $y = x$ and above $y = x^2$.

6.2.41

The area is given by $\int_4^5 (7x - 19 - (x -$

$1)^2)\, dx = (7x^2/2 - 19x - (x-1)^3/3)\Big|_4^5 =$

$(175/2 - 95 - 64/3) - (56 - 76 - 9) = -\frac{173}{6} +$

$29 = \frac{1}{6}.$

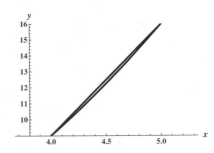

6.2.43

The area is given by $\int_2^5 (5x - 9 - (x-1)^2)\, dx =$

$(5x^2/2 - 9x - (x-1)^3/3)\Big|_2^5 = 125/2 - 45 -$

$64/3 - (10 - 18 - (1/3)) = 4.5.$

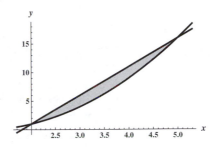

6.2.45 This is given by $\int_0^4 (3y - (y^2 - y))\, dy = \int_0^4 (4y - y^2)\, dy = (2y^2 - y^3/3)\Big|_0^4 = 32 - 64/3 = 32/3.$

6.2.47 The area is given by $\int_0^3 (y/2 + 7.5 - y^2)\, dy = (y^2/4 + 7.5y - y^3/3)\Big|_0^3 = (2.25 + 22.5 - 9) = 15.75.$

6.2.49 On $[\pi/2, 2\pi]$, $1 - \cos x \geq \sin x$, so the area is

$$\int_{\pi/2}^{2\pi} (1 - \cos x - \sin x)\, dx = (x - \sin x + \cos x)\Big|_{\pi/2}^{2\pi} = \frac{3\pi}{2} + 2.$$

6.2.51

a. The area of R_1 is $\int_0^1 (x - x^p)\, dx = \left(x^2/2 - \frac{x^{p+1}}{p+1}\right)\Big|_0^1 = \frac{1}{2} - \frac{1}{p+1} = \frac{p-1}{2p+2}.$

The area of R_2 is $\int_0^1 (x^{1/q} - x)\, dx = \left(\frac{q}{q+1}x^{(q+1)/q} - x^2/2\right)\Big|_0^1 = \frac{q}{q+1} - \frac{1}{2} = \frac{q-1}{2q+2}.$

Clearly, if $q = p$ then $R_1 = R_2$.

b. Using the results above, if $p > q$, then $R_1 - R_2 = \frac{(p-1)(2q+2)-(q-1)(2p+2)}{(2p+2)(2q+2)} = \frac{4p-4q}{(2p+2)(2q+2)} = \frac{p-q}{(p+1)(q+1)} > 0$, so $R_1 > R_2$.

c. If $p < q$, then $R_1 - R_2$ computed above is less than 0, so $R_1 < R_2$.

6.2.53 $y = 8x$ and $y = 9 - x^2$ intersect when $x^2 + 8x - 9 = (x+9)(x-1) = 0$, so at $x = 1$ in the first quadrant. $y = \frac{5}{2}x$ and $y = 9 - x^2$ intersect when $x^2 + \frac{5}{2}x - 9 = 0$, or $2x^2 + 5x - 18 = (2x+9)(x-2) = 0$, so at $x = 2$ in the first quadrant. Thus the area of the region is

$$A = \int_0^1 \left(8x - \frac{5}{2}x\right) dx + \int_1^2 \left(9 - x^2 - \frac{5}{2}x\right) dx$$

$$= \int_0^1 \frac{11}{2}x \, dx + \int_1^2 \left(9 - x^2 - \frac{5}{2}x\right) dx$$

$$= \left(\frac{11}{4}x^2\right)\Big|_0^1 + \left(9x - \frac{x^3}{3} - \frac{5}{4}x^2\right)\Big|_1^2$$

$$= \frac{11}{4} + 18 - \frac{8}{3} - 5 - 9 + \frac{1}{3} + \frac{5}{4} = \frac{17}{3}.$$

6.2.55 $A = \int_{-4}^5 (8 - y - (y-2)^2/3) \, dy = \left(8y - y^2/2 - (y-2)^3/9\right)\Big|_{-4}^5 = 40 - 25/2 - 3 - (-32 - 8 + 24) = 40.5.$

6.2.57 $A_n = \int_0^1 (x^{1/n} - x) \, dx = \left(\frac{nx^{(n+1)/n}}{n+1} - x^2/2\right)\Big|_0^1 = \frac{n}{n+1} - \frac{1}{2} = \frac{n-1}{2n+2}.$

6.2.59 Using the result of the previous problem, $\lim_{n \to \infty} A_n = \lim_{n \to \infty} \frac{n-1}{n+1} = 1$. As $n \to \infty$, the region in question approaches the 1×1 square over the interval $[0, 1]$, which has area 1.

6.2.61 This is a triangle with base 2 and height 1, so it has area 1. Note that for $0 \le x \le 1$ the line which forms the top of the triangle is $y = x$, and for $1 \le x \le 2$, the line is $y = 2 - x$. Integrating with respect to y, we are looking for k so that $\int_0^k (2 - y - y) \, dy = \frac{1}{2}$. Now,

$$\int_0^k (2 - 2y) \, dy = \left(2y - y^2\right)\Big|_0^k = 2k - k^2.$$

Setting $2k - k^2 = \frac{1}{2}$, we have $2k^2 - 4k + 1 = 0$, which has roots $k = \frac{2 \pm \sqrt{2}}{2}$. Only the negative sign choice gives a value between 0 and 1, so we want $k = \frac{2 - \sqrt{2}}{2} = 1 - \frac{\sqrt{2}}{2}$.

6.2.63 Using the result of Exercise 57, the total area bounded by these two curves is $\frac{2-1}{2 \cdot 2 + 2} = \frac{1}{6}$. We are looking for k so that $\int_0^k (y - y^2) \, dy = \frac{1}{12}$. Now,

$$\int_0^k (y - y^2) \, dy = \left(\frac{1}{2}y^2 - \frac{1}{3}y^3\right)\Big|_0^k = \frac{1}{2}k^2 - \frac{1}{3}k^3.$$

Setting this equal to $\frac{1}{12}$ and simplifying gives $4k^3 - 6k^2 + 1 = 0$. The root $k = \frac{1}{2}$ can be found by inspection or through the use of a computer algebra system. By long division, we can see that $4k^3 - 6k^2 + 1 = (2k - 1)(2k^2 - 2k - 1)$, so the other roots are $k = \frac{1 \pm \sqrt{3}}{2}$. The only root in the interval $(0, 1)$ is $k = \frac{1}{2}$.

6.2.65

a. The point (n, n) on the curve $y = x$ would represent the notion that the lowest $p\%$ of the society owns $p\%$ of the wealth, which would represent a form of equality.

b. The function must be increasing and concave up because the poorest $p\%$ cannot own more than $p\%$ of the wealth.

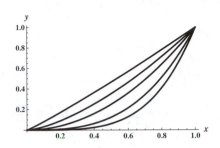

c. $y = x^{1.1}$ is closest to $y = x$, and $y = x^4$ is furthest from $y = x$.

d. Note that $B = \int_0^1 L(x)\,dx$, and $A + B = 1/2$, so $A = \frac{1}{2} - \int_0^1 L(x)\,dx$. Then $G = \frac{A}{A+B} = \frac{A}{1/2} = 2A = 1 - 2\int_0^1 L(x)\,dx$.

e. For $L(x) = x^p$, we have $G = 1 - 2\int_0^1 x^p\,dx = 1 - 2\left(\frac{x^{p+1}}{p+1}\right)\Big|_0^1 = 1 - \frac{2}{p+1} = \frac{p-1}{p+1}$. So we have

p	1.1	1.5	2	3	4
G	1/21	1/5	1/3	1/2	3/5

f. For $p = 1$ we have $G = 0$. Because $\lim_{p\to\infty} \frac{p-1}{p+1} = 1$, the largest possible value of G approaches 1.

g. For $L(x) = 5x^2/6 + x/6$, note that $L(0) = 0$, $L(1) = 1$, $L'(x) = 5x/3 + 1/6 > 0$ on $[0,1]$, and $L''(x) = 5/3 > 0$ as well. The Gini index is

$$G = 1 - 2\int_0^1 (5x^2/6 + x/6)\,dx = 1 - 2\left(5x^3/18 + x^2/12\right)\Big|_0^1 = 1 - 2(5/18 + 1/12) = 1 - 5/9 - 1/6 = 5/18.$$

6.2.67

Some experimental data:

a	-2	-1	0	1
Approximate Area	7.812	6.928	7.812	10.667

The area seems to be minimized for $a = -1$.

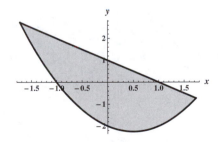

This can be confirmed using an computer algebra system. Note that $y = (x+1)(x-2)$ and $y = ax+1$ intersect at $(1/2)(1 + a \pm \sqrt{a^2 + 2a + 13})$. Then

$$\int_{(1/2)(1+a-\sqrt{a^2+2a+13})}^{(1/2)(1+a+\sqrt{a^2+2a+13})} [ax + 1 - (x+1)(x-2)]\,dx = (1/6)(a^2 + 2a + 13)^{3/2}.$$

Let $f(a) = (1/6)(a^2 + 2a + 13)^{3/2}$. Then $f'(a) = (1/2)(a+1)\sqrt{a^2 + 2a + 13}$, which is zero for $a = -1$. Also, $f''(-1) = \sqrt{3} > 0$, so by the Second Derivative test, there is a minimum for $a = -1$.

6.2.69 $A = 2\int_0^1 y^2\sqrt{1 - y^3}\,dy = \frac{2}{3}\int_0^1 \sqrt{u}\,du$ (where $u = 1 - y^3$). So $A = \frac{2}{3}\left(\frac{2}{3}u^{3/2}\right)\Big|_0^1 = \frac{4}{9}$.

6.2.71

a.

$$F(a) = \int_0^b x(x-a)(x-b)\,dx = \int_0^b (abx - ax^2 - bx^2 + x^3)\,dx$$

$$= \left(abx^2/2 - ax^3/3 - bx^3/3 + x^4/4\right)\Big|_0^b = ab^3/2 - ab^3/3 - b^4/3 + b^4/4 = ab^3/6 - b^4/12.$$

So $F(a) = \frac{b^3}{6}\left(a - \frac{b}{2}\right)$ and $F(a) = 0$ when $a = \frac{b}{2}$ or when $b = 0$. If $b = 0$, then because $0 \le a \le b$, we have $a = 0$ as well. In either case, we have $a = \frac{b}{2}$.

b. Note that $f(x) \ge 0$ for $x \in [0, a]$ and $f(x) \le 0$ for $x \in [a, b]$. So the area under the curve is

$$A(a) = \int_0^a x(x-a)(x-b)\,dx - \int_a^b x(x-a)(x-b)\,dx$$

$$= \left(abx^2/2 - ax^3/3 - bx^3/3 + x^4/4\right)\Big|_0^a - \left(abx^2/2 - ax^3/3 - bx^3/3 + x^4/4\right)\Big|_a^b$$

$$= -\frac{a^4}{6} + \frac{a^3 b}{3} - \frac{ab^3}{6} + \frac{b^4}{12}$$

$$A'(a) = -\frac{2}{3}a^3 + a^2 b - \frac{1}{6}b^3 = -\frac{1}{6}(2a - b)(2a^2 - 2ab - b^2).$$

$$A''(a) = -2a^3 + 2ab.$$

Note that $A'(b/2) = 0$, and $A''(b/2) = -b^2/2 + b^2 > 0$, so there is a minimum at $a = b/2$. The other critical numbers are $a = \frac{2b \pm \sqrt{4b^2 - 4 \cdot 2(-b^2)}}{4} = \frac{b \pm \sqrt{3b^2}}{2} = \left(\frac{1 \pm \sqrt{3}}{2}\right)b$ are not on the interval $(0, b)$. The maximum value of A is $b^4/12$ which occurs at $a = 0$ and $a = b$.

6.2.73

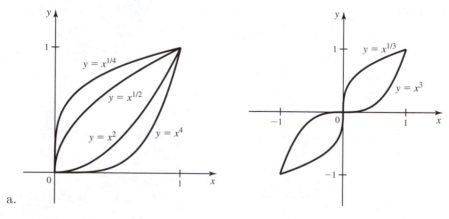

a.

b. $A_n(x)$ is the net area of the region between the graphs of f and g from 0 to x.

c. Note that $\int_0^1 (f(s) - g(s))\,ds = \left(\frac{s^{n+1}}{n+1} - \frac{s^{(1/n)+1}}{(1/n)+1}\right)\Big|_0^1 = \frac{1}{n+1} - \frac{n}{n+1} = \frac{1-n}{n+1} < 0$. So we seek the smallest

c so that $\int_1^c (f(s) - g(s))\,ds = \frac{n-1}{n+1}$. This occurs when $\left(\frac{s^{n+1}}{n+1} - \frac{s^{(1/n)+1}}{(1/n)+1}\right)\Big|_1^c = \frac{n-1}{n+1}$, or $\frac{c^{n+1}}{n+1} = \frac{c^{(n+1)/n}}{(n+1)/n}$,

or $c^{n+1} = nc^{(n+1)/n}$, or $c^{n-(1/n)} = n$, so $c = n^{n/(n^2-1)}$. As n increases, this root increases as well.

6.3 Volume by Slicing

6.3.1 $A(x)$ is the area of the cross section through the solid at the point x.

6.3.3 $V = \pi \int_0^2 (4x^2 - x^4)\,dx$.

6.3.5 The cross sections are disks and $A(x)$ is the area of a disk.

6.3.7 The curves intersect when $2 - x^2 = x^2$, or $x = \pm 1$.
$A(x) = ((2 - x^2) - x^2)^2 = (2(1 - x^2))^2 = 4(1 - 2x^2 + x^4)$. The volume is given by

$$\int_{-1}^1 4(1 - 2x^2 + x^4)\,dx = 4\left(x - 2x^3/3 + x^5/5\right)\Big|_{-1}^1 = 4((1 - 2/3 + 1/5) - (-1 + 2/3 - 1/5)) = 64/15.$$

6.3.9 The curves intersect when $\sqrt{\cos x} = 0$, or $x = \pm\pi/2$. $A(x) = \frac{1}{2}\left(\sqrt{\cos x}\right)^2 = \frac{\cos x}{2}$. The volume is given by

$$\int_{-\pi/2}^{\pi/2} \frac{\cos x}{2}\,dx = \frac{\sin x}{2}\Big|_{-\pi/2}^{\pi/2} = 1/2 - (-1/2) = 1.$$

6.3.11 $A(x) = (2\sqrt{25 - x^2})^2 = 100 - 4x^2$.

$$V = \int_0^5 A(x)\,dx = \int_0^5 (100 - 4x^2)\,dx = \left(100x - 4x^3/3\right)\Big|_0^5 = 500 - \frac{500}{3} = \frac{1000}{3}.$$

6.3.13 For each value of x, the height of the triangle is $2 - x$, which is the diameter of the semicircle, so the area of that semicircle is

$$A(x) = \frac{\pi}{2}\left(\frac{2 - x}{2}\right)^2 = \frac{\pi}{8}(4 - 4x + x^2).$$

Thus the volume is

$$V = \int_0^2 A(x)\,dx = \int_0^2 \frac{\pi}{8}(4 - 4x + x^2)\,dx = \frac{\pi}{8}\left(4x - 2x^2 + x^3/3\right)\Big|_0^2 = \frac{\pi}{3}.$$

6.3.15 The relationship between the height h of tetrahedron and the edge length l is $h = l\sqrt{2/3}$, which can be deduced using triangle geometry and the Pythagorean theorem.

Let z be the distance from the top vertex of the tetrahedron down toward the base perpendicularly. The cross sections perpendicular to this axis are all equilateral triangles with height z, so their side length is $\sqrt{3/2}z$, and their area is

$$A(z) = \frac{\sqrt{3}}{4}\left(z\sqrt{\frac{3}{2}}\right)^2 = \frac{3\sqrt{3}z^2}{8}.$$

The volume is thus

$$\int_0^{4\sqrt{2/3}} \frac{3\sqrt{3}}{8}z^2\,dz = \frac{3\sqrt{3}}{8}\left(z^3/3\right)\Big|_0^{4\sqrt{2/3}} = \frac{16\sqrt{2}}{3}.$$

6.3.17 $V = \pi \int_0^3 4x^2\,dx = 4\pi\left(x^3/3\right)\Big|_0^3 = 36\pi$.

6.3.19 The volume of the solid, using the disk method, is

$$\int_0^2 \pi f(x)^2\,dx = \int_0^2 \pi(4 - x^2)^2\,dx = \pi\left(16x - \frac{8}{3}x^3 + \frac{1}{5}x^5\right)\Big|_0^2 = \frac{256}{15}\pi.$$

6.3.21 $V = \pi \int_0^\pi \sin^2 x\,dx = \frac{\pi}{2}\int_0^\pi (1 - \cos 2x)\,dx = \frac{\pi}{2}\left(x - \frac{\sin 2x}{2}\right)\Big|_0^\pi = \frac{\pi^2}{2}$.

6.3.23 $V = \pi \int_1^4 \left(\frac{1}{x^2}\right)^2 dx = \pi \int_1^4 x^{-4} dx = \pi \left(-\frac{1}{3x^3}\right)\Big|_1^4 = \pi\left(-\frac{1}{3\cdot 64} + \frac{1}{3}\right) = \frac{21\pi}{64}.$

6.3.25 $V = \pi \int_0^4 ((2\sqrt{x})^2 - x^2) dx = \pi \left(2x^2 - x^3/3\right)\Big|_0^4 = \pi(32 - 64/3) = \frac{32\pi}{3}.$

6.3.27

$$V = \int_{\pi/6}^{5\pi/6} \pi(f(x)^2 - g(x)^2) dx = \int_{\pi/6}^{5\pi/6} \pi(2\sin x - 1) dx = \pi(-2\cos x - x)\Big|_{\pi/6}^{5\pi/6} = \frac{2\pi}{3}(3\sqrt{3} - \pi).$$

6.3.29 $V = \pi \int_{-1}^2 ((x+3)^2 - (x^2+1)^2) dx = \pi \int_{-1}^2 (x^2 + 6x + 9 - x^4 - 2x^2 - 1) dx = \pi \int_{-1}^2 (-x^4 - x^2 + 6x + 8) dx =$
$\pi \left(-x^5/5 - x^3/3 + 3x^2 + 8x\right)\Big|_{-1}^2 = \pi(-32/5 - 8/3 + 12 + 16 - (1/5 + 1/3 + 3 - 8)) = \frac{117\pi}{5}.$

6.3.31 $V = \pi \int_0^{\pi/2} (\sin x - \sin^2 x) dx = \pi \int_0^{\pi/2} (\sin x - \frac{1}{2}(1 - \cos(2x))) dx = \pi \left(\sin(2x)/4 - \cos x - x/2\right)\Big|_0^{\pi/2} =$
$\pi(1 - \pi/4) = \frac{4\pi - \pi^2}{4}.$

6.3.33 $V = \pi \int_0^6 (y^2 - y^2/4) dy = \frac{3\pi}{4}\left(y^3/3\right)\Big|_0^6 = 54\pi.$

6.3.35 $V = \pi \int_0^8 (4 - y^{2/3}) dy = \pi \left(4y - (3/5)y^{5/3}\right)\Big|_0^8 = \frac{64\pi}{5}.$

6.3.37 $V = \pi \int_{-2}^2 (\sqrt{4 - y^2})^2 dy = \pi \left(4y - y^3/3\right)\Big|_{-2}^2 = \pi((8 - 8/3) - (-8 + 8/3)) = \frac{32\pi}{3}.$

6.3.39 About the x-axis: $V_x = \pi \int_0^5 4x^2 dx = \pi \left(4x^3/3\right)\Big|_0^5 = \frac{500\pi}{3}.$

About the y-axis: $V_y = \pi \int_0^{10} (25 - y^2/4) dy = \pi \left(25y - y^3/12\right)\Big|_0^{10} = \frac{500\pi}{3}.$
The volumes are the same.

6.3.41 About the x-axis: $V_x = \pi \int_0^1 (1 - x^3)^2 dx = \pi \int_0^1 (1 - 2x^3 + x^6) dx = \pi \left(x - x^4/2 + x^7/7\right)\Big|_0^1 = \frac{9\pi}{14}.$

About the y-axis: $V_y = \pi \int_0^1 (\sqrt[3]{1 - y})^2 dy = -\pi \left(\frac{3}{5}(1-y)^{5/3}\right)\Big|_0^1 = -\pi(0 - 3/5) = \frac{3\pi}{5}.$ The volume V_x is
bigger.

6.3.43 Using the disk method, a disk located at x has a radius of $1 - \sqrt{x}$ when revolved about the line $y = 1$, so the volume is

$$V = \pi \int_0^1 (1 - \sqrt{x})^2 dx = \pi \int_0^1 (1 - 2\sqrt{x} + x) dx = \pi \left(x - \frac{4}{3}x^{3/2} + \frac{1}{2}x^2\right)\Big|_0^1 = \pi\left(1 - \frac{4}{3} + \frac{1}{2}\right) = \frac{\pi}{6}.$$

6.3.45 We use the washer method. For each x, the washer has outer radius $2 + 2\sin x$ and inner radius 2, so the volume is

$$V = \pi \int_0^\pi \left((2 + 2\sin x)^2 - 2^2\right) dx = \pi \int_0^\pi (8\sin x + 4\sin^2 x) dx$$

$$= \pi \int_0^\pi (8\sin x + 2(1 - \cos 2x)) dx = \pi(-8\cos x + 2x - \sin 2x)\Big|_0^\pi$$

$$= \pi(8 + 2\pi + 8) = 2\pi(\pi + 8).$$

6.3.47 We use the washer method. For each x, the outer radius is $1 + \sin x$ and the inner radius is $1 + (1 - \sin x) = 2 - \sin x$. Thus the volume is

$$V = \pi \int_{\pi/6}^{5\pi/6} \left((1 + \sin x)^2 - (2 - \sin x)^2 \right) dx = \pi \int_{\pi/6}^{5\pi/6} (6 \sin x - 3) \, dx$$

$$= \pi(-6 \cos x - 3x) \Big|_{\pi/6}^{5\pi/6} = \pi \left(3\sqrt{3} - \frac{5\pi}{2} + 3\sqrt{3} + \frac{\pi}{2} \right) = \pi \left(6\sqrt{3} - 2\pi \right).$$

6.3.49 The lines $x = 2 - y$ and $x = 1 - \frac{y}{2}$ meet at the point $(0, 2)$. For each y, we obtain a washer whose outer radius is $3 - (1 - \frac{y}{2}) = 2 + \frac{y}{2}$ and whose inner radius is $3 - (2 - y) = y + 1$. Thus the volume is

$$V = \pi \int_0^2 \left(\left(2 + \frac{y}{2} \right)^2 - (y + 1)^2 \right) dy = \pi \int_0^2 \left(-\frac{3}{4}y^2 + 3 \right) dy = \pi \left(-\frac{1}{4}y^3 + 3y \right) \Big|_0^2 = 4\pi.$$

6.3.51

a. False. The cross sections are not disks or washers.

b. True. It is given by $V = \pi \int_0^R (R^2 - x^2) \, dx$.

c. True. This is because if we shift the sine function horizontally by $\pi/2$ units, we obtain the cosine function.

6.3.53 Using the washer method, the volume is

$$\int_1^6 \pi(1 - x^{-3}) \, dx = \left(\pi x + \frac{\pi}{2x^2} \right) \Big|_1^6 = \frac{325}{72} \pi.$$

6.3.55 This region is symmetric about the line $y = 2$, so we find the volume from $y = 0$ to $y = 2$ and double it. Solving for x in $y = x^{1/3}$ gives $x = y^3$, so we have

$$2 \int_0^2 \pi y^6 \, dy = \frac{2\pi}{7} y^7 \Big|_0^2 = \frac{256}{7} \pi.$$

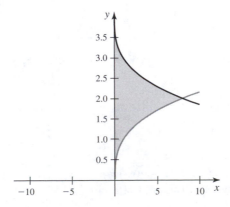

6.3.57 The volume V_S is $\pi \int_0^{\sqrt{a}} (y^2 - a)^2 \, dy = \pi \left(y^5/5 - 2ay^3/3 + a^2 y \right) \Big|_0^{\sqrt{a}} = \frac{8\pi}{15} a^{5/2}$.

The volume V_T is $\frac{1}{3}\pi a^2 \cdot a^{1/2} = \frac{1}{3}\pi a^{5/2}$. The ratio of $V_S/V_T = \frac{8/15}{1/3} = \frac{8}{5}$.

6.3.59

a. This comes from revolving the region in the first quadrant bounded by $\sin x$ and the line $y = 0$ between $x = 0$ and $x = \pi$ around the x axis.

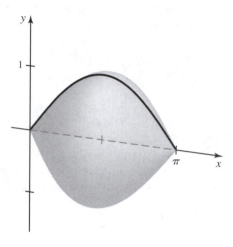

b. This comes from revolving the region in the first quadrant bounded by $y = x + 1$ and the line $y = 0$ between $x = 0$ and $x = 2$ around the x axis.

6.3.61

a. Think of the cone as being obtained by revolving the region under $y = x$ in the first quadrant from $x = 0$ to $x = R$ around the x-axis. The volume is $\pi \int_0^R x^2 \, dx = \pi R^3/3 = \frac{1}{3}V_C$.

b. Think of the hemisphere as being obtained by revolving the region under $y = \sqrt{R^2 - x^2}$ between $x = 0$ and $x = R$ around the x-axis. The volume is $\pi \int_0^R (R^2 - x^2) \, dx = \pi \left(R^2 x - x^3/3 \right) \Big|_0^R = \frac{2\pi R^3}{3} = \frac{2}{3}V_C$.

6.3.63 $V = \pi \int_{-2}^2 ((3 + \sqrt{4 - y^2})^2 - (3 - \sqrt{4 - y^2})^2) \, dy = 24\pi \int_0^2 \sqrt{4 - y^2} \, dy = 24\pi(\pi) = 24\pi^2$. Note that the last integral evaluated represents $1/4$ the area of a circle of radius 2

6.3.65

   a. By the general slicing method, $V(x) = \int_a^b A(x)\,dx$. Because the two figures have the same cross sections $A(x)$, they must therefore have the same volumes.

   b. We are seeking the value of r so that $10\pi r^2 = 40$, so $r^2 = \frac{4}{\pi}$, and $r = \frac{2}{\sqrt{\pi}}$ meters.

6.4   Volume by Shells

6.4.1 $V = 2\pi \int_a^b x(f(x) - g(x))\,dx$.

6.4.3 ...revolved about the x axis. ...using the disk/washer method and integrating with respect to $\underline{x}$ or using the shell method and integrating with respect to $\underline{y}$.

6.4.5 $V = 2\pi \int_0^1 x(x - x^2)\,dx = 2\pi \int_0^1 (x^2 - x^3)\,dx = 2\pi \left(x^3/3 - x^4/4 \right) \Big|_0^1 = 2\pi\,(1/3 - 1/4) = \pi/6$.

6.4.7 $V = 2\pi \int_0^1 \frac{x}{(1+x^2)^2}\,dx$. Let $u = x^2 + 1$. Then $du = 2x\,dx$. Substituting yields $V = \pi \int_1^2 \frac{1}{u^2}\,du = \pi \left(-\frac{1}{u}\right)\Big|_1^2 = \pi \left(1 - \frac{1}{2}\right) = \frac{\pi}{2}$.

6.4.9 $V = 2\pi \int_0^1 x(3 - 3x)\,dx = 2\pi \left(3x^2/2 - x^3 \right) \Big|_0^1 = \pi$.

6.4.11 Note that the line $y = 1$ intersects the curve $y = x^3 - x^8 + 1$ at $x = 0$ and $x = 1$. $V = 2\pi \int_0^1 x(x^3 - x^8 + 1 - 1)\,dx = 2\pi \int_0^1 (x^4 - x^9)\,dx = 2\pi \left(x^5/5 - x^{10}/10 \right) \Big|_0^1 = 2\pi(1/5 - 1/10) = \pi/5$.

6.4.13 $V = 2\pi \int_0^{\sqrt{\pi/2}} x\cos(x^2)\,dx = \pi \left(\sin(x^2) \right) \Big|_0^{\sqrt{\pi/2}} = \pi$.

6.4.15 $V = 2\pi \int_0^2 y(4 - y^2)\,dy = 2\pi \left(2y^2 - y^4/4 \right) \Big|_0^2 = 8\pi$.

6.4.17 $V = 2\pi \int_2^4 y(4 - y)\,dy = 2\pi \left(2y^2 - y^3/3 \right) \Big|_2^4 = \frac{32\pi}{3}$.

6.4.19 Note that the lines intersect at $(0,0)$, $(2,0)$ and $(1,1)$.

$$V = 2\pi \int_0^1 y((2 - y) - y)\,dy = 2\pi \int_0^1 (2y - 2y^2)\,dy = 2\pi \left(y^2 - 2y^3/3 \right) \Big|_0^1 = 2\pi/3.$$

6.4.21 $V = 2\pi \int_0^3 y(y^2)\,dy = 2\pi \int_0^3 y^3\,dy = 2\pi \left(y^4/4 \right) \Big|_0^3 = 81\pi/2$.

6.4.23 $V = 2\pi \int_2^{16} y(2/y)^{2/3}\,dy = 2^{5/3}\pi \int_2^{16} y^{1/3}\,dy = 2^{5/3}\pi \left(3y^{4/3}/4 \right) \Big|_2^{16} = 2^{5/3}\pi(3 \cdot 2^{10/3} - 3 \cdot 2^{-2/3}) = 3\pi(2^5 - 2) = 90\pi$.

6.4.25 Consider the region in the first quadrant bounded by the coordinate axes and the line $y = 8 - (8/3)x$. We can generate the desired cone by revolving this region around the y-axis. We then have $V = 2\pi \int_0^3 x(8 - (8/3)x)\,dx = \frac{16\pi}{3} \int_0^3 (3x - x^2)\,dx = \frac{16\pi}{3} \left(3x^2/2 - x^3/3 \right) \Big|_0^3 = 24\pi$.

6.4.27 Consider the triangle in the first quadrant bounded by $y = 0$, $x = 3$, and $y = 9 - (3/2)x$. The solid in question is formed by revolving this region around the y-axis. The volume is $V = 2\pi \int_3^6 x(9 - (3/2)x)\,dx = 2\pi \left(9x^2/2 - x^3/2\right)\Big|_3^6 = 54\pi$.

6.4.29 Consider the part of the ellipse which lies in the first quadrant. We can obtain half the ellipsoid by revolving this region around the y-axis. So the volume of the whole ellipsoid is $V = 4\pi \int_0^2 x\sqrt{2 - (x^2/2)}\,dx$.

Let $u = 2 - x^2/2$. Then $V = 4\pi \int_0^2 u^{1/2}\,du = 8\pi/3 \left(u^{3/2}\right)\Big|_0^2 = \frac{16\pi\sqrt{2}}{3}$.

6.4.31 $V = 2\pi \int_0^1 (x+2)x^2\,dx = 2\pi \left(x^4/4 + 2x^3/3\right)\Big|_0^1 = \frac{11\pi}{6}$.

6.4.33

$$V = 2\pi \int_0^1 (y+2)(1 - \sqrt{y})\,dy = 2\pi \int_0^1 (y + 2 - y^{3/2} - 2y^{1/2})\,dy = 2\pi \left(y^2/2 + 2y - 2y^{5/2}/5 - 4y^{3/2}/3\right)\Big|_0^1$$

$$= 2\pi(1/2 + 2 - 2/5 - 4/3) = \frac{23\pi}{15}.$$

6.4.35 Using washers, we have

$$V = \pi \int_0^1 (3^2 - (x^2 + 2)^2)\,dx = \pi \int_0^1 (9 - x^4 - 4x^2 - 4)\,dx = \pi \int_0^1 (5 - x^4 - 4x^2)\,dx$$

$$= \pi \left(5x - \frac{x^5}{5} - \frac{4x^3}{3}\right)\Big|_0^1 = \pi \left(5 = \frac{1}{5} - \frac{4}{3}\right) = \frac{52\pi}{15}.$$

6.4.37 Using washers, we have

$$V = \pi \int_0^1 ((6 - x^2)^2 - 5^2)\,dx = \pi \int_0^1 (x^4 - 12x^2 + 11)\,dx = \pi \left(\frac{x^5}{5} - 4x^3 + 11x\right)\Big|_0^1 = \frac{36\pi}{5}.$$

6.4.39 With washers we have

$$V = \pi \int_0^1 (x^{2/3} - x^2)\,dx = \pi \left(\frac{3x^{5/3}}{5} - \frac{x^3}{3}\right)\Big|_0^1 = \frac{4\pi}{15}.$$

With shells we have

$$V = 2\pi \int_0^1 y(y - y^3)\,dy = 2\pi \left(\frac{y^3}{3} - \frac{y^5}{5}\right)\Big|_0^1 = 2\pi \left(\frac{1}{3} - \frac{1}{5}\right) = \frac{4\pi}{15}.$$

The two methods are equally easy to apply.

6.4.41 Using washers we have

$$V = \pi \int_0^2 \left(\left(1 - \frac{x}{3}\right)^2 - \frac{1}{(x+1)^2}\right)\,dx = \pi \left(x - \frac{x^2}{3} + \frac{x^3}{27} + \frac{1}{(x+1)}\right)\Big|_0^2 = \frac{8\pi}{27}.$$

Using shells we have

$$V = 2\pi \int_{1/3}^1 y\left(3 - 3y - \left(\frac{1}{y} - 1\right)\right)\,dy = 2\pi \int_{1/3}^1 (4y - 3y^2 - 1)\,dy = 2\pi \left(2y^2 - y^3 - y\right)\Big|_{1/3}^1$$

$$= 2\pi \left(0 - \left(\frac{2}{9} - \frac{1}{27} - \frac{1}{3}\right)\right) = \frac{8\pi}{27}.$$

The shell method seems a little easier to apply.

6.4.43 The curves intersect when $8 - 2x = 16 - x^2$, or $x^2 - 2x - 8 = 0$, so at $x = -2, 4$.

Washers: The area is

$$\int_{-2}^{4} \pi((16 - x^2)^2 - (8 - 2x)^2)\, dx = \pi \int_{-2}^{4} x^4 - 36x^2 + 32x + 192\, dx = \frac{3456}{5}\pi.$$

Shells: To use the shell method, solve for x to get $x = \pm\sqrt{16 - y}$ and $x = (8 - y)/2$. We need two integrals, one for y from 0 to 12 and one for y from 12 to 16:

$$\int_{0}^{12} 2\pi y \left(\sqrt{16 - y} - \frac{8 - y}{2} \right) dy + \int_{12}^{16} 2\pi y \cdot 2\sqrt{16 - y}\, dy$$

$$= 2\pi \left[\int_{0}^{12} \left(y\sqrt{16 - y} - 4y + \frac{y^2}{2} \right) dy + \int_{12}^{16} y \cdot 2\sqrt{16 - y}\, dy \right]$$

$$= 2\pi \left[\left(-\frac{2}{15}(32 + 3y)(16 - y)^{3/2} - 2y^2 + \frac{y^3}{6} \right) \Big|_{0}^{12} + \left(-\frac{4}{15}(32 + 3y)(16 - y)^{3/2} \right) \Big|_{12}^{16} \right]$$

$$= 2\pi \left[\left(-\frac{2}{15}(68)(8) - 288 + 288 + \frac{2}{15}(32)(64) \right) - \frac{4}{15}(0 - 68(8)) \right]$$

$$= 2\pi \left(\frac{3008}{15} + \frac{2176}{15} \right) = 2\pi \cdot \frac{1728}{5} = \frac{3456}{5}\pi.$$

The washer method is easier to apply.

6.4.45 First, note that the curve and the line intersect at $(0,0)$ and $(1,0)$.

Washers:

$$V = \pi \int_{0}^{1} (x - x^4)^2\, dx = \pi \int_{0}^{1} (x^2 - 2x^5 + x^8)\, dx$$

$$= \pi \left(x^3/3 - x^6/3 + x^9/9 \right) \Big|_{0}^{1} = \pi/9$$

Shells: It isn't practical to even attempt the shell method, as it is not easy to write x in terms of y for the given expression.

6.4.47

a. True. Otherwise, we wouldn't have shells!

b. False. Either method can be used when revolving around either axis.

c. True.

6.4.49

We use the shell method to avoid two separate integrals:

$$\int_{2}^{4} 2\pi x \cdot x^{-3}\, dx = 2\pi \int_{2}^{4} x^{-2}\, dx = \frac{\pi}{2}.$$

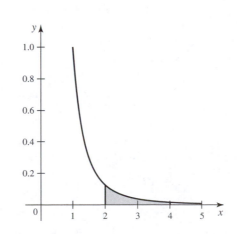

6.4.51 $V = 2\pi \int_0^1 ((2 - x^2)^2 - (x^2)^2)\, dx = 2\pi \int_0^1 (4 - 4x^2)\, dx = 2\pi \left(4x - 4x^3/3 \right) \Big|_0^1 = \frac{16\pi}{3}.$

6.4.53 $V = 2\pi \int_2^6 x(2x + 2 - x)\, dx = 2\pi \int_2^6 (x^2 + 2x)\, dx = 2\pi \left(x^3/3 + x^2 \right) \Big|_2^6 = \frac{608\pi}{3}.$

6.4.55 $V = \int_0^1 A(y)\, dy = \int_0^1 \frac{\pi}{2}(\sqrt{y})^2\, dy = \frac{\pi}{2} \left(y^2/2 \right) \Big|_0^1 = \frac{\pi}{4}.$

6.4.57 By symmetry, $V = 2 \int_0^1 \frac{\pi}{2}(1 - x)^2\, dx = \pi \int_0^1 (1 - 2x + x^2)\, dx = \pi \left(x - x^2 + x^3/3 \right) \Big|_0^1 = \frac{\pi}{3}.$

6.4.59

a. $V_1 = \pi \int_0^1 (ax^2 + 1)^2\, dx = \pi \int_0^1 (a^2 x^4 + 2ax^2 + 1)\, dx = \pi(a^2/5 + 2a/3 + 1).$

   $V_2 = 2\pi \int_0^1 x(ax^2 + 1)\, dx = 2\pi(a/4 + 1/2) = \pi a/2 + \pi.$

b. These are equal when $\pi a/2 + \pi = \pi a^2/5 + 2a\pi/3 + \pi$, or when $\pi a^2/5 + a\pi/6 = 0$, which occurs when $a = 0$ and when $a = -5/6$.

6.4.61

a. $V = \pi \int_0^h (rx/h)^2\, dx = \frac{\pi r^2}{h^2} \left(x^3/3 \right) \Big|_0^h = \frac{\pi r^2 h}{3}.$

b. $V = 2\pi \int_0^r y(h - (h/r)y)\, dy = 2\pi h \left(y^2/2 - y^3/(3r) \right) \Big|_0^r = \frac{\pi r^2 h}{3}.$

6.4.63 The bowl is the surface of revolution when $y = -\sqrt{64 - x^2}$ for $0 \le x \le 8$ is revolved boy the y-axis. The volume of water in the bowl up to height h is the volume integral for the bowl evaluated from -8 (the bottom) to $-8 + h$ (h above the bottom). Using shells, each shell has height $(-8 + h) - (-\sqrt{64 - x^2}) = h - 8 + \sqrt{64 - x^2}$. Further, the quarter-circle meets the line $y = h - 8$ when $h - 8 = -\sqrt{64 - x^2}$, so that $x = \sqrt{16h - h^2}$. Thus the bounds of integration are $x = 0$ to $x = \sqrt{16h - h^2}$. We have

$$V = 2\pi \int_0^{\sqrt{16h - h^2}} x(h - 8 + \sqrt{64 - x^2})\, dx = 2\pi \left(hx^2/2 - 4x^2 - \frac{1}{3}\left(64 - x^2\right)^{3/2} \right) \Bigg|_0^{\sqrt{16h - h^2}}$$

$$= \pi(16h^2 - h^3 - 8(16h - h^2)) - \frac{2}{3}(8 - h)^3 - (0 - 0 - 1024/3)) = \frac{1}{3}(24 - h)\pi h^2.$$

Note that when $h = 0$ the bowl is empty, and in fact the formula above gives $V = 0$. When $h = 8$ the bowl is full, so we have a hemisphere of radius 8, which has volume $\frac{2}{3}\pi \cdot 8^3 = \frac{1024}{3}\pi$; evaluating the expression above at $h = 8$ gives $\frac{1}{3}(24 - 8)\pi \cdot 64 = \frac{1024}{3}\pi$ as well.

6.4.65

$$V = 2\pi \int_0^2 ((3 + \sqrt{4 - x^2})^2 - (3 - \sqrt{4 - x^2})^2)\, dx = 24\pi \int_0^2 \sqrt{4 - x^2}\, dx = 24\pi(\pi) = 24\pi^2.$$

The last integral can be computed with a computer, or one can note that it represents $1/4$ of the area of the circle of radius 2 centered at the origin, so its value is $(1/4)\pi \cdot 4 = \pi$.

6.4.67

a. The cross sections of such an object are washers, with inner radius given b $g(x) - y_0$ and outer radius given by $f(x) - y_0$, so the volume is given by $\pi \int_a^b ((f(x) - y_0)^2 - (g(x) - y_0)^2)\, dx$.

b. The cross section of this object are also washers, and this time the inner radius is $y_0 - f(x)$ and the outer radius is $y_0 - g(x)$, so the volume is given by $\pi \int_a^b ((y_0 - g(x))^2 - (y_0 - f(x))^2)\, dx$.

6.4.69 $V = 2\pi \int_0^2 x f(x^2)\, dx = \pi \int_0^4 f(u)\, du$, where $u = x^2$. Thus, $V = 10\pi$.

6.4.71

a. The longest diagonal of the cube is equal to the diameter of the sphere, which is $\sqrt{r^2 + r^2 + r^2} = \sqrt{3}r$. Thus, if R is the radius of the sphere, we must have $R = \frac{\sqrt{3}}{2}r$.

Now analyzing the cone which contains the sphere, we see that the height h of the cone is $3R$, so $h = \frac{3\sqrt{3}}{2}r$.

Now the volume of the cylinder is $V = \pi \left(\frac{3r}{2}\right)^2 \left(\frac{3\sqrt{3}r}{2}\right) = \frac{27\sqrt{3}\pi r^3}{8}$.

b. Imagine the cone with its vertex up. Consider the plane which contains the vertex of the cone and two non-adjacent vertices of the cube's bottom face. The cross section of this plane with the cone and cube consists of an $r \times \sqrt{2}r$ rectangle (where r is the side length of the cube) and an isosceles triangle of base 2 and height 3, with the rectangle inscribed in the triangle, and the longer side of the rectangle lying on the base of the triangle. Using similar triangles, we have $\frac{r}{3} = \frac{1 - (r/\sqrt{2})}{1}$, so $r = \frac{3\sqrt{2}}{3+\sqrt{2}}$, and the volume of the cube is $\frac{54\sqrt{2}}{(3+\sqrt{2})^3}$.

c. Imagine the sphere with the hole as being obtained by revolving the region pictured around the x-axis, where the relevant curves are $y = r$ and $y = \sqrt{R^2 - x^2}$ where r is the radius of the hole and R is the radius of the sphere. Note that if you draw the triangle with vertices $(0,0)$, $(5,0)$, and $(5,r)$, you have a right triangle with legs of length 5 and r, and hypotenuse of length R, so $25 + r^2 = R^2$.

The volume we are interested in is

$$V = 2\pi \int_0^5 ((R^2 - x^2) - r^2)\, dx = 2\pi \int_0^5 (25 - x^2)\, dx = 2\pi \left(25x - x^3/3\right)\Big|_0^5 = \frac{500\pi}{3}.$$

Note that (surprisingly!), the result doesn't depend on the radius of the original sphere.

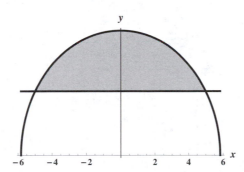

6.5 Length of Curves

6.5.1 Given $f(x)$ and a and b, compute $f'(x)$ and then compute $\int_a^b \sqrt{1 + f'(x)^2}\, dx$.

6.5.3 Because $f'(x) = 3x^2$, the arc length is $\int_{-2}^5 \sqrt{1 + f'(x)^2}\, dx = \int_{-2}^5 \sqrt{1 + 9x^4}\, dx$.

6.5.5 Because $f'(x) = -\frac{2}{x^3}$, the arc length is $\int_1^2 \sqrt{1 + f'(x)^2}\, dx = \int_1^2 \sqrt{1 + \frac{4}{x^6}}\, dx$.

6.5.7 $L = \int_1^5 \sqrt{1 + (y')^2}\, dx = \int_1^5 \sqrt{1 + 2^2}\, dx = 4\sqrt{5}$.

6.5.9 $L = \int_{-2}^{6} \sqrt{1 + (y')^2}\, dx = \int_{-2}^{6} \sqrt{1 + (-8)^2}\, dx = \int_{-2}^{6} \sqrt{65}\, dx = 8\sqrt{65}.$

6.5.11 $y' = \sqrt{x}/2$, so $1 + (y')^2 = 1 + x/4$. Thus,

$$L = \int_{0}^{60} \sqrt{1 + (x/4)}\, dx = \int_{0}^{60} \frac{1}{2}\sqrt{x+4}\, dx = \left(\frac{1}{3}(x+4)^{3/2} \right)\Big|_{0}^{60} = \frac{504}{3} = 168.$$

6.5.13 $y' = x(x^2 + 2)^{1/2}$, so $1 + (y')^2 = 1 + x^2(x^2 + 2) = x^4 + 2x^2 + 1 = (x^2 + 1)^2$. Thus,

$$L = \int_{0}^{1} (x^2 + 1)\, dx = \left(x^3/3 + x \right)\Big|_{0}^{1} = \frac{4}{3}.$$

6.5.15 $y' = x^3 - \frac{1}{4x^3}$, so $1 + (y')^2 = 1 + x^6 - \frac{1}{2} + \frac{1}{16x^6} = \left(x^3 + \frac{1}{4x^3} \right)^2$. Thus,

$$L = \int_{1}^{2} (x^3 + \frac{1}{4x^3})\, dx = \left(x^4/4 + \frac{-1}{8}x^{-2} \right)\Big|_{1}^{2} = 4 - (1/32) - (1/4 - (1/8)) = \frac{123}{32}.$$

6.5.17

a. $y' = 2x$, so $1 + (y')^2 = 1 + 4x^2$, so $L = \int_{-1}^{1} \sqrt{1 + 4x^2}\, dx$.

b. $L = \int_{-1}^{1} \sqrt{1 + 4x^2}\, dx \approx 2.958.$

6.5.19

a. $f'(x) = \sec^2 x$, so $L = \int_{0}^{\pi/4} \sqrt{1 + \sec^4 x}\, dx$

b. Using technology, the value of this integral is approximately 1.277978.

6.5.21

a. $y' = \frac{1}{2\sqrt{x-2}}$, so $1 + (y')^2 = 1 + \frac{1}{4(x-2)} = \frac{4x-7}{4x-8}$, so $L = \int_{3}^{4} \sqrt{\frac{4x-7}{4x-8}}\, dx$.

b. $L = \int_{3}^{4} \sqrt{\frac{4x-7}{4x-8}}\, dx \approx 1.083.$

6.5.23

a. $y' = -2\sin(2x)$, so $1 + (y')^2 = 1 + 4\sin^2(2x)$, so $L = \int_{0}^{\pi} \sqrt{1 + 4\sin^2(2x)}\, dx$.

b. $L = \int_{0}^{\pi} \sqrt{1 + 4\sin^2(2x)}\, dx \approx 5.270.$

6.5.25

a. $y' = -1/x^2$, so $1 + (y')^2 = 1 + \frac{1}{x^4} = \frac{x^4+1}{x^4}$. Thus, $L = \int_{1}^{10} \frac{\sqrt{x^4+1}}{x^2}\, dx$.

b. $L = \int_{1}^{10} \frac{\sqrt{x^4+1}}{x^2}\, dx \approx 9.153.$

6.5.27 $\frac{dx}{dy} = 2$, so $1 + \left(\frac{dx}{dy} \right)^2 = 1 + 4 = 5$, so $L = \int_{-3}^{4} \sqrt{5}\, dy = 7\sqrt{5}.$

6.5.29 $\frac{dx}{dy} = y^3 - 1/(4y^3)$, so $1 + \left(\frac{dx}{dy} \right)^2 = 1 + y^6 - 1/2 + 1/(16y^6) = \left(y^3 + \frac{1}{4y^3} \right)^2$. So $L = \int_{1}^{2} (y^3 + \frac{1}{4} \cdot y^{-3})\, dy = \left(y^4/4 - \frac{1}{8y^2} \right)\Big|_{1}^{2} = 4 - (1/32) - (1/4 - 1/8) = \frac{123}{32}.$

6.5.31

a. False. For example, if $f(x) = x^2$, the first integrand is $\sqrt{1 + 4x^2}$ and the second is $1 + 2x$, which clearly yield different values for (for example) $a = 0$ and $b = 1$.

b. True. They are both equal to $\int_a^b \sqrt{1 + f'(x)^2}\, dx$.

c. False. Because $\sqrt{1 + f'(x)^2} > 0$, arc length can't be negative.

6.5.33

a. We are seeking functions $f(x)$ so that $f'(x) = \pm 4x^2$, so any function of the form $f(x) = \pm 4x^3/3 + C$ will work.

b. We are seeking functions $f(x)$ so that $f'(x) = \pm 6\cos(2x)$, so any function of the form $f(x) = \pm 3\sin(2x) + C$ will work.

6.5.35 The length of the parabola is $\int_{-1}^1 \sqrt{1 + 4x^2}\, dx \approx 2.9597$.

The length of the given cosine function is $\int_{-1}^1 \sqrt{1 + \frac{\pi^2}{4}\sin^2(\pi x/2)}\, dx \approx 2.924$, so the parabola is longer.

6.5.37 $f'(x) = 0.00074x$, so $L = \int_{-640}^{640} \sqrt{1 + (0.00074x)^2}\, dx \approx 1326.4$ meters.

6.5.39

a. Let $u = 2x$. Then the given integral is equal to $\frac{1}{2}\int_a^b \sqrt{1 + f'(u)^2}\, du = \frac{L}{2}$.

b. Let $u = cx$. Then the given integral is equal to $\frac{1}{c}\int_a^b \sqrt{1 + f'(u)^2}\, du = \frac{L}{c}$.

6.5.41

a. $f'(x) = nax^{n-1} - \frac{1}{4anx^{n-1}}$, so that

$$1 + f'(x)^2 = 1 + n^2a^2x^{2n-2} + \frac{1}{16a^2n^2x^{2n-2}} - \frac{1}{2} = (nax^{n-1})^2 + \frac{1}{2} + \frac{1}{(4nax^{n-1})^2}$$

$$= \left(nax^{n-1} + \frac{1}{4nax^{n-1}}\right)^2.$$

Then the arc length is

$$\int_c^d \sqrt{1 + f'(x)^2}\, dx = \int_c^d \left(nax^{n-1} + \frac{1}{4na}x^{-(n-1)}\right) dx$$

which is a rational function that we know how to integrate.

b.

$$\int_1^2 anx^{n-1} + \frac{1}{4an}x^{1-n}\, dx = ax^n + \frac{1}{4an(2-n)}x^{2-n}\Big|_1^2 = a(2^n - 1) + \frac{2^{2-n} - 1}{4an(2-n)}.$$

6.6 Surface Area

6.6.1 $S = \pi r\sqrt{r^2 + h^2} = 3\pi\sqrt{9 + 16} = 15\pi$.

6.6.3 Evaluate $\int_a^b 2\pi f(x)\sqrt{1 + f'(x)^2}\, dx$.

6.6.5 $S = 2\pi \int_0^6 (3x + 4)\sqrt{1 + 9}\, dx = 2\sqrt{10}\pi\left(3x^2/2 + 4x\right)\Big|_0^6 = 2\sqrt{10}\pi(54 + 24) = 156\sqrt{10}\pi$.

6.6.7 $S = 2\pi \int_9^{20} 8\sqrt{x}\sqrt{1 + (16/x)}\, dx = 16\pi \int_9^{20} \sqrt{x + 16}\, dx = 16\pi \left(\frac{2(x+16)^{3/2}}{3} \right) \Big|_9^{20} = \frac{32\pi}{3}(216 - 125) = \frac{2912\pi}{3}$.

6.6.9 $S = 2\pi \int_1^2 \left(x^{3/2} - \frac{\sqrt{x}}{3} \right)\sqrt{1 + (3\sqrt{x}/2 - x^{-1/2}/6)^2}\, dx = 2\pi \int_1^2 (x^{3/2} - \sqrt{x}/3)\sqrt{1 + \left(\frac{9x-1}{6\sqrt{x}} \right)^2}\, dx =$

$2\pi \int_1^2 (x^{3/2} - \sqrt{x}/3)\sqrt{\frac{36x + 81x^2 - 18x + 1}{36x}}\, dx = 2\pi \int_1^2 (x^{3/2} - \sqrt{x}/3)\sqrt{\left(\frac{(1+9x)^2}{36x} \right)}\, dx =$

$2\pi \int_1^2 \left(\frac{(1+9x)x}{6} - \frac{1+9x}{18} \right) dx = \frac{2\pi}{18} \int_1^2 (3x + 27x^2 - 1 - 9x)\, dx = \frac{2\pi}{18} \int_1^2 (27x^2 - 6x - 1)\, dx = \frac{\pi}{9}\left(9x^3 - 3x^2 - x \right) \Big|_1^2 = \frac{\pi}{9}(72 - 12 - 2 - (9 - 3 - 1)) = \frac{53\pi}{9}$.

6.6.11 $S = 2\pi \int_2^3 \sqrt{-x^2 + 6x - 5}\sqrt{1 + \left(\frac{6-2x}{2\sqrt{-x^2+6x-5}} \right)^2}\, dx$. This can be written as $2\pi \int_2^3 \sqrt{-x^2 + 6x - 5} \cdot$

$\sqrt{\frac{4(-x^2+6x-5)+36-24x+4x^2}{4(-x^2+6x-5)}}\, dx = 2\pi \int_2^3 \sqrt{-x^2 + 6x - 5} \cdot \frac{2}{\sqrt{-x^2+6x-5}}\, dx = 4\pi$.

6.6.13 $S = 2\pi \int_{1/2}^2 (x^3/3 + x^{-1}/4)\sqrt{1 + (x^2 - x^{-2}/4)^2}\, dx =$

$2\pi \int_{1/2}^2 (x^3/3 + x^{-1}/4)\sqrt{1 + x^4 - 1/2 + x^{-4}/16}\, dx = 2\pi \int_{1/2}^2 (x^3/3 + x^{-1}/4)\sqrt{x^4 + 1/2 + x^{-4}/16}\, dx =$

$2\pi \int_{1/2}^2 (x^3/3 + x^{-1}/4)\sqrt{(x^2 + x^{-2}/4)^2}\, dx = 2\pi \int_{1/2}^2 (x^3/3 + x^{-1}/4)(x^2 + x^{-2}/4)\, dx = 2\pi \int_{1/2}^2 (x^5/3 + x/12 +$

$x/4 + x^{-3}/16)\, dx = 2\pi \int_{1/2}^2 (x^5/3 + x/3 + x^{-3}/16)\, dx = 2\pi \left(x^6/18 + x^2/6 - x^{-2}/32 \right) \Big|_{1/2}^2 = 2\pi(32/9 + 2/3 -$

$1/128 - (1/1152 + 1/24 - 1/8)) = \frac{275\pi}{32}$.

6.6.15 $S = 2\pi \int_1^7 \sqrt{8x - x^2}\sqrt{1 + \frac{(8-2x)^2}{4(8x-x^2)}}\, dx = 2\pi \int_1^7 \sqrt{8x - x^2}\sqrt{\frac{32x - 4x^2 + 64 - 32x + 4x^2}{4(8x-x^2)}}\, dx =$

$2\pi \int_1^7 \sqrt{8x - x^2}\sqrt{\frac{64}{4(8x-x^2)}}\, dx = 2\pi \int_1^7 4\, dx = 8\pi(7 - 1) = 48\pi$. Because the surface area is 48π square meters, the volume of paint required to cover the surface to a thickness of $.0015$ meters is $48\pi(.0015) \approx .226$ cubic meters. This is about 59.75 gallons.

6.6.17 Note that $x = y^3/3$ for $0 \le y \le 2$. The surface area is $S = 2\pi \int_0^2 (y^3/3)\sqrt{1 + y^4}\, dy$. Let $u = 1 + y^4$ so that $du = 4y^3\, dy$. Substituting yields $S = 2\pi \int_1^{17} \frac{1}{12} u^{1/2}\, du = \frac{\pi}{9}\left(u^{3/2} \right) \Big|_1^{17} = \frac{\pi}{9}(17\sqrt{17} - 1)$.

6.6.19 Note that $x = \frac{y+1}{4}$ for $3 \le y \le 15$. The surface area is $S = 2\pi \int_3^{15} \frac{y+1}{4}\sqrt{1 + \frac{1}{16}}\, dy = \frac{\pi\sqrt{17}}{8} \int_3^{15}(y + 1)\, dy = \frac{\pi\sqrt{17}}{8}\left(y^2/2 + y \right) \Big|_3^{15} = \frac{\pi\sqrt{17}}{8}(225/2 + 15 - (9/2 + 3)) = \frac{\pi\sqrt{17}}{8}(120) = 15\sqrt{17}\pi$.

6.6.21

a. False. One would need to find $x = f^{-1}(y)$ and compute the corresponding integral using f^{-1}.

b. False. For example, the curve given in number 14 above isn't one-to-one on the given interval, but the surface is still defined.

c. True. Because the curve is symmetric about the y-axis, the surface generated by revolving half the curve is half the surface generated by revolving the whole curve.

d. False. This curve is symmetric about the y-axis, so the surface generated by revolving the whole curve is the same as the surface generated by revolving the portion over $[0, 4]$.

6.6.23 $S = 2\pi \int_1^4 (4y^{3/2} - y^{1/2}/12)\sqrt{1 + (6y^{1/2} - y^{-1/2}/24)^2}\, dy =$
$2\pi \int_1^4 (4y^{3/2} - y^{1/2}/12)\sqrt{1 + 36y - 1/2 + 1/(576y)}\, dy = 2\pi \int_1^4 (4y^{3/2} - y^{1/2}/12)\sqrt{36y + 1/2 + 1/(576y)}\, dy =$
$2\pi \int_1^4 (4y^{3/2} - y^{1/2}/12)\sqrt{(6y^{1/2} + y^{-1/2}/24)^2}\, dy = 2\pi \int_1^4 (4y^{3/2} - y^{1/2}/12)(6y^{1/2} + y^{-1/2}/24)\, dy =$
$2\pi \int_1^4 (24y^2 - y/3 - 1/288)\, dy = 2\pi \left(8y^3 - y^2/6 - y/288 \right) \Big|_1^4 = 2\pi(512 - 8/3 - 1/72 - (8 - 1/6 - 1/288)) = \frac{48143\pi}{48}$.

6.6.25

$$S = 2\pi \int_1^8 (9x^{2/3} - x^{4/3}/32)\sqrt{1 + (6x^{-1/3} - x^{1/3}/24)^2}\, dx$$

$$= 2\pi \int_1^8 (9x^{2/3} - x^{4/3}/32)\sqrt{1 + 36x^{-2/3} - 1/2 + x^{2/3}/576}\, dx$$

$$= 2\pi \int_1^8 (9x^{2/3} - x^{4/3}/32)\sqrt{36x^{-2/3} + 1/2 + x^{2/3}/576}\, dx$$

$$= 2\pi \int_1^8 (9x^{2/3} - x^{4/3}/32)\sqrt{(6x^{-1/3} + x^{1/3}/24)^2}\, dx = 2\pi \int_1^8 (9x^{2/3} - x^{4/3}/32)(6x^{-1/3} - x^{1/3}/24)\, dx$$

$$= 2\pi \int_1^8 54x^{1/3} + 3x/16 - x^{5/3}/768\, dx = 2\pi \left(\frac{81x^{4/3}}{2} + \frac{3x^2}{32} - \frac{x^{8/3}}{2048}\right)\Bigg|_1^8$$

$$= 2\pi\,(648 + 6 - 1/8 - (81/2 + 3/32 - 1/2048)) = \frac{1256001\pi}{1024}$$

6.6.27

a. $S = 2\pi \int_0^{\pi/2} \cos x \sqrt{1 + \sin^2 x}\, dx$.

b. $S \approx 7.21$.

6.6.29

a. $S = 2\pi \int_0^{\pi/4} \tan x \sqrt{1 + \sec^4 x}\, dx$.

b. $S \approx 3.84$.

6.6.31 Let $y = f(x) = (a^{2/3} - x^{2/3})^{3/2}$. Note that $f'(x) = -x^{-1/3}\sqrt{a^{2/3} - x^{2/3}}$, so

$$\sqrt{1 + f'(x)^2} = \sqrt{1 + \frac{a^{2/3} - x^{2/3}}{x^{2/3}}} = \sqrt{\frac{x^{2/3} + a^{2/3} - x^{2/3}}{x^{2/3}}} = \frac{a^{1/3}}{x^{1/3}}.$$

Thus (using symmetry) $S = 4\pi \int_0^a (a^{2/3} - x^{2/3})^{3/2}(a^{1/3}x^{-1/3})\, dx$. Let $u = (a^{2/3} - x^{2/3})$, so that $du = (-2x^{-1/3}/3)\, dx$. Substituting gives $S = -6\pi a^{1/3} \int_{a^{2/3}}^0 u^{3/2}\, du = 6\pi a^{1/3} \int_0^{a^{2/3}} u^{3/2}\, du = \frac{12\pi}{5} a^{1/3}\left(u^{5/2}\right)\Big|_0^{a^{2/3}} = \frac{12\pi a^2}{5}$.

6.6.33 Let $y = f(x) = \sqrt{r^2 - x^2}$, and imagine the surface obtained by revolving this curve around the x-axis for $a \le x \le a + h$. The surface area is $S = 2\pi \int_a^{a+h} \sqrt{r^2 - x^2}\sqrt{1 + \left(\frac{-x}{\sqrt{r^2-x^2}}\right)^2}\, dx$. This can be written as $2\pi \int_a^{a+h} \sqrt{r^2 - x^2}\sqrt{\frac{r^2 - x^2 + x^2}{r^2 - x^2}}\, dx = 2\pi \int_a^{a+h} \sqrt{r^2 - x^2}\frac{\sqrt{r^2}}{\sqrt{r^2-x^2}}\, dx = 2\pi \int_a^{a+h} r\, dx = 2\pi r(a + h - a) = 2\pi rh$.

6.6.35

a. The ratio is $\frac{6a^2}{a^3} = \frac{6}{a}$.

b. The ratio is $\frac{4\pi a^2}{(4/3)\pi a^3} = \frac{3}{a}$.

c. The volume of a ball of radius r is $\frac{4}{3}\pi r^3$, so the volume of one with the given radius is $\frac{4}{3}\pi \frac{h}{4} = \frac{\pi h}{3}$. The volume of the ellipsoid is $\frac{4}{3}\pi \cdot \sqrt[3]{h} \cdot \frac{\sqrt[3]{h}}{2} \cdot \frac{\sqrt[3]{h}}{2} = \frac{\pi h}{3}$.

	h	Volume	Ball Surface Area	Ellipsoid Surface Area	Ball SAV	Ellipsoid SAV
	1.1	1.15192	5.31412	5.72187	4.61327	4.96724
d.	5	5.23599	14.582	15.7008	2.78495	2.99864
	10	10.472	3.1475	24.9235	2.21041	2.38001
	20	20.944	36.7443	39.5636	1.75441	1.88902

e. More spherical is better for minimizing heat loss.

6.6.37

a. Note that $g(x) = cf(x)$, so that $g'(x) = cf'(x)$.

$$2\pi \int_a^b g(x)\sqrt{c^2 + g'(x)^2}\, dx = 2\pi \int_a^b cf(x)\sqrt{c^2 + c^2 f'(x)^2}\, dx = 2c^2\pi \int_a^b f(x)\sqrt{1 + f'(x)^2}\, dx = c^2 A.$$

b. Note that $h'(x) = cf'(cx)$. Consider

$$2\pi \int_{a/c}^{b/c} f(cx)\sqrt{c^2 + c^2 f'(cx)^2}\, dx = 2c\pi \int_{a/c}^{b/c} f(cx)\sqrt{1 + f'(cx)^2}\, dx.$$

Let $u = cx$ so that $du = c\, dx$. Then our integral is equal to

$$2\pi \int_a^b f(u)\sqrt{1 + f'(u)^2}\, du = A.$$

6.7   Physical Applications

6.7.1 $m = \rho_1 \cdot l_1 + \rho_2 \cdot l_2 = 1\,\text{g/cm} \cdot 50\,\text{cm} + 2\,\text{g/cm} \cdot 50\,\text{cm} = 150$ g.

6.7.3 The work is the product of 5 Newtons and 5 meters, which is 25 J.

6.7.5 Different volumes of water are moved different distances.

6.7.7 $F = \rho g h = 1000 \cdot 9.8 \cdot 4 = 39,200$ N/m^2.

6.7.9 $m = \int_0^\pi (1 + \sin x)\, dx = (x - \cos x)\Big|_0^\pi = \pi - (-1) - (0 - 1) = \pi + 2.$

6.7.11 $= \int_0^2 (2 - x/2)\, dx = \left(2x - x^2/4\right)\Big|_0^2 = 3.$

6.7.13 $m = \int_0^1 x\sqrt{2 - x^2}\, dx = \left(-\frac{1}{3}(2 - x^2)^{3/2}\right)\Big|_0^1 = (-1/3) + (2\sqrt{2}/3) = \frac{2\sqrt{2}-1}{3}.$

6.7.15 $m = \int_0^2 1\, dx + \int_2^4 (1 + x)\, dx = 2 + \left(x + x^2/2\right)\Big|_2^4 = 2 + (4 + 8) - (2 + 2) = 10.$

6.7.17 $W = \int_0^3 2x\, dx = x^2\Big|_0^3 = 9$ J.

6.7.19

a. Because $f(.2) = .2k = 30$, we have $k = 150$.

b. $W = \int_0^{-.4} 150x\, dx = 75x^2\Big|_0^{-.4} = 75 \cdot 0.16 = 12$ J.

c. $W = \int_0^{.3} 150x\, dx = 75x^2\Big|_0^{.3} = 75 \cdot 0.09 = 6.75$ J.

d. $W = \int_{.2}^{.4} 150x\, dx = 75x^2\Big|_{.2}^{.4} = 75(0.16 - 0.04) = 9$ J.

6.7.21

a. $f(x) = kx$, and $f(.5) = 50$, so $k(.5) = 50$, so $k = 100$.

Therefore $W = \int_0^{1.5} 100x\,dx = \left(50x^2\right)\Big|_0^{1.5} = 112.5$ J.

b. $\int_0^{-.5} 100x\,dx = \left(50x^2\right)\Big|_0^{-.5} = 12.5$ J.

6.7.23

a. $f(.2) = .2k = 50$, so $k = 250$. $W = \int_0^{.5} 250x\,dx = 125x^2\Big|_0^{.5} = 125 \cdot 0.25 = 31.25$ J.

b. $\int_0^{.2} kx\,dx = kx^2/2\Big|_0^{.2} = .02k = 50$, so $k = 2500$. $W = \int_0^{.5} 2500x\,dx = 1250x^2\Big|_0^{.5} = 1250 \cdot 0.25 = 312.5$ J.

6.7.25

a. We have $\int_0^{0.5} kx\,dx = \frac{k}{2}x^2\Big|_0^{0.5} = 0.125k = 100$, so that $k = 800$. Then

$$W = \int_0^{1.25} 800x\,dx = 400x^2\Big|_0^{1.25} = 625\,\text{J}.$$

b. $f(0.5) = 0.5k = 250$, so $k = 500$ and thus

$$W = \int_0^{1.25} 500x\,dx = 250x^2\Big|_0^{1.25} = 390.625\,\text{J}.$$

6.7.27 $W = \int_0^{2.5} \rho g A(y)(2.5 - y)\,dy = 1000 \cdot 9.8 \cdot 25 \cdot 15 \int_0^{2.5}(2.5 - y)\,dy = 3675000\left(2.5y - y^2/2\right)\Big|_0^{2.5} = 3675000 \cdot (2.5)^2/2 = 11,484,375$ J.

6.7.29 $W = \int_0^4 \rho g \pi 2^2(10 - y)\,dy = 4\pi\rho g\left(10y - y^2/2\right)\Big|_0^4 = 4\pi\rho g(40 - 8) = 128\pi\rho g \approx 3.941 \times 10^6$ J.

6.7.31

a. Let the vertex of the cone be at $(0,0)$, with the y-axis vertically oriented. Note that the area of a horizontal slice at height y is $\pi y^2/16$, and it must move $6 - y$ meters to get to the top.

$$W = \int_0^6 \rho g \pi \frac{y^2}{16}(6-y)\,dy = \frac{\pi\rho g}{16}\int_0^6 (6y^2 - y^3)\,dy = \frac{\pi\rho g}{16}\left(2y^3 - y^4/4\right)\Big|_0^6 = (\pi\rho g/16)(108) = 66,150\pi\,\text{J}.$$

b. Not true.

$$\int_0^3 \rho g \pi \frac{y^2}{16}(6 - y)\,dy = \frac{\pi\rho g}{16}\int_0^3 (6y^2 - y^3)\,dy = \frac{\pi\rho g}{16}\left(2y^3 - y^4/4\right)\Big|_0^3 = (\pi\rho g/16) \cdot 33.75 \approx 20672\pi\,\text{J},$$

less than half the amount from part (a). Note that while the water must be raised further than water in the top half, due to the shape of the tank, there is far less water in the bottom half than in the top.

6.7.33

a. Orient the axes so that the south pole of the tank is at $(0,0)$ and the north pole is at $(0,16)$. The cross section of the tank which contains the xy plane intersects the tank in the circle centered at $(0,8)$ with radius 8, so the curve is $x^2 + (y-8)^2 = 8^2$. A slice at height y has area $\pi x^2 = \pi(16y - y^2)$.

$$W = \int_0^{16} \rho g \pi (16y - y^2)\, dy = \pi \rho g \left(8y^2 - y^3/3\right) \Big|_0^{16} = \pi \rho g \frac{16^3}{6} = 200704000 \cdot \pi/3 \approx 2.102 \times 10^8 \text{ J}.$$

b. The total weight of the water lifted up for 18 meters is

$$W = \frac{4\pi}{3}R^3 \rho g h = \frac{4\pi}{3} 8^3 \cdot 1000 \cdot 9.8 \cdot 18 = 120422400\pi \approx 3.783 \times 10^8 \text{ J}.$$

6.7.35

a. Orient the axes so that the lower corners of the trough are at $(-0.25, 0)$ and at $(.25, 0)$. Then the upper corners are at $(-.5, 1)$ and at $(.5, 1)$. Note that the line between $(.25, 0)$ and $(.5, 1)$ is given by $y = 4x - 1$. The area of a slice at height y is $2x \cdot 10 = 20 \cdot \frac{1}{4}(y + 1) = 5(y + 1)$. Thus, $W =$

$$\rho g \int_0^1 5(y+1)(1-y)\, dy = 5\rho g \int_0^1 (1 - y^2)\, dy = 5\rho g \left(y - y^3/3\right) \Big|_0^1 = \frac{10\rho g}{3} \approx 32,667 \text{ J}.$$

b. Yes. If the length is doubled, the area of each slice is doubled, so the work integral is doubled as well.

6.7.37 Let the vertex of the cone be at $(0,0)$, with the y-axis vertically oriented. Note that the area of a horizontal slice at height y is $\pi y^2/16$, and it must move $3 - y$ meters to get to the point 1 meter above the top. $W = \int_0^2 \rho g \pi \frac{y^2}{16}(3 - y)\, dy = (\pi \rho g/16) \int_0^2 (3y^2 - y^3)\, dy = (\pi \rho g/16)\left(y^3 - y^4/4\right)\Big|_0^2 = (\pi \rho g/16)(4) = \frac{\pi \rho g}{4} \approx$ 7696.9 J.

6.7.39 Orient the axes so that the lower corners of the trapezoid are at $(5, 0)$ and $(-5, 0)$, and the upper corners are at $(10, 15)$ and $(-10, 15)$. Note that the line between the corners for $x > 0$ is given by $y = 3(x-5)$, so at level y, we have a width of $2x = \frac{2y}{3} + 10$.

$$F = \rho g \int_0^{15} (15 - y)\left(\frac{2y + 30}{3}\right) dy = \frac{2\rho g}{3} \int_0^{15} (225 - y^2)\, dy = \frac{2\rho g}{3}\left(225y - y^3/3\right)\Big|_0^{15} = \frac{2\rho g}{3} \cdot 2250 = 1500\rho g =$$
1.470×10^7 N.

6.7.41 Orient the axes so that the bottom vertex is at $(0, 0)$. The other vertices are at $(\pm 10, 30)$, and the line between $(0, 0)$ and $(10, 30)$ is given by $y = 3x$. Thus, a slice at height y has width $2x = 2y/3$.

$$F = \int_0^{30} \rho g(30 - y)\frac{2y}{3}\, dy = \frac{2\rho g}{3}\left(15y^2 - y^3/3\right)\Big|_0^{30} = \frac{2\rho g}{3}(4500) = 3000\rho g = 2.940 \times 10^7 \text{ N}.$$

6.7.43 The width of the plate at depth y is $2 - y$, so the force on the plate is

$$F = \int_1^2 \rho g(2 - y)y\, dy = \rho g\left(y^2 - \frac{y^3}{3}\right)\Big|_1^2 = \rho g \cdot \frac{2}{3} \approx 6533 \text{ N}.$$

6.7.45 $F = \int_0^{50} (150 + 2y) \cdot 80\, dy = 80\left(150y + y^2\right)\Big|_0^{50} = 8 \times 10^5$ N.

6.7.47 $F = \int_1^{1.5} \rho g(4 - y) \cdot 0.5\, dy = \rho g\left(2y - y^2/4\right)\Big|_1^{1.5} = \frac{11\rho g}{16} = 6737.5$ N.

6.7.49

a. True. $m = \int_a^b \rho(x)\, dx = \frac{1}{b-a}\int_a^b \rho(x)\, dx \cdot (b - a) = \overline{\rho} \cdot L$.

b. True. $\int_0^L kx\, dx = \frac{kL^2}{2} = \int_0^{-L} kx\, dx$.

c. True. This follows because work is force times distance.

d. False. Although they have the same geometry, they are placed at different depths of the water, so the force is different.

6.7.51

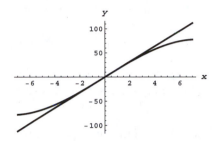

a. Compared to the linear spring $F(x) = 16x$, the restoring force is less for large displacements.

b. $W = \int_0^{1.5}(16x - 0.1x^3)\,dx = \left(8x^2 - .025x^4\right)\Big|_0^{1.5} = 17.87$ J.

c. $W = \int_0^{-2}(16x - 0.1x^3)\,dx = \left(8x^2 - .025x^4\right)\Big|_0^{-2} = 31.6$ J.

6.7.53 Orient the axes so that $(0,0)$ is in the middle of the bottom of the cup. Note that the line between $(0.02, 0)$ and $(0.025, 0.15)$ is given by $y = 30x - 3/5$, so $x = \frac{y}{30} + \frac{1}{50}$. The area of a cross section at height y is given by $\pi x^2 = \pi\left(\frac{5y+3}{150}\right)^2$. Note that the distance the slice must travel is $0.2 - y$, because it must go 0.05 above the top of the glass. Thus, $W = \pi\rho g \int_0^{0.15}\left(\frac{5y+3}{150}\right)^2(.2 - y)\,dy = \frac{\pi\rho g}{5\cdot150^2}\int_0^{0.15}(5y+3)^2(1 - 5y)\,dy = \frac{\pi\rho g}{5\cdot150^2}\int_0^{0.15}(-125y^3 - 125y^2 - 15y + 9)\,dy = \frac{\pi\rho g}{5\cdot150^2}\left(-125y^4/4 - 125y^3/3 - 15y^2/2 + 9y\right)\Big|_0^{0.15} = \frac{\pi\rho g}{5\cdot150^2}\cdot1.0248 \approx 0.280$ J.

6.7.55

a.
$$W = \int_0^{2500000}\frac{GMm}{(x+R)^2}\,dx = GMm\left(\frac{-1}{x+R}\right)\Big|_0^{2500000}$$
$$= GMm\left(\frac{1}{R} - \frac{1}{R+2500000}\right) = \frac{GMm2500000}{R(R+2500000)} \approx 8.87435\times10^9 \text{ J.}$$

b. $W(x) = \int_0^x \frac{GMm}{(t+R)^2}\,dt = GMm\left(\frac{-1}{t+R}\right)\Big|_0^x = GMm\left(\frac{1}{R} - \frac{1}{R+x}\right) = \frac{GMmx}{R(R+x)} = \frac{500GMx}{R(R+x)}.$

c. $\displaystyle\lim_{x\to\infty}\frac{GMmx}{R(R+x)} = \lim_{x\to\infty}\frac{GMm}{R\left(\left(\frac{R}{x}\right)+1\right)} = \frac{GMm}{R}.$

d. Suppose $\frac{GMmx}{R(R+x)} = \frac{1}{2}mv^2$, then $v^2 = \frac{2GMx}{R(R+x)}$, and as $x\to\infty$ we have $v^2 = \frac{2GM}{R}$, so $v = \sqrt{\frac{2GM}{R}}$.

6.7.57

a. $W_1 = \int_0^{30}\rho g(30 - y)\,dy = 5g\left(30y - y^2/2\right)\Big|_0^{30} = 2250g$ J.

b. $W = W_1 + W_2$, where W_2 is the work to just lift the block. $W_2 = 50g\cdot30 = 1500g$, so $W = 2250g + 1500g = 3750g$ J.

6.7.59

a. The acceleration due to gravity is $F = mg$, and is in the vertical direction. The tangent direction to the curve is perpendicular to the normal, which makes an angle of θ with the the vertical. If we form a right triangle with hypotenuse of length mg, and angle θ, then the two legs must have lengths $mg \sin \theta$ (parallel to the curve) and $mg \cos \theta$ (normal to the curve).

b. Note that the angle t is $t = S/L$, where S is arc length. Then $S = Lt$ and $dS = L\,dt$. So $W = \int_0^\theta F\,ds = \int_0^\theta mg \sin \theta \cdot L\,d\theta = mgL\left(-\cos\theta\right)\Big|_0^\theta = mg(L - L\cos\theta) = mgh.$

6.7.61   The plate on the left has more than half of its area below the horizontal line which is $3/2$ below the surface, while the plate on the right has exactly half its area below that line, so the plate on the left should have more force than the plate on the right.

a. Note that the line in the first quadrant form $(0,0)$ to $(\sqrt{2}/2, \sqrt{2}/2)$ is given by $y = x$, while the line from $(\sqrt{2}/2, \sqrt{2}/2)$ to $(0, \sqrt{2})$ is given by $y = -x + \sqrt{2}$. So the width of a slice at height y in the lower part of the region is $2y$, and in the upper part of the region is $2(\sqrt{2} - y)$.

$F = \rho g \int_0^{\sqrt{2}/2} (\sqrt{2} + 1 - y)(2y)\,dy + \rho g \int_{\sqrt{2}/2}^{\sqrt{2}} (\sqrt{2} + 1 - y)2(\sqrt{2} - y)\,dy = \rho g \int_0^{\sqrt{2}/2} (-2y^2 + 2\sqrt{2}y + 2y)\,dy +$

$\rho g \int_{\sqrt{2}/2}^{\sqrt{2}} (2y^2 - 4\sqrt{2}y - 2y + 2\sqrt{2} + 4)\,dy = \rho g\left(-\frac{2y^3}{3} + \sqrt{2}y^2 + y^2\right)\Big|_0^{\sqrt{2}/2}$

$+ \rho g \left(2\left(\frac{y^3}{3} - \frac{1}{2}(1 + 2\sqrt{2})y^2 + \sqrt{2}y + 2y\right)\right)\Big|_{\sqrt{2}/2}^{\sqrt{2}} = \rho g\left(\frac{1}{2} + \frac{\sqrt{2}}{3} + \frac{1}{2} + \frac{1}{3\sqrt{2}}\right) = \rho g\left(1 + \frac{\sqrt{2}}{2}\right)$ N. This is 16,730 N.

b. $F = \int_0^1 \rho g(2 - y) \cdot 1\,dy = \rho g\left(2y - y^2/2\right)\Big|_0^1 = \frac{3\rho g}{2}$ N. This is 14,700 N.

6.7.63

a. $F = \int_1^3 \rho g(4 - y) \cdot 2\,dy = \rho g\left(8y - y^2\right)\Big|_1^3 = 8\rho g = 78,400$ N. This is less than 90,000 N., so the window can withstand the force.

b. $F(h) = \int_1^3 \rho g(h - y) \cdot 2\,dy = \rho g\left(2hy - y^2\right)\Big|_1^3 = \rho g(6h - 9 - (2h - 1)) = 4\rho g(h - 2)$. This is less than or equal to 90,000 Newtons when $h \le 4.296$ meters.

Chapter Six Review

1

a. True. A vertical slice would lead to shells, while a horizontal slice would lead to either disks or washers.

b. True. In order to find position, you would also need to know either its initial position, or at least its position at some time.

c. True. If dV/dt is constant, then V is a linear function of time.

3   The position $s(t)$ and the displacement are the same, because the projective started on the ground at position 0. $s(t) = \int_0^t v(x)\,dx = \int_0^t (20 - 10x)\,dx = \left(20x - 5x^2\right)\Big|_0^t = 20t - 5t^2.$ Note that the projectile is moving up for $0 < t < 2$ and down for $2 < t < 4$. Thus the distance traveled is equal to the position for $0 \le t < 2$, but for $2 \le t < 4$ the distance traveled is $20 + (20 - (20t - 5t^2)) = 40 - 20t + 5t^2$.

5

a. $v(t) = \int a(t)\,dt = \int 2\sin(\pi t/4)\,dt = -\frac{8}{\pi}\cos(\pi t/4) + C$, and because $v(0) = -\frac{8}{\pi}$, we have $C = 0$. Thus, $v(t) = -\frac{8}{\pi}\cos(\pi t/4)$.

$s(t) = \int v(t)\,dt = \int -\frac{8}{\pi}\cos(\pi t/4)\,dt = -\frac{32}{\pi^2}\sin(\pi t/4) + D$, and because $s(0) = 0$ we have $D = 0$. Thus, $s(t) = -\frac{32}{\pi^2}\sin(\pi t/4)$.

b. s is periodic with period 8, so we only consider $0 \le t \le 8$. There are critical numbers for s at $t = 2$ and $t = 6$. There is a maximum for s of $\frac{32}{\pi^2}$ at $t = 6$ and a minimum of $-\frac{32}{\pi^2}$ at $t = 2$.

c. The average velocity is $\frac{1}{8}\int_0^8 -\frac{8}{\pi}\cos(\pi t/4)\,dt = -\frac{4}{\pi^2}\left(\sin(\pi t/4)\right)\Big|_0^8 = 0.$

The average position is $\frac{1}{8}\int_0^8 -\frac{32}{\pi^2}\sin(\pi t/4)\,dt = \frac{16}{\pi^3}\left(\cos(\pi t/4)\right)\Big|_0^8 = 0.$

7

a. For $0 \le t \le 8$ we have $R'(t) = 4t^{1/3}$, so $R(t) = 3t^{4/3} + C$, but $C = 0$, so $R(t) = 3t^{4/3}$.

b. Note that $R(8) = 48$, so for $t > 8$, $R(t) = 48 + \int_8^t 2\,dx = 48 + 2(t - 8)$.

We have $R(t) = \begin{cases} 3t^{4/3} & \text{if } 0 \le t \le 8; \\ 2t + 32 & \text{if } t > 8. \end{cases}$

c. The fuel runs out when $150 = 48 + 2(t - 8)$, which occurs for $t = 59$.

9 $\int_0^1 (-x - (x^2 - 2))\,dx = \int_0^1 (2 - x - x^2)\,dx =$
$(2x - x^2/2 - x^3/3)\big|_0^1 = 2 - \frac{1}{2} - \frac{1}{3} = \frac{7}{6}.$

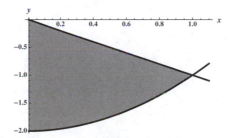

11 These two curves meet when $4x = x\sqrt{25 - x^2}$, so at $x = 0$ and when $4 = \sqrt{25 - x^2}$, which occurs for $x = 3$. A plot of the region is

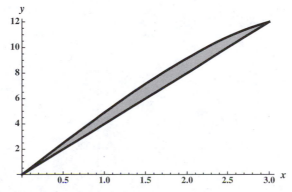

On $[0, 3]$, we have $x\sqrt{25 - x^2} > 4x$ (for example check at $x = 1$), so the area is

$$A = \int_0^3 \left(x\sqrt{25 - x^2} - 4x \right)\,dx.$$

Note that $\int_0^3 -4x\,dx = -2x^2 \Big|_0^3 = -18$. For the first term, use the substitution $u = 25 - x^2$, so that $du = -2x\,dx$ and the integration bounds become $u = 25$ to $u = 16$. Then we have

$$A = -\frac{1}{2}\int_{25}^{16} u^{1/2}\,du - 18 = -\frac{1}{2}\left(\frac{2}{3}u^{3/2}\right)\Bigg|_{25}^{16} - 18 = -\frac{64}{3} + \frac{125}{3} - 18 = \frac{7}{3}.$$

13 Note that (by inspection followed by an easy calculation) $2\sqrt{x}$ and $x(x-3)$ meet at $(0,0)$ and at $(4,4)$. Also, $2\sqrt{x}$ and $3-x$ meet at $(1,2)$ and $x(x-3)$ and $3-x$ meet at $(-1,4)$ and $(3,0)$. So the area of R_1 is

$$\int_{-1}^0 ((3-x) - (x^2 - 3x))\,dx + \int_0^1 (3 - x - 2\sqrt{x})\,dx$$

$$= \int_{-1}^0 (3 + 2x - x^2)\,dx + \int_0^1 (3 - x - 2\sqrt{x})\,dx$$

$$= \left(3x + x^2 - x^3/3\right)\Big|_{-1}^0 + \left(3x - x^2/2 - 4x^{3/2}/3\right)\Big|_0^1$$

$$= 0 - \left(-3 + 1 + \frac{1}{3}\right) + \left(3 - \frac{1}{2} - \frac{4}{3}\right) - 0 = \frac{17}{6}.$$

The area of R_2 is

$$\int_0^1 (2\sqrt{x} - (x^2 - 3x))\,dx + \int_1^3 ((3-x) - (x^2 - 3x))\,dx$$

$$= \int_0^1 (2\sqrt{x} - x^2 + 3x)\,dx + \int_1^3 (3 + 2x - x^2)\,dx$$

$$= \left(4x^{3/2}/3 - x^3/3 + 3x^2/2\right)\Big|_0^1 + \left(3x + x^2 - x^3/3\right)\Big|_1^3$$

$$= \left(\frac{4}{3} - \frac{1}{3} + \frac{3}{2}\right) - 0 + (9 + 9 - 9) - \left(3 + 1 - \frac{1}{3}\right) = \frac{47}{6}.$$

The area of R_3 is

$$\int_1^3 (2\sqrt{x} - (3-x))\,dx + \int_3^4 (2\sqrt{x} - (x^2 - 3x))\,dx$$

$$= \left(4x^{3/2}/3 - 3x + x^2/2\right)\Big|_1^3 + \left(4x^{3/2}/3 - x^3/3 + 3x^2/2\right)\Big|_3^4$$

$$= \left(4\sqrt{3} - 9 + \frac{9}{2}\right) - \left(\frac{4}{3} - 3 + \frac{1}{2}\right) + \left(\frac{32}{3} - \frac{64}{3} + 24\right) - \left(4\sqrt{3} - 9 + \frac{27}{2}\right) = \frac{11}{2}.$$

15

$$\int_0^2 x^2\,dx + \int_2^4 (x^2 - 2x^2 + 4x)\,dx$$

$$= \left(x^3/3\right)\Big|_0^2 + \left(-x^3/3 + 2x^2\right)\Big|_2^4$$

$$= (8/3) + -64/3 + 32 - (-8/3 + 8) = 8.$$

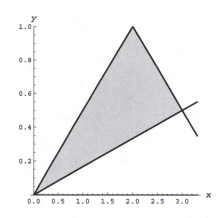

$$\int_0^2 (x/2 - x/6)\,dx + \int_2^3 (2 - x/2 - x/6)\,dx$$

17 $= \left(x^2/6 \right)\Big|_0^2 + \left(2x - x^2/3 \right)\Big|_2^3$

$$= \frac{2}{3} + (6 - 3) - (4 - 4/3) = 1.$$

19 $A(a) = \int_0^{\sqrt[3]{a}} \left(\sqrt{x/a} - x^2/a \right) dx = \left(\frac{2}{3}\frac{x^{3/2}}{\sqrt{a}} - \frac{x^3}{3a} \right)\Big|_0^{\sqrt[3]{a}} = \frac{1}{3}.$

21 From $x = 0$ to $x = 1$ the area function is $A(x) = (\sqrt{x})^2 = x$, while from $x = 1$ to $x = 2$ it is $A(x) = (2 - x)^2$.

$$\int_0^1 x\,dx + \int_1^2 (2 - x)^2\,dx = \frac{1}{2}x^2\Big|_0^1 - \frac{1}{3}(2 - x)^3\Big|_1^2 = \frac{1}{2} + \frac{1}{3} = \frac{5}{6}.$$

23 Because the cross sections are parallel to the x-axis, solve for x to get $x = y^2$ and $x = 2 - y$. Then from $y = 0$ to $y = 1$ the area function is $A(y) = (y^2)^2 = y^4$ and from $y = 1$ to $y = 2$ it is $A(y) = (2 - y)^2$. The by the general slicing method, the volume is

$$\int_0^1 y^4\,dy + \int_1^2 (2 - y)^2\,dy = \frac{y^5}{5}\Big|_0^1 - \frac{1}{3}(2 - y)^3\Big|_1^2 = \frac{1}{5} + \frac{1}{3} = \frac{8}{15}.$$

25

a. $V = 2\pi \int_0^1 x((1 + \sqrt{x}) - (1 - \sqrt{x}))\,dx = 4\pi \int_0^1 x^{3/2}\,dx = 4\pi \cdot \frac{2x^{5/2}}{5}\Big|_0^1 = \frac{8\pi}{5}.$

b. $V = \pi \int_0^2 (1^2 - ((1 - y)^2)^2)\,dy = \pi \int_0^2 (1 - (y^4 - 4y^3 + 6y^2 - 4y + 1))\,dy = \pi \int_0^2 (-y^4 + 4y^3 - 6y^2 + 4y)\,dy = \pi \left(-\frac{y^5}{5} + y^4 - 2y^3 + 2y^2 \right)\Big|_0^2 = \frac{8\pi}{5}.$

27 $V = \pi \int_0^h y^2\,dx = \pi r^2 \int_0^h (1 - x/h)^2\,dx = \pi r^2 \int_0^h (1 - 2x/h + x^2/(h^2))\,dx = \pi r^2 \left(x - x^2/h + x^3/(3h^2) \right)\Big|_0^h = \frac{\pi r^2 h}{3}.$

29 We use the shell method. $V = 2\pi \int_0^{\sqrt{3}/2} \frac{x}{\sqrt{1 - x^2}}\,dx.$ Let $u = 1 - x^2$ so that $du = -2x\,dx.$ Substituting gives $V = 2\pi \int_1^{1/4} \frac{-1}{2\sqrt{u}}\,du = \pi \int_{1/4}^1 u^{-1/2}\,du = 2\pi \sqrt{u}\Big|_{1/4}^1 = 2\pi \left(1 - \frac{1}{2} \right) = \pi.$

31 We use the disk method. The radius of each disk is $4 - (x - 2)^2 = 4x - x^2$, so the volume is

$$V = \pi \int_0^4 (4x - x^2)^2\,dx = \pi \int_0^4 (x^4 - 8x^3 + 16x^2)\,dx$$

$$= \pi \left(\frac{x^5}{5} - 2x^4 + \frac{16}{3}x^3 \right)\Big|_0^4 = \pi \left(\frac{1024}{5} - 512 + \frac{1024}{3} \right) = \frac{512\pi}{15}.$$

33 The two non-horizontal lines intersect at $(2, 4)$. To revolve about $y = -2$, we use the shell method. First solve for x to obtain $x = \frac{y}{2}$ and $x = 6 - y$; then the height of each shell is $6 - y - \frac{y}{2} = 6 - \frac{3}{2}y$, and the radius at y is $y - (-2) = y + 2$. The volume is

$$2\pi \int_0^4 (y+2)\left(6 - \frac{3}{2}y\right) dy = 2\pi \int_0^4 \left(12 + 3y - \frac{3}{2}y^2\right) dy = 2\pi \left(12y + \frac{3}{2}y^3 - \frac{1}{2}y^3\right)\bigg|_0^4 = 80\pi.$$

To revolve about $x = -2$, we use the washer method. With the equations above, the outer radius of each washer is $6 - y - (-2) = 8 - y$ and the inner radius is $\frac{y}{2} - (-2) = 2 + \frac{y}{2}$. The volume is

$$V = \pi \int_0^4 \left((8-y)^2 - \left(2 + \frac{y}{2}\right)^2\right) dy = \pi \int_0^4 \left(\frac{3}{4}y^2 - 18y + 60\right) dy$$
$$= \pi \left(\frac{1}{4}y^3 - 9y^2 + 60y\right)\bigg|_0^4 = \pi(16 - 144 + 240) = 112\pi.$$

35

a. We use washers for both calculations.

Around the x-axis we have $\pi \int_0^1 ((4-x^2)^2 - 9x^2) \, dx = \pi \int_0^1 (16 - 17x^2 + x^4) \, dx =$

$\pi \left(16x - 17x^3/3 + x^5/5\right)\bigg|_0^1 = \pi(16 - 17/3 + 1/5) = 158\pi/15.$

Around the line $y = 5$ we have $\pi \int_0^1 ((5 - 3x)^2 - (5 - (4 - x^2))^2) \, dx = \pi \int_0^1 (24 - 30x + 7x^2 - x^4) \, dx =$

$\pi \left(24x - 15x^2 + 7x^3/3 - x^5/5\right)\bigg|_0^1 = \pi(24 - 15 + 7/3 - 1/5) = 167\pi/15.$ The solid obtained by revolving R about the line $y = 5$ is bigger.

b. We use shells for both calculations.

Around the y-axis we have $2\pi \int_0^1 x(4 - x^2 - 3x) \, dx = 2\pi \int_0^1 (4x - x^3 - 3x^2) \, dx$

$= 2\pi \left(2x^2 - x^4/4 - x^3\right)\bigg|_0^1 = 2\pi(2 - 1/4 - 1) = 3\pi/2.$

Around the line $x = 1$ we have $2\pi \int_0^1 (1 - x)(4 - x^2 - 3x) \, dx = 2\pi \int_0^1 (4 - 7x + 2x^2 + x^3) \, dx =$

$2\pi \left(4x - 7x^2/2 + 2x^3/3 + x^4/4\right)\bigg|_0^1 = 2\pi(4 - 7/2 + 2/3 + 1/4) = 17\pi/6.$ The volume obtained by revolving R about the line $x = 1$ is bigger.

37 We use disks when revolving about the x-axis. The volume is

$$\pi \int_0^1 c^2 x^2 (1-x)^2 \, dx = \pi c^2 \int_0^1 (x^2 - 2x^3 + x^4) \, dx = \pi c^2 \left(\frac{1}{3}x^3 - \frac{1}{2}x^4 + \frac{1}{5}x^5\right)\bigg|_0^1 = \pi c^2 \left(\frac{1}{3} - \frac{1}{2} + \frac{1}{5}\right) = \frac{\pi c^2}{30}.$$

We use shells when revolving about the y-axis. The volume is

$$2\pi \int_0^1 x \cdot cx(1-x) \, dx = 2\pi c \int_0^1 (x^2 - x^3) \, dx = 2\pi c \left(\frac{1}{3}x^3 - \frac{1}{4}x^4\right)\bigg|_0^1 = \frac{\pi c}{6}.$$

These are equal when $\frac{\pi c^2}{30} = \frac{\pi c}{6}$, which for $c \neq 0$ occurs when $c = 5$.

39 $L = \int_0^2 \sqrt{1 + \pi^2 \cos^2 \pi x} \, dx \approx 4.60979.$ This estimate was obtained with a computer algebra system.

41 Note that $y' = 1/(2\sqrt{x}) - \sqrt{x}/2$, so $1 + y'^2 = (1/(2\sqrt{x}) + \sqrt{x}/2)^2.$ $L = \int_1^3 \sqrt{1 + y'^2} \, dy = \int_1^3 1/(2\sqrt{x}) + \sqrt{x}/2 \, dx = \left(\sqrt{x} + x^{3/2}/3\right)\bigg|_1^3 = 2\sqrt{3} - 4/3.$

43 Note that $y' = \frac{2x^3+x}{\sqrt{x^4+x^2}}$, so $1+y'^2 = \frac{x^4+x^2+(2x^3+x)^2}{x^4+x^2} = \frac{4x^6+5x^4+2x^2}{x^4+x^2} = \frac{4x^4+5x^2+2}{x^2+1}$.

So $L = \int_0^1 \sqrt{\frac{4x^4+5x^2+2}{x^2+1}}\,dx \approx 1.74306$. This estimate was obtained with a computer algebra system.

45

a. $S = 2\pi \int_0^3 \sqrt{3x-x^2}\sqrt{1+\frac{(3-2x)^2}{4(3x-x^2)}}\,dx = 2\pi \int_0^3 \sqrt{3x-x^2}\sqrt{\frac{12x-4x^2+9-12x+4x^2}{4(3x-x^2)}}\,dx = \pi \int_0^3 3\,dx = 3\pi(3-0) = 9\pi$.

b. $V = \pi \int_0^3 (3x-x^2)\,dx = \pi \left(3x^2/2 - x^3/3\right)\Big|_0^3 = \pi(27/2-9) = 9\pi/2$.

47

a. $S = 2\pi \int_1^2 \left(\frac{x^4}{2}+\frac{1}{16x^2}\right)\sqrt{1+(2x^3-(1/8x^3))^2}\,dx =$

$2\pi \int_1^2 \left(\frac{x^4}{2}+\frac{1}{16x^2}\right)\sqrt{1+(4x^6-1/2+(64/x^6))}\,dx = 2\pi \int_1^2 \left(\frac{x^4}{2}+\frac{1}{16x^2}\right)\sqrt{4x^6+1/2+64/x^6}\,dx =$

$2\pi \int_1^2 \left(\frac{x^4}{2}+\frac{1}{16x^2}\right)\sqrt{(2x^3+(1/8x^3))^2}\,dx = 2\pi \int_1^2 \left(\frac{x^4}{2}+\frac{1}{16x^2}\right)\left(2x^3+\frac{1}{8x^3}\right)\,dx = 2\pi \int_1^2 (x^7 + x/16 +$

$x/8 + x^{-5}/128)\,dx = 2\pi \int_1^2 (x^7 + 3x/16 + x^{-5}/128)\,dx = 2\pi \left(x^8/8 + 3x^2/32 - x^{-4}/512\right)\Big|_1^2 = 2\pi((32 +$

$3/8 - 1/8192) - (1/8 + 3/32 - 1/512)) = \frac{263439\pi}{4096}$.

b. Using the fact that $\sqrt{1+f'(x)^2} = 2x^3+1/(8x^3)$ which was discovered during the previous calculation,

we have $L = \int_1^2 (2x^3 + x^{-3}/8)\,dx = \left(x^4/2 - x^{-2}/16\right)\Big|_1^2 = (8-1/64)-(1/2-1/16) = \frac{483}{64}$.

c. $V = \pi \int_1^2 (x^4/2 + x^{-2}/16)^2\,dx = \pi \int_1^2 (x^8/4 + x^2/16 + x^{-4}/256)\,dx = \pi \left(x^9/36 + x^3/48 - x^{-3}/768\right)\Big|_1^2 =$

$\pi(512/36 + 8/48 - 1/6144 - (1/36 + 1/48 - 1/768)) = \frac{264341\pi}{18432}$.

49 $m = \int_0^3 (150 - \sqrt{3x})\,dx = \left(150x - \frac{2\sqrt{3}x^{3/2}}{3}\right)\Big|_0^3 = 450 - 6 = 444$ gm.

51

a. Because $50 = \int_0^{.2} kx\,dx = \frac{kx^2}{2}\Big|_0^{0.2} = \frac{k}{50}$, we must have $k = 2500$.

$$W = \int_{0.2}^{0.7} 2500x\,dx = 1250x^2\Big|_{0.2}^{0.7} = 562.5\text{ J}.$$

b. $f(0.2) = 0.2k = 50$, so $k = 250$, and thus

$$W = \int_{0.2}^{0.7} 250x\,dx = 125x^2\Big|_{0.2}^{0.7} = 56.25\text{ J}.$$

53 Orient the semicircle so that the center is at the point $(0, 20)$. Then the force is given by $\int_0^{20} \rho g(20 - y)(2)\sqrt{40y - y^2}\,dy$. Let $u = 40y - y^2$ so that $du = (40 - 2y)\,dy$. Then we have $\rho g \int_0^{400} u^{1/2}\,du = \rho g \left(\frac{2}{3}u^{3/2}\right)\Big|_0^{400} \approx 5.2 \times 10^7$.

Chapter 7

Logarithmic and Exponential Functions

7.1   Inverse Functions

7.1.1 If a function f is not one-to-one, then there are domain values $x_1 \neq x_2$ with $f(x_1) = f(x_2)$. If f^{-1} were to exist, then $f^{-1}(f(x_1)) = f^{-1}(f(x_2))$ which would imply that $x_1 = x_2$, a contradiction.

7.1.3

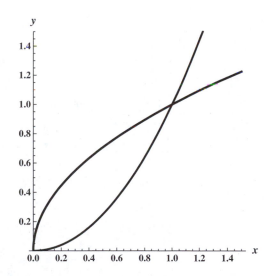

7.1.5 $y = -\sqrt{x}$. Note that since we are assuming $x \leq 0$, we must take the negative of the square root.

7.1.7 By Theorem 7.3, and since f is one-to-one, we have

$$(f^{-1})'(8) = \frac{1}{f'(f^{-1}(8))} = \frac{1}{f'(2)} = \frac{1}{4}$$

7.1.9 f is one-to-one on $(-\infty, -1]$, on $[-1, 1]$, and on $[1, \infty)$.

7.1.11

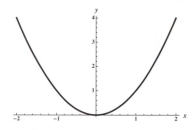

7.1.13 f is one-to-one on $\mathbb{R}$, so it has an inverse on $\mathbb{R}$.

7.1.15 f is one-to-one on its domain, which is $(-\infty, 5) \cup (5, \infty)$, so it has an inverse on that set.

7.1.17 f is one-to-one on the interval $(0, \infty)$, so it has an inverse on that interval. (Alternatively, it is one-to-one on the interval $(-\infty, 0)$, so that interval could be used as well.)

7.1.19

   a. Switching x and y, we have $x = 2y$, so $y = \frac{1}{2}x$. Thus $y = f^{-1}(x) = \frac{1}{2}x$.

   b. $f(f^{-1}(x)) = f\left(\frac{1}{2}x\right) = 2\left(\frac{1}{2}x\right) = x$. Also, $f^{-1}(f(x)) = f^{-1}(2x) = \frac{1}{2} \cdot (2x) = x$.

7.1.21

   a. Switching the x and y, we have $x = 6 - 4y$. Solving for y in terms of x we have $4y = 6 - x$, so $y = f^{-1}(x) = \frac{6-x}{4}$.

   b. $f(f^{-1}(x)) = f\left(\frac{6-x}{4}\right) = 6 - 4 \cdot \left(\frac{6-x}{4}\right) = 6 - (6 - x) = x$.
      $f^{-1}(f(x)) = f^{-1}(6 - 4x) = \frac{6-(6-4x)}{4} = \frac{4x}{4} = x$.

7.1.23

   a. Switching the x and y, we have $x = 3y + 5$. Solving for y in terms of x we have $y = \frac{x-5}{3}$, so $y = f^{-1}(x) = \frac{x-5}{3}$.

   b. $f(f^{-1}(x)) = f\left(\frac{x-5}{3}\right) = 3\left(\frac{x-5}{3}\right) + 5 = (x - 5) + 5 = x$.
      $f^{-1}(f(x)) = f^{-1}(3x + 5) = \frac{(3x+5)-5}{3} = \frac{3x}{3} = x$.

7.1.25

   a. Switching the x and y, we have $x = \sqrt{y + 2}$. Solving for y in terms of x we have $y = x^2 - 2$. Note that since range of f is $[0, \infty)$, that is the domain of f^{-1}.

   b. $f(f^{-1}(x)) = f(x^2 - 2) = \sqrt{x^2 - 2 + 2} = |x| = x$, since x is in the domain of f^{-1} and is thus positive.
      $f^{-1}(f(x)) = f^{-1}(\sqrt{x + 2}) = \sqrt{x+2}^2 - 2 = x + 2 - 2 = x$.

7.1.27 First note that since the expression is symmetric, switching x and y doesn't change the expression. Solving for y gives $|y| = \sqrt{1 - x^2}$. To get the four one-to-one functions, we restrict the domain and choose either the upper part or lower part of the circle as follows:

   a. $f_1(x) = \sqrt{1 - x^2},\ 0 \leq x \leq 1$ $f_2(x) = \sqrt{1 - x^2},\ -1 \leq x \leq 0$ $f_3(x) = -\sqrt{1 - x^2},\ -1 \leq x \leq 0$
      $f_4(x) = -\sqrt{1 - x^2},\ 0 \leq x \leq 1$

   b. Reflecting these functions across the line $y = x$ yields the following: $f_1^{-1}(x) = \sqrt{1 - x^2},\ 0 \leq x \leq 1$
      $f_2^{-1}(x) = -\sqrt{1 - x^2},\ 0 \leq x \leq 1$ $f_3^{-1}(x) = -\sqrt{1 - x^2},\ -1 \leq x \leq 0$ $f_4^{-1}(x) = \sqrt{1 - x^2},\ -1 \leq x \leq 0$

7.1.29

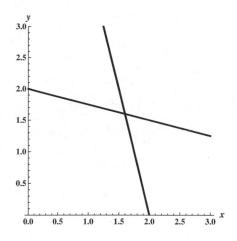

Switching x and y gives $x = 8 - 4y$. Solving this for y yields $y = \frac{8-x}{4}$.

7.1.31

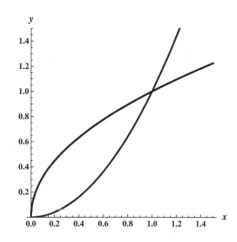

Switching x and y gives $x = \sqrt{y}$. Solving this for y yields $y = x^2$, but note that the range of f is $[0, \infty)$ so that is the domain of f^{-1}.

7.1.33

Switching x and y gives $x = y^4 + 4$. Solving this for y yields $y = \sqrt[4]{x - 4}$.

7.1.35

Begin by completing the square: $f(x) = x^2 - 2x + 6 = (x^2 - 2x + 1) + 5 = (x - 1)^2 + 5$. Switching x and y yields $x = (y - 1)^2 + 5$. Solving for y gives $|y - 1| = \sqrt{x - 5}$. Choosing the principle square root (because the original given interval has x positive) gives $y = f^{-1}(x) = \sqrt{x - 5} + 1$, $x \geq 5$.

7.1.37

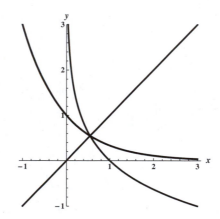

7.1.39 $f(4) = 16$ and $(f^{-1})'(16) = \frac{1}{f'(4)} = \frac{1}{3}$.

7.1.41 $f(1) = -1$ so $(f^{-1})'(-1) = \frac{1}{f'(1)} = \frac{1}{-5} = \frac{-1}{5}$.

7.1.43 $f\left(\frac{\pi}{4}\right) = 1$ and $(f^{-1})'(1) = \frac{1}{f'\left(\frac{\pi}{4}\right)} = \frac{1}{\sec^2\left(\frac{\pi}{4}\right)} = \frac{1}{2}$.

7.1.45 $f(4) = 2$ and $(f^{-1})'(2) = \frac{1}{f'(4)} = \frac{1}{(1/2\sqrt{4})} = 4$.

7.1.47 $f(4) = 36$ and $(f^{-1})'(36) = \frac{1}{f'(4)} = \frac{1}{2(4+2)} = \frac{1}{12}$.

7.1.49 Note that $f(1) = 3$. So $(f^{-1})'(3) = \frac{1}{f'(1)} = \frac{1}{4}$.

7.1.51 $(f^{-1})'(4) = \frac{1}{f'(7)} = \frac{4}{5}$, so $f'(7) = \frac{5}{4}$.

7.1.53

a. Note that $f(0) = 4$, so $f^{-1}(4) = 0$. $(f^{-1})'(4) = \frac{1}{f'(0)} = \frac{1}{2}$.

b. Note that $f(1) = 6$, so $f^{-1}(6) = 1$. $(f^{-1})'(6) = \frac{1}{f'(1)} = \frac{1}{3/2} = \frac{2}{3}$.

c. Note that there is no given x so that $f(x) = 1$, so the desired derivative cannot be determined.

d. From the table directly, $f'(1) = \frac{3}{2}$.

7.1.55

a. False. For example $f(0) = 1$, but the alleged inverse function evaluated at 1 is not 0 (rather, it has value $1/2$.)

b. True. f is its own inverse because $f(f(x)) = f(1/x) = \frac{1}{1/x} = x$.

c. False. In fact, $f(x) = x^3 + x$ is one-to-one, so has one inverse on all of $(-\infty, \infty)$.

d. True. A tenth-degree polynomial could have nine local maxima or minima, giving ten segments on which the function is one-to-one.

e. True. By part (b), $f^{-1}(x) = 1/x$, so that $(f^{-1}(x))' = -1/x^2$.

7.1.57 Note that f is one-to-one, so there is only one inverse. Switching x and y gives $x = (y+1)^3$. Then $\sqrt[3]{x} = y + 1$, so $y = f^{-1}(x) = \sqrt[3]{x} - 1$. The domain of f^{-1} is $\mathbb{R}$.

7.1.59 Note that to get a one-to-one function, we should restrict the domain to either $[0, \infty)$ or $(-\infty, 0]$. Switching x and y yields $x = \frac{2}{y^2+2}$, so $y^2 + 2 = (2/x)$. So $y = \pm\sqrt{(2/x) - 2}$. So the inverse of f when the domain of f is restricted to $[0, \infty)$ is $f^{-1}(x) = \sqrt{(2/x) - 2}$, while if the domain of f is restricted to $(-\infty, 0]$ the inverse is $f^{-1}(x) = -\sqrt{(2/x) - 2}$. In either case, the domain of f^{-1} is $(0, 1]$.

7.1.61 Let $f(y) = 3y - 4$. Then $f'(y) = 3$ for all y in the domain of f. Let $y = f^{-1}(x)$. $(f^{-1})'(x) = \dfrac{1}{f'(y)} = \dfrac{1}{3}$.

7.1.63 Let $x = f(y) = y^2 - 4$ for $y > 0$. Note that this means that $y = \sqrt{x + 4}$. Then $f'(y) = 2y$. So $(f^{-1})'(x) = \dfrac{1}{f'(y)} = \dfrac{1}{2y} = \dfrac{1}{2\sqrt{x+4}}$.

7.1.65 For $y \geq -2$, let $x = \sqrt{y + 2}$. Note that it then follows that $x \geq 0$. Then $1 = \frac{y'}{2\sqrt{y+2}}$, and $x^2 = y + 2$, so $y = x^2 - 2$. Thus we have $(f^{-1})'(x) = y' = 2\sqrt{x^2 - 2 + 2} = 2|x| = 2x$, since $x \geq 0$.

7.1.67 For $y > 0$, let $x = y^{-1/2}$. Then $1 = \frac{-1}{2}y^{-3/2}y'$, so $y' = -2y^{3/2} = -2(x^{-2})^{3/2} = -2x^{-3}$ where $x > 0$.

7.1.69 Solve for r to get $r^3 = \frac{3}{4\pi}V$, so that

$$r = \sqrt[3]{\frac{3}{4\pi}V}$$

7.1.71 Solve for r (remembering that we are assuming $r \geq 0$) to get $r^2 = \frac{V}{10\pi}$, so that

$$r = \sqrt{\frac{V}{10\pi}}$$

7.1.73

a. f is one-to-one on $(-\infty, -\sqrt{2}/2]$, on $[-\sqrt{2}/2, 0]$, on $[0, \sqrt{2}/2]$, and on $[\sqrt{2}/2, \infty)$.

b. If $u = x^2$, then our function becomes $y = u^2 - u$. Completing the square gives $y + (1/4) = u^2 - u + (1/4) = (u - (1/2))^2$. Thus $|u - (1/2)| = \sqrt{y + (1/4)}$, so $u = (1/2) \pm \sqrt{y + (1/4)}$, with the "+" applying for $u = x^2 > (1/2)$ and the "−" applying when $u = x^2 < (1/2)$. Now letting $u = x^2$, we have $x^2 = (1/2) \pm \sqrt{y + (1/4)}$, so $x = \pm\sqrt{(1/2) \pm \sqrt{y + (1/4)}}$.

c. Switching x and y gives the following inverses:

Domain of f	$(-\infty, -\sqrt{2}/2]$	$[-\sqrt{2}/2, 0]$	$[0, \sqrt{2}/2]$	$[\sqrt{2}/2, \infty)$
Range of f	$[-1/4, \infty)$	$[-1/4, 0]$	$[-1/4, 0]$	$[-1/4, \infty)$
Inverse of f	$-\sqrt{(1/2) + \sqrt{x + (1/4)}}$	$-\sqrt{(1/2) - \sqrt{x + (1/4)}}$	$\sqrt{(1/2) - \sqrt{x + (1/4)}}$	$\sqrt{(1/2) + \sqrt{x + (1/4)}}$

7.1.75 Let $y = x^3 + 2x$. This function is one-to-one, so it has an inverse. Making the suggested substitution yields $y = (z - 2/(3z))^3 + 2(z - 2/(3z))$. Expanding gives $y = z^3 - 2z + 4/(3z) - 8/(27z^3) + 2z - 4/(3z) = z^3 - 8/(27z^3)$. Thus we have $y = z^3 - 8/(27z^3)$, so $27z^3y = 27(z^3)^2 - 8$, or $27(z^3)^2 - 27y(z^3) - 8 = 0$.

Applying the quadratic formula gives $z^3 = \dfrac{y}{2} \pm \dfrac{\sqrt{3}\sqrt{32 + 27y^2}}{18}$. We will take the "+" part and finish solving to obtain:

$$z = \sqrt[3]{\frac{y}{2} + \frac{\sqrt{3}\sqrt{32 + 27y^2}}{18}}$$

Now

$$x = z - (2/(3z)) = \frac{3z^2 - 2}{3z} = \frac{3\left(\sqrt[3]{\frac{y}{2} + \frac{\sqrt{3}\sqrt{32 + 27y^2}}{18}}\right)^2 - 2}{3\sqrt[3]{\frac{y}{2} + \frac{\sqrt{3}\sqrt{32 + 27y^2}}{18}}}.$$

So the inverse function $f^{-1}(x)$ is now obtained by switching y and x.

7.1.77   a. Because $y = ax + b$, we have $f'(x_0) = a$, so that $b = y_0 - f'(x_0)x_0$.

b. $(f^{-1})'(y_0) = c$, so

$$c = \frac{1}{f'(x_0)}, \qquad d = x_0 - \frac{y_0}{f'(x_0)}$$

c. We must show that $L(M(x)) = x$. Note that from parts (a) and (b), $ac = 1$; then

$$L(M(x)) = a(cx + d) + b = acx + ad + b = x + a(x_0 - cy_0) + (y_0 - ax_0) = x - acy_0 + y_0 = x$$

7.2   The Natural Logarithmic and Exponential Functions

7.2.1 The domain is $(0, \infty)$ and the range is $(-\infty, \infty)$.

7.2.3 $y = \ln x$ if and only if $x = e^y$. Differentiating implicitly yields $1 = e^y \cdot y'$, so $y' = \frac{1}{e^y} = \frac{1}{e^{\ln x}} = \frac{1}{x}$ for $x > 0$.

7.2.5 $\frac{d}{dx} \ln(kx) = \frac{1}{kx} \cdot k = \frac{1}{x}$. This is valid for $x > 0$ if $k > 0$ and $x < 0$ if $k < 0$. Also, we can write $\ln(kx) = \ln(k) + \ln(x)$, so its derivative is $0 + \frac{1}{x} = \frac{1}{x}$.

7.2.7 $\frac{d}{dx} \ln(7x) = \frac{1}{7x} \cdot 7 = \frac{1}{x}$.

7.2.9 $\frac{d}{dx} \ln(x^2) = \frac{1}{x^2} \cdot (2x) = \frac{2}{x}$. This is valid on $(-\infty, 0) \cup (0, \infty)$.

7.2.11 $\frac{d}{dx} (\ln |\sin x|) = \frac{1}{\sin x} \cdot (\cos x) = \cot x$.

7.2.13 $\frac{d}{dx} \left[\ln \left(\frac{x+1}{x-1}\right)\right] = \frac{x-1}{x+1} \left(\frac{(x-1) - (x+1)}{(x-1)^2}\right) = \frac{-2}{(x+1)(x-1)}$. This is valid on $(-\infty, -1) \cup (1, \infty)$.

7.2.15 $\frac{d}{dx} (x^2 + 1) \ln x = 2x \ln x + \frac{x^2 + 1}{x}$. This is valid on $(0, \infty)$.

7.2.17 $\frac{d}{dx}(\ln(\ln(x))) = \frac{1}{\ln x} \cdot \frac{1}{x}$. This is valid on $(1, \infty)$, which is the domain of $\ln(\ln x)$.

7.2.19 $\frac{d}{dx}\left(\frac{\ln x}{\ln x + 1}\right) = \frac{(\ln x + 1)(1/x) - (\ln x)(1/x)}{(\ln x + 1)^2} = \frac{1}{x(\ln x + 1)^2}$.

7.2.21 $\int \frac{3}{x - 10}\, dx = 3 \ln|x - 10| + C$

7.2.23 $\int \left(\frac{2}{x - 4} - \frac{3}{2x + 1}\right) dx = 2 \ln|x - 4| - \frac{3}{2} \ln|2x + 1| + C$

7.2.25 $\int_0^3 \frac{2x - 1}{x + 1}\, dx = \int_0^3 \left(2 - \frac{3}{x + 1}\right) dx = (2x - 3\ln(x + 1))\Big|_0^3 = 6(1 - \ln 2)$.

7.2.27 Let $u = \ln(\ln x)$, so that $du = \frac{1}{x \ln x}\, dx$. Then,

$$\int_3^4 \frac{dx}{2x \ln x \ln^3(\ln x)} = \int_{\ln(\ln 3)}^{\ln(\ln 4)} \frac{1}{2u^3}\, du = \left(-\frac{1}{4u^2}\right)\Bigg|_{\ln(\ln 3)}^{\ln(\ln 4)} = \frac{1}{4}\left(\frac{1}{(\ln(\ln 3))^2} - \frac{1}{(\ln(\ln 4))^2}\right).$$

7.2.29 Let $u = \ln(\ln x)$ so that $du = \frac{1}{x \ln x}\, dx$. Substituting yields $\int \frac{1}{u}\, du = \ln|u| + C = \ln|\ln(\ln x)| + C$.

7.2.31 $f'(x) = \frac{(e^x + 1)e^x - e^x(e^x)}{(e^x + 1)^2} = \frac{e^x}{(e^x + 1)^2}$.

7.2.33 $f'(x) = -9e^{-x} - 10e^{2x} - 6e^x$.

7.2.35 $f'(x) = \frac{(e^{-x} + 2)(2e^{2x}) - e^{2x}(-e^{-x})}{(e^{-x} + 2)^2} = \frac{2e^x + 4e^{2x} + e^x}{(e^{-x} + 2)^2} = \frac{3e^x + 4e^{2x}}{(e^{-x} + 2)^2}$

7.2.37 $f'(x) = (\sin 2x)' \cdot e^{\sin 2x} = 2\cos(2x)e^{\sin 2x}$, so that $f'(\pi/4) = 2\cos(\pi/2)e^{\sin \pi/2} = 0$.

7.2.39 $y' = \frac{e^x}{4} - 1$, so that at $a = 0$ the slope is $\frac{1}{4} - 1 = -\frac{3}{4}$. At $a = 0$, $f(a) = \frac{1}{4}$, so that the tangent line is

$$y - \frac{1}{4} = -\frac{3}{4}(x - 0), \qquad \text{or } y = \frac{1}{4} - \frac{3}{4}x.$$

7.2.41 $\int \frac{e^{2x} - e^{-2x}}{2}\, dx = \frac{1}{2}\left(\frac{e^{2x}}{2} - \frac{e^{-2x}}{-2}\right) + C = \frac{e^{2x} + e^{-2x}}{4} + C$. Check: $\frac{d}{dx}\left(\frac{e^{2x} + e^{-2x}}{4} + C\right) = \frac{e^{2x} - e^{-2x}}{2}$.

7.2.43 $\int(e^{2x} + 1)\, dx = \frac{1}{2}e^{2x} + x + C$.

7.2.45

$$\int_0^{\ln 3} e^x(e^{3x} + e^{2x} + e^x)\, dx = \int_0^{\ln 3} (e^{4x} + e^{3x} + e^{2x})\, dx = \left(\frac{1}{4}e^{4x} + \frac{1}{3}e^{3x} + \frac{1}{2}e^{2x}\right)\Bigg|_0^{\ln 3}$$

$$= \frac{81}{4} + 9 + \frac{9}{2} - \left(\frac{1}{4} + \frac{1}{3} + \frac{1}{2}\right) = \frac{98}{3}$$

7.2.47 Let $u = e^x - e^{-x}$ so that $du = e^x + e^{-x}\, dx$. Then $\int \frac{e^x + e^{-x}}{e^x - e^{-x}}\, dx = \int \frac{1}{u}\, du = \ln|u| + C = \ln|e^x - e^{-x}| + C$.

7.2.49 Let $u = \sqrt{x}$ so that $du = \frac{1}{2\sqrt{x}}\, dx$. Then $\int \frac{e^{\sqrt{x}}}{\sqrt{x}}\, dx = 2\int e^u\, du = 2e^u + C = 2e^{\sqrt{x}} + C$.

7.2.51 Let $y = \frac{(x + 1)^{10}}{(2x - 4)^8}$, so $\ln y = \ln\left(\frac{(x + 1)^{10}}{(2x - 4)^8}\right) = 10 \ln(x + 1) - 8(\ln(2x - 4))$. Then

$$\frac{1}{y}y' = \frac{10}{x + 1} - \frac{8}{2x - 4} \cdot 2,$$

$$y' = \frac{(x + 1)^{10}}{(2x - 4)^8} \cdot \left(\frac{10}{x + 1} - \frac{8}{x - 2}\right).$$

7.2.53 Let $y = x^{\ln x}$. Then $\ln y = (\ln x)^2$. Thus $\frac{1}{y} y' = 2 \ln x \cdot \frac{1}{x}$, so $y' = x^{\ln x} \left(\frac{2 \ln x}{x} \right)$.

7.2.55 Let $y = \dfrac{(x+1)^{3/2}(x-4)^{5/2}}{(5x+3)^{2/3}}$. Then

$$\ln y = \ln \left(\frac{(x+1)^{3/2}(x-4)^{5/2}}{(5x+3)^{2/3}} \right) = \frac{3}{2} \ln(x+1) + \frac{5}{2} \ln(x-4) - \frac{2}{3} \ln(5x+3)$$

so that

$$\frac{1}{y} y' = \frac{3}{2(x+1)} + \frac{5}{2(x-4)} - \frac{10}{3(5x+3)},$$

$$y' = \frac{(x+1)^{3/2}(x-4)^{5/2}}{(5x+3)^{2/3}} \cdot \left(\frac{3}{2(x+1)} + \frac{5}{2(x-4)} - \frac{10}{3(5x+3)} \right).$$

7.2.57 Let $y = (\sin x)^{\tan x}$, and assume $0 < x < \pi$, $x \neq \frac{\pi}{2}$. Then $\ln y = (\tan x) \ln(\sin x)$. Then

$$\frac{1}{y} y' = (\sec^2 x) \ln(\sin x) + \frac{\tan x \cos x}{\sin x},$$

$$y' = (\sin x)^{\tan x} \left((\sec^2 x) \ln(\sin x) + 1 \right).$$

7.2.59

a. True. This follows because $e^{\ln xy} = xy = e^{\ln x} e^{\ln y} = e^{\ln x + \ln y}$. Because the exponential function is one-to-one, it follows that $\ln xy = \ln x + \ln y$.

b. False. Zero is not in the domain of the natural logarithm function.

c. False. For example, $\ln(1+1) = \ln(2) \neq \ln(1) + \ln(1) = 0$.

d. False. e^3 is a constant, so $\frac{d}{dx}(e^3) = 0$.

e. False. $\frac{d}{dx}(e^x) = e^x$.

f. True, by induction. $f'(x) = f^{(1)}(x) = 3e^{3x}$, so it is true for $n = 1$. If it is true for n, then

$$f^{(n+1)}(x) = f^{(n)'}(x) = (3^n e^{3x})' = 3^{n+1} e^{3x}.$$

g. True. $\int_1^e \frac{1}{x} \, dx = \ln x \Big|_1^e = \ln e - \ln 1 = 1.$

7.2.61

h	$(1+2h)^{1/h}$	h	$(1+2h)^{1/h}$
10^{-1}	6.1917	-10^{-1}	9.3132
10^{-2}	7.2446	-10^{-2}	7.5404
10^{-3}	7.3743	-10^{-3}	7.4039
10^{-4}	7.3876	-10^{-4}	7.3905
10^{-5}	7.3889	-10^{-5}	7.3892
10^{-6}	7.3890	-10^{-6}	7.3891

Let $y = (1+2h)^{1/h}$. Then

$$\ln y = \frac{\ln(1+2h)}{h}.$$

$$\lim_{h \to 0} \ln y = \lim_{h \to 0} \frac{\ln(1+2h)}{h} = \lim_{h \to 0} \frac{2}{1+2h} = 2,$$

so the limit of y as $h \to 0$ is e^2.

7.2.63

x	$\frac{2^x-1}{x}$	x	$\frac{2^x-1}{x}$
10^{-1}	0.71773	-10^{-1}	0.66967
10^{-2}	0.69556	-10^{-2}	0.69075
10^{-3}	0.69339	-10^{-3}	0.69291
10^{-4}	0.69317	-10^{-4}	0.69312
10^{-5}	0.69315	-10^{-5}	0.69314
10^{-6}	0.69315	-10^{-6}	0.69315

$$\lim_{x\to 0}\frac{2^x-1}{x} = \lim_{x\to 0}\frac{2^x\ln(2)}{1} = \ln(2).$$

7.2.65

a. Use the change of variables $u = \cos x$; then $du = -\sin x\, dx$, and

$$\int \tan x\, dx = \int \frac{\sin x}{\cos x}\, dx = -\int \frac{1}{u}\, du = -\ln|u| + C = -\ln|\cos x| + C$$

Because $|\sec x| = \frac{1}{|\cos x|}$, we have $\ln|\sec x| = -\ln|\cos x|$, so that the integral is also equal to $\ln|\sec x| + C$.

b. Use the change of variables $u = \sin x$; then $du = \cos x\, dx$, and

$$\int \cot x\, dx = \int \frac{\cos x}{\sin x}\, dx = \int \frac{1}{u}\, du = \ln|u| + C = \ln|\sin x| + C$$

7.2.67 The average value is given by $\frac{1}{p-1}\int_1^p \frac{1}{x}\, dx = \frac{1}{p-1}\left(\ln x\right)\Big|_1^p = \frac{\ln p}{p-1}$. Note that

$$\lim_{p\to\infty}\frac{\ln p}{p-1} = \lim_{p\to\infty}\frac{1/p}{1} = 0.$$

7.2.69 $\frac{d}{dx}e^{-10x^2} = e^{-10x^2}\cdot -20x$.

7.2.71 $\frac{d}{dx}\ln(\sqrt{10x}) = \frac{d}{dx}\ln((10x)^{1/2}) = \frac{1}{2}\frac{d}{dx}\ln(10x) = \frac{1}{2x}$

7.2.73 If $f(x) = \frac{(x^2+1)(x-3)}{(x+2)^3}$, then $\ln f(x) = \ln(x^2+1) + \ln(x-3) - 3\ln(x+2)$, so that using logarithmic differentiation,

$$f'(x) = f(x)\frac{d}{dx}(\ln f(x)) = f(x)\left(\frac{2x}{x^2+1} + \frac{1}{x-3} - \frac{3}{x+2}\right) = \frac{9x^2 - 14x + 11}{(x+2)^4}$$

7.2.75

$$\frac{d}{dx}\sin(\sin(e^x)) = \cos(\sin(e^x))\frac{d}{dx}\sin(e^x)$$
$$= \cos(\sin(e^x))\cdot\cos(e^x)\cdot e^x$$

7.2.77 $y' = \frac{(x+1)((1)e^x + xe^x) - xe^x(1)}{(x+1)^2} = \frac{xe^x + x^2e^x + e^x + xe^x - xe^x}{(x+1)^2} = \frac{e^x(x^2 + x + 1)}{(x+1)^2}.$

7.2.79 Let $u = x^3$, so that $du = 3x^2\, dx$. Then $\int x^2 e^{x^3}\, dx = \int \frac{1}{3}e^u\, du = \frac{1}{3}e^u + C = \frac{1}{3}e^{x^3} + C$.

7.2.81 $e^{\ln x} = x$, so

$$\int_1^{2e} \frac{e^{\ln x}}{x}\, dx = \int_1^{2e} 1\, dx = 2e - 1$$

7.2.83 Let $u = \ln x$ so that $du = \frac{1}{x}\,dx$. Then we have $\int_0^2 u^5\,du = \left(\frac{u^6}{6}\right)\Big|_0^2 = \frac{32}{3}$.

7.2.85 Let $u = x - 2$, so that $u + 2 = x$. Then $du = dx$. Substituting yields $\int \frac{u+2}{u}\,du = \int \left(1 + \frac{2}{u}\right)\,du = u + 2\ln|u| + C = x - 2 + 2\ln|x-2| + C$.

7.2.87 Let $u = \sin^2 y + 2$ so that $du = 2\sin y \cos y\,dy = \sin(2y)\,dy$. Substituting yields $\int_2^{9/4} \frac{1}{u}\,du = (\ln|u|)\Big|_2^{9/4} = \ln(9/4) - \ln(2) = \ln(9/8)$.

7.2.89 The numerator of f is defined for $-3 < x < 3$. The denominator is defined everywhere, but is zero when $2e^x = e^{-x}$. Simplifying gives $2e^{2x} = 1$, or $e^{2x} = 1/2$. This has the solution $x = (-\ln 2)/2 \approx -0.347$, which lies in the domain of the numerator. So the domain of f is $\{x : -3 < x < 3, x \neq (-\ln 2)/2\}$, and any questions about horizontal asymptotes are moot. As $x \to 3^-$ or as $x \to -3^+$, the numerator approaches $-\infty$, which follows because $\lim_{t\to 0^+} \ln t = -\infty$, and the denominator is approaching a positive constant as $x \to 3^-$ and a negative constant as $x \to -3^+$. Thus $\lim_{x\to 3^-} f(x) = -\infty$ and $\lim_{x\to -3^+} f(x) = \infty$, and there are vertical asymptotes at $x = 3$ and $x = -3$.

As $x \to -(\ln 2)/2$, the numerator is nonzero because $9 - x^2$ is not approaching 1. Thus, there is a vertical asymptote at $x = -(\ln 2)/2$. A graph of the function, with the vertical asymptotes shown in gray, verifies this analysis. Note that the vertical asymptotes at ± 3 require a different viewing window.

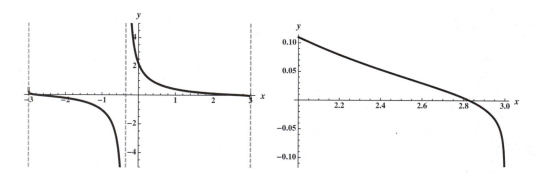

7.2.91

a. We have $f'(x) = e^{-x^2/2} + xe^{-x^2/2} \cdot (-x) = (1 - x^2)e^{-x^2/2}$. This is zero only when $x = \pm 1$, so those are the two critical numbers.

b. Note that $f'(-2) = -3e^{-2} < 0$, $f'(0) = 1 > 0$, and $f'(2) = -3e^{-2} < 0$. Thus f is decreasing on $(-\infty, -1)$ and on $(1, \infty)$, and is increasing on $(-1, 1)$. There is a local minimum at $x = -1$ and a local maximum at $x = 1$.

c. $f''(x) = -2xe^{-x^2/2} + (1 - x^2)e^{-x^2/2}(-x) = (x^3 - 3x)e^{-x^2/2} = x(x^2 - 3)e^{-x^2/2}$. Note that f'' is zero for $x = 0$ and for $x = \pm\sqrt{3}$. f is concave down on $(-\infty, -\sqrt{3})$ and on $(0, \sqrt{3})$, and it is concave up on $(-\sqrt{3}, 0)$ and on $(\sqrt{3}, \infty)$. There are inflection points at $x = 0$ and at $x = \pm\sqrt{3}$.

7.2.93

a. With $f(x) = e^{-x}$ and $a = \ln 2$, we have $f(a) = \frac{1}{2}$ and $f'(a) = -e^{-a} = -\frac{1}{2}$. Thus the linear approximation to f at $x = \ln 2$ is $L(x) = f(\ln 2) + f'(\ln 2)(x - \ln 2) = \frac{1}{2} + -\frac{1}{2}(x - \ln 2) = -\frac{1}{2}x + \frac{1}{2}(1 + \ln 2)$.

b. A plot of f with L in gray:

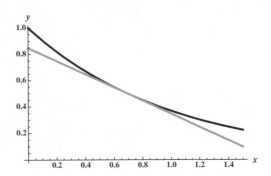

c. $1/e = f(1) \approx L(1) = \frac{\ln 2}{2}$.

7.2.95 Note that the area of the triangle described is $\frac{1}{2}xy$, so if $xy \geq \frac{1}{2}$, then $\frac{1}{2}xy \geq \frac{1}{4}$, which is not what we are seeking. However, if $xy < 1/2$, then the triangle formed will give us what we want. Note that for any choice of x with $0 < x < 1/2$ and any choice of possible y, we have $xy < 1/2$ since $0 < y < 1$. Thus since there is a probability of $\frac{1}{2}$ of choosing $0 < x < 1/2$, the probability we seek is at least $1/2$. In addition, for $1/2 < x < 1$, if $y < \frac{1}{2x}$, then $xy < \frac{1}{2}$. So we should add $\int_{1/2}^{1} \frac{1}{2x}\, dx = \left(\frac{1}{2}\ln x\right)\Big|_{1/2}^{1} = (\ln 2)/2$. Thus the probability we seek is $\frac{1}{2}(1 + \ln 2)$.

7.2.97

a.

We used a graphing rectangle of $[0, 25] \times [0, 8000]$.

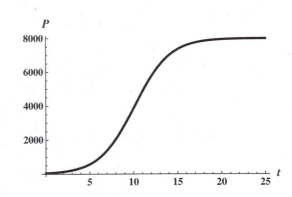

b. To find when $P(t)$ hits 5000, we solve $5000 = \dfrac{400000}{50 + 7950e^{-0.5t}}$, or $50 + 7950e^{-0.5t} = 80$. This leads to $7950e^{-0.5t} = 30$, or $-0.5t = \ln\left(\dfrac{30}{7950}\right)$. So we have $t = 2\ln(265) \approx 11.16$ years. The carrying capacity is $\displaystyle\lim_{t \to \infty} P(t) = \dfrac{400,000}{50} = 8000$. Ninety percent of 8000 is 7200, so we seek the time when $P(t) = 7200$. We have $7200 = \dfrac{400000}{50 + 7950e^{-0.5t}}$, or $50 + 7950e^{-0.5t} = \dfrac{500}{9}$. This leads to $7950e^{-0.5t} = \dfrac{50}{9}$, or $-0.5t = \ln\left(\dfrac{50}{71550}\right)$. So we have $t = 2\ln\left(\dfrac{71550}{50}\right) \approx 14.53$ years.

c. $\dfrac{dP}{dt} = \dfrac{-400000}{(50 + 7950e^{-0.5t})^2} \cdot (7950)(-0.5)e^{-0.5t}$. At $t = 0$ we have $P'(0) = \dfrac{400,000 \cdot 7950 \cdot .5}{8000^2} = \dfrac{1,590,000,000}{8000^2} \approx 25$ fish per year. At $t = 5$ we have $P'(5) = \dfrac{1,590,000,000e^{-5/2}}{(50 + 7950e^{-5/2})^2} \approx 264$ fish per year.

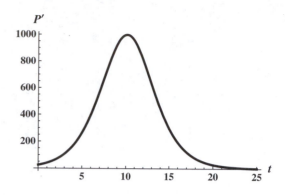

d. The maximum is at about $t = 10$ years.

7.2.99

a. $\ln(P(t)) = \ln(3 \cdot 10^{10}) - \ln(2 + 3e^{-0.025t})$. $\frac{d}{dt} \ln(P(t)) = \frac{P'(t)}{P(t)} = r(t) = \frac{0.075 \cdot e^{-0.025t}}{2 + 3e^{-0.025t}}$. $r(0) = \frac{0.075}{5} = 0.015$, so the population is growing at 1.5% per year in 1999.

b. $r(11) = \frac{0.075 e^{-0.275}}{2 + 3e^{-0.275}} \approx 0.0133$. $r(21) = \frac{0.075 e^{-0.5}}{2 + 3e^{-0.5}} \approx 0.0118$. The relative growth rate decreases over time.

c. $\lim_{t \to \infty} r(t) = \lim_{t \to \infty} \frac{0.075}{3 + 2e^{0.025t}} = 0$, since the denominator increases without bound. The relative growth rate becomes smaller and smaller as the population nears the carrying capacity.

7.2.101

a. $Q(t) = \int r(t) \, dt = \int 10^7 e^{-kt} \, dt = \frac{10^7}{-k} e^{-kt} + C$. When $t = 0$ we have $Q(0) = 10^7$, so $C + \frac{10^7}{-k} = 0$, so $C = \frac{10^7}{k}$. Thus $Q(t) = \frac{10^7(1 - e^{-kt})}{k}$.

b. $\lim_{t \to \infty} Q(t) = \lim_{t \to \infty} \frac{10^7(1 - e^{-kt})}{k} = \frac{10^7}{k}$. This represents the total number of barrels extracted if the nation extracts the oil indefinitely where it is assumed that the nation has at least $\frac{10^7}{k}$ barrels of oil in reserve.

c. We seek k so that $\lim_{b \to \infty} \int_0^b 10^7 e^{-kt} \, dt = 2 \times 10^9$. Thus we require $\lim_{b \to \infty} \left(\frac{e^{-kt}}{-k} \right) \Big|_0^b = 200$, or $\lim_{b \to \infty} \left(\frac{1}{-ke^{bt}} - \frac{1}{-k} \right) = 200$, so $k = \frac{1}{200} = .005$.

d. We want T so that $(2 \times 10^7) \int_0^T e^{-.005t} \, dt = 2 \times 10^9$, so $\left(-200 e^{-.005t} \right) \Big|_0^T = 100$, so $1 - e^{-T/200} = 1/2$, so $T = 200 \ln(2) \approx 138.63$ years.

7.2.103

a. Let $\exp(0) = z$. Then $\ln z = 0$, so $z = 1$, and thus $\exp(0) = 1$.

b. $\ln \left(\frac{\exp(x)}{\exp(y)} \right) = \ln(\exp(x)) - \ln(\exp(y)) = x - y$. Thus, $\exp(x - y) = \frac{\exp(x)}{\exp(y)}$.

c. $\ln((\exp(x))^y) = y \ln(\exp(x)) = yx$, so $\exp(xy) = (\exp(x))^y$.

7.2.105 Because $1/x$ is decreasing, we know that the left Riemann sum is an overestimate. Using $n = 2$ subintervals, we have $\int_1^2 1/x \, dx = \ln 2 < \frac{1}{1} \cdot \frac{1}{2} + \frac{1}{3/2} \cdot \frac{1}{2} = \frac{5}{6} < 1$. Because $1/x$ is decreasing, we know that the right Riemann sum is an underestimate. We have

$$\int_1^3 1/x \, dx = \ln 3 > \frac{2}{7} \left(\frac{1}{8/7} + \frac{1}{10/7} + \frac{1}{12/7} + \frac{1}{14/7} + \frac{1}{16/7} + \frac{1}{18/7} + \frac{1}{20/7} \right) = \frac{2761}{2520} > 1.$$

7.2.107 Because $1/x$ is decreasing, the left Riemann sum is an overestimate to the integral. Over the interval $[1, n+1]$ with n subdivisions, we must have $\int_1^{n+1} \frac{1}{x}\, dx = \ln(n+1) < \frac{n+1-1}{n}(1 + \frac{1}{2} + \frac{1}{3} + \cdots + \frac{1}{n}) = \sum_{i=1}^{n} \frac{1}{i}$. Because $\ln(n+1) \to \infty$ as $n \to \infty$, it must follow that the harmonic partial sum $(1 + \frac{1}{2} + \frac{1}{3} + \cdots + \frac{1}{n}) \to \infty$ as well.

7.2.109 Let $z = \left(\frac{a}{b}\right)^x$. Then $\ln z = x \ln(a/b) = x(\ln a - \ln b) = x \ln a - x \ln b = \ln a^x - \ln b^x = \ln\left(\frac{a^x}{b^x}\right)$. Thus $z = \left(\frac{a}{b}\right)^x = \left(\frac{a^x}{b^x}\right)$.

7.3 Logarithmic and Exponential Functions with Other Bases

7.3.1 $\frac{d}{dx} b^x = b^x \ln b$, for $b > 0$ and all x. In the case $b = e$, we rule states that $\frac{d}{dx} e^x = e^x \ln e = e^x$ because $\ln e = 1$.

7.3.3 $b^x = (e^{\ln b})^x = e^{(\ln b) \cdot x} = e^{x \ln b}$.

7.3.5 $\int 4^x\, dx = \int e^{x \ln 4}\, dx = \frac{1}{\ln 4} e^{x \ln 4} + C = \frac{4^x}{\ln 4} + C$.

7.3.7 $3^x = e^{x \ln 3}$. $x^\pi = e^{\pi \ln x}$. $x^{\sin x} = e^{\sin x \cdot \ln x}$.

7.3.9 If $\log_{10} x = 3$, then $10^3 = x$, so $x = 1000$.

7.3.11 If $\log_8 x = 1/3$, then $x = 8^{1/3} = 2$.

7.3.13 Because $7^x = 21$, we have that $\ln 7^x = \ln 21$, so $x \ln 7 = \ln 21$, and $x = \frac{\ln 21}{\ln 7}$.

7.3.15 Because $3^{3x-4} = 15$, we have that $\ln 3^{3x-4} = \ln 15$, so $(3x - 4) \ln 3 = \ln 15$. Thus, $3x - 4 = \frac{\ln 15}{\ln 3}$, so
$$x = \frac{(\ln 15)/(\ln 3) + 4}{3} = \frac{\ln 15 + 4 \ln 3}{3 \ln 3} = \frac{\ln 5 + \ln 3 + 4 \ln 3}{3 \ln 3} = \frac{\ln 5 + 5 \ln 3}{3 \ln 3} = \frac{\ln 5}{3 \ln 3} + \frac{5}{3}.$$

7.3.17 Because $10^{x^2 - 4} = 1 = 10^0$, we must have $x^2 - 4 = 0$. So $x = \pm 2$.

7.3.19 Because $2^{|x|} = 16 = 2^4$, we must have $|x| = 4$, so $x = \pm 4$.

7.3.21 $\frac{dy}{dx} = 8^x \ln 8$.

7.3.23 $y' = 5 \cdot \frac{d}{dx} 4^x = 5 \cdot \ln 4 \cdot 4^x$.

7.3.25 $y' = 3x^2 3^x + x^3 3^x \ln 3$.

7.3.27 $\frac{dA}{dt} = 250(1.045)^{4t} \cdot \ln(1.045) \cdot 4 = 1000 \cdot \ln(1.045) \cdot (1.045^{4t})$.

7.3.29

a. $T = 10 \cdot 2^{-0.274 \cdot 16}$ minutes ≈ 28.7 seconds.

b. $\frac{\Delta T}{\Delta a} = \frac{10 \cdot 2^{-0.274 \cdot 8} - 10 \cdot 2^{-0.274 \cdot 2}}{8 - 2} \approx -0.78$ minutes per 1000 feet, which is about -46.512 seconds per 1000 feet.

c. $\frac{dT}{da} = -2.74 \cdot 2^{-0.274 \cdot a} \cdot \ln 2$. At $a = 8$ we have $\frac{dT}{da} = -2.74 \cdot 2^{-0.274 \cdot 8} \cdot \ln 2 \approx -0.42$ minutes per 1000 feet. Every 1000 feet the airplane climbs, leaves about .42 minutes less time of consciousness, which corresponds to about 24.94 seconds.

7.3.31

a. At $Q = 10 \mu \text{Ci}$ we have $10 = 350 \cdot \left(\frac{1}{2}\right)^{t/13.1}$, so $\ln(1/35) = \frac{t}{13.1} \ln(1/2)$, so $t = 13.1 \cdot \frac{\ln 35}{\ln 2} \approx 67.19$ hours.

b. $\frac{dQ}{dt} = \frac{350}{13.1} \cdot \ln\left(\frac{1}{2}\right) \cdot \left(\frac{1}{2}\right)^{t/13.1}$. We have $Q'(12) \approx -9.81453$, $Q'(24) \approx -5.20136$, and $Q'(48) \approx -1.46087$. The rate at which the iodine decreases is decreasing in absolute value as time increases.

7.3.33 $\displaystyle\int_{-1}^{1} 10^x \, dx = \left(\frac{10^x}{\ln 10}\right)\Big|_{-1}^{1} = \frac{10 - 10^{-1}}{\ln 10} = \frac{99}{10\ln 10}$.

7.3.35 $\int_1^2 (1 + \ln x)x^x \, dx = (x^x)\big|_1^2 = 4 - 1 = 3$.

7.3.37 Let $u = x^3 + 8$ so that $du = 3x^2 \, dx$. Substituting gives $\frac{1}{3}\int 6^u \, du = \frac{6^u}{3\ln 6} + C = \frac{6^{x^3+8}}{3\ln 6} + C$.

7.3.39 $g'(y) = e^y y^e + e^{y+1}y^{e-1}$.

7.3.41 $s'(t) = -\sin(2^t) \cdot 2^t \ln 2$.

7.3.43 $f'(x) = 2x^{3/2} + \frac{3}{2}(2x - 3)x^{1/2} = 5x^{3/2} - \frac{9}{2}x^{1/2} = \frac{10x\sqrt{x} - 9\sqrt{x}}{2} = \frac{\sqrt{x}}{2}(10x - 9)$.

7.3.45 Let $y = x^{\cos x}$. Then $\ln y = \cos x \ln x$. Differentiating both sides gives

$$\frac{1}{y}y' = (-\sin x)\ln x + \cos x \cdot \frac{1}{x}.$$

Therefore,

$$y' = x^{\cos x}\left(\frac{\cos x}{x} - \sin x \ln x\right).$$

At $\pi/2$ we have $y'(\pi/2) = (\pi/2)^0(0/(\pi/2) - \ln(\pi/2)) = -\ln(\pi/2)$.

7.3.47 Let $y = x^{\sqrt{x}}$. Then $\ln y = \sqrt{x} \cdot \ln x$. Differentiating both sides gives

$$\frac{1}{y}y' = \frac{1}{2\sqrt{x}}\ln x + \frac{\sqrt{x}}{x}.$$

Therefore,

$$y' = x^{\sqrt{x}}\left(\frac{\ln x + 2}{2\sqrt{x}}\right).$$

At 4 we have $y'(4) = 4^2\left(\frac{\ln 4 + 2}{4}\right) = 4\ln 4 + 8$.

7.3.49 Because $f(x) = \sin^{\ln x} x$, we have $\ln f(x) = \ln x \ln \sin x$. Differentiating both sides gives

$$\frac{1}{f(x)}f'(x) = \frac{1}{x} \cdot \ln \sin x + \ln x \cdot \frac{1}{\sin x}\cos x.$$

Therefore,

$$f'(x) = \sin^{\ln x} x \left(\frac{\ln \sin x + x \ln x \cot x}{x}\right).$$

We have $f'(\pi/2) = 0$ because $\cot \pi/2 = 0$ and $\ln \sin(\pi/2) = \ln 1 = 0$.

7.3.51 $f(x) = e^{4x\ln(2x)}$ so $f'(x) = e^{4x\ln(2x)}\left(4x \cdot (1/x) + \ln(2x) \cdot 4\right) = (2x)^{4x}(4 + 4\ln(2x))$.

7.3.53 $y' = 2^{x^2} \cdot \ln 2 \cdot (2x) = 2^{x^2+1}x \ln 2$.

7.3.55 $\ln(H(x)) = 2x\ln(x+1)$, so $\frac{1}{H(x)}H'(x) = \frac{2x}{x+1} + 2\ln(x+1)$, and thus $H'(x) = (x+1)^{2x}\left(\frac{2x}{x+1} + 2\ln(x+1)\right)$.

7.3.57 Let $y = x^{\sin x}$. Then $\ln y = \sin x \ln x$, so $\frac{1}{y}y' = \cos x \ln x + \frac{\sin x}{x}$. At the point $(1, 1)$ we have $y' = \sin 1$, so the tangent line is given by $y - 1 = (\sin 1)(x - 1)$.

7.3.59 Let $y = (x^2)^x = x^{2x}$. Then $\ln y = x \ln x^2$ and $\frac{1}{y} y' = \ln x^2 + 2$, so $y' = x^{2x}(\ln x^2 + 2)$. This quantity is zero when $\ln x^2 = -2$, or $x^2 = e^{-2}$. Thus there are horizontal tangents at $|x| = e^{-1}$, so for $x = \pm \frac{1}{e}$. The two tangent lines are given by $y = \frac{1}{e^{2/e}}$ (at $\left(\frac{1}{e}, \frac{1}{e^{2/e}}\right)$) and $y = e^{2/e}$ (at $\left(-\frac{1}{e}, e^{2/e}\right)$.)

7.3.61 $y' = 4 \cdot \dfrac{2x}{(x^2 - 1) \cdot \ln 3} = \dfrac{8x}{(x^2 - 1) \cdot \ln 3}$.

7.3.63 $y' = -\sin x(\ln(\cos^2 x)) + \cos x \cdot \left(\dfrac{2 \cos x(-\sin x)}{\cos^2 x}\right) = (-\sin x)(\ln(\cos^2 x) + 2)$.

7.3.65 $y' = \dfrac{d}{dx}(\log_4 x)^{-1} = -(\log_4 x)^{-2} \cdot \dfrac{1}{x \ln 4} = -\dfrac{1}{x(\ln 4)(\log_4 x)^2} = -\dfrac{\ln 4}{x \ln^2 x}$.

7.3.67

a. False. $\log_2 9$ is a constant, so its derivative is 0.

b. False. If $x < -1$, the right-hand side is defined while the left-hand side isn't.

c. False. The correct way to write that function would be $e^{(x+1)\ln 2}$.

d. False. $\frac{d}{dx}(\sqrt{2})^x = (\sqrt{2})^x \ln(\sqrt{2})$.

e. True. This follows from the generalized power rule.

7.3.69

7.3.71

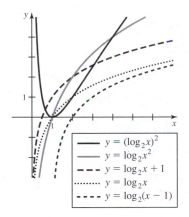

——	$y = (\log_2 x)^2$
——	$y = \log_2 x^2$
– – –	$y = \log_2 x + 1$
········	$y = \log_2 x$
– · – ·	$y = \log_2(x - 1)$

7.3.73 $\dfrac{d^2}{dx^2}(\log x) = \dfrac{d}{dx}\left(\dfrac{1}{x \ln 10}\right) = -\dfrac{1}{x^2 \ln 10}$.

7.3.75 $\dfrac{d^3}{dx^3}(x^2 \ln x) = \dfrac{d^2}{dx^2}(2x \ln x + x) = \dfrac{d}{dx}(2 \ln x + 2 + 1) = \dfrac{2}{x}$.

7.3.77

(i) $y' = \dfrac{d}{dx}\left(e^{x \ln 3}\right) = (e^{x \ln 3}) \cdot \ln 3 = 3^x \ln 3$.

(ii) Let $y = 3^x$. Then $\ln y = x \ln 3$. So $\frac{1}{y} y' = \ln 3$, and $y' = 3^x \ln 3$.

7.3.79 $f'(x) = \dfrac{d}{dx}(4 \ln(3x + 1)) = \dfrac{4}{3x+1} \cdot 3 = \dfrac{12}{3x+1}$.

7.3.81 $f'(x) = \dfrac{d}{dx}\left(\frac{1}{2} \log 10x\right) = \dfrac{d}{dx}\frac{1}{2}[\log 10 + \log x] = \dfrac{1}{2x \ln 10}$.

7.3.83 $f'(x) = \dfrac{d}{dx}(\ln(2x - 1) + 3 \ln(x + 2) - 2 \ln(1 - 4x)) = \dfrac{2}{2x-1} + \dfrac{3}{x+2} + \dfrac{8}{1-4x}$.

7.3.85

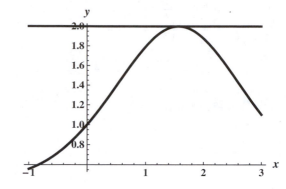

$y' = \frac{d}{dx}e^{\sin x \ln 2} = (\cos x)(\ln 2)2^{\sin x}$. At $x = \pi/2$ we have $y' = 0$, so the tangent line is given by $y = 2$.

7.3.87 Let $y = x^{10x}$. Then $\ln y = 10x \ln x$, so $\frac{1}{y}y' = 10 \ln x + 10$, and $y' = x^{10x}(10)(\ln x + 1)$.

7.3.89 Let $y = x^{\cos x}$. Then $\ln y = \cos x \ln x$, and $\frac{1}{y}y' = (-\sin x)(\ln x) + \frac{\cos x}{x}$. Thus,

$$y' = x^{\cos x}\left(\frac{\cos x}{x} - \sin x \ln x\right).$$

7.3.91 Let $y = \left(1 + \frac{1}{x}\right)^x$. Then $\ln y = x \ln\left(1 + \frac{1}{x}\right)$, so

$$\frac{1}{y}y' = \ln\left(1 + \frac{1}{x}\right) + x\left(\frac{-1/x^2}{1 + 1/x}\right) = \ln\left(1 + \frac{1}{x}\right) - \frac{1}{x+1}.$$

Thus,

$$y' = \left(1 + \frac{1}{x}\right)^x\left(\ln\left(1 + \frac{1}{x}\right) - \frac{1}{x+1}\right).$$

7.3.93 Let $y = x^{x^{10}}$. Then $\ln y = x^{10} \ln x$, so $\frac{1}{y}y' = 10x^9 \ln x + \frac{x^{10}}{x} = x^9(10 \ln x + 1)$. Thus $y' = x^{x^{10}} \cdot x^9(10 \ln x + 1)$.

7.3.95 $\int 3^{-2x}\, dx = \int e^{-2x \ln 3}\, dx = e^{-2x \ln 3} \cdot \frac{1}{-2 \ln 3} + C = \frac{-3^{-2x}}{2 \ln 3} + C$.

7.3.97 Let $u = x^3$, so that $du = 3x^2\, dx$. Then we have $\int \frac{1}{3}10^u\, du = \frac{10^u}{3 \ln 10} + C = \frac{10^{x^3}}{3 \ln 10} + C$.

7.3.99 Let $u = \ln x$ so that $du = \frac{1}{x}\, dx$. Then we have $\int_0^{\ln(2)+1} 3^u\, du = \left(\frac{3^u}{\ln 3}\right)\Big|_0^{\ln(2)+1} = \frac{3 \cdot 3^{\ln 2} - 1}{\ln 3}$.

7.3.101 Let $f(x) = \ln x$ and $a = e$. Then $f'(e) = \lim\limits_{x \to e} \frac{\ln x - 1}{x - e} = \frac{1}{e}$.

7.3.103 Let $f(x) = x^x$ and $a = 3$. Then $f'(3) = \lim\limits_{h \to 0} \frac{(3 + h)^{3+h} - 27}{h} = 3^3 \cdot (\ln 3 + 1) = 27(1 + \ln 3)$.

7.3.105 Let $y = u(x)^{v(x)}$. Then $\ln y = v(x) \ln(u(x))$, so $\frac{1}{y}y' = v'(x) \ln(u(x)) + v(x) \cdot \frac{u'(x)}{u(x)}$. Thus we have $y' = u(x)^{v(x)} \cdot \left(v'(x) \ln(u(x)) + v(x) \cdot \frac{u'(x)}{u(x)}\right) = u(x)^{v(x)} \cdot \left(\frac{v(x)}{u(x)}\frac{du}{dx} + \ln(u(x))\frac{dv}{dx}\right)$.

7.3.107 Using the change of base formulas $\log_b c = \frac{\ln c}{\ln b}$ and $\log_c b = \frac{\ln b}{\ln c}$ we have

$$(\log_b c) \cdot (\log_c b) = \frac{\ln c}{\ln b} \cdot \frac{\ln b}{\ln c} = 1.$$

7.4   Exponential Models

7.4.1 Exponential growth occurs for a constant relative growth rate.

7.4.3 It is the time it takes a function to double in value.

7.4.5 For exponential growth modeled by $y = y_0 e^{kt}$, the doubling time T_2 is $T_2 = \frac{\ln 2}{k}$, where $k > 0$ is the growth constant.

7.4.7 Compound interest and world population growth.

7.4.9 For $f(t)$, $\frac{df}{dt} = 10.5$, so the absolute growth rate is constant. For $g(t)$, $\frac{dg}{dt} = 100e^{t/10} \cdot \frac{1}{10} = 10e^{t/10}$, so the growth rate is not constant but the relative growth rate $\frac{1}{g(t)} g'(t) = \frac{1}{10}$ is constant.

7.4.11 The growth is modeled by $p(t) = 90,000e^{kt}$, with $t = 0$ corresponding to 2010, and time measured in years. Because $(1.024)(90,000) = 90,000e^k$, we have $k = \ln(1.024)$. The doubling time is $T_2 = \frac{\ln 2}{\ln(1.024)} \approx 29.2263$, so it should reach 180,000 around the year 2039.

7.4.13 The growth is modeled by $p(t) = 50000e^{kt}$. When $t = 10$, we have $p(10) = 50000e^{10k} = 55000$, so $k = \ln(11/10)/10$. Thus in 20 years, the population should be about $50000e^{2\ln(11/10)} = 50000(1.1)^2 = 60,500$.

7.4.15 The price of a cart of groceries is modeled by $p(t) = 100e^{kt}$ where $t = 0$ corresponds to 2005, and t is measured in years and $k = \ln(1.03)$. The price of a the groceries in 2018 when $t = 13$ should be about $p(10) = 100e^{13\ln(1.03)} \approx \$146.85$.

7.4.17

a. The population is modeled by $P(t) = 309e^{t\ln(1.008)}$. The doubling time is $\frac{\ln 2}{\ln 1.008} \approx 87$ years. The population in 2050 will be about $P(40) = 309e^{40\ln(1.008)} \approx 425$ million.

b. If the growth rate is .6%, the doubling time is $\frac{\ln 2}{\ln 1.006} \approx 115.9$ years and the population in 2050 will be about $309e^{40\ln(1.006)} \approx 392.5$ million. If the growth rate is 1%, the doubling time will is $\frac{\ln 2}{\ln 1.01} \approx 69.7$ and the population in 2050 will be about $309e^{40\ln(1.01)} \approx 460.1$ million.

c. A growth rate swing of just .2% produces large differences in population growth.

7.4.19 The growth is modeled by $p(t) = 20.9e^{kt}$. When $t = 10$, we have $p(10) = 20.9e^{10k} = 25.1$, so $k = \ln(25.1/20.9)/10 \approx .0183$. Thus $p(25) = 20.9e^{.0183(25)} \approx 33$ million people.

7.4.21 The homicide rate is modeled by $H(t) = 800e^{-kt}$, where $t = 0$ corresponds to 2010, and time is measured in years, and $-k = \ln(.97) \approx -.03$. The rate will reach 600 when $6/8 = e^{-.03t}$, or $t = \frac{\ln(6/8)}{-.03} \approx 9.6$ years. So it should achieve this rate in 2019.

7.4.23 The amount of Valium in the bloodstream is modeled by $a(t) = 20e^{-kt}$. If the half-life is 36 hours, then $36 = \frac{\ln 2}{k}$, so $k = \frac{\ln 2}{36}$. So $a(12) = 20e^{-12\ln(2)/36} = 20e^{-\ln(2)/3} \approx 15.87$ mg.

The concentration of Valium will reach 2 mg when $0.1 = e^{-\ln(2)t/36}$, which is when $t = \frac{-36\ln(0.1)}{\ln 2} \approx 119.59$ hours.

7.4.25 The population is modeled by $p(t) = 9.94e^{kt}$, and $p(10) = 9.94e^{10k} = 9.88$, so $k = \ln(9.88/9.94)/10 \approx -0.0006$. We have $p(20) = 9.94e^{(-0.0006)(20)} \approx 9.82$ million people. The population decline is likely due to economic factors which may change if the economy (especially manufacturing) improves.

7.4.27 Let $y = 1000e^{-kx}$ model the pressure at x feet above sea level. We know that $\frac{1}{3} = e^{-30000k}$, so $k = \frac{\ln(1/3)}{-30000}$. We want to know for what x does $\frac{1}{2} = e^{-kx}$, so we are seeking $x = \frac{\ln 2}{k} = \frac{30000\ln(2)}{\ln 3} \approx 18,928$ feet above sea level.

The pressure is 1/100th of the sea-level pressure when $x = \frac{30000\ln(100)}{\ln 3} \approx 125,754$ feet.

7.4.29 The amount of U-238 in the rock is modeled by $a(t) = a_0 e^{-kt}$ with $k = \frac{\ln 2}{4.5}$, where time is measured in billions of years. We seek t so that $a_0 = a_0 e^{-kt}$, so $t = \frac{\ln(.85)}{-k} = \frac{\ln(.85)(-4.5)}{\ln 2} \approx 1.055$. So the cloth was painted about 1.055 billion years ago.

7.4.31

a. False. If that was the correct formula, then $y(1) = y_0 e^{.06} = (1.06184)y_0 \neq (1.06)y_0$.

b. False. If it increases by ten precent per year, then after 3 years it increases by a factor of $(1.1)^3 = 1.331$ which corresponds to 33.1 percent.

c. True. The relative decay rate is constant, so the decay is exponential.

d. True. This follows because the doubling time is related to k by the equation $T_2 = \frac{\ln 2}{k}$.

e. True. This time would be the constant $\frac{\ln 10}{k}$.

7.4.33 As in the previous problem, the doubling time is $\frac{\ln 2}{k}$ which is constant as a function of t.

7.4.35

a. After 5 hours, Abe has run $\int_0^5 \frac{4}{t+1}\, dt = (4\ln(t+1))\big|_0^5 = 4\ln 6 \approx 7.17$ miles. After 5 hours, Bob has run $\int_0^5 4e^{-t/2}\, dt = \left(8e^{-t/2}\right)\big|_5^0 = 8(1 - e^{-5/2}) \approx 7.34$. So after 5 hours, Bob is ahead.

After 10 hours, Abe has run $\int_0^{10} \frac{4}{t+1}\, dt = (4\ln(t+1))\big|_0^{10} = 4\ln 11 \approx 9.59$ miles. After 10 hours, Bob has run $\int_0^{10} 4e^{-t/2}\, dt = \left(8e^{-t/2}\right)\big|_{10}^0 = 8(1 - e^{-5}) \approx 7.95$. So after 10 hours, Abe is ahead.

b. Bob's distance function is given by $8(1-e^{t/2})$ and is bounded above by 8. Abe's distance function is given by $4\ln(t + 1)$ and is unbounded.

7.4.37 $(1 + .008)^{12} - 1 \approx 10.034\%$, which is more than $12 \cdot .008 \approx 9.6\%$.

7.4.39 As in the previous problem, $s(t) = \frac{v_0}{k}(1 - e^{-kt})$. Consider the equations $s(t_2) - s(t_1) = 1200 = \frac{v_0}{k}(e^{-kt_1} - e^{-kt_2})$ and $v(t_2) - v(t_1) = -100 = v_0(e^{-kt_2} - e^{-kt_1})$. Dividing these two equations yields $k = \frac{1}{12}$.
   Now $v(t_1) = 1000 = v_0 e^{-t_1/12}$ and $v(t_2) = 900 = v_0 e^{-t_2/12}$, so $\frac{1000}{900} = e^{(-1/12)(t_1-t_2)}$, so $t_1 - t_2 = -12\ln(10/9) \approx -1.2643$ seconds. The deceleration takes about 1.2643 seconds to occur.

7.4.41 The initial volume is $V_0 = \frac{4\pi}{3}\left(\frac{5}{10000}\right)^3 = \frac{\pi}{6\times 10^9}$ cubic cm. $k = \frac{\ln 2}{35}$. Suppose $0.5 = V_0 e^{kt} = \frac{\pi}{6\times 10^9} e^{(t\ln 2)/35}$. Then $\frac{t\ln 2}{35} = \ln\left(\frac{3\times 10^9}{\pi}\right)$, and $t = \frac{35}{\ln 2}\ln\left(\frac{3\times 10^9}{\pi}\right) \approx 1044$ days.

7.4.43 Revenue is given by $R(x) = 40xe^{-x/50}$. So $R'(x) = 40xe^{-x/50} \cdot \left(\frac{-1}{50}\right) + 40e^{-x/50} = 40e^{-x/50}(1 - (x/50))$. The critical number is $x = 50$, and this number yields a maximum, because $R'(x) > 0$ on $(0, 50)$, while $R'(x) < 0$ for $x > 50$. Thus, she should charge $50.00.

7.4.45 If $y_0 e^{kt} = y_0(1 + r)^t$, then $(e^k)^t = (1 + r)^t$, so $e^k = 1 + r$, and $k = \ln(1 + r)$. If $y_0 e^{kt} = y_0 2^{t/T_2}$, then $(e^k)^t = (2^{1/T_2})^t$, so $e^k = 2^{1/T_2}$, and $k = \frac{\ln 2}{T_2}$, or $T_2 = \frac{\ln 2}{k}$. Also, we have $T_2 = \frac{\ln 2}{k} = \frac{\ln 2}{\ln(1+r)}$, so $T_2 \ln(1 + r) = \ln 2$. Then $(1 + r)^{T_2} = 2$, so $r = 2^{1/T_2} - 1$.

7.5 Inverse Trigonometric Functions

7.5.1 The sine function is not one-to-one over its whole domain, so in order to define an inverse, it must be restricted to an interval on which it is one-to-one.

7.5.3 $\tan(\tan^{-1}(x)) = x$ for all real numbers x. (Note that the domain of the inverse tangent is $\mathbb{R}$). However, it is not always true that $\tan^{-1}(\tan x) = x$. For example, $\tan 27\pi = 0$, and $\tan^{-1}(0) = 0$. Thus $\tan^{-1}(\tan(27\pi)) \neq 27\pi$.

7.5.5 The numbers $\pm\pi/2$ are not in the range of $\tan^{-1} x$. The range is $(-\pi/2, \pi/2)$. However, it is true that as x increases without bound, the values of $\tan^{-1} x$ get close to $\pi/2$, and as x decreases without bound, the values of $\tan^{-1} x$ get close to $-\pi/2$.

7.5.7

$$\frac{d}{dx}\sin^{-1}(x) = \frac{1}{\sqrt{1-x^2}}, \quad -1 < x < 1$$

$$\frac{d}{dx}\tan^{-1}(x) = \frac{1}{1+x^2}, \quad -\infty < x < \infty$$

$$\frac{d}{dx}\sec^{-1}(x) = \frac{1}{|x|\sqrt{x^2-1}}, \quad |x| > 1.$$

7.5.9 $y' = \dfrac{1}{1+x^2}$. At $x = -2$ we have $y'(-2) = \frac{1}{1+4} = \frac{1}{5}$.

7.5.11 Let $z = \sin^{-1}(1)$. Then $\sin z = 1$, and because $\sin \pi/2 = 1$, and $\pi/2$ is in the desired interval, $z = \pi/2$.

7.5.13 $\sin^{-1}(-1/2) = -\pi/6$, because $\sin(-\pi/6) = -1/2$.

7.5.15 $\cos^{-1}(-1/2) = 2\pi/3$, because $\cos(2\pi/3) = -1/2$.

7.5.17 $\cos(\cos^{-1}(-1)) = \cos(\pi) = -1$.

7.5.19

Let $\theta = \sin^{-1} x$. Then $\sin\theta = x$. $\cos\theta = \frac{\text{side adjacent to }\theta}{\text{hypotenuse}} = \frac{\sqrt{1-x^2}}{1} = \sqrt{1-x^2}$.

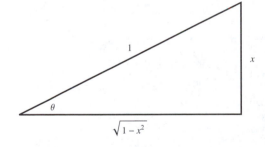

7.5.21

Let $\theta = \cos^{-1} x/2$. Then $\cos\theta = x/2$. $\sin\theta = \frac{\text{side opposite of }\theta}{\text{hypotenuse}} = \frac{\sqrt{4-x^2}}{2}$.

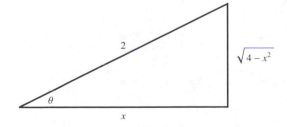

7.5.23

Let $\theta = \cos^{-1} x$.   Then $\cos\theta = x$.   Using the identity given, we have $\sin 2\theta = 2\sin\theta\cos\theta = 2x\sin\theta = 2x\sqrt{1-x^2}$.

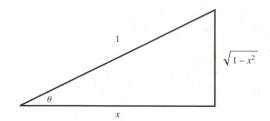

7.5.25

Let $\theta = \cos^{-1}(x)$, and note from the diagram that it then follows that $\cos^{-1}(-x) = \pi - \theta$.   So $\cos^{-1}(x) + \cos^{-1}(-x) = \theta + \pi - \theta = \pi$.

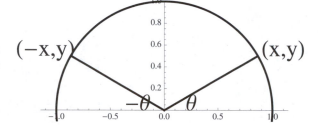

7.5.27 $\tan^{-1}(\sqrt{3}) = \tan^{-1}\left(\frac{\sqrt{3}/2}{1/2}\right) = \pi/3$, because $\sin(\pi/3) = \sqrt{3}/2$ and $\cos(\pi/3) = 1/2$.

7.5.29 $\sec^{-1}(2) = \sec^{-1}\left(\frac{1}{1/2}\right) = \pi/3$, because $\sec(\pi/3) = \frac{1}{\cos(\pi/3)} = \frac{1}{1/2} = 2$.

7.5.31 $\tan^{-1}(\tan(\pi/4)) = \tan^{-1}(1) = \pi/4$.

7.5.33 Let $\csc^{-1}(\sec 2) = z$. Then $\csc z = \sec 2$, so $\sin z = \cos 2$. Now by applying the result of problem 22, we see that $z = \sin^{-1}(\cos 2) = \pi/2 - 2 = \frac{\pi-4}{2}$.

7.5.35

Let $\theta = \tan^{-1} x$. Then $\tan\theta = x$.

$$\cos\theta = \frac{\text{side adjacent to }\theta}{\text{hypotenuse}} = \frac{1}{\sqrt{x^2+1}}.$$

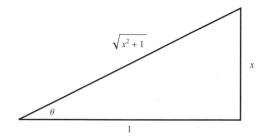

7.5.37

Let $\theta = \tan^{-1} x$. Then $\tan\theta = x$.

$$\sec\theta = \frac{\text{hypotenuse}}{\text{side adjacent to }\theta} = \sqrt{x^2+1}.$$

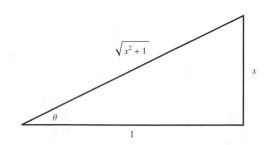

7.5.39

Let $\theta = \csc^{-1} 2x$. Then $\csc\theta = 2x$. We have

$$\sec\theta = \frac{\text{hypotenuse}}{\text{side adjacent to } \theta} = \frac{2x}{\sqrt{4x^2 - 1}}.$$

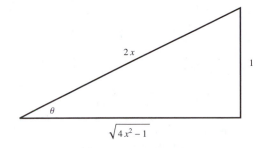

7.5.41 Because $\sin\theta = \frac{x}{6}$, $\theta = \sin^{-1}(x/6)$. Also, $\theta = \tan^{-1}\left(\frac{x}{\sqrt{36-x^2}}\right) = \sec^{-1}\left(\frac{6}{\sqrt{36-x^2}}\right)$.

7.5.43 $\dfrac{d}{dx}\sin^{-1}(2x) = \dfrac{2}{\sqrt{1-4x^2}}.$

7.5.45 $\dfrac{d}{dw}\cos(\sin^{-1}(2w)) = (-\sin(\sin^{-1}(2w)))\cdot\dfrac{d}{dw}\left(\sin^{-1}(2w)\right) = -2w\cdot\dfrac{2}{\sqrt{1-4w^2}} = -\dfrac{4w}{\sqrt{1-4w^2}}.$

7.5.47 $\dfrac{d}{dx}\sin^{-1}(e^{-2x}) = \dfrac{1}{\sqrt{1-e^{-4x}}}\cdot\dfrac{d}{dx}e^{-2x} = -\dfrac{2e^{-2x}}{\sqrt{1-e^{-4x}}}.$

7.5.49 $\dfrac{d}{dy}\tan^{-1}(2y^2-4) = \dfrac{1}{1+(2y^2-4)^2}\cdot\dfrac{d}{dy}(2y^2-4) = \dfrac{4y}{1+(2y^2-4)^2}.$

7.5.51 $\dfrac{d}{dz}\cot^{-1}(\sqrt{z}) = -\dfrac{1}{1+\sqrt{z}^2}\cdot\dfrac{d}{dz}\sqrt{z} = -\dfrac{1}{1+z}\cdot\dfrac{1}{2\sqrt{z}} = -\dfrac{1}{2\sqrt{z}(1+z)}.$

7.5.53 $\dfrac{d}{dx}\cos^{-1}\left(\dfrac{1}{x}\right) = -\dfrac{1}{\sqrt{1-\frac{1}{x^2}}}\cdot\left(-\dfrac{1}{x^2}\right) = \dfrac{1}{x^2\sqrt{\frac{x^2-1}{x^2}}} = \dfrac{|x|}{x^2\sqrt{x^2-1}} = \dfrac{1}{|x|\sqrt{x^2-1}}$, for $|x| > 1$.

7.5.55 $\dfrac{d}{du}\csc^{-1}(2u+1) = -\dfrac{1}{|2u+1|\sqrt{(2u+1)^2-1}}\cdot 2 = -\dfrac{2}{|2u+1|\sqrt{(2u+1)^2-1}} = -\dfrac{2}{|2u+1|\sqrt{4u^2+4u}} =$

$-\dfrac{1}{|2u+1|\sqrt{u^2+u}}.$

7.5.57 $\dfrac{d}{dy}\cot^{-1}\left(\dfrac{1}{1+y^2}\right) = \left(-\dfrac{1}{1+\left(\frac{1}{1+y^2}\right)^2}\right)\cdot\left(-\dfrac{2y}{(1+y^2)^2}\right) = \dfrac{2y}{(1+y^2)^2+1}.$

7.5.59 $\dfrac{d}{dx}\sec^{-1}(\ln x) = \dfrac{1}{|\ln x|\sqrt{(\ln x)^2-1}}\cdot\dfrac{1}{x} = \dfrac{1}{x|\ln x|\sqrt{(\ln x)^2-1}}.$

7.5.61 $\dfrac{d}{dx}\csc^{-1}(\tan(e^x)) = -\dfrac{1}{|\tan e^x|\sqrt{(\tan(e^x))^2-1}}\cdot\sec^2(e^x)\cdot e^x.$

7.5.63 $\dfrac{d}{ds}\cot^{-1}(e^s) = -\dfrac{1}{1+e^{2s}}\cdot e^s = -\dfrac{e^s}{1+e^{2s}}.$

7.5.65

a. $\frac{x}{150} = \cot\theta$, so $\theta = \cot^{-1}\left(\frac{x}{150}\right)$. Then $\frac{d\theta}{dx} = -\frac{1}{1+\left(\frac{x}{150}\right)^2} \cdot \frac{1}{150} = -\frac{150}{(150)^2+x^2}$. When $x = 500$, we have $\frac{d\theta}{dx} = -\frac{150}{150^2+500^2} \approx -0.00055$ radians per meter.

b. The most rapid change is at $x = 0$ where $\frac{d\theta}{dx} = -\frac{1}{150} \approx -0.0067$ radians per meter.

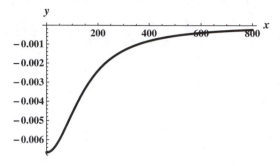

7.5.67 Using Table 4.6 (formula 9, $a = 5$), $\int \frac{6}{\sqrt{25-x^2}}\, dx = 6\sin^{-1}\left(\frac{x}{5}\right) + C$. Check: $\frac{d}{dx}\left(6\sin^{-1}\left(\frac{x}{5}\right) + C\right) = \frac{6}{\sqrt{1-(x^2/25)}} \cdot \frac{1}{5} = \frac{6}{\sqrt{25-x^2}}$.

7.5.69 Using Table 4.6 (formula 11, $a = 10$), $\int \frac{1}{x\sqrt{x^2-100}}\, dx = \frac{1}{10}\sec^{-1}\left|\frac{x}{10}\right| + C$. Check: $\frac{d}{dx}\left(\frac{1}{10}\sec^{-1}\left|\frac{x}{10}\right| + C\right) = \frac{1}{10} \cdot \frac{1}{(x/10)\cdot\sqrt{(x^2/100)-1}} \cdot \frac{1}{10} = \frac{1}{x\sqrt{x^2-100}}$.

7.5.71
$$\int_0^3 \frac{dx}{\sqrt{36-x^2}} = \sin^{-1}\left(\frac{x}{6}\right)\Big|_0^3 = \frac{\pi}{6}$$

7.5.73
$$\int_0^{3/2} \frac{dx}{\sqrt{36-4x^2}} = \frac{1}{2}\int_0^{3/2} \frac{dx}{\sqrt{9-x^2}} = \frac{1}{2}\sin^{-1}\left(\frac{x}{3}\right)\Big|_0^{3/2} = \frac{\pi}{12}$$

7.5.75 Let $u = e^x$. Then $du = e^x\, dx$. We have
$$\int_{-\ln\sqrt{3}}^0 \frac{e^x}{1+e^{2x}}\, dx = \int_{1/\sqrt{3}}^1 \frac{1}{1+u^2}\, du = \left(\tan^{-1} u\right)\Big|_{1/\sqrt{3}}^1 = \frac{\pi}{4} - \frac{\pi}{6} = \frac{\pi}{12}.$$

7.5.77

a. False. In general, the two sides bear no relationship to one another. For example, if $x = \sqrt{3}$, $\tan^{-1} x = \pi/3$ while $\sin^{-1} x$ and $\cos^{-1} x$ are not even defined.

b. True. The range of $\cos^{-1}$ is $[0, \pi]$, and $15\pi/16$ is in this range.

c. False. For example if $x = \pi/2$, the right-hand side evaluates to 1 while the left-hand side is not defined.

d. True, because $\frac{d}{dx}\sin^{-1}(x) = -\frac{d}{dx}\cos^{-1}(x)$.

e. False. $\frac{d}{dx}\tan^{-1}(x) = \frac{1}{1+x^2}$ for all x, and this doesn't equal $\sec^2 x$ anywhere except at the origin (one curve is always less than or equal to one, and the other is always greater than or equal to one.)

f. True. $\frac{d}{dx}\sin^{-1}(x) = \frac{1}{\sqrt{1-x^2}}$, and this is minimal when its denominator is as big as possible, which occurs when $x = 0$. So the smallest possible slope of a tangent line for this function on $(-1, 1)$ is $\frac{1}{\sqrt{1-0^2}} = 1$.

g. True. $\frac{d}{dx}\sin x = \cos x$ and $\cos x = 1$ for $x = 0$ and $-1 \le \cos x \le 1$ for all x. Thus 1 is the largest possible slope for a tangent line to the sine function.

7.5.79 If $\cos\theta = 5/13$, then the Pythagorean identity gives $|\sin\theta| = 12/13$. But if $0 < \theta < \pi/2$, then the sine of θ is positive, so $\sin\theta = 12/13$. Thus $\tan\theta = 12/5$, $\cot\theta = 5/12$, $\sec\theta = 13/5$, and $\csc\theta = 13/12$.

7.5.81 If $\csc\theta = 13/12$, then $\sin\theta = 12/13$, and the Pythagorean identity gives $|\cos\theta| = 5/13$. But if $0 < \theta < \pi/2$, then the cosine of θ is positive, so $\cos\theta = 5/13$. Thus $\tan\theta = 12/5$, $\cot\theta = 5/12$, and $\sec\theta = 13/5$.

7.5.83

a.

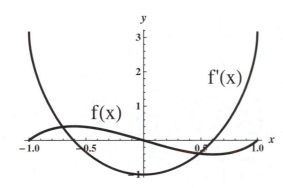

b. $f'(x) = 2x\sin^{-1}(x) + \frac{x^2-1}{\sqrt{1-x^2}}$.

c. Note that f' is zero and f has a horizontal tangent line at about $x = -0.61$ and at about $x = 0.61$.

7.5.85

a.

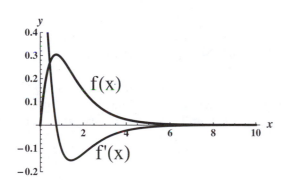

b. $f'(x) = -e^{-x}\tan^{-1}(x) + e^{-x} \cdot \frac{1}{1+x^2}$.

c. Note that f' is zero and f has a horizontal tangent line at about $x = 0.75$.

7.5.87 $y^2 - 4y + 5 = (y-2)^2 + 1$, so

$$\int \frac{dy}{y^2 - 4y + 5} = \int \frac{dy}{(y-2)^2 + 1} = \tan^{-1}(y-2) + C$$

7.5.89 Let $u = e^x$; then $du = e^x\,dx$ so that

$$\int \frac{e^x}{e^{2x} + 4}\,dx = \int \frac{du}{u^2 + 4} = \frac{1}{2}\tan^{-1}\left(\frac{u}{2}\right) = \frac{1}{2}\tan^{-1}\left(\frac{e^x}{2}\right) + C$$

7.5.91

a. Because $\frac{l}{10} = \csc(\theta)$, $\theta = \csc^{-1}\left(\frac{l}{10}\right)$, and $\frac{d\theta}{dl} = -\frac{1}{(l/10)\sqrt{(l/10)^2 - 1}} \cdot \frac{1}{10} = -\frac{10}{l\sqrt{l^2 - 100}}$.

b. $\frac{d\theta}{dl}\big|_{l=50} = -\frac{10}{50\sqrt{2500-100}} \approx -0.0041$ radians per foot. $\frac{d\theta}{dl}\big|_{l=20} = -\frac{10}{20\sqrt{400-100}} \approx -0.029$ radians per foot. $\frac{d\theta}{dl}\big|_{l=11} = -\frac{10}{11\sqrt{121-100}} \approx -0.198$ radians per foot.

c. $\lim\limits_{l \to 10^+} -\dfrac{10}{l\sqrt{l^2 - 100}} = -\infty$. The angle changes very quickly as we approach the dock.

d. $\frac{d\theta}{dl}$ is negative because this measures the change in θ as l increases – but when the boat is approaching the dock, l is decreasing.

7.5.93

a. $\sin\theta = \frac{c}{D}$, so $\theta = \sin^{-1}\left(\frac{c}{D}\right)$. Thus $\dfrac{d\theta}{dc} = \dfrac{1/D}{\sqrt{1 - \left(\frac{c}{D}\right)^2}} = \dfrac{1}{\sqrt{D^2 - c^2}}$.

b. $\dfrac{d\theta}{dc}\bigg|_{c=0} = \dfrac{1}{\sqrt{D^2}} = \dfrac{1}{D}$.

7.5.95 $(f^{-1})'(y_0) = \frac{1}{f'(x_0)}$ where $y_0 = f(x_0)$.

$$\frac{d}{dx}\sin^{-1}(x) = \frac{1}{\cos(\sin^{-1}(x))} = \frac{1}{\sqrt{1 - \sin^2(\sin^{-1}(x))}} = \frac{1}{\sqrt{1 - x^2}}.$$

7.5.97 Using the identity $\cot^{-1}(x) + \tan^{-1}(x) = \frac{\pi}{2}$, we have the $\frac{d}{dx}\cot^{-1}(x) + \frac{d}{dx}\tan^{-1}(x) = 0$, so $\frac{d}{dx}\cot^{-1}(x) = -\frac{d}{dx}\tan^{-1}(x)$. Likewise, because $\csc^{-1}(x) + \sec^{-1}(x) = \frac{\pi}{2}$, we have $\frac{d}{dx}\csc^{-1}(x) + \frac{d}{dx}\sec^{-1}(x) = 0$, so $\frac{d}{dx}\csc^{-1}(x) = -\frac{d}{dx}\sec^{-1}(x)$.

7.5.99 $\cos(\sin^{-1}(x)) = \sqrt{1 - \sin^2(\sin^{-1}(x))} = \sqrt{1 - x^2}$ for $-1 \le x \le 1$.

7.5.101 $\tan(2\tan^{-1}(x)) = \frac{2\tan(\tan^{-1}(x))}{1 - \tan^2(\tan^{-1}(x))} = \frac{2x}{1 - x^2}$ for $-1 < x < 1$.

7.6   L'Hôpital's Rule and Growth Rates of Functions

7.6.1 If $\lim\limits_{x \to a} f(x) = 1$ and $\lim\limits_{x \to a} g(x) = \infty$, then $f(x)^{g(x)} \to 1^\infty$ as $x \to a$, which is meaningless; so direct substitution does not work. Note that the fact that the base has limit 1 competes with the fact that the exponent has limit ∞; the former makes the expression tend to 1 while the latter makes the expression tend to ∞.

7.6.3 This means $\lim\limits_{x \to \infty} \dfrac{g(x)}{f(x)} = 0$.

7.6.5 By Theorem 7.14, we have $\ln x, x^3, 2^x, x^x$ in order of increasing growth rates.

7.6.7 L'Hôpital's rule gives $\lim\limits_{x \to 1} \dfrac{\ln x}{4x - x^2 - 3} = \lim\limits_{x \to 1} \dfrac{1/x}{4 - 2x} = \dfrac{1}{2}$.

7.6.9 L'Hôpital's rule gives $\lim\limits_{x \to e} \dfrac{\ln x - 1}{x - e} = \lim\limits_{x \to e} \dfrac{1/x}{1} = \dfrac{1}{e}$.

7.6.11 Apply l'Hôpital's rule twice gives:

$$\lim_{x \to 0} \frac{e^x - \sin x - 1}{x^4 + 8x^3 + 12x^2} = \lim_{x \to 0} \frac{e^x - \cos x}{4x^3 + 24x^2 + 24x} = \lim_{x \to 0} \frac{e^x + \sin x}{12x^2 + 48x + 24} = \frac{1}{24}.$$

7.6.13 Applying l'Hôpital's rule twice gives:

$$\lim_{x \to \infty} \frac{\ln(3x + 5)}{\ln(7x + 3) + 1} = \lim_{x \to \infty} \frac{3/(3x + 5)}{7/(7x + 3)} = \frac{3}{7} \lim_{x \to \infty} \frac{7x + 3}{3x + 5} = \frac{3}{7} \lim_{x \to \infty} \frac{7}{3} = 1.$$

7.6.15 Applying l'Hôpital's rule twice gives:

$$\lim_{x\to\infty}\frac{x^2-\ln(2/x)}{3x^2+2x}=\lim_{x\to\infty}\frac{2x+(1/x)}{6x+2}=\lim_{x\to\infty}\frac{2-1/x^2}{6}=\frac{2}{6}=\frac{1}{3}.$$

7.6.17 Applying l'Hôpital's rule gives:

$$\lim_{x\to0^+}\sin^{-1}x\cot x=\lim_{x\to0^+}\frac{\sin^{-1}x}{\tan x}=\lim_{x\to0^+}\frac{1/\sqrt{1-x^2}}{\sec^2 x}=\lim_{x\to0^+}\frac{\cos^2 x}{\sqrt{1-x^2}}=1.$$

7.6.19 Note that $\ln(1+2x)^{3/x}=\frac{3\ln(1+2x)}{x}$, so we evaluate $L=\lim_{x\to0}\frac{3\ln(1+2x)}{x}=\lim_{x\to0}\frac{\frac{6}{1+2x}}{1}=6$. Thus, $\lim_{x\to0}(1+2x)^{3/x}=e^L=e^6$.

7.6.21 Note that $\ln(\tan\theta)^{\cos\theta}=\cos\theta\ln\tan\theta$, so we evaluate $L=\lim_{\theta\to\pi/2^-}\cos\theta\ln\tan\theta=\lim_{\theta\to\pi/2^-}\frac{\ln\tan\theta}{\sec\theta}$.

L'Hôpital's rule gives $\lim_{\theta\to\pi/2^-}\frac{\ln\tan\theta}{\sec\theta}=\lim_{\theta\to\pi/2^-}\frac{\sec^2\theta/\tan\theta}{\sec\theta\tan\theta}=\lim_{\theta\to\pi/2^-}\frac{\sec\theta}{\tan^2\theta}=\lim_{\theta\to\pi/2^-}\frac{\cos\theta}{\sin^2\theta}=0$, so $\lim_{\theta\to\pi/2^-}(\tan\theta)^{\cos\theta}=e^L=1$.

7.6.23 Note that $\ln(1+x)^{\cot x}=\cot x\ln(1+x)$, so we evaluate $L=\lim_{x\to0^+}\cot x\ln(1+x)=\lim_{x\to0^+}\frac{\ln(1+x)}{\tan x}=\lim_{x\to0^+}\frac{1/(1+x)}{\sec^2 x}=\lim_{x\to0^+}\frac{\cos^2 x}{1+x}=1$ by l'Hôpital's rule. Therefore $\lim_{x\to0^+}(1+x)^{\cot x}=e^L=e$.

7.6.25 Note that $\ln(\tan x)^x=x\ln\tan x$, so we evaluate

$$L=\lim_{x\to0+}x\ln\tan x=\lim_{x\to0+}\frac{\ln\tan x}{1/x}=\lim_{x\to0+}\frac{\sec^2 x/\tan x}{-1/x^2}=-\lim_{x\to0+}\frac{x^2}{\sin x\cos x}$$

by l'Hôpital's rule. Next, observe that $\lim_{x\to0+}\frac{x^2}{\sin x\cos x}=\lim_{x\to0+}\frac{x}{\sin x}\cdot\lim_{x\to0+}\frac{x}{\cos x}=1\cdot0=0$. Therefore $L=0$ and $\lim_{x\to0+}(\tan x)^x=e^L=1$.

7.6.27 Note that $\ln(x+\cos x)^{1/x}=(\ln(x+\cos x))/x$, so we evaluate

$$L=\lim_{x\to0}\frac{\ln(x+\cos x)}{x}=\lim_{x\to0}\frac{(x+\cos x)^{-1}(1-\sin x)}{1}=1$$

by l'Hôpital's rule. Therefore $\lim_{x\to0}(x+\cos x)^{1/x}=e^L=e$.

7.6.29 By Theorem 7.14, $e^{0.01x}$ grows faster than x^{10} as $x\to\infty$.

7.6.31 Note that $\ln x^{20}=20\ln x$, so $\ln x^{20}$ and $\ln x$ have comparable growth rates as $x\to\infty$.

7.6.33 By Theorem 7.14, x^x grows faster than 100^x as $x\to\infty$.

7.6.35 By Theorem 7.14, 1.00001^x grows faster than x^{20} as $x\to\infty$.

7.6.37 Observe that $\lim_{x\to\infty}\frac{(x/2)^x}{x^x}=\lim_{x\to\infty}2^{-x}=0$, so x^x grows faster than $(x/2)^x$ as $x\to\infty$.

7.6.39 Note that $\lim_{x\to\infty}\frac{e^{x^2}}{e^{10x}}=\lim_{x\to\infty}e^{x^2-10x}=\infty$, so e^{x^2} grows faster than e^{10x} as $x\to\infty$.

7.6.41

   a. False. This limit has the form 0^∞ so the limit is 0.

   b. False. This limit has the indeterminate form 1^∞ which is not always 1.

   c. True. $\ln x^{100} = 100 \ln x$ so that $\displaystyle\lim_{x\to\infty} \frac{\ln x^{100}}{\ln x} = 100 < \infty$.

   d. True. Note that $\displaystyle\lim_{x\to\infty} \frac{e^x}{2^x} = \lim_{x\to\infty} \left(\frac{e}{2}\right)^x = \infty$ because $e/2 > 1$.

7.6.43 Note that $\ln\left(\dfrac{\sin x}{x}\right)^{1/x^2} = \dfrac{\ln \sin x - \ln x}{x^2}$, so we evaluate

$$L = \lim_{x\to 0} \frac{\ln \sin x - \ln x}{x^2} = \lim_{x\to 0} \frac{\frac{\cos x}{\sin x} - \frac{1}{x}}{2x} = \lim_{x\to 0} \frac{x \cos x - \sin x}{2x^2 \sin x}$$

by l'Hôpital's rule. Next, observe that

$$\lim_{x\to 0} \frac{x\cos x - \sin x}{2x^2 \sin x} = \lim_{x\to 0} \frac{\cos x - x \sin x - \cos x}{4x \sin x + 2x^2 \cos x} = \lim_{x\to 0} \frac{-\frac{\sin x}{x}}{\frac{4 \sin x}{x} + 2\cos x} = -\frac{1}{6}$$

using l'Hôpital's rule and $\displaystyle\lim_{x\to 0} \sin x/x = 1$. Therefore $\displaystyle\lim_{x\to 0}\left(\frac{\sin x}{x}\right)^{1/x^2} = e^L = e^{-1/6}$.

7.6.45 $x^{1/x} = e^{\ln x/x}$. Now, $\lim_{x\to\infty} \ln x/x$ has the indeterminate form ∞/∞, so using L'Hôpital's rule,

$$\lim_{x\to\infty} \frac{\ln x}{x} = \lim_{x\to\infty} \frac{1}{x} = 0$$

so that $\lim_{x\to\infty} x^{1/x} = e^{\lim_{x\to\infty} \ln x/x} = e^0 = 1$.

7.6.47 $(\sqrt{x-1})^{\sin \pi x} = e^{(\sin \pi x)\ln \sqrt{x-1}} = e^{(\sin \pi x)\ln(x-1)/2}$. As $x \to 1^+$ the exponent has the indeterminate form $0 \cdot \infty$. Using L'Hôpital's rule twice,

$$\lim_{x\to 1^+} \frac{(\sin \pi x)\ln(x-1)}{2} = \lim_{x\to 1^+} \frac{\ln(x-1)}{2/\sin \pi x} = \lim_{x\to 1^+} \frac{-\sin^2 \pi x}{2(x-1)\pi \cos \pi x}$$

$$= -\lim_{x\to 1^+} \frac{2\pi \sin \pi x \cos \pi x}{2\pi(\cos \pi x - (x-1)\sin \pi x)} = -\lim_{x\to 1^+} \frac{\sin \pi x \cos \pi x}{\cos \pi x - (x-1)\sin \pi x} = 0$$

so that $\lim_{x\to 1^+}(\sqrt{x-1})^{\sin \pi x} = e^0 = 1$.

7.6.49

   a. Approximately 3.44 x 10^{15}.

   b. Approximately 3536.

   c. We can explicitly solve for x in this case: $x^{x/100} = e^x \implies x^{1/100} = e \implies x = e^{100}$.

   d. Approximately 163.

7.6.51 Note that $\ln(a^x - b^x)^x = x\ln(a^x - b^x)$, so we evaluate $L = \displaystyle\lim_{x\to 0^+} x\ln(a^x - b^x) = \lim_{x\to 0^+} \frac{\ln(a^x - b^x)}{1/x} = $

$\displaystyle\lim_{x\to 0^+} -x^2\left(\frac{(\ln a)a^x - (\ln b)b^x}{a^x - b^x}\right)$ by l'Hôpital's rule. We have

$$\lim_{x\to 0^+} -x^2\left(\frac{(\ln a)a^x - (\ln b)b^x}{a^x - b^x}\right) = -\lim_{x\to 0^+} x\left((\ln a)a^x - (\ln b)b^x\right)\frac{x}{a^x - b^x}$$

and one more application of l'Hôpital's rule gives $\displaystyle\lim_{x\to 0^+} \frac{x}{a^x - b^x} = \lim_{x\to 0^+} \frac{1}{(\ln a)a^x - (\ln b)b^x} = \frac{1}{\ln a - \ln b}$, so $L = 0$ and therefore $\displaystyle\lim_{x\to 0^+}(a^x - b^x)^x = e^L = 1$.

7.6.53 Apply l'Hôpital's rule: $\lim\limits_{x \to 0} \dfrac{a^x - b^x}{x} = \lim\limits_{x \to 0} \dfrac{(\ln a)a^x - (\ln b)b^x}{1} = \ln a - \ln b.$

7.6.55

    a. Note that if $f(x) = 3^x$ and $a = 0$, then $f(0+h) = f(h) = 3^h$ and $f(a) = f(0) = 3^0 = 1$. Thus, the given expression is $\lim\limits_{h \to 0} \dfrac{f(0+h) - f(0)}{h}$ which is the derivative of $f(x) = 3^x$ at 0. We have $f'(x) = 3^x \ln 3$ so $f'(0) = \ln 3$.

    b. By l'Hôpital's rule, we have
$$\lim\limits_{h \to 0} \frac{3^h - 1}{h} = \lim\limits_{h \to 0} \frac{3^h \ln 3}{1} = \ln 3.$$

7.6.57

    a. After each year the balance increases by the factor $1 + r$; therefore the balance after t years is $B(t) = P(1 + r)^t$.

    b. Observe that $\lim\limits_{m \to \infty} (1 + r/m)^m = \lim\limits_{m \to \infty} \left(1 + \dfrac{1}{m/r}\right)^{(m/r)r} = e^r$, since $\lim\limits_{n \to \infty} \left(1 + \dfrac{1}{n}\right)^n = e$. So with continuous compounding the balance after t years is $B(t) = Pe^{rt}$.

7.6.59 Let $t = b^x$, as in Example 8; then $x = \ln t / \ln b$ and we have $\lim\limits_{x \to \infty} \dfrac{x^p}{b^x} = \lim\limits_{t \to \infty} \dfrac{\ln^p t}{t \ln^p b} = 0$, by Theorem 7.14.

7.6.61 Note that $\log_a x = \ln x / \ln a$, so $\dfrac{\log_a x}{\log_b x} = \dfrac{\ln b}{\ln a}$, and therefore $\log_a x$ and $\log_b x$ grow at a comparable rate as $x \to \infty$.

7.6.63 Note that $\ln\left(1 + \dfrac{a}{x}\right)^x = \dfrac{\ln(1+a/x)}{1/x}$ so we evaluate $L = \lim\limits_{x \to \infty} \dfrac{\ln(1 + a/x)}{1/x} = \lim\limits_{x \to \infty} \dfrac{1}{1 + a/x} \cdot \dfrac{-a}{x^2} \cdot$
$\dfrac{1}{-1/x^2} = \lim\limits_{x \to \infty} \dfrac{a}{1 + a/x} = a$ by l'Hôpital's rule. Therefore $\lim\limits_{x \to \infty} \left(1 + \dfrac{a}{x}\right)^x = e^L = e^a.$

7.6.65

    a. Observe that $\lim\limits_{x \to \infty} \dfrac{b^x}{e^x} = \lim\limits_{x \to \infty} \left(\dfrac{b}{e}\right)^x$. This limit is ∞ exactly when $b > e$.

    b. Observe that $\lim\limits_{x \to \infty} \dfrac{e^{ax}}{e^x} = \lim\limits_{x \to \infty} e^{(a-1)x}$. This limit is ∞ when $a > 1$, is equal to 1 when $a = 1$, and is equal to 0 when $0 < a < 1$.

7.7   Hyperbolic Functions

7.7.1 $\cosh x = \dfrac{e^x + e^{-x}}{2}$; $\sinh x = \dfrac{e^x - e^{-x}}{2}$.

7.7.3 $\cosh^2 x - \sinh^2 x = 1$.

7.7.5 $\sinh^{-1} x = \ln(x + \sqrt{x^2 + 1})$.

7.7.7 Evaluate $\sinh^{-1}(1/5)$, which has the same value.

7.7.9 $\int \dfrac{dx}{16 - x^2} = \dfrac{1}{4}\coth^{-1}(x/4) + C$ for $|x| > 4$; in this case, the values in the interval of integration satisfy $|x| > 4$.

7.7.11 $\tanh x = \dfrac{\sinh x}{\cosh x} = \dfrac{\frac{e^x - e^{-x}}{2}}{\frac{e^x + e^{-x}}{2}} = \dfrac{e^x - e^{-x}}{e^x + e^{-x}} \cdot \dfrac{e^x}{e^x} = \dfrac{e^{2x} - 1}{e^{2x} + 1}.$

7.7.13 $\cosh^2 x + \sinh^2 x = \left(\frac{e^x + e^{-x}}{2}\right)^2 + \left(\frac{e^x - e^{-x}}{2}\right)^2 = \frac{e^{2x} + 2 + e^{-2x}}{4} + \frac{e^{2x} - 2 + e^{-2x}}{4} = \frac{2e^{2x} + 2e^{-2x}}{4} = \frac{e^{2x} + e^{-2x}}{2} = \cosh(2x)$.

7.7.15 $\cosh x + \sinh x = \frac{e^x + e^{-x}}{2} + \frac{e^x - e^{-x}}{2} = \frac{2e^x}{2} = e^x$.

7.7.17 $\frac{1 + \cosh 2x}{2} = \frac{1 + \cosh^2 x + \sinh^2 x}{2} = \frac{\cosh^2 x + (\sinh^2 x + 1)}{2} = \frac{\cosh^2 x + \cosh^2 x}{2} = \cosh^2 x$.

$\frac{\cosh 2x - 1}{2} = \frac{\cosh^2 x + \sinh^2 x - 1}{2} = \frac{\sinh^2 x + (\cosh^2 x - 1)}{2} = \frac{\sinh^2 x + \sinh^2 x}{2} = \sinh^2 x$.

7.7.19 $\frac{d}{dx}(\coth x) = \frac{d}{dx} \frac{\cosh x}{\sinh x} = \frac{\sinh^2 x - \cosh^2 x}{\sinh^2 x} = \frac{-1}{\sinh^2 x} = -\operatorname{csch}^2 x$.

7.7.21 $\frac{d}{dx}(\operatorname{csch} x) = \frac{d}{dx} \frac{1}{\sinh x} = \frac{\sinh x \cdot 0 - \cosh x}{\sinh^2 x} = -\operatorname{csch} x \coth x$.

7.7.23 $y' = 2(\cosh x)(\sinh x)$.

7.7.25 $y' = 2(\tanh x)(\operatorname{sech}^2 x)$.

7.7.27 $y' = \frac{1}{\operatorname{sech} 2x}(-\operatorname{sech} 2x \tanh 2x)(2) = -2 \tanh 2x$.

7.7.29 $y' = 2x \cosh^2 3x + x^2 (2 \cosh(3x) \sinh(3x)(3)) = 2x \cosh(3x)(\cosh(3x) + 3x \sinh(3x))$.

7.7.31 $\int \cosh 2x \, dx = \frac{1}{2} \sinh 2x + C$.

7.7.33 Let $u = 1 + \cosh x$ so that $du = \sinh x \, dx$. Substituting gives $\int \frac{1}{u} du = \ln|u| + C = \ln|1 + \cosh x| + C$.

7.7.35 Recall that $\tanh^2 x = 1 - \operatorname{sech}^2 x$. So we have $\int (1 - \operatorname{sech}^2 x) \, dx = x - \tanh x + C$.

7.7.37 Let $u = \cosh 3x$ so that $du = 3 \sinh 3x \, dx$. Substituting gives $\frac{1}{3} \int_1^{\cosh 3} u^3 \, du = \left. (u^4/12) \right|_1^{\cosh 3} = \cosh^4 3/12 - 1/12 \approx 856.034$.

7.7.39 $\int_0^{\ln 2} \tanh x \, dx = \int_0^{\ln 2} \frac{\sinh x}{\cosh x} \, dx$. Let $u = \cosh x$ so that $du = \sinh x \, dx$. Substituting gives $\int_1^{5/4} \frac{1}{u} du = \left. (\ln u) \right|_1^{5/4} = \ln(5/4)$.

7.7.41

a. Note that $\sinh \ln x = \frac{e^{\ln x} - e^{-\ln x}}{2} = \frac{x - 1/x}{2} = x/2 - 1/(2x)$. So $\int \frac{\sinh \ln x}{x} = \int (1/2 - x^{-2}/2) \, dx = x/2 + x^{-1}/2 + C = \frac{x^2 + 1}{2x} + C$.

b. Let $u = \ln x$, so that $du = \frac{1}{x} dx$. Substituting gives $\int \sinh u \, du = \cosh u + C = \cosh \ln x + C = \frac{e^{\ln x} + e^{-\ln x}}{2} + C = \frac{x + 1/x}{2} + C = \frac{x^2 + 1}{2x} + C$.

7.7.43

a. The values of $\coth x$ on the interval $[5, 10]$ are very close to 1. So the area under the curve is very close to that of a 5×1 rectangle.

b. $\int_5^{10} \coth x \, dx = \int_5^{10} \frac{\cosh x}{\sinh x} \, dx = \left. (\ln|\sinh x|) \right|_5^{10} = \ln \sinh 10 - \ln \sinh 5 \approx 5.0000454$. The absolute value of the error is about 0.0000454.

7.7.45

a. The curves intersect when $\frac{1}{\cosh x} = \frac{\sinh x}{\cosh x}$, or when $\sinh x = 1$, so at $x = \sinh^{-1}(1) = \ln(1 + \sqrt{2})$.

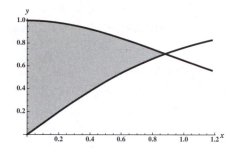

b. We need to compute $A = \int_0^{\ln(1+\sqrt{2})} \operatorname{sech} x - \tanh x\, dx$. Recall that $\int \operatorname{sech} x\, dx = \tan^{-1}|\sinh x| + C$ and $\int \tanh x\, dx = \ln \cosh x + C$. Thus, $A = \left(\tan^{-1}\sinh x - \ln \cosh x\right)\big|_0^{\ln(1+\sqrt{2})} = \tan^{-1}\sinh(\ln(1 + \sqrt{2})) - \ln \cosh(\ln(1 + \sqrt{2})) - (0 - 0) \approx 0.44$.

7.7.47 $f'(x) = \frac{1}{\sqrt{16x^2-1}} \cdot 4 = \frac{4}{\sqrt{16x^2-1}}$.

7.7.49 $f'(v) = \frac{1}{\sqrt{v^4+1}} \cdot (2v) = \frac{2v}{\sqrt{v^4+1}}$.

7.7.51 $f'(x) = \sinh^{-1}(x) + \frac{x}{\sqrt{x^2+1}} - \frac{x}{\sqrt{x^2+1}} = \sinh^{-1}(x)$.

7.7.53 We have $\frac{1}{8}\int \frac{dx}{1-(x/\sqrt{8})^2}\, dx$. Let $u = x/\sqrt{8}$ so that $du = \frac{dx}{\sqrt{8}}$. Substituting gives $\frac{1}{\sqrt{8}}\int \frac{du}{1-u^2} = \frac{1}{\sqrt{8}}\coth^{-1}(u) + C = \frac{1}{\sqrt{8}}\coth^{-1}(x/\sqrt{8}) + C$.

7.7.55 We have $\frac{1}{36}\int \frac{e^x}{1-(e^x/6)^2}\, dx$.

Let $u = e^x/6$ so that $du = e^x/6\, dx$. Substituting gives $\frac{1}{6}\int \frac{du}{1-u^2} = \frac{1}{6}\tanh^{-1}(u) + C = \frac{1}{6}\tanh^{-1}(e^x/6) + C$.

7.7.57 We have $\frac{1}{2}\int \frac{dx}{x\sqrt{1-(x^4/2)^2}}$. Let $u = x^4/2$ so that $du = 2x^3\, dx$. Substituting gives $\frac{1}{2}\int \frac{du}{4u\sqrt{1-u^2}} = \frac{-1}{8}\operatorname{sech}^{-1}u + C = \frac{-1}{8}\operatorname{sech}^{-1}(x^4/2) + C$.

7.7.59 Let $u = \ln x$ so that $du = \frac{dx}{x}$. Substituting gives $\int_0^2 \frac{du}{\sqrt{u^2+1}} = \left(\sinh^{-1} u\right)\big|_0^2 = \sinh^{-1}(2) - 0 = \ln(2 + \sqrt{5})$.

7.7.61 $\int_{-2}^2 \frac{dt}{t^2-3^2} = \frac{-1}{3}\left(\tanh^{-1}(x/3)\right)\big|_{-2}^2 = \frac{-1}{3}\left(\tanh^{-1}(2/3) - \tanh^{-1}(-2/3)\right) = \frac{-1}{3}\left(\frac{1}{2}\ln((5/3)/(1/3)) - \frac{1}{2}\ln((1/3)/(5/3))\right) = \frac{-1}{3}\ln 5 \approx -0.54$.

7.7.63 Let $u = x^{1/3}$ so that $du = \frac{1}{3}x^{-2/3}\, dx$. Substituting gives $\int_{1/2}^1 \frac{3u^2\, du}{u^3\sqrt{1+u^2}} = 3\int_{1/2}^1 \frac{du}{u\sqrt{1+u^2}} = -3\left(\operatorname{csch}^{-1}x\right)\big|_{1/2}^1 = 3(\operatorname{csch}^{-1}(1/2) - \operatorname{csch}^{-1}(1)) = 3(\sinh^{-1}(2) - \sinh^{-1}(1)) = 3(\ln(2+\sqrt{5}) - \ln(1+\sqrt{2})) \approx 1.69$.

7.7.65 By symmetry, we compute $I = 2\int_0^{\cosh^{-1}(17/15)}(17/15 - \cosh x)\, dx$ and divide by $2\cosh^{-1}(17/15)$. The value of I is $2\left((17/15)x - \sinh x\right)\big|_0^{\cosh^{-1}(17/15)} = 2\left((17/15)\cosh^{-1}(17/15) - \sinh(\cosh^{-1}(17/15))\right) = (34/15)\cosh^{-1}(17/15) - 16/15$. When we divide by $2\cosh^{-1}(17/15)$ we obtain $17/15 - \frac{8}{15\cosh^{-1}(17/15)} \approx 0.09$.

7.7.67

a. Let a be the value that produces a sag of 10 feet. Note that $f(0) + \text{sag} = f(50)$, so $a + \text{sag} = a\cosh(50/a)$, and dividing by a yields $1 + \frac{\text{sag}}{a} = \cosh(50/a)$, which can be written as $\cosh(50/a) - 1 = 10/a$.

b. If $t = 10/a$, we have $\cosh(5t) - 1 = t$. Solving for t yields $t \approx 0.08$.

c. If $0.08 = 10/a$, then $a = 10/0.08 = 125$. The length of the power line is $2\int_0^{50} \sqrt{1 + \sinh^2(x/125)}\, dx = 2\int_0^{50}\cosh(x/125)\, dx = 250\left(\sinh(x/125)\right)\big|_0^{50} = 250\sinh(50/125) = 250\sinh(2/5) \approx 102.7$ feet.

7.7.69 We are seeking λ so that $7 = \sqrt{\frac{9.8\lambda}{2\pi} \tanh\left(\frac{20\pi}{\lambda}\right)}$. A computer algebra system finds the approximate root to be 32.81 meters.

7.7.71

a. $f'(x) = \operatorname{sech}^2(x)$, so $f'(0) = 1$. The point of tangency is $(0, \tanh 0) = (0, 0)$, so the linearization is $L(x) = 0 + 1(x - 0) = x$.

b. When $d/\lambda < 0.05$, $2\pi d/\lambda$ is small. Because $\tanh x \approx x$ for small values of x, $\tanh(2\pi d/\lambda) \approx 2\pi d/\lambda$. Thus,

$$v = \sqrt{\frac{g\lambda}{2\pi} \tanh\left(\frac{2\pi d}{\lambda}\right)} \approx \sqrt{\frac{g\lambda}{2\pi} \cdot \frac{2\pi d}{\lambda}} = \sqrt{gd}.$$

c. $v = \sqrt{gd}$ is a function of depth only. As $d \to 0$, $v \to 0$ as well.

7.7.73

a. False. $\sinh(\ln 3)$ is a constant, so its derivative is zero.

b. False. $\frac{d}{dx} \cosh x = \sinh x$, not $-\sinh x$.

c. False. This velocity equation is not a function of time.

d. True. Note that $-\ln(\sqrt{2} - 1) = \ln\left(\frac{1}{\sqrt{2}-1}\right) = \ln\left(\frac{1}{\sqrt{2}-1} \cdot \frac{\sqrt{2}+1}{\sqrt{2}+1}\right) = \ln(\sqrt{2} + 1)$.

e. False. $\int \frac{dx}{2^2 - x^2} = \frac{1}{2} \tanh^{-1}(x/2)$ for $|x| < 2$.

7.7.75

a. $\cosh(0) = 1$

b. $\tanh(0) = 0$

c. Does not exist

d. $\operatorname{sech}(0) = 1$

e. $\coth(\ln 5) = \frac{5 + 1/5}{5 - 1/5} = \frac{13}{12}$

f. $\sinh(\ln 9) = \frac{9 - 1/9}{2} = \frac{40}{9}$

g. $\left(\frac{e + (1/e)}{2}\right)^2 = \left(\frac{e^2 + 1}{2e}\right)^2$

h. Does not exist

i. $\ln(17/8 + \sqrt{\frac{17^2 - 8^2}{8^2}}) = \ln 4$

j. $\sinh^{-1}\left(\frac{e^1 - e^{-1}}{2}\right) = 1$

7.7.77 $f'(x) = 2\sinh x \cosh x \cosh x + \sinh^2 x \sinh x = \sinh x(2\cosh^2 x + \sinh^2 x)$. Note that $\sinh x = 0$ for $x = 0$, while $2\cosh^2 x + \sinh^2 x > 0$ for all x. So the only critical point is $x = 0$.

7.7.79 $f'(x) = 2\tanh x \operatorname{sech}^2 x = \frac{2\sinh x}{\cosh^3 x}$. $f''(x) = \frac{2\cosh^4 x - 2(\sinh x)(3\cosh^2 x)(\sinh x)}{\cosh^6 x} = \frac{2\cosh^2 x - 6\sinh^2 x}{\cosh^4 x}$. This is zero for $\tanh^2 x = 1/3$, or $x = \tanh^{-1}(\pm\sqrt{1/3}) = \pm\tanh^{-1}(\sqrt{1/3})$.

7.7.81 This area would be the area under $\operatorname{sech} x$ between 0 and 1, minus the area of a quarter circle. Thus we have $\int_0^1 \operatorname{sech} x\, dx - \frac{\pi}{4} = \left(\tan^{-1}|\sinh x|\right)\big|_0^1 - \frac{\pi}{4} = \tan^{-1}(\sinh(1)) - \frac{\pi}{4} \approx 0.08$.

7.7.83 Applying L'Hôpital's rule twice brings you back to the initial limit. If we write $\frac{\sinh x}{\cosh x} = \frac{(\sinh x)e^x}{(\cosh x)e^x} = \frac{e^{2x} - 1}{e^{2x} + 1}$, then applying L'Hôpital's rule once shows that the limit is 1.

7.7.85 $\lim_{x \to 0} \frac{\tanh^{-1} x}{\tan(\pi x/2)} = \lim_{x \to 0} \frac{\frac{1}{1 - x^2}}{(\pi/2)\sec^2(\pi x/2)} = \frac{2}{\pi} \lim_{x \to 0} \frac{\cos^2(\pi x/2)}{1 - x^2} = \frac{2}{\pi}$.

7.7.87 Let $y = (\tanh x)^x$. Then $\ln y = x\ln(\tanh x) = \frac{\ln \tanh x}{1/x}$. Note that $\lim_{x \to 0} \frac{\ln \tanh x}{1/x} = \lim_{x \to 0} \frac{1/(\sinh x \cosh x)}{-1/x^2} = \lim_{x \to 0} \frac{-x^2}{\sinh x \cosh x} = \lim_{x \to 0} \frac{-2x}{\sinh^2 x + \cosh^2 x} = 0$. Thus, $\lim_{x \to 0} y = \lim_{x \to 0}(\tanh x)^x = e^0 = 1$.

7.7.89 $\int \frac{\cosh z}{\sinh^2 z}\,dz = \int \coth z \csch z \, dx = -\csch z + C.$

7.7.91 Let $u = \sinh^{-1} x$ so that $du = \frac{dx}{\sqrt{x^2+1}}$. Substituting gives $\int_{\sinh^{-1}(5/12)}^{\sinh^{-1}(3/4)} u\,du = \left. \left(u^2/2 \right) \right|_{\sinh^{-1}(5/12)}^{\sinh^{-1}(3/4)} = $
$\frac{1}{2}\left(\left(\sinh^{-1}(3/4) \right)^2 - \left(\sinh^{-1}(5/12) \right)^2 \right) \approx 0.158.$

7.7.93 Let A be the area under $3 - \cosh x$ between $\cosh^{-1}(-3)$ and $\cosh^{-1}(3)$. Then the volume of the kiln will be $6A$. By symmetry, $A = 2 \int_0^{\cosh^{-1}(3)} (3 - \cosh x)\,dx = 2\left. (3x - \sinh x) \right|_0^{\cosh^{-1}(3)} = 6\cosh^{-1}(3) - 2\sinh(\cosh^{-1}(3)) \approx 4.92$. So the volume of the kiln is about $6 \cdot 4.92 \approx 29.5$

7.7.95

a. $d(10) = \frac{75}{.2} \ln\left(\cosh\left(\sqrt{\frac{(.2)(9.8)}{75}} \cdot 10 \right) \right) \approx 360.8.$

b. We need to solve $100 = \frac{75}{.2} \ln\left(\cosh\left(\sqrt{\frac{(.2)(9.8)}{75}} \cdot t \right) \right)$ for t. We have $t = \cosh^{-1}(e^{\frac{100 \cdot .2}{75}}) \cdot \sqrt{\frac{75}{(.2)(9.8)}} \approx$
   4.72 seconds. The first 200 feet take $t = \cosh^{-1}(e^{\frac{200 \cdot .2}{75}}) \cdot \sqrt{\frac{75}{(.2)(9.8)}} \approx 6.97$ seconds, which means that the second 100 feet took about $6.97 - 4.72 = 2.25$ seconds. The average velocity over the first 100 feet is $\frac{100}{4.72} \approx 21.2$ meters per second, while over the second 100 feet is $\frac{100}{2.25} \approx 44.5$ meters per second.

7.7.97

a. $\sqrt{\frac{mg}{k}} \lim_{t \to \infty} \tanh\left(\sqrt{\frac{kg}{m}} t \right) = \sqrt{\frac{mg}{k}} \cdot 1 = \sqrt{\frac{mg}{k}}.$

b. $\sqrt{\frac{(75)(9.8)}{.2}} \approx 60.6$ meters per second.

c. We seek t so that $v(t) = \sqrt{\frac{mg}{k}} \tanh\left(\sqrt{\frac{kg}{m}} t \right) = .95 \sqrt{\frac{mg}{k}}.$

   So $\tanh\left(\sqrt{\frac{kg}{m}} t \right) = .95$, and $t = \tanh^{-1}(.95) \sqrt{\frac{m}{kg}}.$

d. It takes $\tanh^{-1}(.95) \sqrt{\frac{75}{(.2)(9.8)}} \approx 11.33$ seconds to achieve 95 percent of terminal velocity. It takes $d(11.33) \approx 436.5$ feet to achieve this. So the cliff should be at least $436.5 + 300 = 736.5$ feet high.

7.7.99 For $y = A \sinh kx$, we have $y' = kA \cosh kx$ and $y'' = k^2 A \sinh kx$. So $y'' - k^2 y = k^2 A \sinh kx - k^2 A \sinh kx = 0$.
   For $y = B \cosh kx$, we have $y' = kB \sinh kx$ and $y'' = k^2 B \cosh kx$. So $y'' - k^2 y = k^2 B \cosh kx - k^2 B \cosh kx = 0$.

7.7.101

$$\sinh(\cosh^{-1}(x)) = \sinh(\ln(x + \sqrt{x^2 - 1})) = \frac{e^{\ln(x+\sqrt{x^2-1})} - \frac{1}{e^{\ln(x+\sqrt{x^2-1})}}}{2}$$

$$= \frac{1}{2}\left(x + \sqrt{x^2 - 1} - \frac{1}{x + \sqrt{x^2 - 1}} \right).$$

Multiplying the last term by $\frac{x - \sqrt{x^2-1}}{x - \sqrt{x^2-1}}$ gives

$$\frac{1}{2}\left((x + \sqrt{x^2 - 1}) - (x - \sqrt{x^2 - 1}) \right) = \sqrt{x^2 - 1}.$$

7.7.103 $\cosh(x+y) = \frac{e^{x+y}+e^{-(x+y)}}{2} = \frac{2e^{x+y}+2e^{-(x+y)}}{4}$. This can be written as

$$\frac{1}{4}\left(e^{x+y} + e^{x-y} + e^{y-x} + e^{-(x+y)} + e^{x+y} - e^{x-y} - e^{y-x} + e^{-(x+y)}\right)$$

$$= \left(\frac{e^x + e^{-x}}{2}\right)\left(\frac{e^y + e^{-y}}{2}\right) + \left(\frac{e^x - e^{-x}}{2}\right)\left(\frac{e^y - e^{-y}}{2}\right)$$

$$= \cosh x \cosh y + \sinh x \sinh y$$

7.7.105 First suppose $x \geq 0$. Then $\cosh^{-1}(\cosh x) = \ln(\cosh x + \sqrt{1+\cosh^2 x}) = \ln(\cosh x + \sqrt{\sinh^2 x}) = \ln(\cosh x + \sinh x) = \ln e^x = x$.

Now suppose $x < 0$. Then $\cosh^{-1}(\cosh x) = \ln(\cosh x + \sqrt{1+\cosh^2 x}) = \ln(\cosh x + \sqrt{\sinh^2 x}) = \ln(\cosh x - \sinh x) = \ln e^{-x} = -x$.

Thus $\cosh^{-1}(\cosh x) = |x|$.

7.7.107

a. $\int \operatorname{sech} x \, dx = \int \frac{\cosh x}{1+\sinh^2 x} \, dx$. Let $u = \sinh x$ so that $du = \cosh x \, dx$. Then we have $\int \frac{1}{1+u^2} \, du = \tan^{-1}(u) + C = \tan^{-1}(\sinh x) + C$.

b. Note that $\operatorname{sech} x = \frac{\operatorname{sech}^2 x}{\sqrt{\operatorname{sech}^2 x}} = \frac{\operatorname{sech}^2 x}{\sqrt{1-\tanh^2 x}}$. So $\int \operatorname{sech} x \, dx = \int \frac{\operatorname{sech}^2 x}{\sqrt{1-\tanh^2 x}} \, dx$. Let $u = \tanh x$ so that $du = \operatorname{sech}^2 x \, dx$. Substituting gives $\int \frac{1}{\sqrt{1-u^2}} \, du = \sin^{-1}(u) + C = \sin^{-1}(\tanh x) + C$.

c. $\frac{d}{dx}(2\tan^{-1}(e^x) + C) = 2\left(\frac{1}{1+e^{2x}}\right)e^x + 0 = \frac{2e^x}{1+e^{2x}} \cdot \frac{e^{-x}}{e^{-x}} = \frac{2}{e^{-x}+e^x} = \frac{1}{\cosh x} = \operatorname{sech} x$.

7.7.109 $L = \int_{\ln 2}^{\ln 8} \sqrt{1 + \operatorname{csch}^x} \, dx = \int_{\ln 2}^{\ln 8} \sqrt{\coth^2 x} \, dx = \int_{\ln 2}^{\ln 8} \coth x \, dx = \int_{\ln 2}^{\ln 8} \frac{\cosh x}{\sinh x} \, dx$. Let $u = \sinh x$ so that $du = \cosh x \, dx$. Substituting gives $\int_{3/4}^{63/16} \frac{1}{u} \, du = (\ln u)\big|_{3/4}^{63/16} = \ln(63/16) - \ln(3/4) = \ln(21/4) \approx 1.66$.

7.7.111 Suppose $r > 0$ and $0 < x < 1$. If $u = x^r$, then $du = rx^{r-1} \, dx$. Substituting gives $\int \frac{du}{ru\sqrt{1-u^2}} = \frac{-1}{r}(\operatorname{sech}^{-1}(u)) + C = \frac{-1}{r}(\operatorname{sech}^{-1}(x^r)) + C$.

Chapter Seven Review

1

a. True. $y = 2/x$; solving for x gives $x = 2/y$ so that $f^{-1}(x) = 2/x$.

b. False. $\ln(xy) = \ln x + \ln y$.

c. True. $\frac{d}{dx}(e^x) = e^x$, which is never zero.

d. False, but it's close. $\dfrac{Ae^{0.1(t+1)}}{Ae^{0.1t}} = e^{0.1} \approx 1.1052 \neq 1.1$.

e. True. $\dfrac{d}{dx}(b^x) = b^x \ln b$, so $b^x = b^x \ln b$ only when $\ln b = 1$ so that $b = e$.

f. False. It is of the form 0^∞, which is not indeterminate. Looking at the limit, the base approaches zero while the exponent grows large, so clearly the power approaches zero.

g. False. That is the *range*. The domain of $\tan^{-1} x$ is $\mathbb{R}$.

h. False. $\pi^{3x} = e^{3x \ln \pi}$.

i. True. $\sinh(\ln x) = \frac{e^{\ln x} - e^{-\ln x}}{2} = \frac{x - (1/x)}{2} = \frac{x^2 - 1}{2x}$.

3 If $\log x^2 + 3\log x = \log 32$, then $\log(x^2 \cdot x^3) = \log(32)$, so $x^5 = 32$ and $x = 2$. The answer does not depend on the base of the log.

5

By graphing, it is clear that this function is not one-to-one on its whole domain, but it is one-to-one on the interval $(-\infty, 0]$, on the interval $[0, 2]$, and on the interval $[2, \infty)$, so it would have an inverse if we restricted it to any of these particular intervals.

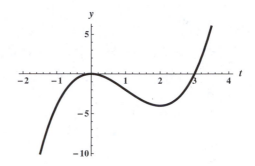

7

Completing the square gives $f(x) = x^2 - 4x + 4 + 1 = (x-2)^2 + 1$. Switching the x and y and solving for y yields $(y-2)^2 = x - 1$, so $|y-2| = \sqrt{x-1}$, and thus $y = 2 + \sqrt{x-1}$ (we choose the "+" rather than the "−" because the domain of f is $x > 2$, so the range of f^{-1} must also consist of numbers greater than 2.)

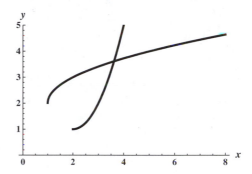

9 $\left. (f^{-1}(x))' \right|_{x=f(\pi/4)} = \dfrac{1}{f'(\pi/4)} = \dfrac{1}{-\sin(\pi/4)} = -\sqrt{2}$

11 Note that $f(2) = 3$. $\left. (f^{-1}(x))' \right|_{y=3} = \dfrac{1}{f'(2)} = \dfrac{2\sqrt{2^3 + 2 - 1}}{3(2)^2 + 1} = \dfrac{6}{13}$.

13 Let $x = 6 - 3y$. then $3y = 6 - x$ and $y = 2 - \frac{x}{3}$. We have $f^{-1}(x) = 2 - \frac{x}{3}$ and $(f^{-1})'(x) = -\frac{1}{3}$.

15 Note that $y \geq 0$ in the given function when $x \geq \frac{1}{4}$. Let $x = \sqrt{4y-1}$. Then $4y - 1 = x^2$, $4y = x^2 + 1$, and $y = \frac{x^2+1}{4}$. We have $f^{-1}(x) = \frac{x^2+1}{4}$, $x \geq 0$, and $(f^{-1})'(x) = \frac{x}{2}$, $x \geq 0$.

17

  a. $(f^{-1})'\left(\frac{1}{\sqrt{2}}\right) = \dfrac{1}{f'\left(\frac{\pi}{4}\right)} = \dfrac{1}{\cos\left(\frac{\pi}{4}\right)} = \sqrt{2}$.

  b. $\left. \dfrac{d}{dx}\sin^{-1}(x) \right|_{x=1/\sqrt{2}} = \dfrac{1}{\sqrt{1-(1/2)}} = \dfrac{1}{\sqrt{1/2}} = \sqrt{2}$.

19 $f'(x) = \ln^2 x + x \cdot 2\ln x \cdot \left(\frac{1}{x}\right) = \ln x \cdot (\ln x + 2)$.

21 $f'(x) = 2^{x^2-x} \cdot \ln 2 \cdot (2x - 1)$.

23 $f'(x) = \dfrac{1}{\sqrt{1 - \left(\frac{1}{x}\right)^2}} \cdot -\dfrac{1}{x^2} = -\dfrac{1}{|x|\sqrt{x^2-1}}$.

25 $\frac{d}{dx}x^{1/x} = \frac{d}{dx}e^{\frac{\ln x}{x}} = e^{\frac{\ln x}{x}} \cdot \left(\frac{1-\ln x}{x^2}\right) = x^{1/x}\left(\frac{1-\ln x}{x^2}\right)$. So $\left. \frac{d}{dx}x^{1/x} \right|_{x=1} = 1 \cdot \frac{1-0}{1^2} = 1$.

27 $f'(x) = \sec^{-1} x + \frac{1}{\sqrt{x^2-1}}$. So $f'(2/\sqrt{3}) = \frac{\pi}{6} + \sqrt{3}$.

29 $\frac{d}{dx}(x^2 \sinh x) = 2x \sinh x + x^2 \cosh x$.

31 Since $\cos(\pi/6) = \sqrt{3}/2$, $\cos^{-1}(\sqrt{3}/2) = \pi/6$.

33 Since $\sin(-\pi/2) = -1$, $\sin^{-1}(-1) = -\pi/2$.

35 $\sin(\sin^{-1}(x)) = x$, for all x in the domain of the inverse sine function.

37 If $\theta = \sin^{-1}(12/13)$, then $0 < \theta < \pi/2$, and $\sin \theta = 12/13$. Then (using the Pythagorean identity) we can deduce that $\cos \theta = 5/13$. It must follow that $\tan \theta = 12/5$, $\cot \theta = 5/12$, $\sec \theta = 13/5$, and $\csc \theta = 13/12$.

39

Note that
$$\tan \theta = \frac{a}{b} = \cot(\pi/2 - \theta).$$

Thus, $\cot^{-1}(\tan \theta) = \cot^{-1}(\cot(\pi/2 - \theta)) = \pi/2 - \theta$.

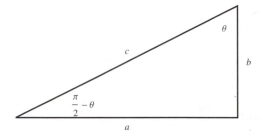

41 Let $\theta = \sin^{-1}(x)$. Then $\sin \theta = x$ and note that then $\sin(-\theta) = -\sin \theta = -x$, so $-\theta = \sin^{-1}(-x)$. Then $\sin^{-1}(x) + \sin^{-1}(-x) = \theta + -\theta = 0$.

43

Let $\theta = \sin^{-1}(2x - 1)$. Then $\sin \theta = 2x - 1$ and

$$\cot \theta = \frac{\text{side adjacent to } \theta}{\text{side opposite } \theta} = \frac{\sqrt{4x - 4x^2}}{2x - 1} = \frac{2\sqrt{x(1-x)}}{2x - 1}.$$

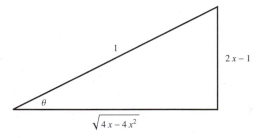

45 Let $u = \ln x$ so that $du = \frac{1}{x} dx$. Then we have $\int_2^8 \frac{1}{u} du = (\ln u)|_2^8 = \ln 4$.

47 Let $u = x^2 + 8x + 25$ so that $du = (2x + 8) dx = 2(x + 4) dx$. Then we have $\frac{1}{2} \int \frac{1}{u} du = \frac{1}{2} \ln|u| + C = \frac{1}{2} \ln|x^2 + 8x + 25| + C$.

49
$$\int \frac{3}{2x^2 + 1} dx = \frac{3}{2} \int \frac{1}{x^2 + 1/2} dx = \frac{3\sqrt{2}}{2} \tan^{-1}(\sqrt{2}x) + C$$

51
$$\int_{-1}^{2\sqrt{3}-1} \frac{dx}{x^2 + 2x + 5} = \int_{-1}^{2\sqrt{3}-1} \frac{dx}{(x+1)^2 + 4} = \frac{1}{2} \tan^{-1}\left(\frac{x+1}{2}\right)\Bigg|_{-1}^{2\sqrt{3}-1} = \frac{\pi}{6}$$

53
$$\int_{-2/3}^{2/\sqrt{3}} \frac{dx}{\sqrt{16 - 9x^2}} = \frac{1}{3} \int_{-2/3}^{2/\sqrt{3}} \frac{dx}{\sqrt{16/9 - x^2}} = \frac{1}{3} \sin^{-1}\left(\frac{3x}{4}\right)\Bigg|_{-2/3}^{2/\sqrt{3}}$$
$$= \frac{1}{3}\left(\sin^{-1}(\sqrt{3}/2) - \sin^{-1}(-1/2)\right) = \frac{\pi}{6}$$

55 $\int \frac{dx}{\sqrt{x^2-9}} = \cosh^{-1}(x/3) + C = \ln(x + \sqrt{x^2-9}) + C.$

57 Let $u = x^3$ so that $du = 3x^2\,dx$. Substituting gives $\frac{1}{3}\int_0^1 \frac{du}{3^2 - u^2} = \frac{1}{9}\left(\tanh^{-1}(u/3)\right)\big|_0^1 = \frac{1}{9}\tanh^{-1}(1/3) \approx 0.0385.$

59 $f'(x) = \dfrac{3}{x} - \dfrac{x}{12}$; then

$$1 + f'(x)^2 = 1 + \left(\frac{3}{x} - \frac{x}{12}\right)^2 = 1 + \frac{9}{x^2} + \frac{x^2}{144} - \frac{1}{2} = \left(\frac{3}{x} + \frac{x}{12}\right)^2$$

so that

$$\int_1^6 \sqrt{1 + \left(\frac{3}{x} - \frac{x}{12}\right)^2}\,dx = \int_1^6 \frac{3}{x} + \frac{x}{12}\,dx = \left(3\ln x + \frac{x^2}{24}\right)\bigg|_1^6 = 3\ln 6 + \frac{35}{24}$$

61 $f'(y) = 2\sqrt{2}e^{\sqrt{2}y} - \frac{\sqrt{2}}{16}e^{-\sqrt{2}y}$; then

$$1 + f'(y)^2 = 1 + 8e^{2\sqrt{2}y} - \frac{1}{2} + \frac{1}{128}e^{-2\sqrt{2}y} = \left(2\sqrt{2}e^{\sqrt{2}y} + \frac{\sqrt{2}}{16}e^{-\sqrt{2}y}\right)^2$$

so that

$$\int_0^{(\ln 2)/\sqrt{2}} \sqrt{1 + f'(y)^2}\,dy = \int_0^{(\ln 2)/\sqrt{2}} \left(2\sqrt{2}e^{\sqrt{2}y} + \frac{\sqrt{2}}{16}e^{-\sqrt{2}y}\right)\,dy$$

$$= 2e^{\sqrt{2}y} - \frac{1}{16}e^{-\sqrt{2}y}\bigg|_0^{(\ln 2)/\sqrt{2}} = \frac{65}{32}$$

63 Using the shell method, the volume is

$$\int_0^4 \frac{4\pi x}{x^2 + 1}\,dx = 2\pi \ln(x^2 + 1)\bigg|_0^4 = 2\pi \ln 17$$

65

   a.

Tom's position function is given by

$$\int_0^t 20e^{-2x}\,dx = \left(-10e^{-2x}\right)\big|_0^t$$

$$= 10(1 - e^{-2t}).$$

Sue's position is given by

$$\int_0^t 15e^{-x}\,dx = \left(-15e^{-x}\right)\big|_0^t$$

$$= 15(1 - e^{-t}).$$

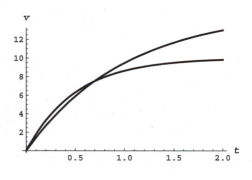

   b. We are looking for t so that $10(1 - e^{-2t}) = 15(1 - e^{-t})$, which occurs when $10(e^{-t})^2 - 15e^{-t} + 5 = 0$, or $2u^2 - 3u + 1 = 0$ where $u = e^{-t}$. This quadratic factors as $(2u - 1)(u - 1)$, so if $e^{-t} = 1$ or $e^{-t} = 1/2$, so $t = 0$ or $t = \ln 2$.

   c. Sue takes the lead and doesn't relinquish it at $t = \ln 2$.

67 Growth is modeled by $p(t) = 150,000e^{kt}$ where $k = \ln(1.04)$. The population reaches 1,000,000 when $1,000,000 = 150,000e^{\ln(1.04)t}$, or when $\ln(20/3) = \ln(1.04)t$, so when $t = \frac{\ln(20/3)}{\ln(1.04)} \approx 48.37$ years.

69

The domain of f is the set of all real numbers. Note that $\lim_{x\to\infty} f(x) = \infty$ and $\lim_{x\to-\infty} f(x) = 0$. $f'(x) = e^x(2x-1)+(x^2-x)e^x = e^x(x^2+x-1)$, which is 0 for $x = \frac{-1\pm\sqrt{5}}{2}$. Note that $f'(x) > 0$ on the interval $(-\infty, (-1-\sqrt{5})/2)$ and on $((-1+\sqrt{5})/2, \infty)$, so f is increasing on those intervals, while $f'(x) < 0$ (and so f is decreasing) on $((-1-\sqrt{5})/2, (-1+\sqrt{5})/2)$. There is a local maximum at $(-1-\sqrt{5})/2$ and a local minimum at $(-1+\sqrt{5})/2$. Note that $f''(x) = e^x(2x+1)+(x^2+x-1)e^x = e^x(x^2+3x)$, which is 0 for $x = 0$ and $x = -3$. $f''(x) > 0$ on $(-\infty, -3)$ and on $(0, \infty)$, so f is concave up on those intervals, while $f''(x) < 0$ on $(-3, 0)$, so f is concave down there. There are inflection points at $x = -3$ and at $x = 0$.

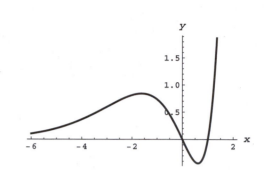

71 Note that $y' = 1/x$, so $1 + y'^2 = \frac{x^2+1}{x^2}$, and $\sqrt{1+y'^2} = \frac{\sqrt{x^2+1}}{x}$. So

$$L = \int_1^b \frac{\sqrt{x^2+1}}{x}\,dx$$

$$= \left(\sqrt{x^2+1} - \ln\left(\frac{1+\sqrt{x^2+1}}{x}\right)\right)\Bigg|_1^b = \sqrt{b^2+1} - \sqrt{2} + \ln\left(\frac{(\sqrt{b^2+1}-1)(1+\sqrt{2})}{b}\right).$$

Using a computer algebra system, we see that this has value 2 for $b \approx 2.715$.

73 Note that $\ln(\sin x)^{\tan x} = \tan x \ln\sin x$, so we evaluate $L = \lim_{x\to\pi/2^-} \tan x \ln\sin x = \lim_{x\to\pi/2^-} \frac{\ln\sin x}{\cot x} = \lim_{x\to\pi/2^-} \frac{\cot x}{-\csc^2 x} = \lim_{x\to\pi/2^-} -\cos x \sin x = 0$ by l'Hôpital's rule. Therefore $\lim_{x\to\pi/2^-} (\sin x)^{\tan x} = e^L = 1$.

75 Observe that $\lim_{x\to\infty} \frac{x+1}{x-1} = 1$, either by l'Hôpital's rule or Theorem 2.7. Hence $\lim_{x\to\infty} \ln\left(\frac{x+1}{x-1}\right) = \ln 1 = 0$.

77 Note that $\ln\left(1-\frac{3}{x}\right)^x = x\ln\left(1-\frac{3}{x}\right) = \frac{\ln(1-3/x)}{1/x}$, so we evaluate

$$L = \lim_{x\to\infty} \frac{\ln(1-3/x)}{1/x} = \lim_{x\to\infty} \frac{(1-3/x)^{-1}(3/x^2)}{-1/x^2} = -3$$

by l'Hôpital's rule. Therefore $\lim_{x\to\infty}\left(1-\frac{3}{x}\right)^x = e^L = e^{-3}$.

79 Observe that $\lim_{x\to\infty} \frac{x^{1/2}}{x^{1/3}} = \lim_{x\to\infty} x^{1/6} = \infty$, so $x^{1/2}$ grows faster than $x^{1/3}$ as $x \to \infty$.

81 By Theorem 7.14, $\sqrt{x}$ grows faster than $\ln^{10} x$ as $x \to \infty$.

83 Observe that $\lim_{x\to\infty} \frac{e^x}{3^x} = \lim_{x\to\infty}\left(\frac{e}{3}\right)^x = 0$ since $e/3 < 1$. Therefore 3^x grows faster than e^x as $x \to \infty$.

85 Observe that $4^{x/2} = (4^{1/2})^x = 2^x$, so these functions are identical and hence have comparable growth rates as $x \to \infty$.

87

For the second limit, let $y = x^2$:

$$\lim_{x \to 0} \frac{x^2}{1 - e^{-x^2}} = \lim_{y \to 0} \frac{y}{1 - e^{-y}}$$

$$= \lim_{y \to 0} \frac{1}{e^{-y}} = 1$$

by l'Hôpital's rule. For the first, observe that

for $x > 0$ $\dfrac{x}{\sqrt{1 - e^{-x^2}}} = \left(\dfrac{x^2}{1 - e^{-x^2}}\right)^{1/2}$, so

$\lim\limits_{x \to 0^+} \dfrac{x}{\sqrt{1 - e^{-x^2}}} = 1$ as well.

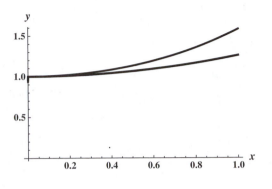

89 Note that $\lim\limits_{x \to 0+} x^x = 1$, as shown in Example 6 a. Therefore $\lim\limits_{x \to 0^+} f(x) = \lim\limits_{x \to 0^+} (x^x)^x = 1^0 = 1$, and $\lim\limits_{x \to 0} g(x) = \lim\limits_{x \to 0} x^{(x^x)} = 0^1 = 0$.

91 First, observe that $\lim\limits_{x \to \infty} \left(1 + \dfrac{1}{x}\right)^{x+a} = \lim\limits_{x \to \infty} \left(1 + \dfrac{1}{x}\right)^{x} \left(1 + \dfrac{1}{x}\right)^{a} = e \cdot 1 = e$. Thus $\lim\limits_{x \to \infty} \ln g(x) = 1$. It suffices to determine whether $\ln g(x) - 1$ is positive or negative as $x \to \infty$. To do this, consider $\lim\limits_{x \to \infty} x(\ln g(x) - 1) = \lim\limits_{t \to 0} \dfrac{(1 + at)\ln(1 + t) - t}{t^2}$, where we make the change of variables $t = 1/x$. This limit can be evaluated by using l'Hôpital's rule twice: $\lim\limits_{t \to 0} \dfrac{(1 + at)\ln(1 + t) - t}{t^2} = a - \dfrac{1}{2}$. Therefore when $a > 1/2$ we have $g(x) > e$ as $x \to \infty$, and when $0 < a < 1/2$, $g(x) < e$ as $x \to \infty$. In the case $a = 1/2$ we consider the limit $\lim\limits_{x \to \infty} x^2(\ln g(x) - 1) = \lim\limits_{t \to 0} \dfrac{(1 + at)\ln(1 + t) - t}{t^3}$, which can be evaluated by using l'Hôpital's three times: $\lim\limits_{t \to 0} \dfrac{(1 + t/2)\ln(1 + t) - t}{t^3} = \dfrac{1}{12}$. Therefore $g(x) > e$ as $x \to \infty$ in this case as well.

93

a.

It appears that $\lim_{x \to 0^+} f(x) = 0$.

b. Let $x = e^y$, so that $y = \ln x$ and as $x \to 0^+$ we have $y \to -\infty$. Then $\lim_{x \to 0^+} f(x) = \lim_{y \to -\infty} \frac{e^{-y^2/2\sigma^2}}{\sigma\sqrt{2\pi}\,e^y}$
$= \frac{1}{\sigma\sqrt{2\pi}} \lim_{y \to -\infty} \frac{1}{e^y \cdot e^{y^2/2\sigma^2}} = 0$.

c. Write f as $f(x) = \frac{1}{\sigma\sqrt{2\pi}} \left(\frac{1}{x e^{(\ln^2 x)/2\sigma^2}} \right)$. Then

$$f'(x) = -\frac{1}{\sigma\sqrt{2\pi}} \left(\frac{1}{x^2 e^{(\ln^2 x)/\sigma^2}} \right) \left(e^{(\ln^2 x)/2\sigma^2} + x e^{(\ln^2 x)/2\sigma^2} \cdot \frac{1}{\sigma^2} \ln x \cdot \frac{1}{x} \right) =$$

$$-\frac{1}{\sigma\sqrt{2\pi}} \left(\frac{1}{x^2} \right) \left(1 + \frac{\ln x}{\sigma^2} \right).$$

This quantity is 0 only when $1 + \frac{\ln x}{\sigma^2} = 0$, which occurs for $x^* = e^{-\sigma^2}$, and this critical number yields a maximum.

d. $f(e^{-\sigma^2}) = \frac{1}{\sigma\sqrt{2\pi}} \left(\frac{1}{(e^{-\sigma^2} e^{(\ln^2(e^{-\sigma^2}))/2\sigma^2})} \right) = \frac{1}{\sigma\sqrt{2\pi}} \left(\frac{1}{e^{-\sigma^2} e^{\sigma^2/2}} \right) = \frac{e^{\sigma^2/2}}{\sigma\sqrt{2\pi}}$.

e. Let $g(\sigma) = \frac{e^{\sigma^2/2}}{\sigma\sqrt{2\pi}}$. Then $g'(\sigma) = \frac{e^{\frac{\sigma^2}{2}}(\sigma^2 - 1)}{\sqrt{2\pi}\sigma^2}$, so there is a critical point at $\sigma = 1$. Note that g is decreasing on $(0, 1)$ and increasing on $(1, \infty)$, so g has a minimum at $\sigma = 1$.

95 The viewing angle θ is given by $\theta = \cot^{-1}\left(\frac{x}{10}\right) - \cot^{-1}\left(\frac{x}{3}\right)$, and we wish to maximize this function for $x > 0$. The critical points satisfy $\theta'(x) = -\frac{1}{1+\left(\frac{x}{10}\right)^2} \cdot \frac{1}{10} - (-)\frac{1}{1+\left(\frac{x}{3}\right)^2} \cdot \frac{1}{3} = \frac{3}{x^2+3^2} - \frac{10}{x^2+10^2} = 0$ which simplifies to $3(x^2 + 100) = 10(x^2 + 9)$ or $x^2 = 30$. Therefore $x = \sqrt{30} \approx 5.5$ ft is the only critical point in $(0, \infty)$. By the First (or Second) Derivative Test, this critical point corresponds to a local maximum, and by Theorem 4.5, this solitary local maximum must be the absolute maximum on the interval $(0, \infty)$.

97

a. $v(t) = y'(t) = -15e^{-t}\cos t - 15e^{-t}\sin t$, so $v(1) \approx -7.625$ meters per second, and $v(3) \approx 0.63$ meters per second.

b. She is moving down for approximately 2.4 seconds, and then up until about 5.5 seconds, and then down again until about 8.6 seconds, and then up again.

c. The maximum velocity going up appears to be about about .65 meters per second.

99

a. Because $\frac{d}{dx}\cosh x = \sinh x$ and $\frac{d}{dx}\sinh x = \cosh x$, we see that every nth derivative of $\cosh x$ where n is even is equal to $\cosh x$. So $\frac{d^6}{dx^6}\cosh x = \cosh x$.

b. $\frac{d}{dx} x \operatorname{sech} x = \operatorname{sech} x - x \operatorname{sech} x \tanh x = \operatorname{sech} x (1 - x \tanh x)$.

101 $f'(\ln 3) = \sinh(\ln 3) = 4/3$. So the linearization is $L(x) = 5/3 + (4/3)(x - \ln 3)$. Using this approximation, we have $\cosh(1) \approx L(1) = (5/3) + (4/3)(1 - \ln 3) \approx 1.535$.

Chapter 8

Integration Techniques

8.1 Basic Approaches

8.1.1 Let $u = 4 - 7x$. Then $du = -7\,dx$ and we obtain $-\frac{1}{7}\int u^{-6}\,du$.

8.1.3 $\sin^2 x = \frac{1-\cos 2x}{2}$.

8.1.5 Complete the square in the denominator to get $\int \frac{10}{(x-2)^2 + 1}\,dx$. The integral now matches item 12 in Table 7.1.

8.1.7 Let $u = 3 - 5x$ so that $du = -5\,dx$. Substituting gives $\frac{-1}{5}\int u^{-4}\,du = \frac{1}{15}u^{-3} + C = \frac{1}{15(3-5x)^3} + C$.

8.1.9 Let $u = 2x - \pi/4$ so that $du = 2\,dx$. Substituting gives

$$\frac{1}{2}\int_{-\pi/4}^{\pi/2} \sin u\,du = \frac{1}{2}\left(-\cos u\right)\Big|_{-\pi/4}^{\pi/2} = \frac{1}{2}\left(0 + \sqrt{2}/2\right) = \frac{\sqrt{2}}{4}.$$

8.1.11 Let $u = \ln(2x)$ so that $du = \frac{dx}{x}$. Substituting gives $\int u\,du = u^2/2 + C = \frac{1}{2}\ln^2 2x + C$.

8.1.13 Let $u = e^x + 1$ so that $du = e^x\,dx$ Substituting gives $\int \frac{1}{u}\,du = \ln|u| + C = \ln(e^x + 1) + C$.

8.1.15 We rewrite the integral by multiplying the numerator and denominator of the integrand by e^x. We have $\int \frac{e^x}{e^x - 2e^{-x}}\,dx = \int \frac{e^x}{e^x - 2e^{-x}} \cdot \frac{e^x}{e^x}\,dx = \int \frac{e^{2x}}{e^{2x}-2}\,dx$. Now let $u = e^{2x} - 2$ so that $du = 2e^{2x}\,dx$. Substituting gives $\frac{1}{2}\int \frac{du}{u} = \frac{1}{2}\ln|u| + C = \frac{1}{2}\ln|e^{2x} - 2| + C$.

8.1.17 Let $u = \ln x^2 = 2\ln x$. Then $du = \frac{2}{x}\,dx$. Substituting gives $\frac{1}{2}\int_0^4 u^2\,du = \left(u^3/6\right)\Big|_0^4 = 32/3$.

8.1.19 The integral can be written as $\cot^4 x \csc^2 x\,dx$. Let $u = \cot x$ so that $du = -\csc^2 x\,dx$. Substituting gives $-\int u^4\,du = -u^5/5 + C = -(\cot^5 x)/5 + C$.

8.1.21 $\int \frac{dx}{x^{-1}+1} = \int \frac{1}{x^{-1}+1} \cdot \frac{x}{x}\,dx = \int \frac{x}{x+1}\,dx$. Using long division, we have $\frac{x}{x+1} = 1 - \frac{1}{x+1}$. Then $\int \left(1 - \frac{1}{x+1}\right)\,dx = x - \ln|x+1| + C$.

8.1.23 $\int \frac{x}{x^2+4}\,dx + 2\int \frac{1}{x^2+4}\,dx = \frac{1}{2}\ln(x^2 + 4) + \tan^{-1}(x/2) + C$.

8.1.25
$$\int \frac{\sin t}{\cos^2 t}\,dt + \int \frac{\tan t}{\cos^2 t}\,dt = \int \tan t \sec t\,dt + \int \frac{\sin t}{\cos^3 t}\,dt = \sec t + \int \frac{\sin t}{\cos^3 t}\,dt.$$
Now let $u = \cos t$ so that $du = -\sin t\,dt$. We have
$$\sec t - \int u^{-3}\,du = \sec t + u^{-2}/2 + C = \sec t + \frac{1}{2\cos^2 t} + C = \sec t + \frac{1}{2}\sec^2 t + C.$$

8.1.27

$$\int \frac{2}{\sqrt{1-x^2}}\,dx - 3\int \frac{x}{\sqrt{1-x^2}}\,dx = 2\sin^{-1} x - 3\int \frac{x}{\sqrt{1-x^2}}\,dx.$$

Let $u = 1 - x^2$ so that $du = -2x\,dx$. Substituting gives

$$2\sin^{-1} x + \frac{3}{2}\int u^{-1/2}\,du = 2\sin^{-1} x + 3\sqrt{u} + C = 2\sin^{-1} x + 3\sqrt{1-x^2} + C.$$

8.1.29 Note that long division gives $\frac{x+2}{x+4} = 1 - \frac{2}{x+4}$. Thus our integral is equal to

$$\int \left(1 - \frac{2}{x+4}\right) dx = x - 2\ln|x+4| + C.$$

8.1.31 Note that long division gives $\frac{t^3-2}{t+1} = t^2 - t + 1 - \frac{3}{t+1}$. Our integral is therefore

$$\int \left(t^2 - t + 1 - \frac{3}{t+1}\right) dt = t^3/3 - t^2/2 + t - 3\ln|t+1| + C.$$

8.1.33 Note that $x^2 - 2x + 10 = (x^2 - 2x + 1) + 9 = (x-1)^2 + 9$. So we have $\int \frac{dx}{(x-1)^2 + 3^2} = \frac{1}{3}\tan^{-1}((x-1)/3) + C$.

8.1.35 Note that $27 - 6\theta - \theta^2 = -(\theta^2 + 6\theta + 9 - 36) = -((\theta+3)^2 - 36) = 36 - (\theta+3)^2$. Thus our integral is $\int \frac{d\theta}{\sqrt{36-(\theta+3)^2}} = \sin^{-1}((\theta+3)/6) + C$.

8.1.37 $\int \frac{1}{1+\sin\theta} \cdot \frac{1-\sin\theta}{1-\sin\theta}\,d\theta = \int \frac{1-\sin\theta}{1-\sin^2\theta}\,d\theta = \int \frac{1-\sin\theta}{\cos^2\theta}\,d\theta = \int \sec^2\theta\,d\theta - \int \frac{\sin\theta}{\cos^2\theta}\,d\theta = \tan\theta - \int \frac{\sin\theta}{\cos^2\theta}\,d\theta$. Let $u = \cos\theta$ so that $du = -\sin\theta\,d\theta$. Substituting gives $\tan\theta + \int u^{-2}\,du = \tan\theta - \frac{1}{u} + C = \tan\theta - \sec\theta + C$.

8.1.39 $\int \frac{1}{\sec x-1} \cdot \frac{\sec x+1}{\sec x+1}\,dx = \int \frac{\sec x+1}{\sec^2 x-1}\,dx = \int \frac{\sec x+1}{\tan^2 x}\,dx = \int \frac{\sec x}{\tan^2 x}\,dx + \int \cot^2 x\,dx = \int \cot x\csc x\,dx + \int \cot^2 x\,dx = -\csc x + \int (\csc^2 x - 1)\,dx = -\csc x - \cot x - x + C$.

8.1.41

a. False. This seem to use the untrue "identity" that $\frac{a}{b+c} = \frac{a}{b} + \frac{a}{c}$.

b. False. The degree of the numerator is already less than the degree of the denominator, so long division won't help.

c. False. This is false because $\frac{d}{dx}\ln|\sin x + 1| + C \neq \frac{1}{\sin x+1}$. The substitution $u = \sin x + 1$ can't be carried out because $du = \cos x\,dx$ can't be accounted for.

d. False. In fact, $\int e^{-x}\,dx = -e^{-x} + C \neq \ln e^x + C$.

8.1.43 $\int_{-1}^{0} \frac{x}{x^2+2x+2}\,dx = \int_{-1}^{0} \frac{x}{(x+1)^2+1}\,dx$. Let $u = x + 1$ so that $du = dx$. Substituting gives $\int_0^1 \frac{u-1}{u^2+1}\,du = \int_0^1 \frac{u}{u^2+1}\,du - \int_0^1 \frac{1}{u^2+1}\,du = \left((1/2)\ln(u^2+1) - \tan^{-1}(u)\right)\Big|_0^1 = (1/2)\ln 2 - \frac{\pi}{4} = \frac{1}{4}(\ln 4 - \pi)$.

8.1.45 Using the identity $\sin 2x = 2\sin x\cos x$, we have $2\int \sin^2 x\cos x\,dx$. Let $u = \sin x$ so that $du = \cos x\,dx$. We have $2\int u^2\,du = 2u^3/3 + C = 2(\sin^3 x)/3 + C$.

8.1.47 Rewrite the integral as $\int \frac{1}{\sqrt{x}} \cdot \frac{1}{1+(\sqrt{x})^2}\,dx$ and let $u = \sqrt{x}$. Then $du = \frac{1}{2\sqrt{x}}\,dx$, and substituting gives

$$2\int \frac{1}{1+u^2}\,du = 2\tan^{-1} u + C = 2\tan^{-1}\sqrt{x} + C.$$

8.1.49 Note that $x^2 + 6x + 13 = (x^2 + 6x + 9) + 4 = (x+3)^2 + 4$. Also note that we can write the numerator $x - 2 = x + 3 - 5 = \frac{1}{2}(2x+6) - 5$. We have $\int \frac{\frac{1}{2}(2x+6)}{x^2+6x+13}\,dx - \int \frac{5}{(x+3)^2+4}\,dx$. For the first integral, let $u = x^2 + 6x + 13$ so that $du = (2x+6)\,dx$. We have (for just the first integral) $\frac{1}{2}\int \frac{1}{u}\,du = \frac{1}{2}\ln|u| + C = \frac{1}{2}\ln(x^2 + 6x + 13) + C$. The second integrand has antiderivative equal to $\frac{5}{2}\tan^{-1}((x+3)/2)$, so the original integral is equal to

$$\frac{1}{2}\ln(x^2 + 6x + 13) - \frac{5}{2}\tan^{-1}((x+3)/2) + C.$$

8.1.51 Let $u = e^x$ so that $du = e^x\,dx$. Substituting gives $\int \frac{1}{u^2+2u+1}\,du = \int (u+1)^{-2}\,du = -\frac{1}{u+1} + C = -\frac{1}{e^x+1} + C$.

8.1.53 The denominator factors as $(x+1)^2$. $\int_1^3 \frac{2}{(x+1)^2}\,dx = -2\left(\frac{1}{x+1}\right)\Big|_1^3 = -2(1/4 - 1/2) = 1/2$.

8.1.55

  a. If $u = \tan x$ then $du = \sec^2 x\,dx$. Substituting gives $\int u\,du = u^2/2 + C = (\tan^2 x)/2 + C$.

  b. If $u = \sec x$, then $du = \sec x \tan x\,dx$. Substituting gives $\int u\,du = u^2/2 + C = (\sec^2 x)/2 + C$.

  c. The seemingly different answers are the same, since $(\tan^2 x)/2$ and $(\sec^2 x)/2$ differ by a constant. In fact, $(\tan^2 x)/2 - (\sec^2 x)/2 = \frac{-1}{2}$.

8.1.57

  a. Let $u = x + 1$ so that $du = dx$. Note that $x = u - 1$, so that $x^2 = (u-1)^2$. Substituting gives $\int \frac{u^2 - 2u + 1}{u}\,du = \int (u - 2 + (1/u))\,du = u^2/2 - 2u + \ln|u| + C = (x+1)^2/2 - 2(x+1) + \ln|x+1| + C$.

  b. By long division, $\frac{x^2}{x+1} = x - 1 + \frac{1}{x+1}$. Thus, $\int \frac{x^2}{x+1}\,dx = \int \left(x - 1 + \frac{1}{x+1}\right)dx = x^2/2 - x + \ln|x+1| + C$.

  c. The seemingly different answers are the same, because they differ by a constant. In fact, $(x+1)^2/2 - 2(x+1) + \ln|x+1| - (x^2/2 - x + \ln|x+1|) = -\frac{3}{2}$.

8.1.59 $A = \int_2^4 \frac{x^2-1}{x^3-3x}\,dx$. Let $u = x^3 - 3x$ so that $du = 3x^2 - 3\,dx$. Substituting gives $A = \frac{1}{3}\int_2^{52} \frac{1}{u}\,du = \frac{1}{3}\ln u \Big|_2^{52} = \frac{1}{3}(\ln 52 - \ln 2) = \ln(26)/3$.

8.1.61

  a. $V = \pi \int_0^2 (x^2 + 1)\,dx = \pi \left(x^3/3 + x\right)\Big|_0^2 = \pi(8/3 + 2) = 14\pi/3$.

  b. $V = 2\pi \int_0^2 x\sqrt{x^2+1}\,dx$. Let $u = x^2 + 1$ so that $du = 2x\,dx$. Substituting gives $\pi \int_1^5 u^{1/2}\,du = \pi\left(2u^{3/2}/3\right)\Big|_1^5 = \pi(10\sqrt{5}/3 - 2/3)$.

8.1.63 $L = \int_0^1 \sqrt{1 + \frac{25x^{1/2}}{16}}\,dx$. Let $u^2 = 1 + \frac{25x^{1/2}}{16}$. Then $2u\,du = \frac{25}{32\sqrt{x}}\,dx$. Note that $\sqrt{x} = \frac{16}{25}(u^2 - 1)$, and that $dx = \frac{64\sqrt{x}}{25}u\,du = \frac{1024}{625}(u^3 - u)\,du$. Substituting gives

$$L = \int_1^{\sqrt{41/16}} \frac{1024}{625}(u^4 - u^2)\,du = \frac{1024}{625}\left(u^5/5 - u^3/3\right)\Big|_1^{\sqrt{41/16}}$$

$$= \frac{1024}{625}((\sqrt{41/16})^5/5 - (\sqrt{41/16})^3/3 - (1/5 - 1/3)) = \frac{1024}{625}\left(\frac{2}{15} + \frac{1763\sqrt{41}}{15360}\right)$$

$$= \frac{2048 + 1763\sqrt{41}}{9375} \approx 1.423.$$

8.1.65 $A = 2\pi \int_0^1 \sqrt{x+1}\sqrt{1 + \frac{1}{4(x+1)}}\,dx = 2\pi \int_0^1 \sqrt{x+5/4}\,dx = 2\pi \left((2/3)(x+5/4)^{3/2}\right)\Big|_0^1 = \frac{4\pi}{3}(27/8 - 5\sqrt{5}/8) = \frac{9\pi}{2} - \frac{5\sqrt{5}\pi}{6}.$

8.2   Integration by Parts

8.2.1 It is based on the product rule. In fact, the rule can be obtained by writing down the product rule, then integrating both sides and rearranging the terms in the result.

8.2.3 It is generally a good idea to let u be something easy to differentiate, keeping in mind that whatever is left for dv is something which you will need to be able to integrate. In this case, it would be prudent to let $u = x^n$ and $dv = \cos(ax)\,dx$. Note that differentiating x^n results in something simpler (lower degree,) while integrating it make it more complicated (higher degree.) However, differentiating or integrating $\cos(ax)$ yields essentially the same thing (a constant times the sine function).

8.2.5 Those for which the choice for dv is easily integrated and when the resulting new integral is no more difficult than the original.

8.2.7 Let $u = x$ and $dv = \cos x\,dx$. Then $du = dx$ and $v = \sin x$. Then $\int x \cos x\,dx = x \sin x - \int \sin x\,dx = x \sin x + \cos x + C.$

8.2.9 Let $u = t$ and $dv = e^t\,dt$. Then $du = dt$ and $v = e^t$. Then $\int te^t\,dt = te^t - \int e^t\,dt = te^t - e^t + C.$

8.2.11 Let $u = x$ and $dv = \frac{dx}{\sqrt{x+1}}$. Then $du = dx$ and $v = 2\sqrt{x+1}$. Then $\int \frac{x}{\sqrt{x+1}}\,dx = 2x\sqrt{x+1} - \int 2\sqrt{x+1}\,dx. = 2x\sqrt{x+1} - \frac{4}{3}(x+1)^{3/2} + C = \frac{2}{3}\sqrt{x+1}(x-2) + C.$

8.2.13 Let $u = \ln x^3 = 3\ln x$ and let $dv = x^2\,dx$. Then $du = \frac{3\,dx}{x}$ and $v = x^3/3$. Then $\int x^2 \ln x^3\,dx = x^3 \ln x - \int x^2\,dx = x^3 \ln x - x^3/3 + C.$

8.2.15 Let $u = \ln x$ and $dv = x^2\,dx$. Then $du = \frac{1}{x}\,dx$ and $v = \frac{x^3}{3}$. Then $\int x^2 \ln x\,dx = \frac{x^3}{3}\ln x - \frac{1}{3}\int x^2\,dx = \frac{x^3}{3}\ln x - \frac{x^3}{9} + C = \frac{x^3}{9}(3\ln x - 1) + C.$

8.2.17 Let $u = \ln x$ and $dv = x^{-10}\,dx$. Then $du = \frac{1}{x}\,dx$ and $v = -\frac{1}{9}x^{-9}$. Then $\int \frac{\ln x}{x^{10}}\,dx = -\frac{1}{9x^9}\ln x + \frac{1}{9}\int x^{-10}\,dx = -\frac{1}{9x^9}\ln x + -\frac{1}{81x^9} + C.$

8.2.19 Let $u = \tan^{-1} x$ and $dv = dx$. Then $du = \frac{1}{1+x^2}\,dx$ and $v = x$. Then $\int \tan^{-1} x\,dx = x\tan^{-1} x - \int \frac{x}{1+x^2}\,dx = x\tan^{-1} x - \frac{1}{2}\ln(1+x^2) + C.$ The fact that $-\int \frac{x}{1+x^2}\,dx = -\frac{1}{2}\ln(1+x^2) + C$ follows from the ordinary substitution $u = 1 + x^2$.

8.2.21 $\int x \sin x \cos x\,dx = \frac{1}{2}\int x\cdot(2\sin x \cos x)\,dx = \frac{1}{2}\int x \sin 2x\,dx.$ Now using the result of problem 8, we have
$\int x \sin x \cos x\,dx = \frac{1}{2}\cdot\left(-\frac{1}{2}x\cos 2x + \frac{1}{4}\sin 2x\right) + C = -\frac{1}{4}x\cos 2x + \frac{1}{8}\sin 2x + C.$

8.2.23 Let $u = t^2$ and $dv = e^{-t}\,dt$. Then $du = 2t\,dt$ and $v = -e^{-t}$. We have $\displaystyle\int t^2 e^{-t}\,dt = -t^2 e^{-t} + 2\int te^{-t}\,dt$. To compute this last integral, we let $u = t$ and $dv = e^{-t}\,dt$. Then $\displaystyle\int te^{-t}\,dt = -te^{-t} + \int e^{-t}\,dt = -te^{-t} - e^{-t} + C$.

Putting these results together, we obtain

$$\int t^2 e^{-t}\,dt = -t^2 e^{-t} + 2(-te^{-t} - e^{-t}) + C = -e^{-t}(t^2 + 2t + 2) + C.$$

8.2.25 Let $u = \sin 4x$ and $dv = e^{-x}\,dx$. Then $du = 4\cos 4x\,dx$ and $v = -e^{-x}$. We have $\displaystyle\int e^{-x}\sin 4x\,dx = -e^{-x}\sin 4x + 4\int e^{-x}\cos 4x\,dx$.

Now in order to compute the integral which comprises this last term, we let $u = \cos 4x$ and $dv = e^{-x}\,dx$. Then $du = -4\sin 4x\,dx$ and $v = -e^{-x}$. Thus, $\displaystyle\int e^{-x}\cos 4x\,dx = -e^{-x}\cos 4x - 4\int e^{-x}\sin 4x\,dx$.

Putting these results together gives

$$\int e^{-x}\sin 4x\,dx = -e^{-x}\sin 4x - 4e^{-x}\cos 4x - 16\int e^{-x}\sin 4x\,dx$$

$$17\int e^{-x}\sin 4x\,dx = -e^{-x}\sin 4x - 4e^{-x}\cos 4x + C$$

$$\int e^{-x}\sin 4x\,dx = -\frac{e^{-x}}{17}(\sin 4x + 4\cos 4x) + C.$$

8.2.27 Let $u = \cos x$ and $dv = e^x\,dx$. Then $du = -\sin x\,dx$ and $v = e^x$. We have $\displaystyle\int e^x\cos x\,dx = e^x\cos x + \int e^x\sin x\,dx$.

Now in order to compute the integral which comprises this last term, we let $u = \sin x$ and $dv = e^x\,dx$. Then $du = \cos x\,dx$ and $v = e^x$. Thus, $\displaystyle\int e^x\sin x\,dx = e^x\sin x - \int e^x\cos x\,dx$.

Putting these results together gives

$$\int e^x\cos x\,dx = e^x\cos x + e^x\sin x - \int e^x\cos x\,dx$$

$$2\int e^x\cos x\,dx = e^x(\cos x + \sin x) + C$$

$$\int e^x\cos x\,dx = \frac{e^x}{2}(\cos x + \sin x) + C.$$

8.2.29 Let $u = x^2$ and $dv = \sin 2x\,dx$. Then $du = 2x\,dx$ and $v = -\frac{1}{2}\cos 2x$. Then $\displaystyle\int x^2\sin 2x\,dx = -\frac{1}{2}x^2\cos 2x + \int x\cos 2x\,dx$.

Now we consider computing this last term $\displaystyle\int x\cos 2x\,dx$ as a new problem. Let $u = x$ and $dv = \cos 2x\,dx$. Then $du = dx$ and $v = \frac{1}{2}\sin 2x$. So $\displaystyle\int x\cos 2x\,dx = \frac{1}{2}x\sin 2x - \frac{1}{2}\int \sin 2x\,dx = \frac{1}{2}x\sin 2x + \frac{1}{4}\cos 2x + C$.

Combining these results we have $\displaystyle\int x^2\sin 2x\,dx = -\frac{1}{2}x^2\cos 2x + \frac{1}{2}x\sin 2x + \frac{1}{4}\cos 2x + C$.

8.2.31 Let $u = x$ and $dv = \sin x\,dx$. Then $du = dx$ and $v = -\cos x$. Then $\displaystyle\int_0^\pi x\sin x\,dx = -x\cos x\,\Big|_0^\pi + \int_0^\pi \cos x\,dx = \pi - 0 + \sin x\,\Big|_0^\pi = \pi - 0 + 0 - 0 = \pi$.

8.2.33 Let $u = x$ and $dv = \cos 2x\, dx$. Then $du = dx$ and $v = \frac{1}{2}\sin 2x$. Then $\displaystyle\int_0^{\pi/2} x\cos 2x\, dx =$

$$\frac{1}{2}x\sin 2x\,\Big|_0^{\pi/2} - \frac{1}{2}\int_0^{\pi/2}\sin 2x\, dx = 0 - \left(\frac{1}{2}\cdot\frac{(-\cos 2x)}{2}\right)\Big|_0^{\pi/2} = -\frac{1}{4} - \frac{1}{4} = -\frac{1}{2}.$$

8.2.35 Let $u = \ln x$ and $dv = x^2\, dx$. Then $du = \frac{1}{x}\, dx$ and $v = \frac{x^3}{3}$. Then $\displaystyle\int_1^{e^2} x^2\ln x\, dx = \frac{1}{3}x^3\ln x\,\Big|_1^{e^2} -$

$$\frac{1}{3}\int_1^{e^2} x^2\, dx = \frac{2}{3}e^6 - \frac{1}{9}x^3\,\Big|_1^{e^2} = \frac{2}{3}e^6 - \frac{1}{9}\left(e^6 - 1\right) = \frac{5}{9}e^6 + \frac{1}{9}.$$

8.2.37 By problem 18, $\displaystyle\int \sin^{-1} x\, dx = x\sin^{-1} x + \sqrt{1 - x^2}$. Thus,

$$\int_{1/2}^{\sqrt{3}/2}\sin^{-1} x\, dx = \left(x\sin^{-1} x + \sqrt{1 - x^2}\right)\Big|_{1/2}^{\sqrt{3}/2}$$

$$= \left(\frac{\sqrt{3}}{2}\cdot\frac{\pi}{3} + \frac{1}{2}\right) - \left(\frac{1}{2}\cdot\frac{\pi}{6} + \frac{\sqrt{3}}{2}\right)$$

$$= \frac{\pi}{6}\left(\sqrt{3} - \frac{1}{2}\right) + \frac{1}{2}\left(1 - \sqrt{3}\right).$$

8.2.39 Using shells, we have $\dfrac{V}{2\pi} = \displaystyle\int_0^{\ln 2} xe^{-x}\, dx$. Let $u = x$ and $dv = e^{-x}\, dx$, so that $du = dx$ and $v = -e^{-x}$. Then

$$\frac{V}{2\pi} = -xe^{-x}\,\Big|_0^{\ln 2} + \int_0^{\ln 2} e^{-x}\, dx = -\frac{1}{2}\ln 2 - e^{-x}\,\Big|_0^{\ln 2} = -\frac{\ln 2}{2} - \left(\frac{1}{2} - 1\right) = \frac{1}{2}\left(1 - \ln 2\right).$$

Thus $V = \pi(1 - \ln 2)$.

8.2.41 Using disks, we have $\dfrac{V}{\pi} = \displaystyle\int_1^{e^2} x^2\ln^2 x\, dx$. By problem 26, we have $\displaystyle\int x^2\ln^2 x\, dx = \frac{1}{3}x^3\ln^2 x -$ $\dfrac{2}{9}x^3\ln x + \dfrac{2}{27}x^3 + C$. Thus,

$$\frac{V}{\pi} = \left(\frac{1}{3}x^3\ln^2 x - \frac{2}{9}x^3\ln x + \frac{2}{27}x^3\right)\Big|_1^{e^2} = \left(\frac{4}{3}e^6 - \frac{4}{9}e^6 + \frac{2}{27}e^6\right) - \left(\frac{2}{27}\right) = \frac{26}{27}e^6 - \frac{2}{27}.$$

Thus, $V = \dfrac{\pi}{27}\left(26e^6 - 2\right)$.

8.2.43

a. False. For example, suppose $u = x$ and $dv = x\, dx$. Then $\displaystyle\int uv'\, dx = \int x^2\, dx = \frac{x^3}{3} + C$, but

$$\int u\, dx\int v'\, dx = \left(\int x\, dx\right)^2 = \left(\frac{x^2}{2} + C\right)^2.$$

b. True. This is one way to write the integration by parts formula.

c. True. This is the integration by parts formula with the roles of u and v reversed.

8.2.45 Let $u = x^n$ and $dv = \cos ax\, dx$. Then $du = nx^{n-1}\, dx$ and $v = \frac{\sin ax}{a}$. Then $\displaystyle\int x^n\cos ax\, dx =$ $\dfrac{x^n\sin ax}{a} - \dfrac{n}{a}\displaystyle\int x^{n-1}\sin ax\, dx$.

8.2.47 Let $u = \ln^n x$ and $dv = dx$. Then $du = \frac{n \ln^{n-1}(x)}{x} dx$ and $v = x$. Then $\int \ln^n(x)\, dx = x \ln^n x - n \int \ln^{n-1}(x)\, dx$.

8.2.49

$$\int x^2 \cos 5x\, dx = \frac{x^2 \sin 5x}{5} - \frac{2}{5} \int x \sin 5x\, dx$$

$$= \frac{x^2 \sin 5x}{5} - \frac{2}{5} \left(-\frac{x \cos 5x}{5} + \frac{1}{5} \int \cos 5x\, dx \right)$$

$$= \frac{1}{5} \left(x^2 \sin 5x + \frac{2}{5} x \cos 5x - \frac{2}{25} \sin 5x \right) + C.$$

8.2.51

$$\int \ln^4 x\, dx = x \ln^4 x - 4 \int \ln^3 x\, dx$$

$$= x \ln^4 x - 4 \left(x \ln^3 x - 3 \int \ln^2 x\, dx \right)$$

$$= x \ln^4 x - 4x \ln^3 x + 12 \left(x \ln^2 x - 2 \int \ln x\, dx \right)$$

$$= x \ln^4 x - 4x \ln^3 x + 12x \ln^2 x - 24 \left(x \ln x - x \right) + C.$$

8.2.53 Let $u = \tan x + 2$, so that $du = \sec^2 x\, dx$. Then $\int \sec^2 x \ln(\tan x + 2)\, dx = \int \ln u\, du = u \ln u - u + C = (\tan x + 2) \ln(\tan x + 2) - \tan x + C.$

8.2.55 Using the change of base formula, we have $\int \log_b x\, dx = \int \frac{\ln x}{\ln b}\, dx = \frac{1}{\ln b} (x \ln x - x) + C.$

8.2.57 Let $z = \sqrt{x}$, so that $dz = \frac{1}{2\sqrt{x}}\, dx$. Substituting yields $2 \int \frac{\sqrt{x} \cos \sqrt{x}}{2\sqrt{x}}\, dx = 2 \int z \cos z\, dz$. Now let $u = z$ and $dv = \cos z\, dz$. then $du = dz$ and $v = \sin z$. Then by Integration by Parts, we have $2 \int z \cos z\, dz = 2 \left(z \sin z - \int \sin z\, dz \right) = 2z \sin z + 2 \cos z + C$. Thus, the original given integral is equal to $2(\sqrt{x} \sin \sqrt{x} + \cos \sqrt{x}) + C.$

8.2.59 By the Fundamental Theorem, $f'(x) = \sqrt{\ln^2 x - 1}$. So the arc length is $\int_e^{e^3} \sqrt{1 + (f'(x))^2}\, dx = \int_e^{e^3} \ln x\, dx = (x \ln x - x) \Big|_e^{e^3} = 3e^3 - e^3 - (e - e) = 2e^3.$

8.2.61 Using shells, we have $\frac{V}{2\pi} = \int_0^{\pi/2} x \cos x\, dx$. Let $u = x$ and $dv = \cos x\, dx$, so that $du = dx$ and $v = \sin x$. We have $\frac{V}{2\pi} = x \sin x \Big|_0^{\pi/2} - \int_0^{\pi/2} \sin x\, dx = \frac{\pi}{2} - 1.$
Thus, $V = \pi(\pi - 2)$.

8.2.63 Let V_1 be the volume generated when R is revolved about the x-axis, and V_2 the volume generated when R is revolved about the y-axis.
Using disks, we have

$$\frac{V_1}{\pi} = \int_0^\pi \sin^2 x\, dx = \frac{1}{2} \int_0^\pi (1 - \cos 2x)\, dx = \frac{1}{2} \left(\pi - \frac{1}{2} \int_0^\pi 2 \cos 2x\, dx \right) = \frac{1}{2} \left(\pi - \frac{1}{2} \int_0^{2\pi} \cos u\, du \right) = \frac{\pi}{2},$$

where the ordinary substitution $u = 2x$ was made. So $V_1 = \frac{\pi^2}{2}$.

Using shells to compute V_2, we have $\dfrac{V_2}{2\pi} = \displaystyle\int_0^\pi x \sin x \, dx$. Letting $u = x$ and $dv = \sin x \, dx$, we have $du = dx$ and $v = -\cos x$. Thus, $\dfrac{V_2}{2\pi} = -x \cos x \Big|_0^\pi + \displaystyle\int_0^\pi \cos x \, dx = \pi$. Thus, $V_2 = 2\pi^2$, and $V_2 > V_1$.

8.2.65

a. Let $u = x$ and $dv = f'(x)\,dx$. Then $du = dx$ and $v = f(x)$. So $\displaystyle\int xf'(x)\,dx = xf(x) - \int f(x)\,dx$.

b. Letting $f'(x) = e^{3x}$ we have $\displaystyle\int xe^{3x}\,dx = \frac{1}{3}xe^{3x} - \frac{1}{9}e^{3x} + C$.

8.2.67 Let $u = \sec x$ and $dv = \sec^2 x\,dx$, so that $du = \sec x \tan x\,dx$ and $v = \tan x$. Then $\displaystyle\int \sec^3 x\,dx = \sec x \tan x - \int \sec x \tan^2 x\,dx = \sec x \tan x - \int \sec x\left(\sec^2 x - 1\right)\,dx = \sec x \tan x - \int \sec^3 x\,dx + \int \sec x\,dx$. Thus $2\displaystyle\int \sec^3 x\,dx = \sec x \tan x + \int \sec x\,dx$, so $\displaystyle\int \sec^3 x\,dx = \frac{1}{2}\sec x \tan x + \frac{1}{2}\int \sec x\,dx$.

8.2.69

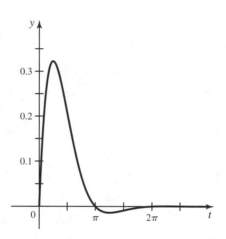

a. We have $s(t) = 0$ when $\sin t = 0$, which occurs for $t = k\pi$, where k is an integer.

b. This is given by $\dfrac{1}{\pi}\displaystyle\int_0^\pi e^{-t}\sin t\,dt$. Using the previous problem, this is equal to

$$\frac{1}{\pi}e^{-t} \cdot \frac{-\sin t - \cos t}{2}\bigg|_0^\pi = \frac{1}{2\pi}(e^{-\pi} + 1).$$

c. This is given by $\dfrac{1}{\pi}\displaystyle\int_{n\pi}^{(n+1)\pi} e^{-t}\sin t\,dt$. Using the previous problem, this is equal to

$$\frac{1}{\pi}e^{-t} \cdot \frac{-\sin t - \cos t}{2}\bigg|_{n\pi}^{(n+1)\pi} = \frac{1}{2\pi}\left(e^{-(n+1)\pi}\left(-\sin(n+1)\pi - \cos(n+1)\pi\right) - e^{-n\pi}\left(-\sin n\pi - \cos n\pi\right)\right)$$

$$= \frac{1}{2\pi}\left(-e^{-(n+1)\pi}\cos(n+1)\pi + e^{-n\pi}\cos n\pi\right) = \frac{e^{-n\pi}}{2\pi}\left(\cos n\pi - e^{-\pi}\cos(n+1)\pi\right)$$

$$= \frac{e^{-n\pi}}{2\pi}\left((-1)^n - e^{-\pi}(-1)^{n+1}\right) = (-1)^n\frac{e^{-n\pi}}{2\pi}\left(1 + e^{-\pi}\right).$$

d. Each a_n is $e^{-\pi}$ times a_{n-1}.

8.2.71

a. If $u = f(x)$ then $du = f'(x)\,dx$, and if $dv = g(x)\,dx$, then $v = G_1(x)$. The integration by parts formula gives

$$\int f(x)g(x)\,dx = f(x)G_1(x) - \int f'(x)G_1(x)\,dx.$$

b. Letting $u = f'(x)$ yields $du = f''(x)\,dx$ and letting $dv = G_1(x)\,dx$ yields $v = G_2(x)$. Then continuing from part (a) we have

$$\int f(x)g(x)\,dx = f(x)G_1(x) - \left(f'(x)G_2(x) - \int f''(x)G_2(x)\,dx \right)$$

$$= f(x)G_1(x) - f'(x)G_2(x) + \int f''(x)G_2(x)\,dx.$$

Note that we end up with three terms which consist of products as indicated by the arrows, with the last product inside an integral sign. The signs alternate as indicated on the chart.

c. For the trailing integral from part (b), we let $u = f''(x)$ so that $du = f'''(x)\,dx$, and let $dv = G_2(x)\,dx$ so that $v = G_3(x)$. Then we have

$$\int f(x)g(x)\,dx = f(x)G_1(x) - f'(x)G_2(x) + \int f''(x)G_2(x)\,dx$$

$$= f(x)G_1(x) - f'(x)G_2(x) + f''(x)G_3(x) - \int f'''(x)G_3(x)\,dx.$$

f and its derivatives	g and its integrals
$f(x)$ $\quad$ +	$g(x)$
$f'(x)$ $\quad$ −	$G_1(x)$
$f''(x)$ $\quad$ +	$G_2(x)$
$f'''(x)$ $\quad$ −	$G_3(x)$

d. The table is as follows:

f and its derivatives	g and its integrals
x^2 $\quad$ +	$e^{0.5x}$
$2x$ $\quad$ −	$2e^{0.5x}$
2 $\quad$ +	$4e^{0.5x}$
0 $\quad$ −	$8e^{0.5x}$

So

$$\int x^2 e^{0.5x}\,dx = 2x^2 e^{0.5x} - 8x e^{0.5x} + 16 e^{0.5x} + C.$$

Note that the last integral is $\int 0 \cdot 8 e^{0.5x}\,dx = \int 0\,dx = C$.

e. We have:

f and its derivatives	g and its integrals
x^3 $\quad$ +	$\cos x$
$3x^2$ $\quad$ −	$\sin x$
$6x$ $\quad$ +	$-\cos x$
6 $\quad$ −	$-\sin x$
0 $\quad$ +	$\cos x$

So $\int x^3 \cos x \, dx = x^3 \sin x + 3x^2 \cos x - 6x \sin x - 6 \cos x + C$.

f. $\frac{d^k}{dx^k}(p_n(x)) = 0$ for $k \geq n + 1$, so the process will end.

8.2.73

a. $\int e^x \cos x \, dx = e^x \sin x + e^x \cos x - \int e^x \cos x \, dx$.

b. Adding $\int e^x \cos x \, dx$ to both sides of the previous equation gives $2 \int e^x \cos x \, dx = e^x \sin x + e^x \cos x$, so

$$\int e^x \cos x \, dx = \frac{e^x \sin x + e^x \cos x}{2} + C.$$

c. The chart looks like this:

f and its derivatives		g and its integrals
e^{-2x}	$+$	$\sin 3x$
$-2e^{-2x}$	$-$	$-\frac{1}{3} \cos 3x$
$4e^{-2x}$	$+$	$-\frac{1}{9} \sin 3x$

So we have

$$\int e^{-2x} \sin 3x = -\frac{1}{3} e^{-2x} \cos 3x - \frac{2}{9} e^{-2x} \sin 3x - \frac{4}{9} e^{-2x} \sin 3x.$$

Adding $\frac{4}{9} \int e^{-2x} \sin 3x \, dx$ to both sides gives

$$\frac{13}{9} \int e^{-2x} \sin 3x \, dx = -\frac{1}{3} e^{-2x} \cos 3x - \frac{2}{9} e^{-2x} \sin 3x + D,$$

where D is an arbitrary constant so

$$\int e^{-2x} \sin 3x \, dx = -\frac{3}{13} e^{-2x} \cos 3x - \frac{2}{13} e^{-2x} \sin 3x + C.$$

8.2.75 Let $u = x$ and $dv = f''(x) \, dx$. Then $du = dx$ and $v = f'(x)$. We have $\displaystyle\int_a^b x f''(x) \, dx = x f'(x) \Big|_a^b - \displaystyle\int_a^b f'(x) \, dx = (0 - 0) - f(x) \Big|_a^b = -(f(b) - f(a)) = f(a) - f(b)$.

8.2.77

a. To compute I_1, we let $u = -x^2$, so that $du = -2x \, dx$. Then an ordinary substitution yields

$$-\frac{1}{2} \int e^u \, du = -\frac{1}{2} e^u + C = -\frac{1}{2} e^{-x^2} + C.$$

b. To compute I_3, we let $u = x^2$ and $dv = xe^{-x^2} \, dx$, so that $du = 2x \, dx$ and $v = -\frac{1}{2} e^{-x^2}$ (by part (a)). Then $I_3 = -\frac{1}{2} x^2 e^{-x^2} + \int xe^{-x^2} \, dx = -\frac{1}{2} x^2 e^{-x^2} - \frac{1}{2} e^{-x^2} + C = -\frac{1}{2} e^{-x^2}(x^2 + 1) + C$.

c. To compute I_5, we let $u = x^4$ and $dv = xe^{-x^2} \, dx$. Then $du = 4x^3 \, dx$ and $v = -\frac{1}{2} e^{-x^2}$. Then $I_5 = -\frac{1}{2} x^4 e^{-x^2} + 2 \int x^3 e^{-x^2} \, dx = -\frac{1}{2} x^4 e^{-x^2} + 2I_3 = -\frac{1}{2} e^{-x^2}(x^4 + 2x^2 + 2)$.

d. Suppose that n is odd and that it is true that $I_n = -\frac{1}{2}e^{-x^2}p_{n-1}(x)$ where $p_{n-1}(x)$ is an even polynomial of degree $n-1$. We will show that I_{n+2} also has this property, so the result will follow by induction. Let $u = x^{n+1}$ and let $dv = xe^{-x^2}\,dx$, so that $du = (n+1)x^n\,dx$ and $v = -\frac{1}{2}e^{-x^2}$. Then $I_{n+2} = -\frac{1}{2}x^{n+1}e^{-x^2} + \frac{n+1}{2}I_n = -\frac{1}{2}e^{-x^2}(x^{n+1} + \frac{n+1}{2}p_{n-1}(x)) + C = -\frac{1}{2}e^{-x^2}p_{n+1}(x) + C$. Note that if $p_{n-1}(x)$ is an even polynomial of degree $n-1$ then $\frac{n+1}{2}p_{n-1}(x) + x^{n+1}$ is an even polynomial of degree $n+1$.

e. Note that $I_2 = -\frac{1}{2}xe^{-x^2} + \frac{1}{2}I_0$. (Using a similar technique to that above.) Now if I_2 were expressible in terms of elementary functions, then I_0 would be as well, but we are given that it isn't. Similarly, we can express I_{2k} in terms of I_{2k-2} using Integration by Parts, and if any of these were expressible in terms of elementary functions, then the even numbered one below it would be. So the inability to express I_0 that way implies the inability to express I_2 that way, which implies the inability to express I_4 that way, and so on.

8.3 Trigonometric Integrals

8.3.1 The half-angle identities for sine and cosine:

$$\sin^2 x = \frac{1 - \cos 2x}{2} \quad \text{and} \quad \cos^2 x = \frac{1 + \cos 2x}{2},$$

which are conjugates.

8.3.3 To integrate $\sin^3 x$, write $\sin^3 x = \sin x \sin^2 x = \sin x(1 - \cos^2 x)$, and let $u = \cos x$ so that $du = -\sin x\,dx$.

8.3.5 A reduction formula is a recursive formula involving integrals. Using it, one can rewrite an integral of a certain type in a simpler form – which can then perhaps be evaluated or further reduced.

8.3.7 One would compute this integral by letting $u = \tan x$, so that $du = \sec^2 x\,dx$. This substitution leads to the integral $\int u^{10}\,du$, which can easily be evaluated.

8.3.9 $\int \sin^2 x\,dx = \frac{1}{2}\int (1 - \cos 2x)\,dx = \frac{1}{2}\left(x - \frac{1}{2}\sin 2x\right) + C.$

8.3.11 $\int \cos^3 x\,dx = \int \cos x(1 - \sin^2 x)\,dx = \int \cos x\,dx - \int \cos x \sin^2 x\,dx = \sin x - \int \cos x \sin^2 x\,dx.$ Let $u = \sin x$ so that $du = \cos x\,dx$. Substituting gives $\sin x - \int u^2\,du = \sin x - u^3/3 + C = \sin x - (\sin^3 x)/3 + C.$

8.3.13 $\int \sin^5 x\,dx = \int (\sin^2 x)^2 \sin x\,dx = \int (1 - \cos^2 x)^2 \sin x\,dx.$ Let $u = \cos x$ so that $du = -\sin x\,dx$. Substituting yields $-\int (1 - u^2)^2\,du = \int (-u^4 + 2u^2 - 1)\,du = \frac{-u^5}{5} + \frac{2u^3}{3} - u + C = \frac{-\cos^5 x}{5} + \frac{2\cos^3 x}{3} - \cos x + C.$

8.3.15

$$\int \sin^2 x \cos^2 x\,dx = \int \left(\frac{1 - \cos 2x}{2}\right)\left(\frac{1 + \cos 2x}{2}\right)\,dx = \frac{1}{4}\int 1 - \cos^2 2x\,dx$$

$$= \frac{1}{4}\int \left(1 - \frac{1 + \cos 4x}{2}\right)\,dx = \frac{1}{4}\int \left(\frac{1}{2} - \frac{1}{2}\cos 4x\right)\,dx = \frac{1}{4}\left(\frac{x}{2} - \frac{\sin 4x}{8}\right) + C.$$

8.3.17 $\int \sin^3 x \cos^2 x\,dx = \int \sin x(1 - \cos^2 x)(\cos^2 x)\,dx.$ Let $u = \cos x$ so that $du = -\sin x\,dx$. Substituting gives $\int (u^4 - u^2)\,du = u^5/5 - u^3/3 + C = (\cos^5 x)/5 - (\cos^3 x)/3 + C.$

8.3.19 $\displaystyle\int \cos^3 x \sqrt{\sin x}\,dx = \int \cos x(1 - \sin^2 x)\sqrt{\sin x}\,dx.$ Let $u = \sin x$ so that $du = \cos x\,dx.$ Substituting

gives $\displaystyle\int (1 - u^2)u^{1/2}\,du = \int (u^{1/2} - u^{5/2})\,du = 2u^{3/2}/3 - 2u^{7/2}/7 + C = 2(\sin^{3/2} x)/3 - 2(\sin^{7/2} x)/7 + C.$

8.3.21 $\displaystyle\int \sin^5 x \cos^{-2} x\,dx = \int (\sin x)\left(\frac{(1 - \cos^2 x)^2}{\cos^2 x}\right)\,dx.$ Let $u = \cos x$ so that $du = -\sin x\,dx.$ Substi-

tuting yields $\displaystyle -\int \frac{(1 - u^2)^2}{u^2}\,du = \int -u^{-2} + 2 - u^2\,du = \frac{1}{u} + 2u - \frac{u^3}{3} + C = \sec x + 2\cos x - \frac{\cos^3 x}{3} + C.$

8.3.23 $\displaystyle\int \sin^2 x \cos^4 x\,dx = \int \left(\frac{1 - \cos 2x}{2}\right)\left(\frac{1 + \cos 2x}{2}\right)^2\,dx = \frac{1}{8}\int (1 - \cos^2 2x)(1 + \cos 2x)\,dx = \frac{1}{8}\int 1 +$

$\displaystyle \cos 2x - \cos^2 2x - \cos^3 2x\,dx = \frac{1}{8}\int 1 + \cos 2x - \frac{1 + \cos 4x}{2} - \cos^3 2x\,dx = \frac{1}{8}\int \frac{1}{2} + \cos 2x - \frac{1}{2}\cos 4x\,dx - \frac{1}{8}\int (1 -$

$\displaystyle \sin^2 2x)\cos 2x\,dx = \frac{x}{16} + \frac{\sin 2x}{16} - \frac{\sin 4x}{64} - \frac{1}{8}\int (1 - \sin^2 2x)\cos 2x\,dx.$ To compute this last integral, we let

$u = \sin 2x,$ so that $du = 2\cos 2x\,dx.$ Then $\displaystyle\int (1 - \sin^2 2x)\cos 2x\,dx = \frac{1}{2}\int (1 - u^2)\,du = \frac{1}{2}\left(u - \frac{u^3}{3}\right) + C =$

$\displaystyle \frac{1}{2}\left(\sin 2x - \frac{\sin^3 2x}{3}\right) + C.$

Thus, our original given integral is equal to $\displaystyle \frac{x}{16} + \frac{\sin 2x}{16} - \frac{\sin 4x}{64} - \frac{1}{8}\left(\frac{1}{2}\left(\sin 2x - \frac{\sin^3 2x}{3}\right)\right) + C =$

$\displaystyle \frac{x}{16} - \frac{\sin 4x}{64} + \frac{\sin^3 2x}{48} + C.$

8.3.25 $\displaystyle\int \tan^2 x\,dx = \int (\sec^2 x - 1)\,dx = \tan x - x + C.$

8.3.27 $\displaystyle\int \cot^4 x\,dx = \int \cot^2 x(\csc^2 x - 1)\,dx = \int (\cot^2 x \csc^2 x - (\csc^2 x - 1))\,dx = \int \cot^2 x \csc^2 x\,dx + \cot x +$

$x.$ Let $u = \cot x$ so that $du = -\csc^2 x\,dx.$ Substituting gives $\displaystyle -\int u^2\,du + \cot x + x = -u^3/3 + \cot x + x + C =$

$-(\cot^3 x)/3 + \cot x + x + C.$

8.3.29

$$\int 20\tan^6 x\,dx = 20\int (\tan^4 x)(\sec^2 x - 1)\,dx = 20\int ((\tan^4 x)\sec^2 x - (\tan^2 x)(\sec^2 x - 1))\,dx$$

$$= 20\int (\tan^4 x \sec^2 x - \tan^2 x \sec^2 x + \sec^2 x - 1)\,dx.$$

Let $u = \tan x$ so that $du = \sec^2 x\,dx.$ We have $\displaystyle 20\left(\int (u^4 - u^2)\,du + \tan x - x\right) + C = 4u^5 - \frac{20u^3}{3} + 20\tan x -$

$\displaystyle 20x + C = 4\tan^5 x - \frac{20\tan^3 x}{3} + 20\tan x - 20x + C.$

8.3.31 Let $u = \tan x$ so that $du = \sec^2 x\,dx.$ Substituting gives $\displaystyle\int 10u^9\,du = u^{10} + C = \tan^{10} x + C.$

8.3.33 Let $u = \sec x$ so that $du = \sec x \tan x\,dx.$ Substituting gives $\displaystyle\int u^2\,du = u^3/3 + C = (\sec^3 x)/3 + C.$

8.3.35

$$\int \tan^3 4x\,dx = \int (\tan 4x)(\sec^2 4x - 1)\,dx = \int (\tan 4x)\sec^2 4x\,dx - \int \tan 4x\,dx$$

$$= \int (\tan 4x)\sec^2 4x\,dx + \frac{\ln|\cos 4x|}{4} + C.$$

Let $u = \tan 4x$ so that $du = 4 \sec^2 4x \, dx$. Substituting gives

$$\frac{1}{4} \int u \, du + \frac{\ln |\cos 4x|}{4} + C = \frac{u^2}{8} + \frac{\ln |\cos 4x|}{4} + C = \frac{\tan^2 4x}{8} + \frac{\ln |\cos 4x|}{4} + C.$$

8.3.37 Let $u = \tan x$ so that $du = \sec^2 x \, dx$. Then

$$\int \sec^2 x \tan^{1/2} x \, dx = \int u^{1/2} \, du = \frac{2}{3} u^{3/2} + C = \frac{2}{3} \tan^{3/2} x + C.$$

8.3.39 $\int \frac{\csc^4 x}{\cot^2 x} \, dx = \int (\csc^2 x) \left(\frac{\cot^2 x + 1}{\cot^2 x} \right) \, dx$. Let $u = \cot x$ so that $du = -\csc^2 x \, dx$. Substituting

gives $- \int \frac{u^2 + 1}{u^2} \, du = \int -1 - u^{-2} \, du = -u + \frac{1}{u} + C = -\cot x + \tan x + C.$

8.3.41 $\int_0^{\pi/4} \sec^4 \theta \, d\theta = \int_0^{\pi/4} (\sec^2 \theta)(1 + \tan^2 \theta) \, d\theta$. Let $u = \tan \theta$ so that $du = \sec^2 \theta \, d\theta$. Note that when

$\theta = 0$ we have $u = 0$ and when $\theta = \frac{\pi}{4}$ we have $u = 1$. So the original integral is equal to $\int_0^1 (1 + u^2) \, du =$

$\left(u + \frac{u^3}{3} \right) \Big|_0^1 = 1 + \frac{1}{3} = \frac{4}{3}.$

8.3.43 $\int_{\pi/6}^{\pi/3} \cot^3 \theta \, d\theta = \int_{\pi/6}^{\pi/3} (\cot \theta)(\csc^2 \theta - 1) \, d\theta = \int_{\pi/6}^{\pi/3} \cot \theta \csc^2 \theta \, d\theta - \int_{\pi/6}^{\pi/3} \frac{\cos \theta}{\sin \theta} \, d\theta$. For the first

integral, let $u = \cot \theta$ so that $du = -\csc^2 \theta \, d\theta$. For the second integral, let $w = \sin \theta$ so that $dw = \cos \theta \, d\theta$.
Substituting gives

$$-\int_{\sqrt{3}}^{1/\sqrt{3}} u \, du - \int_{1/2}^{\sqrt{3}/2} \frac{1}{w} \, dw = -\frac{u^2}{2} \Big|_{\sqrt{3}}^{1/\sqrt{3}} - \ln w \Big|_{1/2}^{\sqrt{3}/2}$$

$$= -\frac{1}{2} \left(\frac{1}{3} - 3 \right) - \left(\ln \sqrt{3} - \ln 2 - \ln 1 + \ln 2 \right) = \frac{4}{3} - \frac{\ln 3}{2}.$$

8.3.45

a. True. We have $\int_0^\pi \cos^{2m+1} x \, dx = \int_0^\pi (\cos^2 x)^m \cos x \, dx = \int_0^\pi (1 - \sin^2 x)^m \cos x \, dx$. Let $u = \sin x$ so

that $du = \cos x \, dx$. Substituting yields $\int_0^0 (1 - u^2)^m \, du = 0.$

b. False. For example, suppose $m = 1$. Then $\int_0^\pi \sin x \, dx = -\cos x \Big|_0^\pi = -(-1 - 1) = 2 \neq 0.$

8.3.47 $\int \csc x \, dx = \int (\csc x) \left(\frac{\csc x + \cot x}{\csc x + \cot x} \right) \, dx = \int \frac{\csc^2 x + \csc x \cot x}{\csc x + \cot x} \, dx$. Let $u = \csc x + \cot x$ so that

$du = -\csc^2 x - \csc x \cot x \, dx$. Substituting then yields $-\int \frac{1}{u} \, du = -\ln |u| + C = -\ln |\csc x + \cot x| + C.$

8.3.49 $A = \int_0^{\pi/4} \sec x - \tan x \, dx = \ln |\sec x + \tan x| + \ln |\cos x| \Big|_0^{\pi/4} = \ln \left(\sqrt{2} + 1 \right) + \ln \left(\frac{\sqrt{2}}{2} \right) - (0 + 0) =$

$\ln \left(1 + \sqrt{2}/2 \right).$

8.3.51 Let $u = \ln \theta$ so that $du = \frac{1}{\theta} \, d\theta$. Substituting yields $\int \sec^4 u \, du = \int (\sec^2 u)(1 + \tan^2 u) \, du$. Let

$w = \tan u$ so that $dw = \sec^2 u \, du$. Substituting again gives $\int (1 + w^2) \, dw = w + \frac{w^3}{3} + C = \tan(\ln(\theta)) +$

$\frac{\tan^3(\ln(\theta))}{3} + C.$

8.3.53

$$\int_{-\pi/3}^{\pi/3} \sqrt{\sec^2\theta - 1}\, d\theta = 2\int_0^{\pi/3}\sqrt{\sec^2\theta - 1}\, d\theta = 2\int_0^{\pi/3}\tan\theta\, d\theta$$

$$= -2\ln|\cos\theta|\ \Big|_0^{\pi/3} = -2\ln(1/2) + 2\ln(1) = 2\ln 2.$$

8.3.55 $\int_0^{\pi}(1 - \cos 2x)^{3/2}\, dx = \int_0^{\pi}(2\sin^2 x)^{3/2}\, dx = 2\sqrt{2}\int_0^{\pi}\sin^3 x\, dx = 2\sqrt{2}\int_0^{\pi}(\sin x)(1 - \cos^2 x)\, dx.$ Let $u = \cos x$ so that $du = -\sin x\, dx.$ Substituting yields $-2\sqrt{2}\int_1^{-1}(1 - u^2)\, du = 2\sqrt{2}\int_{-1}^{1}(1 - u^2)\, du = 4\sqrt{2}\int_0^1(1 - u^2)\, du = 4\sqrt{2}\left(u - \dfrac{u^3}{3}\right)\Big|_0^1 = \dfrac{8\sqrt{2}}{3}.$

8.3.57 Let $u = e^x + 1$ so that $du = e^x\, dx.$ Substituting gives $\int\sec u\, du = \ln|\sec u + \tan u| + C = \ln|\sec(e^x + 1) + \tan(e^x + 1)| + C.$

8.3.59 $\int_0^{\pi/2}\sqrt{1 - \cos 2x}\, dx = \sqrt{2}\int_0^{\pi/2}\sin x\, dx = -\sqrt{2}\cos x\ \Big|_0^{\pi/2} = \sqrt{2}.$

8.3.61 $\int_0^{\pi/4}(1 + \cos 4x)^{3/2}\, dx = \int_0^{\pi/4}(2\cos^2 2x)^{3/2}\, dx = 2\sqrt{2}\int_0^{\pi/4}\cos^3 2x\, dx = 2\sqrt{2}\int_0^{\pi/4}(\cos 2x)(1 - \sin^2 2x)\, dx.$ Let $u = \sin 2x$ so that $du = 2\cos 2x\, dx.$ Substituting gives

$$\sqrt{2}\int_0^1(1 - u^2)\, du = \sqrt{2}\left(u - \frac{u^3}{3}\right)\Big|_0^1 = \frac{2\sqrt{2}}{3}.$$

8.3.63 If $y = \ln(\sec x)$ then $\frac{dy}{dx} = \tan x.$ Thus, $L = \int_0^{\pi/4}\sqrt{1 + \tan^2 x}\, dx = \int_0^{\pi/4}\sec x\, dx$

$= \ln|\sec x + \tan x|\ \Big|_0^{\pi/4} = \ln(\sqrt{2} + 1).$

8.3.65 For $n \neq 1,$ $\int\tan^n x\, dx = \int(\tan^{n-2} x)(\sec^2 x - 1)\, dx = \int\tan^{n-2} x\sec^2 x\, dx - \int\tan^{n-2} x\, dx.$ Let $u = \tan x$ so that $du = \sec^2 x\, dx.$ Then substituting in the first of these last two integrals yields $\int u^{n-2}\, du - \int\tan^{n-2} x\, dx = \dfrac{u^{n-1}}{n-1} - \int\tan^{n-2} x\, dx = \dfrac{\tan^{n-1} x}{n-1} - \int\tan^{n-2} x\, dx.$

Thus $\int_0^{\pi/4}\tan^3 x\, dx = \dfrac{\tan^2 x}{2}\Big|_0^{\pi/4} - \int_0^{\pi/4}\tan x\, dx = \dfrac{1}{2} + \ln|\cos x|\ \Big|_0^{\pi/4} = \dfrac{1}{2} - \dfrac{\ln 2}{2}.$

8.3.67 $\int\sin 3x\cos 7x\, dx = \dfrac{1}{2}\left(\int\sin(-4x)\, dx + \int\sin 10x\, dx\right) = \dfrac{1}{2}\left(\dfrac{\cos(-4x)}{4} - \dfrac{\cos 10x}{10}\right) + C = \dfrac{\cos 4x}{8} - \dfrac{\cos 10x}{20} + C.$

8.3.69 $\int\sin 3x\sin 2x\, dx = \dfrac{1}{2}\left(\int\cos x\, dx - \int\cos 5x\, dx\right) = \dfrac{\sin x}{2} - \dfrac{\sin 5x}{10} + C.$

8.3.71

a. $\int_0^{\pi}\sin mx\sin nx\, dx = \dfrac{1}{2}\left(\int_0^{\pi}\cos(m - n)x\, dx - \int_0^{\pi}\cos(m + n)x\, dx\right) = \dfrac{1}{2}\left(\dfrac{1}{m - n}\int_0^{(m-n)\pi}\cos u\, du - \dfrac{1}{m + n}\int_0^{(m+n)\pi}\cos v\, dv\right)$ where $u = (m - n)x$ and $v = (m + n)x.$ But

this yields $\dfrac{1}{2}\left(\dfrac{1}{m - n}\sin u\ \Big|_0^{(m-n)\pi} - \dfrac{1}{m + n}\sin v\ \Big|_0^{(m+n)\pi}\right) = \dfrac{1}{2}(0 - 0) = 0.$

b. $\displaystyle\int_0^\pi \cos mx \cos nx\, dx = \frac{1}{2}\left(\int_0^\pi \cos(m-n)x\, dx + \int_0^\pi \cos(m+n)x\, dx\right) = 0$ by the previous part of this problem,

c. $\displaystyle\int_0^\pi \sin mx \cos nx\, dx = \frac{1}{2}\left(\int_0^\pi \sin(m-n)x\, dx + \int_0^\pi \sin(m+n)x\, dx\right) =$

$\displaystyle\frac{1}{2}\left(\frac{1}{m-n}\int_0^{(m-n)\pi}\sin u\, du + \frac{1}{m+n}\int_0^{(m+n)\pi}\sin v\, dv\right)$ where $u = (m-n)x$ and $v = (m+n)x$. This

quantity is equal to $\displaystyle\frac{-1}{2}\left(\frac{1}{m-n}\cos u\Big|_0^{(m-n)\pi} + \frac{1}{m+n}\cos v\Big|_0^{(m+n)\pi}\right) =$

$\displaystyle\frac{-1}{2}\left(\frac{1}{m-n}\left(\cos(m-n)\pi - 1\right) + \frac{1}{m+n}\left(\cos(m+n)\pi - 1\right)\right) =$

$\begin{cases} 0 & \text{if } m \text{ and } n \text{ are both even or both odd;} \\ \frac{1}{m-n} + \frac{1}{m+n} = \frac{2m}{m^2-n^2} & \text{otherwise.} \end{cases}$

8.3.73

a. $\displaystyle\int_0^\pi \sin^2 x\, dx = \frac{1}{2}\int_0^\pi (1 - \cos 2x)\, dx =$
$\displaystyle\frac{1}{2}\left(x - \frac{\sin 2x}{2}\right)\Big|_0^\pi = \frac{\pi}{2}.$
$\displaystyle\int_0^\pi \sin^2 2x\, dx = \frac{1}{2}\int_0^\pi (1 - \cos 4x)\, dx =$
$\displaystyle\frac{1}{2}\left(x - \frac{\sin 4x}{4}\right)\Big|_0^\pi = \frac{\pi}{2}.$

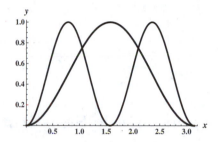

b. $\displaystyle\int_0^\pi \sin^2 3x\, dx = \frac{1}{2}\int_0^\pi (1 - \cos 6x)\, dx =$
$\displaystyle\frac{1}{2}\left(x - \frac{\sin 6x}{6}\right)\Big|_0^\pi = \frac{\pi}{2}.$
$\displaystyle\int_0^\pi \sin^2 4x\, dx = \frac{1}{2}\int_0^\pi (1 - \cos 8x)\, dx =$
$\displaystyle\frac{1}{2}\left(x - \frac{\sin 8x}{8}\right)\Big|_0^\pi = \frac{\pi}{2}.$

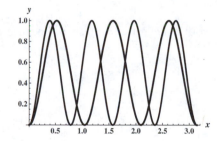

c. $\displaystyle\int_0^\pi \sin^2 nx\, dx = \frac{1}{2}\int_0^\pi (1 - \cos 2nx)\, dx = \frac{1}{2}\left(x - \frac{\sin 2nx}{2n}\right)\Big|_0^\pi = \frac{\pi}{2}.$

d. Yes. $\displaystyle\int_0^\pi \cos^2 nx\, dx = \frac{1}{2}\int_0^\pi (1 + \cos 2nx)\, dx = \frac{1}{2}\left(x + \frac{\sin 2nx}{2n}\right)\Big|_0^\pi = \frac{\pi}{2}.$

e. Claim: The corresponding integrals are all equal to $\frac{3\pi}{8}$. Proof:

$\displaystyle\int_0^\pi \sin^4 nx\, dx = \int_0^\pi \left(\frac{1 - \cos 2nx}{2}\right)^2 dx$

$\displaystyle = \int_0^\pi \frac{1 - 2\cos 2nx + \cos^2 2nx}{4}\, dx = \int_0^\pi \frac{1}{4}\, dx - \frac{1}{2}\int_0^\pi \cos 2nx\, dx + \frac{1}{4}\int_0^\pi \cos^2 2nx\, dx$

$\displaystyle = \frac{\pi}{4} - \frac{1}{2}\left(\frac{\sin 2nx}{2n}\right)\Big|_0^\pi + \frac{1}{4}\cdot\frac{\pi}{2} = \frac{\pi}{4} - 0 + \frac{\pi}{8} = \frac{3\pi}{8}.$

f. The proof is by strong induction. The base step ($m = 1$) follows by part (a) of this problem. Now suppose the result holds for all positive integer values less than m. Now

$$\int_0^\pi \sin^{2m} x\, dx = -\frac{\sin^{2m-1} x \cos x}{2m}\Big|_0^\pi + \frac{2m-1}{2m}\int_0^\pi \sin^{2m-2} x\, dx$$

$$= \frac{2m-1}{2m}\int_0^\pi \sin^{2(m-1)} x\, dx$$

$$= \frac{2m-1}{2m}\cdot \pi \cdot \frac{1\cdot 3\cdot 5\cdots(2m-3)}{2\cdot 4\cdot 6\cdots 2(m-1)} = \pi\cdot\frac{1\cdot 3\cdot 5\cdots(2m-1)}{2\cdot 4\cdot 6\cdots 2m}.$$

Thus the result holds for all positive numbers m. Note the use of the reduction formula from problem number 64.

To see that the same result holds for $\int_0^\pi \cos^{2m} x\, dx$ first note that $\int_{\pi/2}^\pi \sin^{2m} x\, dx = -\int_{\pi/2}^0 \sin^{2m}(\pi - u)\, du = \int_0^{\pi/2} \sin^{2m} u\, du$ (using the substitution $x = \pi - u$.)

Thus, $\int_0^\pi \cos^{2m} x\, dx = -\int_{\pi/2}^{-\pi/2} \cos^{2m}(\pi/2 - u)\, du$. (Via the substitution $x = \pi/2 - u$.) This last integral can be written as $\int_{-\pi/2}^{\pi/2} \sin^{2m} u\, du = 2\int_0^{\pi/2} \sin^{2m} u\, du = \int_0^{\pi/2} \sin^{2m} u\, du + \int_{\pi/2}^\pi \sin^{2m} u\, du = \int_0^\pi \sin^{2m} x\, dx$.

8.4   Trigonometric Substitutions

8.4.1 This would suggest $x = 3\sec\theta$, because then $\sqrt{x^2 - 9} = 3\sqrt{\sec^2\theta - 1} = 3\sqrt{\tan^2\theta} = 3\tan\theta$, for $\theta \in [0, \pi/2)$.

8.4.3 This would suggest $x = 10\sin\theta$, because then $\sqrt{100 - x^2} = 10\sqrt{1 - \sin^2\theta} = 10\sqrt{\cos^2\theta} = 10\cos\theta$, for $|\theta| \le \frac{\pi}{2}$.

8.4.5

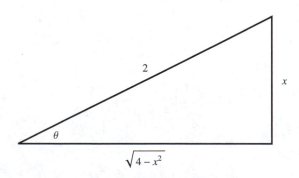

If $x = 2\sin\theta$ then $\frac{x^2}{4} = \sin^2\theta = \frac{1}{\csc^2\theta}$. Then $\cot^2\theta = \csc^2\theta - 1 = \frac{4}{x^2} - 1 = \frac{4-x^2}{x^2}$. So $\cot\theta = \frac{\sqrt{4-x^2}}{x}$ for $0 < |\theta| \le \frac{\pi}{2}$.

8.4.7 Let $x = 5\sin\theta$, so that $dx = 5\cos\theta\, d\theta$. Note that $\sqrt{25 - x^2} = 5\cos\theta$. Then $\int_0^{5/2} \frac{1}{\sqrt{25 - x^2}}\, dx = \int_0^{\pi/6} \frac{5\cos\theta}{5\cos\theta}\, d\theta = \frac{\pi}{6}$.

8.4.9 Let $x = 10 \sin \theta$ so that $dx = 10 \cos \theta \, d\theta$. Note that $\sqrt{100 - x^2} = 10 \cos \theta$. Then $\displaystyle\int_5^{10} \sqrt{100 - x^2} \, dx =$

$100 \displaystyle\int_{\pi/6}^{\pi/2} \cos^2 \theta \, d\theta = 50 \int_{\pi/6}^{\pi/2} (1 + \cos 2\theta) \, d\theta = 50 \left(\theta + \frac{\sin 2\theta}{2} \right) \Big|_{\pi/6}^{\pi/2} = 50 \left(\frac{\pi}{2} + 0 - \left(\frac{\pi}{6} + \frac{\sqrt{3}}{4} \right) \right) = \frac{50\pi}{3} - \frac{25\sqrt{3}}{2}$.

8.4.11 Let $x = \sin \theta$ so that $dx = \cos \theta \, d\theta$. Note that $\sqrt{1 - x^2} = \sqrt{1 - \sin^2 \theta} = \cos \theta$. Substituting gives
$\displaystyle\int_0^{\pi/6} \sin^2 \theta \, d\theta = \int_0^{\pi/6} \frac{1 - \cos 2\theta}{2} \, d\theta = \frac{\theta}{2} \Big|_0^{\pi/6} - \frac{\sin 2\theta}{4} \Big|_0^{\pi/6} = \frac{\pi}{12} - \frac{\sqrt{3}}{8}$.

8.4.13 Let $x = 4 \sin \theta$ so that $dx = 4 \cos \theta \, d\theta$. Note that $\sqrt{16 - x^2} = 4 \cos \theta$. Thus, $\displaystyle\int \frac{1}{\sqrt{16 - x^2}} \, dx = \int \frac{4 \cos \theta}{4 \cos \theta} \, d\theta = \theta + C = \sin^{-1} \left(\frac{x}{4} \right) + C$.

8.4.15 Let $x = 3 \sin \theta$ so that $dx = 3 \cos \theta \, d\theta$ and $\sqrt{9 - x^2} = 3 \cos \theta$. Then

$$\int \frac{\sqrt{9 - x^2}}{x} \, dx = \int \frac{3 \cos \theta \cdot 3 \cos \theta}{3 \sin \theta} \, d\theta = 3 \int \frac{1 - \sin^2 \theta}{\sin \theta} \, d\theta$$

$$= 3 \left(\int \csc \theta \, d\theta - \int \sin \theta \, d\theta \right) \, d\theta = 3 \left(-\ln|\csc \theta + \cot \theta| + \cos \theta \right)$$

$$= -3 \ln \left| \frac{3}{x} + \frac{\sqrt{9 - x^2}}{x} \right| + \sqrt{9 - x^2} + C.$$

8.4.17 Let $x = 8 \sin \theta$ so that $dx = 8 \cos \theta \, d\theta$ and $\sqrt{64 - x^2} = 8 \cos \theta$. Then,

$$\int \sqrt{64 - x^2} \, dx = \int 64 \cos^2 \theta \, d\theta = 32 \int (1 + \cos 2\theta) \, d\theta = 32\theta + 16 \sin 2\theta + C$$

$$= 32\theta + 32 \sin \theta \cos \theta + C = 32 \sin^{-1} \left(\frac{x}{8} \right) + \frac{x\sqrt{64 - x^2}}{2} + C.$$

8.4.19 Let $x = \sin \theta$ so that $dx = \cos \theta \, d\theta$. Note that $\sqrt{1 - x^2} = \sqrt{1 - \sin^2 \theta} = \cos \theta$. Substituting gives
$\displaystyle\int \sec^2 \theta \, d\theta = \tan \theta + C = \tan(\sin^{-1}(x)) + C = \frac{x}{\sqrt{1 - x^2}} + C$.

8.4.21 Let $x = 3 \tan \theta$ so that $dx = 3 \sec^2 \theta \, d\theta$. Note that $\sqrt{x^2 + 9} = \sqrt{9(\tan^2 \theta + 1)} = 3 \sec \theta$. Substituting gives $\displaystyle\int \frac{1}{9} \cot \theta \csc \theta \, d\theta = -\frac{1}{9} \csc \theta + C = -\frac{1}{9} \csc(\tan^{-1}(x/3)) + C = \frac{-\sqrt{x^2 + 9}}{9x} + C$.

8.4.23 Let $x = 6 \sin \theta$ so that $dx = 6 \cos \theta \, d\theta$ and $\sqrt{36 - x^2} = 6 \cos \theta$. Then $\displaystyle\int \frac{1}{\sqrt{36 - x^2}} \, dx = \int \frac{6 \cos \theta}{6 \cos \theta} \, d\theta = \theta + C = \sin^{-1} \left(\frac{x}{6} \right) + C$.

8.4.25 Let $x = 9 \sec \theta$ with $\theta \in (0, \pi/2)$. Then $dx = 9 \sec \theta \tan \theta \, d\theta$ and $\sqrt{x^2 - 81} = 9 \tan \theta$. Then $\displaystyle\int \frac{1}{\sqrt{x^2 - 81}} \, dx = \int \frac{9 \sec \theta \tan \theta}{9 \tan \theta} \, d\theta = \int \sec \theta \, d\theta = \ln|\sec \theta + \tan \theta| + C = \ln \left| \frac{x}{9} + \frac{\sqrt{x^2 - 81}}{9} \right| + C$. Note that because $x > 9$, the absolute value signs are unnecessary, and the final result can be written as $\ln(\sqrt{x^2 - 81} + x) + C$.

8.4.27 Let $x = \frac{\tan \theta}{2}$ so that $dx = \frac{\sec^2 \theta}{2} \, d\theta$ and $\sqrt{1 + 4x^2} = \sec \theta$. Then $\displaystyle\int \frac{1}{(1 + 4x^2)^{3/2}} \, dx = \frac{1}{2} \int \frac{\sec^2 \theta}{\sec^3 \theta} \, d\theta = \frac{1}{2} \int \cos \theta \, d\theta = \frac{\sin \theta}{2} + C = \frac{x}{\sqrt{1 + 4x^2}} + C$.

8.4.29 Let $x = 4\sin\theta$ so that $dx = 4\cos\theta\, d\theta$. Note that $\sqrt{16 - x^2} = 4\cos\theta$. Then $\displaystyle\int \frac{x^2}{\sqrt{16 - x^2}}\, dx =$

$\displaystyle\int \frac{16\sin^2\theta \cdot 4\cos\theta}{4\cos\theta}\, d\theta = 16\int \sin^2\theta\, d\theta = 8\int (1 - \cos 2\theta)\, d\theta = 8\left(\theta - \frac{\sin 2\theta}{2}\right) + C = 8\theta - 8\sin\theta\cos\theta + C =$

$8\sin^{-1}\left(\dfrac{x}{4}\right) - \dfrac{x\sqrt{16 - x^2}}{2} + C.$

8.4.31 Let $x = 3\sec\theta$ where $\theta \in (0, \pi/2)$. Then $dx = 3\sec\theta\tan\theta$ and $\sqrt{x^2 - 9} = 3\tan\theta$. Thus we

have $\displaystyle\int \frac{\sqrt{x^2 - 9}}{x}\, dx = \int \frac{3\sec\theta\tan\theta \cdot 3\tan\theta}{3\sec\theta}\, d\theta = 3\int \tan^2\theta\, d\theta = 3\int \sec^2\theta - 1\, d\theta = 3\left(\tan\theta - \theta\right) + C =$

$\sqrt{x^2 - 9} - 3\tan^{-1}\left(\dfrac{\sqrt{x^2 - 9}}{3}\right) + C = \sqrt{x^2 - 9} - 3\sec^{-1}(x/3) + C.$

8.4.33 Let $x = 2\tan\theta$ so that $dx = 2\sec^2\theta\, d\theta$. Note that $\sqrt{4 + x^2} = 2\sec\theta$. Then $\displaystyle\int \frac{x^2}{\sqrt{4 + x^2}}\, dx =$

$\displaystyle\int \frac{4\tan^2\theta \cdot 2\sec^2\theta}{2\sec\theta}\, d\theta = 4\int \tan^2\theta\sec\theta\, d\theta = 4\int (\sec^2\theta - 1)\sec\theta\, d\theta = 4\left(\int \sec^3\theta\, d\theta - \int \sec\theta\, d\theta\right) =$

$4\left(\dfrac{1}{2}\left(\sec\theta\tan\theta + \int \sec\theta\, d\theta\right) - \int \sec\theta\, d\theta\right) = 2\sec\theta\tan\theta - 2\int \sec\theta\, d\theta = 2\sec\theta\tan\theta - 2\ln|\sec\theta + \tan\theta| +$

$C = \dfrac{x\sqrt{4 + x^2}}{2} - 2\ln\left|\dfrac{\sqrt{4 + x^2}}{2} + \dfrac{x}{2}\right| + C.$ This can be written as $\dfrac{x\sqrt{4+x^2}}{2} - 2\ln(x + \sqrt{4 + x^2}) + C.$

8.4.35 $\displaystyle\int \frac{1}{\sqrt{3 - 2x - x^2}}\, dx = \int \frac{1}{\sqrt{4 - (x + 1)^2}}\, dx = \int \frac{1}{\sqrt{4 - u^2}}\, du$ where $u = x + 1$. Then let $u = 2\sin\theta$

so that $du = 2\cos\theta\, d\theta$. We have $\displaystyle\int \frac{2\cos\theta}{2\cos\theta}\, d\theta = \theta + C = \sin^{-1}\left(\dfrac{x + 1}{2}\right) + C.$

8.4.37 Let $x = \frac{5}{3}\sec\theta$ where $\theta \in [0, \pi/2)$. Then $dx = \frac{5}{3}\sec\theta\tan\theta\, d\theta$ and $\sqrt{9x^2 - 25} = 5\tan\theta$. Thus,

$\displaystyle\int \frac{\sqrt{9x^2 - 25}}{x^3}\, dx = \int \frac{5\tan\theta \cdot \frac{5}{3}\cdot\sec\theta\tan\theta}{\frac{125}{27}\sec^3\theta}\, d\theta = \frac{9}{5}\int \frac{\tan^2\theta}{\sec^2\theta}\, d\theta = \frac{9}{5}\int \frac{\sec^2\theta - 1}{\sec^2\theta}\, d\theta = \frac{9}{5}\int 1 - \cos^2\theta\, d\theta =$

$\dfrac{9}{5}\int \sin^2\theta\, d\theta = \dfrac{9}{10}\int 1 - \cos 2\theta\, d\theta = \dfrac{9\theta}{10} - \dfrac{9\sin 2\theta}{20} + C = \dfrac{9\theta}{10} - \dfrac{9\sin\theta\cos\theta}{10} = \dfrac{9\cos^{-1}(5/3x)}{10} - \dfrac{\sqrt{9x^2 - 25}}{2x^2} + C.$

8.4.39 Let $x = 5\tan\theta$ so that $dx = 5\sec^2\theta\, d\theta$. Note that $25 + x^2 = 25\sec^2\theta$. Thus, $\displaystyle\int \frac{x^2}{(25 + x^2)^2}\, dx =$

$\displaystyle\int \frac{25\tan^2\theta \cdot 5\sec^2\theta}{25^2\sec^4\theta}\, d\theta = \frac{1}{5}\int \frac{\tan^2\theta}{\sec^2\theta}\, d\theta = \frac{1}{5}\int \frac{\sec^2\theta - 1}{\sec^2\theta}\, d\theta = \frac{1}{5}\int (1 - \cos^2\theta)\, d\theta = \frac{1}{5}\int \sin^2\theta\, d\theta =$

$\dfrac{1}{10}\int 1 - \cos 2\theta\, d\theta = \dfrac{1}{10}\left(\theta - \dfrac{\sin 2\theta}{2}\right) + C = \dfrac{1}{10}\left(\theta - \sin\theta\cos\theta\right) + C = \dfrac{1}{10}\left(\tan^{-1}(x/5) - \dfrac{5x}{25 + x^2}\right) + C.$

8.4.41 Let $x = 10\sin\theta$ so that $dx = 10\cos\theta\, d\theta$. Note that $\sqrt{100 - x^2} = 10\cos\theta$. Thus,

$$\int \frac{x^2}{(100 - x^2)^{3/2}}\, dx = \int \frac{100\sin^2\theta \cdot 10\cos\theta}{1000\cos^3\theta}\, d\theta = \int \tan^2\theta\, d\theta$$

$$= \int (\sec^2\theta - 1)\, d\theta = \tan\theta - \theta + C = \frac{x}{\sqrt{100 - x^2}} - \sin^{-1}(x/10) + C.$$

8.4.43 Let $x = 9\sin\theta$ so that $dx = 9\cos\theta\, d\theta$. Note that $81 - x^2 = 81\cos^2\theta$. Thus,

$$\int \frac{x^3}{(81 - x^2)^2}\, dx = \int \frac{9^3\sin^3\theta \cdot 9\cos\theta}{9^4\cos^4\theta}\, d\theta = \int \tan^3\theta\, d\theta$$

$$= \int (\tan\theta)(\sec^2\theta - 1)\, d\theta = \int \sec^2\theta\tan\theta\, d\theta - \int \tan\theta\, d\theta$$

$$= \frac{\sec^2\theta}{2} + \ln|\cos\theta| + C = \frac{81}{2(81 - x^2)} + \ln\left|\frac{\sqrt{81 - x^2}}{9}\right| + C.$$

This can be written as $\frac{81}{2(81-x^2)} + \ln\sqrt{81-x^2} + C$.

8.4.45 Let $x = \sec\theta$ where $\theta \in (0, \pi/2)$. Then $dx = \sec\theta\tan\theta\,d\theta$ and $\sqrt{x^2-1} = \tan\theta$. Then

$$\int \frac{1}{x(x^2-1)^{3/2}}\,dx = \int \frac{\sec\theta\tan\theta}{\sec\theta\tan^3\theta}\,d\theta = \int \cot^2\theta\,d\theta$$

$$= \int \csc^2\theta - 1\,d\theta = -\cot\theta - \theta + C = -\frac{1}{\sqrt{x^2-1}} - \sec^{-1}x + C.$$

8.4.47 Let $x = 4\tan\theta$ so that $dx = 4\sec^2\theta\,d\theta$. Note that $\sqrt{x^2+16} = 4\sec\theta$. Thus, $\displaystyle\int_0^1 \frac{1}{\sqrt{x^2+16}}\,dx =$

$$\int_0^{\tan^{-1}(1/4)} \frac{4\sec^2\theta}{4\sec\theta}\,d\theta = \int_0^{\tan^{-1}(1/4)} \sec\theta\,d\theta = \ln|\sec\theta + \tan\theta|\,\Big|_0^{\tan^{-1}(1/4)} = \ln\left(\frac{\sqrt{17}+1}{4}\right).$$

8.4.49 Let $x = \tan\theta$ so that $dx = \sec^2\theta\,d\theta$ and $\sqrt{1+x^2} = \sec\theta$. Substituting gives $\int_{\pi/6}^{\pi/4} \cot\theta\csc\theta\,d\theta =$

$$(-\csc\theta)\,\Big|_{\pi/6}^{\pi/4} = -(\sqrt{2} - 2) = 2 - \sqrt{2}.$$

8.4.51 Let $x = \tan\theta$ so that $dx = \sec^2\theta\,d\theta$. Note that $\sqrt{x^2+1} = \sqrt{\tan^2\theta+1} = \sqrt{\sec^2\theta} = \sec\theta$. Substituting gives $\int_0^{\pi/6} \sec^3\theta\,d\theta$. Recall from section 7.2 number 67 that $\int \sec^3\theta\,d\theta = \frac{1}{2}\sec\theta\tan\theta + \frac{1}{2}\ln|\sec\theta + \tan\theta|$.

Thus the original integral is equal to $\left(\frac{1}{2}\sec\theta\tan\theta + \frac{1}{2}\ln|\sec\theta + \tan\theta|\right)\Big|_0^{\pi/6} = \frac{1\cdot 2\cdot 1}{2\cdot\sqrt{3}\cdot\sqrt{3}} + \frac{1}{2}\ln\left(\frac{2}{\sqrt{3}} + \frac{1}{\sqrt{3}}\right) = $

$\frac{1}{3} + \frac{\ln 3}{4}$.

8.4.53 Let $x = \frac{1}{3}\tan\theta$ so that $dx = \frac{1}{3}\sec^2\theta\,d\theta$. Note that $\sqrt{9x^2+1} = \sec\theta$. Thus $\displaystyle\int_0^{1/3} \frac{1}{(9x^2+1)^{3/2}}\,dx =$

$$\int_0^{\pi/4} \frac{\frac{1}{3}\sec^2\theta}{\sec^3\theta}\,d\theta = \frac{1}{3}\int_0^{\pi/4} \cos\theta\,d\theta = \frac{1}{3}\sin\theta\,\Big|_0^{\pi/4} = \frac{\sqrt{2}}{6}.$$

8.4.55 Let $x = 2\sec\theta$ so that $dx = 2\sec\theta\tan\theta\,d\theta$ and $x^2 - 4 = 4\tan^2\theta$. Thus, $\displaystyle\int_{4/\sqrt{3}}^4 \frac{1}{x^2(x^2-4)}\,dx =$

$$\int_{\pi/6}^{\pi/3} \frac{2\sec\theta\tan\theta}{4\sec^2\theta\cdot 4\tan^2\theta}\,d\theta = \frac{1}{8}\int_{\pi/6}^{\pi/3} \frac{\cos^2\theta}{\sin\theta}\,d\theta = \frac{1}{8}\int_{\pi/6}^{\pi/3} \frac{1-\sin^2\theta}{\sin\theta}\,d\theta = \frac{1}{8}\int_{\pi/6}^{\pi/3} \csc\theta - \sin\theta\,d\theta =$$

$$\frac{1}{8}\left(-\ln|\csc\theta + \cot\theta| + \cos\theta\right)\Big|_{\pi/6}^{\pi/3} = \frac{1}{8}\left(-\ln(\sqrt{3}(2-\sqrt{3})) + \frac{1-\sqrt{3}}{2}\right) = \frac{1}{16}[1 - \sqrt{3} - \ln(21 - 12\sqrt{3})].$$

8.4.57

a. False. In fact, we would have $\csc\theta = \frac{\sqrt{x^2+16}}{x}$.

b. True. Almost every number in the interval $[1, 2]$ is not in the domain of $\sqrt{1-x^2}$, so this integral isn't defined.

c. False. It does represent a finite real number, because $\sqrt{x^2-1}$ is continuous on the interval $[1, 2]$.

d. False. It can be so evaluated. The integral is equivalent to $\displaystyle\int \frac{1}{(x+2)^2 + 5}\,dx$, and this can be evaluated by the substitution $x + 2 = \sqrt{5}\tan\theta$.

8.4.59 Note that the given integral can be written $\displaystyle\int \frac{1}{(x+3)^2 + 9}\,dx = \int \frac{1}{u^2 + 9}\,du$ where $u = x + 3$. Now let $u = 3\tan\theta$ so that $du = 3\sec^2\theta\,d\theta$ and $u^2 + 9 = 9\sec^2\theta$. Thus we have

$$\int \frac{3\sec^2\theta}{9\sec^2\theta}\,d\theta = \frac{\theta}{3} + C = \frac{\tan^{-1}((x+3)/3)}{3} + C.$$

8.4.61 Note that the given integral can be written as $\displaystyle\int \frac{(x-1)^2}{\sqrt{(x-1)^2+9}}\,dx = \int \frac{u^2}{\sqrt{u^2+9}}\,du$ where $u = x-1$.
Now let $u = 3\tan\theta$ so that $du = 3\sec^2\theta\,d\theta$ and $u^2 + 9 = 9\sec^2\theta$. Thus we have

$$\int \frac{9\tan^2\theta \cdot 3\sec^2\theta}{3\sec\theta}\,d\theta = 9\int \tan^2\theta\sec\theta\,d\theta = 9\int (\sec^2\theta - 1)(\sec\theta)\,d\theta$$

$$= 9\int (\sec^3\theta - \sec\theta)\,d\theta = 9\left(\frac{1}{2}\sec\theta\tan\theta + \frac{1}{2}\int \sec\theta\,d\theta - \int \sec\theta\,d\theta\right)$$

$$= \frac{9}{2}\left(\sec\theta\tan\theta - \int \sec\theta\,d\theta\right)$$

$$= \frac{9}{2}(\sec\theta\tan\theta - \ln|\sec\theta + \tan\theta|) + C$$

$$= \frac{9}{2}\left(\frac{(x-1)\sqrt{(x-1)^2+9}}{9} - \ln\left|\frac{\sqrt{(x-1)^2+9}}{3} + \frac{x-1}{3}\right|\right) + C.$$

This can be written as $\frac{x-1}{2}\sqrt{x^2 - 2x + 10} - \frac{9}{2}\ln(x-1+\sqrt{x^2-2x+10}) + C$. Note that in the middle of
this derivation we used the reduction formula for $\displaystyle\int \sec^3\theta\,d\theta$ given in problem 54 in the previous section.

8.4.63 Note that the given integral can be written as $\displaystyle\int \frac{(x-4)^2}{(25-(x-4)^2)^{3/2}}\,dx$. Let $u = x-4$, and note that

we have $\displaystyle\int \frac{u^2}{(25-u^2)^{3/2}}\,du$. Now let $u = 5\sin\theta$ so that $du = 5\cos\theta\,d\theta$, and note that $\sqrt{25-u^2} = 5\cos\theta$.

Thus we have $\displaystyle\int \frac{25\sin^2\theta \cdot 5\cos\theta}{5^3\cos^3\theta}\,d\theta = \int \tan^2\theta\,d\theta = \int \sec^2\theta - 1\,d\theta = \tan\theta - \theta + C = \frac{x-4}{\sqrt{25-(x-4)^2}} - $

$\sin^{-1}\left(\dfrac{x-4}{5}\right) + C.$

8.4.65 $\displaystyle\int_{1/2}^{(\sqrt{2}+3)/2\sqrt{2}} \frac{1}{8x^2 - 8x + 11}\,dx = \int_{1/2}^{(\sqrt{2}+3)/2\sqrt{2}} \frac{1}{8(x-1/2)^2 + 9}\,dx$. Let $u = x - 1/2$, so that our

integral becomes $\displaystyle\int_0^{3/2\sqrt{2}} \frac{1}{8u^2 + 9}\,du$. Now let $u = \frac{3}{\sqrt{8}}\tan\theta$ so that $du = \frac{3}{\sqrt{8}}\sec^2\theta\,d\theta$. Substituting gives

$\displaystyle\int_0^{\pi/4} \frac{\frac{3}{\sqrt{8}}\sec^2\theta}{9\sec^2\theta}\,d\theta = \frac{1}{6\sqrt{2}}\int d\theta = \frac{1}{6\sqrt{2}}\theta\Big|_0^{\pi/4} = \frac{\pi\sqrt{2}}{48}.$

8.4.67

a. Recall that the area of a circular sector subtended by an angle θ is given by $\frac{\theta r^2}{2}$. So the area of the
    cap is this area minus the area of the isosceles triangle with two sides of length r and angle between
    them θ. So $A_{\text{cap}} = A_{\text{sector}} - A_{\text{triangle}} = \frac{\theta r^2}{2} - \frac{r^2\sin\theta}{2} = \frac{r^2}{2}(\theta - \sin\theta).$

b. For a cap we have $0 \le \theta \le \pi$ so $0 \le \theta/2 \le \pi/2$. By symmetry, $\dfrac{A_{\text{cap}}}{2} = \displaystyle\int_{r\cos\theta/2}^{r} \sqrt{r^2 - x^2}\,dx$. Let

    $x = r\cos\alpha/2$ so that $dx = -\frac{r}{2}\sin\alpha/2\,d\alpha$. Then we have $\dfrac{A_{\text{cap}}}{2} = \displaystyle\int_\theta^0 r\sin(\alpha/2) \cdot -\frac{r}{2}\sin(\alpha/2)\,d\alpha = $

    $\dfrac{r^2}{2}\displaystyle\int_0^\theta \sin^2(\alpha/2)\,d\alpha = \frac{r^2}{4}\int_0^\theta (1 - \cos\alpha)\,d\alpha = \frac{r^2}{4}(\alpha - \sin\alpha)\Big|_0^\theta = \frac{r^2}{4}(\theta - \sin\theta).$

    Thus $A_{\text{cap}} = \frac{r^2}{2}(\theta - \sin\theta).$

8.4.69

a. The area is given by $\int_0^4 \frac{1}{\sqrt{9+x^2}}\,dx$. Let $x = 3\tan\theta$, so that $dx = 3\sec^2\theta\,d\theta$. Substituting yields

$$\int_0^{\tan^{-1}(4/3)} \frac{3\sec^2\theta}{3\sec\theta}\,d\theta = \int_0^{\tan^{-1}(4/3)} \sec\theta\,d\theta = \ln|\sec\theta + \tan\theta|\,\Big|_0^{\tan^{-1}(4/3)} = \ln\left(\frac{5}{3}+\frac{4}{3}\right) = \ln 3.$$

b. Using disks, we have $\dfrac{V}{\pi} = \int_0^4 \dfrac{1}{9+x^2}\,dx$. Let $x = 3\tan\theta$ so that $dx = 3\sec^2\theta\,d\theta$. Substituting yields

$$\int_0^{\tan^{-1}(4/3)} \frac{3\sec^2\theta}{9\sec^2\theta}\,d\theta = \frac{1}{3}\int_0^{\tan^{-1}(4/3)} 1\,d\theta = \frac{1}{3}\theta\,\Big|_0^{\tan^{-1}(4/3)} = \frac{1}{3}\tan^{-1}(4/3). \text{ Thus } V = \frac{\pi}{3}\tan^{-1}(4/3).$$

c. Using shells, we have $\dfrac{V}{2\pi} = \int_0^4 \dfrac{x}{(9+x^2)^{1/2}}\,dx$. Let $u = 9 + x^2$, so that $du = 2x\,dx$. Then we have

$$\frac{1}{2}\int_9^{25} u^{-1/2}\,du = \sqrt{u}\,\Big|_9^{25} = 5 - 3 = 2. \text{ So } V = 4\pi.$$

8.4.71 Because $y = ax^2$, we have $1 + \left(\dfrac{dy}{dx}\right)^2 = 1 + 4a^2x^2$, so the arc length is given by $\int_0^{10} \sqrt{1 + 4a^2x^2}\,dx$.
Let $x = \frac{1}{2a}\tan\theta$ so that $dx = \frac{1}{2a}\sec^2\theta\,d\theta$. Then we have

$$\frac{1}{2a}\int_0^{\tan^{-1}(20a)} \sec^2\theta\sec\theta\,d\theta = \frac{1}{2a}\int_0^{\tan^{-1}(20a)} \sec^3\theta\,d\theta = \frac{1}{4a}\left(\sec\theta\tan\theta + \ln|\sec\theta + \tan\theta|\right)\Big|_0^{\tan^{-1}(20a)}$$

$$= \frac{1}{4a}\left(\sqrt{1 + 400a^2}\,(20a) + \ln(\sqrt{1 + 400a^2} + 20a)\right).$$

8.4.73

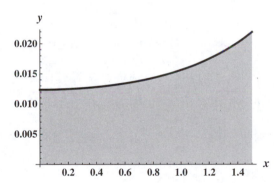

The area is given by $\int_0^{3/2} \dfrac{1}{(9-x^2)^2}\,dx$. Let $x = 3\sin\theta$ so that $dx = 3\cos\theta\,d\theta$. Then we have

$$\int_0^{\pi/6} \frac{3\cos\theta}{9^2\cos^4\theta}\,d\theta = \frac{1}{27}\int_0^{\pi/6} \sec^3\theta\,d\theta = \frac{1}{54}\left(\sec\theta\tan\theta + \ln|\sec\theta + \tan\theta|\right)\Big|_0^{\pi/6}$$

$$= \frac{1}{54}\left(\frac{2}{\sqrt{3}}\cdot\frac{1}{\sqrt{3}} + \ln\left(\frac{2}{\sqrt{3}} + \frac{1}{\sqrt{3}}\right)\right) = \frac{1}{81} + \frac{\ln(\sqrt{3})}{54}.$$

8.4.75

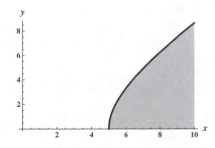

The area is given by $\int_5^{10} \sqrt{x^2 - 25}\, dx$. Let $x = 5\sec\theta$ so that $dx = 5\sec\theta\tan\theta\, d\theta$. Then we have

$$\int_0^{\pi/3} 5\sec\theta\tan\theta \cdot 5\tan\theta\, d\theta = 25\int_0^{\pi/3} \sec\theta\tan^2\theta\, d\theta = 25\int_0^{\pi/3}\sec\theta(\sec^2\theta - 1)\, d\theta$$

$$= 25\int_0^{\pi/3}(\sec^3\theta - \sec\theta)\, d\theta = \frac{25}{2}\left(\sec\theta\tan\theta - \ln|\sec\theta + \tan\theta|\right)\Big|_0^{\pi/3}$$

$$= \frac{25}{2}\left(2\sqrt{3} - \ln(2 + \sqrt{3})\right) = 25\sqrt{3} - \frac{25}{2}\cdot\ln(2 + \sqrt{3}).$$

8.4.77 $\int_{2+\sqrt{2}}^4 \dfrac{1}{\sqrt{(x-1)(x-3)}}\, dx = \int_{2+\sqrt{2}}^4 \dfrac{1}{\sqrt{(x-2)^2 - 1}}\, dx = \int_{\sqrt{2}}^2 \dfrac{1}{\sqrt{u^2 - 1}}\, du$, where $u = x - 2$. Now let

$u = \sec\theta$, so that $du = \sec\theta\tan\theta\, d\theta$. Then $\int_{\pi/4}^{\pi/3} \dfrac{\sec\theta\tan\theta}{\tan\theta}\, d\theta = \int_{\pi/4}^{\pi/3}\sec\theta\, d\theta = \ln(\sec\theta + \tan\theta)\Big|_{\pi/4}^{\pi/3} =$

$\ln(2 + \sqrt{3}) - \ln(\sqrt{2} + 1) = \ln\left(\dfrac{2 + \sqrt{3}}{\sqrt{2} + 1}\right) = \ln((2 + \sqrt{3})(\sqrt{2} - 1))$.

8.4.79 Using washers, $V = \int_{-4}^4 A(x)\, dx$. Now $A(x) = \pi((f(x))^2 - (g(x))^2) = \pi((6 + \sqrt{16 - x^2})^2 -$

$(6 - \sqrt{16 - x^2})^2) = 24\pi\sqrt{16 - x^2}$. Thus our integral is $\int_{-4}^4 24\pi\sqrt{16 - x^2}\, dx = 48\pi\int_0^4 \sqrt{16 - x^2}\, dx =$

$48\pi\left(\dfrac{\pi\cdot 4^2}{4}\right) = 192\pi^2$.

8.4.81

a. $E_x(a) = \dfrac{kQa}{2L}\int_{-L}^L \dfrac{dy}{(a^2 + y^2)^{3/2}}$. Let $y = a\tan\theta$ so that $dy = a\sec^2\theta\, d\theta$. Then note that

$$\int_{-L}^L \frac{dy}{(a^2 + y^2)^{3/2}} = 2\int_0^L \frac{dy}{(a^2 + y^2)^{3/2}} = 2\int_0^{\tan^{-1}(L/a)} \frac{a\sec^2\theta}{a^3\sec^3\theta}\, d\theta = \frac{2}{a^2}\int_0^{\tan^{-1}(L/a)}\cos\theta\, d\theta$$

$$= \frac{2}{a^2}\sin\theta\Big|_0^{\tan^{-1}(L/a)} = \frac{2}{a^2}\cdot\frac{L}{\sqrt{a^2 + L^2}} = \frac{2L}{a^2\sqrt{a^2 + L^2}}.$$

Thus, $E_x(a) = \dfrac{kQ}{a\sqrt{a^2 + L^2}}$.

b. Set $\rho = Q/(2L)$. Then because $\displaystyle\lim_{L\to\infty}\dfrac{2L}{a^2\sqrt{a^2 + L^2}} = \dfrac{2}{a^2}$, we have

$$E_x(a, 0) \approx \frac{kQa}{2L}\lim_{L\to\infty}\int_{-L}^L \frac{dy}{(a^2 + y^2)^{3/2}} = \frac{kQa}{2L}\lim_{L\to\infty}\frac{2L}{a^2\sqrt{a^2 + L^2}} = \frac{kQa}{2L}\left(\frac{2}{a^2}\right) = \frac{2kQ}{2aL} = \frac{2k\rho}{a}.$$

8.4.83

a. Because $t \in [0, \pi]$ so that $\sin t \geq 0$, we have

$$\int_a^b \sqrt{\frac{1 - \cos t}{g(\cos a - \cos t)}}\, dt = \int_a^b \sqrt{\frac{(1 - \cos t)(1 + \cos t)}{g(1 + \cos t)(\cos a - \cos t)}}\, dt = \int_a^b \sin t\sqrt{\frac{1}{g(1 + \cos t)(\cos a - \cos t)}}\, dt.$$

Let $u = \cos t$ so that $du = -\sin t\, dt$. Then the given integral is equal to

$$-\frac{1}{\sqrt{g}}\int_{\cos a}^{\cos b}\sqrt{\frac{1}{(1 + u)(\cos a - u)}}\, du.$$

Now we complete the square:

$$(1+u)(\cos a - u) = \cos a + (\cos a - 1)u - u^2$$

$$= -\left(u^2 - (\cos a - 1)u + \left(\frac{\cos a - 1}{2}\right)^2 - \left(\frac{\cos a - 1}{2}\right)^2\right) + \cos a$$

$$= \cos a + \left(\frac{\cos a - 1}{2}\right)^2 - \left(u - \frac{\cos a - 1}{2}\right)^2 = \left(\frac{\cos a + 1}{2}\right)^2 - \left(u - \frac{\cos a - 1}{2}\right)^2.$$

Thus, setting $v = u - \frac{\cos a - 1}{2}$ we have that the original integral is equal to

$$-\frac{1}{\sqrt{g}}\int_{(\cos a + 1)/2}^{\cos b - \frac{\cos a - 1}{2}} \frac{1}{\sqrt{k^2 - v^2}}\, dv \text{ where } k = \frac{(\cos a + 1)}{2}.$$

Now, $\int \frac{1}{\sqrt{k^2 - v^2}}\, dv = \int \frac{k\cos\theta}{k\cos\theta}\, d\theta = \theta + C = \sin^{-1}(v/k) + C$ where $v = k\sin\theta$.

Therefore, the original integral is equal to

$$-\frac{1}{\sqrt{g}}\sin^{-1}\left(\frac{2v}{\cos a + 1}\right)\Bigg|_{(\cos a + 1)/2}^{\cos b - (\cos a - 1)/2} = \frac{1}{\sqrt{g}}\left(\sin^{-1}\left(\frac{\cos a + 1}{\cos a + 1}\right) - \sin^{-1}\left(\frac{2\cos b - \cos a + 1}{\cos a + 1}\right)\right)$$

$$= \frac{1}{\sqrt{g}}\left(\frac{\pi}{2} - \sin^{-1}\left(\frac{2\cos b - \cos a + 1}{\cos a + 1}\right)\right).$$

b. Letting $b = \pi$, we have that the integral is equal to

$$\frac{1}{\sqrt{g}}\left(\frac{\pi}{2} - \sin^{-1}\left(\frac{-2 - \cos a + 1}{\cos a + 1}\right)\right)$$

$$= \frac{1}{\sqrt{g}}\left(\frac{\pi}{2} - \sin^{-1}(-1)\right) = \frac{1}{\sqrt{g}}\left(\frac{\pi}{2} - \left(-\frac{\pi}{2}\right)\right) = \frac{\pi}{\sqrt{g}}.$$

8.4.85 If $1 < x = \sec\theta$ then we need $\theta \in (0, \pi/2)$. Alternatively, if $-1 > x = \sec\theta$, then we need $\theta \in (\pi/2, \pi)$. So, in the former, $\int \frac{1}{x\sqrt{x^2 - 1}}\, dx = \int \frac{\sec\theta\tan\theta}{\sec\theta\tan\theta}\, d\theta = \theta + C = \sec^{-1} x + C = \tan^{-1}\sqrt{x^2 - 1} + C$. In the latter case, $\int \frac{1}{x\sqrt{x^2 - 1}}\, dx = \int \frac{\sec\theta}{\sec\theta(-\tan\theta)}\, d\theta = -\theta + C = -\sec^{-1} x + C = -\tan^{-1}\sqrt{x^2 - 1} + C.$

8.4.87

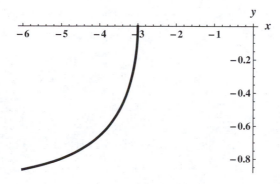

Let $x = 3\sec\theta$ for $x \in (\pi/2, \pi)$. Then $dx = 3\sec\theta\tan\theta\, d\theta$, and note that $|\tan\theta| = -\tan\theta$. We have

$$\int_{-6}^{-3} \frac{\sqrt{x^2 - 9}}{x}\, dx = \int_{2\pi/3}^{\pi} \frac{3\sec\theta\tan\theta(-3\tan\theta)}{3\sec\theta}\, d\theta = -3\int_{2\pi/3}^{\pi} \tan^2\theta\, d\theta$$

$$= -3\left(\tan\theta - \theta\right)\Bigg|_{2\pi/3}^{\pi} = -3\left(-\pi - (-\sqrt{3} - 2\pi/3)\right) = \pi - 3\sqrt{3}.$$

Note that this number is less than zero, as suggested by the graph.

8.4.89

a. The area of sector OAB is given by the formula $\frac{\theta}{2}a^2$ where $\sin\theta = x/a$. The area of triangle OBC is $\frac{x}{2}\sqrt{a^2 - x^2}$. Thus,

$$F(x) = \frac{a^2\sin^{-1}(x/a)}{2} + \frac{x\sqrt{a^2 - x^2}}{2}.$$

b. By the first fundamental theorem of calculus, $F(x)$ is an antiderivative of $\sqrt{a^2 - x^2}$, because $F(x) = \int_0^x \sqrt{a^2 - t^2}\, dt$. Thus, any other antiderivative differs from this by a constant, so

$$\int \sqrt{a^2 - x^2}\, dx = \frac{a^2\sin^{-1}(x/a)}{2} + \frac{x\sqrt{a^2 - x^2}}{2} + C.$$

8.5   Partial Fractions

8.5.1 Proper rational functions can be integrated using partial fraction decomposition.

8.5.3

a. $\dfrac{A}{x - 3}$.

b. $\dfrac{A_1}{x - 4}, \dfrac{A_2}{(x - 4)^2}, \dfrac{A_3}{(x - 4)^3}$.

c $\dfrac{Ax + B}{x^2 + 2x + 6}$.

8.5.5 $\dfrac{2}{x^2 - 2x - 8} = \dfrac{2}{(x - 4)(x + 2)} = \dfrac{A}{x - 4} + \dfrac{B}{x + 2}$. Thus, $2 = A(x + 2) + B(x - 4)$. Equating coefficients gives $A + B = 0$ and $2A - 4B = 2$. Solving this system yields $A = 1/3$ and $B = -1/3$. Thus,

$$\frac{2}{x^2 - 2x - 8} = \frac{1/3}{x - 4} - \frac{1/3}{x + 2}.$$

8.5.7 $\dfrac{5x - 7}{x^2 - 3x + 2} = \dfrac{5x - 7}{(x - 1)(x - 2)} = \dfrac{A}{x - 1} + \dfrac{B}{x - 2}$. Thus, $A(x - 2) + B(x - 1) = 5x - 7$. Equating coefficients gives $A + B = 5$ and $-2A - B = -7$. Solving this system yields $A = 2$, $B = 3$. Thus,

$$\frac{5x - 7}{x^2 - 3x + 2} = \frac{2}{x - 1} + \frac{3}{x - 2}.$$

8.5.9 $\dfrac{x^2}{x^3 - 16x} = \dfrac{x}{(x - 4)(x + 4)} = \dfrac{A}{x - 4} + \dfrac{B}{x + 4}$. Thus, $x = A(x + 4) + B(x - 4)$. Equating coefficients gives $A + B = 1$, $4A - 4B = 0$. Solving this system yields $A = B = 1/2$. Thus,

$$\frac{x^2}{x^3 - 16x} = \frac{1/2}{x - 4} + \frac{1/2}{x + 4}.$$

8.5.11 $\dfrac{x + 2}{x^3 - 3x^2 + 2x} = \dfrac{x + 2}{(x - 1) \cdot x \cdot (x - 2)} = \dfrac{A}{x - 1} + \dfrac{B}{x} + \dfrac{C}{x - 2}$. Thus $x + 2 = Ax(x - 2) + B(x - 1)(x - 2) + Cx(x - 1)$. Letting $x = 1$ gives $A = -3$, letting $x = 0$ gives $B = 1$, and letting $x = 2$ gives $C = 2$. Therefore,

$$\frac{x + 2}{x^3 - 3x^2 + 2x} = \frac{-3}{x - 1} + \frac{1}{x} + \frac{2}{x - 2}.$$

8.5.13 If we write $\dfrac{3}{(x-1)(x+2)} = \dfrac{A}{x-1} + \dfrac{B}{x+2}$, we have $3 = A(x+2) + B(x-1)$. Letting $x = -2$ yields $B = -1$ and letting $x = 1$ yields $A = 1$. Thus, the original integral is equal to $\displaystyle\int \left(\dfrac{1}{x-1} - \dfrac{1}{x+2} \right) dx = \ln|x-1| - \ln|x+2| + C = \ln\left| \dfrac{x-1}{x+2} \right| + C.$

8.5.15 If we write $\dfrac{6}{x^2-1} = \dfrac{6}{(x-1)(x+1)} = \dfrac{A}{x-1} + \dfrac{B}{x+1}$, then we have $6 = A(x+1) + B(x-1)$. Letting $x = -1$ yields $B = -3$ and letting $x = 1$ yields $A = 3$. Thus, the original integral is equal to $\displaystyle\int \left(\dfrac{3}{x-1} - \dfrac{3}{x+1} \right) dx = 3\left(\ln|x-1| - \ln|x+1|\right) + C = 3\ln\left| \dfrac{x-1}{x+1} \right| + C.$

8.5.17 If we write $\dfrac{5x}{x^2-x-6} = \dfrac{A}{x-3} + \dfrac{B}{x+2}$, then we have $5x = A(x+2) + B(x-3)$. Letting $x = -2$ yields $B = 2$ and letting $x = 3$ yields $A = 3$. Thus the original integral is equal to $\displaystyle\int_{-1}^{2} \left(\dfrac{3}{x-3} + \dfrac{2}{x+2} \right) dx = \left. \left(3\ln|x-3| + 2\ln|x+2|\right)\right|_{-1}^{2} = \ln(16) - \ln(64) = -\ln 4.$

8.5.19 If we write $\dfrac{10x}{x^2-2x-24} = \dfrac{10x}{(x-6)(x+4)} = \dfrac{A}{x-6} + \dfrac{B}{x+4}$, then we have $10x = A(x+4) + B(x-6)$. Letting $x = -4$ yields $B = 4$ and letting $x = 6$ yields $A = 6$. Thus the original integral is equal to $\displaystyle\int \left(\dfrac{6}{x-6} + \dfrac{4}{x+4} \right) dx = \ln\left|(x-6)^6(x+4)^4\right| + C.$

8.5.21 Let $\dfrac{6x^2}{x^4-5x^2+4} = \dfrac{6x^2}{(x-2)(x+2)(x-1)(x+1)} = \dfrac{A}{x-2} + \dfrac{B}{x+2} + \dfrac{C}{x-1} + \dfrac{D}{x+1}$. Then $6x^2 = A(x+2)(x-1)(x+1) + B(x-2)(x-1)(x+1) + C(x-2)(x+2)(x+1) + D(x-2)(x+2)(x-1)$. Letting $x = 2$ gives $A = 2$, letting $x = -2$ gives $B = -2$, letting $x = 1$ gives $C = -1$, and letting $x = -1$ gives $D = 1$. Thus, the original integral is equal to $\displaystyle\int \left(\dfrac{2}{x-2} - \dfrac{2}{x+2} - \dfrac{1}{x-1} + \dfrac{1}{x+1} \right) dx = \ln\left| \dfrac{(x-2)^2(x+1)}{(x+2)^2(x-1)} \right| + C$

8.5.23 Let $\dfrac{x^2+12x-4}{x(x-2)(x+2)} = \dfrac{A}{x} + \dfrac{B}{x-2} + \dfrac{C}{x+2}$. Then $x^2+12x-4 = A(x-2)(x+2) + Bx(x+2) + Cx(x-2)$. Letting $x = 0$ gives $A = 1$, letting $x = 2$ gives $B = 3$, and letting $x = -2$ gives $C = -3$. Thus, the original integral is equal to $\displaystyle\int \left(\dfrac{1}{x} + \dfrac{3}{x-2} - \dfrac{3}{x+2} \right) dx = \ln\left| \dfrac{x(x-2)^3}{(x+2)^3} \right| + C.$

8.5.25 If we write $\dfrac{1}{x^4-10x^2+9} = \dfrac{1}{(x-1)(x+1)(x-3)(x+3)} = \dfrac{A}{x-1} + \dfrac{B}{x+1} + \dfrac{C}{x-3} + \dfrac{D}{x+3}$ then $1 = A(x+1)(x-3)(x+3) + B(x-1)(x-3)(x+3) + C(x-1)(x+1)(x+3) + D(x-1)(x+1)(x-3)$. Letting $x = -1$ yields $B = 1/16$. Letting $x = 3$ yields $C = 1/48$. Letting $x = -3$ yields $D = -1/48$, and letting $x = 1$ yields $A = -1/16$. Thus the original integral is equal to

$$\int \left(-\frac{1/16}{x-1} + \frac{1/16}{x+1} + \frac{1/48}{x-3} - \frac{1/48}{x+3} \right) dx$$

$$= -\frac{1}{16}\ln|x-1| + \frac{1}{16}\ln|x+1| + \frac{1}{48}\ln|x-3| - \frac{1}{48}\ln|x+3| + C$$

$$= \ln\left| \frac{(x+1)^3(x-3)}{(x-1)^3(x+3)} \right|^{1/48} + C.$$

8.5.27 If we write $\dfrac{81}{x^3-9x^2} = \dfrac{A}{x} + \dfrac{B}{x^2} + \dfrac{C}{x-9}$, then $81 = Ax(x-9) + B(x-9) + C(x^2)$. Letting $x = 0$ yields $B = -9$. Letting $x = 9$ yields $C = 1$. If we let $x = 10$, then we have $81 = 10A + B + 100C = 10A - 9 + 100$, so $A = -1$. Thus, the original integral is equal to $\displaystyle\int \left(-\dfrac{1}{x} - \dfrac{9}{x^2} + \dfrac{1}{x-9} \right) dx = \ln\left| \dfrac{(x-9)}{x} \right| + \dfrac{9}{x} + C.$

8.5.29 If we write $\dfrac{x}{(x+3)^2} = \dfrac{A}{x+3} + \dfrac{B}{(x+3)^2}$, then we have $x = A(x+3) + B$. Letting $x = -3$ yields $B = -3$, and then letting $x = -2$ yields $A = 1$. Thus the original integral is equal to

$$\int_{-1}^{1} \left(\frac{1}{x+3} - \frac{3}{(x+3)^2} \right) dx = \left(\ln|x+3| + \frac{3}{x+3} \right) \Bigg|_{-1}^{1} = \ln 4 + \frac{3}{4} - \left(\ln 2 + \frac{3}{2} \right) = \ln 2 - \frac{3}{4}.$$

8.5.31 If we write $\dfrac{2}{x^3 + x^2} = \dfrac{A}{x} + \dfrac{B}{x^2} + \dfrac{C}{x+1}$, then $2 = Ax(x+1) + B(x+1) + Cx^2$. Letting $x = 0$ yields $B = 2$, and letting $x = -1$ yields $C = 2$. Then letting $x = 1$ yields $A = -2$. So the original integral is equal to

$$\int \left(-\frac{2}{x} + \frac{2}{x^2} + \frac{2}{x+1} \right) dx = 2\left(\ln|x+1| - \ln|x| \right) - \frac{2}{x} + C.$$

8.5.33 If we write $\dfrac{x-5}{x^2(x+1)} = \dfrac{A}{x} + \dfrac{B}{x^2} + \dfrac{C}{x+1}$, then we have $x - 5 = Ax(x+1) + B(x+1) + Cx^2$. Letting $x = 0$ yields $B = -5$, and letting $x = -1$ yields $C = -6$. Then letting $x = 1$ yields $-4 = 2A - 10 - 6$, so $A = 6$. The original integral is thus equal to

$$\int \left(\frac{6}{x} - \frac{5}{x^2} - \frac{6}{x+1} \right) dx = 6\left(\ln|x| - \ln|x+1| \right) + \frac{5}{x} + C.$$

8.5.35 Let $\dfrac{x^2 - x}{(x-2)(x-3)^2} = \dfrac{A}{x-2} - \dfrac{B}{x-3} + \dfrac{C}{(x-3)^2}$. Then $x^2 - x = A(x-3)^2 + B(x-2)(x-3) + C(x-2)$. Letting $x = 2$ gives $A = 2$. Letting $x = 3$ gives $C = 6$. Letting $x = 0$ gives $0 = 18 + 6B - 12$, so $B = -1$. The original integral is thus

$$\int \left(\frac{2}{x-2} - \frac{1}{x-3} + \frac{6}{(x-3)^2} \right) dx = \ln \left| \frac{(x-2)^2}{x-3} \right| + \frac{-6}{x-3} + C.$$

8.5.37 Let $\dfrac{x^2 - 4}{x^3 - 2x^2 + x} = \dfrac{A}{x} + \dfrac{B}{x-1} + \dfrac{C}{(x-1)^2}$. Then $x^2 - 4 = A(x-1)^2 + Bx(x-1) + Cx$. Letting $x = 0$ gives $A = -4$. Letting $x = 1$ gives $C = -3$. Letting $x = 2$ gives $0 = -4 + 2B - 6$, so $B = 5$. The given integral is thus equal to

$$\int \left(-\frac{4}{x} + \frac{5}{x-1} - \frac{3}{(x-1)^2} \right) dx = \ln \left| \frac{(x-1)^5}{x^4} \right| + \frac{3}{x-1} + C.$$

8.5.39 $\dfrac{20x}{(x-1)^2(x^2+1)} = \dfrac{A}{x-1} + \dfrac{B}{(x-1)^2} + \dfrac{Cx+D}{x^2+1}$.

8.5.41 $\dfrac{2x^2 + 3}{(x^2 - 8x + 16)(x^2 + 3x + 4)} = \dfrac{2x^2 + 3}{(x-4)^2(x^2 + 3x + 4)} = \dfrac{A}{x-4} + \dfrac{B}{(x-4)^2} + \dfrac{Cx+D}{x^2 + 3x + 4}$.

8.5.43 Let $\dfrac{x^2 + x + 2}{(x+1)(x^2+1)} = \dfrac{Ax+B}{x^2+1} + \dfrac{C}{x+1}$. Then $x^2 + x + 2 = (Ax+B)(x+1) + C(x^2+1)$. Letting $x = -1$ gives $C = 1$. Letting $x = 0$ gives $2 = B + 1$, so $B = 1$. Letting $x = 1$ gives $4 = 2A + 2 + 2$, so $A = 0$. The original integral is therefore equal to

$$\int \left(\frac{1}{x^2+1} + \frac{1}{x+1} \right) dx = \tan^{-1}(x) + \ln|x+1| + C.$$

8.5.45 Let $\dfrac{2x^2 + 5x + 5}{(x+1)(x^2 + 2x + 2)} = \dfrac{Ax+B}{x^2 + 2x + 2} + \dfrac{C}{x+1}$. Then $2x^2 + 5x + 5 = (Ax+B)(x+1) + C(x^2 + 2x + 2)$. Letting $x = -1$ gives $C = 2$. Letting $x = 0$ gives $5 = B + 4$, so $B = 1$. Letting $x = 1$ gives $12 = 2A + 2 + 2(5)$, so $A = 0$. The original integral is therefore equal to

$$\int \left(\frac{1}{x^2 + 2x + 2} + \frac{2}{x+1} \right) dx = \int \left(\frac{1}{(x+1)^2 + 1} + \frac{2}{x+1} \right) dx = \tan^{-1}(x+1) + \ln((x+1)^2) + C.$$

8.5.47 If we write $\dfrac{20x}{(x-1)(x^2+4x+5)} = \dfrac{A}{x-1} + \dfrac{Bx+C}{x^2+4x+5}$, then $20x = A(x^2+4x+5)+(Bx+C)(x-1)$.
Letting $x=1$ yields $A=2$. Letting $x=0$ yields $0=10-C$, so $C=10$. Letting $x=2$ yields $40=34+2B+10$,
so $B=-2$. The original integral is thus $\displaystyle\int\left(\dfrac{2}{x-1} - \dfrac{2x-10}{x^2+4x+5}\right)dx = \int\left(\dfrac{2}{x-1} - \dfrac{(2x+4)-14}{x^2+4x+5}\right)dx =$
$\displaystyle\int\left(\dfrac{2}{x-1} - \dfrac{(2x+4)}{x^2+4x+5} + \dfrac{14}{(x+2)^2+1}\right)dx = \ln\left|\dfrac{(x-1)^2}{x^2+4x+5}\right| + 14\tan^{-1}(x+2)+C$

8.5.49 If we write $\dfrac{x^2}{x^3-x^2+4x-4} = \dfrac{x^2}{(x-1)(x^2+4)} = \dfrac{A}{x-1} + \dfrac{Bx+C}{x^2+4}$, then $x^2 = A(x^2+4)+(Bx+$
$C)(x-1)$. Letting $x=1$ yields $A=1/5$. Letting $x=0$ yields $C=4/5$, and then letting $x=2$ and solving
for B yields $B=4/5$. Thus the original integral is equal to $\displaystyle\int\dfrac{1/5}{x-1}dx + \dfrac{4}{5}\int\dfrac{x}{x^2+4}dx + \dfrac{4}{5}\int\dfrac{1}{x^2+4}dx$.
Thus, the original integral is equal to $\frac{1}{5}\ln|x-1| + \frac{2}{5}\ln(x^2+4) + \frac{2}{5}\tan^{-1}(x/2)+C$.

8.5.51

    a. False. Because the given integrand is improper, the first step would be to use long division to write
the integrand as the sum of a polynomial and a proper rational function.

    b. False. This is easy to evaluate via the substitution $u = 3x^2 + x$.

    c. False. The discriminant of the denominator is $b^2-4ac = 169-168 = 1 > 0$, so the denominator factors
into linear factors of the real numbers.

    d. True. The discriminant of the denominator is $b^2 - 4ac = 169 - 172 = -3 < 0$, so the given quadratic
expression is irreducible.

8.5.53

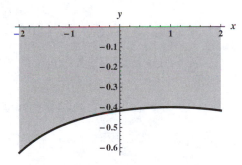

If we write $\dfrac{10}{x^2-2x-24} = \dfrac{10}{(x-6)(x+4)} = \dfrac{A}{x-6} + \dfrac{B}{x+4}$, then $10 = A(x+4)+B(x-6)$. Letting $x=-4$
gives $B=-1$ and letting $x=6$ gives $A=1$. Thus the area in question is given by

$$-\int_{-2}^{2}\left(-\dfrac{1}{x+4} + \dfrac{1}{x-6}\right)dx = (\ln(x+4) - \ln|x-6|)\Big|_{-2}^{2} = \ln 6 - \ln 4 - (\ln 2 - \ln 8) = \ln 6.$$

8.5.55

The curve intersects the x-axis at $x = 2 \pm 2\sqrt{2}$. We can write the integrand as $1 + \frac{1}{x^2 - 4x - 5} = 1 + \frac{1}{(x-5)(x+1)}$. If we write this second term in the form $\frac{A}{x-5} + \frac{B}{x+1}$, then $1 = A(x+1) + B(x-5)$. Letting $x = -1$ yields $B = -1/6$ and letting $x = 5$ yields $A = 1/6$. Thus the area in question is given by

$$\int_{2-2\sqrt{2}}^{2+2\sqrt{2}} \left(1 + \frac{1/6}{x-5} - \frac{1/6}{x+1} \right) dx = x + \frac{1}{6} \ln|x-5| - \frac{1}{6} \ln|x+1| \Big|_{2-2\sqrt{2}}^{2+2\sqrt{2}} = 4\sqrt{2} + \frac{1}{6} \ln \left(\frac{3 - 2\sqrt{2}}{3 + 2\sqrt{2}} \right)^2.$$

8.5.57 Using disks, we have $\frac{V}{\pi} = \int_0^4 \frac{x^2}{(x+1)^2}\, dx$. Now we can perform long division to rewrite this integral as $\int_0^4 \left(1 - \frac{2x+2}{x^2 + 2x + 1} + \frac{1}{(x+1)^2} \right) dx$. Thus we have $V = \pi \left(x - \ln|x^2 + 2x + 1| - \frac{1}{x+1} \Big|_0^4 \right) = \pi \left(4 - \ln(25) - \frac{1}{5} - (-1) \right) = \pi \left(\frac{24}{5} - \ln(25) \right)$.

8.5.59 Using disks, we have $\frac{V}{\pi} = \int_1^2 \frac{1}{x(3-x)}\, dx$. Now if we write $\frac{1}{x(3-x)} = \frac{A}{x} + \frac{B}{3-x}$, then we have $1 = A(3-x) + Bx$. Letting $x = 0$ yields $A = 1/3$ and letting $x = 3$ yields $B = 1/3$. Thus we have $V = \frac{\pi}{3} \int_1^2 \left(\frac{1}{x} - \frac{1}{x-3} \right) dx = \frac{\pi}{3} \left(\ln x - \ln|x-3| \Big|_1^2 \right) = \frac{\pi}{3} \cdot 2 \ln 2$.

8.5.61 Using shells, we have $\frac{V}{2\pi} = \int_0^3 \frac{x+1}{x+2}\, dx = \int_0^3 \left(1 - \frac{1}{x+2} \right) dx = x - \ln|x+2| \Big|_0^3 = 3 + \ln(2/5)$. Thus, $V = 2\pi(3 + \ln(2/5))$.

8.5.63 Let $u = e^x$, so that $du = e^x\, dx$. Then $\int \frac{1}{1 + e^x} \cdot \frac{e^x}{e^x}\, dx = \int \frac{1}{u(1+u)}\, du$. If we write $\frac{1}{u(1+u)} = \frac{A}{u} + \frac{B}{1+u}$, then $1 = A(1+u) + Bu$. Letting $u = 0$ yields $A = 1$ and letting $u = -1$ yields $B = -1$, so the original integral is equal to $\int \left(\frac{1}{u} - \frac{1}{1+u} \right) du = \ln|u| - \ln|1+u| + C = x - \ln(1 + e^x) + C$.

8.5.65 After performing long division, we have that the original integrand is equal to $3 + \frac{13x - 12}{(x-1)(x-2)}$, and if we write $\frac{13x - 12}{(x-1)(x-2)} = \frac{A}{x-1} + \frac{B}{x-2}$, then $13x - 12 = A(x-2) + B(x-1)$. Letting $x = 1$ yields $A = -1$ and letting $x = 2$ yields $B = 14$. Thus the original integral is equal to $3x - \int \frac{1}{x-1}\, dx + 14 \int \frac{1}{x-2}\, dx = 3x - \ln|x-1| + 14 \ln|x-2| + C$.

8.5.67 $\int \frac{1}{2 + e^t}\, dt = \int \frac{e^t}{(e^t + 2)e^t}\, dt$. Let $u = e^t$, so that $du = e^t\, dt$. Then we have $\int \frac{1}{(u+2)u}\, du$. If we write $\frac{1}{(u+2)u} = \frac{A}{u} + \frac{B}{u+2}$, then $A(u+2) + Bu = 1$, and we find that $A = 1/2$ and $B = -1/2$. Then we integrate $\int \left(\frac{1/2}{u} - \frac{1/2}{u+2} \right) du = \frac{1}{2} (\ln|u| - \ln|u+2|) + C = \frac{1}{2} (t - \ln(2 + e^t)) + C$.

8.5.69 This can be written as

$$\int \frac{dt}{\cos t(1 + \sin t)} = \int \frac{\cos t}{\cos^2 t(1 + \sin t)}\, dt = \int \frac{\cos t}{(1 - \sin^2 t)(1 + \sin t)}\, dt.$$

Now let $u = \sin t$ so that $du = \cos t\, dt$. Substituting gives $\int \frac{du}{(1 - u^2)(1 + u)} = \int \frac{du}{(1+u)^2(1-u)}$. If we write $\frac{1}{(1+u)^2(1-u)} = \frac{A}{1+u} + \frac{B}{(1+u)^2} + \frac{C}{1-u}$ then $1 = A(1+u)(1-u) + B(1-u) + C(1+u)^2$. Letting

$u = -1$ yields $B = 1/2$. Letting $u = 1$ yields $C = 1/4$. Letting $u = 0$ then yields $A = 1/4$. Then our integral is equal to

$$\int \left(\frac{1/4}{1+u} + \frac{1/2}{(1+u)^2} + \frac{1/4}{1-u} \right) du = \frac{1}{4} \left(\frac{-2}{1+u} + \ln \left| \frac{1+u}{1-u} \right| \right) + C = \frac{1}{4} \left(\frac{-2}{1+\sin t} + \ln \left(\frac{1+\sin t}{1-\sin t} \right) \right) + C.$$

8.5.71 Let $u = e^x$ so that $du = e^x \, dx$. Then the original integral is equal to $\displaystyle\int \frac{1}{(u-1)(u+2)} \, du$. If we

write $\dfrac{1}{(u-1)(u+2)} = \dfrac{A}{u-1} + \dfrac{B}{u+2}$, then $1 = A(u+2) + B(u-1)$. Letting $u = -2$ yields $B = -1/3$

and letting $u = 1$ yields $A = 1/3$. Thus we have $\displaystyle\int \left(\frac{1/3}{u-1} - \frac{1/3}{u+2} \right) du = \frac{1}{3} \left(\ln |u-1| - \ln |u+2| \right) + C =$

$\dfrac{1}{3} \ln \left| \dfrac{e^x - 1}{e^x + 2} \right| + C.$

8.5.73 $\displaystyle\int \frac{1}{(e^x + e^{-x})^2} \cdot \frac{(e^x)^2}{(e^x)^2} \, dx = \int \frac{e^{2x}}{(e^{2x} + 1)^2} \, dx$. Let $u = e^x$ so that $du = e^x \, dx$. Then we have

$\displaystyle\int \frac{u}{(u^2+1)^2} \, du$. If we let $w = u^2+1$, then $dw = 2u \, du$, so we have $\dfrac{1}{2} \displaystyle\int w^{-2} \, dw = -\dfrac{1}{2w} + C = -\dfrac{1}{2u^2+2} + C =$

$-\dfrac{1}{2e^{2x}+2} + C.$

8.5.75

a. $\sec x = \dfrac{1}{\cos x} \cdot \dfrac{\cos x}{\cos x} = \dfrac{\cos x}{\cos^2 x} = \dfrac{\cos x}{1 - \sin^2 x}.$

b. $\displaystyle\int \sec x \, dx = \int \frac{\cos x}{1 - \sin^2 x} \, dx$. Let $u = \sin x$ so that $du = \cos x \, dx$. Then our integral becomes

$\displaystyle\int \frac{du}{1 - u^2}$. If we write $\dfrac{1}{1-u^2} = \dfrac{1}{(1-u)(1+u)} = \dfrac{A}{1-u} + \dfrac{B}{1+u}$, we have $1 = A(1+u) + B(1-u)$, so

$A = \frac{1}{2}$ and $B = \frac{1}{2}$. We have

$$\int \frac{du}{1-u^2} = \frac{1}{2} \int \left(\frac{1}{1-u} + \frac{1}{1+u} \right) du = \frac{1}{2} (\ln |1+u| - \ln |1-u|) + C = \frac{1}{2} \ln \left| \frac{1+\sin x}{1-\sin x} \right| + C.$$

8.5.77 If we let $u^4 = x + 2$, then $4u^3 \, du = dx$. Substituting yields

$$\int \frac{4u^3}{u+1} \, du = \int \left(4u^2 - 4u + 4 - \frac{4}{u+1} \right) du = \frac{4}{3} u^3 - 2u^2 + 4u - 4 \ln |u+1| + C$$

$$= \frac{4}{3} (x+2)^{3/4} - 2(x+2)^{1/2} + 4(x+2)^{1/4} - 4 \ln((x+2)^{1/4} + 1) + C.$$

8.5.79 If we let $u^6 = x$, then $6u^5 \, du = dx$. Substituting yields

$$\int \frac{6u^5}{u^3 + u^2} \, du = 6 \int \frac{u^3}{u+1} \, du = 6 \int \left(u^2 - u + 1 - \frac{1}{u+1} \right) du$$

$$= 2u^3 - 3u^2 + 6u - 6 \ln |u+1| + C = 2\sqrt{x} - 3\sqrt[3]{x} + 6\sqrt[6]{x} - 6 \ln(\sqrt[6]{x} + 1) + C.$$

8.5.81 If we let $(u^2 - 1)^2 = x$, then $2(u^2 - 1) \cdot 2u \, du = dx$. Substituting yields $\displaystyle\int \frac{4u(u^2-1)}{u} \, du = 4 \int (u^2 - $

$1) \, du = \dfrac{4u^3}{3} - 4u + C = \dfrac{4}{3} u(u^2 - 3) + C = \dfrac{4}{3} \sqrt{1 + \sqrt{x}} (\sqrt{x} - 2) + C.$

8.5.83 If we write $\dfrac{2}{x(x^2+1)^2} = \dfrac{A}{x} + \dfrac{Bx+C}{x^2+1} + \dfrac{Dx+E}{(x^2+1)^2}$, then $2 = A(x^2+1)^2 + (Bx+C)x(x^2+1) + (Dx+E)x$.
Letting $x = 0$ yields $A = 2$. Expanding the right-hand side yields $2 = (2+B)x^4 + Cx^3 + (4+B+D)x^2 +$

$(C+E)x+2$. Equating coefficients gives us the equations $2+B=0$, $C=0$, $4+B+D=0$, and $C+E=0$, from which we can deduce that $B=-2$, $C=0$, $D=-2$ and $E=0$. The original integral is thus equal to

$$\int \left(\frac{2}{x} - \frac{2x}{x^2+1} - \frac{2x}{(x^2+1)^2} \right) dx = 2\ln|x| - \ln|x^2+1| + \frac{1}{x^2+1} + C.$$

8.5.85 If we write $\dfrac{x}{(x-1)(x^2+2x+2)^2} = \dfrac{A}{x-1} + \dfrac{Bx+C}{x^2+2x+2} + \dfrac{Dx+E}{(x^2+2x+2)^2}$, then $x = A(x^2+2x+2)^2 + (Bx+C)(x-1)(x^2+2x+2) + (Dx+E)(x-1)$. Letting $x=1$ yields $A = 1/25$. Then expanding the polynomial on the right-hand side gives $25x = (1+25B)x^4 + (4+25(B+C))x^3 + (8+25D+25C)x^2 + (8+25E-25D-50B)x + (4-25E-50C)$. Equating coefficients and then solving for the unknowns yields $B = -1/25$, $C = -3/25$, $D = -5/25$, and $E = 10/25$. The original integral is thus equal to

$$\frac{1}{25}\int \left(\frac{1}{x-1} - \frac{x+3}{x^2+2x+2} - \frac{5x-10}{(x^2+2x+2)^2} \right) dx$$

$$= \frac{1}{25}\int \left(\frac{1}{x-1} - \frac{x+1}{x^2+2x+2} - 5\frac{x+1}{(x^2+2x+2)^2} - 2\frac{1}{(x+1)^2+1} + 15\frac{1}{((x+1)^2+1)^2} \right) dx$$

$$= \frac{1}{25}\left(\ln|x-1| - \frac{1}{2}\ln|x^2+2x+2| + \frac{5}{2}\cdot\frac{1}{x^2+2x+2} - 2\tan^{-1}(x+1) \right) + \frac{3}{5}\int \frac{1}{(v^2+1)^2}\,dv,$$

where $v = x+1$. To compute this last term, we let $v = \tan\theta$ so that $dv = \sec^2\theta\,d\theta$. Then the last term is equal to

$$\frac{3}{5}\int \frac{\sec^2\theta}{\sec^4\theta}\,d\theta = \frac{3}{5}\int \cos^2\theta\,d\theta = \frac{3}{10}\int (1+\cos 2\theta)\,d\theta$$

$$= \frac{3}{10}\left(\theta + \frac{\sin 2\theta}{2} \right) + C = \frac{3}{10}\left(\tan^{-1}(x+1) + \frac{x+1}{(x+1)^2+1} \right) + C.$$

Putting this all together yields

$$\frac{1}{25}\left(\ln|x-1| - \frac{1}{2}\ln|x^2+2x+2| \right) + \frac{11}{50}\tan^{-1}(x+1) + \frac{1}{10}\left(\frac{3x+4}{(x+1)^2+1} \right) + C.$$

8.5.87 If we write $\dfrac{1}{x^2-1} = \dfrac{A}{x-1} + \dfrac{B}{x+1}$, then $1 = A(x+1) + B(x-1)$, so $A = 1/2$ and $B = -1/2$. Thus we have $\dfrac{1}{2}\int \dfrac{1}{x-1} - \dfrac{1}{x+1}\,dx = \dfrac{1}{2}\left(\ln|x-1| - \ln|x+1| \right) + C$.

Now let $x = \sec\theta$, so that $dx = \sec\theta\tan\theta\,d\theta$. Then the original integral is equal to $\displaystyle\int \csc\theta\,d\theta =$

$-\ln|\csc\theta + \cot\theta| + C = -\ln\left| \dfrac{x}{\sqrt{x^2-1}} + \dfrac{1}{\sqrt{x^2-1}} \right| + C = -\ln\left(\dfrac{|x+1|}{\sqrt{x^2-1}} \right) + C = -\ln\left(\sqrt{\left|\dfrac{x+1}{x-1}\right|} \right) + C =$

$\ln\left(\sqrt{\left|\dfrac{x-1}{x+1}\right|} \right) + C$. The two answers are equivalent.

8.5.89 Using the substitution $x = 2\tan^{-1}u$ yields

$$\int \frac{1}{1+\sin x}\,dx = \int \frac{2}{1+u^2}\cdot\frac{1}{1+\frac{2u}{1+u^2}}\,du = \int \frac{2}{u^2+2u+1}\,du$$

$$= 2\int \frac{1}{(u+1)^2}\,du = -\frac{2}{u+1} + C = -\frac{2}{\tan(x/2)+1} + C.$$

8.5.91 Using the substitution $x = 2\tan^{-1}u$ yields $\displaystyle\int \frac{1}{1-\frac{1-u^2}{1+u^2}}\cdot\frac{2}{1+u^2}\,du = \int u^{-2}\,du = -\frac{1}{u} + C =$

$-\cot(x/2) + C$.

8.5.93 Using the substitution $\theta = 2\tan^{-1}u$ yields $\int_0^1 \frac{1}{\frac{1-u^2}{1+u^2} + \frac{2u}{1+u^2}} \cdot \frac{2}{1+u^2}\,du = \int_0^1 \frac{2}{1+2u-u^2}\,du =$

$-2\int_0^1 \frac{1}{u^2-2u+1-2}\,du = -2\int_0^1 \frac{1}{(u-1)^2-2}\,du$. If we factor the denominator as the difference of squares we have $(u-1)^2-2 = (u-1-\sqrt{2})(u-1+\sqrt{2})$, and using a partial fractions decomposition yields

$\frac{1}{\sqrt{2}}\int_0^1 \left(\frac{1}{u+\sqrt{2}-1} - \frac{1}{u-\sqrt{2}-1}\right)du = \frac{1}{\sqrt{2}}\ln\left|\frac{u+\sqrt{2}-1}{u-\sqrt{2}-1}\right|\bigg|_0^1 = \frac{1}{\sqrt{2}}\ln\left(\frac{\sqrt{2}+1}{\sqrt{2}-1}\right)$.

8.5.95 $s_A(t) = \int v_A(t)\,dt = 88\int \frac{t}{t+1}\,dt = 88\int\left(1-\frac{1}{1+t}\right)dt = 88t - 88\ln(t+1)+C$. Because $s_A(0)=0$, we see that $C=0$, so $s_A(t) = 88t - 88\ln(1+t)$.

$s_B(t) = \int v_B(t)\,dt = 88\int \frac{t^2}{(t+1)^2}\,dt = 88\int\left(1-\frac{2t+2}{t^2+2t+1}+\frac{1}{(t+1)^2}\right)dt = 88(t-\ln(t^2+2t+1)-\frac{1}{t+1})+D$. Because $s_B(0)=0$, we see that $D=88$, so $s_B(t) = 88(t-\ln(t^2+2t+1)-\frac{1}{t+1}+1)$.

$s_C(t) = \int v_C(t)\,dt = 88\int \frac{t^2}{t^2+1}\,dt = 88\int\left(1-\frac{1}{t^2+1}\right)dt = 88(t-\tan^{-1}(t))+E$. Because $s_C(0)=0$, we have that $E=0$. Thus $s_C(t) = 88t - 88\tan^{-1}(t)$.

a. $s_A(1) = 88(1-\ln(2)) \approx 27$. $s_B(1) = 88(1-\ln 4 - (1/2)+1) \approx 88(3/2-\ln 4) \approx 10$. $s_C(1) = 88(1-\tan^{-1}(1)) \approx 18.9$. So car A travels farthest.

b. $s_A(5) = 88(5-\ln 6) \approx 282$. $s_B(5) = 88(5-\ln 36 - (1/6)+1) \approx 198$. $s_C(5) = 88(5-\tan^{-1}(5)) \approx 319$. So car C travels farthest.

c. See the development above.

d. Ultimately car C gains the lead. This can be seen by the fact that car C's velocity function is greater than that of the other cars.

8.5.97 First note that the numerator of the given integrand can be written as $x^8 - 4x^7 + 6x^6 - 4x^5 + x^4$, and this quantity when divided by x^2+1 yields $x^6 - 4x^5 + 5x^4 - 4x^2 + 4 - \frac{4}{1+x^2}$. Thus the given integral is equal to $\left(\frac{x^7}{7} - \frac{2x^6}{3} + x^5 - \frac{4x^3}{3} + 4x\right)\bigg|_0^1 - \left(4\tan^{-1}(x)\right)\bigg|_0^1 = \frac{1}{7} - \frac{2}{3} + 1 - \frac{4}{3} + 4 - \pi = \frac{22}{7} - \pi$. Because the given integrand is positive on the interval $(0,1)$, we know that this integral is positive. Thus,

$$0 < \int_0^1 \frac{x^4(1-x)^4}{1+x^2}\,dx = \frac{22}{7} - \pi.$$

Adding π to both sides of this inequality yields $\pi < \frac{22}{7}$.

8.6 Other Integration Strategies

8.6.1 The power rule, substitution, integration by parts, and partial fraction decomposition are examples of analytical methods.

8.6.3 The computer algebra system may use a different algorithm than whoever prepared the table, so the results may look different – however, they should differ by a constant.

8.6.5 Using table entry 17, we have $\int \cos^{-1}(x)\,dx = x\cos^{-1}x - \sqrt{1-x^2} + C$.

8.6.7 Using table entry 77 for $\int \frac{1}{\sqrt{x^2+a^2}}\,dx$, we see that $\int \frac{1}{\sqrt{x^2+16}}\,dx = \ln(x+\sqrt{x^2+16})+C$.

8.6.9 Using table entry 90 for $\int \frac{x}{ax+b}\,dx$, we see that $\int \frac{3u}{2u+7}\,du = \frac{3u}{2} - \frac{21}{4}\ln|2u+7| + C$.

8.6.11 Using table entry 47, we have $\int \dfrac{dx}{1 - \cos 4x} = -\dfrac{1}{4}\cot(2x) + C$.

8.6.13 Using table entry 94 for $\int \dfrac{x}{\sqrt{ax + b}}\,dx$, we have that $\int \dfrac{x\,dx}{\sqrt{4x + 1}} = \dfrac{1}{24}(4x - 2)\sqrt{4x + 1} + C = \dfrac{1}{12}(2x - 1)\sqrt{4x + 1} + C$.

8.6.15 Using table entry 69 for $\int \dfrac{1}{\sqrt{x^2 - a^2}}\,dx$ we have $\int \dfrac{1}{\sqrt{9x^2 - 100}}\,du = \dfrac{1}{3}\int \dfrac{1}{\sqrt{u^2 - 100}}\,du$ where $u = 3x$. This is then equal to $\frac{1}{3}\ln|u + \sqrt{u^2 - 100}| + C = \frac{1}{3}\ln|3x + \sqrt{9x^2 - 100}| + C$. This can be written as $\frac{1}{3}\ln|x + \sqrt{x^2 - (100/9)}| + C$.

8.6.17 Using table entry 84 for $\int \dfrac{1}{(a^2 + x^2)^{3/2}}\,dx$, we have $\int \dfrac{1}{(16 + 9x^2)^{3/2}}\,dx = \dfrac{1}{3}\int \dfrac{1}{(16 + u^2)^{3/2}}\,du = \dfrac{1}{3}\cdot\dfrac{3x}{16\sqrt{16 + 9x^2}} + C = \dfrac{x}{16\sqrt{16 + 9x^2}} + C$.

8.6.19 Using table entry 62 for $\int \dfrac{1}{x\sqrt{a^2 - x^2}}\,dx$, we have $\int \dfrac{1}{x\sqrt{(12)^2 - x^2}}\,dx = \dfrac{-1}{12}\ln\left|\dfrac{12 + \sqrt{144 - x^2}}{x}\right| + C$.

8.6.21 Using table entry 103 for $\int \ln^n x\,dx$ we have that $\int \ln^2 x\,dx = x\ln^2 x - 2\int \ln x\,dx = x\ln^2 x - 2(x\ln x - x) + C = 2x + x\ln^2 x - 2x\ln x + C$.

8.6.23 Note that $\sqrt{x^2 + 10x} = \sqrt{x^2 + 10x + 25 - 25} = \sqrt{(x + 5)^2 - 5^2}$. Thus, the given integral is equal to $\int \sqrt{u^2 - 5^2}\,du$ where $u = x + 5$. Using table entry 68 we have that this is equal to $\frac{u}{2}\sqrt{u^2 - 25} - \frac{25}{2}\ln|u + \sqrt{u^2 - 25}| + C = \frac{x+5}{2}\sqrt{(x + 5)^2 - 25} - \frac{25}{2}\ln|x + 5 + \sqrt{(x + 5)^2 - 25}| + C$.

8.6.25 $\int \dfrac{1}{x^2 + 2x + 10}\,dx = \int \dfrac{1}{(x + 1)^2 + 9}\,dx$. Using table entry 14 this is equal to $\frac{1}{3}\tan^{-1}\left(\frac{x+1}{3}\right) + C$.

8.6.27 $\int \dfrac{1}{x(x^{10} + 1)}\,dx = \int \dfrac{10x^9}{10x^{10}(x^{10} + 1)}\,dx$. Let $u = x^{10} + 1$ so that $du = 10x^9\,dx$. Substituting yields $\int \dfrac{1}{10}\cdot\dfrac{1}{(u - 1)u}\,du$. Using table entry 96 for $\int \dfrac{dx}{x(ax + b)}$ we have $\dfrac{1}{10}\ln\left|\dfrac{u - 1}{u}\right| + C = \dfrac{1}{10}\ln\left|\dfrac{x^{10}}{x^{10} + 1}\right| + C$.

8.6.29 $\int \dfrac{1}{\sqrt{x^2 - 6x}}\,dx = \int \dfrac{1}{\sqrt{(x - 3)^2 - 9}}\,dx$. Using table entry 69, this is equal to $\ln|x - 3 + \sqrt{(x - 3)^2 - 9}| + C$. This can also be written as $2\ln(\sqrt{x - 6} + \sqrt{x}) + C$.

8.6.31 $\int \dfrac{e^x}{\sqrt{e^{2x} + 4}}\,dx = \int \dfrac{1}{\sqrt{u^2 + 4}}\,du$ where $u = e^x$. Then using table entry 77, we have $\ln(u + \sqrt{u^2 + 4}) + C = \ln(e^x + \sqrt{e^{2x} + 4}) + C$.

8.6.33 $\int \dfrac{\cos x}{\sin^2 x + 2\sin x}\,dx = \int \dfrac{1}{u^2 + 2u}\,du = \int \dfrac{1}{(u + 1)^2 - 1}\,du$ where $u = \sin x$. Then using table entry 74, we have

$$-\dfrac{1}{2}\ln\left|\dfrac{u + 2}{u}\right| + C = -\dfrac{1}{2}\ln\left|\dfrac{\sin x + 2}{\sin x}\right| + C.$$

8.6.35 Let $u = x^3$, so that $du = 3x^2\,dx$. Substituting yields

$$\dfrac{1}{3}\int \dfrac{\tan^{-1}(u)}{u^2}\,du = -\dfrac{1}{3}\left(\dfrac{1}{u}\tan^{-1}(u) - \int \dfrac{1}{u(1 + u^2)}\,du\right) = -\dfrac{1}{3}\left(\dfrac{\tan^{-1}(x^3)}{x^3}\right) + \dfrac{1}{3}\int \dfrac{1}{u(1 + u^2)}\,du,$$

where we used table entry 18. Now let $w = u^2 + 1$ so that $dw = 2u\,du$. This last integral is thus equal to $\dfrac{1}{3}\displaystyle\int \dfrac{2u\,du}{2u^2(1+u^2)} = \dfrac{1}{6}\displaystyle\int \dfrac{dw}{(w-1)w} = -\dfrac{1}{6}\ln\left|\dfrac{w}{w-1}\right| + C = -\dfrac{1}{6}\ln\left|\dfrac{u^2+1}{u^2}\right| + C = -\dfrac{1}{6}\ln\left|\dfrac{x^6+1}{x^6}\right| + C,$

where we used table entry 96. Thus the original integral is equal to $-\dfrac{1}{3}\left(\dfrac{\tan^{-1}(x^3)}{x^3}\right) - \dfrac{1}{6}\ln\left|\dfrac{x^6+1}{x^6}\right| + C.$

This can be written as $-\dfrac{\tan^{-1}(x^3)}{3x^3} + \ln\left|\dfrac{x}{(x^6+1)^{1/6}}\right| + C.$

8.6.37 Let $u = \ln x$, so that $du = \frac{1}{x}\,dx$. Substituting yields $\displaystyle\int u\sin^{-1}(u)\,du = \dfrac{2u^2 - 1}{4}\sin^{-1}(u) + \dfrac{u\sqrt{1-u^2}}{4} +$

$C = \dfrac{2\ln^2 x - 1}{4}\sin^{-1}(\ln x) + \dfrac{\ln x\sqrt{1 - \ln^2 x}}{4} + C$, where we used table entry 105.

8.6.39 The integral which gives the length of the curve is $\dfrac{1}{2}\displaystyle\int_0^8 \sqrt{4 + x^2}\,dx.$ Using table entry 76, we have

$\frac{1}{2}\left(\frac{x}{2}\sqrt{4+x^2} + \frac{4}{2}\ln(x + \sqrt{4+x^2})\right)\Big|_0^8 = 4\sqrt{17} + \ln(8 + 2\sqrt{17}) - \ln 2 = 4\sqrt{17} + \ln(4 + \sqrt{17}) \approx 18.59.$

8.6.41 The integral which gives the length of the curve is $\displaystyle\int_0^{\ln 2} \sqrt{1 + e^{2x}}\,dx = \displaystyle\int_1^2 \dfrac{\sqrt{1+u^2}}{u}\,du$ where $u = e^x$.

This is equal to $\left(\sqrt{1+u^2} - \ln\left|\dfrac{1+\sqrt{1+u^2}}{u}\right|\right)\Big|_1^2 = \sqrt{5} - \sqrt{2} + \ln(1+\sqrt{2}) - \ln\left(\dfrac{1+\sqrt{5}}{2}\right) \approx 1.22.$ Note

that we used table entry 83.

8.6.43 Using the method of shells, we have $\dfrac{V}{2\pi} = \displaystyle\int_0^{12} \dfrac{x}{\sqrt{x+4}}\,dx = \dfrac{2}{3}(x-8)\sqrt{x+4}\,\Big|_0^{12} = \dfrac{32}{3} + \dfrac{32}{3} = \dfrac{64}{3}.$

Thus $V = \dfrac{128\pi}{3}.$ We used table entry 94 to compute the integral.

8.6.45 Using the method of disks, we have $\dfrac{V}{\pi} = \displaystyle\int_0^{\pi/2} \sin^2 y\,dy = \dfrac{y}{2} - \dfrac{\sin 2y}{4}\,\Big|_0^{\pi/2} = \dfrac{\pi}{4}.$ Thus $V = \dfrac{\pi^2}{4}.$ We

used table entry 31 to compute the antiderivative.

8.6.47 $\displaystyle\int \dfrac{x}{\sqrt{2x+3}}\,dx = \dfrac{1}{3}(x-3)\sqrt{3+2x} + C.$

8.6.49 $\displaystyle\int \tan^2 3x\,dx = \dfrac{1}{3}\tan 3x - x + C.$

8.6.51 $\displaystyle\int \dfrac{(x^2-a^2)^{3/2}}{x}\,dx = \dfrac{1}{3}(x^2-a^2)^{3/2} - a^2\sqrt{x^2-a^2} + a^3\cos^{-1}(a/x) + C.$

8.6.53 $\displaystyle\int (a^2-x^2)^{3/2}\,dx = \dfrac{-x}{8}(2x^2-5a^2)\sqrt{a^2-x^2} + \dfrac{3a^4}{8}\sin^{-1}(x/a) + C.$

8.6.55 $\displaystyle\int_{2/3}^{4/5} x^8\,dx = \dfrac{(4/5)^9 - (2/3)^9}{9} \approx 0.012.$

8.6.57 $\displaystyle\int_0^4 (9+x^2)^{3/2}\,dx = \dfrac{1540 + 243\ln 3}{8} \approx 225.870.$

8.6.59 $\displaystyle\int_0^{\pi/2} \dfrac{1}{1+\tan^2 t}\,dt = \dfrac{\pi}{4} \approx 0.785.$

8.6.61 $\displaystyle\int_0^1 \ln x\ln(1+x)\,dx = 2 - \dfrac{\pi^2}{12} - \ln 4 \approx -0.209.$

8.6.63

a. Yes, it is possible, because these are equal for $x > 1$.

b. Yes, because $\frac{1}{9} = 0.\overline{11}$.

8.6.65 The two answers differ by a constant, namely the constant one. This can be seen as follows:

$$\frac{2\sin(x/2)}{\cos(x/2) + \sin(x/2)} = \frac{2\tan(x/2)}{1 + \tan(x/2)} = 2\frac{\sin x}{1 + \cos x} \cdot \frac{1}{1 + \frac{\sin x}{1 + \cos x}}$$

$$= \frac{2\sin x}{1 + \cos x + \sin x} = \frac{2\sin x}{1 + \cos x + \sin x} \cdot \frac{1 - \cos x - \sin x}{1 - \cos x - \sin x}$$

$$= \frac{2\sin x(1 - \cos x - \sin x)}{-2\sin x \cos x} = \frac{\sin x + \cos x - 1}{\cos x} = \frac{\sin x - 1}{\cos x} + 1.$$

8.6.67 $\int x^3 e^{2x}\, dx = \frac{1}{2}x^3 e^{2x} - \frac{3}{2}\int x^2 e^{2x}\, dx = \frac{1}{2}x^3 e^{2x} - \frac{3}{2}\left(\frac{1}{2}x^2 e^{2x} - \int x e^{2x}\, dx\right) = \frac{1}{2}x^3 e^{2x} - \frac{3}{4}x^2 e^{2x} +$

$\frac{3}{2}\left(\frac{1}{2}x e^{2x} - \frac{1}{2}\int e^{2x}\, dx\right) = \frac{1}{2}x^3 e^{2x} - \frac{3}{4}x^2 e^{2x} + \frac{3}{4}x e^{2x} - \frac{3}{8}e^{2x} + C.$

8.6.69 Let $u = 3y$. Then $\int \tan^4 3y\, dy = \frac{1}{3}\int \tan^4 u\, du = \frac{1}{3}\left(\frac{1}{3}\tan^3 u - \int \tan^2 u\, du\right) = \frac{1}{9}\tan^3 u -$

$\frac{1}{3}\left(\tan u - \int du\right) = \frac{1}{9}\tan^3 3y - \frac{1}{3}\tan 3y + y + C.$

8.6.71 Let $u = 2x$. Then $\int x\sin^{-1} 2x\, dx = \frac{1}{4}\int u\sin^{-1} u\, du = \frac{2u^2 - 1}{16}\sin^{-1} u + \frac{u\sqrt{1 - u^2}}{16} + C =$

$\frac{8x^2 - 1}{16}\sin^{-1} 2x + \frac{x\sqrt{1 - 4x^2}}{8} + C.$

8.6.73 $\int x^{-2}\tan^{-1} x\, dx = -(1/x)\tan^{-1} x + \int \frac{1}{x(1 + x^2)}\, dx = \frac{-\tan^{-1} x}{x} + \ln\left(\frac{|x|}{\sqrt{1 + x^2}}\right) + C.$

8.6.75

a. Note that $\int_0^b f(x)\, dx = \int_0^{b/2} f(x)\, dx + \int_{b/2}^b f(x)\, dx$. For the second integral, let $u = b - x$. Then

$du = -dx$. The second integral is equal to $-\int_{b-(b/2)}^0 f(b - u)\, du = \int_0^{b/2} f(b - u)\, du$. Because this can

be written as $\int_0^{b/2} f(b - x)\, dx$, we have $\int_0^b f(x)\, dx = \int_0^{b/2} (f(x) + f(b - x))\, dx.$

b.

$$\int_0^{\pi/4} \ln(1 + \tan x)\, dx = \int_0^{\pi/8} (\ln(1 + \tan x) + \ln(1 + \tan(\pi/4 - x)))\, dx$$

$$= \int_0^{\pi/8} \left(\ln(1 + \tan x) + \ln\left(1 + \frac{\tan \pi/4 - \tan x}{1 + \tan \pi/4 \tan x}\right)\right)\, dx$$

$$= \int_0^{\pi/8} \left(\ln(1 + \tan x) + \ln\left(1 + \frac{1 - \tan x}{1 + \tan x}\right)\right)\, dx$$

$$= \int_0^{\pi/8} \left(\ln(1 + \tan x) + \ln\left(\frac{2}{1 + \tan x}\right)\right) = \int_0^{\pi/8} \ln(2)\, dx = \frac{\pi \ln 2}{8}.$$

8.6.77

	T	Relative Error
θ_0		
0.1	6.27927	0.000623603
0.2	6.26762	0.0024778
0.3	6.24854	0.0051388
0.4	6.22253	0.00965413
0.5	6.19021	0.0147967
0.6	6.15236	0.0208215
0.7	6.10979	0.0275963
0.8	6.06338	0.0349831
0.9	6.01399	0.0428433
1.0	5.96247	0.0510427

a. Using a computer algebra system, we have

b. All of these values are within 10 percent of 2π.

8.6.79 Let $u = ax + b$, so that $du = a\,dx$. Then we have $\dfrac{1}{a}\displaystyle\int \dfrac{ax+b-b}{ax+b}\,dx = \dfrac{1}{a}\displaystyle\int \left(1 - \dfrac{b}{ax+b}\right)\,dx =$
$\dfrac{1}{a}\left(x - \dfrac{b}{a}\displaystyle\int \dfrac{1}{u}\,du\right) = \dfrac{1}{a}\left(x - \dfrac{b}{a}\ln|u|\right) + C = \dfrac{1}{a}\left(x - \dfrac{b}{a}\ln|ax+b|\right) + C.$

8.6.81 Let $u = ax + b$, so that $du = a\,dx$. Then we have $\dfrac{1}{a}\displaystyle\int \dfrac{u-b}{a}u^n\,du = \dfrac{1}{a^2}\displaystyle\int (u^{n+1} - bu^n)\,du =$
$\dfrac{1}{a^2}\left(\dfrac{u^{n+2}}{n+2} - \dfrac{bu^{n+1}}{n+1}\right) + C = \dfrac{1}{a^2}\left(\dfrac{(ax+b)^{n+2}}{n+2} - \dfrac{b(ax+b)^{n+1}}{n+1}\right) + C.$

8.6.83

a. The result holds.

b. First note that $\displaystyle\int_0^{\pi/2} \cos^n x\,dx = -\int_{\pi/2}^0 \cos^n(\pi/2 - \theta)\,d\theta = \int_0^{\pi/2} \sin^n\theta\,d\theta$. So we only need to show

the result for $\displaystyle\int_0^{\pi/2} \sin^{10} x\,dx$. Repeatedly using the reduction formula we have $\displaystyle\int_0^{\pi/2} \sin^{10} x\,dx =$
$\dfrac{9}{10}\displaystyle\int_0^{\pi/2} \sin^8 x\,dx = \dfrac{9\cdot 7}{10\cdot 8}\displaystyle\int_0^{\pi/2} \sin^6 x\,dx = \dfrac{9\cdot 7\cdot 5}{10\cdot 8\cdot 6}\displaystyle\int_0^{\pi/2} \sin^4 x\,dx = \dfrac{9\cdot 7\cdot 5\cdot 3}{10\cdot 8\cdot 6\cdot 4}\displaystyle\int_0^{\pi/2} \sin^2 x\,dx =$
$\dfrac{9\cdot 7\cdot 5\cdot 3}{10\cdot 8\cdot 6\cdot 4\cdot 2}\displaystyle\int_0^{\pi/2} dx = \dfrac{63\pi}{2^9}.$

c. The values decrease as n increases.

8.7 Numerical Integration

8.7.1 $\Delta x = \dfrac{18-4}{28} = \dfrac{1}{2}$.

8.7.3 The Trapezoidal Rule approximates the definite integral by using a trapezoid over each subinterval rather than a rectangle.

8.7.5 The endpoints of the subintervals are -1, 1, 3, 5, 7, and 9. The trapezoidal rule uses the value of f at each of these endpoints.

8.7.7 The absolute error is $|\pi - 3.14| \approx 0.0015926536$. The relative error is $\dfrac{|\pi - 3.14|}{\pi} \approx 5 \times 10^{-4}$.

8.7.9 The absolute error is $|e - 2.72| \approx 0.0017181715$. The relative error is $\frac{|e-2.72|}{e} \approx 6.32 \times 10^{-4}$.

8.7.11

For $n = 1$, we have $f(6) \cdot 8 = 72 \cdot 8 = 576$.

For $n = 2$ we have $f(4) \cdot 4 + f(8) \cdot 4 = 32 \cdot 4 + 128 \cdot 4 = 640$.

For $n = 4$, we have $f(3) \cdot 2 + f(5) \cdot 2 + f(7) \cdot 2 + f(9) \cdot 2 = 18 \cdot 2 + 50 \cdot 2 + 98 \cdot 2 + 162 \cdot 2 = 656$.

8.7.13 We have

$$\frac{1}{6} \left(\sin(\pi/12) + \sin(\pi/4) + \sin(5\pi/12) + \sin(7\pi/12) + \sin(3\pi/4) + \sin(11\pi/12) \right) \approx 0.6439505509.$$

8.7.15 For $n = 2$ we have $T(2) = \frac{4}{2}(f(2) + 2f(6) + f(10)) = 2(8 + 2(72) + 200) = 704$.

Using the results of number 11: For $n = 4$, note that $T(4) = \frac{T(2)+M(2)}{2} = \frac{704+640}{2} = 672$.

Using the results of number 11: For $n = 8$ we have that $T(8) = \frac{T(4)+M(4)}{2} = \frac{672+656}{2} = 664$.

8.7.17 We have $T(6) = \frac{1}{12}(\sin 0 + 2\sin \pi/6 + 2\sin \pi/3 + 2\sin \pi/2 + 2\sin 2\pi/3 + 2\sin 5\pi/6 + \sin \pi) = \frac{1}{6}(\frac{1}{2} + \frac{\sqrt{3}}{2} + 1 + \frac{\sqrt{3}}{2} + \frac{1}{2}) = \frac{1}{6}(2 + \sqrt{3})$.

8.7.19 The width of each subinterval is $1/25$, so $M(25) = \frac{1}{25}(\sin \pi/50 + \sin 3\pi/50 + \sin 5\pi/50 + \cdots + \sin 49\pi/50) \approx .6370388444$. Since $\int_0^1 \sin \pi x \, dx = \frac{2}{\pi}$, the absolute error is $|2/\pi - M(25)| \approx 4.19 \times 10^{-4}$ and the relative error is this number divided by $2/\pi$ which is approximately 6.58×10^{-4}. The Trapezoidal Rule yields approximately $.6357817937$, with a relative error of ≈ 0.001316.

8.7.21

n	$M(n)$	Absolute Error	$T(n)$	Absolute Error
4	99	1	102	2
8	99.75	0.250	100.5	0.5
16	99.9375	0.0625	100.125	0.125
32	99.984375	0.0156	100.03125	0.03125

8.7.23

n	$M(n)$	Absolute Error	$T(n)$	Absolute Error
4	1.50968181	9.7×10^{-3}	1.48067370	1.9×10^{-2}
8	1.50241228	2.4×10^{-3}	1.49517776	4.8×10^{-3}
16	1.50060256	6.0×10^{-4}	1.49879502	1.2×10^{-3}
32	1.50015061	1.5×10^{-4}	1.49969879	3.0×10^{-4}

8.7.25 Because the given function has odd symmetry about the midpoint of the interval $[0, \pi]$, the midpoint rule calculates to be zero for all even values of n, as does the trapezoidal rule.

8.7.27 Answers may vary.

$$\overline{T} = \frac{1}{12}\int_0^{12} T(t)\, dt \approx \frac{1}{12}\text{Trapezoid}(12) \approx \frac{1}{24}(47+2(50+46+45+48+52+54+61+62+63+63+59)+55) = 54.5.$$

8.7.29 Answers may vary.

$$\overline{T} = \frac{1}{12}\int_0^{12} T(t)\, dt \approx \frac{1}{12}\text{Trapezoid}(12) \approx \frac{1}{24}(35 + 2(34 + 34 + 36 + 36 + 37 + 37 + 36 + 35 + 35 + 34 + 33) + 32)) \approx 35.0.$$

8.7.31

a. Left Riemann sum: $\frac{1}{120}(70 \cdot 20 + 130 \cdot 25 + 200 \cdot 15 + 239 \cdot 30 + 311 \cdot 20 + 355 \cdot 10) = 204.917$.
   Right Riemann sum: $\frac{1}{120}(130 \cdot 20 + 200 \cdot 25 + 239 \cdot 15 + 311 \cdot 30 + 355 \cdot 20 + 375 \cdot 10) = 261.375$.
   Trapezoidal rule:

$$\frac{(70+130)\cdot 20}{2 \cdot 120} + \frac{(130+200)\cdot 25}{2 \cdot 120} + \frac{(200+239)\cdot 15}{2 \cdot 120} + \frac{(239+311)\cdot 30}{2 \cdot 120}$$
$$+ \frac{(311+355)\cdot 20}{2 \cdot 120} + \frac{(355+375)\cdot 10}{2 \cdot 120} = 233.146$$

These are approximations to the average temperature of the curling iron on the interval $[0, 120]$.

b. Because the function is increasing, the left Riemann sum is an underestimate and the right Riemann sum is an overestimate. The Trapezoidal rule appears to be a slight underestimate. The followings is a plot of the points with straight line segments connecting them. The shape suggests that the actual function is concave down on the interval, so the trapezoidal rule will be an underestimate.

c. The change in temperature over the time interval is

$$\int_0^{120} T'(t)\, dt = T(120) - T(0) = 375 - 70 = 305 \text{ degrees}.$$

8.7.33

a. The net change in elevation is $\int_0^5 v(t)\, dt$, which can be approximated by the trapezoidal rule to give

$$\frac{(0+100)\cdot 1}{2} + \frac{(100+120)\cdot 0.5}{2} + \frac{(120+150)\cdot 1.5}{2}$$
$$+ \frac{(150+110)\cdot 0.5}{2} + \frac{(110+90)\cdot 0.5}{2} + \frac{(90+80)\cdot 1}{2} = 507.5.$$

If we add the net change to the original elevation of 5400 feet, we have an elevation of approximately 5907.5 feet.

b. The right Riemann sum is

$$100 \cdot 1 + 120 \cdot 0.5 + 150 \cdot 1.5 + 110 \cdot 0.5 + 90 \cdot 0.5 + 80 \cdot 1 = 565.$$

If we add the net change to the original elevation of 5400 feet, we have an elevation of approximately 5965 feet.

c. The elevation can be estimated by

$$5400 + \int_0^5 g(t)\, dt = 5400 + \int_0^5 (3.49t^3 - 43.21t^2 + 142.43t - 1.75)\, dt \approx 5916.52,$$

so the elevation of the balloon is about 5917 feet.

8.7.35

n	$T(n)$	Absolute Error	$S(n)$	Absolute Error
25	3.19623162	—	—	—
50	3.19495398	4.3×10^{-4}	3.19452809	4.5×10^{-8}

8.7.37

n	$T(n)$	Absolute Error	$S(n)$	Absolute Error
50	1.00008509	—	—	—
100	1.00002127	2.1×10^{-5}	1.00000000	4.6×10^{-9}

8.7.39

n	$T(n)$	Absolute Error	$S(n)$	Absolute Error
4	1820	284	—	—
8	1607.75	71.8	1537	1
16	1553.9844	18	1536.0625	6.3×10^{-2}
32	1540.4990	4.5	1536.0039	3.9×10^{-3}

8.7.41

n	$T(n)$	Absolute Error	$S(n)$	Absolute Error
4	0.46911538	5.3×10^{-2}	—	—
8	0.50826998	1.3×10^{-2}	0.52132152	2.9×10^{-4}
16	0.51825968	3.4×10^{-3}	0.52158957	1.7×10^{-5}
32	0.52076933	8.4×10^{-4}	0.52160588	1.1×10^{-6}

8.7.43

a. True. In the case of a linear function, the region under the curve and over each subinterval is a trapezoid, so the trapezoidal rule gives the exact area.

b. False. Since $E_M(n) \leq \frac{k(b-a)^3}{24n^2}$, we have $E_M(3n) \leq \frac{k(b-a)^3}{24(9n^2)}$, so the error decreases by a factor of about 9.

c. True. Since $E_T(n) \leq \frac{k(b-a)^3}{12n^2}$, we have $E_T(4n) \leq \frac{k(b-a)^3}{24(16n^2)}$, so the error decreases by a factor of about 16.

8.7.45 $\int_0^{\pi/2} \cos^9 x \, dx = \frac{128}{315}$.

n	$M(n)$	Absolute Error	$T(n)$	Absolute Error
4	0.40635058	1.4×10^{-6}	0.40634783	1.4×10^{-6}
8	0.40634921	7.6×10^{-10}	0.40634921	7.6×10^{-9}
16	0.40634921	6.6×10^{-13}	0.40634921	6.6×10^{-13}
32	0.40634921	8.9×10^{-16}	0.40634921	7.8×10^{-16}

8.7.47 $\int_0^{\pi} \ln(5 + 3\cos x) x \, dx = \pi \ln(9/2)$.

n	$M(n)$	Absolute Error	$T(n)$	Absolute Error
4	4.72531820	1.2×10^{-4}	4.72507878	1.2×10^{-4}
8	4.72519851	9.1×10^{-9}	4.72519849	9.1×10^{-9}
16	4.72519850	0.	4.72519850	8.9×10^{-16}
32	4.72519850	0.	4.72519850	8.9×10^{-19}

8.7.49 $\displaystyle\int_0^\pi \frac{\cos x}{(5/4) - \cos x}\, dx = \frac{2\pi}{3} \approx 2.094395102.$

n	$S(n)$	Absolute Error
4	1.916439963	0.178
8	2.080919302	0.0135
16	2.094341841	5.3×10^{-5}

8.7.51 $\displaystyle\int_0^\pi \sin 6x \cos 3x\, dx = \frac{4}{9} = 0.\overline{4}.$

n	$S(n)$	Absolute Error
8	0.0305049084	0.475
16	0.4540112289	0.0096
32	0.44487	0.00087

8.7.53 $\displaystyle\int_0^{2\pi} \sqrt{a^2 \cos^2 t + b^2 \sin^2 t}\, dt,\ a = 4,\ b = 8.$

n	$S(n)$
4	41.88790205
8	39.05860599

8.7.55 We are computing $\displaystyle\frac{1}{3\sqrt{2\pi}} \int_{66}^{72} e^{-(x-69)^2/18}\, dx.$ If we use Simpson's rule we obtain

n	$S(n)$
4	0.683
8	0.683

So about 68.3%.

8.7.57

a. For even n we have $S(n) = \frac{20}{n}(365)(f(a) + 4f(x_1) + 4f(x_2) + \ldots + 4f(x_{n-1}) + f(b))$. For $n = 6$ since there are 6 decades between 1940 and 2000, we have $S(6) \approx 160,000$ millions of barrels produced.

b. Following part (a) with $n = 6$ we have $S(n) \approx 68,000$ millions of barrels imported.

8.7.59

a. $T(40) = \frac{1}{80}(\sin 1 + 2\sum_{i=1}^{39} \sin e^{i/40} + \sin e) \approx .8748.$

b. $f(x) = \sin e^x$, so $f'(x) = e^x \cos e^x$. $f''(x) = -e^{2x} \sin e^x + e^x \cos e^x = e^x(\cos e^x - e^x \sin e^x)$.

c. $|f''(x)| = |e^x| \cdot |\cos e^x - e^x \sin e^x| \leq |e^x|(|\cos e^x| + |e^x \sin e^x|) \leq e(1 + e) < 10.11$. However, the graph of the absolute value of $f''(x)$ reveals that is is actually bounded by 5.75 on the interval $[0, 1]$.

d. Since $E_T(n) \leq k(b-a)^3/12n^2 = 6/(12(40^2)) = .0003125$, $T(40)$ is accurate to at least 3 decimal places.

8.7.61

a. The exact value is $\int_0^4 x^3\, dx = \frac{x^4}{4}\Big|_0^4 = 64$. Simpson's rule gives $S(2) = (0 + 4 \cdot 8 + 64)\frac{2}{3} = 64$. The values match exactly.

b. $S(4) = (0 + 4 \cdot 1 + 2 \cdot 8 + 4 \cdot 27 + 64)\frac{1}{3} = 64$. This also matches the exact value exactly.

c. The 4th derivative of x^3 is 0, so the value of K in the theorem can be taken to be 0, so the error is 0.

d. Any polynomial of degree 3 or less has 4th derivative equal to 0, so the value of K in the theorem can be taken to be 0, so the error is 0.

8.7.63 The trapezoidal rule will be an overestimate in this case. This is because of the fact that if the function is above the axis and concave up on the given interval, then each trapezoid on each subinterval lies over the area under the curve for that corresponding subinterval.

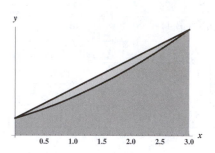

8.7.65 Using the previous results, we have

$$S(2n) = \frac{4T(2n) - T(n)}{3} = \frac{2(T(n) + M(n)) - T(n)}{3} = \frac{2M(n) + T(n)}{3}.$$

So $S(50) = \frac{1}{3}(2M(25) + T(25)) \approx .8298251909$.

8.8   Improper Integrals

8.8.1 The interval of integration is infinite or the integrand is unbounded on the interval of integration.

8.8.3 Compute $\displaystyle\int_0^1 \frac{1}{\sqrt{x}}\,dx = \lim_{b \to 0^+} \int_b^1 \frac{1}{\sqrt{x}}\,dx$.

8.8.5 $\displaystyle\int_1^\infty x^{-2}\,dx = \lim_{b\to\infty}\int_1^b x^{-2}\,dx = \lim_{b\to\infty}\left(-\frac{1}{x}\right)\Big|_1^b = \lim_{b\to\infty}\left(1 - \frac{1}{b}\right) = 1.$

8.8.7 $\displaystyle\int_{-\infty}^0 e^x\,dx = \lim_{b\to -\infty}\int_b^0 e^x\,dx = \lim_{b\to -\infty} e^x\Big|_b^0 = \lim_{b\to -\infty}\left(1 - e^b\right) = 1.$

8.8.9 $\displaystyle\int_2^\infty \frac{dx}{\sqrt{x}} = \lim_{b\to\infty}\int_2^b \frac{dx}{\sqrt{x}} = \lim_{b\to\infty} 2\sqrt{x}\Big|_2^b = \lim_{b\to\infty} 2(\sqrt{b} - \sqrt{2}) = \infty$, so the integral diverges.

8.8.11 $\displaystyle\int_0^\infty e^{-2x}\,dx = \lim_{b\to\infty}\int_0^b e^{-2x}\,dx = \lim_{b\to\infty}\left(-\frac{1}{2}e^{-2x}\right)\Big|_0^b = \lim_{b\to\infty}\frac{1}{2}\left(1 - e^{-2b}\right) = \frac{1}{2}.$

8.8.13 $\displaystyle\int_0^\infty e^{-ax}\,dx = \lim_{b\to\infty}\int_0^b e^{-ax}\,dx = \lim_{b\to\infty}\left(-e^{-x}/a\right)\Big|_0^b = \lim_{b\to\infty}\left(-\frac{1}{ae^b} + \frac{1}{a}\right) = \frac{1}{a}.$

8.8.15

$$\int_{e^2}^\infty \frac{dx}{x\ln^p x} = \lim_{b\to\infty}\int_{e^2}^b \frac{dx}{x\ln^p x} = \lim_{b\to\infty}\left(\frac{1}{1-p}\ln^{1-p} x\right)\Big|_{e^2}^b = \lim_{b\to\infty}\frac{1}{p-1}\left(2^{1-p} - \ln^{1-p} b\right) = \frac{1}{(p-1)2^{p-1}}.$$

8.8.17

$$\int_{-\infty}^\infty xe^{-x^2}\,dx = \lim_{b\to -\infty}\int_b^0 xe^{-x^2} + \lim_{b\to\infty}\int_0^b xe^{-x^2}$$

$$= \lim_{b\to -\infty}\left(-\frac{1}{2}e^{-x^2}\right)\Big|_b^0 + \lim_{b\to\infty}\left(-\frac{1}{2}e^{-x^2}\right)\Big|_0^b$$

$$= \lim_{b\to -\infty}\left(-\frac{1}{2} + \frac{1}{2}e^{-b^2}\right) + \lim_{b\to\infty}\left(-\frac{1}{2}e^{-b^2} + \frac{1}{2}\right) = -\frac{1}{2} + \frac{1}{2} = 0.$$

8.8.19 $\displaystyle\int_2^\infty \frac{\cos(\pi/x)}{x^2}\,dx = \lim_{b\to\infty}\int_2^b \frac{\cos(\pi/x)}{x^2}\,dx = \lim_{b\to\infty}\left(-\frac{1}{\pi}\sin(\pi/x)\right)\bigg|_2^b = \lim_{b\to\infty}\frac{1}{\pi}(1-\sin(\pi/b)) = \frac{1}{\pi}.$

8.8.21 $\displaystyle\int_0^\infty \frac{e^u}{e^{2u}+1}\,du = \lim_{b\to\infty}\int_0^b \frac{e^u}{e^{2u}+1}\,du.$ Let $u = e^u$ so that $du = e^u\,du.$ After substitution we have

$\displaystyle\lim_{b\to\infty}\int_1^{e^b}\frac{1}{u^2+1}\,du = \lim_{b\to\infty}\left(\tan^{-1}(u)\right)\bigg|_1^{e^b} = \lim_{b\to\infty}\left(\tan^{-1}(e^b)-\tan^{-1}(1)\right) = \pi/2 - \pi/4 = \pi/4.$

8.8.23 $\displaystyle\int_1^\infty \frac{1}{v(v+1)}\,dv = \lim_{b\to\infty}\int_1^b\left(\frac{1}{v}-\frac{1}{v+1}\right)dv = \lim_{b\to\infty}\ln\left|\frac{v}{v+1}\right|\bigg|_1^b = \lim_{b\to\infty}\ln\left|\frac{b}{b+1}\right| - \ln(1/2) =$
$0 - \ln(1/2) = \ln 2.$

8.8.25 Let $u = x^3 + x$ so that $du = (3x^2+1)\,dx$ Substituting gives

$$\int_2^\infty \frac{1}{u}\,du = \lim_{b\to\infty}\int_2^b \frac{1}{u}\,du = \lim_{b\to\infty}\ln u\bigg|_2^b = \lim_{b\to\infty}(\ln b - \ln 2) = \infty.$$

The given integral diverges.

8.8.27 $\displaystyle\int_2^\infty \frac{dx}{(x+2)^2} = \lim_{b\to\infty}\int_2^b \frac{dx}{(x+2)^2} = \lim_{b\to\infty}-\frac{1}{x+2}\bigg|_2^b = \lim_{b\to\infty}-\frac{1}{b+2}+\frac{1}{4} = \frac{1}{4}.$

8.8.29 Using the result from Example 2, we see that the volume is given by $V = \pi\displaystyle\int_1^\infty x^{-4}\,dx = \frac{\pi}{4-1} = \frac{\pi}{3}.$

8.8.31 Using the result from Example 2, we see that the volume is given by

$$V = \pi\int_1^\infty\left(\frac{1}{x^2}+\frac{1}{x^3}\right)dx = \frac{\pi}{2-1}+\frac{\pi}{3-1} = \frac{3\pi}{2}.$$

8.8.33 $V = \pi\displaystyle\int_2^\infty \frac{dx}{x(\ln x)^2} = \pi\lim_{b\to\infty}\int_2^b \frac{dx}{x(\ln x)^2} = \pi\lim_{b\to\infty}\left(-\frac{1}{\ln x}\right)\bigg|_2^b = \pi\lim_{b\to\infty}\left(\frac{1}{\ln 2}-\frac{1}{\ln b}\right) = \frac{\pi}{\ln 2}.$

8.8.35 $\displaystyle\int_0^8 \frac{dx}{\sqrt[3]{x}} = \lim_{c\to 0^+}\int_c^8 x^{-1/3}\,dx = \lim_{c\to 0^+}\left(\frac{3}{2}x^{2/3}\right)\bigg|_c^8 = \frac{3}{2}\lim_{c\to 0^+}(4-c^{2/3}) = 6.$

8.8.37 $\displaystyle\lim_{c\to 1^+}\int_c^2 \frac{1}{\sqrt{x-1}}\,dx = \lim_{c\to 1^+}\left(2\sqrt{x-1}\right)\bigg|_c^2 = \lim_{c\to 1^+}\left(2-2\sqrt{c-1}\right) = 2.$

8.8.39 $\displaystyle\lim_{b\to(\pi/2)^-}\int_0^b \sec x\tan x\,dx = \lim_{b\to(\pi/2)^-}(\sec x)\bigg|_0^b = \lim_{b\to(\pi/2)^-}(\sec b - 1) = \infty.$ The given integral diverges.

8.8.41 Note that $\displaystyle\int \frac{e^{\sqrt{x}}}{\sqrt{x}}\,dx = 2e^{\sqrt{x}} + C.$ Thus,

$$\lim_{c\to 0^+}\int_c^1 \frac{e^{\sqrt{x}}}{\sqrt{x}}\,dx = \lim_{c\to 0^+}\left(2e^{\sqrt{x}}\right)\bigg|_c^1 = \lim_{c\to 0^+}\left(2e - 2e^{\sqrt{c}}\right) = 2e - 2.$$

8.8.43 $\displaystyle\int_0^1 \frac{x^3}{x^4-1}\,dx = \lim_{c\to 1^-}\int_0^c \frac{x^3}{x^4-1}\,dx = \lim_{c\to 1^-}\left(\frac{1}{4}\ln|x^4-1|\right)\bigg|_0^c = \frac{1}{4}\lim_{c\to 1^-}\ln|c^4-1| = -\infty,$ so the integral diverges.

8.8.45

$$\int_0^{10} \frac{dx}{\sqrt[4]{10-x}} = \lim_{c \to 10^-} \int_0^c (10-x)^{-1/4}\, dx = \lim_{c \to 10^-} \left(-\frac{4}{3}(10-x)^{3/4} \right) \Big|_0^c$$

$$= \frac{4}{3} \lim_{c \to 10^-} \left(10^{3/4} - (10-c)^{3/4} \right) = \frac{4}{3} 10^{3/4}.$$

8.8.47 By the even symmetry of the integrand, this should be equal to $2 \int_0^1 \ln y^2\, dy$. Thus we have

$$2 \int_0^1 \ln y^2\, dy = 4 \lim_{c \to 0^+} \int_c^1 \ln y\, dx = 4 \lim_{c \to 0^+} (y \ln y - y) \Big|_c^1 = 4 \lim_{c \to 0^+} (-1 - c \ln c + c) = -4.$$ (The fact that $\lim_{c \to 0^+} c \ln c = 0$ can be derived using L'Hôpital's rule, or from the result $\lim_{c \to 0^+} c^c = 1$).

8.8.49 By symmetry,

$$\int_{-2}^2 \frac{dp}{\sqrt{4-p^2}} = 2 \int_0^2 \frac{dp}{\sqrt{4-p^2}} = 2 \lim_{c \to 2^-} \int_0^c \frac{dp}{\sqrt{4-p^2}}$$

$$= 2 \lim_{c \to 2^-} \left(\sin^{-1}(p/2) \right) \Big|_0^c = 2 \lim_{c \to 2^-} \left(\sin^{-1}(c/2) - \sin^{-1} 0 \right)$$

$$= 2(\sin^{-1} 1 - 0) = \pi.$$

8.8.51 Using disks, we have

$$V = \pi \int_1^2 (x-1)^{-1/2}\, dx = \pi \lim_{c \to 1^+} \int_c^2 (x-1)^{-1/2}\, dx = \pi \lim_{c \to 1^+} \left(2(x-1)^{1/2} \right) \Big|_c^2 = 2\pi \lim_{c \to 1^+} (1 - \sqrt{c-1}) = 2\pi.$$

8.8.53 Using shells, we have

$$V = 2\pi \int_0^4 x(4-x)^{-1/3}\, dx = 2\pi \int_0^4 (4-u)u^{-1/3}\, du = 2\pi \int_0^4 (4u^{-1/3} - u^{2/3})\, du$$

via the substitution $u = 4 - x$. Therefore

$$V = 2\pi \lim_{c \to 0^+} \int_c^4 (4u^{-1/3} - u^{2/3})\, du = 2\pi \lim_{c \to 0^+} \left(6u^{2/3} - \frac{3}{5}u^{5/3} \right) \Big|_c^4 = \frac{72 \cdot 2^{1/3}\pi}{5}.$$

8.8.55 $V = \pi \int_0^{\pi/2} \tan^2 x\, dx = \lim_{a \to \frac{\pi}{2}^-} \pi \int_0^a \tan^2 x\, dx = \lim_{a \to \frac{\pi}{2}^-} \pi \int_0^a (\sec^2 x - 1)\, dx = \lim_{a \to \frac{\pi}{2}^-} \pi (\tan x - x) \Big|_0^a = \lim_{a \to \frac{\pi}{2}^-} \pi (\tan a - a) = \infty$, so the volume doesn't exist.

8.8.57 As in Example 7, we have

$$\text{AUC}_i = \int_0^\infty C_i(t)\, dt = 250 \int_0^\infty e^{-0.08t}\, dt = \frac{250}{0.08} = 3125$$

and

$$\text{AUC}_o = \int_0^\infty C_0(t)\, dt = 200 \int_0^\infty (e^{-0.08t} - e^{-1.8t})\, dt = 200 \left(\frac{1}{0.08} - \frac{1}{1.8} \right) = \frac{21,500}{9} \approx 2389,$$

(here we use the fact that $\int_0^\infty e^{-ax}\, dx = \frac{1}{a}$ for $a > 0$). Therefore the bioavailability of the drug is

$$F = \frac{\text{AUC}_o}{\text{AUC}_i} = \frac{21,500}{9 \cdot 3125} \approx 0.764.$$

8.8.59 The maximum distance is

$$D = 10 \int_0^\infty (t+1)^{-2}\, dt = 10 \lim_{b\to\infty} \int_0^b (t+1)^{-2}\, dt = 10 \lim_{b\to\infty} \left(-(t+1)^{-1}\right)\Big|_0^b = 10\, \text{mi}.$$

8.8.61

a. True. The area under the curve $y = f(x)$ from 0 to ∞ is less than the area under $y = g(x)$ on this interval, which by assumption is finite.

b. False. For example, take $f(x) = 1$; then $\int_0^\infty f(x)\, dx = \infty$.

c. False. For example, take $p = 1/2$ and $q = 1$.

d. True. The area under the curve $y = x^{-q}$ from 1 to ∞ is less than the area under $y = x^{-p}$ on this interval, which by assumption is finite.

e. True. Using the result in Example 2, we see that this integral exists if and only if $3p + 2 > 1$, which is equivalent to $p > -1/3$.

8.8.63

a. The function $e^{-|x|}$ is even, so $\int_{-\infty}^\infty e^{-|x|}\, dx = 2\int_0^\infty e^{-x}\, dx = 2$. (Note that here we used the fact that $\int_0^\infty e^{-ax}\, dx = \frac{1}{a}$ for $a > 0$).

b. The function $x^3/(1 + x^8)$ is odd, so $\int_{-\infty}^\infty \frac{x^3}{1+x^8}\, dx = \int_0^\infty \frac{x^3}{1+x^8}\, dx + \int_{-\infty}^0 \frac{x^3}{1+x^8}\, dx = \int_0^\infty \frac{x^3}{1+x^8}\, dx - \int_0^\infty \frac{x^3}{1+x^8}\, dx = 0$ assuming $\int_0^\infty \frac{x^3}{1+x^8}\, dx$ exists, which is true because $0 < \frac{x^3}{1+x^8} < \frac{1}{x^5}$ on $[1, \infty)$ and $\int_1^\infty x^{-5}\, dx$ exists.

8.8.65

n	$T_2(n)$	$T_4(n)$	$T_8(n)$
4	.880619	.886319	1.036632
8	.881704	.886227	.886319
16	.881986	.886227	.886227
32	.882058	.886227	.886227

Based on these results, we conclude that $\int_0^\infty e^{-x^2}\, dx \approx 0.886$.

8.8.67 Integration by parts gives $\int x \ln x\, dx = \frac{x^2}{4}(2 \ln x - 1) + C$, so $\int_0^1 x \ln x\, dx = \lim_{c\to 0^+} \int_c^1 x \ln x\, dx =$

$\lim_{c\to 0^+} \left(\frac{x^2}{4}(2 \ln x - 1)\right)\Big|_c^1 = \frac{1}{4} \lim_{c\to 0^+} (c^2(-2 \ln c + 1) - 1) = -\frac{1}{4}$. (The fact that $\lim_{c\to 0^+} c^2 \ln c = 0$ can be derived using L'Hôpital's rule, or from the result $\lim_{c\to 0^+} c^c = 1$).

8.8.69 Let $u = x^2$ so that $2x\, dx = du$. The first integral is then equal to $\frac{1}{2}\int_0^\infty e^{-u}\, du = \frac{1}{2}$, using the fact that $\int_0^\infty e^{-ax}\, dx = \frac{1}{a}$ for $a > 0$. The second integral cannot be evaluated by finding an antiderivative for $x^2 e^{-x^2}$; however using more advanced methods it can be shown that $\int_0^\infty x^2 e^{-x^2}\, dx = \frac{\sqrt{\pi}}{4} \approx 0.443$.

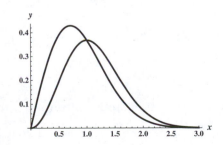

8.8.71 The region R has area $A = \int_0^\infty e^{-bx}\,dx - \int_0^\infty e^{-ax}\,dx = \frac{1}{b} - \frac{1}{a}$.

8.8.73

a. We have

$$A(a,b) = \int_b^\infty e^{-ax}\,dx = \lim_{c\to\infty}\int_b^c e^{-ax}\,dx = \lim_{c\to\infty}\left(-\frac{1}{a}e^{-ax}\right)\Big|_b^c = \frac{1}{a}\lim_{c\to\infty}(e^{-ab} - e^{-ac}) = \frac{e^{-ab}}{a}.$$

b. Solving $e^{-ab} = 2a$ for b gives $b = g(a) = -\frac{1}{a}\ln(2a)$.

c. The function g has $g'(x) = \frac{1}{x^2}\ln 2x - \frac{1}{x^2} = \frac{\ln 2x - 1}{x^2}$, so g has a critical point at $x = e/2$, and the first derivative test shows that g takes a minimum at this point. Hence $b^* = g(e/2) = -\frac{2}{e}$.

8.8.75

a. The solid has volume $V = \pi\int_0^1 x^{-2p}\,dx$, which by the result in problem 74 is finite if and only if $2p < 1$, or $p < 1/2$.

b. The solid has volume $V = 2\pi\int_0^1 x^{1-p}\,dx$, which by the result in problem 74 is finite if and only if $p - 1 < 1$, or $p < 2$.

8.8.77 These integrals cannot be evaluated by finding an antiderivative of their integrands. However, if we make the substitution $u = \pi/2 - x$ we find that $\int_0^{\pi/2}\ln\sin x\,dx = -\int_{\pi/2}^0 \ln\cos u\,du = \int_0^{\pi/2}\ln\cos x\,dx$. We also have

$$\int_{\pi/2}^\pi \ln\sin x\,dx = \int_0^{\pi/2}\ln\sin(x + \pi/2)\,dx = \int_0^{\pi/2}\ln\cos x\,dx,$$

and therefore

$$\int_0^{\pi/2}\ln\sin x\,dx = \frac{1}{2}\int_0^\pi \ln\sin x\,dx = \int_0^{\pi/2}\ln\sin 2y\,dy = \int_0^{\pi/2}(\ln\sin y + \ln\cos y + \ln 2)\,dy.$$

This implies $\int_0^{\pi/2}\ln\cos x\,dx = -\int_0^{\pi/2}\ln 2\,dx = -\frac{\pi\ln 2}{2}$.

8.8.79 This integral cannot be evaluated by finding an antiderivative of the integrand; however the result may be verified by numerical approximation.

8.8.81 We have the relation $B = I\int_0^\infty e^{-rt}\,dt = \frac{I}{r}$, using the result that $\int_0^\infty e^{-ax}\,dx = \frac{1}{a}$ for $a > 0$. Therefore $B = 5000/0.12 = \$41{,}666.67$.

8.8.83

a. We have $\int_0^\infty e^{-ax}\cos bx\,dx = \lim_{c\to\infty}\int_0^c e^{-ax}\cos bx\,dx = \lim_{c\to\infty}\left(\frac{e^{-ax}(b\sin bx - a\cos bx)}{a^2 + b^2}\right)\Big|_0^c =$

$\lim_{c\to\infty}\frac{a + e^{-ac}(b\sin bc - a\cos bc)}{a^2 + b^2} = \frac{a}{a^2 + b^2}$.

b. We have $\int_0^\infty e^{-ax}\sin bx\,dx = \lim_{c\to\infty}\int_0^c e^{-ax}\sin bx\,dx = \lim_{c\to\infty}\left(-\frac{e^{-ax}(a\sin bx + b\cos bx)}{a^2 + b^2}\right)\Big|_0^c =$

$\lim_{c\to\infty}\frac{b - e^{-ac}(a\sin bc + b\cos bc)}{a^2 + b^2} = \frac{b}{a^2 + b^2}$.

8.8.85 Evaluate the improper integral $\int_0^\infty te^{-at}\,dt = \lim_{b\to\infty}\int_0^b te^{-at}\,dt = \lim_{b\to\infty}\left(-\frac{e^{-at}(at+1)}{a^2}\right)\Big|_0^b =$

$\frac{1}{a^2}\lim_{b\to\infty}\left(1 - e^{-ab}(ab+1)\right) = \frac{1}{a^2}$, provided $a > 0$. Therefore $0.00005\int_0^\infty te^{-0.00005t}\,dt = \frac{0.00005}{0.00005^2} = 20{,}000\,\text{hrs}$.

8.8.87

a. We have $W = GMm \int_R^\infty x^{-2}\, dx = GMm \lim_{b\to\infty} \left(-\frac{1}{x}\right)\big|_R^b = \frac{GMm}{R} \approx 6.279 \times 10^7\, \text{m J}.$

b. Solve $\frac{1}{2}v_e^2 = 6.279 \times 10^7$ to obtain $v_e \approx 11.207\, \text{km/s}.$

c. We need $\frac{GM}{R} \geq \frac{1}{2}c^2 \iff R \leq \frac{2GM}{c^2} \approx 9\, \text{mm}.$

8.8.89

a.

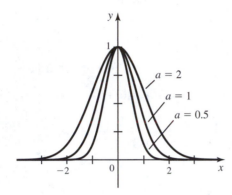

b. The areas are $\sqrt{2\pi}, \sqrt{\pi}, \sqrt{\pi/2}$ respectively.

c. Completing the square gives

$$ax^2 + bx + c = a\left(x + \frac{b}{2a}\right)^2 + \frac{4ac - b^2}{4a},$$

and therefore

$$\int_{-\infty}^{\infty} e^{-(ax^2 + bx + c)}\, dx = e^{(b^2 - 4ac)/(4a)} \int_{-\infty}^{\infty} e^{-a(x + (b/2a))^2}\, dx.$$

Make the substitution $y = x + b/(2a)$ to obtain

$$\int_{-\infty}^{\infty} e^{-(ax^2 + bx + c)}\, dx = e^{(b^2 - 4ac)/(4a)} \int_{-\infty}^{\infty} e^{-ay^2}\, dy = e^{(b^2 - 4ac)/(4a)} \sqrt{\frac{\pi}{a}}.$$

8.8.91 The Laplace transform of $f(t) = e^{at}$ is given by $F(s) = \int_0^\infty e^{-st}e^{at}\, dt = \int_0^\infty e^{-(s-a)t}\, dt = \frac{1}{s-a}$, using the formula $\int_0^\infty e^{-cx}\, dx = \frac{1}{c}$ for $c > 0$.

8.8.93 The Laplace transform of $f(t) = \sin at$ is given by $F(s) = \int_0^\infty e^{-st} \sin at\, dt = \frac{a}{s^2 + a^2}$ (this formula is derived in the solution to problem 83 b).

8.8.95

a. Make the substitution $x = y + 2$; then

$$\int_1^3 \frac{dx}{\sqrt{(x-1)(3-x)}} = \int_{-1}^1 \frac{dy}{\sqrt{1 - y^2}} = 2 \int_0^1 \frac{dy}{\sqrt{1 - y^2}} = 2 \lim_{c\to 1^-} \left(\sin^{-1} x\right)\Big|_0^c = \pi.$$

b. The substitution $y = e^x$ gives

$$\int_1^\infty \frac{dx}{e^{x+1} + e^{3-x}} = \frac{1}{e} \int_1^\infty \frac{e^x\, dx}{e^{2x} + e^2} = \frac{1}{e} \int_e^\infty \frac{dy}{y^2 + e^2} = \frac{1}{e} \lim_{b\to\infty} \left(\frac{1}{e} \tan^{-1}\left(\frac{y}{e}\right)\right)\Big|_e^b = \frac{\pi}{4e^2}.$$

8.8.97 Because $p > 0$, the functions $1/(x^p + x^{-p})$ and $1/x^p$ have the same growth rate as $x \to \infty$; therefore by the result in Example 2, we have $\int_0^\infty \frac{dx}{x^p + x^{-p}} < \infty \iff p > 1$.

8.8.99

a. Integrate by parts with $u = \sqrt{x}\ln x$ and $v = -1/(1+x)$: $\int_0^\infty \frac{\sqrt{x}\ln x}{(1+x)^2}\,dx = \frac{1}{2}\int_0^\infty \frac{\ln x + 2}{\sqrt{x}(x+1)}\,dx = \frac{1}{2}\int_0^\infty \frac{\ln x}{\sqrt{x}(x+1)}\,dx + \int_0^\infty \frac{dx}{\sqrt{x}(x+1)}$ (the integration by parts is legitimate for this improper integral because the product uv has limit 0 as $x \to \infty$ and as $x \to 0^+$).

b. Let $y = 1/x$. Then $dy = -\frac{1}{x^2}\,dx$.

c. We have $\int_0^1 \frac{\ln x}{\sqrt{x}(x+1)}\,dx = \int_\infty^1 \frac{-\ln y}{\frac{1}{\sqrt{y}}(\frac{1}{y}+1)}\left(-\frac{dy}{y^2}\right) = -\int_1^\infty \frac{\ln y}{\sqrt{y}(1+y)}\,dy$, and hence $\int_0^\infty \frac{\ln x}{\sqrt{x}(x+1)}\,dx = 0$.

d. The change of variables $z = \sqrt{x}$ gives $\int_0^\infty \frac{dx}{\sqrt{x}(x+1)} = 2\int_0^\infty \frac{dz}{z^2+1} = \pi$.

8.8.101 Recall that $\frac{d}{dx}x^x = x^x(1 + \ln x)$, which can be deduced from logarithmic differentiation. Thus we have $\int_0^a x^x(\ln x + 1)\,dx = \lim_{c \to 0^+} x^x\Big|_c^a = \lim_{c \to 0^+}(a^a - c^c) = (a^a - 1)$. This last limit is deduced as follows; let $z = c^c$, then $\ln z = c\ln c = \frac{\ln c}{1/c}$, and by L'hôpital's rule, we have $\lim_{c \to 0^+}\frac{\ln c}{1/c} = \lim_{c \to 0^+}\frac{1/c}{-1/c^2} = \lim_{c \to 0^+} -c = 0$. Thus $\lim_{c \to 0^+} z = e^0 = 1$.

8.9   Introduction to Differential Equations

8.9.1 Second-order, because the highest-order derivative appearing in the equation is second order.

8.9.3 The equation is second-order, so we expect two arbitrary constants in the general solution.

8.9.5 A separable first-order differential equation is one that can be written in the form $g(y)y'(t) = h(t)$.

8.9.7 Integrate both sides with respect to t and convert the integral on the left side to an integral with respect to y.

8.9.9 $y = Ce^{-5t}$ so $y' = -5Ce^{-5t}$. Then $y' + 5y = -5Ce^{-5t} + 5Ce^{-5t} = 0$.

8.9.11 $y = C_1 \sin 4t + C_2 \cos 4t$ so $y' = 4C_1 \cos 4t - 4C_2 \sin 4t$ and then $y'' = -16C_1 \sin 4t - 16C_2 \cos 4t$. Then $y'' + 16y = -16C_1 \sin 4t - 16C_2 \cos 4t + 16C_1 \sin 4t + 16C_2 \cos 4t = 0$.

8.9.13 $y = 16e^{2t} - 10$, so $y' = 32e^{2t}$. So $y' - 2y = 32e^{2t} - 32e^{2t} + 20 = 20$. Also $y(0) = 16 - 10 = 6$.

8.9.15 $y = -3\cos 3t$ so $y' = 9\sin 3t$ and $y'' = 27\cos 3t$. Then $y'' + 9y = 27\cos 3t - 27\cos 3t = 0$. Also $y(0) = -3\cos 0 = -3$ and $y'(0) = 9\sin 0 = 0$.

8.9.17 Integrate both sides with respect to t: $\int y'(t)\,dt = \int (3t^2 - 4t + 10)\,dt$ $y(t) = t^3 - 2t^2 + 10t + C$; then substitute $y(0) = 20$ to obtain $y(0) = (t^3 - 2t^2 + 10t + C)\big|_{t=0} = C = 20$, so $y(t) = t^3 - 2t^2 + 10t + 20$.

8.9.19 Integrate both sides with respect to t: $\int y'(t)\,dt = \int \frac{2t^2+4}{t}\,dt = \int \left(2t + \frac{4}{t}\right)\,dt$ $y(t) = t^2 + 4\ln t + C$; then substitute $y(1) = 2$ to obtain $y(1) = (t^2 + 4\ln t + C)\big|_{t=1} = 1 + C = 2$, so $C = 1$ and thus $y(t) = t^2 + 4\ln t + 1$.

8.9.21 The general solution is $y(t) = Ce^{3t} + \frac{4}{3}$.

8.9.23 The general solution is $y(x) = Ce^{-2x} - 2$.

8.9.25 The general solution is $y(t) = Ce^{3t} + 2$; substitute $y(0) = 9$ to obtain $C + 2 = 9$, so $C = 7$; hence $y(t) = 7e^{3t} + 2$.

8.9.27 The general solution is $y(t) = Ce^{-2t} - 2$; substitute $y(0) = 0$ to obtain $C - 2 = 0$, so $C = 2$; hence $y(t) = 2e^{-2t} - 2$.

8.9.29

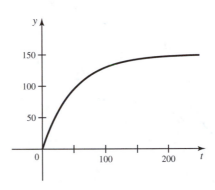

a. The general solution is $y(t) = Ce^{-0.02t} + 150$; substitute $y(0) = 0$ to obtain $C + 150 = 0$, so $C = -150$; hence $y(t) = 150(1 - e^{-0.02t})$.

b. The steady-state level is $\lim_{t \to \infty} 150(1 - e^{-0.02t}) = 150$ mg.

c. We have $150(1 - e^{-0.02t}) = 0.9 \cdot 150$, so $e^{-0.02t} = 0.1$, and thus $t = \frac{\ln 10}{0.02} \approx 115$ hrs.

8.9.31 The equation is separable, so we have $\int y \, dy = \int 3t^2 \, dt$. So $\frac{y^2}{2} = t^3 + C$, and thus $y = \pm\sqrt{2t^3 + C}$.

8.9.33 The equation is separable, so we have $\int e^{-y/2} \, dy = \int \sin t \, dt$, and so $-2e^{-y/2} = -\cos t + C$. Thus, $y = -2 \ln\left(\frac{1}{2}\cos t + C\right)$.

8.9.35 This equation is not separable.

8.9.37 This equation is separable, so we have $\int 2y \, dy = \int e^t \, dt$, so $y^2 = e^t + C$, and thus $y = \pm\sqrt{e^t + C}$. Substituting $y(\ln 2) = 1$ gives $1 = 2 + C$ so $C = -1$, and the solution to this initial value problem is $y = \sqrt{e^t - 1}$.

8.9.39 This equation is separable, so we have $\int e^y \, dy = \int e^x \, dx$, and thus $e^y = e^x + C$. Therefore, $y = \ln(e^x + C)$. Substituting $y(0) = \ln 3$ gives $\ln 3 = \ln(1 + C)$, so $C = 2$ and the solution to this initial value problem is $y = \ln(e^x + 2)$.

8.9.41

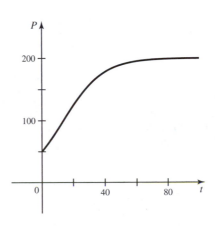

a. This equation is separable, so we have $\int \frac{200}{P(200-P)} \, dP = \int 0.08 \, dt$, so $\int \left(\frac{1}{P} + \frac{1}{200-P}\right) dP = 0.08t + C$. Therefore, $\ln\left|\frac{P}{200-P}\right| = 0.08t + C$. Substituting $P(0) = 50$ gives $-\ln 3 = C$, and solving for P gives $P(t) = \frac{200}{3e^{-0.08t}+1}$.

b. The steady-state population is $\lim_{t \to \infty} P(t) = 200$.

8.9.43

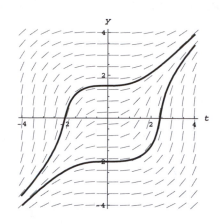

8.9.45

a. This matches with D.

b. This matches with B.

c. This matches with A.

d. This matches with C.

8.9.47

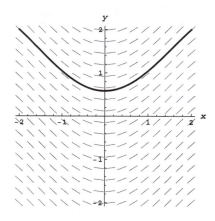

8.9.49

a. False. The general solution is $y = Ce^{20t}$.

b. False. They both do satisfy this differential equation.

c. False. Rewrite the equation as $y'(t) = (t + 2)(y + 2)$.

d. True. $y'(t) = 2(t + 1) = 2\sqrt{(t + 1)^2} = 2\sqrt{y}$.

8.9.51

a. $y = 0$ is an equilibrium solution.

b and c.

8.9.53

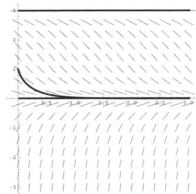

a. Solve $y(y-3) = 0$ to get equilibrium solutions $y = 0$ and $y = 3$.

b and c.

8.9.55

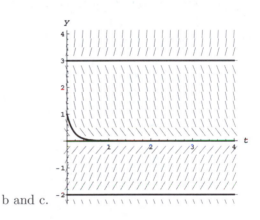

a. The equilibrium solutions are $y = 0$, $y = -2$ and $y = 3$.

b and c.

8.9.57 This equation is separable, so we have $\int \frac{dp}{p+1} = \int \frac{dt}{t^2}$, so $\ln|p+1| = -\frac{1}{t} + D$, and thus $p = Ce^{-1/t} - 1$. Substituting $p(1) = 3$ gives $3 = Ce^{-1} - 1$, so $C = 4e$ and the solution to this initial value problem is $p = 4e^{1-1/t} - 1$.

8.9.59 This equation is separable, so we have $\int \sec^2 w \, dw = \int 2t \, dt$, so $\tan w = t^2 + C$. Thus, $w = \tan^{-1}(t^2 + C)$. Substituting $w(0) = \pi/4$ gives $\pi/4 = \tan^{-1} C$, so $C = \tan(\pi/4) = 1$ and the solution to this initial value problem is $w = \tan^{-1}(t^2 + 1)$.

8.9.61

a. This equation is separable, so we have $\int \frac{1}{y(1-y)} \, dy = \int k \, dt$, so $\int \left(\frac{1}{y} + \frac{1}{1-y} \right) dy = kt + D$. Therefore, $\ln \left| \frac{y}{1-y} \right| = kt + D$, which is equivalent to $\frac{y}{1-y} = Ce^{kt}$. Substituting $y(0) = y_0$ gives $C = y_0/(1 - y_0)$, and thus $y = \frac{y_0}{(1-y_0)e^{-kt} + y_0}$.

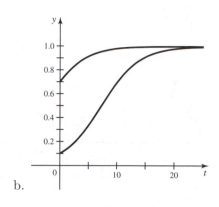

b.

c. The denominator in $y(t)$ above is positive for all $t \geq 0$ when $0 < y_0 < 1$, so $y(t)$ is defined for all $t \geq 0$; we have $\lim_{t \to \infty} y(t) = 1$, which is the steady-state solution.

8.9.63

a. We have $mv'(t) = mg - Rv$, so $v'(t) = g - bv$ with $b = R/m$.

b. Solve $bv = g$ to obtain terminal velocity $\tilde{v} = g/b = mg/R$.

c. The equation $v' = g - bv$ is first-order linear, with general solution $v = Ce^{-bv} + \tilde{v}$. The initial condition $v(0) = 0$ gives $C = -\tilde{v}$, which gives $v = \tilde{v}(1 - e^{-bt})$.

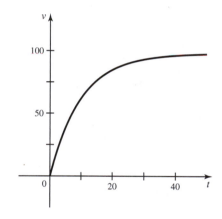

d. We have $b = 0.1$, $\tilde{v} = 98$ m/s.

8.9.65

a. The general solution to $y' = -ky$ is $y = Ce^{-kt}$.

b. The equation $y' = -ky^2$ is separable, so we have $-\int \frac{dy}{y^2} = \int k\,dt$, so $\frac{1}{y} = kt + C$. The initial condition $y(0) = y_0$ gives $C = 1/y_0$, and solving for y gives $y = \frac{1}{kt + 1/y_0} = \frac{y_0}{1 + ky_0 t}$.

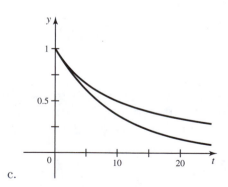

c.

8.9.67

a. The equation $B' = aB - m$ is first-order linear, with general solution $B = Ce^{at} + m/a$; in this case $a = 0.05$, $m/a = 20{,}000$, and the initial condition $B_0 = 15{,}000$ gives $C = -5000$, so $B = 20{,}000 - 5000e^{0.05t}$. The balance decreases.

b. The steady-state (constant balance) solution is $B = m/a = \$50,000$, which gives $m = 0.05 \cdot 50,000 = \$2500$.

8.9.69

a. Solving $y' = 0$ gives the equilibrium solution $y = -b/a$, which is a horizontal line.

b.

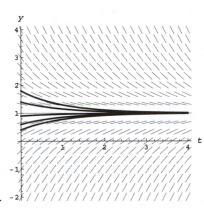

c.

Note that the general solutions $y = (y_0 + \frac{b}{a})e^{at} - \frac{b}{a}$ increases without bound if $a > 0$ and $y_0 > \frac{-b}{a}$, and decreases without bound if $a > 0$ but $y_0 < \frac{-b}{a}$. But if $a < 0$, the general solutions have limit $\frac{-b}{a}$, but must increase to it if $y_0 < \frac{-b}{a}$ and decrease to it if $y_0 > \frac{-b}{a}$.

Chapter Eight Review

1

a. True. Two applications of integration by parts are needed to reduce to $\int x^2 e^{2x}\, dx$.

b. False. This integral can be done using a trigonometric substitution.

c. False. Both are correct, since $-\cos^2 x = \sin^2 x - 1$, so $\sin^2 x$ and $-\cos^2 x$ differ by a constant.

d. True. Recall that $2\sin x \cos x = \sin 2x$.

e. False. Use long division to write the integrand as the sum of a polynomial and a proper rational function.

3 Let $u = x + 4$ so that $du = dx$. Note that $x = u - 4$. Substituting gives

$$\int \frac{3(u-4)}{\sqrt{u}}\, du = 3\int (u^{1/2} - 4u^{-1/2})\, du = 3((2/3)u^{3/2} - 8u^{1/2}) + C$$

$$= 2(x+4)^{3/2} - 24\sqrt{x+4} + C = 2\sqrt{x+4}(x+4-12) + C = 2\sqrt{x+4}(x-8) + C.$$

5

$$\int_{-2}^{1} \frac{3}{(x+2)^2 + 9}\, dx = \left(\tan^{-1}((x+2)/3)\right)\Big|_{-2}^{1} = \pi/4 - 0 = \pi/4.$$

7 Let $u = \sqrt{t-1}$. Then $u^2 + 1 = t$, and $2u\, du = dt$. Substituting gives $\int \frac{u^2}{u^2+1}\, du = \int \left(1 - \frac{1}{u^2+1}\right) du = u - \tan^{-1} u + C = \sqrt{t-1} - \tan^{-1}\sqrt{t-1} + C.$

9 Let $u = x$ and $dv = (1/2)(x+2)^{-1/2}\,dx$. Then $du = dx$ and $v = (x+2)^{1/2}$. We have

$$\int \frac{x}{2\sqrt{x+2}}\,dx = x\sqrt{x+2} - \int \sqrt{x+2}\,dx = x\sqrt{x+2} - \frac{2}{3}(x+2)^{3/2} + C$$

$$= \frac{1}{3}\sqrt{x+2}(3x - 2x - 4) + C = \frac{1}{3}\sqrt{x+2}(x-4) + C.$$

11 Let $u = x$ and $dv = \sinh x\,dx$. Then $du = dx$ and $v = \cosh x$. We have $\int x \sinh x\,dx = x \cosh x - \int \cosh x\,dx = x \cosh x - \sinh x + C$.

13 $\int_0^{\pi/4} \cos^5 2x \sin^2 2x\,dx = \int_0^{\pi/4} \cos(2x)(1 - \sin^2 2x)^2 \sin^2 2x\,dx$. Let $u = \sin 2x$ so that $du = 2\cos 2x\,dx$. Substituting gives

$$\frac{1}{2}\int_0^1 (1-u^2)^2 u^2\,du = \frac{1}{2}\int_0^1 (u^2 - 2u^4 + u^6)\,du = \frac{1}{2}\left. \left(u^3/3 - 2u^5/5 + u^7/7\right)\right|_0^1 = \frac{1}{2}(1/3 - 2/5 + 1/7) - 0 = \frac{4}{105}.$$

15 $\int \tan^4 t \sec^2 t\,dt$. Let $u = \tan t$ so that $du = \sec^2 t\,dt$. Substituting gives

$$\int u^4\,du = u^5/5 + C = \frac{1}{5}\tan^5 t + C.$$

17 $\int \tan^2 \theta \sec^2 \theta (\sec \theta \tan \theta)\,d\theta = \int (\sec^2 \theta - 1)\sec^2 \theta(\sec \theta \tan \theta)\,d\theta$. Let $u = \sec \theta$ so that $du = \sec \theta \tan \theta\,d\theta$. Substituting gives

$$\int (u^2 - 1)u^2\,du = \int (u^4 - u^2)\,du = u^5/5 - u^3/3 + C = \frac{1}{5}\sec^5 \theta - \frac{1}{3}\sec^3 \theta + C.$$

19 Let $x = \sec \theta$ so that $dx = \sec \theta \tan \theta\,d\theta$ and $\sqrt{x^2 - 1} = \tan \theta$. Substituting gives

$$\int_{\pi/4}^{\pi/3} \tan^2 \theta\,d\theta = \int_{\pi/4}^{\pi/3} (\sec^2 \theta - 1)\,d\theta = \left.(\tan \theta - \theta)\right|_{\pi/4}^{\pi/3} = \sqrt{3} - \pi/3 - (1 - \pi/4) = \sqrt{3} - 1 - \frac{\pi}{12}.$$

21 Let $x = 2\tan \theta$ so that $dx = 2\sec^2 \theta\,d\theta$ and $\sqrt{x^2 + 4} = 2\sec \theta$. Substituting gives

$$\int \frac{8\tan^3 \theta}{2\sec \theta} 2\sec^2 \theta\,d\theta = 8\int \tan \theta \sec \theta(\sec^2 \theta - 1)\,d\theta.$$

Let $u = \sec \theta$ so that $du = \sec \theta \tan \theta\,d\theta$. Substituting again gives

$$8\int (u^2 - 1)\,du = \frac{8}{3}u^3 - 8u + C = \frac{8}{3}\sec^3 \theta - 8\sec \theta + C$$

$$= \frac{8}{3}(\sec(\tan^{-1}(x/2)))^3 - 8\sec \tan^{-1}(x/2) + C = \frac{1}{3}\sqrt{x^2 + 4}(x^2 + 4) - 4\sqrt{x^2 + 4} + C$$

$$= \frac{1}{3}\sqrt{x^2 + 4}(x^2 + 4 - 12) + C = \frac{1}{3}\sqrt{x^2 + 4}(x^2 - 8) + C.$$

23 Write $\frac{2x^2 + 7x + 4}{x^3 + 2x^2 + 2x} = \frac{Ax + B}{x^2 + 2x + 2} + \frac{C}{x}$. Then $2x^2 + 7x + 4 = Ax^2 + Bx + C(x^2 + 2x + 2)$. Letting $x = 0$ gives $C = 2$. Letting $x = 1$ gives $13 = A + B + 10$, so $A + B = 3$. Letting $x = -1$ gives $-1 = A - B + 2$, so $A - B = -3$. Solving this system of linear equations gives $A = 0$ and $B = 3$. Thus our integral is equal to

$$\int \left(\frac{3}{x^2 + 2x + 2} + \frac{2}{x}\right)dx = \int \left(\frac{3}{(x+1)^2 + 1} + \frac{2}{x}\right)dx = 3\tan^{-1}(x+1) + 2\ln|x| + C.$$

25 Write $\frac{3x^3 + 4x^2 + 6x}{(x+1)^2(x^2+4)} = \frac{Ax + B}{x^2 + 4} + \frac{C}{x+1} + \frac{D}{(x+1)^2}$. Then $3x^3 + 4x^2 + 6x = (Ax + B)(x+1)^2 + C(x+1)(x^2 + 4) + D(x^2 + 4)$. Letting $x = -1$ gives $D = -1$. Letting $x = 0$ gives $0 = B + 4C - 4$, so $B + 4C = 4$. Letting $x = 1$ gives $13 = 4A + 4B + 10C - 5$, so $9 = 2A + 2B + 5C$. Letting $x = 2$ gives $52 = 18A + 9B + 24C - 8$, so $60 = 18A + 9B + 24C$, so $20 = 6A + 3B + 8C$. Solving the system of linear equations gives $A = 2$, $B = 0$, and $C = 1$. Our integral is thus equal to

$$\int \left(\frac{2x}{x^2 + 4} + \frac{1}{x+1} - \frac{1}{(x+1)^2}\right)dx = \ln\left|(x^2 + 4)(x+1)\right| + \frac{1}{x+1} + C.$$

27 Using the table of integrals entry 77, we find that

$$\int \frac{dx}{x\sqrt{ax-b}} = \frac{2}{\sqrt{b}} \tan^{-1}\left(\sqrt{\frac{ax-b}{b}}\right) + C.$$

Therefore

$$\int \frac{dx}{x\sqrt{4x-6}} = \frac{\sqrt{6}}{3} \tan^{-1}\left(\sqrt{\frac{2x-3}{3}}\right) + C.$$

29 Using the table of integrals reduction formula (number 45) twice, we have

$$\int \sec^5 x\, dx = \frac{\sec^3 x \tan x}{4} + \frac{3}{4}\int \sec^3 x\, dx$$

$$= \frac{\sec^3 x \tan x}{4} + \frac{3}{4}\left(\frac{\sec x \tan x}{2} + \frac{1}{2}\int \sec x\, dx\right)$$

$$= \frac{\sec^3 x \tan x}{4} + \frac{3}{4}\left(\frac{\sec x \tan x}{2} + \frac{1}{2}(\ln|\sec x + \tan x|)\right) + C$$

$$= \frac{\sec^3 x \tan x}{4} + \frac{3 \sec x \tan x}{8} + \frac{3}{8}(\ln|\sec x + \tan x| + C.$$

31 Using a CAS, we find that

$$\int_{-1}^{1} e^{-2x^2}\, dx \approx 1.196.$$

33

a. Using a calculator program for the trapezoid and midpoint rules, we find that $T_6 = 9.125$, $M_6 = 8.937$.

b. Similarly, $T_{12} = 9.031$, $M_{12} = 8.984$.

35 First evaluate

$$\int_0^b xe^{-x}\, dx = -e^{-x}(x+1)\Big|_0^b = 1 - (b+1)e^{-b}.$$

Then

$$\int_0^\infty xe^{-x}\, dx = \lim_{b\to\infty}\left(1 - (b+1)e^{-b}\right) = 1.$$

37 First take $0 < c < 3$ and evaluate

$$\int_0^c \frac{dx}{\sqrt{9-x^2}} = \sin^{-1}\left(\frac{x}{3}\right)\Big|_0^c = \sin^{-1}\left(\frac{c}{3}\right).$$

Then

$$\int_0^3 \frac{dx}{\sqrt{9-x^2}} = \lim_{c\to 3^-} \sin^{-1}\left(\frac{c}{3}\right) = \frac{\pi}{2}.$$

39

$$\int \frac{1}{1+\cos\theta}\cdot\frac{1-\cos\theta}{1-\cos\theta}\, d\theta = \int \frac{1-\cos\theta}{\sin^2\theta}\, d\theta = \int (\csc^2\theta - \csc\theta\cot\theta)\, d\theta = -\cot\theta + \csc\theta + C.$$

41 Two applications of integration by parts gives

$$\int e^x \sin x\, dx = e^x \sin x - \int e^x \cos x\, dx = e^x \sin x - e^x \cos x - \int e^x \sin x\, dx.$$

Thus,

$$2\int e^x \sin x\, dx = e^x \sin x - e^x \cos x,$$

so

$$\int e^x \sin x\, dx = \frac{1}{2}(e^x \sin x - e^x \cos x) + C.$$

43 Let $u = 4\theta$, Then

$$\int \cos^2 \theta \, d\theta = \frac{1}{4} \int \cos^2 u \, du = \frac{1}{4} \int \frac{1 + \cos 2u}{2} \, du = \frac{u}{8} + \frac{\sin 2u}{16} + C = \frac{\theta}{2} + \frac{\sin 8\theta}{16} + C.$$

45 Let $u = \sec z$. Then $du = \sec z \tan z \, dz$ and

$$\int \sec^5 z \tan z \, dz = \int u^4 \, du = \frac{u^5}{5} + C = \frac{\sec^5 z}{5} + C.$$

47 Let $u = \cos x$ so that $du = -\sin x \, dx$. Then

$$\int_0^{\pi/6} \sin^5 \theta \, d\theta = \int_0^{\pi/6} \sin^4 \theta \cdot \sin \theta \, d\theta = -\int_1^{\sqrt{3}/2} (1 - u^2)^2 \, du = \int_{\sqrt{3}/2}^1 (1 - u^2)^2 \, du$$

$$= \int_{\sqrt{3}/2}^1 (1 - 2u^2 + u^4) \, du = \left(u - \frac{2}{3} u^3 + \frac{1}{5} u^5 \right) \Big|_{\sqrt{3}/2}^1$$

$$= \frac{8}{15} - \frac{49\sqrt{3}}{160} = \frac{256 - 147\sqrt{3}}{480}.$$

49 Let $x = 2 \sin \theta$ so that $dx = 2 \cos \theta \, d\theta$. Then

$$\int \frac{dx}{\sqrt{4 - x^2}} = \int \frac{2 \cos \theta \, d\theta}{2 \cos \theta} = \theta + C = \sin^{-1}(x/2) + C.$$

51 Let $y = 3 \sin \theta$ so that $dy = 3 \cos \theta \, d\theta$. Then

$$\int \frac{dy}{y^2 \sqrt{9 - y^2}} = \int \frac{3 \cos \theta \, d\theta}{9 \sin^2 \theta \cdot 3 \cos \theta} = \frac{1}{9} \int \csc^2 \theta \, d\theta = -\frac{1}{9} \cot \theta + C.$$

Now use $\sin \theta = y/3$, $\cos \theta = (1/3)\sqrt{9 - y^2}$ to obtain

$$\int \frac{dy}{y^2 \sqrt{9 - y^2}} = -\frac{1}{9y} \sqrt{9 - y^2} + C.$$

53 Let $x = (3/2) \tan \theta$ so that $dx = (3/2) \sec^2 \theta \, d\theta$, and note that $(3/2) \tan(\pi/6) = \sqrt{3}/2$. Then

$$\int_0^{\sqrt{3}/2} \frac{4}{9 + 4x^2} \, dx = \int_0^{\pi/6} \frac{4}{9 \sec^2 \theta} \cdot \frac{3}{2} \sec^2 \theta \, d\theta = \frac{2}{3} \int_0^{\pi/6} d\theta = \frac{\pi}{9}.$$

55 Let $u = \cosh x$ so that $du = \sinh x \, dx$. Then

$$\int \frac{\sinh x}{\cosh^2 x} \, dx = \int u^{-2} \, du = \frac{-1}{u} + C = \frac{-1}{\cosh x} + C = -\operatorname{sech} x + C.$$

57 Let $u = \sinh x$ so that $du = \cosh x \, dx$. Substituting gives

$$\int_0^{\sqrt{3}} \frac{1}{\sqrt{4 - u^2}} \, du = \left(\sin^{-1}(u/2) \right) \Big|_0^{\sqrt{3}} = \frac{\pi}{3}.$$

59 Using the method of partial fractions, we express $\frac{1}{x^2 - 2x - 15} = \frac{1}{(x-5)(x+3)} = \frac{A}{x-5} + \frac{B}{x+3}$. Clearing denominators gives $1 = A(x + 3) + B(x - 5)$ and comparing coefficients gives $A + B = 0$, $3A - 5B = 1$ which has solution $A = 1/8$, $B = -1/8$. Hence

$$\int \frac{dx}{x^2 - 2x - 15} = \frac{1}{8} \int \left(\frac{1}{x-5} - \frac{1}{x+3} \right) dx = \frac{1}{8} \ln \left| \frac{x-5}{x+3} \right| + C.$$

61 Using the method of partial fractions, we express $\frac{1}{(y+1)(y^2+1)} = \frac{A}{y+1} + \frac{By+C}{y^2+1}$. Clearing denominators gives $1 = A(y^2+1) + (By+C)(y+1)$ and comparing coefficients gives $A + B = 0$, $B + C = 0$ and $A + C = 1$, which has solution $A = 1/2$, $B = -1/2$, $C = 1/2$. Hence

$$\int_0^1 \frac{dy}{(y+1)(y^2+1)} = \frac{1}{2}\int_0^1 \left(\frac{1}{y+1} + \frac{1-y}{y^2+1}\right) dy = \frac{1}{2}\left(\ln(y+1) + \tan^{-1} y - \frac{1}{2}\ln(y^2+1)\right)\Big|_0^1 = \frac{1}{4}\ln 2 + \frac{\pi}{8}.$$

63 First consider $\int_0^1 \frac{dx}{\sqrt[3]{|x-1|}} = \lim_{a\to 1^-}\int \frac{dx}{\sqrt[3]{1-x}} = \lim_{a\to 1^-} -\frac{3}{2}(1-x)^{2/3}\Big|_0^a = \lim_{a\to 1^-} -\frac{3}{2}\left((1-a)^{2/3} - 1\right) = \frac{3}{2}$.

Now consider $\int_1^2 \frac{dx}{\sqrt[3]{|x-1|}} = \lim_{b\to 1^+}\int_b^2 \frac{dx}{\sqrt[3]{x-1}} = \lim_{b\to 1^+} \frac{3}{2}(x-1)^{2/3}\Big|_b^2 = \lim_{b\to 1^+} \frac{3}{2}\left(1 - (b-1)^{2/3}\right) = \frac{3}{2}$.

Thus, $\int_0^2 \frac{dx}{\sqrt[3]{|x-1|}} = \frac{3}{2} + \frac{3}{2} = 3$.

65 Factor $x^2 - x - 2 = (x-2)(x+1)$ and use the method of partial fractions: $\frac{1}{(x-2)(x+1)} = \frac{A}{x-2} + \frac{B}{x+1}$. Clearing denominators gives $1 = A(x+1) + B(x-2)$ and equating coefficients gives $A + B = 0$, $A - 2B = 1$, so $A = 1/3$, $B = -1/3$. Therefore

$$\int \frac{dx}{x^2 - x - 2} = \frac{1}{3}\int \left(\frac{1}{x-2} - \frac{1}{x+1}\right) dx = \frac{1}{3}\ln\left|\frac{x-2}{x+1}\right| + C.$$

67 As a preliminary step, observe that

$$\frac{2x^2 - 4x}{x^2 - 4} = \frac{2x(x-2)}{(x-2)(x+2)} = \frac{2x}{x+2} = \frac{2(x+2-2)}{x+2} = 2 - \frac{4}{x+2}.$$

Hence

$$\int \frac{2x^2 - 4x}{x^2 - 4}\, dx = 2x - 4\ln|x+2| + C = 2(x - 2\ln|x+2|) + C.$$

69 Make the preliminary substitution $x = e^{2t}$, $dx = 2e^{2t}$. Then we have

$$\int \frac{e^{2t}}{(1+e^{4t})^{3/2}}\, dt = \frac{1}{2}\int \frac{dx}{(1+x^2)^{3/2}}.$$

Now let $x = \tan\theta$, $dx = \sec^2\theta\, d\theta$ and

$$\frac{1}{2}\int \frac{dx}{(1+x^2)^{3/2}} = \frac{1}{2}\int \frac{\sec^2\theta}{\sec^3\theta}\, d\theta = \frac{1}{2}\int \cos\theta\, d\theta = \frac{1}{2}\sin\theta + C.$$

Now

$$\sin\theta = \frac{\tan\theta}{\sec\theta} = \frac{x}{\sqrt{1+x^2}} = \frac{e^{2t}}{\sqrt{1+e^{4t}}}.$$

Therefore

$$\int \frac{e^{2t}}{(1+e^{4t})^{3/2}}\, dt = \frac{1}{2}\frac{e^{2t}}{\sqrt{1+e^{4t}}} + C.$$

71 The volume is

$$V = \pi\int_1^e (\ln x)^2\, dx = \pi x((\ln x)^2 - 2\ln x + 2)\Big|_1^e = \pi(e - 2).$$

73 The volume is

$$V = 2\pi\int_1^e (x-1)\ln x\, dx = \frac{\pi}{2}x\left(2(x-2)\ln x - x + 4\right)\Big|_1^e = \frac{\pi}{2}\left(e^2 - 3\right).$$

75 The volume generated by revolving around the x-axis is

$$V_x = \pi \int_0^\pi \sin^2 x \, dx = \pi \left(\frac{x}{2} - \frac{\sin 2x}{4} \right) \Big|_0^\pi = \frac{\pi^2}{2},$$

and the volume generated by revolving around the y-axis is

$$V_y = 2\pi \int_0^\pi x \sin x \, dx = 2\pi (\sin x - x \cos x) \Big|_0^\pi = 2\pi^2,$$

so the greater volume is obtained by revolving around the y-axis.

77

a. Observe that

$$\int_{1/2}^b \ln x \, dx = (x \ln x - x) \Big|_{1/2}^b \approx b \ln b - b + 0.847;$$

solve $b \ln b - b + 0.847 = 0$ numerically to obtain $b \approx 1.603$.

b. Similarly, we have

$$\int_{1/3}^b \ln x \, dx = (x \ln x - x) \Big|_{1/3}^b \approx b \ln b - b + 0.700;$$

solve $b \ln b - b + 0.700 = 0$ numerically to obtain $b \approx 1.870$.

c. In general, the pair (a, b) must satisfy the equation

$$\int_a^b \ln x \, dx = (b \ln b - b) - (a \ln a - a) = 0,$$

which gives $b \ln b - b = a \ln a - a$.

d. As a increases there is less negative area to the right of $x = 1$, so $b = g(a)$ is a decreasing function of a.

79 The average velocity is

$$\bar{v} = \frac{1}{\pi} \int_0^\pi 10 \sin 3t \, dt = -\frac{10}{3\pi} \cos 3t \Big|_0^\pi = \frac{20}{3\pi}.$$

81 The number of cars is given by the integral

$$\int_0^4 800 t e^{-t/2} \, dt = -1600(t+2)e^{-t/2} \Big|_0^4 = 3200(1 - 3e^{-2}) \approx 1901.$$

83

a. Using integration by parts, we find that

$$I(p) = \int_1^e \frac{\ln x}{x^p} \, dx = -\frac{x^{1-p}}{(p-1)^2} \left((p-1) \ln x + 1 \right) \Big|_1^e = \frac{1}{(p-1)^2} (1 - pe^{1-p})$$

for $p \neq 1$, and using the substitution $u = \ln x$ gives

$$I(1) = \int_1^e \frac{\ln x}{x} \, dx = \frac{(\ln x)^2}{2} \Big|_1^e = \frac{1}{2}.$$

b. We have

$$\lim_{p \to \infty} I(p) = \lim_{p \to \infty} \frac{1}{(p-1)^2}(1 - pe^{1-p}) = \lim_{p \to \infty} \left(\frac{1}{(p-1)^2} - \frac{pe}{(p-1)^2}e^{-p} \right) = 0,$$

and

$$\lim_{p \to -\infty} I(p) = \lim_{p \to -\infty} \left(\frac{1}{(p-1)^2} - \frac{pe}{(p-1)^2}e^{-p} \right) = \infty,$$

since e^{-p} grows much faster than $(p-1)^2/p$ as $p \to -\infty$.

c. By inspection we see that $I(0) = 1$.

85 Use a calculator program for, say, Simpson's rule with $n = 100$, but replace the limits with 0.00000001 and 0.99999999 to avoid errors coming from trying to evaluate the function at $x = 0$ or $x = 1$; we obtain

$$\int_0^1 \frac{x^2 - x}{\ln x}\, dx \approx 0.4054651.$$

87 Numerically approximate the integral to obtain

$$\int_0^1 \frac{\sin^{-1} x}{x}\, dx \approx 1.0889;$$

therefore $n = \pi \ln 2/1.0889 = 2$. (When approximating the integral, replace the lower limit 0 with a small positive number like 0.00001 to avoid an error from evaluating the integrand at $x = 0$.)

89

a. Using integration by parts or a CAS, we find that the volume is $V_1(a) = \pi \int_1^a (\ln x)^2\, dx = \pi x((\ln x)^2 - 2\ln x + 2)\Big|_1^a = \pi[(a\ln^2 a - 2a\ln a + 2(a-1)]$.

b. Using integration by parts or a CAS, we find that the volume is $V_2(a) = 2\pi \int_1^a x\ln x\, dx = \frac{\pi}{2}x^2(2\ln x - 1)\Big|_1^a = \frac{\pi}{2}(2a^2\ln a - a^2 + 1)$.

c. As shown by the graph, $V_2(a) > V_1(a)$ for all $a > 1$.

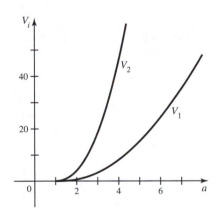

91 We have

$$V_1 = \pi \int_0^b e^{-2ax}\, dx = -\frac{\pi}{2a}e^{-2ax}\Big|_0^b = \frac{\pi}{2a}\left(1 - e^{-2ab}\right),$$

and

$$V_2 = \pi \int_b^\infty e^{-2ax}\, dx = \lim_{c \to \infty}\left(-\frac{\pi}{2a}e^{-2ax}\Big|_b^c\right) = \frac{\pi}{2a}e^{-2ab}.$$

Equating and solving gives $e^{-2ab} = 1/2$, which is equivalent to $ab = (1/2)\ln 2$.

93 This first-order linear equation has general solution $y = Ce^{2t} - 2$; the initial condition gives $C - 2 = 8$, so $C = 10$ and hence $y = 10e^{2t} - 2$.

95 This equation is separable. We have $\int 2y \, dy = \int \left(1 + \frac{1}{t}\right) dt$, so $y^2 = t + \ln t + C$. The initial condition gives $16 = 1 + C$, so $C = 15$ and solving for y gives $y = \sqrt{t + \ln t + 15}$.

97 This equation is separable. We have $\int \cos y \, dy = \int \frac{dt}{t^2}$, so $\sin y = -\frac{1}{t} + C$. The initial condition gives $0 = -1 + C$, so $C = 1$ and solving for y gives $y = \sin^{-1}\left(1 - \frac{1}{t}\right)$. Therefore $\lim_{t \to \infty} y(t) = \lim_{t \to \infty} \sin^{-1}\left(1 - \frac{1}{t}\right) = \sin^{-1} 1 = \frac{\pi}{2}$.

99

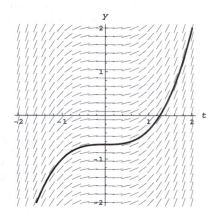

101 This equation is separable. We have

$$\int dt = -\frac{1}{10} \int \left(1 + \frac{5}{s}\right) ds \implies t = -\frac{s + 5\ln s}{10} + C.$$

The initial condition gives $s(0) = 50$ gives $C = (50 + 5\ln 50)/10$, so the solution is given implicitly by

$$t = \frac{-s - 5\ln s + 50 + 5\ln 50}{10}.$$

As $t \to \infty$ we must have $s + 5\ln s \to -\infty$, which implies that $\lim_{t \to \infty} s(t) = 0$.

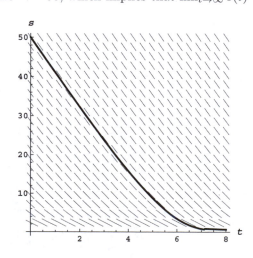

Chapter 9

Sequences and Infinite Series

9.1 An Overview

9.1.1 A *sequence* is an ordered list of numbers $a_1, a_2, a_3, \ldots$, often written $\{a_1, a_2, \ldots\}$ or $\{a_n\}$. For example, the natural numbers $\{1, 2, 3, \ldots\}$ are a sequence where $a_n = n$ for every n.

9.1.3 $a_1 = 1$ (given); $a_2 = 1 \cdot a_1 = 1$; $a_3 = 2 \cdot a_2 = 2$; $a_4 = 3 \cdot a_3 = 6$; $a_5 = 4 \cdot a_4 = 24$.

9.1.5 An *infinite series* is an infinite sum of numbers. Thus if $\{a_n\}$ is a sequence, then $a_1 + a_2 + \cdots = \sum_{k=1}^{\infty} a_k$ is an infinite series. For example, if $a_k = \frac{1}{k}$, then $\sum_{k=1}^{\infty} a_k = \sum_{k=1}^{\infty} \frac{1}{k}$ is an infinite series.

9.1.7 $S_1 = \sum_{k=1}^{1} k^2 = 1$; $S_2 = \sum_{k=1}^{2} k^2 = 1 + 4 = 5$; $S_3 = \sum_{k=1}^{3} k^2 = 1 + 4 + 9 = 14$; $S_4 = \sum_{k=1}^{4} k^2 = 1 + 4 + 9 + 16 = 30$.

9.1.9 $a_1 = \dfrac{1}{10}$; $a_2 = \dfrac{1}{100}$; $a_3 = \dfrac{1}{1000}$; $a_4 = \dfrac{1}{10000}$.

9.1.11 $a_1 = \frac{-1}{2}$, $a_2 = \frac{1}{2^2} = \frac{1}{4}$. $a_3 = \frac{-2}{2^3} = \frac{-1}{8}$, $a_4 = \frac{1}{2^4} = \frac{1}{16}$.

9.1.13 $a_1 = \frac{2^2}{2+1} = \frac{4}{3}$. $a_2 = \frac{2^3}{2^2+1} = \frac{8}{5}$. $a_3 = \frac{2^4}{2^3+1} = \frac{16}{9}$. $a_4 = \frac{2^5}{2^4+1} = \frac{32}{17}$.

9.1.15 $a_1 = 1 + \sin(\pi/2) = 2$; $a_2 = 1 + \sin(2\pi/2) = 1 + \sin \pi = 1$; $a_3 = 1 + \sin(3\pi/2) = 0$; $a_4 = 1 + \sin(4\pi/2) = 1 + \sin 2\pi = 1$.

9.1.17 $a_1 = 2$, $a_2 = 2 \cdot 2 = 4$, $a_3 = 2(4) = 8$, $a_4 = 2 \cdot 8 = 16$.

9.1.19 $a_1 = 10$ (given); $a_2 = 3 \cdot a_1 - 12 = 30 - 12 = 18$; $a_3 = 3 \cdot a_2 - 12 = 54 - 12 = 42$; $a_4 = 3 \cdot a_3 - 12 = 126 - 12 = 114$.

9.1.21 $a_1 = 0$ (given); $a_2 = 3 \cdot a_1^2 + 1 + 1 = 2$; $a_3 = 3 \cdot a_2^2 + 2 + 1 = 15$; $a_4 = 3 \cdot a_3^2 + 3 + 1 = 679$.

9.1.23

    a. $\frac{1}{32}, \frac{1}{64}$.

    b. $a_1 = 1$; $a_{n+1} = \frac{a_n}{2}$.

    c. $a_n = \frac{1}{2^{n-1}}$.

9.1.25

    a. $-5, 5$.

    b. $a_1 = -5$, $a_{n+1} = -a_n$.

c. $a_n = (-1)^n \cdot 5$.

9.1.27

a. 32, 64.

b. $a_1 = 1$; $a_{n+1} = 2a_n$.

c. $a_n = 2^{n-1}$.

9.1.29

a. 243, 729.

b. $a_1 = 1$; $a_{n+1} = 3a_n$.

c. $a_n = 3^{n-1}$.

9.1.31 $a_1 = 9$, $a_2 = 99$, $a_3 = 999$, $a_4 = 9999$. This sequence diverges, because the terms get larger without bound.

9.1.33 $a_1 = \frac{1}{10}$, $a_2 = \frac{1}{100}$, $a_3 = \frac{1}{1000}$, $a_4 = \frac{1}{10,000}$. This sequence converges to zero.

9.1.35 $a_1 = -\frac{1}{2}$, $a_2 = \frac{1}{4}$, $a_3 = -\frac{1}{8}$, $a_4 = \frac{1}{16}$. This sequence converges to 0 because each term is smaller in absolute value than the preceding term and they get arbitrarily close to zero.

9.1.37 $a_1 = 1 + 1 = 2$, $a_2 = 1 + 1 = 2$, $a_3 = 2$, $a_4 = 2$. This constant sequence converges to 2.

9.1.39 $a_1 = \frac{50}{11} + 50 \approx 54.545$, $a_2 = \frac{54.545}{11} + 50 \approx 54.959$, $a_3 = \frac{54.959}{11} + 50 \approx 54.996$, $a_4 = \frac{54.996}{11} + 50 \approx 55.000$. This sequence converges to 55.

9.1.41

n	1	2	3	4	4	6	7	8	9	10
a_n	0.4636	0.2450	0.1244	0.0624	0.0312	0.0156	0.0078	0.0039	0.0020	0.0010

This sequence appears to converge to 0.

9.1.43

n	1	2	3	4	5	6	7	8	9	10
a_n	0	2	6	12	20	30	42	56	72	90

This sequence appears to diverge.

9.1.45

n	1	2	3	4	5	6	7	8	9	10
a_n	0.83333	0.96154	0.99206	0.99840	0.99968	0.99994	0.99999	1.0000	1.0000	1.0000

This sequence appears to converge to 1.

9.1.47

a. 2.5, 2.25, 2.125, 2.0625.

b. The limit is 2.

9.1.49

n	0	1	2	3	4	5	6	7	8	9	10
a_n	3	3.500	3.750	3.875	3.938	3.969	3.984	3.992	3.996	3.998	3.999

This sequence converges to 4.

9.1.51

n	0	1	2	3	4	5	6	7	8	9	10
a_n	0	1	3	7	15	31	63	127	255	511	1023

This sequence diverges.

9.1.53

n	0	1	2	3	4	5	6	7	8	9
a_n	1000	18.811	5.1686	4.1367	4.0169	4.0021	4.0003	4.0000	4.0000	4.0000

This sequence converges to 4.

9.1.55

a. 20, 10, 5, 2.5.

b. $h_n = 20(0.5)^n$.

9.1.57

a. 30, 7.5, 1.875, 0.46875.

b. $h_n = 30(0.25)^n$.

9.1.59 $S_1 = 0.3$, $S_2 = 0.33$, $S_3 = 0.333$, $S_4 = 0.3333$. It appears that the infinite series has a value of $0.3333\ldots = \frac{1}{3}$.

9.1.61 $S_1 = 4$, $S_2 = 4.9$, $S_3 = 4.99$, $S_4 = 4.999$. The infinite series has a value of $4.999\cdots = 5$.

9.1.63

a. $S_1 = \frac{2}{3}$, $S_2 = \frac{4}{5}$, $S_3 = \frac{6}{7}$, $S_4 = \frac{8}{9}$.

b. It appears that $S_n = \frac{2n}{2n+1}$.

c. The series has a value of 1 (the partial sums converge to 1).

9.1.65

a. $S_1 = \frac{1}{3}$, $S_2 = \frac{2}{5}$, $S_3 = \frac{3}{7}$, $S_4 = \frac{4}{9}$.

b. $S_n = \frac{n}{2n+1}$.

c. The partial sums converge to $\frac{1}{2}$, which is the value of the series.

9.1.67

a. True. For example, $S_2 = 1 + 2 = 3$, and $S_4 = a_1 + a_2 + a_3 + a_4 = 1 + 2 + 3 + 4 = 10$.

b. False. For example, $\frac{1}{2}, \frac{3}{4}, \frac{7}{8}, \cdots$ where $a_n = 1 - \frac{1}{2^n}$ converges to 1, but each term is greater than the previous one.

c. True. In order for the partial sums to converge, they must get closer and closer together. In order for this to happen, the difference between successive partial sums, which is just the value of a_n, must approach zero.

9.1.69 Using the work from the previous problem:

a. Here $h_0 = 20$, $r = 0.75$, so $S_1 = 40$, $S_2 = 40 + 40 \cdot 0.75 = 70$, $S_3 = S_2 + 40 \cdot (0.75)^2 = 92.5$, $S_4 = S_3 + 40 \cdot (0.75)^3 = 109.375$, $S_5 = S_4 + 40 \cdot (0.75)^4 = 122.03125$

b.

n	1	2	3	4	5	6
a_n	40	70	92.5	109.375	122.031	131.523
n	7	8	9	10	11	12
a_n	138.643	143.982	147.986	150.990	153.242	154.932
n	13	14	15	16	17	18
a_n	156.199	157.149	157.862	158.396	158.797	159.098
n	19	20	21	22	23	24
a_n	159.323	159.493	159.619	159.715	159.786	159.839

The sequence converges to 160.

9.1.71

a. 0.9, 0.99, 0.999, .9999.

b. The limit is 1.

9.1.73

a. $\frac{1}{3}, \frac{4}{9}, \frac{13}{27}, \frac{40}{81}$.

b. The limit is $1/2$.

9.1.75

a. $-1, 0, -1, 0$.

b. The limit does not exist.

9.1.77

a. $\frac{3}{10} = 0.3$, $\frac{33}{100} = 0.33$, $\frac{333}{1000} = 0.333$, $\frac{3333}{10000} = 0.3333$.

b. The limit is $1/3$.

9.1.79

a. $M_0 = 20$, $M_1 = 20 \cdot 0.5 = 10$, $M_2 = 20 \cdot 0.5^2 = 5$, $M_3 = 20 \cdot 0.5^3 = 2.5$, $M_4 = 20 \cdot 0.5^4 = 1.25$

b. $M_n = 20 \cdot 0.5^n$.

c. The initial mass is $M_0 = 20$. We are given that 50% of the mass is gone after each decade, so that $M_{n+1} = 0.5 \cdot M_n$, $n \geq 0$.

d. The amount of material goes to 0.

9.1.81

a. $d_0 = 200$, $d_1 = 200 \cdot .95 = 190$, $d_2 = 200 \cdot .95^2 = 180.5$, $d_3 = 200 \cdot .95^3 = 171.475$, $d_4 = 200 \cdot .95^4 = 162.90125$.

b. $d_n = 200(0.95)^n$, $n \geq 0$.

c. We are given $d_0 = 200$; because 5% of the drug is washed out every hour, that means that 95% of the preceding amount is left every hour, so that $d_{n+1} = 0.95 \cdot d_n$.

d. The sequence converges to 0.

9.2 Sequences

9.2.1 There are many examples; one is $a_n = \frac{1}{n}$. This sequence is nonincreasing (in fact, it is decreasing) and has a limit of 0.

9.2.3 There are many examples; one is $a_n = \frac{1}{n}$. This sequence is nonincreasing (in fact, it is decreasing), is bounded above by 1 and below by 0, and has a limit of 0.

9.2.5 $\{r^n\}$ converges for $-1 < r \leq 1$. It diverges for all other values of r (see Theorem 8.3).

9.2.7 $\{e^{n/100}\}$ grows faster than $\{n^{100}\}$ as $n \to \infty$.

9.2.9 Divide numerator and denominator by n^4 to get $\lim\limits_{n \to \infty} \frac{1/n}{1 + \frac{1}{n^4}} = 0$.

9.2.11 Divide numerator and denominator by n^3 to get $\lim\limits_{n \to \infty} \frac{3 - n^{-3}}{2 + n^{-3}} = \frac{3}{2}$.

9.2.13 Divide numerator and denominator by 3^n to get $\lim\limits_{n \to \infty} \frac{3 + (1/3^{n-1})}{1} = 3$.

9.2.15 $\lim\limits_{n \to \infty} \tan^{-1} n = \frac{\pi}{2}$.

9.2.17 Because $\lim\limits_{n \to \infty} \tan^{-1} n = \frac{\pi}{2}$, $\lim\limits_{n \to \infty} \frac{\tan^{-1} n}{n} = 0$.

9.2.19 Find the limit of the logarithm of the expression, which is $n \ln\left(1 + \frac{2}{n}\right)$. Using L'Hôpital's rule:

$$\lim\limits_{n \to \infty} n \ln\left(1 + \frac{2}{n}\right) = \lim\limits_{n \to \infty} \frac{\ln\left(1 + \frac{2}{n}\right)}{1/n} = \lim\limits_{n \to \infty} \frac{\frac{1}{1 + (2/n)}\left(\frac{-2}{n^2}\right)}{-1/n^2} = \lim\limits_{n \to \infty} \frac{2}{1 + (2/n)} = 2.$$

Thus the limit of the original expression is e^2.

9.2.21 Take the logarithm of the expression and use L'Hôpital's rule:

$$\lim\limits_{n \to \infty} \frac{n}{2} \ln\left(1 + \frac{1}{2n}\right) = \lim\limits_{n \to \infty} \frac{\ln(1 + (1/2n))}{2/n} = \lim\limits_{n \to \infty} \frac{\frac{1}{1 + (1/2n)} \cdot \frac{-1}{2n^2}}{-2/n^2} = \lim\limits_{n \to \infty} \frac{1}{4(1 + (1/2n))} = \frac{1}{4}.$$

Thus the original limit is $e^{1/4}$.

9.2.23 Using L'Hôpital's rule: $\lim\limits_{n \to \infty} \frac{n}{e^n + 3n} = \lim\limits_{n \to \infty} \frac{1}{e^n + 3} = 0$.

9.2.25 Taking logs, we have $\lim\limits_{n \to \infty} \frac{1}{n} \ln(1/n) = \lim\limits_{n \to \infty} -\frac{\ln n}{n} = \lim\limits_{n \to \infty} \frac{-1}{n} = 0$ by L'Hôpital's rule. Thus the original sequence has limit $e^0 = 1$.

9.2.27 Except for a finite number of terms, this sequence is just $a_n = ne^{-n}$, so it has the same limit as this sequence. Note that $\lim\limits_{n \to \infty} \frac{n}{e^n} = \lim\limits_{n \to \infty} \frac{1}{e^n} = 0$, by L'Hôpital's rule.

9.2.29 $\ln(\sin(1/n)) + \ln n = \ln(n \sin(1/n)) = \ln\left(\frac{\sin(1/n)}{1/n}\right)$. As $n \to \infty$, $\sin(1/n)/(1/n) \to 1$, so the limit of the original sequence is $\ln 1 = 0$.

9.2.31 $\lim\limits_{n \to \infty} n \sin(6/n) = \lim\limits_{n \to \infty} \frac{\sin(6/n)}{1/n} = \lim\limits_{n \to \infty} \frac{\frac{-6 \cos(6/n)}{n^2}}{(-1/n^2)} = \lim\limits_{n \to \infty} 6 \cos(6/n) = 6 \cdot \cos 0 = 6$.

9.2.33 The terms with odd-numbered subscripts have the form $-\frac{n}{n+1}$, so they approach -1, while the terms with even-numbered subscripts have the form $\frac{n}{n+1}$ so they approach 1. Thus, the sequence has no limit.

9.2.35 When n is an integer, $\sin\left(\frac{n\pi}{2}\right)$ oscillates between the values ± 1 and 0, so this sequence does not converge.

9.2.37 The numerator is bounded in absolute value by 1, while the denominator goes to ∞, so the limit of this sequence is 0.

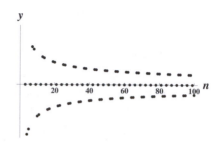

9.2.39 $\displaystyle\lim_{n\to\infty}(1+\cos(1/n)) = 1 + \cos(0) = 2.$

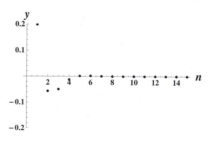

9.2.41 This is the sequence $\frac{\cos n}{e^n}$; the numerator is bounded in absolute value by 1 and the denominator increases without bound, so the limit is zero.

9.2.43 Ignoring the factor of $(-1)^n$ for the moment, we see, taking logs, that $\displaystyle\lim_{n\to\infty}\frac{\ln n}{n} = 0$, so that $\displaystyle\lim_{n\to\infty}\sqrt[n]{n} = e^0 = 1$. Taking the sign into account, the odd terms converge to -1 while the even terms converge to 1. Thus the sequence does not converge.

9.2.45 Because $0.2 < 1$, this sequence converges to 0. Because $0.2 > 0$, the convergence is monotone.

9.2.47 Because $|-0.7| < 1$, the sequence converges to 0; because $-0.7 < 0$, it does not do so monotonically. The sequence converges by oscillation.

9.2.49 Because $1.00001 > 1$, the sequence diverges; because $1.00001 > 0$, the divergence is monotone.

9.2.51 Because $|-2.5| > 1$, the sequence diverges; because $-2.5 < 0$, the divergence is not monotone. The sequence diverges by oscillation.

9.2.53 Because $-1 \leq \cos n \leq 1$, we have $\frac{-1}{n} \leq \frac{\cos n}{n} \leq \frac{1}{n}$. Because both $\frac{-1}{n}$ and $\frac{1}{n}$ have limit 0 as $n \to \infty$, the given sequence does as well.

9.2.55 Because $-1 \leq \sin n \leq 1$ for all n, the given sequence satisfies $-\frac{1}{2^n} \leq \frac{\sin n}{2^n} \leq \frac{1}{2^n}$, and because both $\pm \frac{1}{2^n} \to 0$ as $n \to \infty$, the given sequence converges to zero as well by the Squeeze Theorem.

9.2.57 The inverse tangent function takes values between $-\pi/2$ and $\pi/2$, so the numerator is always between $-\pi$ and π. Thus $\frac{-\pi}{n^3+4} \leq \frac{2\tan^{-1} n}{n^3+4} \leq \frac{\pi}{n^3+4}$, and by the Squeeze Theorem, the given sequence converges to zero.

9.2.59

a. After the nth dose is given, the amount of drug in the bloodstream is $d_n = 0.5 \cdot d_{n-1} + 80$, because the half-life is one day. The initial condition is $d_1 = 80$.

b. The limit of this sequence is 160 mg.

c. Let $L = \lim_{n \to \infty} d_n$. Then from the recurrence relation, we have $d_n = 0.5 \cdot d_{n-1} + 80$, and thus $\lim_{n \to \infty} d_n = 0.5 \cdot \lim_{n \to \infty} d_{n-1} + 80$, so $L = 0.5 \cdot L + 80$, and therefore $L = 160$.

9.2.61

a.

$$B_0 = 0$$
$$B_1 = 1.0075 \cdot B_0 + \$100 = \$100$$
$$B_2 = 1.0075 \cdot B_1 + \$100 = \$200.75$$
$$B_3 = 1.0075 \cdot B_2 + \$100 = \$302.26$$
$$B_4 = 1.0075 \cdot B_3 + \$100 = \$404.52$$
$$B_5 = 1.0075 \cdot B_4 + \$100 = \$507.56$$

b. $B_n = 1.0075 \cdot B_{n-1} + \$100$.

c. Using a calculator or computer program, $B_n > \$5,000$ during the 43rd month.

9.2.63 Because $n! \ll n^n$ by Theorem 8.6, we have $\lim_{n \to \infty} \frac{n!}{n^n} = 0$.

9.2.65 Theorem 8.6 indicates that $\ln^q n \ll n^p$, so $\ln^{20} n \ll n^{10}$, so $\lim_{n \to \infty} \frac{n^{10}}{\ln^{20} n} = \infty$.

9.2.67 By Theorem 8.6, $n^p \ll b^n$, so $n^{1000} \ll 2^n$, and thus $\lim_{n \to \infty} \frac{n^{1000}}{2^n} = 0$.

9.2.69 Let $\varepsilon > 0$ be given and let N be an integer with $N > \frac{1}{\varepsilon}$. Then if $n > N$, we have $\left| \frac{1}{n} - 0 \right| = \frac{1}{n} < \frac{1}{N} < \varepsilon$.

9.2.71 Let $\varepsilon > 0$ be given. We wish to find N such that for $n > N$, $\left| \frac{3n^2}{4n^2+1} - \frac{3}{4} \right| = \left| \frac{-3}{4(4n^2+1)} \right| = \frac{3}{4(4n^2+1)} < \varepsilon$. But this means that $3 < 4\varepsilon(4n^2 + 1)$, or $16\varepsilon n^2 + (4\varepsilon - 3) > 0$. Solving the quadratic, we get $n > \frac{1}{4}\sqrt{\frac{3}{\varepsilon} - 4}$, provided $\varepsilon < 3/4$. So let $N = \frac{1}{4}\sqrt{\frac{3}{\varepsilon}}$ if $\epsilon < 3/4$ and let $N = 1$ otherwise.

9.2.73 Let $\varepsilon > 0$ be given. We wish to find N such that for $n > N$, $\left|\frac{cn}{bn+1} - \frac{c}{b}\right| = \left|\frac{-c}{b(bn+1)}\right| = \frac{c}{b(bn+1)} < \varepsilon$. But this means that $\varepsilon b^2 n + (b\varepsilon - c) > 0$, so that $N > \frac{c}{b^2\varepsilon}$ will work.

9.2.75

a. True. See Theorem 8.2 part 4.

b. False. For example, if $a_n = 1/n$ and $b_n = e^n$, then $\lim\limits_{n\to\infty} a_n b_n = \infty$.

c. True. The definition of the limit of a sequence involves only the behavior of the nth term of a sequence as n gets large (see the Definition of Limit of a Sequence). Thus suppose a_n, b_n differ in only finitely many terms, and that M is large enough so that $a_n = b_n$ for $n > M$. Suppose a_n has limit L. Then for $\varepsilon > 0$, if N is such that $|a_n - L| < \varepsilon$ for $n > N$, first increase N if required so that $N > M$ as well. Then we also have $|b_n - L| < \varepsilon$ for $n > N$. Thus a_n and b_n have the same limit. A similar argument applies if a_n has no limit.

d. True. Note that a_n converges to zero. Intuitively, the nonzero terms of b_n are those of a_n, which converge to zero. More formally, given ϵ, choose N_1 such that for $n > N_1$, $a_n < \epsilon$. Let $N = 2N_1 + 1$. Then for $n > N$, consider b_n. If n is even, then $b_n = 0$ so certainly $b_n < \epsilon$. If n is odd, then $b_n = a_{(n-1)/2}$, and $(n-1)/2 > ((2N_1 + 1) - 1)/2 = N_1$ so that $a_{(n-1)/2} < \epsilon$. Thus b_n converges to zero as well.

e. False. If $\{a_n\}$ happens to converge to zero, the statement is true. But consider for example $a_n = 2 + \frac{1}{n}$. Then $\lim\limits_{n\to\infty} a_n = 2$, but $(-1)^n a_n$ does not converge (it oscillates between positive and negative values increasingly close to ± 2).

f. True. Suppose $\{0.000001 a_n\}$ converged to L, and let $\epsilon > 0$ be given. Choose N such that for $n > N$, $|0.000001 a_n - L| < \epsilon \cdot 0.000001$. Dividing through by 0.000001, we get that for $n > N$, $|a_n - 1000000L| < \epsilon$, so that a_n converges as well (to $1000000L$).

9.2.77 $\{(n-2)^2 + 6(n-2) - 9\}_{n=3}^{\infty} = \{n^2 + 2n - 17\}_{n=3}^{\infty}$.

9.2.79 Evaluate the limit of each term separately: $\lim\limits_{n\to\infty} \frac{75^{n-1}}{99^n} = \frac{1}{99} \lim\limits_{n\to\infty} \left(\frac{75}{99}\right)^{n-1} = 0$, while $\frac{-5^n}{8^n} \le \frac{5^n \sin n}{8^n} \le \frac{5^n}{8^n}$, so by the Squeeze Theorem, this second term converges to 0 as well. Thus the sum of the terms converges to zero.

9.2.81 Because $\lim\limits_{n\to\infty} 0.99^n = 0$, and because cosine is continuous, the first term converges to $\cos 0 = 1$. The limit of the second term is $\lim\limits_{n\to\infty} \frac{7^n + 9^n}{63^n} = \lim\limits_{n\to\infty} \left(\frac{7}{63}\right)^n + \lim\limits_{n\to\infty} \left(\frac{9}{63}\right)^n = 0$. Thus the sum converges to 1.

9.2.83 Dividing the numerator and denominator by 6^n gives $a_n = \frac{1 + (1/2)^n}{1 + (n^{100}/6^n)}$. By Theorem 8.6, $n^{100} \ll 6^n$. Thus $\lim\limits_{n\to\infty} a_n = \frac{1+0}{1+0} = 1$.

9.2.85 We can write $a_n = \frac{(7/5)^n}{n^7}$. Theorem 8.6 indicates that $n^7 \ll b^n$ for $b > 1$, so $\lim\limits_{n\to\infty} a_n = \infty$.

9.2.87 A graph shows that the sequence appears to converge. Let its supposed limit be L, then $\lim\limits_{n\to\infty} a_{n+1} = \lim\limits_{n\to\infty} (2a_n(1-a_n)) = 2(\lim\limits_{n\to\infty} a_n)(1 - \lim\limits_{n\to\infty} a_n)$, so $L = 2L(1-L) = 2L - 2L^2$, and thus $2L^2 - L = 0$, so $L = 0, \frac{1}{2}$. Thus the limit appears to be either 0 or 1/2; with the given initial condition, doing a few iterations by hand confirms that the sequence converges to 1/2: $a_0 = 0.3$; $a_1 = 2 \cdot 0.3 \cdot 0.7 = .42$; $a_2 = 2 \cdot 0.42 \cdot 0.58 = 0.4872$.

9.2.89 Computing three terms gives $a_0 = 0.5, a_1 = 4 \cdot 0.5 \cdot 0.5 = 1, a_2 = 4 \cdot 1 \cdot (1 - 1) = 0$. All successive terms are obviously zero, so the sequence converges to 0.

9.2.91 For $b = 2$, $2^3 > 3!$ but $16 = 2^4 < 4! = 24$, so the crossover point is $n = 4$. For e, $e^5 \approx 148.41 > 5! = 120$ while $e^6 \approx 403.4 < 6! = 720$, so the crossover point is $n = 6$. For 10, $24! \approx 6.2 \times 10^{23} < 10^{24}$, while $25! \approx 1.55 \times 10^{25} > 10^{25}$, so the crossover point is $n = 25$.

9.2.93

a. The profits for each of the first ten days, in dollars are:

n	0	1	2	3	4	5	6	7	8	9	10
h_n	130.00	130.75	131.40	131.95	132.40	132.75	133.00	133.15	133.20	133.15	133.00

b. The profit on an item is revenue minus cost. The total cost of keeping the heifer for n days is $.45n$, and the revenue for selling the heifer on the nth day is $(200 + 5n) \cdot (.65 - .01n)$, because the heifer gains 5 pounds per day but is worth a penny less per pound each day. Thus the total profit on the nth day is $h_n = (200 + 5n) \cdot (.65 - .01n) - .45n = 130 + 0.8n - 0.05n^2$. The maximum profit occurs when $-.1n + .8 = 0$, which occurs when $n = 8$. The maximum profit is achieved by selling the heifer on the 8th day.

9.2.95 The approximate first few values of this sequence are:

n	0	1	2	3	4	5	6
c_n	.7071	.6325	.6136	.6088	.6076	.6074	.6073

The value of the constant appears to be around 0.607.

9.2.97

a. If we "cut off" the expression after n square roots, we get a_n from the recurrence given. We can thus *define* the infinite expression to be the limit of a_n as $n \to \infty$.

b. $a_0 = 1$, $a_1 = \sqrt{2}$, $a_2 = \sqrt{1 + \sqrt{2}} \approx 1.5538$, $a_3 \approx 1.5981$, $a_4 \approx 1.6118$, and $a_5 \approx 1.6161$.

c. $a_{10} \approx 1.618$, which differs from $\frac{1+\sqrt{5}}{2} \approx 1.61803394$ by less than $.001$.

d. Assume $\lim\limits_{n \to \infty} a_n = L$. Then $\lim\limits_{n \to \infty} a_{n+1} = \lim\limits_{n \to \infty} \sqrt{1 + a_n} = \sqrt{1 + \lim\limits_{n \to \infty} a_n}$, so $L = \sqrt{1 + L}$, and thus $L^2 = 1 + L$. Therefore we have $L^2 - L - 1 = 0$, so $L = \frac{1 \pm \sqrt{5}}{2}$.

Because clearly the limit is positive, it must be the positive square root.

e. Letting $a_{n+1} = \sqrt{p + \sqrt{a_n}}$ with $a_0 = p$ and assuming a limit exists we have $\lim\limits_{n \to \infty} a_{n+1} = \lim\limits_{n \to \infty} \sqrt{p + a_n}$ $= \sqrt{p + \lim\limits_{n \to \infty} a_n}$, so $L = \sqrt{p + L}$, and thus $L^2 = p + L$. Therefore, $L^2 - L - p = 0$, so $L = \frac{1 \pm \sqrt{1+4p}}{2}$, and because we know that L is positive, we have $L = \frac{1 + \sqrt{4p+1}}{2}$. The limit exists for all positive p.

9.2.99

a. Define a_n as given in the problem statement. Then we can *define* the value of the continued fraction to be $\lim\limits_{n \to \infty} a_n$.

b. $a_0 = 1$, $a_1 = 1 + \frac{1}{a_0} = 2$, $a_2 = 1 + \frac{1}{a_1} = \frac{3}{2} = 1.5$, $a_3 = 1 + \frac{1}{a_2} = \frac{5}{3} \approx 1.667$, $a_4 = 1 + \frac{1}{a_3} = \frac{8}{5} = 1.6$, $a_5 = 1 + \frac{1}{a_4} = \frac{13}{8} = 1.625$.

c. From the list above, the values of the sequence alternately decrease and increase, so we would expect that the limit is somewhere between 1.6 and 1.625.

d. Assume that the limit is equal to L. Then from $a_{n+1} = 1 + \frac{1}{a_n}$, we have $\lim\limits_{n \to \infty} a_{n+1} = 1 + \frac{1}{\lim\limits_{n \to \infty} a_n}$, so $L = 1 + \frac{1}{L}$, and thus $L^2 - L - 1 = 0$. Therefore, $L = \frac{1 \pm \sqrt{5}}{2}$, and because L is clearly positive, it must be equal to $\frac{1+\sqrt{5}}{2} \approx 1.618$.

e. Here $a_0 = a$ and $a_{n+1} = a + \frac{b}{a_n}$. Assuming that $\lim\limits_{n \to \infty} a_n = L$ we have $L = a + \frac{b}{L}$, so $L^2 = aL + b$, and thus $L^2 - aL - b = 0$. Therefore, $L = \frac{a \pm \sqrt{a^2 + 4b}}{2}$, and because $L > 0$ we have $L = \frac{a + \sqrt{a^2 + 4b}}{2}$.

9.2.101

a. $f_0 = f_1 = 1, f_2 = 2, f_3 = 3, f_4 = 5, f_5 = 8, f_6 = 13, f_7 = 21, f_8 = 34, f_9 = 55, f_{10} = 89$.

b. The sequence is clearly not bounded.

c. $\frac{f_{10}}{f_9} \approx 1.61818$

d. We use induction. Note that $\frac{1}{\sqrt{5}}\left(\varphi + \frac{1}{\varphi}\right) = \frac{1}{\sqrt{5}}\left(\frac{1+\sqrt{5}}{2} + \frac{2}{1+\sqrt{5}}\right) = \frac{1}{\sqrt{5}}\left(\frac{1+2\sqrt{5}+5+4}{2(1+\sqrt{5})}\right) = 1 = f_1$. Also note that $\frac{1}{\sqrt{5}}\left(\varphi^2 - \frac{1}{\varphi^2}\right) = \frac{1}{\sqrt{5}}\left(\frac{3+\sqrt{5}}{2} - \frac{2}{3+\sqrt{5}}\right) = \frac{1}{\sqrt{5}}\left(\frac{9+6\sqrt{5}+5-4}{2(3+\sqrt{5})}\right) = 1 = f_2$. Now note that

$$f_{n-1} + f_{n-2} = \frac{1}{\sqrt{5}}\left(\varphi^{n-1} - (-1)^{n-1}\varphi^{1-n} + \varphi^{n-2} - (-1)^{n-2}\varphi^{2-n}\right)$$

$$= \frac{1}{\sqrt{5}}\left((\varphi^{n-1} + \varphi^{n-2}) - (-1)^n(\varphi^{2-n} - \varphi^{1-n})\right).$$

Now, note that $\varphi - 1 = \frac{1}{\varphi}$, so that

$$\varphi^{n-1} + \varphi^{n-2} = \varphi^{n-1}\left(1 + \frac{1}{\varphi}\right) = \varphi^{n-1} \cdot \varphi = \varphi^n$$

and

$$\varphi^{2-n} - \varphi^{1-n} = \varphi^{-n}(\varphi^2 - \varphi) = \varphi^{-n}(\varphi(\varphi - 1)) = \varphi^{-n}.$$

Making these substitutions, we get

$$f_n = f_{n-1} + f_{n-2} = \frac{1}{\sqrt{5}}(\varphi^n - (-1)^n\varphi^{-n})$$

9.2.103

a.

$$
\begin{array}{rl}
2: & 1 \\
3: & 10, 5, 16, 8, 4, 2, 1 \\
4: & 2, 1 \\
5: & 16, 8, 4, 2, 1 \\
6: & 3, 10, 5, 16, 8, 4, 2, 1 \\
7: & 22, 11, 34, 17, 52, 26, 13, 40, 20, 10, 5, 16, 8, 4, 2, 1 \\
8: & 4, 2, 1 \\
9: & 28, 14, 7, 22, 11, 34, 17, 52, 26, 13, 40, 20, 10, 5, 16, 8, 4, 2, 1 \\
10: & 5, 16, 8, 4, 2, 1
\end{array}
$$

b. From the above, $H_2 = 1, H_3 = 7$, and $H_4 = 2$.

c. This plot is for $1 \leq n \leq 100$. Like hailstones, the numbers in the sequence a_n rise and fall but eventually crash to the earth. The conjecture appears to be true.

9.2.105

a. Note that $a_2 = \sqrt{3a_1} = \sqrt{3\sqrt{3}} > \sqrt{3} = a_1$. Now assume that $\sqrt{3} = a_1 < a_2 < \ldots a_{k-1} < a_k$. Then

$$a_{k+1} = \sqrt{3a_k} > \sqrt{3a_{k-1}} = a_k.$$

Thus $\{a_n\}$ is increasing.

b. Clearly because $a_1 = \sqrt{3} > 0$ and $\{a_n\}$ is increasing, the sequence is bounded below by $\sqrt{3} > 0$. Further, $a_1 = \sqrt{3} < 3$; assume that $a_k < 3$. Then $a_{k+1} = \sqrt{3a_k} < \sqrt{3 \cdot 3} = 3$, so that $a_{k+1} < 3$. So by induction, $\{a_k\}$ is bounded above by 3.

c. Because $\{a_n\}$ is bounded and monotonically increasing, $\lim\limits_{n\to\infty} a_n$ exists by Theorem 8.5.

d. Because the limit exists, we have

$$\lim_{n\to\infty} a_{n+1} = \lim_{n\to\infty} \sqrt{3a_n} = \sqrt{3} \lim_{n\to\infty} \sqrt{a_n} = \sqrt{3}\sqrt{\lim_{n\to\infty} a_n}.$$

Let $L = \lim\limits_{n\to\infty} a_{n+1} = \lim\limits_{n\to\infty} a_n$; then $L = \sqrt{3}\sqrt{L}$, so that $L = 3$.

9.2.107 By Theorem 8.6,

$$\lim_{n\to\infty} \frac{n^5}{e^{n/2}} = 2^5 \lim_{n\to\infty} \frac{(n/2)^5}{e^{n/2}} = 0,$$

so that $e^{n/2}$ has the larger growth rate. Using computational software we see that $e^{35/2} \approx 3.982 \times 10^7 < 35^5 \approx 5.252 \times 10^7$, while $e^{36/2} \approx 6.566 \times 10^7 > 36^5 \approx 6.047 \times 10^7$.

9.2.109 Experiment with a few widely separated values of n:

n	$n!$	$n^{0.7n}$
1	1	1
10	3.63×10^6	10^7
100	9.33×10^{157}	10^{140}
1000	4.02×10^{2567}	10^{2100}

It appears that $n^{0.7n}$ starts out larger, but is overtaken by the factorial somewhere between $n = 10$ and $n = 100$, and that the gap grows wider as n increases. Looking between $n = 10$ and $n = 100$ revels that for $n = 18$, we have $n! \approx 6.402 \times 10^{15} < n^{0.7n} \approx 6.553 \times 10^{15}$ while for $n = 19$ we have $n! \approx 1.216 \times 10^{17} > n^{0.7n} \approx 1.017 \times 10^{17}$.

9.2.111 First note that for $a = 1$ we already know that $\{n^n\}$ grows fast than $\{n!\}$. So if $a > 1$, then $n^{an} \geq n^n$, so that $\{n^{an}\}$ grows faster than $\{n!\}$ for $a > 1$ as well. To settle the case $a < 1$, recall Stirling's formula which states that for large values of n,

$$n! \sim \sqrt{2\pi n}\, n^n e^{-n}.$$

Thus

$$\lim_{n\to\infty} \frac{n!}{n^{an}} = \lim_{n\to\infty} \frac{\sqrt{2\pi n}\, n^n e^{-n}}{n^{an}}$$
$$= \sqrt{2\pi} \lim_{n\to\infty} n^{\frac{1}{2}+(1-a)n} e^{-n}$$
$$\geq \sqrt{2\pi} \lim_{n\to\infty} n^{(1-a)n} e^{-n}$$
$$= \sqrt{2\pi} \lim_{n\to\infty} e^{(1-a)n\ln n} e^{-n}$$
$$= \sqrt{2\pi} \lim_{n\to\infty} e^{((1-a)\ln n - 1)n}.$$

If $a < 1$ then $(1-a)\ln n - 1 > 0$ for large values of n because $1 - a > 0$, so that this limit is infinite. Hence $\{n!\}$ grows faster than $\{n^{an}\}$ exactly when $a < 1$.

9.3 Infinite Series

9.3.1 A geometric series is a series in which the ratio of successive terms in the underlying sequence is a constant. Thus a geometric series has the form $\sum ar^k$ where r is the constant. One example is $3 + 6 + 12 + 24 + 48 + \cdots$ in which $a = 3$ and $r = 2$.

9.3.3 The ratio is the common ratio between successive terms in the sum.

9.3.5 No. For example, the geometric series with $a_n = 3 \cdot 2^n$ does not have a finite sum.

9.3.7 $S = 1 \cdot \dfrac{1 - 3^9}{1 - 3} = \dfrac{19682}{2} = 9841.$

9.3.9 $S = 1 \cdot \dfrac{1 - (4/25)^{21}}{1 - 4/25} = \dfrac{25^{21} - 4^{21}}{25^{21} - 4 \cdot 25^{20}} \approx 1.1905.$

9.3.11 $S = 1 \cdot \dfrac{1 - (-3/4)^{10}}{1 + 3/4} = \dfrac{4^{10} - 3^{10}}{4^{10} + 3 \cdot 4^9} = \dfrac{141361}{262144} \approx 0.5392.$

9.3.13 $S = 1 \cdot \dfrac{1 - \pi^7}{1 - \pi} = \dfrac{\pi^7 - 1}{\pi - 1} \approx 1409.84.$

9.3.15 $S = 1 \cdot \dfrac{1 - (-1)^{21}}{2} = 1.$

9.3.17 $\dfrac{1093}{2916}.$

9.3.19 $\dfrac{1}{1 - 1/4} = \dfrac{4}{3}.$

9.3.21 $\dfrac{1}{1 - 0.9} = 10.$

9.3.23 Divergent, because $r > 1$.

9.3.25 $\dfrac{e^{-2}}{1 - e^{-2}} = \dfrac{1}{e^2 - 1}.$

9.3.27 $\dfrac{2^{-3}}{1 - 2^{-3}} = \dfrac{1}{7}.$

9.3.29 $\dfrac{1/625}{1 - 1/5} = \dfrac{1}{500}.$

9.3.31 $\dfrac{1}{1 - e/\pi} = \dfrac{\pi}{\pi - e}.$ (Note that $e < \pi$, so $r < 1$ for this series.)

9.3.33 $\displaystyle\sum_{k=0}^{\infty} \left(\dfrac{1}{4}\right)^k 5^{3-k} = 5^3 \sum_{k=0}^{\infty} \left(\dfrac{1}{20}\right)^k = 5^3 \cdot \dfrac{1}{1 - 1/20} = \dfrac{5^3 \cdot 20}{19} = \dfrac{2500}{19}.$

9.3.35 $\dfrac{1}{1 + 9/10} = \dfrac{10}{19}.$

9.3.37 $3 \cdot \dfrac{1}{1 + 1/\pi} = \dfrac{3\pi}{\pi + 1}.$

9.3.39 $\dfrac{0.15^2}{1.15} = \dfrac{9}{460} \approx 0.0196.$

9.3.41

  a. $0.\overline{3} = 0.333\ldots = \sum_{k=1}^{\infty} 3(0.1)^k$.

  b. The limit of the sequence of partial sums is $1/3$.

9.3.43

  a. $0.\overline{1} = 0.111\ldots = \sum_{k=1}^{\infty} (0.1)^k$.

  b. The limit of the sequence of partial sums is $1/9$.

9.3.45

  a. $0.\overline{09} = 0.0909\ldots = \sum_{k=1}^{\infty} 9(0.01)^k$.

  b. The limit of the sequence of partial sums is $1/11$.

9.3.47

  a. $0.\overline{037} = 0.037037037\ldots = \sum_{k=1}^{\infty} 37(0.001)^k$.

  b. The limit of the sequence of partial sums is $37/999 = 1/27$.

9.3.49 $0.\overline{12} = 0.121212\ldots = \sum_{k=0}^{\infty} .12 \cdot 10^{-2k} = \dfrac{.12}{1 - 1/100} = \dfrac{12}{99} = \dfrac{4}{33}$.

9.3.51 $0.\overline{456} = 0.456456456\ldots = \sum_{k=0}^{\infty} .456 \cdot 10^{-3k} = \dfrac{.456}{1 - 1/1000} = \dfrac{456}{999} = \dfrac{152}{333}$.

9.3.53 $0.00\overline{952} = 0.00952952\ldots = \sum_{k=0}^{\infty} .00952 \cdot 10^{-3k} = \dfrac{.00952}{1 - 1/1000} = \dfrac{9.52}{999} = \dfrac{952}{99900} = \dfrac{238}{24975}$.

9.3.55 The second part of each term cancels with the first part of the succeeding term, so $S_n = \frac{1}{1+1} - \frac{1}{n+2} = \frac{n}{2n+4}$, and $\lim\limits_{n \to \infty} \frac{n}{2n+4} = \frac{1}{2}$.

9.3.57 $\dfrac{1}{(k+6)(k+7)} = \dfrac{1}{k+6} - \dfrac{1}{k+7}$, so the series given is the same as $\sum_{k=1}^{\infty} \left(\frac{1}{k+6} - \frac{1}{k+7} \right)$. In that series, the second part of each term cancels with the first part of the succeeding term, so $S_n = \frac{1}{1+6} - \frac{1}{n+7}$. Thus $\lim\limits_{n \to \infty} S_n = \frac{1}{7}$.

9.3.59 Note that $\frac{4}{(4k-3)(4k+1)} = \frac{1}{4k-3} - \frac{1}{4k+1}$. Thus the given series is the same as $\sum_{k=3}^{\infty} \left(\dfrac{1}{4k-3} - \dfrac{1}{4k+1} \right)$. In that series, the second part of each term cancels with the first part of the succeeding term (because $4(k+1) - 3 = 4k+1$), so we have $S_n = \frac{1}{9} - \frac{1}{4n+1}$, and thus $\lim\limits_{n \to \infty} S_n = \dfrac{1}{9}$.

9.3.61 $\ln\left(\dfrac{k+1}{k} \right) = \ln(k+1) - \ln k$, so the series given is the same as $\sum_{k=1}^{\infty} (\ln(k+1) - \ln k)$, in which the first part of each term cancels with the second part of the next term, so we have $S_n = \ln(n+1) - \ln 1 = \ln(n+1)$, and thus the series diverges.

9.3.63 $\dfrac{1}{(k+p)(k+p+1)} = \dfrac{1}{k+p} - \dfrac{1}{k+p+1}$, so that $\sum_{k=1}^{\infty} \dfrac{1}{(k+p)(k+p+1)} = \sum_{k=1}^{\infty} \left(\dfrac{1}{k+p} - \dfrac{1}{k+p+1} \right)$ and this series telescopes to give $S_n = \frac{1}{p+1} - \frac{1}{n+p+1} = \frac{n}{n(p+1)+(p+1)^2}$ so that $\lim\limits_{n \to \infty} S_n = \frac{1}{p+1}$.

9.3.65 Let $a_n = \dfrac{1}{\sqrt{n+1}} - \dfrac{1}{\sqrt{n+3}}$. Then the second term of a_n cancels with the first term of a_{n+2}, so the series telescopes and $S_n = \frac{1}{\sqrt{2}} + \frac{1}{\sqrt{3}} - \frac{1}{\sqrt{n-1+3}} - \frac{1}{\sqrt{n+3}}$ and thus the sum of the series is the limit of S_n, which is $\dfrac{1}{\sqrt{2}} + \dfrac{1}{\sqrt{3}}$.

9.3.67 $16k^2 + 8k - 3 = (4k+3)(4k-1)$, so $\frac{1}{16k^2+8k-3} = \frac{1}{(4k+3)(4k-1)} = \frac{1}{4}\left(\frac{1}{4k-1} - \frac{1}{4k+3}\right)$. Thus the series given is equal to $\dfrac{1}{4}\displaystyle\sum_{k=0}^{\infty}\left(\frac{1}{4k-1} - \frac{1}{4k+3}\right)$. This series telescopes, so $S_n = \frac{1}{4}\left(-1 - \frac{1}{4n+3}\right)$, so the sum of the series is equal to $\displaystyle\lim_{n\to\infty} S_n = -\frac{1}{4}$.

9.3.69

a. True. $\left(\dfrac{\pi}{e}\right)^{-k} = \left(\dfrac{e}{\pi}\right)^{k}$; because $e < \pi$, this is a geometric series with ratio less than 1.

b. True. If $\displaystyle\sum_{k=12}^{\infty} a^k = L$, then $\displaystyle\sum_{k=0}^{\infty} a^k = \left(\displaystyle\sum_{k=0}^{11} a^k\right) + L$.

c. False. For example, let $0 < a < 1$ and $b > 1$.

d. True. Suppose $a > \frac{1}{2}$. Then we want $a = \sum_{k=0}^{\infty} r^k = \frac{1}{1-r}$. Solving for r gives $r = 1 - \frac{1}{a}$. Because $a > 0$ we have $r < 1$; because $a > \frac{1}{2}$ we have $r > 1 - \frac{1}{1/2} = -1$. Thus $|r| < 1$ so that $\sum_{k=0}^{\infty} r^k$ converges, and it converges to a.

e. True. Suppose $a > -\frac{1}{2}$. Then we want $a = \sum_{k=1}^{\infty} r^k = \frac{r}{1-r}$. Solving for r gives $r = \frac{a}{a+1}$. For $a \geq 0$, clearly $0 \leq r < 1$ so that $\sum_{k=1}^{\infty} r^k$ converges to a. For $-\frac{1}{2} < a < 0$, clearly $r < 0$, but $|a| < |a+1|$, so that $|r| < 1$. Thus in this case $\sum_{k=1}^{\infty} r^k$ also converges to a.

9.3.71 This can be written as $\dfrac{1}{3}\displaystyle\sum_{k=1}^{\infty}\left(-\frac{2}{3}\right)^{k}$. This is a geometric series with ratio $r = -\frac{2}{3}$ so the sum is $\frac{1}{3} \cdot \frac{-2/3}{1-(-2/3)} = \frac{1}{3} \cdot \left(-\frac{2}{5}\right) = -\frac{2}{15}$.

9.3.73 Note that

$$\frac{\ln((k+1)k^{-1})}{(\ln k)\ln(k+1)} = \frac{\ln(k+1)}{(\ln k)\ln(k+1)} - \frac{\ln k}{(\ln k)\ln(k+1)} = \frac{1}{\ln k} - \frac{1}{\ln(k+1)}.$$

In the partial sum S_n, the first part of each term cancels the second part of the preceding term, so we have $S_n = \frac{1}{\ln 2} - \frac{1}{\ln(n+1)}$. Thus we have $\displaystyle\lim_{n\to\infty} S_n = \dfrac{1}{\ln 2}$.

9.3.75

a. Because the first part of each term cancels the second part of the previous term, the nth partial sum telescopes to be $S_n = \frac{4}{3} - \frac{4}{3^{n+1}}$. Thus, the sum of the series is $\displaystyle\lim_{n\to\infty} S_n = \dfrac{4}{3}$.

b. Note that $\frac{4}{3^k} - \frac{4}{3^{k+1}} = \frac{4 \cdot 3^{k+1} - 4 \cdot 3^k}{3^k 3^{k+1}} = \frac{8}{3^{k+1}}$. Thus, the original series can be written as $\displaystyle\sum_{k=1}^{\infty} \frac{8}{3^{k+1}}$ which is geometric with $r = 1/3$ and $a = 8/9$, so the sum is $\frac{8/9}{1-1/3} = \frac{8}{9} \cdot \frac{3}{2} = \frac{4}{3}$.

9.3.77 At the nth stage, there are 2^{n-1} triangles of area $A_n = \frac{1}{8}A_{n-1} = \frac{1}{8^{n-1}}A_1$, so the total area of the triangles formed at the nth stage is $\dfrac{2^{n-1}}{8^{n-1}}A_1 = \left(\dfrac{1}{4}\right)^{n-1} A_1$. Thus the total area under the parabola is

$$\sum_{n=1}^{\infty}\left(\frac{1}{4}\right)^{n-1} A_1 = A_1 \sum_{n=1}^{\infty}\left(\frac{1}{4}\right)^{n-1} = A_1 \frac{1}{1-1/4} = \frac{4}{3}A_1.$$

9.3.79 It appears that the loan is paid off after about 470 months. Let B_n be the loan balance after n months. Then $B_0 = 180000$ and $B_n = 1.005 \cdot B_{n-1} - 1000$. Then $B_n = 1.005 \cdot B_{n-1} - 1000 = 1.005(1.005 \cdot B_{n-2} - 1000) - 1000 = (1.005)^2 \cdot B_{n-2} - 1000(1 + 1.005) = (1.005)^2 \cdot (1.005 \cdot B_{n-3} - 1000) - 1000(1 + 1.005) = (1.005)^3 \cdot B_{n-3} - 1000(1 + 1.005 + (1.005)^2) = \cdots = (1.005)^n B_0 - 1000(1 + 1.005 + (1.005)^2 + \cdots + (1.005)^{n-1}) = (1.005)^n \cdot 180000 - 1000\left(\frac{(1.005)^n - 1}{1.005 - 1}\right)$. Solving this equation for $B_n = 0$ gives $n \approx 461.667$ months, so the loan is paid off after 462 months.

9.3.81 $F_n = (1.015)F_{n-1} - 120 = (1.015)((1.015)F_{n-2} - 120) - 120 = (1.015)((1.015)((1.015)F_{n-3} - 120) - 120) - 120 = \cdots = (1.015)^n(4000) - 120(1 + (1.015) + (1.015)^2 + \cdots + (1.015)^{n-1})$. This is equal to

$$(1.015)^n(4000) - 120\left(\frac{(1.015)^n - 1}{1.015 - 1}\right) = (-4000)(1.015)^n + 8000.$$

The long term population of the fish is 0.

9.3.83 Under the one-child policy, each couple will have one child. Under the one-son policy, we compute the expected number of children as follows: with probability $1/2$ the first child will be a son; with probability $(1/2)^2$, the first child will be a daughter and the second child will be a son; in general, with probability $(1/2)^n$, the first $n - 1$ children will be girls and the nth a boy. Thus the expected number of children is the sum $\sum_{i=1}^{\infty} i \cdot \left(\frac{1}{2}\right)^i$. To evaluate this series, use the following "trick": Let $f(x) = \sum_{i=1}^{\infty} ix^i$. Then

$f(x) + \sum_{i=1}^{\infty} x^i = \sum_{i=1}^{\infty} (i+1)x^i$. Now, let

$$g(x) = \sum_{i=1}^{\infty} x^{i+1} = -1 - x + \sum_{i=0}^{\infty} x^i = -1 - x + \frac{1}{1-x}$$

and

$$g'(x) = f(x) + \sum_{i=1}^{\infty} x^i = f(x) - 1 + \sum_{i=0}^{\infty} x^i = f(x) - 1 + \frac{1}{1-x}.$$

Evaluate $g'(x) = -1 - \frac{1}{(1-x)^2}$; then

$$f(x) = 1 - \frac{1}{1-x} - 1 - \frac{1}{(1-x)^2} = \frac{-1+x+1}{(1-x)^2} = \frac{x}{(1-x)^2}$$

Finally, evaluate at $x = \frac{1}{2}$ to get $f\left(\frac{1}{2}\right) = \sum_{i=1}^{\infty} i \cdot \left(\frac{1}{2}\right)^i = \frac{1/2}{(1-1/2)^2} = 2$. There will thus be twice as many children under the one-son policy as under the one-child policy.

9.3.85 Ignoring the initial drop for the moment, the height after the nth bounce is $10p^n$, so the total time spent in that bounce is $2 \cdot \sqrt{2 \cdot 10p^n/g}$ seconds. The total time before the ball comes to rest (now including the time for the initial drop) is then $\sqrt{20/g} + \sum_{i=1}^{\infty} 2 \cdot \sqrt{2 \cdot 10p^n/g} = \sqrt{\frac{20}{g}} + 2\sqrt{\frac{20}{g}} \sum_{i=1}^{\infty} (\sqrt{p})^n = \sqrt{\frac{20}{g}} + 2\sqrt{\frac{20}{g}} \frac{\sqrt{p}}{1-\sqrt{p}} = \sqrt{\frac{20}{g}}\left(1 + \frac{2\sqrt{p}}{1-\sqrt{p}}\right) = \sqrt{\frac{20}{g}}\left(\frac{1+\sqrt{p}}{1-\sqrt{p}}\right)$ seconds.

9.3.87

a. I_{n+1} is obtained by I_n by dividing each edge into three equal parts, removing the middle part, and adding two parts equal to it. Thus 3 equal parts turn into 4, so $L_{n+1} = \frac{4}{3}L_n$. This is a geometric sequence with a ratio greater than 1, so the nth term grows without bound.

b. As the result of part (a), I_n has $3 \cdot 4^n$ sides of length $\frac{1}{3^n}$; each of those sides turns into an added triangle in I_{n+1} of side length 3^{-n-1}. Thus the added area in I_{n+1} consists of $3 \cdot 4^n$ equilateral triangles with side 3^{-n-1}. The area of an equilateral triangle with side x is $\dfrac{x^2\sqrt{3}}{4}$. Thus $A_{n+1} = A_n + 3 \cdot 4^n \cdot \dfrac{3^{-2n-2}\sqrt{3}}{4} = A_n + \frac{\sqrt{3}}{12} \cdot \left(\frac{4}{9}\right)^n$, and $A_0 = \frac{\sqrt{3}}{4}$. Thus $A_{n+1} = A_0 + \sum_{i=0}^{n} \frac{\sqrt{3}}{12} \cdot \left(\frac{4}{9}\right)^i$, so that

$$A_\infty = A_0 + \frac{\sqrt{3}}{12}\sum_{i=0}^{\infty}\left(\frac{4}{9}\right)^i = \frac{\sqrt{3}}{4} + \frac{\sqrt{3}}{12}\frac{1}{1-4/9} = \frac{\sqrt{3}}{4}\left(1+\frac{3}{5}\right) = \frac{2}{5}\sqrt{3}.$$

9.3.89 $|S - S_n| = \left|\displaystyle\sum_{i=n}^{\infty} r^k\right| = \left|\dfrac{r^n}{1-r}\right|$ because the latter sum is simply a geometric series with first term r^n and ratio r.

9.3.91

a. Solve $\left|\dfrac{(-0.8)^n}{1.8}\right| = \dfrac{0.8^n}{1.8} < 10^{-6}$ for n to get $n = 60$.

b. Solve $\dfrac{0.2^n}{0.8} < 10^{-6}$ for n to get $n = 9$.

9.3.93

a. Solve $\dfrac{1/\pi^n}{1-1/\pi} < 10^{-6}$ for n to get $n = 13$.

b. Solve $\dfrac{1/e^n}{1-1/e} < 10^{-6}$ for n to get $n = 15$.

9.3.95

a. $f(x) = \sum_{k=0}^{\infty}(-1)^k x^k = \frac{1}{1+x}$; because f is a geometric series, $f(x)$ exists only when the ratio, $-x$, is such that $|-x| = |x| < 1$. Then $f(0) = 1$, $f(0.2) = \frac{1}{1.2} = \frac{5}{6}$, $f(0.5) = \frac{1}{1+.05} = \frac{2}{3}$. Neither $f(1)$ nor $f(1.5)$ exists.

b. The domain of f is $\{x : |x| < 1\}$.

9.3.97 $f(x)$ is a geometric series with ratio $\frac{1}{1+x}$; thus $f(x)$ converges when $\left|\frac{1}{1+x}\right| < 1$. For $x > -1$, $\left|\dfrac{1}{1+x}\right| = \dfrac{1}{1+x}$ and $\dfrac{1}{1+x} < 1$ when $1 < 1 + x$, $x > 0$. For $x < -1$, $\left|\dfrac{1}{1+x}\right| = \dfrac{1}{-1-x}$, and this is less than 1 when $1 < -1 - x$, i.e. $x < -2$. So $f(x)$ converges for $x > 0$ and for $x < -2$. When $f(x)$ converges, its value is $\dfrac{1}{1-\frac{1}{1+x}} = \dfrac{1+x}{x}$, so $f(x) = 3$ when $1 + x = 3x$, $x = \frac{1}{2}$.

9.3.99

a. Using Theorem 8.7 in each case except for $r = 0$ gives

r	$f(r)$
-0.9	0.526
-0.7	0.588
-0.5	0.667
-0.2	0.833
0	1
0.2	1.250
0.5	2
0.7	3.333
0.9	10

b. A plot of f is

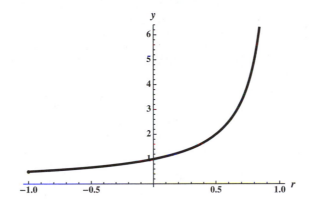

c. For $-1 < r < 1$ we have $f(r) = \frac{1}{1-r}$, so that

$$\lim_{r \to -1^+} f(r) = \lim_{r \to -1^+} \frac{1}{1-r} = \frac{1}{2}, \qquad \lim_{r \to 1^-} f(r) = \lim_{r \to 1^-} \frac{1}{1-r} = \infty.$$

9.4 The Divergence and Integral Tests

9.4.1 If the sequence of terms has limit 1, then the corresponding series diverges. It is necessary (but not sufficient) that the sequence of terms has limit 0 in order for the corresponding series to be convergent.

9.4.3 Yes. Either the series and the integral both converge, or both diverge, if the terms are positive and decreasing.

9.4.5 For the same values of p as in the previous problem – it converges for $p > 1$, and diverges for all other values of p.

9.4.7 The remainder of an infinite series is the error in approximating a convergent infinite series by a finite number of terms.

9.4.9 $a_k = \frac{k}{2k+1}$ and $\lim_{k \to \infty} a_k = \frac{1}{2}$, so the series diverges.

9.4.11 $a_k = \frac{k}{\ln k}$ and $\lim_{k \to \infty} a_k = \infty$, so the series diverges.

9.4.13 $a_k = \frac{1}{1000+k}$ and $\lim_{k \to \infty} a_k = 0$, so the divergence test is inconclusive.

9.4.15 $a_k = \frac{\sqrt{k}}{\ln^{10} k}$ and $\lim\limits_{k \to \infty} a_k = \infty$, so the series diverges.

9.4.17 $a_k = k^{1/k}$. In order to compute $\lim_{k \to \infty} a_k$, we let $y_k = \ln a_k = \frac{\ln k}{k}$. By Theorem 9.6, (or by L'Hôpital's rule), $\lim_{k \to \infty} y_k = 0$, so $\lim_{k \to \infty} a_k = e^0 = 1$. The given series thus diverges.

9.4.19 Clearly $\frac{1}{e^x} = e^{-x}$ is continuous, positive, and decreasing for $x \geq 2$ (in fact, for all x), so the integral test applies. Because

$$\int_2^\infty e^{-x}\, dx = \lim_{c \to \infty} \int_2^c e^{-x}\, dx = \lim_{c \to \infty} \left(-e^{-x} \right)\Big|_2^c = \lim_{c \to \infty} \left(e^{-2} - e^{-c} \right) = e^{-2},$$

the Integral Test tells us that the original series converges as well.

9.4.21 Let $f(x) = x \cdot e^{-2x^2}$. This function is continuous for $x \geq 1$. Its derivative is $e^{-2x^2}(1 - 4x^2) < 0$ for $x \geq 1$, so $f(x)$ is decreasing. Because $\int_1^\infty x \cdot e^{-2x^2}\, dx = \frac{1}{4e^2}$, the series converges.

9.4.23 Let $f(x) = \frac{1}{\sqrt{x+8}}$. $f(x)$ is obviously continuous and decreasing for $x \geq 1$. Because $\int_1^\infty \frac{1}{\sqrt{x+8}}\, dx = \infty$, the series diverges.

9.4.25 Let $f(x) = \frac{x}{e^x}$. $f(x)$ is clearly continuous for $x > 1$, and its derivative, $f'(x) = \frac{e^x - xe^x}{e^{2x}} = (1 - x)\frac{e^x}{e^{2x}}$, is negative for $x > 1$ so that $f(x)$ is decreasing. Because $\int_1^\infty f(x)\, dx = 2e^{-1}$, the series converges.

9.4.27 The integral test does not apply, because the sequence of terms is not decreasing.

9.4.29 This is a p-series with $p = 10$, so this series converges.

9.4.31 $\sum_{k=3}^\infty \frac{1}{(k-2)^4} = \sum_{k=1}^\infty \frac{1}{k^4}$, which is a p-series with $p = 4$, thus convergent.

9.4.33 $\sum_{k=1}^\infty \frac{1}{\sqrt[3]{k}} = \sum_{k=1}^\infty \frac{1}{k^{1/3}}$ is a p-series with $p = 1/3$, thus divergent.

9.4.35

a. The remainder R_n is bounded by $\int_n^\infty \frac{1}{x^6}\, dx = \frac{1}{5n^5}$.

b. We solve $\frac{1}{5n^5} < 10^{-3}$ to get $n = 3$.

c. $L_n = S_n + \int_{n+1}^\infty \frac{1}{x^6}\, dx = S_n + \frac{1}{5(n+1)^5}$, and $U_n = S_n + \int_n^\infty \frac{1}{x^6}\, dx = S_n + \frac{1}{5n^5}$.

d. $S_{10} \approx 1.017341512$, so $L_{10} \approx 1.017341512 + \frac{1}{5 \cdot 11^5} \approx 1.017342754$, and $U_{10} \approx 1.017341512 + \frac{1}{5 \cdot 10^5} \approx 1.017343512$.

9.4.37

a. The remainder R_n is bounded by $\int_n^\infty \frac{1}{3^x}\, dx = \frac{1}{3^n \ln 3}$.

b. We solve $\frac{1}{3^n \ln 3} < 10^{-3}$ to obtain $n = 7$.

c. $L_n = S_n + \int_{n+1}^\infty \frac{1}{3^x}\, dx = S_n + \frac{1}{3^{n+1} \ln 3}$, and $U_n = S_n + \int_n^\infty \frac{1}{3^x}\, dx = S_n + \frac{1}{3^n \ln 3}$.

d. $S_{10} \approx 0.4999915325$, so $L_{10} \approx 0.4999915325 + \frac{1}{3^{11} \ln 3} \approx 0.4999966708$, and $U_{10} \approx 0.4999915325 + \frac{1}{3^{10} \ln 3} \approx 0.5000069475$.

9.4.39

a. The remainder R_n is bounded by $\int_n^\infty \frac{1}{x^{3/2}}\, dx = 2n^{-1/2}$.

b. We solve $2n^{-1/2} < 10^{-3}$ to get $n > 4 \times 10^6$, so let $n = 4 \times 10^6 + 1$.

c. $L_n = S_n + \int_{n+1}^\infty \frac{1}{x^{3/2}}\, dx = S_n + 2(n+1)^{-1/2}$, and $U_n = S_n + \int_n^\infty \frac{1}{x^{3/2}}\, dx = S_n + 2n^{-1/2}$.

d. $S_{10} = \sum_{k=1}^{10} \frac{1}{k^{3/2}} \approx 1.995336493$, so $L_{10} \approx 1.995336493 + 2 \cdot 11^{-1/2} \approx 2.598359182$, and $U_{10} \approx$ 1.995336493 + 2 \cdot 10^{-1/2} \approx 2.627792025$.

9.4.41

a. The remainder R_n is bounded by $\int_n^\infty \frac{1}{x^3}\, dx = \frac{1}{2n^2}$.

b. We solve $\frac{1}{2n^2} < 10^{-3}$ to get $n = 23$.

c. $L_n = S_n + \int_{n+1}^\infty \frac{1}{x^3}\, dx = S_n + \frac{1}{2(n+1)^2}$, and $U_n = S_n + \int_n^\infty \frac{1}{x^3}\, dx = S_n + \frac{1}{2n^2}$.

d. $S_{10} \approx 1.197531986$, so $L_{10} \approx 1.197531986 + \frac{1}{2 \cdot 11^2} \approx 1.201664217$, and $U_{10} \approx 1.197531986 + \frac{1}{2 \cdot 10^2} \approx$ 1.202531986.

9.4.43 This is a geometric series with $a = \frac{1}{3}$ and $r = \frac{1}{12}$, so $\sum_{k=1}^\infty \frac{4}{12^k} = \frac{1/3}{1 - 1/12} = \frac{1/3}{11/12} = \frac{4}{11}$.

9.4.45 $\displaystyle\sum_{k=0}^\infty \left(3\left(\frac{2}{5}\right)^k - 2\left(\frac{5}{7}\right)^k \right) = 3\sum_{k=0}^\infty \left(\frac{2}{5}\right)^k - 2\sum_{k=0}^\infty \left(\frac{5}{7}\right)^k = 3\left(\frac{1}{3/5}\right) - 2\left(\frac{1}{2/7}\right) = 5 - 7 = -2.$

9.4.47 $\displaystyle\sum_{k=1}^\infty \left(\frac{1}{3}\left(\frac{5}{6}\right)^k + \frac{3}{5}\left(\frac{7}{9}\right)^k \right) = \frac{1}{3}\sum_{k=1}^\infty \left(\frac{5}{6}\right)^k + \frac{3}{5}\sum_{k=1}^\infty \left(\frac{7}{9}\right)^k = \frac{1}{3}\left(\frac{5/6}{1/6}\right) + \frac{3}{5}\left(\frac{7/9}{2/9}\right) = \frac{5}{3} + \frac{21}{10} = \frac{113}{30}.$

9.4.49 $\displaystyle\sum_{k=1}^\infty \left(\left(\frac{1}{6}\right)^k + \left(\frac{1}{3}\right)^{k-1} \right) = \sum_{k=1}^\infty \left(\frac{1}{6}\right)^k + \sum_{k=1}^\infty \left(\frac{1}{3}\right)^{k-1} = \frac{1/6}{5/6} + \frac{1}{2/3} = \frac{17}{10}.$

9.4.51

a. True. The two series differ by a finite amount ($\sum_{k=1}^9 a_k$), so if one converges, so does the other.

b. True. The same argument applies as in part (a).

c. False. If $\sum a_k$ converges, then $a_k \to 0$ as $k \to \infty$, so that $a_k + 0.0001 \to 0.0001$ as $k \to \infty$, so that $\sum (a_k + 0.0001)$ cannot converge.

d. False. Suppose $p = -1.0001$. Then $\sum p^k$ diverges but $p + 0.001 = -0.9991$ so that $\sum (p + .0001)^k$ converges.

e. False. Let $p = 1.0005$; then $-p + .001 = -(p - .001) = -.9995$, so that $\sum k^{-p}$ converges (p-series) but $\sum k^{-p+.001}$ diverges.

f. False. Let $a_k = \frac{1}{k}$, the harmonic series.

9.4.53 Converges by the Integral Test because $\displaystyle\int_1^\infty \frac{1}{(3x+1)(3x+4)}\, dx = \int_1^\infty \frac{1}{3(3x+1)} - \frac{1}{3(3x+4)}\, dx =$

$\displaystyle\lim_{b\to\infty} \int_1^b \left(\frac{1}{3(3x+1)} - \frac{1}{3(3x+4)} \right) dx = \lim_{b\to\infty} \frac{1}{9}\left(\ln\left(\frac{3x+1}{3x+4}\right) \right)\Big|_1^b = \lim_{b\to\infty} = -\frac{1}{9} \cdot \ln(4/7) \approx 0.06217 < \infty.$

Alternatively, this is a telescoping series with nth partial sum equal to $S_n = \frac{1}{3}\left(\frac{1}{4} - \frac{1}{3n+4}\right)$ which converges to $\frac{1}{12}$.

9.4.55 Diverges by the Divergence Test because $\displaystyle\lim_{k\to\infty} a_k = \lim_{k\to\infty} \frac{k}{\sqrt{k^2+1}} = 1 \neq 0.$

9.4.57 Converges by the Integral Test because $\displaystyle\int_2^\infty \frac{4}{x\ln^2 x}\, dx = \lim_{b\to\infty} \left(\frac{-4}{\ln x}\Big|_2^b \right) = \frac{4}{\ln 2} < \infty.$

9.4.59

   a. Note that $\int \frac{1}{x \ln x (\ln \ln x)^p}\, dx = \frac{1}{1-p}(\ln \ln x)^{1-p}$, and thus the improper integral with bounds n and ∞ exists only if $p > 1$ because $\ln \ln x > 0$ for $x > e$. So this series converges for $p > 1$.

   b. For large values of z, clearly $\sqrt{z} > \ln z$, so that $z > (\ln z)^2$. Write $z = \ln x$; then for large x, $\ln x > (\ln \ln x)^2$; multiplying both sides by $x \ln x$ we have that $x \ln^2 x > x \ln x (\ln \ln x)^2$, so that the first series converges faster because the terms get smaller faster.

9.4.61 Let $S_n = \sum_{k=1}^n \frac{1}{\sqrt{k}}$. Then this looks like a left Riemann sum for the function $y = \frac{1}{\sqrt{x}}$ on $[1, n+1]$. Because each rectangle lies above the curve itself, we see that S_n is bounded below by the integral of $\frac{1}{\sqrt{x}}$ on $[1, n+1]$. Now,

$$\int_1^{n+1} \frac{1}{\sqrt{x}}\, dx = \int_1^{n+1} x^{-1/2}\, dx = 2\sqrt{x}\Big|_1^{n+1} = 2\sqrt{n+1} - 2.$$

This integral diverges as $n \to \infty$, so the series does as well by the bound above.

9.4.63 $\sum_{k=1}^\infty c a_k = \lim_{n \to \infty} \sum_{k=1}^n c a_k = \lim_{n \to \infty} c \sum_{k=1}^n a_k = c \lim_{n \to \infty} \sum_{k=1}^n a_k$, so that one sum diverges if and only if the other one does.

9.4.65 To approximate the sequence for $\zeta(m)$, note that the remainder R_n after n terms is bounded by

$$\int_n^\infty \frac{1}{x^m}\, dx = \frac{1}{m-1} n^{1-m}.$$

For $m = 3$, if we wish to approximate the value to within 10^{-3}, we must solve $\frac{1}{2} n^{-2} < 10^{-3}$, so that $n = 23$,

and $\sum_{k=1}^{23} \frac{1}{k^3} \approx 1.201151926$. The true value is ≈ 1.202056903.

For $m = 5$, if we wish to approximate the value to within 10^{-3}, we must solve $\frac{1}{4} n^{-4} < 10^{-3}$, so that $n = 4$,

and $\sum_{k=1}^{4} \frac{1}{k^5} \approx 1.036341789$. The true value is ≈ 1.036927755.

9.4.67 $\sum_{k=1}^\infty \frac{1}{k^2} = \sum_{k=1}^\infty \frac{1}{(2k)^2} + \sum_{k=1}^\infty \frac{1}{(2k-1)^2}$, splitting the series into even and odd terms. But $\sum_{k=1}^\infty \frac{1}{(2k)^2} = \frac{1}{4} \sum_{k=1}^\infty \frac{1}{k^2}$. Thus $\frac{\pi^2}{6} = \frac{1}{4}\frac{\pi^2}{6} + \sum_{k=1}^\infty \frac{1}{(2k-1)^2}$, so that the sum in question is $\frac{3\pi^2}{24} = \frac{\pi^2}{8}$.

9.4.69

   a. $x_1 = \sum_{k=2}^2 \frac{1}{k} = \frac{1}{2}$, $x_2 = \sum_{k=3}^4 \frac{1}{k} = \frac{1}{3} + \frac{1}{4} = \frac{7}{12}$, $x_3 = \sum_{k=4}^6 \frac{1}{k} = \frac{1}{4} + \frac{1}{5} + \frac{1}{6} = \frac{37}{60}$.

   b. x_n has n terms. Each term is bounded below by $\frac{1}{2n}$ and bounded above by $\frac{1}{n+1}$. Thus $x_n \geq n \cdot \frac{1}{2n} = \frac{1}{2}$, and $x_n \leq n \cdot \frac{1}{n+1} < n \cdot \frac{1}{n} = 1$.

   c. The right Riemann sum for $\int_1^2 \frac{dx}{x}$ using n subintervals has n rectangles of width $\frac{1}{n}$; the right edges of those rectangles are at $1 + \frac{i}{n} = \frac{n+i}{n}$ for $i = 1, 2, \ldots, n$. The height of such a rectangle is the value of $\frac{1}{x}$ at the right endpoint, which is $\frac{n}{n+i}$. Thus the area of the rectangle is $\frac{1}{n} \cdot \frac{n}{n+i} = \frac{1}{n+i}$. Adding up over all the rectangles gives x_n.

   d. The limit $\lim_{n \to \infty} x_n$ is the limit of the right Riemann sum as the width of the rectangles approaches zero.

      This is precisely $\int_1^2 \frac{dx}{x} = \ln x\Big|_1^2 = \ln 2$.

9.4.71

a. Note that the center of gravity of any stack of dominoes is the average of the locations of their centers. Define the midpoint of the zeroth (top) domino to be $x = 0$, and stack additional dominoes down and to its right (to increasingly positive x-coordinates). Let $m(n)$ be the x-coordinate of the midpoint of the nth domino. Then in order for the stack not to fall over, the left edge of the nth domino must be placed directly under the center of gravity of dominos 0 through $n - 1$, which is $\frac{1}{n} \sum_{i=0}^{n-1} m(i)$, so that $m(n) = 1 + \frac{1}{n} \sum_{i=0}^{n-1} m(i)$. We claim that in fact $m(n) = \sum_{k=1}^{n} \frac{1}{k}$. Use induction. This is certainly true for $n = 1$. Note first that $m(0) = 0$, so we can start the sum at 1 rather than at 0. Now, $m(n) = 1 + \frac{1}{n} \sum_{i=1}^{n-1} m(i) = 1 + \frac{1}{n} \sum_{i=1}^{n-1} \sum_{j=1}^{i} \frac{1}{j}$. Now, 1 appears $n - 1$ times in the double sum, 2 appears $n - 2$ times, and so forth, so we can rewrite this sum as $m(n) = 1 + \frac{1}{n} \sum_{i=1}^{n-1} \frac{n-i}{i} = 1 + \frac{1}{n} \sum_{i=1}^{n-1} \left(\frac{n}{i} - 1 \right) = 1 + \frac{1}{n} \left(n \sum_{i=1}^{n-1} \frac{1}{i} - (n-1) \right) = \sum_{i=1}^{n-1} \frac{1}{i} + 1 - \frac{n-1}{n} = \sum_{i=1}^{n} \frac{1}{i}$, and we are done by induction (noting that the statement is clearly true for $n = 0$, $n = 1$). Thus the maximum overhang is $\sum_{k=2}^{n} \frac{1}{k}$.

b. For an infinite number of dominos, because the overhang is the harmonic series, the distance is potentially infinite.

9.4.73

a. Dividing both sides of the recurrence equation by f_n gives $\frac{f_{n+1}}{f_n} = 1 + \frac{f_{n-1}}{f_n}$. Let the limit of the ratio of successive terms be L. Taking the limit of the previous equation gives $L = 1 + \frac{1}{L}$. Thus $L^2 = L + 1$, so $L^2 - L - 1 = 0$. The quadratic formula gives $L = \frac{1 \pm \sqrt{1 - 4 \cdot (-1)}}{2}$, but we know that all the terms are positive, so we must have $L = \frac{1+\sqrt{5}}{2} = \phi \approx 1.618$.

b. Write the recurrence in the form $f_{n-1} = f_{n+1} - f_n$ and divide both sides by f_{n+1}. Then we have $\frac{f_{n-1}}{f_{n+1}} = 1 - \frac{f_n}{f_{n+1}}$. Taking the limit gives $1 - \frac{1}{\phi}$ on the right-hand side.

c. Consider the harmonic series with the given groupings, and compare it with the sum of $\frac{f_{k-1}}{f_{k+1}}$ as shown. The first three terms match exactly. The sum of the next two are $\frac{1}{4} + \frac{1}{5} > \frac{1}{5} + \frac{1}{5} = \frac{2}{5}$. The sum of the next three are $\frac{1}{6} + \frac{1}{7} + \frac{1}{8} > \frac{1}{8} + \frac{1}{8} + \frac{1}{8} = \frac{3}{8}$. The sum of the next five are $\frac{1}{9} + \cdots + \frac{1}{13} > 5 \cdot \frac{1}{13} = \frac{5}{13}$. Thus the harmonic series is bounded below by the series $\sum_{k=1}^{\infty} \frac{f_{k-1}}{f_{k+1}}$.

d. The result above implies that the harmonic series diverges, because the series $\sum_{k=1}^{\infty} \frac{f_{k-1}}{f_{k+1}}$ diverges, since its general term has limit $1 - \frac{1}{\phi} \neq 0$.

9.5   The Ratio, Root, and Comparison Tests

9.5.1 Given a series $\sum a_k$ of positive terms, compute $\lim_{k \to \infty} \frac{a_{k+1}}{a_k}$ and call it r. If $0 \leq r < 1$, the given series converges. If $r > 1$ (including $r = \infty$), the given series diverges. If $r = 1$, the test is inconclusive.

9.5.3 Given a series of positive terms $\sum a_k$ that you suspect converges, find a series $\sum b_k$ that you know converges, for which $\lim_{k \to \infty} \frac{a_k}{b_k} = L$ where $L \geq 0$ is a finite number. If you are successful, you will have shown that the series $\sum a_k$ converges.

Given a series of positive terms $\sum a_k$ that you suspect diverges, find a series $\sum b_k$ that you know diverges, for which $\lim_{k \to \infty} \frac{a_k}{b_k} = L$ where $L > 0$ (including the case $L = \infty$). If you are successful, you will have shown that $\sum a_k$ diverges.

9.5.5 The Ratio Test.

9.5.7 The difference between successive partial sums is a term in the sequence. Because the terms are positive, differences between successive partial sums are as well, so the sequence of partial sums is increasing.

9.5.9 The ratio between successive terms is $\frac{a_{k+1}}{a_k} = \frac{1}{(k+1)!} \cdot \frac{(k)!}{1} = \frac{1}{k+1}$, which goes to zero as $k \to \infty$, so the given series converges by the Ratio Test.

9.5.11 The ratio between successive terms is $\frac{a_{k+1}}{a_k} = \frac{(k+1)^2}{4(k+1)} \cdot \frac{4^k}{(k)^2} = \frac{1}{4}\left(\frac{k+1}{k}\right)^2$. The limit is $1/4$ as $k \to \infty$, so the given series converges by the Ratio Test.

9.5.13 The ratio between successive terms is $\frac{a_{k+1}}{a_k} = \frac{(k+1)e^{-(k+1)}}{(k)e^{-(k)}} = \frac{k+1}{(k)e}$. The limit of this ratio as $k \to \infty$ is $1/e < 1$, so the given series converges by the Ratio Test.

9.5.15 The ratio between successive terms is $\frac{2^{k+1}}{(k+1)^{99}} \cdot \frac{(k)^{99}}{2^k} = 2\left(\frac{k}{k+1}\right)^{99}$; the limit as $k \to \infty$ is 2, so the given series diverges by the Ratio Test.

9.5.17 The ratio between successive terms is $\frac{((k+1)!)^2}{(2(k+1))!} \cdot \frac{(2k)!}{((k)!)^2} = \frac{(k+1)^2}{(2k+2)(2k+1)}$; the limit as $k \to \infty$ is $1/4$, so the given series converges by the Ratio Test.

9.5.19 The kth root of the kth term is $\frac{10k^3+3}{9k^3+k+1}$. The limit of this as $k \to \infty$ is $\frac{10}{9} > 1$, so the given series diverges by the Root Test.

9.5.21 The kth root of the kth term is $\frac{k^{2/k}}{2}$. The limit of this as $k \to \infty$ is $\frac{1}{2} < 1$, so the given series converges by the Root Test.

9.5.23 The kth root of the kth term is $\left(\frac{k}{k+1}\right)^{2k}$. The limit of this as $k \to \infty$ is $e^{-2} < 1$, so the given series converges by the Root Test.

9.5.25 The kth root of the kth term is $\left(\frac{1}{k^k}\right)$. The limit of this as $k \to \infty$ is 0, so the given series converges by the Root Test.

9.5.27 $\frac{1}{k^2+4} < \frac{1}{k^2}$, and $\sum_{k=1}^{\infty} \frac{1}{k^2}$ converges, so $\sum_{k=1}^{\infty} \frac{1}{k^2+4}$ converges as well, by the Comparison Test.

9.5.29 Use the Limit Comparison Test with $\left\{\frac{1}{k}\right\}$. The ratio of the terms of the two series is $\frac{k^3-k}{k^3+4}$ which has limit 1 as $k \to \infty$. Because the comparison series diverges, the given series does as well.

9.5.31 For all k, $\frac{1}{k^{3/2}+1} < \frac{1}{k^{3/2}}$. The series whose terms are $\frac{1}{k^{3/2}}$ is a p-series which converges, so the given series converges as well by the Comparison Test.

9.5.33 $\sin(1/k) > 0$ for $k \geq 1$, so we can apply the Comparison Test with $1/k^2$. $\sin(1/k) < 1$, so $\frac{\sin(1/k)}{k^2} < \frac{1}{k^2}$. Because the comparison series converges, the given series converges as well.

9.5.35 Use the Limit Comparison Test with $\{1/k\}$. The ratio of the terms of the two series is $\frac{k}{2k-\sqrt{k}} = \frac{1}{2-1/\sqrt{k}}$, which has limit $1/2$ as $k \to \infty$. Because the comparison series diverges, the given series does as well.

9.5.37 Use the Limit Comparison Test with $\frac{k^{2/3}}{k^{3/2}}$. The ratio of corresponding terms of the two series is $\frac{\sqrt[3]{k^2+1}}{\sqrt{k^3+1}} \cdot \frac{k^{3/2}}{k^{2/3}} = \frac{\sqrt[3]{k^2+1}}{\sqrt[3]{k^2}} \cdot \frac{\sqrt{k^3}}{\sqrt{k^3+1}}$, which has limit 1 as $k \to \infty$. The comparison series is the series whose terms are $k^{2/3-3/2} = k^{-5/6}$, which is a p-series with $p < 1$, so it, and the given series, both diverge.

9.5.39

    a. False. For example, let $\{a_k\}$ be all zeros, and $\{b_k\}$ be all 1's.

    b. True. This is a result of the Comparison Test.

    c. True. Both of these statements follow from the Comparison Test.

    d. True. The limit of the ratio is always 1 in the case, so the test is inconclusive.

9.5.41 Use the Divergence Test: $\lim_{k\to\infty} a_k = \lim_{k\to\infty} \left(1 + \frac{2}{k}\right)^k = e^2 \neq 0$, so the given series diverges.

9.5.43 Use the Ratio Test: the ratio of successive terms is $\frac{(k+1)^{100}}{(k+2)!} \cdot \frac{(k+1)!}{k^{100}} = \left(\frac{k+1}{k}\right)^{100} \cdot \frac{1}{k+2}$. This has limit $1^{100} \cdot 0 = 0$ as $k \to \infty$, so the given series converges by the Ratio Test.

9.5.45 Use the Root Test. The kth root of the kth term is $(k^{1/k} - 1)^2$, which has limit 0 as $k \to \infty$, so the given series converges by the Root Test.

9.5.47 Use the Divergence Test: $\lim_{k\to\infty} \frac{k^2+2k+1}{3k^2+1} = \frac{1}{3} \neq 0$, so the given series diverges.

9.5.49 Use the Limit Comparison Test with the harmonic series. Note that $\lim_{k\to\infty} \frac{\frac{1}{\ln k}}{\frac{1}{k}} = \lim_{k\to\infty} \frac{k}{\ln k} = \infty$, and because the harmonic series diverges, the given series does as well.

9.5.51 Use the Limit Comparison Test with the series whose kth term is $\frac{1}{k^{3/2}}$. Note that $\lim_{k\to\infty} \frac{\frac{1}{\sqrt{k^3-k+1}}}{\frac{\sqrt{k^3}}{1}} = \lim_{k\to\infty} \sqrt{\frac{k^3}{k^3-k+1}} = \sqrt{1} = 1$, and the series $\sum_{k=1}^{\infty} \frac{1}{k^{3/2}}$ converges because it is a p-series with $p = \frac{3}{2}$. Thus, the given series also converges.

9.5.53 Use the Comparison Test. Each term $\frac{1}{k} + 2^{-k} > \frac{1}{k}$. Because the harmonic series diverges, so does this series.

9.5.55 Use the Ratio Test. $\frac{a_{k+1}}{a_k} = \frac{2^{k+1}(k+1)!}{(k+1)^{k+1}} \cdot \frac{(k)^k}{2^k(k)!} = 2\left(\frac{k}{k+1}\right)^k$, which has limit $\frac{2}{e}$ as $k \to \infty$, so the given series converges.

9.5.57 Use the Limit Comparison Test with $\{1/k^3\}$. The ratio of corresponding terms is $\frac{k^{11}}{k^{11}+3}$, which has limit 1 as $k \to \infty$. Because the comparison series converges, so does the given series.

9.5.59 This is a p-series with exponent greater than 1, so it converges.

9.5.61 $\ln\left(\frac{k+2}{k+1}\right) = \ln(k+2) - \ln(k+1)$, so this series telescopes. We get $\sum_{k=1}^{n} \ln\left(\frac{k+2}{k+1}\right) = \ln(n+2) - \ln 2$. Because $\lim_{n\to\infty} \ln(n+2) - \ln 2 = \infty$, the sequence of partial sums diverges, so the given series is divergent.

9.5.63 For $k > 7$, $\ln k > 2$ so note that $\frac{1}{k^{\ln k}} < \frac{1}{k^2}$. Because $\sum_{k=1}^{\infty} \frac{1}{k^2}$ converges, the given series converges as well.

9.5.65 Use the Limit Comparison Test with the harmonic series. $\frac{\tan(1/k)}{1/k}$ has limit 1 as $k \to \infty$ because $\lim_{x\to 0} \frac{\tan x}{x} = 1$. Thus the original series diverges.

9.5.67 Note that $\frac{1}{(2k+1)\cdot(2k+3)} = \frac{1}{2}\left(\frac{1}{2k+1} - \frac{1}{2k+3}\right)$. Thus this series telescopes.

$$\sum_{k=0}^{n} \frac{1}{(2k+1)(2k+3)} = \frac{1}{2}\sum_{k=0}^{n}\left(\frac{1}{2k+1} - \frac{1}{2k+3}\right) = \frac{1}{2}\left(-\frac{1}{2n+3} + 1\right),$$

so the given series converges to $1/2$, because that is the limit of the sequence of partial sums.

9.5.69 This series is $\sum_{k=1}^{\infty} \frac{k^2}{k!}$. By the Ratio Test, $\frac{a_{k+1}}{a_k} = \frac{(k+1)^2}{(k+1)!} \cdot \frac{k!}{k^2} = \frac{1}{k+1}\left(\frac{k+1}{k}\right)^2$, which has limit 0 as $k \to \infty$, so the given series converges.

9.5.71 For $p \leq 1$ and $k > e$, $\frac{\ln k}{k^p} > \frac{1}{k^p}$. The series $\sum_{k=1}^{\infty} \frac{1}{k^p}$ diverges, so the given series diverges. For $p > 1$, let $q < p - 1$; then for sufficiently large k, $\ln k < k^q$, so that by the Comparison Test, $\frac{\ln k}{k^p} < \frac{k^q}{k^p} = \frac{1}{k^{p-q}}$. But $p - q > 1$, so that $\sum_{k=1}^{\infty} \frac{1}{k^{p-q}}$ is a convergent p-series. Thus the original series is convergent precisely when $p > 1$.

9.5.73 For $p \leq 1$, $\frac{(\ln k)^p}{k^p} > \frac{1}{k^p}$ for $k \geq 3$, and $\sum_{k=1}^{\infty} \frac{1}{k^p}$ diverges for $p \leq 1$, so the original series diverges. For $p > 1$, let $q < p - 1$; then for sufficiently large k, $(\ln k)^p < k^q$. Note that $\frac{(\ln k)^p}{k^p} < \frac{k^q}{k^p} = \frac{1}{k^{p-q}}$. But $p - q > 1$, so $\sum_{k=1}^{\infty} \frac{1}{k^{p-q}}$ converges, so the given series converges. Thus, the given series converges exactly for $p > 1$.

9.5.75 Use the Ratio Test:

$$\lim_{k \to \infty} \frac{a_{k+1}}{a_k} = \lim_{k \to \infty} \frac{(k+1)p^{k+1}}{k+2} \cdot \frac{k+1}{kp^k} = p,$$

so the given series converges for $p < 1$ and diverges for $p > 1$. For $p = 1$ the given series diverges by limit comparison with the harmonic series.

9.5.77 $\lim\limits_{k \to \infty} a_k = \lim\limits_{k \to \infty} \left(1 - \frac{p}{k}\right)^k = e^{-p} \neq 0$, so this sequence diverges for all p by the Divergence Test.

9.5.79 These tests apply only for series with positive terms, so assume $r > 0$. Clearly the series do not converge for $r = 1$, so we assume $r \neq 1$ in what follows. Using the Integral Test, $\sum r^k$ converges if and only if $\int_1^{\infty} r^x dx$ converges. This improper integral has value $\lim\limits_{b \to \infty} \frac{r^x}{\ln r}\Big|_1^b$, which converges only when $\lim\limits_{b \to \infty} r^b$ exists, which occurs only for $r < 1$. Using the Ratio Test, $\frac{a_{k+1}}{a_k} = \frac{r^{k+1}}{r^k} = r$, so by the Ratio Test, the series converges if and only if $r < 1$. Using the Root Test, $\lim\limits_{k \to \infty} \sqrt[k]{a_k} = \lim\limits_{k \to \infty} \sqrt[k]{r^k} = \lim\limits_{k \to \infty} r = r$, so again we have convergence if and only if $r < 1$. By the Divergence Test, we know that a geometric series diverges if $|r| \geq 1$.

9.5.81 To prove case (2), assume $L = 0$ and that $\sum b_k$ converges. Because $L = 0$, for every $\varepsilon > 0$, there is some N such that for all $n > N$, $\left|\frac{a_k}{b_k}\right| < \varepsilon$. Take $\varepsilon = 1$; this then says that there is some N such that for all $n > N$, $0 < a_k < b_k$. By the Comparison Test, because $\sum b_k$ converges, so does $\sum a_k$. To prove case (3), because $L = \infty$, then $\lim\limits_{k \to \infty} \frac{b_k}{a_k} = 0$, so by the argument above, we have $0 < b_k < a_k$ for sufficient large k. But $\sum b_k$ diverges, so by the Comparison Test, $\sum a_k$ does as well.

9.5.83 The series clearly converges for $x = 0$. For $x \neq 0$, we have $\frac{a_{k+1}}{a_k} = \frac{x^{k+1}}{x^k} = x$. This has limit x as $k \to \infty$, so the series converges for $x < 1$. It clearly does not converge for $x = 1$. So the series converges for $x \in [0, 1)$.

9.5.85 The series clearly converges for $x = 0$. For $x \neq 0$, we have $\frac{a_{k+1}}{a_k} = \frac{x^{k+1}}{(k+1)^2} \cdot \frac{k^2}{x^k} = x\left(\frac{k}{k+1}\right)^2$, which has limit x as $k \to \infty$. Thus the series converges for $x < 1$. When $x = 1$, the series is $\frac{1}{k^2}$, which converges. Thus the original series converges for $0 \leq x \leq 1$.

9.5.87 The series clearly converges for $x = 0$. For $x \neq 0$, we have $\frac{a_{k+1}}{a_k} = \frac{x^{k+1}}{2^{k+1}} \cdot \frac{2^k}{x^k} = \frac{x}{2}$, which has limit $x/2$ as $k \to \infty$. Thus the series converges for $0 \leq x < 2$. For $x = 2$, it is obviously divergent.

9.5.89

a. $\ln \prod_{k=0}^{\infty} e^{1/2^k} = \sum_{k=0}^{\infty} \frac{1}{2^k} = 2$, so that the original product converges to e^2.

b. $\ln \prod_{k=2}^{\infty} \left(1 - \frac{1}{k}\right) = \ln \prod_{k=2}^{\infty} \frac{k-1}{k} = \sum_{k=2}^{\infty} \ln \frac{k-1}{k} = \sum_{k=2}^{\infty}(\ln(k-1) - \ln(k))$. This series telescopes to give $S_n = -\ln(n)$, so the original series has limit $\lim\limits_{n \to \infty} P_n = \lim\limits_{n \to \infty} e^{-\ln(n)} = 0$.

9.5.91

a. Use the Ratio Test:

$$\frac{a_{k+1}}{a_k} = \frac{1 \cdot 3 \cdot 5 \cdots (2k+1)}{p^{k+1}(k+1)!} \cdot \frac{p^k(k)!}{1 \cdot 3 \cdot 5 \cdots (2k-1)} = \frac{(2k+1)}{(k+1)p}$$

and this expression has limit $\frac{2}{p}$ as $k \to \infty$. Thus the series converges for $p > 2$.

b. Following the hint, when $p = 2$ we have $\sum_{k=1}^{\infty} \dfrac{(2k)!}{2^k k!(2 \cdot 4 \cdot 6 \cdots 2k)} = \sum_{k=1}^{\infty} \dfrac{(2k)!}{(2^k)^2(k!)^2}$. Using Stirling's formula, the numerator is asymptotic to $2\sqrt{\pi}\sqrt{k}(2k)^{2k}e^{-2k} = 2\sqrt{\pi}\sqrt{k}(2^k)^2(k^k)^2 e^{-2k}$ while the denominator is asymptotic to $(2^k)^2 2\pi k(k^k)^2 e^{-2k}$, so the quotient is asymptotic to $\dfrac{1}{\sqrt{\pi}\sqrt{k}}$. Thus the original series diverges for $p = 2$ by the Limit Comparison Test with the divergent p-series $\sum_{k=1}^{\infty} \dfrac{1}{k^{1/2}}$.

9.6 Alternating Series

9.6.1 Because $S_{n+1} - S_n = (-1)^n a_{n+1}$ alternates signs.

9.6.3 We have

$$S = S_{2n+1} + (a_{2n} - a_{2n+1}) + (a_{2n+2} - a_{2n+3}) + \cdots$$

and each term of the form $a_{2k} - a_{2k+1} > 0$, so that $S_{2n+1} < S$. Also

$$S = S_{2n} + (-a_{2n+1} + a_{2n+2}) + (-a_{2n+3} + a_{2n+4}) + \cdots$$

and each term of the form $-a_{2k+1} + a_{2k+2} < 0$, so that $S < S_{2n}$. Thus the sum of the series is trapped between the odd partial sums and the even partial sums.

9.6.5 The remainder is less than the first neglected term because

$$S - S_n = (-1)^{n+1}(a_{n+1} + (-a_{n+2} + a_{n+3}) + \cdots)$$

so that the sum of the series *after* the first disregarded term has the opposite sign from the first disregarded term.

9.6.7 No. If the terms are positive, then the absolute value of each term is the term itself, so convergence and absolute convergence would mean the same thing in this context.

9.6.9 Yes. For example, $\sum \dfrac{(-1)^k}{k^3}$ converges absolutely and thus not conditionally (see the definition).

9.6.11 The terms of the series decrease in magnitude, and $\lim_{k \to \infty} \dfrac{1}{2k+1} = 0$, so the given series converges.

9.6.13 $\lim_{k \to \infty} \dfrac{k}{3k+2} = \dfrac{1}{3} \neq 0$, so the given series diverges.

9.6.15 The terms of the series decrease in magnitude, and $\lim_{k \to \infty} \dfrac{1}{k^3} = 0$, so the given series converges.

9.6.17 The terms of the series decrease in magnitude, and $\lim_{k \to \infty} \dfrac{k^2}{k^3+1} = \lim_{k \to \infty} \dfrac{1/k}{1+1/k^3} = 0$, so the given series converges.

9.6.19 $\lim_{k \to \infty} \dfrac{k^2-1}{k^2+3} = 1$, so the terms of the series do not tend to zero and thus the given series diverges.

9.6.21 $\lim_{k \to \infty} \left(1 + \dfrac{1}{k}\right) = 1$, so the given series diverges.

9.6.23 The derivative of $f(k) = \dfrac{k^{10}+2k^5+1}{k(k^{10}+1)}$ is $f'(k) = \dfrac{-(k^{20}+2k^{10}+12k^{15}-8k^5+1)}{k^2(k^{10}+1)^2}$. The numerator is negative for large enough values of k, and the denominator is always positive, so the derivative is negative for large enough k. Also, $\lim_{k \to \infty} \dfrac{k^{10}+2k^5+1}{k(k^{10}+1)} = \lim_{k \to \infty} \dfrac{1+2k^{-5}+k^{-10}}{k+k^{-9}} = 0$. Thus the given series converges.

9.6.25 $\lim_{k \to \infty} k^{1/k} = 1$ (for example, take logs and apply L'Hôpital's rule), so the given series diverges by the Divergence Test.

9.6.27 $\dfrac{1}{\sqrt{k^2+4}}$ is decreasing and tends to zero as $k \to \infty$, so the given series converges.

9.6.29 We want $\frac{1}{n+1} < 10^{-4}$, or $n+1 > 10^4$, so $n = 10^4$.

9.6.31 The series starts with $k = 0$, so we want $\frac{1}{2n+1} < 10^{-4}$, or $2n + 1 > 10^4$, $n = 5000$.

9.6.33 We want $\frac{1}{(n+1)^4} < 10^{-4}$, or $(n+1)^4 > 10^4$, so $n = 10$.

9.6.35 The series starts with $k = 0$, so we want $\frac{1}{3n+1} < 10^{-4}$, or $3n + 1 > 10^4$, $n = 3334$.

9.6.37 The series starts with $k = 0$, so we want $\frac{1}{4^n}\left(\frac{2}{4n+1} + \frac{2}{4n+2} + \frac{1}{4n+3}\right) < 10^{-4}$, or $\frac{4^n(4n+1)(4n+2)(4n+3)}{4(20n^2+21n+5)} > 10000$, which occurs first for $n = 6$.

9.6.39 To figure out how many terms we need to sum, we must find n such that $\frac{1}{(n+1)^5} < 10^{-3}$, so that $(n+1)^5 > 1000$; this occurs first for $n = 3$. Thus $\frac{-1}{1} + \frac{1}{2^5} - \frac{1}{3^5} \approx -0.973$.

9.6.41 To figure out how many terms we need to sum, we must find n so that $\frac{n+1}{(n+1)^2+1} < 10^{-3}$, so that $\frac{(n+1)^2+1}{n+1} = n + 1 + \frac{1}{n+1} > 1000$. This occurs first for $n = 999$. We have $\sum_{k=1}^{999} \frac{(-1)^k k}{k^2+1} \approx -0.269$.

9.6.43 To figure how many terms we need to sum, we must find n such that $\frac{1}{(n+1)^{n+1}} < 10^{-3}$, or $(n+1)^{n+1} > 1000$, so $n = 4$ ($5^5 = 3125$). Thus the approximation is $\sum_{k=1}^{4} \frac{(-1)^n}{n^n} \approx -.783$.

9.6.45 The series of absolute values is a p-series with $p = 2/3$, so it diverges. The given alternating series does converge, though, by the Alternating Series Test. Thus, the given series is conditionally convergent.

9.6.47 The series of absolute values is a p-series with $p = 3/2$, so it converges absolutely.

9.6.49 The series of absolute values is $\sum \frac{|\cos(k)|}{k^3}$, which converges by the Comparison Test because $\frac{|\cos(k)|}{k^3} \leq \frac{1}{k^3}$. Thus the series converges absolutely.

9.6.51 The absolute value of the kth term of this series has limit $\pi/2$ as $k \to \infty$, so the given series is divergent by the Divergence Test.

9.6.53 The series of absolute values is $\sum \frac{k}{2k+1}$, but $\lim_{k\to\infty} \frac{k}{2k+1} = \frac{1}{2}$, so by the Divergence Test, this series diverges. The original series does not converge conditionally, either, because $\lim_{k\to\infty} a_k = \frac{1}{2} \neq 0$.

9.6.55 The series of absolute values is $\sum \frac{\tan^{-1}(k)}{k^3}$, which converges by the Comparison Test because $\frac{\tan^{-1}(k)}{k^3} < \frac{\pi}{2}\frac{1}{k^3}$, and $\sum \frac{\pi}{2}\frac{1}{k^3}$ converges because it is a constant multiple of a convergent p-series. So the original series converges absolutely.

9.6.57

  a. False. For example, consider the alternating harmonic series.

  b. True. This is part of Theorem 8.21.

  c. True. This statement is simply saying that a convergent series converges.

  d. True. This is part of Theorem 8.21.

  e. False. Let $a_k = \frac{1}{k}$.

  f. True. Use the Comparison Test: $\lim_{k\to\infty} \frac{a_k^2}{a_k} = \lim_{k\to\infty} a_k = 0$ because $\sum a_k$ converges, so $\sum a_k^2$ and $\sum a_k$ converge or diverge together. Because the latter converges, so does the former.

  g. True, by definition. If $\sum |a_k|$ converged, the original series would converge absolutely, not conditionally.

9.6.59 $\sum_{k=1}^{\infty} \frac{1}{k^2} - \sum_{k=1}^{\infty} \frac{(-1)^{k+1}}{k^2} = 2\sum_{k=1}^{\infty} \frac{1}{(2k)^2} = 2 \cdot \frac{1}{4} \sum_{k=1}^{\infty} \frac{1}{k^2}$, and thus $\sum_{k=1}^{\infty} \frac{(-1)^{k+1}}{k^2} = \frac{\pi^2}{6} - \frac{1}{2} \cdot \frac{\pi^2}{6} = \frac{\pi^2}{12}$.

9.6.61 Write $r = -s$; then $0 < s < 1$ and $\sum r^k = \sum (-1)^k s^k$. Because $|s| < 1$, the terms s^k are nonincreasing and tend to zero, so by the Alternating Series Test, the series $\sum (-1)^k s^k = \sum r^k$ converges.

9.6.63 Let $S = 1 - \frac{1}{2} + \frac{1}{3} - \cdots$. Then

$$
\begin{aligned}
S &= \left(1 - \tfrac{1}{2}\right) + \left(\tfrac{1}{3} - \tfrac{1}{4}\right) + \left(\tfrac{1}{5} - \tfrac{1}{6}\right) + \left(\tfrac{1}{7} - \tfrac{1}{8}\right) + \cdots \\
\tfrac{1}{2}S &= \qquad \tfrac{1}{2} \qquad - \qquad \tfrac{1}{4} \qquad + \qquad \tfrac{1}{6} \qquad - \qquad \tfrac{1}{8} \qquad + \cdots
\end{aligned}
$$

Add these two series together to get

$$\frac{3}{2}S = \frac{3}{2}\ln 2 = 1 + \frac{1}{3} - \frac{1}{2} + \frac{1}{5} + \cdots$$

To see that the results are as desired, consider a collection of four terms:

$$\cdots + \left(\frac{1}{4k+1} - \frac{1}{4k+2}\right) + \left(\frac{1}{4k+3} - \frac{1}{4k+4}\right) + \cdots$$

$$\cdots + \frac{1}{4k+2} - \frac{1}{4k+4} + \cdots$$

Adding these results in the desired sign pattern. This repeats for each group of four elements.

9.6.65 Both series diverge, so comparisons of their values are not meaningful.

Chapter Nine Review

1

    a. False. Let $a_n = 1 - \frac{1}{n}$. This sequence has limit 1.

    b. False. The terms of a sequence tending to zero is necessary but not sufficient for convergence of the series.

    c. True. This is the definition of convergence of a series.

    d. False. If a series converges absolutely, the definition says that it does not converge conditionally.

    e. True. It has limit 1 as $n \to \infty$.

    f. False. The subsequence of the even terms has limit 1 and the subsequence of odd terms has limit -1, so the sequence does not have a limit.

    g. False. It diverges by the Divergence Test because $\lim_{k \to \infty} \frac{k^2}{k^2+1} = 1 \neq 0$.

    h. True. The given series converges by the Limit Comparison Test with the series $\sum_{k=1}^{\infty} \frac{1}{k^2}$, and thus its sequence of partial sums converges.

3 $\lim_{n \to \infty} \frac{8^n}{n!} = 0$ because exponentials grow more slowly than factorials.

5 Take logs and compute $\lim_{n \to \infty} (1/n)\ln n = \lim_{n \to \infty} (\ln n)/n = \lim_{n \to \infty} \frac{1}{n} = 0$ by L'Hôpital's rule. Thus the original limit is $e^0 = 1$.

7 Take logs, and then evaluate $\lim_{n \to \infty} \frac{1}{\ln n}\ln(1/n) = \lim_{n \to \infty} (-1) = -1$, so the original limit is e^{-1}.

9 $a_n = (-1/0.9)^n = (-10/9)^n$. The terms grow without bound so the sequence does not converge.

11

    a. $S_1 = \frac{1}{3}$, $S_2 = \frac{11}{24}$, $S_3 = \frac{21}{40}$, $S_4 = \frac{17}{30}$.

    b. $S_n = \frac{1}{2}\left(\frac{1}{1} + \frac{1}{2} - \frac{1}{n+1} - \frac{1}{n+2}\right)$, because the series telescopes.

    c. From part (b), $\lim\limits_{n\to\infty} S_n = \frac{3}{4}$, which is the sum of the series.

13 $\sum_{k=1}^{\infty} 3(1.001)^k = 3\sum_{k=1}^{\infty}(1.001)^k$. This is a geometric series with ratio greater than 1, so it diverges.

15 $\frac{1}{k(k+1)} = \frac{1}{k} - \frac{1}{k+1}$, so the series telescopes, and $S_n = 1 - \frac{1}{n+1}$. Thus $\lim\limits_{n\to\infty} S_n = 1$, which is the value of the series.

17 This series telescopes. $S_n = 3 - \frac{3}{3n+1}$, so that $\lim\limits_{n\to\infty} S_n = 3$, which is the value of the series.

19 $\displaystyle\sum_{k=1}^{\infty} \frac{2^k}{3^{k+2}} = \frac{1}{9}\sum_{k=1}^{\infty}\left(\frac{2}{3}\right)^k = \frac{1}{9} \cdot \frac{2/3}{1 - 2/3} = \frac{2}{9}$.

21

    a. It appears that the series converges, because the sequence of partial sums appears to converge to 1.5.

    b. The convergence is uncertain.

    c. This series clearly appears to diverge, because the partial sums seem to be growing without bound.

23 The series can be written $\sum \frac{1}{k^{2/3}}$, which is a p-series with $p = 2/3 < 1$, so this series diverges.

25 This is a geometric series with ratio $2/e < 1$, so the series converges.

27 Applying the Ratio Test:

$$\lim_{k\to\infty} \frac{a_{k+1}}{a_k} = \lim_{k\to\infty} \frac{2^{k+1}(k+1)!}{(k+1)^{k+1}} \cdot \frac{k^k}{2^k k!} = \lim_{k\to\infty} 2\left(\frac{k}{k+1}\right)^k = \frac{2}{e} < 1,$$

so the given series converges.

29 Use the Comparison Test: $\frac{3}{2+e^k} < \frac{3}{e^k}$, but $\sum \frac{3}{e^k}$ converges because it is a geometric series with ratio $\frac{1}{e} < 1$. Thus the original series converges as well.

31 $a_k = \frac{k^{1/k}}{k^3} = \frac{1}{k^{3-1/k}}$. For $k \geq 2$, then, $a_k < \frac{1}{k^2}$. Because $\sum \frac{1}{k^2}$ converges, the given series also converges, by the Comparison Test.

33 Use the Ratio Test: $\frac{a_{k+1}}{a_k} = \frac{(k+1)^5}{e^{k+1}} \cdot \frac{e^k}{k^5} = \frac{1}{e} \cdot \left(\frac{k+1}{k}\right)^5$, which has limit $1/e < 1$ as $k \to \infty$. Thus the given series converges.

35 Use the Comparison Test. Because $\lim\limits_{k\to\infty} \frac{\ln k}{k^{1/2}} = 0$, we have that for sufficiently large k, $\ln k < k^{1/2}$, so that $a_k = \frac{2\ln k}{k^2} < \frac{2k^{1/2}}{k^2} = \frac{2}{k^{3/2}}$. Now $\sum \frac{2}{k^{3/2}}$ is convergent, because it is a p-series with $p = 3/2 > 1$. Thus the original series is convergent.

37 Use the Ratio Test. The ratio of successive terms is $\frac{2 \cdot 4^{k+1}}{(2k+3)!} \cdot \frac{(2k+1)!}{2 \cdot 4^k} = \frac{4}{(2k+3)(2k+2)}$. This has limit 0 as $k \to \infty$, so the given series converges.

39 Use the Limit Comparison Test with the harmonic series. Note that $\lim\limits_{k\to\infty} \dfrac{\coth k}{k}\cdot\dfrac{k}{1}=\lim\limits_{k\to\infty}\coth k = 1$. Because the harmonic series diverges, the given series does as well.

41 Use the Divergence Test. $\lim_{k\to\infty}\tanh k = \lim_{k\to\infty}\dfrac{e^k+e^{-k}}{e^k-e^{-k}}=1\neq 0$, so the given series diverges.

43 $|a_k|=\dfrac{1}{k^2-1}$. Use the Limit Comparison Test with the convergent series $\sum\dfrac{1}{k^2}$. Because $\lim\limits_{k\to\infty}\dfrac{\frac{1}{k^2-1}}{\frac{1}{k^2}}=\lim\limits_{k\to\infty}\dfrac{k^2}{k^2-1}=1$, the given series converges absolutely.

45 Use the Ratio Test on the absolute values of the sequence of terms: $\lim\limits_{k\to\infty}\left|\dfrac{a_{k+1}}{a_k}\right|=\lim\limits_{k\to\infty}\dfrac{k+1}{e^{k+1}}\cdot\dfrac{e^k}{k}=\lim\limits_{k\to\infty}\dfrac{1}{e}\cdot\dfrac{k+1}{k}=\dfrac{1}{e}<1$. Thus, the original series is absolutely convergent.

47 Use the Ratio Test on the absolute values of the sequence of terms: $\lim\limits_{k\to\infty}\left|\dfrac{a_{k+1}}{a_k}\right|=\lim\limits_{k\to\infty}\dfrac{10}{k+1}=0$, so the series converges absolutely.

49 Because $k^2\ll 2^k$, $\lim_{k\to\infty}\dfrac{-2\cdot(-2)^k}{k^2}\neq 0$. The given series thus diverges by the Divergence Test.

51

    a. For $|x|<1$, $\lim\limits_{k\to\infty}x^k=0$, so this limit is zero.

    b. This is a geometric series with ratio $-4/5$, so the sum is $\dfrac{1}{1+4/5}=\dfrac{5}{9}$.

53 Consider the constant sequence with $a_k=1$ for all k. The sequence $\{a_k\}$ converges to 1, but the corresponding series $\sum a_k$ diverges by the divergence test.

55

    a. This sequence converges because $\lim_{k\to\infty}\dfrac{k}{k+1}=\lim_{k\to\infty}\dfrac{1}{1+\frac{1}{k}}=\dfrac{1}{1+0}=1$.

    b. Because the sequence of terms has limit 1 (which means its limit isn't zero) this series diverges by the divergence test.

57 Because the series converges, we must have $\lim\limits_{k\to\infty}a_k=0$. Because it converges to 8, the partial sums converge to 8, so that $\lim\limits_{k\to\infty}S_k=8$.

59 The series converges absolutely for $p>1$, conditionally for $0<p\leq 1$ in which case $\{k^{-p}\}$ is decreasing to zero.

61 The sum is 0.2500000000 to ten decimal places. The maximum error is

$$\int_{20}^{\infty}\frac{1}{5^x}\,dx=\lim_{b\to\infty}\left(-\frac{1}{5^x\ln 5}\Big|_{20}^{b}\right)=\frac{1}{5^{20}\ln 5}\approx 6.5\times 10^{-15}.$$

63 The maximum error is a_{n+1}, so we want $a_{n+1}=\dfrac{1}{(k+1)^4}<10^{-8}$, or $(k+1)^4>10^8$, so $k=100$.

65

    a. Let T_n be the amount of additional tunnel dug during week n. Then $T_0=100$ and $T_n=.95\cdot T_{n-1}=(.95)^nT_0=100(0.95)^n$, so the total distance dug in N weeks is

$$S_N=100\sum_{k=0}^{N-1}(0.95)^k=100\left(\frac{1-(0.95)^N}{1-0.95}\right)=2000(1-0.95^N).$$

    Then $S_{10}\approx 802.5$ meters and $S_{20}\approx 1283.03$ meters.

b. The longest possible tunnel is $S_\infty = 100 \sum_{k=0}^{\infty} (0.95)^k = \frac{100}{1-.95} = 2000$ meters.

67

a. The area of a circle of radius r is πr^2. For $r = 2^{1-n}$, this is $2^{2-2n}\pi$. There are 2^{n-1} circles on the nth page, so the total area of circles on the nth page is $2^{n-1} \cdot \pi 2^{2-2n} = 2^{1-n}\pi$.

b. The sum of the areas on all pages is $\sum_{k=1}^{\infty} 2^{1-k}\pi = 2\pi \sum_{k=1}^{\infty} 2^{-k} = 2\pi \cdot \frac{1/2}{1/2} = 2\pi$.

69

a. $B_n = 1.0025 B_{n-1} + 100$ and $B_0 = 100$.

b. $B_n = 100 \cdot 1.0025^n + 100 \cdot \frac{1 - 1.0025^n}{1 - 1.0025} = 100 \cdot 1.0025^n - 40000(1 - 1.0025^n) = 40000(1.0025^{n+1} - 1)$.

71

a. $T_1 = \frac{\sqrt{3}}{16}$ and $T_2 = \frac{7\sqrt{3}}{64}$.

b. At stage n, 3^{n-1} triangles of side length $1/2^n$ are removed. Each of those triangles has an area of $\frac{\sqrt{3}}{4 \cdot 4^n} = \frac{\sqrt{3}}{4^{n+1}}$, so a total of

$$3^{n-1} \cdot \frac{\sqrt{3}}{4^{n+1}} = \frac{\sqrt{3}}{16} \cdot \left(\frac{3}{4}\right)^{n-1}$$

is removed at each stage. Thus

$$T_n = \frac{\sqrt{3}}{16} \sum_{k=1}^{n} \left(\frac{3}{4}\right)^{k-1} = \frac{\sqrt{3}}{16} \sum_{k=0}^{n-1} \left(\frac{3}{4}\right)^{k} = \frac{\sqrt{3}}{4} \left(1 - \left(\frac{3}{4}\right)^{n}\right).$$

c. $\lim_{n \to \infty} T_n = \frac{\sqrt{3}}{4}$ because $\left(\frac{3}{4}\right)^n \to 0$ as $n \to \infty$.

d. The area of the triangle was originally $\frac{\sqrt{3}}{4}$, so none of the original area is left.

Chapter 10

Power Series

10.1 Approximating Functions With Polynomials

10.1.1 Let the polynomial be $p(x)$. Then $p(0) = f(0)$, $p'(0) = f'(0)$, and $p''(0) = f''(0)$.

10.1.3 The approximations are $p_0(0.1) = 1$, $p_1(0.1) = 1 + \frac{0.1}{2} = 1.05$, and $p_2(0.1) = 1 + \frac{0.1}{2} - \frac{.01}{8} = 1.04875$.

10.1.5 The remainder is the difference between the value of the Taylor polynomial at a point and the true value of the function at that point, $R_n(x) = f(x) - p_n(x)$.

10.1.7

a. Note that $f(1) = 8$, and $f'(x) = 12\sqrt{x}$, so $f'(1) = 12$. Thus, $p_1(x) = 8 + 12(x-1)$.

b. $f''(x) = 6/\sqrt{x}$, so $f''(1) = 6$. Thus $p_2(x) = 8 + 12(x-2) + 3(x-1)^2$.

c. $p_1(1.1) = 12 \cdot 0.1 + 8 = 9.2$. $p_2(1.1) = 3(.1)^2 + 12 \cdot 0.1 + 8 = 9.23$.

10.1.9

a. $f'(x) = -e^{-x}$, so $p_1(x) = f(0) + f'(0)x = 1 - x$.

b. $f''(x) = e^{-x}$, so $p_2(x) = f(0) + f'(0)x + \frac{1}{2}f''(0)x^2 = 1 - x + \frac{1}{2}x^2$.

c. $p_1(0.2) = 0.8$, and $p_2(0.2) = 1 - 0.2 + \frac{1}{2}(0.04) = 0.82$.

10.1.11

a. $f'(x) = -\frac{1}{(x+1)^2}$, so $p_1(x) = f(0) + f'(0)x = 1 - x$.

b. $f''(x) = \frac{2}{(x+1)^3}$, so $p_2(x) = f(0) + f'(0)x + \frac{1}{2}f''(0)x^2 = 1 - x + x^2$.

c. $p_1(0.05) = 0.95$, and $p_2(0.05) = 1 - 0.05 + 0.0025 = 0.953$.

10.1.13

a. $f'(x) = (1/3)x^{-2/3}$, so $p_1(x) = f(8) + f'(8)(x-8) = 2 + \frac{1}{12}(x-8)$.

b. $f''(x) = (-2/9)x^{-5/3}$, so $p_2(x) = f(8) + f'(8)(x-8) + \frac{1}{2}f''(8)(x-8)^2 = 2 + \frac{1}{12}(x-8) - \frac{1}{288}(x-8)^2$.

c. $p_1(7.5) \approx 1.958$, $p_2(7.5) \approx 1.957$.

10.1.15 $f(0) = 1$, $f'(0) = -\sin 0 = 0$, $f''(0) = -\cos 0 = -1$, so that $p_0(x) = 1$, $p_1(x) = 1$, $p_2(x) = 1 - \frac{1}{2}x^2$.

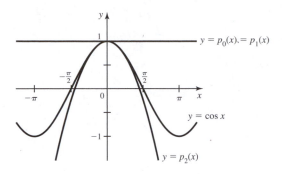

10.1.17 $f(0) = 0$, $f'(0) = -\frac{1}{1-0} = -1$, $f''(0) = -\frac{1}{(1-0)^2} = -1$, so that $p_0(x) = 0$, $p_1(x) = -x$, $p_2(x) = -x - \frac{1}{2}x^2$.

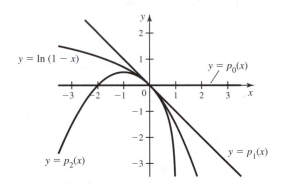

10.1.19 $f(0) = 0$. $f'(x) = \sec^2 x$, $f''(x) = 2\tan x \sec^2 x$, so that $f'(0) = 1$, $f''(0) = 0$. Thus $p_0(x) = 0$, $p_1(x) = x$, $p_2(x) = x$.

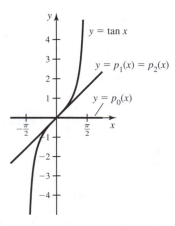

10.1.21 $f(0) = 1$, $f'(0) = -3(1+0)^{-4} = -3$, $f''(0) = 12(1+0)^{-5} = 12$, so that $p_0(x) = 1$, $p_1(x) = 1 - 3x$, $p_2(x) = 1 - 3x + 6x^2$.

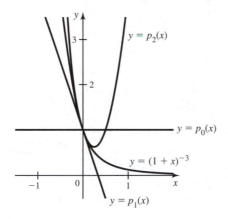

10.1.23

   a. $p_2(0.05) \approx 1.025$.

   b. The absolute error is $\sqrt{1.05} - p_2(0.05) \approx 7.68 \times 10^{-6}$.

10.1.25

   a. $p_2(0.08) \approx 0.962$.

   b. The absolute error is $p_2(0.08) - \frac{1}{\sqrt{1.08}} \approx 1.5 \times 10^{-4}$.

10.1.27

   a. $p_2(0.15) \approx 0.861$.

   b. The absolute error is $p_2(0.15) - e^{-0.15} \approx 5.4 \times 10^{-4}$.

10.1.29

   a. Note that $f(1) = 1$, $f'(1) = 3$, and $f''(1) = 6$. Thus, $p_0(x) = 1$, $p_1(x) = 1 + 3(x - 1)$, and $p_2(x) = 1 + 3(x - 1) + 3(x - 1)^2$.

   b.

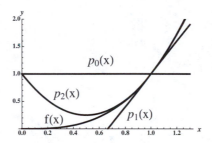

10.1.31

   a. $p_0(x) = \frac{\sqrt{2}}{2}$, $p_1(x) = \frac{\sqrt{2}}{2} + \frac{\sqrt{2}}{2}(x - \frac{\pi}{4})$, $p_2(x) = \frac{\sqrt{2}}{2} + \frac{\sqrt{2}}{2}(x - \frac{\pi}{4}) - \frac{\sqrt{2}}{4}(x - \frac{\pi}{4})^2$.

   b.

10.1.33

   a. $p_0(x) = 3$, $p_1(x) = 3 + \frac{1}{6}(x - 9)$, $p_2(x) = 3 + \frac{1}{6}(x - 9) - \frac{1}{216}(x - 9)^2$.

   b.

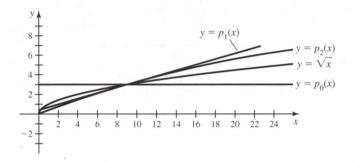

10.1.35

   a. $p_0(x) = 1$, $p_1(x) = 1 + \frac{1}{e}(x - e)$, $p_2(x) = 1 + \frac{1}{e}(x - e) - \frac{1}{2e^2}(x - e)^2$.

   b.

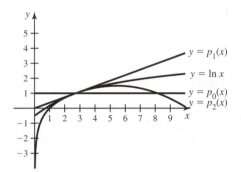

10.1.37

   a. $f(1) = \frac{\pi}{4} + 2$, $f'(1) = \frac{1}{2} + 2 = \frac{5}{2}$. $f''(1) = -\frac{1}{2} + 2 = \frac{3}{2}$. $p_0(x) = 2 + \frac{\pi}{4}$, $p_1(x) = 2 + \frac{\pi}{4} + \frac{5}{2}(x - 1)$,
     $p_2(x) = 2 + \frac{\pi}{4} + \frac{5}{2}(x - 1) + \frac{3}{4}(x - 1)^2$.

   b.

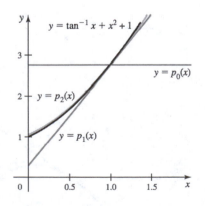

10.1.39

   a. Ue the Taylor polynomial centered at 0 with $f(x) = e^x$. We have $p_3(x) = 1 + x + \frac{1}{2}x^2 + \frac{1}{6}x^3$.
     $p_3(0.12) \approx 1.127$.

b. $|f(0.12) - p_3(0.12)| \approx 8.9 \times 10^{-6}$.

10.1.41

a. Use the Taylor polynomial centered at 0 with $f(x) = \tan(x)$. We have $p_3(x) = x + \frac{1}{3}x^3$. $p_3(-0.1) \approx -0.100$.

b. $|p_3(-0.1) - f(-0.1)| \approx 1.3 \times 10^{-6}$.

10.1.43

a. Use the Taylor polynomial centered at 0 with $f(x) = \sqrt{1+x}$. We have $p_3(x) = 1 + \frac{1}{2}x - \frac{1}{8}x^2 + \frac{1}{16}x^3$. $p_3(0.06) \approx 1.030$.

b. $|f(0.06) - p_3(0.06)| \approx 4.9 \times 10^{-7}$.

10.1.45

a. Use the Taylor polynomial centered at 100 with $f(x) = \sqrt{x}$. We have $p_3(x) = 10 + \frac{1}{20}(x - 100) - \frac{1}{8000}(x - 100)^2 + \frac{1}{1600000}(x - 100)^3$. $p_3(101) \approx 10.050$.

b. $|p_3(101) - f(101)| \approx 3.9 \times 10^{-9}$.

10.1.47

a. Use the Taylor polynomial centered at 0 with $f(x) = \sinh(x)$. Note that $f(0) = 0$, $f'(0) = 1$, $f''(0) = 0$ and $f'''(0) = 1$. Then we have $p_3(x) = x + x^3/6$, so $\sinh(.5) \approx (.5)^3/6 + .5 \approx 0.521$.

b. $|p_3(.5) - \sinh(.5)| \approx 2.6 \times 10^{-4}$.

10.1.49 With $f(x) = \sin x$ we have $R_n(x) = \dfrac{f^{(n+1)}(c)}{(n+1)!}x^{n+1}$ for c between 0 and x.

10.1.51 With $f(x) = e^{-x}$ we have $f^{(n+1)}(x) = (-1)^{n+1}e^{-x}$, so that $R_n(x) = \dfrac{(-1)^{n+1}e^{-c}}{(n+1)!}x^{n+1}$ for c between 0 and x.

10.1.53 With $f(x) = \sin x$ we have $R_n(x) = \dfrac{f^{(n+1)}(c)}{(n+1)!}\left(x - \dfrac{\pi}{2}\right)^{n+1}$ for c between $\frac{\pi}{2}$ and x.

10.1.55 $f(x) = \sin x$, so $f^{(5)}(x) = \cos x$. Because $\cos x$ is bounded in magnitude by 1, the remainder is bounded by $|R_4(x)| \leq \frac{0.3^5}{5!} \approx 2.0 \times 10^{-5}$.

10.1.57 $f(x) = e^x$, so $f^{(5)}(x) = e^x$. Because $e^{0.25}$ is bounded by 2, $|R_4(x)| \leq 2 \cdot \frac{0.25^5}{5!} \approx 1.63 \times 10^{-5}$.

10.1.59 $f(x) = e^{-x}$, so $f^{(5)}(x) = -e^{-x}$. Because $f^{(5)}$ achieves its maximum magnitude in the range at $x = 0$, which has absolute value 1, $|R_4(x)| \leq 1 \cdot \frac{0.5^5}{5!} \approx 2.6 \times 10^{-4}$.

10.1.61 Here $n = 3$ or 4, so use $n = 4$, and $M = 1$ because $f^{(5)}(x) = \cos x$, so that $R_4(x) \leq \frac{(\pi/4)^5}{5!} \approx 2.49 \times 10^{-3}$.

10.1.63 $n = 2$ and $M = e^{1/2} < 2$, so $|R_2(x)| \leq 2 \cdot \frac{(1/2)^3}{3!} \approx 4.2 \times 10^{-2}$.

10.1.65 $n = 2$; $f^{(3)}(x) = \frac{2}{(1+x)^3}$, which achieves its maximum at $x = -0.2$: $|f^{(3)}(x)| = \frac{2}{0.8^3} < 4$. Then $|R_2(x)| \leq 4 \cdot \frac{0.2^3}{3!} \approx 5.4 \times 10^{-3}$.

10.1.67 Use the Taylor series for e^x at $x = 0$. The derivatives of e^x are e^x. On $[-0.5, 0]$, the maximum magnitude of any derivative is thus 1 at $x = 0$, so $|R_n(-0.5)| \leq \frac{0.5^{n+1}}{(n+1)!}$, so for $R_n(-0.5) < 10^{-3}$ we need $n = 4$.

10.1.69 Use the Taylor series for $\cos x$ at $x = 0$. The magnitude of any derivative of $\cos x$ is bounded by 1, so $|R_n(-0.25)| \leq \frac{0.25^{n+1}}{(n+1)!}$, so for $|R_n(-0.25)| < 10^{-3}$ we need $n = 3$.

10.1.71 Use the Taylor series for $f(x) = \sqrt{x}$ at $x = 1$. Then $|f^{(n+1)}(x)| = \frac{1 \cdot 3 \cdots (2n-1)}{2^{n+1}} x^{-(2n+1)/2}$, which achieves its maximum on $[1, 1.06]$ at $x = 1$. Then

$$|R_n(1.06)| \leq \frac{1 \cdot 3 \cdots \cdot (2n-1)}{2^{n+1}} \cdot \frac{(1.06-1)^{n+1}}{(n+1)!},$$

and for $|R_n(0.06)| < 10^{-3}$ we need $n = 1$.

10.1.73

a. False. If $f(x) = e^{-2x}$, then $f^{(n)}(x) = (-1)^n 2^n e^{-2x}$, so that $f^{(n)}(0) \neq 0$ and all powers of x are present in the Taylor series.

b. True. The constant term of the Taylor series is $f(0) = 1$. Higher-order terms all involve derivatives of $f(x) = x^5 - 1$ evaluated at $x = 0$; clearly for $n < 5$, $f^{(n)}(0) = 0$, and for $n > 5$, the derivative itself vanishes. Only for $n = 5$, where $f^{(5)}(x) = 5!$, is the derivative nonzero, so the coefficient of x^5 in the Taylor series is $f^{(5)}(0)/5! = 1$ and the Taylor polynomial of order 10 is in fact $x^5 - 1$. Note that this statement is true of any polynomial of degree at most 10.

c. True. The odd derivatives of $\sqrt{1 + x^2}$ vanish at $x = 0$, while the even ones do not.

d. True. Clearly the second-order Taylor polynomial for f at a has degree at most 2. However, the coefficient of $(x - a)^2$ is $\frac{1}{2} f''(a)$, which is zero because f has an inflection point at a.

10.1.75

a. This matches (C) because for $f(x) = (1 + 2x)^{1/2}$, $f''(x) = -(1 + 2x)^{-3/2}$ so $\frac{f''(0)}{2!} = -\frac{1}{2}$.

b. This matches (E) because for $f(x) = (1 + 2x)^{-1/2}$, $f''(x) = 3(1 + 2x)^{-5/2}$, so $\frac{f''(0)}{2!} = \frac{3}{2}$.

c. This matches (A) because $f^{(n)}(x) = 2^n e^{2x}$, so that $f^{(n)}(0) = 2^n$, which is (A)'s pattern.

d. This matches (D) because $f''(x) = 8(1 + 2x)^{-3}$ and $f''(0) = 8$, so that $f''(0)/2! = 4$

e. This matches (B) because $f'(x) = -6(1 + 2x)^{-4}$ so that $f'(0) = -6$.

f. This matches (F) because $f^{(n)}(x) = (-2)^n e^{-2x}$, so $f^{(n)}(0) = (-2)^n$, which is (F)'s pattern.

10.1.77

a. $p_2(0.1) = 0.1$. The maximum error in the approximation is $1 \cdot \frac{0.1^3}{3!} \approx 1.67 \times 10^{-4}$.

b. $p_2(0.2) = 0.2$. The maximum error in the approximation is $1 \cdot \frac{0.2^3}{3!} \approx 1.33 \times 10^{-3}$.

10.1.79

a. $p_3(0.1) = 1 - .01/2 = 0.995$. The maximum error is $1 \cdot \frac{0.1^4}{4!} \approx 4.2 \times 10^{-6}$.

b. $p_3(0.2) = 1 - .04/2 = 0.98$. The maximum error is $1 \cdot \frac{0.2^4}{4!} \approx 6.7 \times 10^{-5}$.

10.1.81

a. $p_1(0.1) = 1.05$. Because $|f''(x)| = \frac{1}{4}(1 + x)^{-3/2}$ has a maximum value of $1/4$ at $x = 0$, the maximum error is $\frac{1}{4} \cdot \frac{0.1^2}{2} \approx 1.3 \times 10^{-3}$.

b. $p_1(0.2) = 1.1$. The maximum error is $\frac{1}{4} \cdot \frac{0.2^2}{2} = 5 \times 10^{-3}$.

10.1.83

a. $p_1(0.1) = 1.1$. Because $f''(x) = e^x$ is less than 2 on $[0, 0.1]$, the maximum error is less than $2 \cdot \frac{0.1^2}{2!} = 10^{-2}$.

b. $p_1(0.2) = 1.2$. The maximum error is less than $2 \cdot \frac{0.2^2}{2!} = .04 = 4 \times 10^{-2}$.

10.1.85

a.

	$|\sec x - p_2(x)|$	$|\sec x - p_4(x)|$
-0.2	3.4×10^{-4}	5.5×10^{-6}
-0.1	2.1×10^{-5}	8.5×10^{-8}
0.0	0	0
0.1	2.1×10^{-5}	8.5×10^{-8}
0.2	3.4×10^{-4}	5.5×10^{-6}

b. The errors are equal for positive and negative x. This makes sense, because $\sec(-x) = \sec x$ and $p_n(-x) = p_n(x)$ for $n = 2, 4$. The errors appear to get larger as x gets farther from zero.

10.1.87

a.

	$|e^{-x} - p_1(x)|$	$|e^{-x} - p_2(x)|$
-0.2	2.14×10^{-2}	1.40×10^{-3}
-0.1	5.17×10^{-3}	1.71×10^{-4}
0.0	0	0
0.1	4.84×10^{-3}	1.63×10^{-4}
0.2	1.87×10^{-2}	1.27×10^{-3}

b. The errors are different for positive and negative displacements from zero, and appear to get larger as x gets farther from zero.

10.1.89

a.

	$|\tan x - p_1(x)|$	$|\tan x - p_3(x)|$
-0.2	2.71×10^{-3}	4.34×10^{-5}
-0.1	3.35×10^{-4}	1.34×10^{-6}
0.0	0	0
0.1	3.35×10^{-4}	1.34×10^{-6}
0.2	2.71×10^{-3}	4.34×10^{-5}

b. The errors are equal for positive and negative x. This makes sense, because $\tan(-x) = -\tan x$ and $p_n(-x) = -p_n(x)$ for $n = 1, 3$. The errors appear to get larger as x gets farther from zero.

10.1.91 The true value of $e^{0.35} \approx 1.419067549$. The 6th-order Taylor polynomial for e^x centered at $x = 0$ is

$$p_6(x) = 1 + x + \frac{x^2}{2} + \frac{x^3}{6} + \frac{x^4}{24} + \frac{x^5}{120} + \frac{x^6}{720}.$$

Evaluating the polynomials at $x = 0.35$ produces the following table:

n	$p_n(0.35)$	$|p_n(0.35) - e^{0.35}|$
1	1.350000000	6.91×10^{-2}
2	1.411250000	7.82×10^{-3}
3	1.418395833	6.72×10^{-4}
4	1.419021094	4.65×10^{-5}
5	1.419064862	2.69×10^{-6}
6	1.419067415	1.33×10^{-7}

The 6th-order Taylor polynomial for e^x centered at $x = \ln 2$ is

$$p_6(x) = \qquad\qquad 2 + 2(x - \ln 2) + (x - \ln 2)^2 + \frac{1}{3}(x - \ln 2)^3 + \frac{1}{12}(x - \ln 2)^4$$

$$+ \frac{1}{60}(x - \ln 2)^5 + \frac{1}{360}(x - \ln 2)^6.$$

Evaluating the polynomials at $x = 0.35$ produces the following table:

n	$p_n(0.35)$	$\lvert p_n(0.35) - e^{0.35} \rvert$
1	1.313705639	1.05×10^{-1}
2	1.431455626	1.24×10^{-2}
3	1.417987101	1.08×10^{-3}
4	1.419142523	7.50×10^{-5}
5	1.419063227	4.32×10^{-6}
6	1.419067762	2.13×10^{-7}

Comparing the tables shows that using the polynomial centered at $x = 0$ is more accurate for all n. To see why, consider the remainder. Let $f(x) = e^x$. By Theorem 9.2, the magnitude of the remainder when approximating $f(0.35)$ by the polynomial p_n centered at 0 is:

$$\lvert R_n(0.35) \rvert = \frac{\lvert f^{(n+1)}(c) \rvert}{(n+1)!}(0.35)^{n+1} = \frac{e^c}{(n+1)!}(0.35)^{n+1}$$

for some c with $0 < c < 0.35$ while the magnitude of the remainder when approximating $f(0.35)$ by the polynomial p_n centered at $\ln 2$ is:

$$\lvert R_n(0.35) \rvert = \frac{\lvert f^{(n+1)}(c) \rvert}{(n+1)!}\lvert 0.35 - \ln 2 \rvert^{n+1} = \frac{e^c}{(n+1)!}(\ln 2 - 0.35)^{n+1}$$

for some c with $0.35 < c < \ln 2$. Because $\ln 2 - 0.35 \approx 0.35$, the relative size of the magnitudes of the remainders is determined by e^c in each remainder. Because e^x is an increasing function, the remainder in using the polynomial centered at 0 will be less than the remainder in using the polynomial centered at $\ln 2$, and the former polynomial will be more accurate.

10.1.93

a. The slope of the tangent line to $f(x)$ at $x = a$ is by definition $f'(a)$; by the point-slope form for the equation of a line, we have $y - f(a) = f'(a)(x - a)$, or $y = f(a) + f'(a)(x - a)$.

b. The Taylor polynomial centered at a is $p_1(x) = f(a) + f'(a)(x - a)$, which is the tangent line at a.

10.1.95

a. We have

$$
\begin{array}{ll}
f(0) = f^{(4)}(0) = \sin 0 = 0 & f(\pi) = f^{(4)}(\pi) = \sin \pi = 0 \\
f'(0) = f^{(5)}(0) = \cos 0 = 1 & f'(\pi) = f^{(5)}(0) = \cos \pi = -1 \\
f''(0) = -\sin 0 = 0 & f''(\pi) = -\sin \pi = 0 \\
f'''(0) = -\cos 0 = -1 & f'''(\pi) = -\cos \pi = 1.
\end{array}
$$

Thus

$$p_5(x) = x - \frac{x^3}{3!} + \frac{x^5}{5!}$$

$$q_5(x) = -(x - \pi) + \frac{1}{3!}(x - \pi)^3 - \frac{1}{5!}(x - \pi)^5.$$

b. A plot of the three functions, with $\sin x$ the black solid line, $p_5(x)$ the dashed line, and $q_5(x)$ the dotted line is below.

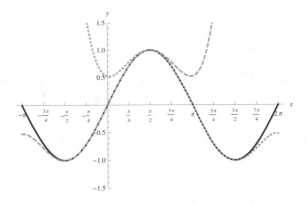

$p_5(x)$ and $\sin x$ are almost indistinguishable on $[-\pi/2, \pi/2]$, after which $p_5(x)$ diverges pretty quickly from $\sin x$. $q_5(x)$ is reasonably close to $\sin x$ over the entire range, but the two are almost indistinguishable on $[\pi/2, 3\pi/2]$. $p_5(x)$ is a better approximation than $q_5(x)$ on about $[-\pi, \pi/2)$, while $q_5(x)$ is better on about $(\pi/2, 2\pi]$.

c. Evaluating the errors gives

x	$|\sin x - p_5(x)|$	$|\sin x - q_5(x)|$
$\frac{\pi}{4}$	3.6×10^{-5}	7.4×10^{-2}
$\frac{\pi}{2}$	4.5×10^{-3}	4.5×10^{-3}
$\frac{3\pi}{4}$	7.4×10^{-2}	3.6×10^{-5}
$\frac{5\pi}{4}$	2.3	3.6×10^{-5}
$\frac{7\pi}{4}$	20.4	7.4×10^{-2}

d. $p_5(x)$ is a better approximation than $q_5(x)$ only at $x = \frac{\pi}{4}$, in accordance with part (b). The two are equal at $x = \frac{\pi}{2}$, after which $q_5(x)$ is a substantially better approximation than $p_5(x)$.

10.1.97

a. We have

$$f(36) = \sqrt{36} = 6 \qquad\qquad f(49) = \sqrt{49} = 7$$
$$f'(36) = \frac{1}{2} \cdot \frac{1}{\sqrt{36}} = \frac{1}{12} \qquad\qquad f'(49) = \frac{1}{2} \cdot \frac{1}{\sqrt{49}} = \frac{1}{14}.$$

Thus

$$p_1(x) = 6 + \frac{1}{12}(x - 36) \qquad\qquad q_1(x) = 7 + \frac{1}{14}(x - 49).$$

b. Evaluating the errors gives

x	$|\sqrt{x} - p_1(x)|$	$|\sqrt{x} - q_1(x)|$
37	5.7×10^{-4}	6.0×10^{-2}
39	5.0×10^{-3}	4.1×10^{-2}
41	1.4×10^{-2}	2.5×10^{-2}
43	2.6×10^{-2}	1.4×10^{-2}
45	4.2×10^{-2}	6.1×10^{-3}
47	6.1×10^{-2}	1.5×10^{-3}

c. $p_1(x)$ is a better approximation than $q_1(x)$ for $x \leq 41$, and $q_1(x)$ is a better approximation for $x \geq 43$. To see why this is true, note that $f''(x) = -\frac{1}{4}x^{-3/2}$, so that on $[36, 49]$ it is bounded in magnitude by $\frac{1}{4} \cdot 36^{-3/2} = \frac{1}{864}$. . Thus (using P_1 for the error term for p_1 and Q_1 for the error term for q_1)

$$P_1(x) \leq \frac{1}{864} \cdot \frac{|x-36|^2}{2!} = \frac{1}{1728}(x-36)^2, \qquad Q_1(x) \leq \frac{1}{864} \cdot \frac{|x-49|^2}{2!} = \frac{1}{1728}(x-49)^2.$$

It follows that the relative sizes of $P_1(x)$ and $Q_1(x)$ are governed by the distance of x from 36 and 49. Looking at the different possibilities for x reveals why the results in part (b) hold.

10.2 Properties of Power Series

10.2.1 $c_0 + c_1 x + c_2 x^2 + c_3 x^3$.

10.2.3 Generally the Ratio Test or Root Test is used.

10.2.5 The radius of convergence does not change, but the interval of convergence may change at the endpoints.

10.2.7 $|x| < \frac{1}{4}$.

10.2.9 Using the Root Test: $\lim_{k \to \infty} \sqrt[k]{|a_k|} = \lim_{k \to \infty} |2x| = |2x|$. So the radius of convergence is $\frac{1}{2}$. At $x = 1/2$ the series is $\sum 1$ which diverges, and at $x = -1/2$ the series is $\sum (-1)^k$ which also diverges. So the interval of convergence is $(-1/2, 1/2)$.

10.2.11 Using the Root Test, $\lim_{k \to \infty} \sqrt[k]{|a_k|} = \lim_{k \to \infty} \frac{|x-1|}{k^{1/k}} = |x-1|$. So the radius of convergence is 1. At $x = 2$, we have the harmonic series (which diverges) and at $x = 0$ we have the alternating harmonic series (which converges). Thus the interval of convergence is $[0, 2)$.

10.2.13 Using the Ratio Test: $\lim_{k \to \infty} \left| \frac{a_{k+1}}{a_k} \right| = \lim_{k \to \infty} \left| \frac{(k+1)^{k+1} x^{k+1}}{k^k x^k} \right| = \lim_{k \to \infty} (k+1) \left(\frac{k+1}{k} \right)^k |x| = \infty$ (for $x \neq 0$) because $\lim_{k \to \infty} \left(\frac{k+1}{k} \right)^k = e$. Thus, the radius of convergence is 0, the series only converges at $x = 0$.

10.2.15 Using the Root Test: $\lim_{k \to \infty} \sqrt[k]{|a_k|} = \lim_{k \to \infty} \sin(1/k)|x| = \sin(0)|x| = 0$. Thus, the radius of convergence is ∞ and the interval of convergence is $(-\infty, \infty)$.

10.2.17 Using the Root Test: $\lim_{k \to \infty} \sqrt[k]{|a_k|} = \lim_{k \to \infty} \frac{|x|}{3} = \frac{|x|}{3}$, so the radius of convergence is 3. At -3, the series is $\sum (-1)^k$, which diverges. At 3, the series is $\sum 1$, which diverges. So the interval of convergence is $(-3, 3)$.

10.2.19 Using the Root Test: $\lim_{k \to \infty} \sqrt[k]{|a_k|} = \lim_{k \to \infty} \frac{|x|}{k} = 0$, so the radius of convergence is infinite and the interval of convergence is $(-\infty, \infty)$.

10.2.21 Using the Ratio Test: $\lim_{k \to \infty} \left| \frac{(k+1)^2 x^{2k+2}}{(k+1)!} \cdot \frac{k!}{k^2 x^{2k}} \right| = \lim_{k \to \infty} \frac{k+1}{k^2} x^2 = 0$, so the radius of convergence is infinite, and the interval of convergence is $(-\infty, \infty)$.

10.2.23 Using the Ratio Test: $\lim_{k \to \infty} \left| \frac{a_{k+1}}{a_k} \right| = \left| \frac{x^{2k+3}}{3^k} \cdot \frac{3^{k-1}}{x^{2k+1}} \right| = \frac{x^2}{3}$ so that the radius of convergence is $\sqrt{3}$. At $x = \sqrt{3}$, the series is $\sum 3\sqrt{3}$, which diverges. At $x = -\sqrt{3}$, the series is $\sum (-3\sqrt{3})$, which also diverges, so the interval of convergence is $(-\sqrt{3}, \sqrt{3})$.

10.2.25 Using the Root Test: $\lim_{k \to \infty} \sqrt[k]{|a_k|} = \lim_{k \to \infty} \frac{(|x-1|)k}{k+1} = |x-1|$, so the series converges when $|x-1| < 1$, so for $0 < x < 2$. The radius of convergence is 1. At $x = 2$, the series diverges by the Divergence Test. At $x = 0$, the series diverges as well by the Divergence Test. Thus the interval of convergence is $(0, 2)$.

10.2.27 Using the Ratio Test: $\lim\limits_{k \to \infty} \left| \frac{a_{k+1}}{a_k} \right| = \left| \frac{(k+1)^{20} x^{k+1}}{(2k+3)!} \cdot \frac{(2k+1)!}{x^k k^{20}} \right| = \lim\limits_{k \to \infty} \left(\frac{k+1}{k} \right)^{20} \frac{|x|}{(2k+2)(2k+3)} = 0$, so the radius of convergence is infinite, and the interval of convergence is $(-\infty, \infty)$.

10.2.29 $f(3x) = \frac{1}{1-3x} = \sum_{k=0}^{\infty} 3^k x^k$, which converges for $|x| < 1/3$, and diverges at the endpoints.

10.2.31 $h(x) = \frac{2x^3}{1-x} = \sum_{k=0}^{\infty} 2x^{k+3}$, which converges for $|x| < 1$ and is divergent at the endpoints.

10.2.33 $p(x) = \frac{4x^{12}}{1-x} = \sum_{k=0}^{\infty} 4x^{k+12} = 4 \sum_{k=0}^{\infty} x^{k+12}$, which converges for $|x| < 1$. It is divergent at the endpoints.

10.2.35 $f(3x) = \ln(1 - 3x) = -\sum_{k=1}^{\infty} \frac{(3x)^k}{k} = -\sum_{k=1}^{\infty} \frac{3^k}{k} x^k$. Using the Ratio Test:

$$\lim_{k \to \infty} \left| \frac{a_{k+1}}{a_k} \right| = \lim_{k \to \infty} \frac{3k}{k+1} |x| = 3 |x|,$$

so the radius of convergence is $1/3$. The series diverges at $1/3$ (harmonic series), and converges at $-1/3$ (alternating harmonic series).

10.2.37 $h(x) = x \ln(1 - x) = -\sum_{k=1}^{\infty} \frac{x^{k+1}}{k}$. Using the Ratio Test: $\lim\limits_{k \to \infty} \left| \frac{a_{k+1}}{a_k} \right| = \lim\limits_{k \to \infty} \frac{k}{k+1} |x| = |x|$, so the radius of convergence is 1, and the series diverges at 1 (harmonic series) but converges at -1 (alternating harmonic series).

10.2.39 $p(x) = 2x^6 \ln(1 - x) = -2 \sum_{k=1}^{\infty} \frac{x^{k+6}}{k}$. Using the Ratio Test: $\lim\limits_{k \to \infty} \left| \frac{a_{k+1}}{a_k} \right| = \lim\limits_{k \to \infty} \frac{k}{k+1} |x| = |x|$, so the radius of convergence is 1. The series diverges at 1 (harmonic series) but converges at -1 (alternating harmonic series).

10.2.41 The power series for $f(x)$ is $\sum_{k=0}^{\infty} (2x)k$, convergent for $-1 < 2x < 1$, so for $-1/2 < x < 1/2$. The power series for $g(x) = f'(x)$ is $\sum_{k=1}^{\infty} k(2x)^{k-1} \cdot 2 = 2 \sum_{k=1}^{\infty} k(2x)^{k-1}$, also convergent on $|x| < 1/2$.

10.2.43 The power series for $f(x)$ is $\sum_{k=0}^{\infty} x^k$, convergent for $-1 < x < 1$, so the power series for $g(x) = \frac{1}{6} f'''(x)$ is $\frac{1}{6} \sum_{k=3}^{\infty} k(k-1)(k-2)x^{k-3} = \frac{1}{6} \sum_{k=0}^{\infty} (k+1)(k+2)(k+3)x^k$, also convergent on $|x| < 1$.

10.2.45 The power series for $\frac{1}{1-3x}$ is $\sum_{k=0}^{\infty} (3x)^k$, convergent on $|x| < 1/3$. Because $g(x) = \ln(1 - 3x) = -3 \int \frac{1}{1-3x} dx$ and because $g(0) = 0$, the power series for $g(x)$ is $-3 \sum_{k=0}^{\infty} 3^k \frac{1}{k+1} x^{k+1} = -\sum_{k=1}^{\infty} \frac{3^k}{k} x^k$, also convergent on $[-1/3, 1/3)$.

10.2.47 Start with $g(x) = \frac{1}{1+x}$. The power series for $g(x)$ is $\sum_{k=0}^{\infty} (-1)^k x^k$. Because $f(x) = g(x^2)$, its power series is $\sum_{k=0}^{\infty} (-1)^k x^{2k}$. The radius of convergence is still 1, and the series is divergent at both endpoints. The interval of convergence is $(-1, 1)$.

10.2.49 Note that $f(x) = \frac{3}{3+x} = \frac{1}{1+(1/3)x}$. Let $g(x) = \frac{1}{1+x}$. The power series for $g(x)$ is $\sum_{k=0}^{\infty} (-1)^k x^k$, so the power series for $f(x) = g((1/3)x)$ is $\sum_{k=0}^{\infty} (-1)^k 3^{-k} x^k = \sum_{k=0}^{\infty} \left(\frac{-x}{3} \right)^k$. Using the Ratio Test: $\lim\limits_{k \to \infty} \left| \frac{a_{k+1}}{a_k} \right| = \lim\limits_{k \to \infty} \left| \frac{3^{-(k+1)} x^{k+1}}{3^{-k} x^k} \right| = \frac{|x|}{3}$, so the radius of convergence is 3. The series diverges at both endpoints. The interval of convergence is $(-3, 3)$.

10.2.51 Note that $f(x) = \ln \sqrt{4 - x^2} = \frac{1}{2} \ln(4 - x^2) = \frac{1}{2} \left(\ln 4 + \ln \left(1 - \frac{x^2}{4} \right) \right) = \ln 2 + \frac{1}{2} \ln \left(1 - \frac{x^2}{4} \right)$. Now, the power series for $g(x) = \ln(1 - x)$ is $-\sum_{k=1}^{\infty} \frac{1}{k} x^k$, so the power series for $f(x)$ is $\ln 2 - \frac{1}{2} \sum_{k=1}^{\infty} \frac{1}{k} \frac{x^{2k}}{4^k} = \ln 2 - \sum_{k=1}^{\infty} \frac{x^{2k}}{k 2^{2k+1}}$. Now, $\lim\limits_{k \to \infty} \left| \frac{a_{k+1}}{a_k} \right| = \lim\limits_{k \to \infty} \left| \frac{x^{2k+2}}{(k+1)2^{2k+3}} \cdot \frac{k 2^{2k+1}}{x^{2k}} \right| = \lim\limits_{k \to \infty} \frac{k}{4(k+1)} x^2 = \frac{x^2}{4}$, so that the radius of convergence is 2. The series diverges at both endpoints, so its interval of convergence is $(-2, 2)$.

Copyright © 2015 Pearson Education, Inc.

10.2.53

    a. True. This power series is centered at $x = 3$, so its interval of convergence will be symmetric about 3.

    b. True. Use the Root Test.

    c. True. Substitute x^2 for x in the series.

    d. True. Because the power series is zero on the interval, all its derivatives are as well, which implies (differentiating the power series) that all the c_k are zero.

10.2.55 Using the Ratio Test: $\lim\limits_{k\to\infty}\left|\dfrac{a_{k+1}}{a_k}\right| = \lim\limits_{k\to\infty}\left|\dfrac{(k+1)!\,x^{k+1}}{(k+1)^{k+1}} \cdot \dfrac{k^k}{k!\,x^k}\right| = \lim\limits_{k\to\infty}\left(\dfrac{k}{k+1}\right)^k |x| = \frac{1}{e}|x|$. The radius of convergence is therefore e.

10.2.57 $\sum_{k=0}^{\infty}(-1)^k \frac{1}{k+1}x^k$

10.2.59 $\sum_{k=1}^{\infty}(-1)^k \frac{x^{2k}}{k!}$

10.2.61 The power series for $f(x-a)$ is $\sum c_k(x-a)^k$. Then $\sum c_k(x-a)^k$ converges if and only if $|x - a| < R$, which happens if and only if $a - R < x < a + R$, so the radius of convergence is the same.

10.2.63 This is a geometric series with ratio $\sqrt{x} - 2$, so its sum is $\frac{1}{1-(\sqrt{x}-2)} = \frac{1}{3-\sqrt{x}}$. Again using the Root Test, $\lim\limits_{k\to\infty}\sqrt[k]{|a_k|} = |\sqrt{x} - 2|$, so the interval of convergence is given by $|\sqrt{x} - 2| < 1$, so $1 < \sqrt{x} < 3$ and $1 < x < 9$. The series diverges at both endpoints.

10.2.65 This is a geometric series with ratio e^{-x}, so its sum is $\frac{1}{1-e^{-x}}$. By the Root Test, $\lim\limits_{k\to\infty}\sqrt[k]{|a_k|} = e^{-x}$, so the power series converges for $x > 0$.

10.2.67 This is a geometric series with ratio $(x^2 - 1)/3$, so its sum is $\frac{1}{1-\frac{x^2-1}{3}} = \frac{3}{3-(x^2-1)} = \frac{3}{4-x^2}$. Using the Root Test, the series converges for $|x^2 - 1| < 3$, so that $-2 < x^2 < 4$ or $-2 < x < 2$. It diverges at both endpoints.

10.2.69 The power series for e^x is $\sum_{k=0}^{\infty}\frac{x^k}{k!}$. Substitute $-x$ for x to get $e^{-x} = \sum_{k=0}^{\infty}(-1)^k\frac{x^k}{k!}$. The series converges for all x.

10.2.71 Substitute $-3x$ for x in the power series for e^x to get $e^{-3x} = \sum_{k=0}^{\infty}\frac{(-3x)^k}{k!} = \sum_{k=0}^{\infty}(-1)^k\frac{3^k}{k!}x^k$. The series converges for all x.

10.2.73 The power series for $x^m f(x)$ is $\sum c_k x^{k+m}$. The radius of convergence of this power series is determined by the limit

$$\lim\limits_{k\to\infty}\left|\frac{c_{k+1}x^{k+1+m}}{c_k x^{k+m}}\right| = \lim\limits_{k\to\infty}\left|\frac{c_{k+1}x^{k+1}}{c_k x^k}\right|,$$

and the right-hand side is the limit used to determine the radius of convergence for the power series for $f(x)$. Thus the two have the same radius of convergence.

10.2.75

    a. $f(x)g(x) = c_0 d_0 + (c_0 d_1 + c_1 d_0)x + (c_0 d_2 + c_1 d_1 + c_2 d_0)x^2 + \ldots$

    b. The coefficient of x^n in $f(x)g(x)$ is $\sum_{i=0}^{n} c_i d_{n-i}$.

10.2.77

a. For both graphs, the difference between the true value and the estimate is greatest at the two ends of the range; the difference at 0.9 is greater than that at -0.9.

b. The difference between $f(x)$ and $S_n(x)$ is greatest for $x = 0.9$; at that point, $f(x) = \frac{1}{(1-0.9)^2} = 100$, so we want to find n such that $S_n(x)$ is within 0.01 of 100. We find that $S_{111} \approx 99.98991435$ and $S_{112} \approx 99.99084790$, so $n = 112$.

10.3   Taylor Series

10.3.1 The nth Taylor Polynomial is the nth sum of the corresponding Taylor Series.

10.3.3 The nth coefficient is $\frac{f^{(n)}(a)}{n!}$.

10.3.5 Substitute x^2 for x in the Taylor series. By theorems proved in the previous section about power series, the interval of convergence does not change except perhaps at the endpoints of the interval.

10.3.7 It means that the limit of the remainder term is zero.

10.3.9

a. Note that $f(0) = 1$, $f'(0) = -1$, $f''(0) = 1$, and $f'''(0) = -1$. So the Maclaurin series is $1 - x + x^2/2 - x^3/6 + \cdots$.

b. $\sum_{k=0}^{\infty} (-1)^k \frac{x^k}{k!}$.

c. The series converges on $(-\infty, \infty)$, as can be seen from the Ratio Test.

10.3.11

a. Because the series for $\frac{1}{1+x}$ is $1 - x + x^2 - x^3 + \cdots$, the series for $\frac{1}{1+x^2}$ is $1 - x^2 + x^4 - x^6 + \cdots$.

b. $\sum_{k=0}^{\infty} (-1)^k x^{2k}$.

c. The absolute value of the ratio of consecutive terms is x^2, so by the Ratio Test, the radius of convergence is 1. The series diverges at the endpoints by the Divergence Test, so the interval of convergence is $(-1, 1)$.

10.3.13

a. Note that $f(0) = 1$, and that $f^{(n)}(0) = 2^n$. Thus, the series is given by $1 + 2x + \frac{4x^2}{2} + \frac{8x^3}{6} + \cdots$.

b. $\sum_{k=0}^{\infty} \frac{(2x)^k}{k!}$.

c. The absolute value of the ratio of consecutive terms is $\frac{2|x|}{n}$, which has limit 0 as $n \to \infty$. So by the Ratio Test, the interval of convergence is $(-\infty, \infty)$.

10.3.15

a. By integrating the Taylor series for $\frac{1}{1+x^2}$ (which is the derivative of $\tan^{-1}(x)$), we obtain the series $x - \frac{x^3}{3} + \frac{x^5}{5} - \frac{x^7}{7} + \cdots$. Then by replacing x by $x/2$ we have $\frac{x}{2} - \frac{x^3}{3 \cdot 2^3} + \frac{x^5}{5 \cdot 2^5} - \frac{x^7}{7 \cdot 2^7} + \cdots$.

b. $\sum_{k=0}^{\infty} (-1)^k \frac{1}{(2k+1) \cdot 2^{2k+1}} x^{2k+1}$.

c. By the Ratio Test (the ratio of consecutive terms has limit $\frac{x^2}{4}$), the radius of convergence is $|x| < 2$. Also, at the endpoints we have convergence by the Alternating Series Test, so the interval of convergence is $[-2, 2]$.

10.3.17

a. Note that $f(0) = 1$, $f'(0) = \ln 3$, $f''(0) = \ln^2 3$, $f''(0) = \ln^3 3$. So the first four terms of the desired series are $1 + (\ln 3)x + \frac{\ln^2 3}{2}x^2 + \frac{\ln^3 3}{6}x^3 + \cdots$.

b. $\sum_{k=0}^{\infty} \frac{(\ln^k 3)x^k}{k!}$.

c. The ratio of successive terms is $\frac{(\ln^{k+1} 3)x^{k+1}}{(k+1)!} \cdot \frac{k!}{(\ln^k 3)x^k} = \frac{\ln 3}{k+1}x$, and the limit as $k \to \infty$ of this quantity is 0, so the interval of convergence is $(-\infty, \infty)$.

10.3.19

a. Note that $f(0) = 1$, $f'(0) = 0$, $f''(0) = 9$, $f'''(0) = 0$, etc. The first terms of the series are $1 + 9x^2/2 + 81x^4/4! + 3^6 x^6/6! + \cdots$.

b. $\sum_{k=0}^{\infty} \frac{(3x)^{2k}}{(2k)!}$.

c. The absolute value of the ratio of successive terms is $\left| \frac{(3x)^{2k+2}}{(2k+2)!} \cdot \frac{(2k)!}{(3x)^{2k}} \right| = \frac{1}{(2k+2)(2k+1)} \cdot 9x^2$, which has limit 0 as $x \to \infty$. The interval of convergence is therefore $(-\infty, \infty)$.

10.3.21

a. Note that $f(\pi/2) = 1$, $f'(\pi/2) = \cos(\pi/2) = 0$, $f''(\pi/2) = -\sin(\pi/2) = -1$, $f'''(\pi/2) = -\cos(\pi/2) = 0$, and so on. Thus the series is given by $1 - \frac{1}{2}\left(x - \frac{\pi}{2}\right)^2 + \frac{1}{24}\left(x - \frac{\pi}{2}\right)^4 - \frac{1}{720}\left(x - \frac{\pi}{2}\right)^6 + \cdots$.

b. $\sum_{k=0}^{\infty} (-1)^k \frac{1}{(2k)!}\left(x - \frac{\pi}{2}\right)^{2k}$.

10.3.23

a. Note that $f^{(k)}(1) = (-1)^k \frac{k!}{1^{k+1}} = (-1)^k \cdot k!$. Thus the series is given by $1 - (x-1) + (x-1)^2 - (x-1)^3 + \cdots$.

b. $\sum_{k=0}^{\infty} (-1)^k (x-1)^k$.

10.3.25

a. Note that $f^{(k)}(3) = (-1)^{k-1} \frac{(k-1)!}{3^k}$. Thus the series is given by $\ln(3) + \frac{x-3}{3} - \frac{1}{18}(x-3)^2 + \frac{1}{81}(x-3)^3 + \cdots$.

b. $\ln 3 + \sum_{k=1}^{\infty} (-1)^{k+1} \frac{1}{k \cdot 3^k}(x-3)^k$.

10.3.27

a. Note that $f(1) = 2$, $f'(1) = 2\ln 2$, $f''(1) = 2\ln^2 2$, $f'''(1) = 2\ln^3 2$. The first terms of the series are
$2 + (2\ln 2)(x - 1) + (\ln^2 2)(x - 1)^2 + \frac{(\ln^3 2)(x-1)^3}{3} + \cdots$.

b. $\sum_{k=0}^{\infty} \frac{2(x-1)^k \ln^k 2}{k!}$.

10.3.29 Because the Taylor series for $\ln(1 + x)$ is $x - \frac{x^2}{2} + \frac{x^3}{3} - \frac{x^4}{4} + \cdots$, the first four terms of the Taylor series for $\ln(1 + x^2)$ are $x^2 - \frac{x^4}{2} + \frac{x^6}{3} - \frac{x^8}{4} + \cdots$, obtained by substituting x^2 for x.

10.3.31 Because the Taylor series for $\frac{1}{1-x} = 1 + x + x^2 + x^3 + \cdots$, the first four terms of the Taylor series for $\frac{1}{1-2x}$ are $1 + 2x + 4x^2 + 8x^3 + \cdots$ obtained by substituting $2x$ for x.

10.3.33 The Taylor series for $e^x - 1$ is the Taylor series for e^x, less the constant term of 1, so it is $x + \frac{x^2}{2} + \frac{x^3}{3!} + \frac{x^4}{4!} + \cdots$. Thus, the first four terms of the Taylor series for $\frac{e^x - 1}{x}$ are $1 + \frac{x}{2!} + \frac{x^2}{3!} + \frac{x^3}{4!} + \cdots$, obtained by dividing the terms of the first series by x.

10.3.35 Because the Taylor series for $(1 + x)^{-1}$ is $1 - x + x^2 - x^3 + \cdots$, if we substitute x^4 for x, we obtain $1 - x^4 + x^8 - x^{12} + \cdots$.

10.3.37 The Taylor series for $\sinh x$ is $x + \frac{x^3}{6} + \frac{x^5}{120} + \frac{x^7}{5040} + \cdots$. Thus, the Taylor series for $\sinh x^2$ is $x^2 + \frac{x^6}{6} + \frac{x^{10}}{120} + \frac{x^{14}}{5040} + \cdots$ obtained by substituting x^2 for x.

10.3.39

a. The binomial coefficients are $\binom{-2}{0} = 1$, $\binom{-2}{1} = \frac{-2}{1!} = -2$, $\binom{-2}{2} = \frac{(-2)(-3)}{2!} = 3$, $\binom{-2}{3} = \frac{(-2)(-3)(-4)}{3!} = -4$.

Thus the first four terms of the series are $1 - 2x + 3x^2 - 4x^3 + \cdots$.

b. $1 - 2 \cdot 0.1 + 3 \cdot 0.01 - 4 \cdot 0.001 = 0.826$

10.3.41

a. The binomial coefficients are $\binom{1/4}{0} = 1$, $\binom{1/4}{1} = \frac{1/4}{1} = \frac{1}{4}$, $\binom{1/4}{2} = \frac{(1/4)(-3/4)}{2!} = -\frac{3}{32}$, $\binom{1/4}{3} = \frac{(1/4)(-3/4)(-7/4)}{3!} = \frac{7}{128}$, so the first four terms of the series are $1 + \frac{1}{4}x - \frac{3}{32}x^2 + \frac{7}{128}x^3 + \cdots$.

b. Substitute $x = 0.12$ to get approximately 1.029.

10.3.43

a. The binomial coefficients are $\binom{-2/3}{0} = 1$, $\binom{-2/3}{1} = -\frac{2}{3}$, $\binom{-2/3}{2} = \frac{(-2/3)(-5/3)}{2!} = \frac{5}{9}$, $\binom{-2/3}{3} = \frac{(-2/3)(-5/3)(-8/3)}{3!} = -\frac{40}{81}$, so the first four terms of the series are $1 - \frac{2}{3}x + \frac{5}{9}x^2 - \frac{40}{81}x^3 + \cdots$.

b. Substitute $x = 0.18$ to get 0.89512.

10.3.45 $\sqrt{1 + x^2} = 1 + \frac{x^2}{2} - \frac{x^4}{8} + \frac{x^6}{16} - \cdots$. By the Ratio Test, the radius of convergence is 1. At the endpoints, the series obtained are convergent by the Alternating Series Test. Thus, the interval of convergence is $[-1, 1]$.

10.3.47 $\sqrt{9 - 9x} = 3\sqrt{1 - x} = 3 - \frac{3}{2}x - \frac{3}{8}x^2 - \frac{3}{16}x^3 - \cdots$. The interval of convergence is $[-1, 1]$.

10.3.49 $\sqrt{a^2 + x^2} = a\sqrt{1 + \frac{x^2}{a^2}} = a + \frac{x^2}{2a} - \frac{x^4}{8a^3} + \frac{x^6}{16a^5} - \cdots$. The series converges when $\frac{x^2}{a^2}$ is less than 1 in magnitude, so the radius of convergence is a. The series given by the endpoints is convergent by the Alternating Series Test, so the interval of convergence is $[-a, a]$.

10.3.51 $(1 + 4x)^{-2} = 1 - 2(4x) + 3(4x)^2 - 4(4x)^3 + \cdots = 1 - 8x + 48x^2 - 256x^3 + \cdots$.

10.3.53 $\frac{1}{(4+x^2)^2} = (4+x^2)^{-2} = \frac{1}{16}(1+(x^2/4))^{-2} = \frac{1}{16}\left(1 - 2\cdot\frac{x^2}{4} + 3\cdot\frac{x^4}{16} - 4\cdot\frac{x^6}{64} + \cdots\right) = \frac{1}{16} - \frac{1}{32}x^2 + \frac{3}{256}x^4 - \frac{1}{256}x^6 + \cdots$

10.3.55 $(3+4x)^{-2} = \frac{1}{9}\left(1 + \frac{4x}{3}\right)^{-2} = \frac{1}{9} - \frac{2}{9}\left(\frac{4x}{3}\right) + \frac{3}{9}\left(\frac{4x}{3}\right)^2 - \frac{4}{9}\left(\frac{4x}{3}\right)^3 + \cdots.$

10.3.57 The interval of convergence for the Taylor series for $f(x) = \sin x$ is $(-\infty, \infty)$. The remainder is $R_n(x) = \frac{f^{(n+1)}(c)}{(n+1)!}x^{n+1}$ for some c. Because $f^{(n+1)}(x)$ is $\pm\sin x$ or $\pm\cos x$, we have

$$\lim_{n\to\infty}|R_n(x)| \le \lim_{n\to\infty}\frac{1}{(n+1)!}\left|x^{n+1}\right| = 0$$

for any x.

10.3.59 The interval of convergence for the Taylor series for e^{-x} is $(-\infty, \infty)$. The remainder is $R_n(x) = \frac{(-1)^{n+1}e^{-c}}{(n+1)!}x^{n+1}$ for some c. Thus $\lim\limits_{n\to\infty}|R_n(x)| = 0$ for any x.

10.3.61

a. False. Not all of its derivatives are defined at zero - in fact, none of them are.

b. True. The derivatives of $\csc x$ involve positive powers of $\csc x$ and $\cot x$, both of which are defined at $\pi/2$, so that $\csc x$ has continuous derivatives at $\pi/2$.

c. False. For example, the Taylor series for $f(x^2)$ doesn't converge at $x = 1.9$, because the Taylor series for $f(x)$ doesn't converge at $1.9^2 = 3.61$.

d. False. The Taylor series centered at 1 involves derivatives of f evaluated at 1, not at 0.

e. True. The follows because the Taylor series must itself be an even function.

10.3.63

a. The relevant Taylor series are: $e^x = 1 + x + \frac{x^2}{2!} + \frac{x^3}{3!} + \frac{x^4}{4!} + \frac{x^5}{5!} + \frac{x^6}{6!} + \cdots$ and $e^{-x} = 1 - x + \frac{x^2}{2!} - \frac{x^3}{3!} + \frac{x^4}{4!} - \frac{x^5}{5!} + \frac{x^6}{6!} + \cdots$. Thus the first four terms of the resulting series are $\frac{1}{2}(e^x + e^{-x}) = 1 + \frac{x^2}{2!} + \frac{x^4}{4!} + \frac{x^6}{6!} + \cdots$.

b. Because each series converges (absolutely) on $(-\infty, \infty)$, so does their sum. The radius of convergence is ∞.

10.3.65

a. Use the binomial theorem. The binomial coefficients are $\binom{-2/3}{0} = 1$, $\binom{-2/3}{1} = -\frac{2}{3}$, $\binom{-2/3}{2} = \frac{(-2/3)(-5/3)}{2!} = \frac{5}{9}$, $\binom{-2/3}{3} = \frac{(-2/3)(-5/3)(-8/3)}{3!} = -\frac{40}{81}$ and then, substituting x^2 for x, we obtain $1 - \frac{2}{3}x^2 + \frac{5}{9}x^4 - \frac{40}{81}x^6 + \cdots$.

b. From Theorem 9.6 the radius of convergence is determined from $\left|x^2\right| < 1$, so it is 1.

10.3.67

a. From the binomial formula, the Taylor series for $(1 - x)^p$ is $\sum \binom{p}{k}(-1)^k x^k$, so the Taylor series for $(1 - x^2)^p$ is $\sum \binom{p}{k}(-1)^k x^{2k}$. Here $p = 1/2$, and the binomial coefficients are $\binom{1/2}{0} = 1$, $\binom{1/2}{1} = \frac{1/2}{1!} = \frac{1}{2}$, $\binom{1/2}{2} = \frac{(1/2)(-1/2)}{2!} = -\frac{1}{8}$, $\binom{1/2}{3} = \frac{(1/2)(-1/2)(-3/2)}{3!} = \frac{1}{16}$ so that $(1 - x^2)^{1/2} = 1 - \frac{1}{2}x^2 - \frac{1}{8}x^4 - \frac{1}{16}x^6 + \cdots$.

b. From Theorem 9.6 the radius of convergence is determined from $\left|x^2\right| < 1$, so it is 1.

10.3.69

a. $f(x) = (1 + x^2)^{-2}$; using the binomial series and substituting x^2 for x we obtain $1 - 2x^2 + 3x^4 - 4x^6 + \cdots$.

b. From Theorem 9.6 the radius of convergence is determined from $\left|x^2\right| < 1$, so it is 1.

10.3.71 Because $f(64) = 4$, and $f'(x) = \frac{1}{3}x^{-2/3}$, $f'(64) = \frac{1}{48}$, $f''(x) = -\frac{2}{9}x^{-5/3}$, $f''(64) = -\frac{1}{4608}$, $f'''(x) = \frac{10}{27}x^{-8/3}$, and $f'''(64) = \frac{10}{1769472} = \frac{5}{884736}$, the first four terms of the Taylor series are $4 + \frac{1}{48}(x - 64) - \frac{1}{4608 \cdot 2!}(x - 64)^2 + \frac{5}{884736 \cdot 3!}(x - 64)^3$. Evaluating at $x = 60$, we get 3.914870274.

10.3.73 Because $f(16) = 2$, and $f'(x) = \frac{1}{4}x^{-3/4}$, $f'(16) = \frac{1}{32}$, $f''(x) = -\frac{3}{16}x^{-7/4}$, $f''(16) = -\frac{3}{2048}$, $f'''(x) = \frac{21}{64}x^{-11/4}$, and $f'''(16) = \frac{21}{131072}$, the first four terms of the Taylor series are $2 + \frac{1}{32}(x - 16) - \frac{3}{2048 \cdot 2!}(x - 16)^2 + \frac{21}{131072 \cdot 3!}(x - 16)^3$. Evaluating at $x = 13$, we get 1.898937225.

10.3.75 Evaluate the binomial coefficient $\binom{1/2}{k} = \frac{(1/2)(-1/2)(-3/2)\cdots(1/2-k+1)}{k!} = \frac{(1/2)(-1/2)\cdots((3-2k)/2)}{k!} = (-1)^{k-1}2^{-k}\frac{1\cdot3\cdots(2k-3)}{k!} = (-1)^{k-1}2^{-k}\frac{(2k-2)!}{2^{k-1}\cdot(k-1)!\cdot k!} = (-1)^{k-1}2^{1-2k}\cdot\frac{1}{k}\binom{2k-2}{k-1}$. This is the coefficient of x^k in the Taylor series for $\sqrt{1+x}$. Substituting $4x$ for x, the Taylor series becomes $\sum_{k=0}^{\infty}(-1)^{k-1}2^{1-2k}\cdot\frac{1}{k}\binom{2k-2}{k-1}(4x)^k = \sum_{k=0}^{\infty}(-1)^{k-1}\frac{2}{k}\binom{2k-2}{k-1}x^k$. If we can show that k divides $\binom{2k-2}{k-1}$, we will be done, for then the coefficient of x^k will be an integer. But $\binom{2k-2}{k-1} - \binom{2k-2}{k-2} = \frac{(2k-2)!}{(k-1)!(k-1)!} - \frac{(2k-2)!}{(k-2)!k!} = \frac{(2k-2)!}{(k-1)!(k-1)!} - \frac{(2k-2)!(k-1)}{(k-1)!(k-1)!k} = \frac{k(2k-2)!-(k-1)(2k-2)!}{k(k-1)!(k-1)!} = \frac{1}{k}\frac{(2k-2)!}{(k-1)!(k-1)!} = \frac{1}{k}\binom{2k-2}{k-1}$ and thus we have shown that k divides $\binom{2k-2}{k-1}$.

10.3.77

a. The Maclaurin series for $\sin x$ is $x - \frac{1}{3!}x^3 + \frac{1}{5!}x^5 - \frac{1}{7!}x^7 + \cdots$. Squaring the first four terms yields

$$\left(x - \frac{1}{3!}x^3 + \frac{1}{5!}x^5 - \frac{1}{7!}x^7\right)^2$$

$$= x^2 - \frac{2}{3!}x^4 + \left(\frac{2}{5!} + \frac{1}{3!3!}\right)x^6 + \left(-2\cdot\frac{1}{7!} - 2\cdot\frac{1}{3!5!}\right)x^8$$

$$= x^2 - \frac{1}{3}x^4 + \frac{2}{45}x^6 - \frac{1}{315}x^8.$$

b. The Maclaurin series for $\cos x$ is $1 - \frac{1}{2}x^2 + \frac{1}{4!}x^4 - \frac{1}{6!}x^6 + \frac{1}{8!}x^8 - \cdots$. Substituting $2x$ for x in the Maclaurin series for $\cos x$ and then computing $(1 - \cos 2x)/2$, we obtain

$$(1 - (1-\frac{1}{2}(2x)^2 + \frac{1}{4!}(2x)^4 - \frac{1}{6!}(2x)^6 + \frac{1}{8!}(2x)^8)/2$$

$$= (2x^2 - \frac{2}{3}x^4 + \frac{4}{45}x^6 - \frac{2}{315}x^8)/2$$

$$= x^2 - \frac{1}{3}x^4 + \frac{2}{45}x^6 - \frac{1}{315}x^8,$$

and the two are the same.

c. If $f(x) = \sin^2 x$, then $f(0) = 0$, $f'(x) = \sin 2x$, so $f'(0) = 0$. $f''(x) = 2\cos 2x$, so $f''(x) = 2$, $f'''(x) = -4\sin 2x$, so $f'''(0) = 0$. Note that from this point $f^{(n)}(0) = 0$ if n is odd and $f^{(n)}(0) = \pm 2^{n-1}$ if n is even, with the signs alternating for every other even n. Thus, the series for $\sin^2 x$ is

$$2x^2/2 - 8x^4/4! + 32x^6/6! - 128x^8/8! + \cdots = x^2 - \frac{1}{3}x^4 + \frac{2}{45}x^6 - \frac{1}{315}x^8 + \cdots.$$

10.3.79 There are many solutions. For example, first find a series that has $(-1, 1)$ as an interval of convergence, say $\frac{1}{1-x} = \sum_{k=0}^{\infty} x^k$. Then the series $\frac{1}{1-x/2} = \sum_{k=0}^{\infty}\left(\frac{x}{2}\right)^k$ has $(-2, 2)$ as its interval of convergence. Now shift the series up so that it is centered at 4. We have $\sum_{k=0}^{\infty}\left(\frac{x-4}{2}\right)^k$, which has interval of convergence $(2, 6)$.

10.3.81 $\frac{1\cdot3\cdot5\cdot7}{2\cdot4\cdot6\cdot8}x^4 - \frac{1\cdot3\cdot5\cdot7\cdot9}{2\cdot4\cdot6\cdot8\cdot10}x^5$.

10.3.83 Use the Taylor series for $\cos x$ centered at $\pi/4$: $\frac{\sqrt{2}}{2}(1 - (x - \pi/4) - \frac{1}{2}(x - \pi/4)^2 + \frac{1}{6}(x - \pi/4)^3 + \cdots)$. The remainder after n terms (because the derivatives of $\cos x$ are bounded by 1 in magnitude) is $|R_n(x)| \leq \frac{1}{(n+1)!} \cdot \left(\frac{\pi}{4} - \frac{2\pi}{9}\right)^{n+1}$.

Solving for $|R_n(x)| < 10^{-4}$, we obtain $n = 3$. Evaluating the first four terms (through $n = 3$) of the series we get 0.7660427050. The true value is ≈ 0.7660444431.

10.3.85 Use the Taylor series for $f(x) = x^{1/3}$ centered at 64: $4 + \frac{1}{48}(x - 64) - \frac{1}{9216}(x - 64)^2 + \cdots$. Because we wish to evaluate this series at $x = 83$, $|R_n(x)| = \frac{|f^{(n+1)}(c)|}{(n+1)!}(83 - 64)^{n+1}$. We compute that $|f^{(n+1)}(c)| = \frac{2 \cdot 5 \cdots (3n-1)}{3^{n+1}c^{(3n+2)/3}}$, which is maximized at $c = 64$. Thus

$$|R_n(x)| \leq \frac{2 \cdot 5 \cdots (3n-1)}{3^{n+1}64^{(3n+2)/3}(n+1)!}19^{n+1}$$

Solving for $|R_n(x)| < 10^{-4}$, we obtain $n = 5$. Evaluating the terms of the series through $n = 5$ gives 4.362122553. The true value is ≈ 4.362070671.

10.3.87

a. Use the Taylor series for $(125 + x)^{1/3}$ centered at $x = 0$. Using the first four terms and evaluating at $x = 3$ gives a result (5.03968) accurate to within 10^{-4}.

b. Use the Taylor series for $x^{1/3}$ centered at $x = 125$. Note that this gives the identical Taylor series except that the exponential terms are $(x - 125)^n$ rather than x^n. Thus we need terms up through $(x - 125)^3$, just as before, evaluated at $x = 128$, and we obtain the identical result.

c. Because the two Taylor series are the same except for the shifting, the results are equivalent.

10.3.89 Consider the remainder after the first term of the Taylor series. Taylor's Theorem indicates that $R_1(x) = \frac{f''(c)}{2}(x - a)^2$ for some c between x and a, so that $f(x) = f(a) + f'(a)(x - a) + \frac{f''(c)}{2}(x - a)^2$. But $f'(a) = 0$, so that for every x in an interval containing a, there is a c between x and a such that $f(x) = f(a) + \frac{f''(c)}{2}(x - a)^2$.

a. If $f''(x) > 0$ on the interval containing a, then for every x in that interval, we have $f(x) = f(a) + \frac{f''(c)}{2}(x - a)^2$ for some c between x and a. But $f''(c) > 0$ and $(x - a)^2 > 0$, so that $f(x) > f(a)$ and a is a local minimum.

b. If $f''(x) < 0$ on the interval containing a, then for every x in that interval, we have $f(x) = f(a) + \frac{f''(c)}{2}(x - a)^2$ for some c between x and a. But $f''(c) < 0$ and $(x - a)^2 > 0$, so that $f(x) < f(a)$ and a is a local maximum.

10.4 Working with Taylor Series

10.4.1 Replace f and g by their Taylor series centered at a, and evaluate the limit.

10.4.3 Substitute -0.6 for x in the Taylor series for e^x centered at 0. Note that this series is an alternating series, so the error can easily be estimated by looking at the magnitude of the first neglected term.

10.4.5 The series is $f'(x) = \sum_{k=1}^{\infty} kc_k x^{k-1}$, which converges for $|x| < b$.

10.4.7 Because $e^x = 1 + x + x^2/2! + x^3/3! + \cdots$, we have $\frac{e^x - 1}{x} = 1 + x/2! + \cdots$, so $\lim_{x \to 0} \frac{e^x - 1}{x} = 1$.

10.4.9 Because $-\ln(1 - x) = x + \frac{x^2}{2} + \frac{x^3}{3} + \frac{x^4}{4} + \frac{x^5}{5} + \cdots$, we have $\frac{-x - \ln(1-x)}{x^2} = \frac{1}{2} + \frac{x}{3} + \frac{x^2}{4} + \cdots$, so $\lim_{x \to 0} \frac{-x - \ln(1-x)}{x^2} = \frac{1}{2}$.

10.4.11 We compute that

$$\frac{e^x - e^{-x}}{x} = \frac{1}{x}\left(\left(1 + x + \frac{x^2}{2} + \frac{x^3}{6} + \cdots\right) - \left(1 - x + \frac{x^2}{2} - \frac{x^3}{6} + \cdots\right)\right)$$

$$= \frac{1}{x}\left(2x + \frac{x^3}{3} + \cdots\right) = 2 + \frac{x^2}{3} + \cdots$$

so the limit of $\dfrac{e^x - e^{-x}}{x}$ as $x \to 0$ is 2.

10.4.13 We compute that

$$\frac{2\cos 2x - 2 + 4x^2}{2x^4} = \frac{1}{2x^4}\left(2\left(1 - \frac{(2x)^2}{2} + \frac{(2x)^4}{24} - \frac{(2x)^6}{720} + \cdots\right) - 2 + 4x^2\right)$$

$$= \frac{1}{2x^4}\left(\frac{(2x)^4}{12} - \frac{(2x)^6}{360} + \cdots\right) = \frac{2}{3} - \frac{4x^2}{45} + \cdots$$

so the limit of $\dfrac{2\cos 2x - 2 + 4x^2}{2x^4}$ as $x \to 0$ is $\dfrac{2}{3}$.

10.4.15 We have $\ln(1 + x) = x - \frac{1}{2}x^2 + \frac{1}{3}x^3 - \frac{1}{4}x^4 + \cdots$, so that

$$\frac{\ln(1 + x) - x + x^2/2}{x^3} = \frac{x^3/3 - x^4/4 + \cdots}{x^3} = \frac{1}{3} - \frac{x}{4} + \cdots$$

so that $\displaystyle\lim_{x\to 0} \frac{\ln(1 + x) - x + x^2/2}{x^3} = \frac{1}{3}$.

10.4.17 We compute that

$$\frac{3\tan^{-1} x - 3x + x^3}{x^5} = \frac{1}{x^5}\left(3\left(x - \frac{x^3}{3} + \frac{x^5}{5} - \frac{x^7}{7} + \cdots\right) - 3x + x^3\right)$$

$$= \frac{1}{x^5}\left(\frac{3x^5}{5} - \frac{3x^7}{7} + \cdots\right) = \frac{3}{5} - \frac{3x^2}{7} + \cdots$$

so the limit of $\dfrac{3\tan^{-1} x - 3x + x^3}{x^5}$ as $x \to 0$ is $\dfrac{3}{5}$.

10.4.19 The Taylor series for $\sin 2x$ centered at 0 is

$$\sin 2x = 2x - \frac{1}{3!}(2x)^3 + \frac{1}{5!}(2x)^5 - \frac{1}{7!}(2x)^7 + \cdots = 2x - \frac{4}{3}x^3 + \frac{4}{15}x^5 - \frac{8}{315}x^7 + \cdots.$$

Thus

$$\frac{12x - 8x^3 - 6\sin 2x}{x^5} = \frac{12 - 8x^3 - (12x - 8x^3 + \frac{8}{5}x^5 - \frac{16}{105}x^7 + \cdots)}{x^5}$$

$$= -\frac{8}{5} + \frac{16}{105}x^2 - \cdots,$$

so $\displaystyle\lim_{x\to 0} \frac{12x - 8x^3 - 6\sin 2x}{x^5} = -\frac{8}{5}$.

10.4.21 The Taylor series for $\ln(x - 1)$ centered at 2 is

$$\ln(x - 1) = (x - 2) - \frac{1}{2}(x - 2)^2 + \cdots.$$

We compute that

$$\frac{x - 2}{\ln(x - 1)} = \frac{x - 2}{(x - 2) - \frac{1}{2}(x - 2)^2 + \cdots} = \frac{1}{1 - \frac{1}{2}(x - 2) + \cdots}$$

so the limit of $\dfrac{x - 2}{\ln(x - 1)}$ as $x \to 2$ is 1.

10.4.23 Computing Taylor series centers at 0 gives

$$e^{-2x} = 1 - 2x + \frac{1}{2!}(-2x)^2 + \frac{1}{3!}(-2x)^3 + \cdots = 1 - 2x + 2x^2 - \frac{4}{3}x^3 + \cdots$$

$$e^{-x/2} = 1 - \frac{x}{2} + \frac{1}{2!}\left(-\frac{x}{2}\right)^2 + \frac{1}{3!}\left(-\frac{x}{2}\right)^3 + \cdots = 1 - \frac{x}{2} + \frac{1}{8}x^2 - \frac{1}{48}x^3 + \cdots.$$

Thus

$$\frac{e^{-2x} - 4e^{-x/2} + 3}{2x^2} = \frac{1 - 2x + 2x^2 - \frac{4}{3}x^3 + \cdots - (4 - 2x + \frac{1}{2}x^2 - \frac{1}{12}x^3 + \cdots) + 3}{2x^2}$$

$$= \frac{\frac{3}{2}x^2 - \frac{5}{4}x^3 + \cdots}{2x^2}$$

$$= \frac{3}{4} - \frac{5}{8}x + \cdots$$

so $\displaystyle\lim_{x\to 0} \frac{e^{-2x} - 4e^{-x/2} + 3}{2x^2} = \frac{3}{4}$.

10.4.25

a. $f'(x) = \frac{d}{dx}\left(\sum_{k=0}^{\infty} \frac{x^k}{k!}\right) = \sum_{k=1}^{\infty} k\frac{x^{k-1}}{k!} = \sum_{k=0}^{\infty} \frac{x^k}{k!} = f(x)$.

b. $f'(x) = e^x$ as well.

c. The series converges on $(-\infty, \infty)$.

10.4.27

a. $f'(x) = \frac{d}{dx}(\ln(1+x)) = \frac{d}{dx}\left(\sum_{k=1}^{\infty}(-1)^{k+1}\frac{1}{k}x^k\right) = \sum_{k=1}^{\infty}(-1)^{k+1}x^{k-1} = \sum_{k=0}^{\infty}(-1)^k x^k$.

b. This is the power series for $\frac{1}{1+x}$.

c. The Taylor series for $\ln(1+x)$ converges on $(-1, 1)$, as does the Taylor series for $\frac{1}{1+x}$.

10.4.29

a.

$$f'(x) = \frac{d}{dx}(e^{-2x}) = \frac{d}{dx}\left(\sum_{k=0}^{\infty} \frac{(-2x)^k}{k!}\right) = \frac{d}{dx}\left(\sum_{k=0}^{\infty}(-2)^k \frac{x^k}{k!}\right) = -2\sum_{k=1}^{\infty}(-2)^{k-1}\frac{x^{k-1}}{(k-1)!} = -2\sum_{k=0}^{\infty} \frac{(-2x)^k}{k!}.$$

b. This is the Taylor series for $-2e^{-2x}$.

c. Because the Taylor series for e^{-2x} converges on $(-\infty, \infty)$, so does this one.

10.4.31

a. $\tan^{-1} x = x - \frac{x^3}{3} + \frac{x^5}{5} - \cdots$, so $\frac{d}{dx}\tan^{-1} x^2 = 1 - x^2 + x^4 - x^6 + \cdots$.

b. This is the series for $\frac{1}{1+x^2}$.

c. Because the series for $\tan^{-1} x$ has a radius of convergence of 1, this series does too. Checking the endpoints shows that the interval of convergence is $(-1, 1)$.

10.4.33

a. Because $y(0) = 2$, we have $0 = y'(0) - y(0) = y'(0) - 2$ so that $y'(0) = 2$. Differentiating the equation gives $y''(0) = y'(0)$, so that $y''(0) = 2$. Successive derivatives also have the value 2 at 0, so the Taylor series is $2\sum_{k=0}^{\infty} \frac{t^k}{k!}$.

b. $2\sum_{k=0}^{\infty} \frac{t^k}{k!} = 2e^t$.

10.4.35

a. $y(0) = 2$, so that $y'(0) = 16$. Differentiating, $y''(t) - 3y'(t) = 0$, so that $y''(0) = 48$, and in general $y^{(k)}(0) = 3y^{(k-1)}(0) = 3^{k-1} \cdot 16$. Thus the power series is $2 + \frac{16}{3}\sum_{k=1}^{\infty} \frac{(3t)^k}{k!} = 2 + \sum_{k=1}^{\infty} \frac{3^{k-1}16}{k!}t^k$.

b. $2 + \frac{16}{3}\sum_{k=1}^{\infty} \frac{(3t)^k}{k!} = 2 + \frac{16}{3}(e^{3t} - 1) = \frac{16}{3}e^{3t} - \frac{10}{3}$.

10.4.37 The Taylor series for e^{-x^2} is $\sum_{k=0}^{\infty}(-1)^k \frac{x^{2k}}{k!}$. Thus, the desired integral is $\int_0^{0.25} \sum_{k=0}^{\infty}(-1)^k \frac{x^{2k}}{k!}\,dx = \sum_{k=0}^{\infty}(-1)^k \frac{x^{2k+1}}{(2k+1)k!}\Big|_0^{0.25} = \sum_{k=0}^{\infty}(-1)^k \frac{1}{(2k+1)k!4^{2k+1}}$. Because this is an alternating series, to approximate it to within 10^{-4}, we must find n such that $a_{n+1} < 10^{-4}$, or $\frac{1}{(2n+3)(n+1)! \cdot 4^{2n+3}} < 10^{-4}$. This occurs for $n = 1$, so $\sum_{k=0}^{1}(-1)^k \frac{1}{(2k+1) \cdot k! \cdot 4^{2k+1}} = \frac{1}{4} - \frac{1}{192} \approx 0.245$.

10.4.39 The Taylor series for $\cos 2x^2$ is $\sum_{k=0}^{\infty}(-1)^k \frac{(2x^2)^{2k}}{(2k)!} = \sum_{k=0}^{\infty}(-1)^k \frac{4^k x^{4k}}{(2k)!}$. Note that $\cos x$ is an even function, so we compute the integral from 0 to 0.35 and double it:

$$2\int_0^{0.35} \sum_{k=0}^{\infty}(-1)^k \frac{4^k x^{4k}}{(2k)!}\,dx = 2\left(\sum_{k=0}^{\infty}(-1)^k \frac{4^k x^{4k+1}}{(4k+1)(2k)!}\right)\Big|_0^{0.35} = 2\left(\sum_{k=0}^{\infty}(-1)^k \frac{4^k(0.35)^{4k+1}}{(4k+1)(2k)!}\right).$$

Because this is an alternating series, to approximate it to within $\frac{1}{2} \cdot 10^{-4}$, we must find n such that $a_{n+1} < \frac{1}{2} \cdot 10^{-4}$, or $\frac{4^{n+1}(0.35)^{4n+5}}{(4n+3)(2n+2)!} < \frac{1}{2} \cdot 10^{-4}$. This occurs first for $n = 1$, and we have $2\left(.35 - \frac{4 \cdot (0.35)^5}{5 \cdot 2!}\right) \approx 0.696$.

10.4.41 $\tan^{-1} x = x - x^3/3 + x^5/5 - x^7/7 + x^9/9 - \cdots$, so $\int \tan^{-1} x\,dx = \int(x - x^3/3 + x^5/5 - x^7/7 + x^9/9 - \cdots)\,dx = C + \frac{x^2}{2} - \frac{x^4}{12} + \frac{x^6}{30} - \frac{x^8}{56} + \cdots$. Thus, $\int_0^{0.35} \tan^{-1} x\,dx = \frac{(0.35)^2}{2} - \frac{(0.35)^4}{12} + \frac{(0.35)^6}{30} - \frac{(0.35)^8}{56} + \cdots$. Note that this series is alternating, and $\frac{(0.35)^6}{30} < 10^{-4}$, so we add the first two terms to approximate the integral to the desired accuracy. Calculating gives approximately 0.060.

10.4.43 The Taylor series for $(1+x^6)^{-1/2}$ is $\sum_{k=0}^{\infty}\binom{-1/2}{k}x^{6k}$, so the desired integral is $\int_0^{0.5} \sum_{k=0}^{\infty}\binom{-1/2}{k}x^{6k}\,dx$ $= \sum_{k=0}^{\infty} \frac{1}{6k+1}\binom{-1/2}{k}x^{6k+1}\Big|_0^{0.5} = \sum_{k=0}^{\infty} \frac{1}{6k+1}\binom{-1/2}{k}(0.5)^{6k+1}$. This is an alternating series because the binomial coefficients alternate in sign, so to approximate it to within 10^{-4}, we must find n such that $a_{n+1} < 10^{-4}$, or $\left|\frac{1}{6n+7}\binom{-1/2}{n+1}(0.5)^{6n+7}\right| < 10^{-4}$. This occurs first for $n = 1$, so we have $\binom{-1/2}{0}0.5 + \frac{1}{7}\binom{-1/2}{1}(0.5)^7 \approx 0.499$.

10.4.45 Use the Taylor series for e^x at 0: $1 + \frac{2}{1!} + \frac{2^2}{2!} + \frac{2^3}{3!}$.

10.4.47 Use the Taylor series for $\cos x$ at 0: $1 - \frac{2^2}{2!} + \frac{2^4}{4!} - \frac{2^6}{6!}$

10.4.49 Use the Taylor series for $\ln(1 + x)$ evaluated at $x = 1/2$: $\frac{1}{2} - \frac{1}{2} \cdot \frac{1}{4} + \frac{1}{3} \cdot \frac{1}{8} - \frac{1}{4} \cdot \frac{1}{16}$.

10.4.51 The Taylor series for f centered at 0 is $\frac{-1 + \sum_{k=0}^{\infty} \frac{x^k}{k!}}{x} = \frac{\sum_{k=1}^{\infty} \frac{x^k}{k!}}{x} = \sum_{k=1}^{\infty} \frac{x^{k-1}}{k!} = \sum_{k=0}^{\infty} \frac{x^k}{(k+1)!}$. Evaluating both sides at $x = 1$, we have $e - 1 = \sum_{k=0}^{\infty} \frac{1}{(k+1)!}$.

10.4.53 The Maclaurin series for $\ln(1+x)$ is $x - \frac{1}{2}x^2 + \frac{1}{3}x^3 - \frac{1}{4}x^4 + \cdots = \sum_{k=1}^{\infty}(-1)^{k+1}\frac{x^k}{k}$. By the Ratio Test, $\lim_{k\to\infty}\left|\frac{a_{k+1}}{a_k}\right| = \lim_{k\to\infty}\left|\frac{x^{k+1}k}{x^k(k+1)}\right| = |x|$, so the radius of convergence is 1. The series diverges at -1 and converges at 1, so the interval of convergence is $(-1, 1]$. Evaluating at 1 gives $\ln 2 = \sum_{k=1}^{\infty}(-1)^{k+1}\frac{1}{k} = 1 - \frac{1}{2} + \frac{1}{3} - \frac{1}{4} + \cdots$.

10.4.55 $\sum_{k=0}^{\infty}\frac{x^k}{2^k} = \sum_{k=0}^{\infty}\left(\frac{x}{2}\right)^k = \frac{1}{1-\frac{x}{2}} = \frac{2}{2-x}$.

10.4.57 $\sum_{k=0}^{\infty}(-1)^k\frac{x^{2k}}{4^k} = \sum_{k=0}^{\infty}\left(\frac{-x^2}{4}\right)^k = \frac{1}{1+\frac{x^2}{4}} = \frac{4}{4+x^2}.$

10.4.59 $\ln(1+x) = -\sum_{k=1}^{\infty}(-1)^k\frac{x^k}{k}$, so $\ln(1-x) = -\sum_{k=1}^{\infty}\frac{x^k}{k}$, and finally $-\ln(1-x) = \sum_{k=1}^{\infty}\frac{x^k}{k}.$

10.4.61

$$
\begin{aligned}
\sum_{k=1}^{\infty}(-1)^k\frac{kx^{k+1}}{3^k} &= \sum_{k=1}^{\infty}(-1)^k\frac{k}{3^k}x^{k+1} = \sum_{k=1}^{\infty}k\left(-\frac{1}{3}\right)^k x^{k+1} \\
&= x^2\sum_{k=1}^{\infty}\left(-\frac{1}{3}\right)^k kx^{k-1} = x^2\sum_{k=1}^{\infty}\left(-\frac{1}{3}\right)^k \frac{d}{dx}(x^k) \\
&= x^2\frac{d}{dx}\left(\sum_{k=1}^{\infty}\left(-\frac{x}{3}\right)^k\right) = x^2\frac{d}{dx}\left(\frac{1}{1+\frac{x}{3}}\right) = -\frac{3x^2}{(x+3)^2}.
\end{aligned}
$$

10.4.63 $\sum_{k=2}^{\infty}\frac{k(k-1)x^k}{3^k} = x^2\sum_{k=2}^{\infty}\frac{k(k-1)x^{k-2}}{3^k} = x^2\frac{d^2}{dx^2}\left(\sum_{k=2}^{\infty}\frac{x^k}{3^k}\right)$
$= x^2\frac{d^2}{dx^2}\left(\sum_{k=2}^{\infty}\left(\frac{x}{3}\right)^k\right) = x^2\frac{d^2}{dx^2}\left(\frac{x^2}{9}\cdot\frac{1}{1-\frac{x}{3}}\right) = x^2\frac{d^2}{dx^2}\left(\frac{x^2}{9-3x}\right) = x^2\frac{-6}{(x-3)^3} = \frac{-6x^2}{(x-3)^3}.$

10.4.65

a. False. This is because $\frac{1}{1-x}$ is not continuous at 1, which is in the interval of integration.

b. False. The Ratio Test shows that the radius of convergence for the Taylor series for $\tan^{-1} x$ centered at 0 is 1.

c. True. $\sum_{k=0}^{\infty}\frac{x^k}{k!} = e^x$. Substitute $x = \ln 2$.

10.4.67 The Taylor series for $\sin x$ centered at 0 is

$$\sin x = x - \frac{x^3}{6} + \frac{x^5}{120} - \cdots.$$

We compute that

$$
\begin{aligned}
\frac{\sin ax}{\sin bx} &= \frac{ax - \frac{(ax)^3}{6} + \frac{(ax)^5}{120} - \cdots}{bx - \frac{(bx)^3}{6} + \frac{(bx)^5}{120} - \cdots} \\
&= \frac{a - \frac{a^3 x^2}{6} + \frac{a^5 x^4}{120} - \cdots}{b - \frac{b^3 x^2}{6} + \frac{b^5 x^4}{120} - \cdots}
\end{aligned}
$$

so the limit of $\frac{\sin ax}{\sin bx}$ as $x \to 0$ is $\frac{a}{b}$.

10.4.69 Compute instead the limit of the log of this expression, $\lim_{x\to 0}\frac{\ln(\sin x/x)}{x^2}$. If the Taylor expansion of $\ln(\sin x/x)$ is $\sum_{k=0}^{\infty}c_k x^k$, then $\lim_{x\to 0}\frac{\ln(\sin x/x)}{x^2} = \lim_{x\to 0}\sum_{k=0}^{\infty}c_k x^{k-2} = \lim_{x\to 0}c_0 x^{-2} + c_1 x^{-1} + c_2$, because the higher-order terms have positive powers of x and thus approach zero as x does. So compute the terms of the Taylor series of $\ln\left(\frac{\sin x}{x}\right)$ up through the quadratic term. The relevant Taylor series are: $\frac{\sin x}{x} = 1 - \frac{1}{6}x^2 + \frac{1}{120}x^4 - \cdots$, $\ln(1+x) = x - \frac{1}{2}x^2 + \frac{1}{3}x^3 - \cdots$ and we substitute the Taylor series for $\frac{\sin x}{x} - 1$ for x in the Taylor series for $\ln(1+x)$. Because the lowest power of x in the first Taylor series is 2, it follows that only the linear term in the series for $\ln(1+x)$ will give any powers of x that are at most quadratic. The only term that results is $-\frac{1}{6}x^2$. Thus $c_0 = c_1 = 0$ in the above, and $c_2 = -\frac{1}{6}$, so that $\lim_{x\to 0}\frac{\ln(\sin x/x)}{x^2} = -\frac{1}{6}$ and thus $\lim_{x\to 0}\left(\frac{\sin x}{x}\right)^{1/x^2} = e^{-1/6}.$

10.4.71 The Taylor series we need are $\cos x = 1 - \frac{1}{2}x^2 + \frac{1}{24}x^4 + \ldots$, $e^t = 1 + t + \frac{1}{2!}t^2 + \frac{1}{3!}t^3 + \frac{1}{4!}t^4 + \ldots$. We are looking for powers of x^3 and x^4 that occur when the first series is substituted for t in the second series. Clearly there will be no odd powers of x, because $\cos x$ has only even powers. Thus the coefficient of x^3 is zero, so that $f^{(3)}(0) = 0$. The coefficient of x^4 comes from the expansion of $1 - \frac{1}{2}x^2 + \frac{1}{24}x^4$ in each term of e^t. Higher powers of x clearly cannot contribute to the coefficient of x^4. Thus consider $\left(1 - \frac{1}{2}x^2 + \frac{1}{24}x^4\right)^k$. The term $-\frac{1}{2}x^2$ generates $\binom{k}{2}$ terms of value $\frac{1}{4}x^4$ for $k \geq 2$, while the other term generates k terms of value $\frac{1}{24}x^4$ for $k \geq 1$. These terms all have to be divided by the $k!$ appearing in the series for e^t. So the total coefficient of x^4 is $\frac{1}{24}\sum_{k=1}^{\infty}\frac{k}{k!} + \frac{1}{4}\sum_{k=2}^{\infty}\binom{k}{2}\frac{1}{k!}$, $= \frac{1}{24}\sum_{k=1}^{\infty}\frac{1}{(k-1)!} + \frac{1}{4}\sum_{k=2}^{\infty}\frac{1}{2\cdot(k-2)!}$, $= \frac{1}{24}\sum_{k=0}^{\infty}\frac{1}{k!} + \frac{1}{8}\sum_{k=0}^{\infty}\frac{1}{k!}$, $= \frac{1}{24}e + \frac{1}{8}e = \frac{e}{6}$ Thus $f^{(4)}(0) = \frac{e}{6}\cdot 4! = 4e$.

10.4.73 The Taylor series for $\sin t^2$ is $\sin t^2 = t^2 - \frac{1}{3!}t^6 + \frac{1}{5!}t^{10} - \ldots$, so that $\int_0^x \sin t^2\, dt = \frac{1}{3}t^3 - \frac{1}{7\cdot 3!}t^7 + \ldots \Big|_0^x = \frac{1}{3}x^3 - \frac{1}{7\cdot 3!}x^7 + \ldots$. Thus $f^{(3)}(0) = \frac{3!}{3} = 2$ and $f^{(4)}(0) = 0$.

10.4.75 Consider the series $\sum_{k=1}^{\infty} x^k = \frac{x}{1-x}$. Differentiating both sides gives $\frac{1}{(1-x)^2} = \sum_{k=0}^{\infty} kx^{k-1} = \frac{1}{x}\sum_{k=0}^{\infty} kx^k$ so that $\frac{x}{(1-x)^2} = \sum_{k=0}^{\infty} kx^k$. Evaluate both sides at $x = 1/2$ to see that the sum of the series is $\frac{1/2}{(1-1/2)^2} = 2$. Thus the expected number of tosses is 2.

10.4.77

a. We look first for a Taylor series for $(1 - k^2\sin^2\theta)^{-1/2}$. Because $(1 - k^2x^2)^{-1/2} = (1 - (kx)^2)^{-1/2} = \sum_{i=0}^{\infty}\binom{-1/2}{i}(kx)^{2i}$, and $\sin\theta = \theta - \frac{1}{3!}\theta^3 + \frac{1}{5!}\theta^5 - \ldots$, substituting the second series into the first gives $\frac{1}{\sqrt{1-k^2\sin^2\theta}} = 1 + \frac{1}{2}k^2\theta^2 + \left(-\frac{1}{6}k^2 + \frac{3}{8}k^4\right)\theta^4 + \left(\frac{1}{45}k^2 - \frac{1}{4}k^4 + \frac{5}{16}k^6\right)\theta^6 + \left(\frac{-1}{630}k^2 + \frac{3}{40}k^4 - \frac{5}{16}k^6 + \frac{35}{128}k^8\right)\theta^8 + \ldots$.
   Integrating with respect to θ and evaluating at $\pi/2$ (the value of the antiderivative is 0 at 0) gives $\frac{1}{2}\pi + \frac{1}{48}k^2\pi^3 + \frac{1}{160}\left(-\frac{1}{6}k^2 + \frac{3}{8}k^4\right)\pi^5 + \frac{1}{896}\left(\frac{1}{45}k^2 - \frac{1}{4}k^4 + \frac{5}{16}k^6\right)\pi^7 + \frac{1}{4608}\left(-\frac{1}{630}k^2 + \frac{3}{40}k^4 - \frac{5}{16}k^6 + \frac{35}{128}k^8\right)\pi^9$.
   Evaluating these terms for $k = 0.1$ gives $F(0.1) \approx 1.574749680$. (The true value is approximately 1.574745562.)

b. The terms above, with coefficients of k^n converted to decimal approximations, is $1.5707 + .3918\cdot k^2 + .3597\cdot k^4 - .9682\cdot k^6 + 1.7689\cdot k^8$. The coefficients are all less than 2 and do not appear to be increasing very much if at all, so if we want the result to be accurate to within 10^{-3} we should probably take n such that $k^n < \frac{1}{2}\times 10^{-3} = .0005$, so $n = 4$ for this value of k.

c. By the above analysis, we would need a larger n because $0.2^n > 0.1^n$ for a given value of n.

10.4.79

a. By the Fundamental Theorem, $S'(x) = \sin x^2$, $C'(x) = \cos x^2$.

b. The relevant Taylor series are $\sin t^2 = t^2 - \frac{1}{3!}t^6 + \frac{1}{5!}t^{10} - \frac{1}{7!}t^{14} + \ldots$, and $\cos t^2 = 1 - \frac{1}{2!}t^4 + \frac{1}{4!}t^8 - \frac{1}{6!}t^{12} + \ldots$. Integrating, we have $S(x) = \frac{1}{3}x^3 - \frac{1}{7\cdot 3!}x^7 + \frac{1}{11\cdot 5!}x^{11} - \frac{1}{15\cdot 7!}x^{15} + \ldots$, and $C(x) = x - \frac{1}{5\cdot 2!}x^5 + \frac{1}{9\cdot 4!}x^9 - \frac{1}{13\cdot 6!}x^{13} + \ldots$.

c. $S(0.05) \approx \frac{1}{3}(0.05)^3 - \frac{1}{42}(0.05)^7 + \frac{1}{1320}(0.05)^{11} - \frac{1}{75600}(0.05)^{15} \approx 4.166664807 \times 10^{-5}$. $C(-0.25) \approx (-0.25) - \frac{1}{10}(-0.25)^5 + \frac{1}{216}(-0.25)^9 - \frac{1}{9360}(-0.25)^{13} \approx -.2499023616$.

d. The series is alternating. Because $a_{n+1} = \frac{1}{(4n+7)(2n+3)!}(0.05)^{4n+7}$, and this is less than 10^{-4} for $n = 0$, only one term is required.

e. The series is alternating. Because $a_{n+1} = \frac{1}{(4n+5)(2n+2)!}(0.25)^{4n+5}$, and this is less than 10^{-6} for $n = 1$, two terms are required.

10.4.81

a. $J_0(x) = 1 - \frac{1}{4}x^2 + \frac{1}{16\cdot 2!^2}x^4 - \frac{1}{2^6\cdot 3!^2}x^6 + \ldots$.

b. Using the Ratio Test: $\left|\dfrac{a_{k+1}}{a_k}\right| = \dfrac{x^{2k+2}}{2^{2k+2}((k+1)!)^2} \cdot \dfrac{2^{2k}(k!)^2}{x^{2k}} = \dfrac{x^2}{4(k+1)^2}$, which has limit 0 as $k \to \infty$ for any x. Thus the radius of convergence is infinite and the interval of convergence is $(-\infty, \infty)$.

c. Starting only with terms up through x^8, we have $J_0(x) = 1 - \frac{1}{4}x^2 + \frac{1}{64}x^4 - \frac{1}{2304}x^6 + \frac{1}{147456}x^8 + \cdots$, $J_0'(x) = -\frac{1}{2}x + \frac{1}{16}x^3 - \frac{1}{384}x^5 + \frac{1}{18432}x^7 + \cdots$, $J_0''(x) = -\frac{1}{2} + \frac{3}{16}x^2 - \frac{5}{384}x^4 + \frac{7}{18432}x^6 + \cdots$ so that $x^2 J_0(x) = x^2 - \frac{1}{4}x^4 + \frac{1}{64}x^6 - \frac{1}{2304}x^8 + \frac{1}{147456}x^{10} + \cdots$, $x J_0'(x) = -\frac{1}{2}x^2 + \frac{1}{16}x^4 - \frac{1}{384}x^6 + \frac{1}{18432}x^8 + \cdots$, $x^2 J_0''(x) = -\frac{1}{2}x^2 + \frac{3}{16}x^4 - \frac{5}{384}x^6 + \frac{7}{18432}x^8 + \cdots$, and $x^2 J_0''(x) + x J_0'(x) + x^2 J_0(x) = 0$.

10.4.83

a. The power series for $\cos x$ has only even powers of x, so that the power series has the same value evaluated at $-x$ as it does at x.

b. The power series for $\sin x$ has only odd powers of x, so that evaluating it at $-x$ gives the opposite of its value at x.

10.4.85

a. Because $f(a) = g(a) = 0$, we use the Taylor series for $f(x)$ and $g(x)$ centered at a to compute that

$$
\begin{aligned}
\lim_{x \to a} \frac{f(x)}{g(x)} &= \lim_{x \to a} \frac{f(a) + f'(a)(x-a) + \frac{1}{2}f''(a)(x-a)^2 + \cdots}{g(a) + g'(a)(x-a) + \frac{1}{2}g''(a)(x-a)^2 + \cdots} \\
&= \lim_{x \to a} \frac{f'(a)(x-a) + \frac{1}{2}f''(a)(x-a)^2 + \cdots}{g'(a)(x-a) + \frac{1}{2}g''(a)(x-a)^2 + \cdots} \\
&= \lim_{x \to a} \frac{f'(a) + \frac{1}{2}f''(a)(x-a) + \cdots}{g'(a) + \frac{1}{2}g''(a)(x-a) + \cdots} = \frac{f'(a)}{g'(a)}.
\end{aligned}
$$

Because $f'(x)$ and $g'(x)$ are assumed to be continuous at a and $g'(a) \neq 0$,

$$
\frac{f'(a)}{g'(a)} = \lim_{x \to a} \frac{f'(x)}{g'(x)}
$$

and we have that

$$
\lim_{x \to a} \frac{f(x)}{g(x)} = \lim_{x \to a} \frac{f'(x)}{g'(x)}
$$

which is one form of L'Hôpital's Rule.

b. Because $f(a) = g(a) = f'(a) = g'(a) = 0$, we use the Taylor series for $f(x)$ and $g(x)$ centered at a to compute that

$$
\begin{aligned}
\lim_{x \to a} \frac{f(x)}{g(x)} &= \lim_{x \to a} \frac{f(a) + f'(a)(x-a) + \frac{1}{2}f''(a)(x-a)^2 + \frac{1}{6}f'''(a)(x-a)^3 + \cdots}{g(a) + g'(a)(x-a) + \frac{1}{2}g''(a)(x-a)^2 + \frac{1}{6}g'''(a)(x-a)^3 + \cdots} \\
&= \lim_{x \to a} \frac{\frac{1}{2}f''(a)(x-a)^2 + \frac{1}{6}f'''(a)(x-a)^3 + \cdots}{\frac{1}{2}g''(a)(x-a)^2 + \frac{1}{6}g'''(a)(x-a)^3 + \cdots} \\
&= \lim_{x \to a} \frac{\frac{1}{2}f''(a) + \frac{1}{6}f'''(a)(x-a) + \cdots}{\frac{1}{2}g''(a) + \frac{1}{6}g'''(a)(x-a) + \cdots} = \frac{f''(a)}{g''(a)}.
\end{aligned}
$$

Because $f''(x)$ and $g''(x)$ are assumed to be continuous at a and $g''(a) \neq 0$,

$$
\frac{f''(a)}{g''(a)} = \lim_{x \to a} \frac{f''(x)}{g''(x)}
$$

and we have that

$$
\lim_{x \to a} \frac{f(x)}{g(x)} = \lim_{x \to a} \frac{f''(x)}{g''(x)}
$$

which is consistent with two applications of L'Hôpital's Rule.

Chapter Ten Review

1

   a. True. The approximations tend to get better as n increases in size, and also when the value being approximated is closer to the center of the series. Because 2.1 is closer to 2 than 2.2 is, and because $3 > 2$, we should have $|p_3(2.1) - f(2.1)| < |p_2(2.2) - f(2.2)|$.

   b. False. The interval of convergence may or may not include the endpoints.

   c. True. The interval of convergence is an interval centered at 0, and the endpoints may or may not be included.

   d. True. Because $f(x)$ is a polynomial, all its derivatives vanish after a certain point (in this case, $f^{(12)}(x)$ is the last nonzero derivative).

3 $p_2(x) = 1$.

5 $p_3(x) = x - \frac{x^2}{2} + \frac{x^3}{3}$.

7 $p_2(x) = x - 1 - \frac{1}{2}(x - 1)^2$.

9 $p_3(x) = \frac{5}{4} + \frac{3(x - \ln 2)}{4} + \frac{5(x - \ln 2)^2}{8} + \frac{(x - \ln 2)^3}{8}$.

11

   a. $p_0(x) = 1$, $p_1(x) = 1 + x$, and $p_2(x) = 1 + x + \frac{x^2}{2}$.

   b.

n	$p_n(-0.08)$	$\left| p_n(-0.08) - e^{-0.08} \right|$
0	1	7.7×10^{-2}
1	0.92	3.1×10^{-3}
2	0.923	8.4×10^{-5}

13

   a. $p_0(x) = \frac{\sqrt{2}}{2}$, $p_1(x) = \frac{\sqrt{2}}{2}(1 + (x - \pi/4))$, and $p_2(x) = \frac{\sqrt{2}}{2}\left(1 + (x - \pi/4) - \frac{1}{2}(x - \pi/4)^2\right)$.

   b.

n	$p_n(\pi/5)$	$\left| p_n(\pi/5) - \sin(\pi/5) \right|$
0	0.707	1.2×10^{-1}
1	0.596	8.2×10^{-3}
2	0.587	4.7×10^{-4}

15 The derivatives of $\sin x$ are bounded in magnitude by 1, so $|R_n(x)| \le M \frac{|x|^{n+1}}{(n+1)!} \le \frac{|x|^{n+1}}{(n+1)!}$. But $|x| < \pi$, so $|R_3(x)| \le \frac{\pi^4}{24}$.

17 Using the Ratio Test, $\lim\limits_{k \to \infty} \left| \frac{a_{k+1}}{a_k} \right| = \lim\limits_{k \to \infty} \left| \frac{(k+1)^2 x^{k+1}}{(k+1)!} \cdot \frac{k!}{k^2 x^k} \right| = \lim\limits_{k \to \infty} \left(\frac{k+1}{k} \right)^2 \frac{|x|}{k+1} = 0$, so the interval of convergence is $(-\infty, \infty)$.

19 Using the Ratio Test, $\lim\limits_{k \to \infty} \frac{a_{k+1}}{a_k} = \lim\limits_{k \to \infty} \left| \frac{(x+1)^{2k+2}}{(k+1)!} \cdot \frac{k!}{(x+1)^{2k}} \right| = \lim\limits_{k \to \infty} \frac{1}{k+1}(x+1)^2 = 0$, so the interval of convergence is $(-\infty, \infty)$.

21 By the Root Test, $\lim\limits_{k\to\infty}\sqrt[k]{|a_k|}=\lim\limits_{k\to\infty}\left(\frac{|x|}{9}\right)^3=\frac{|x^3|}{729}$, so the series converges for $|x|<9$. The series given by letting $x=\pm 9$ are both divergent by the Divergence Test. Thus, $(-9,9)$ is the interval of convergence.

23 By the Ratio Test, $\lim\limits_{k\to\infty}\left|\frac{(x+2)^{k+1}}{2^{k+1}\ln(k+1)}\cdot\frac{2^k\ln k}{(x+2)^k}\right|=\lim\limits_{k\to\infty}\frac{\ln k}{2\ln(k+1)}|x+2|=\frac{|x+2|}{2}$. The radius of convergence is thus 2, and a check of the endpoints gives the divergent series $\sum\frac{1}{\ln k}$ at $x=0$ and the convergent alternating series $\sum\frac{(-1)^k}{\ln k}$ at $x=-4$. The interval of convergence is therefore $[-4,0)$.

25 The Maclaurin series for $f(x)$ is $\sum_{k=0}^{\infty}x^{2k}$. By the Root Test, this converges for $|x^2|<1$, so $-1<x<1$. It diverges at both endpoints, so the interval of convergence is $(-1,1)$.

27 The Maclaurin series for $f(x)$ is $\sum_{k=0}^{\infty}(-5x)^k=\sum_{k=0}^{\infty}(-5)^k x^k$. By the Root Test, this has radius of convergence $1/5$. Checking the endpoints, we obtain an interval of convergence of $(-1/5,1/5)$.

29 Note that $\frac{1}{1-10x}=\sum_{k=0}^{\infty}(10x)^k$, so $\frac{1}{10}\cdot\frac{1}{1-10x}=\frac{1}{10}\sum_{k=0}^{\infty}(10x)^k$. Taking the derivative of $\frac{1}{10}\cdot\frac{1}{1-10x}$ gives $f(x)$. Thus, the Maclaurin series for $f(x)$ is $\frac{1}{10}\sum_{k=1}^{\infty}10k(10x)^{k-1}=\sum_{k=1}^{\infty}k(10x)^{k-1}$. Using the Ratio Test, we see that the radius of convergence is $1/10$, and checking endpoints we obtain an interval of convergence of $(-1/10,1/10)$.

31 The first three terms are $1+3x+\frac{9x^2}{2}$. The series is $\sum_{k=0}^{\infty}\frac{(3x)^k}{k!}$.

33 The first three terms are $-(x-\pi/2)+\frac{1}{6}(x-\pi/2)^3-\frac{1}{120}(x-\pi/2)^5$. The series is

$$\sum_{k=0}^{\infty}(-1)^{k+1}\frac{1}{(2k+1)!}\left(x-\frac{\pi}{2}\right)^{2k+1}.$$

35 The first three terms are $4x-\frac{1}{3}(4x)^3+\frac{1}{5}(4x)^5$. The series is $\sum_{k=0}^{\infty}(-1)^k\frac{(4x)^{2k+1}}{2k+1}$.

37 The nth derivative of $\cosh 3x$ at $x=0$ is 0 if n is odd and is 3^n if n is even. The first 3 terms of the series are thus $1+\frac{9x^2}{2!}+\frac{81x^4}{4!}$. The whole series can be written as $\sum_{k=0}^{\infty}\frac{(3x)^{2k}}{(2k)!}$.

39 $f(x)=\binom{1/3}{0}+\binom{1/3}{1}x+\binom{1/3}{2}x^2+\cdots=1+\frac{1}{3}x-\frac{1}{9}x^2+\cdots$.

41 $f(x)=\binom{-3}{0}+\binom{-3}{1}\frac{x}{2}+\binom{-3}{2}\frac{x^2}{4}+\cdots=1-\frac{3}{2}x+\frac{3}{2}x^2+\cdots$.

43 $R_n(x)=\frac{(-1)^{n+1}e^{-c}}{(n+1)!}x^{n+1}$ for some c between 0 and x, and $\lim\limits_{n\to\infty}|R_n(x)|\le e^{-|x|}\lim\limits_{n\to\infty}\frac{|x|^{n+1}}{(n+1)!}=0$, because $n!$ grows faster than $|x|^n$ as $n\to\infty$ for all x.

45 $R_n(x)=\frac{f^{(n+1)}(c)}{(n+1)!}x^{n+1}$ for some c in $(-1/2,1/2)$. Now, $\left|f^{(n+1)}(c)\right|=\frac{n!}{(1+c)^{n+1}}$, so $\lim\limits_{n\to\infty}|R_n(x)|\le\lim\limits_{n\to\infty}(2|x|)^{n+1}\cdot\frac{1}{n+1}\le\lim\limits_{n\to\infty}1^{n+1}\frac{1}{n+1}=0$.

47 The Taylor series for $\cos x$ centered at 0 is

$$\cos x=1-\frac{x^2}{2}+\frac{x^4}{24}-\frac{x^6}{720}+\cdots.$$

We compute that

$$\frac{x^2/2-1+\cos x}{x^4}=\frac{1}{x^4}\left(x^2/2-1+\left(1-\frac{x^2}{2}+\frac{x^4}{24}-\frac{x^6}{720}+\cdots\right)\right)$$

$$=\frac{1}{x^4}\left(\frac{x^4}{24}-\frac{x^6}{720}+\cdots\right)=\frac{1}{24}-\frac{x^2}{720}+\cdots$$

so the limit of $\dfrac{x^2/2-1+\cos x}{x^4}$ as $x\to 0$ is $\dfrac{1}{24}$.

49 The Taylor series for $\ln(x-3)$ centered at 4 is

$$\ln(x-3) = (x-4) - \frac{1}{2}(x-4)^2 + \frac{1}{3}(x-4)^3 - \cdots.$$

We compute that

$$\begin{aligned}
\frac{\ln(x-3)}{x^2-16} &= \frac{1}{(x-4)(x+4)}\left((x-4) - \frac{1}{2}(x-4)^2 + \frac{1}{3}(x-4)^3 - \cdots\right)\\
&= \frac{1}{(x-4)(x+4)}\left((x-4)\left(1 - \frac{1}{2}(x-4) + \frac{1}{3}(x-4)^2 - \cdots\right)\right)\\
&= \frac{1}{x+4}\left(1 - \frac{1}{2}(x-4) + \frac{1}{3}(x-4)^2 - \cdots\right)
\end{aligned}$$

so the limit of $\dfrac{\ln(x-3)}{x^2-16}$ as $x \to 4$ is $\dfrac{1}{8}$.

51 The Taylor series for $\sec x$ centered at 0 is

$$\sec x = 1 + \frac{x^2}{2} + \frac{5x^4}{24} + \frac{61x^6}{720} + \cdots$$

and the Taylor series for $\cos x$ centered at 0 is

$$\cos x = 1 - \frac{x^2}{2} + \frac{x^4}{24} - \frac{x^6}{720} + \cdots.$$

We compute that

$$\begin{aligned}
&\frac{\sec x - \cos x - x^2}{x^4}\\
&= \frac{1}{x^4}\left(\left(1 + \frac{x^2}{2} + \frac{5x^4}{24} + \frac{61x^6}{720} + \cdots\right) - \left(1 - \frac{x^2}{2} + \frac{x^4}{24} - \frac{x^6}{720} + \cdots\right) - x^2\right)\\
&= \frac{1}{x^4}\left(\frac{x^4}{6} + \frac{31x^6}{360} + \cdots\right) = \frac{1}{6} + \frac{31x^2}{360} + \cdots
\end{aligned}$$

so the limit of $\dfrac{\sec x - \cos x - x^2}{x^4}$ as $x \to 0$ is $\dfrac{1}{6}$.

53 We have $e^{-x^2} = 1 - x^2 + \frac{x^4}{2} - \frac{x^6}{6} + \frac{x^8}{24} - \cdots$, so $\int e^{-x^2}\,dx = \int(1 - x^2 + \frac{x^4}{2} - \frac{x^6}{6} + \frac{x^8}{24} - \cdots)\,dx = C + x - \frac{x^3}{3} + \frac{x^5}{10} - \frac{x^7}{42} + \cdots$. Thus, $\int_0^{1/2} e^{-x^2}\,dx = (0.5) - \frac{(0.5)^3}{3} + \frac{(0.5)^5}{10} - \frac{(0.5)^7}{42} + \cdots$. Because $(0.5)^7/42 < .001$, we can calculate the approximation using the first three numbers shown, arriving at approximately 0.461.

55 $x\cos x = x - \frac{x^3}{2} + \frac{x^5}{24} - \frac{x^7}{720} + \cdots$, so $\int x\cos x\,dx = \int(x - \frac{x^3}{2} + \frac{x^5}{24} - \frac{x^7}{720} + \cdots)\,dx = C + \frac{x^2}{2} - \frac{x^4}{8} + \frac{x^6}{144} - \frac{x^8}{5760} + \frac{x^{10}}{403200} - \cdots$. Thus $\int_0^1 x\cos x\,dx = \frac{1}{2} - \frac{1}{8} + \frac{1}{144} - \frac{1}{5760} + \cdots$. Because $\frac{1}{5760} < .001$, we add the first three terms to approximate to the desired accuracy. Calculating gives $\int_0^1 x\cos x\,dx \approx 0.382$.

57 The series for $f(x) = \sqrt{x}$ centered at $a = 121$ is $11 + \frac{x-121}{22} - \frac{(x-121)^2}{10648} + \frac{(x-121)^3}{2576816} + \cdots$. Letting $x = 119$ gives $\sqrt{119} \approx 11 - \frac{1}{11} - \frac{1}{2\cdot 11^3} - \frac{1}{2\cdot 11^5}$.

59 $\tan^{-1} x = x - x^3/3 + x^5/5 - x^7/7 + x^9/9 + \cdots$, so $\tan^{-1}(-1/3) \approx \frac{-1}{3} + \frac{1}{3\cdot 3^3} - \frac{1}{5\cdot 3^5} + \frac{1}{7\cdot 3^7}$.

61 Because $y(0) = 4$, we have $y'(0) - 16 + 12 = 0$, so $y'(0) = 4$. Differentiating the equation $n-1$ times and evaluating at 0 we obtain $y^{(n)}(0) = 4y^{(n-1)}(0)$, so that $y^{(n)}(0) = 4^n$. The Taylor series for $y(x)$ is thus $y(x) = 4 + 4x + \frac{4^2 x^2}{2!} + \frac{4^3 x^3}{3!} + \cdots$, or $y(x) = 3 + e^{4x}$.

63

a. The Taylor series for $\ln(1+x)$ is $\sum_{k=1}^{\infty}(-1)^{k+1}\frac{x^k}{k}$. Evaluating at $x=1$ gives $\ln 2 = \sum_{k=1}^{\infty}(-1)^{k+1}\frac{1}{k}$.

b. The Taylor series for $\ln(1-x)$ is $-\sum_{k=1}^{\infty}\frac{x^k}{k}$. Evaluating at $x=1/2$ gives $\ln(1/2) = -\sum_{k=1}^{\infty}\frac{1}{k2^k}$, so that $\ln 2 = \sum_{k=1}^{\infty}\frac{1}{k2^k}$.

c. $f(x) = \ln\left(\frac{1+x}{1-x}\right) = \ln(1+x) - \ln(1-x)$. Using the two Taylor series above we have $f(x) = \sum_{k=1}^{\infty}(-1)^{k+1}\frac{x^k}{k} - \left(-\sum_{k=1}^{\infty}\frac{x^k}{k}\right) = \sum_{k=1}^{\infty}(1+(-1)^{k+1})\frac{x^k}{k} = 2\sum_{k=0}^{\infty}\frac{x^{2k+1}}{2k+1}$.

d. Because $\frac{1+x}{1-x} = 2$ when $x = \frac{1}{3}$, the resulting infinite series for $\ln 2$ is $2\sum_{k=0}^{\infty}\frac{1}{3^{2k+1}(2k+1)}$.

e. The first four terms of each series are: $1 - \frac{1}{2} + \frac{1}{3} - \frac{1}{4} \approx 0.5833333333$, $\frac{1}{2} + \frac{1}{8} + \frac{1}{24} + \frac{1}{64} \approx 0.6822916667$, $\frac{2}{3} + \frac{2}{81} + \frac{2}{1215} + \frac{2}{15309} \approx 0.6931347573$ The true value is $\ln 2 \approx 0.6931471806$. The third series converges the fastest, because it has 3^{k+1} in the denominator as opposed to 2^k, so its terms get small faster.

Chapter 11

Parametric and Polar Curves

11.1   Parametric Equations

11.1.1 Given an input value of t, the point $(x(t), y(t))$ can be plotted in the xy-plane, generating a curve.

11.1.3 Let $x = R\cos(\pi t/5)$ and $y = -R\sin(\pi t/5)$. Note that as t ranges from 0 to 10, $\pi t/5$ ranges from 0 to 2π. Because $x^2 + y^2 = R^2$, this curve represents a circle of radius R. Note also that for $t = 0$ the initial point is $(R, 0)$, and for small values of t the plotted points are in the third quadrant — so the curve is being traced with clockwise orientation.

11.1.5 Let $x = t$ and $y = t^2$ for $t \in (-\infty, \infty)$.

11.1.7 Solving the first equation for t gives $t = \frac{1-x}{2}$. Substitute that value for t in the second equation to get $y = 3\left(\frac{1-x}{2}\right)^2$; simplifying gives $y = \frac{3}{4}x^2 - \frac{3}{2}x + \frac{3}{4}$.

11.1.9 The slope of the tangent line is $\frac{dy}{dx} = \frac{dy/dt}{dx/dt}$, so at $t = a$ the slope is given by $\frac{g'(a)}{f'(a)}$, $f'(a) \neq 0$.

11.1.11

a.

t	x	y
-10	-20	-34
-5	-10	-19
0	0	-4
5	10	11
10	20	26

b.

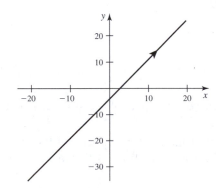

c. Solving $x = 2t$ for t yields $t = x/2$, so $y = 3t - 4 = 3x/2 - 4$.

d. The curve is the line segment from $(-20, -34)$ to $(20, 26)$.

11.1.13

b.

a.

t	x	y
-5	11	-18
-3	9	-12
0	6	-3
3	3	6
5	1	12

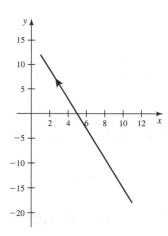

c. Solving $x = -t + 6$ for t yields $t = 6 - x$, so $y = 3t - 3 = 18 - 3x - 3 = 15 - 3x$.

d. The curve is the line segment from $(11, -18)$ to $(1, 12)$.

11.1.15

a. Solving $x = \sqrt{t} + 4$ for t yields $t = (x - 4)^2$. Thus, $y = 3\sqrt{t} = 3(x - 4)$, where x ranges from 4 to 8. Note that all $t \geq 0$, $x > 0$, and $y > 0$.

b. The curve is the line segment from $(4, 0)$ to $(8, 12)$.

11.1.17

a. Because $\cos^2 t + \sin^2 t = 1$, we have $x^2 + y = 1$, so $y = 1 - x^2$, $-1 \leq x \leq 1$.

b. This is a parabola opening downward with a vertex at $(0, 1)$, and starting at $(1, 0)$ and ending at $(-1, 0)$.

11.1.19

a. Solving $x = r - 1$ for r yields $r = x + 1$. Thus, $y = r^3 = (x + 1)^3$, where $-5 \leq x \leq 3$.

b. The curve is the part of the standard cubic curve, shifted one unit to the left, from $(-5, -64)$ to $(3, 64)$.

11.1.21 Note that $x^2 + y^2 = 9\cos^2 t + 9\sin^2 t = 9$, so this represents an arc of the circle of radius 3 centered at the origin from $(-3, 0)$ to $(3, 0)$ traversed counterclockwise.

11.1.23 Note that $x^2 + (y - 1)^2 = \cos^2 t + \sin^2 t = 1$, so we have a circle of radius 1 centered at $(0, 1)$, traversed counterclockwise starting at $(1, 1)$.

11.1.25 Note that $x^2 + y^2 = 49\cos^2 2t + 49\sin^2 2t = 49$, so this represents an arc of the circle of radius 7 centered at the origin from $(-7, 0)$ to $(-7, 0)$ traversed counterclockwise. (So the whole circle is represented.)

11.1.27

Let $x = 4\cos t$ and $y = 4\sin t$ for $0 \le t \le 2\pi$.
Then $x^2 + y^2 = 16\cos^2 t + 16\sin^2 t = 16$.

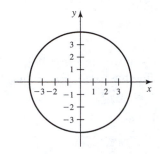

11.1.29

Let $x = \cos t + 2$ and $y = \sin t + 3$ for $0 \le t \le 2\pi$. Then $(x-2)^2 + (y-3)^2 = 1$, which is a circle with the desired center and radius and orientation.

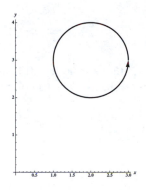

11.1.31

Let $x = -2 + 8\sin t$ and $y = -3 + 8\cos t$ for $0 \le t \le 2\pi$. Then $(x+2)^2 + (y+3)^2 = 64\sin^2 t + 64\cos^2 t = 64$.

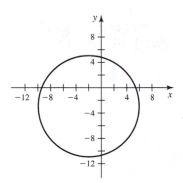

11.1.33 Let t be time in minutes, so $0 \le t \le 1.5$ Let $x = 400\cos(4\pi/3)t$ and $y = 400\sin(4\pi/3)t$. Then because $x^2 + y^2 = 400^2$, the path is a circle of radius 400. Note that the values of x and y are the same at $t = 0$ and $t = 1.5$, and that the circle is traversed counterclockwise.

11.1.35 Let t be time in seconds, so $0 \le t \le 24$ Let $x = 50\cos(\pi/12)t$ and $y = 50\sin(\pi/12)t$. Then because $x^2 + y^2 = 50^2$, the path is a circle of radius 50. Note that the values of x and y are the same at $t = 0$ and $t = 24$, and that the circle is traversed counterclockwise.

11.1.37

Because $t = x - 3$, we have $y = 1 - (x - 3) = 4 - x$, so the line has slope -1. When $t = 0$, we have the point $(3, 1)$.

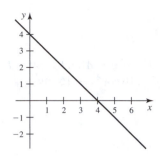

11.1.39

Because $y = 1$, this is a horizontal line with slope 0. When $t = 0$, we have the point $(8, 1)$.

11.1.41 Let $x = x_0 + at$ and $y = y_0 + bt$. Letting $(x_0, y_0) = (0, 0)$, and then finding a and b so that the curve is at the point Q when $t = 1$ yields $x = 2t$, $y = 8t$ for $0 \le t \le 1$.

11.1.43 Let $x = x_0 + at$ and $y = y_0 + bt$. Letting $(x_0, y_0) = (-1, -3)$, and then finding a and b so that the curve is at the point Q when $t = 1$ yields $x = -1 + 7t$, $y = -3 - 13t$ for $0 \le t \le 1$.

11.1.45

Let $x = t$ and $y = 2t^2 - 4$, $-1 \le t \le 5$.

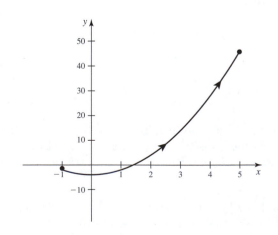

11.1.47

Let $x = -2 + 4t$ and $y = 3 - 6t$, $0 \le t \le 1$,
and $x = t + 1$, $y = 8t - 11$ for $1 \le t \le 2$.

11.1.49

11.1.51

11.1.53

11.1.55

11.1.57

11.1.59

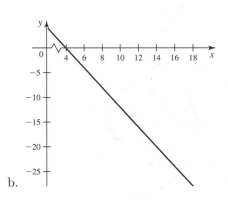

a. $\frac{dy}{dx} = \frac{dy/dt}{dx/dt} = -\frac{8}{4} = -2$ for all t. Because the curve is a line, the tangent line to the curve at the given point is the line itself.

b.

11.1.61

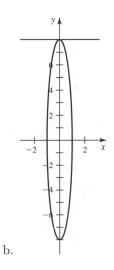

a. $\frac{dy}{dx} = \frac{dy/dt}{dx/dt} = \frac{8\cos t}{-\sin t} = -8\cot t$. At the given value of t, the value of $\frac{dy}{dx}$ is $-8\cot \pi/2 = 0$. The tangent line at the point $(0,8)$ is thus the horizontal line $y = 8$.

b.

11.1.63

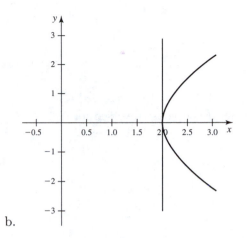

a. $\frac{dy}{dx} = \frac{dy/dt}{dx/dt} = \frac{1+\frac{1}{t^2}}{1-\frac{1}{t^2}} = \frac{t^2+1}{t^2-1}$. At the given value of t, the derivative doesn't exist, and the tangent line is the vertical line $x = 2$, tangent at the point $(2, 0)$.

b.

11.1.65

a. False. This generates a circle in the counterclockwise direction.

b. True. Note that when t is increased by one, the value of $2\pi t$ is increased by 2π, which is the period of both the sine and the cosine functions.

c. False. This generates only the portion of the parabola in the first quadrant, omitting the portion in the second quadrant.

d. True. They describe the portion of the unit circle in the 4th and 1st quadrants.

e. True. This ellipse has vertical tangents at $t = 0$ and $t = \pi$.

11.1.67 The point corresponding to $t = 2$ is $(3, 10)$. $\frac{dy}{dx} = \frac{dy/dt}{dx/dt} = \frac{3t^2+1}{2t}$, so the slope at $t = 2$ is $\frac{13}{4}$. The equation of the tangent line is therefore $y - 10 = \frac{13}{4}(x - 3)$, or $y = \frac{13}{4}x + \frac{1}{4}$.

11.1.69 The point corresponding to $t = \pi/4$ is $\left(\frac{4\sqrt{2}+\pi\sqrt{2}}{8}, \frac{4\sqrt{2}-\pi\sqrt{2}}{8}\right)$. $\frac{dy}{dx} = \frac{dy/dt}{dx/dt} = \frac{\cos t-(\cos t-t\sin t)}{-\sin t+(\sin t+t\cos t)} = \tan t$. At $t = \pi/4$, we have a slope of 1. The equation of the tangent line is thus $y - \frac{4\sqrt{2}-\pi\sqrt{2}}{8} = 1\left(x - \frac{4\sqrt{2}+\pi\sqrt{2}}{8}\right)$, or $y = x - \frac{\pi\sqrt{2}}{4}$.

11.1.71 Let $x = 1 + 2t$ and $y = 1 + 4t$, for $-\infty < t < \infty$. Note that $y = 2(1 + 2t) - 1$, so $y = 2x - 1$.

11.1.73 Let $x = t^2$ and $y = t$, for $0 \le t < \infty$. Note that $x = t^2 = y^2$, and that the starting point is $(0, 0)$.

11.1.75

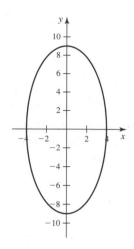

The entire curve is traversed for $0 \le t \le 2\pi$.

11.1.77

Let $x = 3\cos t$ and $y = \frac{3}{2}\sin t$ for $0 \le t \le 2\pi$.
Then the major axis on the x-axis has length
$2 \cdot 3 = 6$ and the minor axis on the y-axis has
length $2 \cdot \frac{3}{2} = 3$. Note that $\left(\frac{x}{3}\right)^2 + \left(\frac{2y}{3}\right)^2 = \cos^2 t + \sin^2 t = 1$.

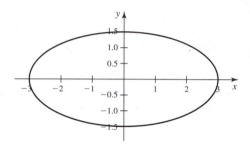

11.1.79

Let $x = 15\cos t - 2$ and $y = 10\sin t - 3$ for
$0 \le t \le 2\pi$. Note that $\left(\frac{x+2}{15}\right)^2 + \left(\frac{y+3}{10}\right)^2 = \cos^2 t + \sin^2 t = 1$.
Then the major axis has length 30 and the
minor axis has length 20.

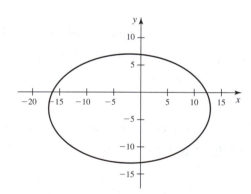

11.1.81

a. For $(1+s, 2s) = (1+2t, 3t)$, we must have $1 + s = 1 + 2t$ and $2s = 3t$, so that $s = 2t$ and $2s = 3t$. The
   only solution to this pair of equations is $s = t = 0$, so these two lines intersect when $s = t = 0$, at the
   point $(1, 0)$.

b. For $(2+5s, 1+s) = (4+10t, 3+2t)$, we must have $2 + 5s = 4 + 10t$ and $1 + s = 3 + 2t$, so that $s = 2 + 2t$
   and $s = 2 + 2t$. This pair of equations has no solutions, so the lines are parallel.

c. For $(1+3s, 4+2s) = (4?3t, 6+4t)$, we must have $1 + 3s = 4?3t$ and $4 + 2s = 6 + 4t$, so that $s = 1?t$ and
   $s = 1 + 2t$. The only solution to this pair of equations is $s = 1$ and $t = 0$, so these two lines intersect
   for these values of s and t, at the point $(4, 6)$.

11.1.83 Note that $x^2 + y^2 = 4\sin^2 8t + 4\cos^2 8t = 4$, so the curve is the circle $x^2 + y^2 = 4$.

11.1.85 Note that because $t = x$, we have $y = \sqrt{4 - t^2} = \sqrt{4 - x^2}$.

11.1.87 Because $\sec^2 t - 1 = \tan^2 t$, we have $y = x^2$.

11.1.89 $\frac{dy}{dx} = \frac{dy/dt}{dx/dt} = \frac{4\cos t}{-4\sin t} = -\cot t$. We seek t so that $\cot t = -1/2$, so $t = \cot^{-1}(-1/2)$. The corresponding points on the curve are $\left(-\frac{4\sqrt{5}}{5}, \frac{8\sqrt{5}}{5}\right)$ and $\left(\frac{4\sqrt{5}}{5}, -\frac{8\sqrt{5}}{5}\right)$.

11.1.91 $\frac{dy}{dx} = \frac{dy/dt}{dx/dt} = \frac{1+(1/t^2)}{1-(1/t^2)} = \frac{t^2+1}{t^2-1}$. We seek t so that $\frac{t^2+1}{t^2-1} = 1$, which never occurs. Thus, there are no points on this curve with slope 1.

11.1.93 Note that in equation B, the parameter is scaled by a factor of 3. Thus, the curves are the same when the corresponding interval for t is scaled by a factor of $1/3$, so for $a = 0$ and $b = \frac{2\pi}{3}$. In fact, the same curve will be generated for $a = p$, $b = p + 2\pi/3$ where p is any real number.

11.1.95

a. $\frac{dy}{dx} = \frac{dy/dt}{dx/dt} = \frac{2\cos t}{2\cos 2t}$. This is zero when $\cos t = 0$ but $\cos 2t \neq 0$, which occurs for $t = \pi/2$ and $t = 3\pi/2$. The corresponding points on the graph are $(0, 2)$ and $(0, -2)$.

b. Using the derivative obtained above, we seek points where $\cos 2t = 0$ but $\cos t \neq 0$. This occurs for $t = \pi/4$, $3\pi/4$, $5\pi/4$, and $7\pi/4$. The corresponding points on the curve are $(1, \sqrt{2})$, $(-1, \sqrt{2})$, $(-1, -\sqrt{2})$, and $(1, -\sqrt{2})$.

11.1.97

a. Let $\text{sgn}(x) = \begin{cases} 1 & \text{if } x \geq 0 \\ -1 & \text{if } x < 0. \end{cases}$ Let $x = a \cdot \text{sgn}(\cos t)\,|\cos(t)|^{2/n}$ and $y = b \cdot \text{sgn}(\sin(t))\,|\sin(t)|^{2/n}$.

b.

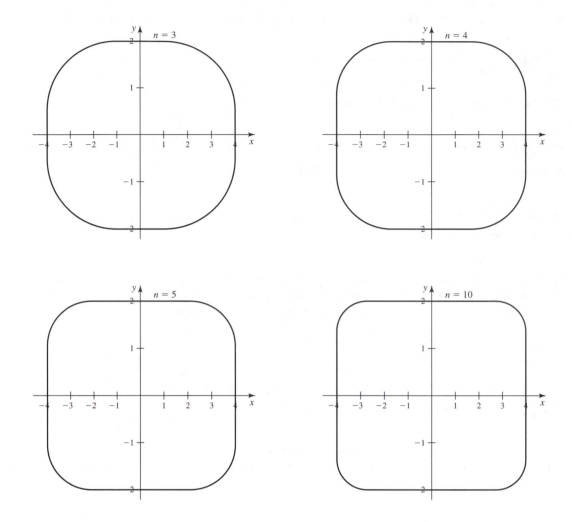

c. As n increases from near 0 to near 1, the curves change from star-shaped to a rectangular shape with corners at $(\pm a, 0)$ and $(0, \pm b)$. As n increases from 1 on, the curves become more rectangular with corners at $(\pm a, \pm b)$.

11.1.99 The first graphic shown is for $a = 1$ and $b = 1$. The second is for $a = 2$, $b = 1$, and the third is for $a = 1$, $b = 2$.

11.1.101

$a = 1$, $b = 2$.

$a = 3.5$, $b = 2$.

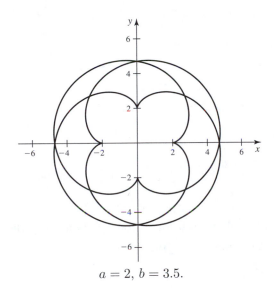

$a = 2$, $b = 3.5$.

Note that for $a < b$, we have cusps pointing inward, while for $a > b$, the cusps point outward.

11.1.103

a.

b.

c.

11.1.105 The package lands when $y = 0$, so when $-4.9t^2 + 4000 = 0$ for $t > 0$. This occurs when $t = \sqrt{\frac{4000}{4.9}} \approx 28.571$ seconds. At that time, $x \approx 100 \cdot 28.57 = 2857$ meters.

11.1.107 Let $x = 1 + \cos^2 t - \sin^2 t$ and $y = t$, for $-\infty < t < \infty$. Note that because $1 - \sin^2 t = \cos^2 t$, we have $x = 2\cos^2 t$, $y = t$.

11.1.109 Suppose that $a^2 + c^2 = b^2 + d^2$, and that $ab + cd = 0$. Note that

$$x^2 + y^2 = a^2 \cos^2 t + 2ab \sin t \cos t + b^2 \sin^2 t + c^2 \cos^2 t + 2cd \sin t \cos t + d^2 \sin^2 t,$$

which can be rewritten as

$$(a^2 + c^2)\cos^2 t + (b^2 + d^2)\sin^2 t + (2ab + 2cd)\sin t \cos t.$$

Because $b^2 + d^2 = a^2 + c^2$ and because $2ab + 2cd = 0$, we can write this as

$$(a^2 + c^2)(\cos^2 t + \sin^2 t) = R^2,$$

so we have the circle $x^2 + y^2 = R^2$, as desired.

11.2 Polar Coordinates

11.2.1 The coordinates $(2, \pi/6)$, $(2, -11\pi/6)$, and $(-2, 7\pi/6)$ all give rise to the same point. Also, the coordinates $(-3, -\pi/2)$, $(3, \pi/2)$ and $(-3, 3\pi/2)$ give rise to the same point.

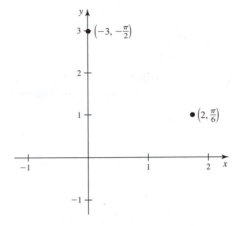

11.2.3 If a point has Cartesian coordinates (x, y) then $r^2 = x^2 + y^2$ and $\tan\theta = y/x$ for $x \neq 0$. If $x = 0$, then $\theta = \pi/2$ and $r = y$.

11.2.5 Because $x = r\cos\theta$, we have that the vertical line $x = 5$ has polar equation $r = 5\sec\theta$.

11.2.7 x-axis symmetry occurs if (r, θ) on the graph implies $(r, -\theta)$ is on the graph. y-axis symmetry occurs if (r, θ) on the graph implies $(r, \pi - \theta) = (-r, -\theta)$ is on the graph. Symmetry about the origin occurs if (r, θ) on the graph implies $(-r, \theta) = (r, \theta + \pi)$ is on the graph.

11.2.9 The coordinates $(2, \pi/4)$, $(-2, 5\pi/4)$, and $(2, 9\pi/4)$ represent the same point.

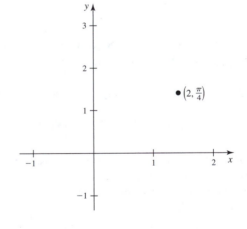

11.2.11 The coordinates $(-1, -\pi/3)$, $(1, 2\pi/3)$ and $(1, -4\pi/3)$ represent the same point.

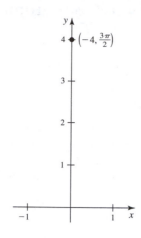

11.2.13 The coordinates $(-4, 3\pi/2)$, $(4, \pi/2)$ and $(-4, -\pi/2)$ represent the same point.

11.2.15 $x = 3\cos(\pi/4) = \frac{3\sqrt{2}}{2}$. $y = 3\sin(\pi/4) = \frac{3\sqrt{2}}{2}$.

11.2.17 $x = \cos(-\pi/3) = \frac{1}{2}$. $y = \sin(-\pi/3) = -\frac{\sqrt{3}}{2}$.

11.2.19 $x = -4\cos(3\pi/4) = 2\sqrt{2}$. $y = -4\sin(3\pi/4) = -2\sqrt{2}$.

11.2.21 $r^2 = x^2 + y^2 = 4 + 4 = 8$, so $r = \sqrt{8}$. $\tan\theta = 1$, so $\theta = \pi/4$, so $(2\sqrt{2}, \pi/4)$ is one representation of this point, and $(-2\sqrt{2}, -3\pi/4)$ is another.

11.2.23 $r^2 = x^2 + y^2 = 1 + 3 = 4$, so $r = \pm 2$. $\tan\theta = \sqrt{3}$, so $\theta = \pi/3, 4\pi/3$. $(2, \pi/3)$ is one representation of this point, and $(-2, -2\pi/3)$ is another.

11.2.25 $r^2 = 64$, so $r = \pm 8$. $\tan\theta = -\sqrt{3}$, so $\theta = -\pi/3, 2\pi/3$. One representation of the given point is $(8, 2\pi/3)$, and $(-8, -\pi/3)$ is another.

11.2.27 $x = r\cos\theta = -4$, so this is the vertical line $x = -4$ through $(-4, 0)$.

11.2.29 Because $x^2 + y^2 = r^2 = 4$, this is a circle of radius 2 centered at the origin.

11.2.31 Note that $x^2 + y^2 = r^2 = 4\sin^2\theta + 8\sin\theta\cos\theta + 4\cos^2\theta = 4 + 8\sin\theta\cos\theta$. Also note that $x = r\cos\theta = 2\sin\theta\cos\theta + 2\cos^2\theta$ and $y = r\sin\theta = 2\sin^2\theta + 2\sin\theta\cos\theta$. Thus, $2x + 2y = 4 + 8\sin\theta\cos\theta$. If we combine these, we see that $x^2 + y^2 - (2x + 2y) = 0$. Thus $(x^2 - 2x + 1) + (y^2 - 2y + 1) = 2$, so we have the circle $(x-1)^2 + (y-1)^2 = 2$. This is a circle of radius $\sqrt{2}$ centered at $(1,1)$.

11.2.33 $r\cos\theta = \sin 2\theta = 2\sin\theta\cos\theta$. Note that if $\cos\theta = 0$, then r can be any real number, and the equation is satisfied. For $\cos\theta \neq 0$, we have $x = r\cos\theta = 2\sin\theta\cos\theta$, so $r = 2\sin\theta$, and thus $y = r\sin\theta = 2\sin^2\theta$. Thus $x^2 + y^2 - 2y = 4\sin^2\theta\cos^2\theta + 4\sin^2\theta\sin^2\theta - 4\sin^2\theta = 4\sin^2\theta(\sin^2\theta + \cos^2\theta) - 4\sin^2\theta = 4\sin^2\theta - 4\sin^2\theta = 0$. Note also that $x^2 + y^2 - 2y = 0$ is equivalent to $x^2 + (y-1)^2 = 1$, so we have a circle of radius one centered at $(0,1)$, as well as the line $x = 0$ which is the y-axis.

11.2.35 $r = 8\sin\theta$, so $r^2 = 8r\sin\theta$, so $x^2 + y^2 = 8y$. This can be written $x^2 + (y-4)^2 = 16$, which represents a circle of radius 4 centered at $(0,4)$.

11.2.37

θ	0	$\pi/6$	$\pi/4$	$\pi/3$	$\pi/2$	$2\pi/3$	$3\pi/4$	$5\pi/6$	π
r	8	$4\sqrt{3}$	$4\sqrt{2}$	4	0	-4	$-4\sqrt{2}$	$-4\sqrt{3}$	-8

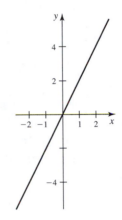

11.2.39 $r(\sin\theta - 2\cos\theta) = 0$ when $r = 0$ or when $\tan\theta = 2$, so the curve is a straight line through the origin of slope 2.

11.2.41

11.2.43

11.2.45

11.2.47

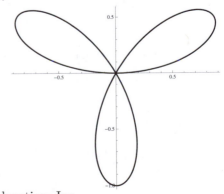

11.2.49 Points B, D, F, H, J and L have y-coordinate 0, so the graph is at the pole for each of these points. Points E, I, and M have maximal radius, so these correspond to the points at the tips of the outer loops. The points C, G and K correspond to the tips of the smaller loops. Point A corresponds to the polar point $(1, 0)$.

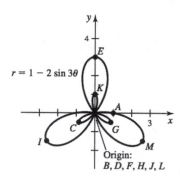

11.2.51 Points B, D, F, H, J, L, N and P are at the origin. C, G, K and O are on the ends of the long loops, while A, E, I and M are at the ends of the smaller loops.

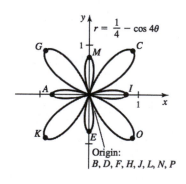

11.2.53   The interval $[0, 8\pi]$ generates the entire graph.

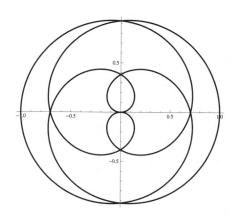

11.2.55 The interval $[0, 2\pi]$ generates the entire graph.

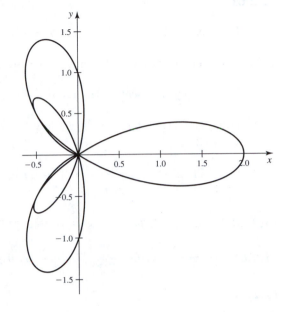

11.2.57 The interval $[0, 5\pi]$ generates the entire graph.

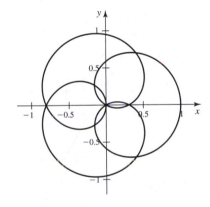

11.2.59 The interval $[0, 2\pi]$ generates the entire graph.

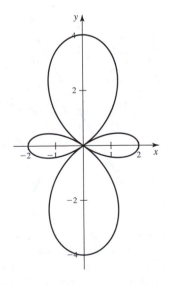

11.2.61

    a. True. Note that $r^2 = 8$ and $\tan\theta = -1$.

    b. True. Their intersection point (in Cartesian coordinates) is $(4, -2)$.

    c. False. They intersect at the polar coordinates $(2, \pi/4)$ and $(2, 5\pi/4)$.

    d. True. Note that for $\theta = \frac{3\pi}{2}$ we have $r = -3$. But the polar point $\left(-3, \frac{3\pi}{2}\right)$ is the same as the polar point $\left(3, \frac{\pi}{2}\right)$.

    e. True. The first is the line $x = 2$ because $x = r\cos\theta = 2\sec\theta\cos\theta = 2$, and the second is $y = 3$ because $y = r\sin\theta = 3\csc\theta\sin\theta = 3$.

11.2.63 We have $r\sin\theta = r^2\cos^2\theta$, so $r = \frac{\sin\theta}{\cos^2\theta} = \tan\theta\sec\theta$.

11.2.65 We have $r\sin\theta = \frac{1}{r\cos\theta}$, so $r^2 = \sec\theta\csc\theta$.

11.2.67   **11.2.69**

11.2.71   **11.2.73**

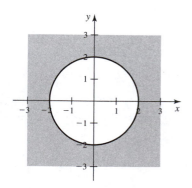

11.2.75 Consider the circle with center $C(r_0, \theta_0)$, and let A be the origin and $B(r, \theta)$ be a point on the circle not collinear with A and C. Note that the length of side BC is R, and that the angle CAB has measure $\theta - \theta_0$. Applying the law of cosines to triangle CAB yields the equation $R^2 = r^2 + r_0^2 - 2rr_0\cos(\theta - \theta_0)$, which is equivalent to the given equation.

11.2.77 In relation to number 75, we have $r_0 = 2$ and $\theta_0 = \pi/3$, and $R^2 - 4 = 12$, so $R^2 = 16$. Thus this is a circle with polar center $(2, \pi/3)$ and radius 4.

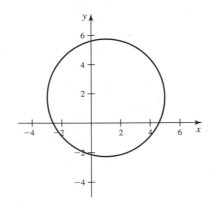

11.2.79 In relation to number 66, we have $a = 2$ and $b = 3$. So $R^2 - a^2 - b^2 = R^2 - 13 = 3$, and thus $R^2 = 16$. Thus we have a circle centered at $(2, 3)$ with radius 4.

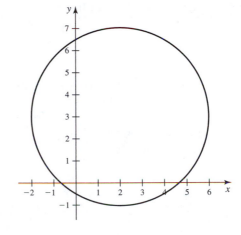

11.2.81 In relation to number 66, we have $a = -1$ and $b = 2$. So $R^2 - a^2 - b^2 = R^2 - 5 = 4$, and thus $R^2 = 9$. Thus we have a circle centered at $(-1, 2)$ with radius 3.

11.2.83

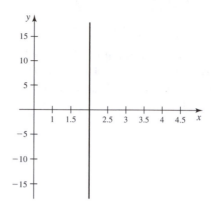

a. On all three intervals, the graph is the same vertical line, oriented upward.

b. For $\theta \neq \frac{2m+1}{2}\pi$ where m is an integer, we have $\cos\theta \neq 0$, so the equation is equivalent to $x = r\cos\theta = 2$. So the graph is a vertical line.

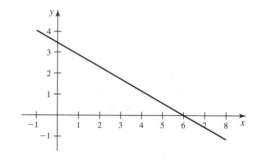

11.2.85 Using problem 84b, this is the line with $r_0 = 3$ and $\theta_0 = \frac{\pi}{3}$. So it is the line through the polar point $(3, \pi/3)$ in the direction of angle $\pi/3 + \pi/2 = 5\pi/6$. The Cartesian equation is $y = -\frac{x}{\sqrt{3}} + 2\sqrt{3}$.

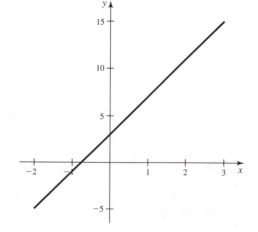

11.2.87 Using problem 84a, this is the line with $b = 3$ and $m = 4$, so $y = 4x + 3$.

11.2.89

a. This matches (A), because we have $|a| = 1 = |b|$, and the graph is a cardioid.

b. This matches (C). This has an inner loop because $|a| = 1 < 2 = |b|$. Note that $r = 1$ when $\theta = 0$, so it can't be (D).

c. This matches (B). This has $|a| = 2 > 1 = |b|$, so it has an oval-like shape.

d. This matches (D). This has an inner loop because $|a| = 1 < 2 = |b|$. Note that $r = -1$ when $\theta = 0$, so this can't be (C).

e. This matches (E). Note that there is an inner loop because $|a| = 1 < 2 = |b|$, and that $r = 3$ when $\theta = \pi/2$.

f. This matches (F).

11.2.91

11.2.93

11.2.95

11.2.97

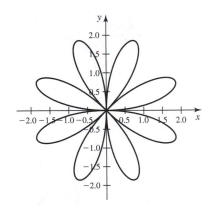

11.2.99 Note that $a \sin m\theta = 0$ for $\theta = \frac{k\pi}{m}$, $k = 1, 2, \ldots, 2m$. Thus the graph is back at the pole $r = 0$ for each of these values, and each of these gives rise to a distinct petal of the rose if m is odd. If m is even, then by symmetry, each petal for $k = 1, 2, \ldots \frac{m}{2}$ is equivalent to one for $k = \frac{m}{2} + 1, \frac{m}{2} + 2, \ldots, m$. (Note that this follows because the sine function is odd.) A similar result holds for the rose $r = a \cos \theta$.

11.2.101 For $a = 1$, the spiral winds outward counterclockwise. For $a = -1$, the spiral winds inward counterclockwise.

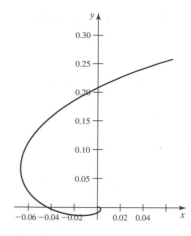

11.2.103 Suppose $2\cos\theta = 1 + \cos\theta$. Then $\cos\theta = 1$, so this occurs for $\theta = 0$ and $\theta = 2\pi$. At those values, $r = 2$, so the curves intersect at the polar point $(2, 0)$. The curves also intersect when $r = 0$, which occurs for $\theta = \pi/2$ and $\theta = 3\pi/2$ for the first curve and $\theta = \pi$ for the second.

11.2.105 Suppose $1 - \sin\theta = 1 + \cos\theta$, or $\tan\theta = -1$. Then $\theta = 3\pi/4$ or $\theta = 7\pi/4$. So the curves intersect at the polar points $(1 + \sqrt{2}/2, 7\pi/4)$ and $(1 - \sqrt{2}/2, 3\pi/4)$. They also intersect at the pole $(0, 0)$, which occurs for the first curve at $\pi/2$ and for the second curve at π.

11.2.107

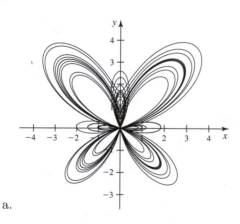

b. It adds multiple layers of the same type of curve as $\sin^5\theta/12$ oscillates between -1 and 1 for $0 \le \theta \le 24\pi$.

a.

11.2.109

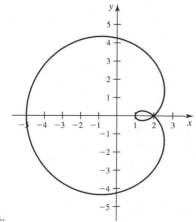

b. $r = 3 - 4\cos\pi t$ is a limaçon, and $x - 2 = r\cos\pi t$ and $y = r\sin\pi t$ is a circle, and the composition of a limaçon and a circle is a limaçon.

a.

11.2.111 With $r = a\cos\theta + b\sin\theta$, we have $r^2 = ar\cos\theta + br\sin\theta$, or $x^2 + y^2 = ax + by$, so $\left(x - \frac{a}{2}\right)^2 + \left(y - \frac{b}{2}\right)^2 = \frac{a^2+b^2}{4}$. Thus, the center is $(a/2, b/2)$ and $r = \frac{\sqrt{a^2+b^2}}{2}$.

11.2.113 Because $\sin(\theta/2) = \sin(\pi - \theta/2) = \sin((2\pi - \theta)/2)$, we have that the graph is symmetric with respect to the x-axis.

11.3   Calculus in Polar Coordinates

11.3.1 Because $x = r\cos\theta$ and $y = r\sin\theta$, we have $x = f(\theta)\cos\theta$ and $y = f(\theta)\sin(\theta)$.

11.3.3 Because slope is given relative to the horizontal and vertical coordinates, it is given by $\frac{dy}{dx}$, not by $\frac{dr}{d\theta}$.

11.3.5 $\frac{dy}{dx} = \frac{-\cos\theta\sin\theta + (1-\sin\theta)\cos\theta}{-\cos^2\theta - (1-\sin\theta)\sin\theta}$. At $(1/2, \pi/6)$, we have $\frac{dy}{dx} = \frac{0}{-1} = 0$. The given curve intersects the origin $r = 0$ for $\theta = \pi/2$. At this point, $\frac{dy}{dx}$ does not exist, and the tangent line is vertical. (It is the line $\theta = \pi/2$.)

11.3.7 $\frac{dy}{dx} = \frac{16\cos\theta\sin\theta}{-8\sin^2\theta + 8\cos^2\theta}$. At $(4, 5\pi/6)$ we have $\frac{dy}{dx} = \frac{-4\sqrt{3}}{4} = -\sqrt{3}$. The given curve intersects the origin $r = 0$ for $\theta = 0$ and $\theta = \pi$. At these points, the derivative is 0, and the tangent line is horizontal, so $\theta = 0$ is the tangent line.

11.3.9 $\frac{dy}{dx} = \frac{-3\sin^2\theta + (6+3\cos\theta)\cos\theta}{-3\cos\theta\sin\theta - (6+3\cos\theta)\sin\theta}$. At both $(3, \pi)$ and $(9, 0)$, this doesn't exist. The given curve does not intersect the origin, because $r \geq 3$ for all θ.

11.3.11 $\frac{dy}{dx} = \frac{-8\sin(2\theta)\sin\theta + 4\cos(2\theta)\cos\theta}{-8\sin(2\theta)\cos\theta - 4\cos(2\theta)\sin\theta}$. The tips of the leaves occur at $\theta = 0$, $\pi/2$, π and $3\pi/2$. At 0 and at π, we have that $\frac{dy}{dx}$ doesn't exist. At $\pi/2$ and $3\pi/2$ we have $\frac{dy}{dx} = 0$. The graph intersects the origin for $\theta = \pi/4$, $\theta = 3\pi/4$, $\theta = 5\pi/4$ and $\theta = 7\pi/4$, and thus the two distinct tangent lines are $\theta = \pi/4$ and $\theta = 3\pi/4$.

11.3.13 The curve hits the origin at $\pm\pi/4$, where the tangent lines are given by $\theta = \pi/4$ and $\theta = -\pi/4$. The slopes of those lines are given by $\tan(\pi/4) = 1$ and $\tan(-\pi/4) = -1$.

11.3.15 Note that the curve is at the origin at $\pi/2$, so there is vertical tangent at $(0, \pi/2)$. Also, $\frac{dy}{dx} = \frac{-4\sin^2\theta + 4\cos^2\theta}{-8\sin\theta\cos\theta} = \frac{1-2\sin^2\theta}{\sin(2\theta)}$. Thus, there are horizontal tangents at $\pi/4$ and $3\pi/4$ (at the polar points $(2\sqrt{2}, \pi/4)$ and $(-2\sqrt{2}, 3\pi/4)$). There is also a vertical tangent where $\theta = 0$, at the point $(4, 0)$.

11.3.17 Using the double angle identities somewhat liberally:

$$\frac{dy}{dx} = \frac{2\cos(2\theta)\sin\theta + \sin(2\theta)\cos\theta}{2\cos(2\theta)\cos\theta - \sin(2\theta)\sin\theta} = \frac{\sin\theta(\cos 2\theta + \cos^2\theta)}{\cos\theta(\cos(2\theta) - \sin^2\theta)} = \frac{\sin\theta(3\cos^2\theta - 1)}{\cos\theta(1 - 3\sin^2\theta)} = \frac{\sin\theta(3\cos^2\theta - 1)}{\cos\theta(3\cos^2\theta - 2)}.$$

The numerator is 0 for $\theta = 0$ and for $\theta = \pm\cos^{-1}(\pm\sqrt{3}/3)$, so there are horizontal tangents at the corresponding points $(0, 0)$, $(0.943, 0.955)$, $(-0.943, 2.186)$, $(0.943, 4.097)$, and $(-0.943, 5.328)$. The denominator is 0 for $\theta = \pi/2$ and $3\pi/2$, and for $\theta = \pm\cos^{-1}(\pm\sqrt{6}/3)$, so there are vertical tangents at $(0, 0)$, $(0.943, 0.615)$, $(-0.943, 2.526)$, $(0.943, 3.757)$, and $(-0.943, 5.668)$.

11.3.19 The curve intersects the origin at $\theta = \pi/2$, and there is a vertical tangent at $(0, \pi/2)$. $\frac{dy}{dx} = \frac{-\cos\theta\sin\theta + (1-\sin\theta)\cos\theta}{-\cos^2\theta - (1-\sin\theta)\sin\theta} = \frac{\cos\theta(1-2\sin\theta)}{\sin^2\theta - \cos^2\theta - \sin\theta} = \frac{\cos\theta(1-2\sin\theta)}{2\sin^2\theta - \sin\theta - 1}$. There are horizontal tangents when $\sin\theta = 1/2$, which occurs for $\theta = \pi/6, 5\pi/6$, and when $\cos\theta = 0$ (but not $\sin\theta = 1$) which occurs at $\theta = 3\pi/2$. So the horizontal tangents are at $(1/2, \pi/6)$, $((1/2, 5\pi/6)$, and $(2, 3\pi/2)$. There are vertical tangents when $2\sin^2\theta - \sin\theta - 1 = (2\sin\theta + 1)(\sin\theta - 1) = 0$, or $\theta = 7\pi/6$ and $\theta = 11\pi/6$. The vertical tangents are thus at $(3/2, 7\pi/6)$, $(3/2, 11\pi/6)$, and $(0, \pi/2)$, as well as the aforementioned $(0, \pi/2)$.

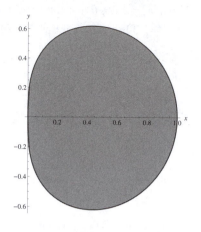

11.3.21 $\quad A = 2 \cdot \frac{1}{2} \int_0^{\pi/2} \cos\theta \, d\theta = \sin\theta \Big|_0^{\pi/2} = 1.$

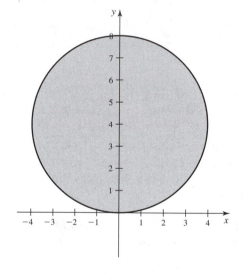

11.3.23 $\quad A = \frac{1}{2} \int_0^{\pi} (8\sin\theta)^2 \, d\theta = 32 \int_0^{\pi} \sin^2\theta \, d\theta = 32 \int_0^{\pi} \frac{1-\cos 2\theta}{2} \, d\theta = 32 \left(\frac{1}{2}\theta - \frac{\sin\theta\cos\theta}{2} \right) \Big|_0^{\pi} = 16\pi.$

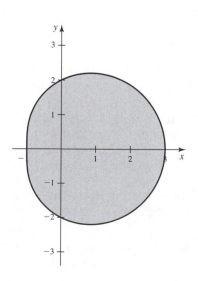

11.3.25 $\quad$ Using symmetry, we have $\frac{1}{2} \cdot 2 \int_0^{\pi} (2 + \cos\theta)^2 \, d\theta = \int_0^{\pi} (4 + 4\cos\theta + \cos^2\theta) \, d\theta = \left(4\theta + 4\sin\theta + \frac{1}{2}\theta + \frac{\sin\theta\cos\theta}{2} \right) \Big|_0^{\pi} = \frac{9\pi}{2}.$

11.3.27

$$2 \cdot \tfrac{1}{2} \int_0^{\pi/6} \cos^2 3\theta \, d\theta = \int_0^{\pi/6} \tfrac{1+\cos 6\theta}{2} \, d\theta =$$

$$\left(\theta/2 + \tfrac{\sin 6\theta}{6}\right)\Big|_0^{\pi/6} = \tfrac{\pi}{12}.$$

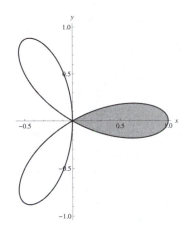

11.3.29

The area is given by $2 \cdot \tfrac{1}{2} \int_0^{\pi/3} (\cos^2 \theta - (1/2)^2) \, d\theta = \int_0^{\pi/3} (\cos(2\theta)/2 + \tfrac{1}{4}) \, d\theta =$

$(\sin(2\theta)/4 + \theta/4)\Big|_0^{\pi/3} = \tfrac{\sqrt{3}}{8} + \tfrac{\pi}{12}.$

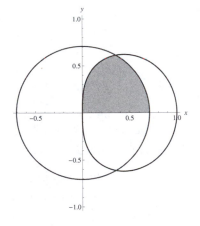

11.3.31

The region inside the circle between 0 and $\pi/3$ is $1/6$ the area of a circle of radius $1/\sqrt{2}$ so it has area $(1/6)\pi(1/2) = \pi/12$. The rest of the area is represented by $\tfrac{1}{2} \int_{\pi/3}^{\pi} \cos \theta \, d\theta =$

$\tfrac{1}{2} (\sin \theta) \Big|_{\pi/3}^{\pi} = \tfrac{1}{2}\left(1 - \sqrt{3}/2\right).$ The total area

is therefore $\tfrac{\pi}{12} + \tfrac{1}{2} - \tfrac{\sqrt{3}}{4}.$

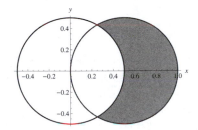

11.3.33

Using symmetry, we compute the area of $1/2$ of one leaf, and then double it. We have $A = \tfrac{1}{2} \int_0^{\pi/10} \cos^2(5\theta) \, d\theta = \tfrac{1}{10} \int_0^{\pi/2} \cos^2 u \, du =$

$\tfrac{1}{10} \left(\tfrac{1}{2}u + \tfrac{\cos u \sin u}{2}\right)\Big|_0^{\pi/2} = \tfrac{\pi}{40}.$ So the area

of one leaf is $2 \cdot \tfrac{\pi}{40} = \tfrac{\pi}{20}.$

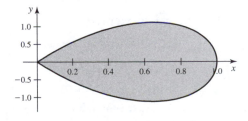

Note that the area inside one leaf of the rose but outside the circle is given by $\frac{1}{2}\int_{\pi/12}^{5\pi/12}(16\sin^2(2\theta) - 4)\,d\theta =$ $(2\theta - \sin(4\theta))\Big|_{\pi/12}^{5\pi/2} = \sqrt{3} + \frac{2\pi}{3}$. Also, the area inside one leaf of the rose is

11.3.35 $\frac{1}{2}\int_0^{\pi/2}16\sin^2(2\theta)\,d\theta = (4\theta - \sin(4\theta))\Big|_0^{\pi/2} = 2\pi$. Thus the area inside one leaf of the rose and inside the circle must be $2\pi - (\sqrt{3} + \frac{2\pi}{3}) = \frac{4\pi}{3} - \sqrt{3}$, and the total area inside the rose and inside the circle must be $4(\frac{4\pi}{3} - \sqrt{3}) = \frac{16\pi}{3} - 4\sqrt{3}$.

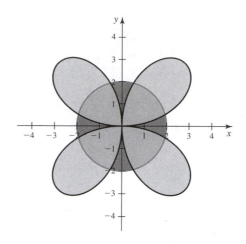

These curves intersect when $\sin\theta = \cos\theta$, which occurs at $\theta = \pi/4$ and $\theta = 5\pi/4$, and when $r = 0$ which occurs for $\theta = 0$ and $\theta = \pi$

11.3.37 for the first curve and $\theta = \pi/2$ and $\theta = 3\pi/2$ for the second curve. Only two of these intersection points are unique: the origin and the point $(3\sqrt{2}/2, \pi/4) = (-3\sqrt{2}/2, 5\pi/4)$.

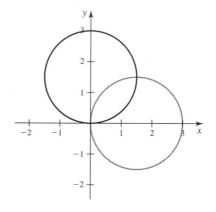

The curves intersect when $\sin\theta = \cos\theta$, which occurs for $\theta = \pi/4$ and $\theta = 5\pi/4$. The corresponding points are $\left(\frac{2+\sqrt{2}}{2}, \frac{\pi}{4}\right)$ and

11.3.39 $\left(\frac{2-\sqrt{2}}{2}, \frac{5\pi}{4}\right)$. They also intersect at the pole: the first curve is at the pole at $(0, \pi)$ and the other at $(0, 3\pi/2)$.

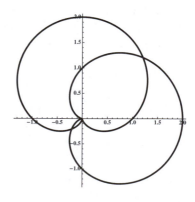

11.3.41 By symmetry, we need to compute the area inside $r = 3\sin\theta$ between 0 and $\pi/4$ and then double that result. We have $2 \cdot \frac{1}{2}\int_0^{\pi/4}9\sin^2\theta\,d\theta = \frac{9}{2}\int_0^{\pi/4}(1 - \cos(2\theta))\,d\theta = \frac{9}{2}\left(\theta - (1/2)\sin(2\theta)\right)\Big|_0^{\pi/4} = \frac{9}{2}\left(\frac{\pi}{4} - \frac{1}{2}\right) = \frac{9}{8}(\pi - 2)$.

11.3.43 By symmetry, we can compute the area between $\pi/4$ and $5\pi/4$ inside $r = 1 + \cos\theta$ and then double it. This will include both the bigger and smaller enclosed regions. We have $2 \cdot \frac{1}{2}\int_{\pi/4}^{5\pi/4}(1 + \cos\theta)^2\,d\theta = \int_{\pi/4}^{5\pi/4}(1 +$

$2\cos\theta+(1/2)(1+\cos(2\theta)))\,d\theta = \int_{\pi/4}^{5\pi/4}((3/2)+2\cos\theta+(1/2)(\cos 2\theta))\,d\theta = (3\theta/2+2\sin\theta+(1/4)\sin 2\theta)\Big|_{\pi/4}^{5\pi/4}$

$= \left(\frac{15\pi}{8}-\sqrt{2}+\frac{1}{4}\right)-\left(\frac{3\pi}{8}+\sqrt{2}+\frac{1}{4}\right)=\frac{3\pi}{2}-2\sqrt{2}.$

11.3.45

a. False. The area is given by $\frac{1}{2}\int_{\alpha}^{\beta}f(\theta)^2\,d\theta$.

b. False. The slope is given by $\frac{dy}{dx}$, which can be computed using the formula

$$\frac{dy}{dx}=\frac{dy/d\theta}{dx/d\theta}=\frac{f'(\theta)\sin\theta+f(\theta)\cos\theta}{f'(\theta)\cos\theta-f(\theta)\sin\theta}.$$

11.3.47 The circles intersect for $\theta=\pi/6$ and $\theta=5\pi/6$.

The area inside $r=2\sin\theta$ but outside of $r=1$ would be given by $\frac{1}{2}\int_{\pi/6}^{5\pi/6}(4\sin^2\theta-1)\,d\theta=$

$\frac{1}{2}\left(x-\sin(2x)\right)\Big|_{\pi/6}^{5\pi/6}=\frac{\pi}{3}+\frac{\sqrt{3}}{2}$. The total area of $r=2\sin\theta$ is π. Thus, the area inside both circles is

$\pi-\left(\frac{\pi}{3}+\frac{\sqrt{3}}{2}\right)=\frac{2\pi}{3}-\frac{\sqrt{3}}{2}.$

11.3.49 The inner loop is traced out between $\theta=\pi/6$ and $\theta=5\pi/6$, so its area is given by $\frac{1}{2}\int_{\pi/6}^{5\pi/6}(3-$

$6\sin\theta)^2\,d\theta=\frac{1}{2}\int_{\pi/6}^{5\pi/6}(9-36\sin\theta+36\sin^2\theta)\,d\theta=\frac{3}{2}(3\theta+12\cos\theta+6\theta-3\sin(2\theta))\Big|_{\pi/6}^{5\pi/6}=9\pi-\frac{27\sqrt{3}}{2}.$

We can determine the area inside the outer loop by using symmetry and doubling the area of the region traced out between $5\pi/6$ and $3\pi/2$. Thus the area inside the outer region is $2\cdot\frac{1}{2}\int_{5\pi/6}^{3\pi/2}(3-6\sin\theta)^2\,d\theta=$

$3(3\theta+12\cos\theta+6\theta-3\sin(2\theta))\Big|_{5\pi/6}^{3\pi/2}=18\pi+\frac{27\sqrt{3}}{2}$. So the area outside the inner loop and inside the outer

loop is $18\pi+\frac{27\sqrt{3}}{2}-\left(9\pi-\frac{27\sqrt{3}}{2}\right)=9\pi+27\sqrt{3}.$

11.3.51 The first horizontal tangent line is at the origin. The next is at approximately $(4.0576, 2.0288)$, and the third at approximately $(9.8262, 4.9131)$. The first vertical tangent line is at approximately $(1.7206, 0.8603)$, the next is at about $(6.8512, 3.4256)$, and the next at approximately $(12.8746, 6.4373)$.

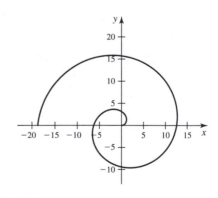

11.3.53

a. $A_n=\frac{1}{2}\int_{(2n-2)\pi}^{(2n-1)\pi}e^{-2\theta}\,d\theta-\frac{1}{2}\int_{2n\pi}^{(2n+1)\pi}e^{-2\theta}\,d\theta=-\frac{1}{4}e^{-(4n-2)\pi}+\frac{1}{4}e^{-(4n-4)\pi}+\frac{1}{4}e^{-(4n+2)\pi}-\frac{1}{4}e^{-4n\pi}.$

b. Each term tends to 0 as $n\to\infty$ so $\lim_{n\to\infty}A_n=0$.

c. $\frac{A_{n+1}}{A_n}=\frac{e^{-(4n+2)\pi}+e^{-(4n)\pi}+e^{-(4n+6)\pi}-e^{-(4n+4)\pi}}{e^{-(4n-2)\pi}+e^{-(4n-4)\pi}+e^{-(4n+2)\pi}-e^{-4n\pi}}=e^{-4\pi}$, so $\lim_{n\to\infty}\frac{A_{n+1}}{A_n}=e^{-4\pi}.$

11.3.55 One half of the area is given by $\frac{1}{2}\int_0^{\pi/2}6\sin 2\theta\,d\theta=-\frac{3}{2}\cos 2\theta\Big|_0^{\pi/2}=3$, so the total area is 6.

11.3.57 The area is given by

$$\frac{1}{2}\int_0^{2\pi}(4-2\cos\theta)^2\,d\theta = \int_0^{2\pi}(8-8\cos\theta+2\cos^2\theta)\,d\theta = \left(8\theta - 8\sin\theta + \theta + \frac{1}{2}\sin(2\theta)\right)\Big|_0^{2\pi} = 18\pi.$$

11.3.59 Suppose that the goat is tethered at the origin, and that the center of the corral is $(1,\pi)$. The circle that the goat can graze is $r=a$, and the corral is given by $r=-2\cos\theta$. The intersection occurs for $\theta=\cos^{-1}(-a/2)$.

The area grazed by the goat is twice the area of the sector of the circle $r=a$ between $\cos^{-1}(-a/2)$ and π, plus twice the area of the circle $r=-2\cos\theta$ between $\pi/2$ and $\cos^{-1}(-a/2)$. Thus we need to compute

$$A = \int_{\cos^{-1}(-a/2)}^{\pi}a^2\,d\theta + \int_{\pi/2}^{\cos^{-1}(-a/2)}4\cos^2\theta\,d\theta = a^2\pi - a^2\cos^{-1}(-a/2) + (2\cos\theta\sin\theta + 2\theta)\Big|_{\pi/2}^{\cos^{-1}(-a/2)} =$$

$a^2(\pi - \cos^{-1}(-a/2)) - \pi - \frac{1}{2}a\sqrt{4-a^2} + 2\cos^{-1}(-a/2)$. Note that $\pi - \cos^{-1}(-a/2) = \cos^{-1}(a/2)$, so this can be written as $(a^2-2)\cos^{-1}(a/2) + \pi - \frac{1}{2}a\sqrt{4-a^2}$. Note that for $a=0$ this is 0, and for $a=2$, this is π, as desired.

11.3.61 Again, suppose that the goat is tethered at the origin, and that the center of the corral is $(1,\pi)$. The equation of the corral fence is given by $r=-2\cos\theta$. Note that to the right of the vertical line $\theta=\pi/2$, the goat can graze a half-circle of area $\pi a^2/2$. Also, there is a region in the 2nd quadrant and one in the 3rd quadrant of equal size that can also be grazed. Let this region have area A, so that the total area grazed will then be $\frac{\pi a^2}{2} + 2A$.

Imagine that the goat is walking "west" from the polar point $(a,\pi/2)$, and is keeping the rope taut until his whole rope is along the fence in the third quadrant. Let ϕ be the central angle angle from the origin to the polar point $(1,\pi)$ to the point on the fence that the goat's rope is touching as he makes this walk. When the goat is at $(a,\pi/2)$, we have $\phi=0$. When the goat is all the way to the fence, we have $\phi=a$. Then length of the rope not along the fence is $a-\phi$. Thus, the value of A is $\frac{1}{2}\int_0^a(a-\phi)^2d\phi = \frac{1}{2}\left(a^2\phi - a\phi^2 + \frac{\phi^3}{3}\right)\Big|_0^a = \frac{a^3}{6}$.

Thus, the goat can graze a region of area $\frac{\pi a^2}{2} + \frac{a^3}{3}$.

11.3.63

a. If $\cot\phi = \frac{f'(\theta)}{f(\theta)}$ is constant for all θ, then $\phi = \cot^{-1}\left(\frac{f'(\theta)}{f(\theta)}\right)$ is constant. Then $\frac{d}{d\theta}\ln(f(\theta)) = \frac{1}{f(\theta)}\cdot f'(\theta) = \cot\phi$ is constant.

b. If $f(\theta) = Ce^{k\theta}$, then $\cot\phi = \frac{f'(\theta)}{f(\theta)} = \frac{kCe^{k\theta}}{Ce^{k\theta}} = k$.

c.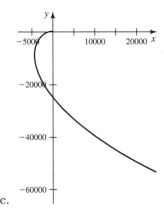

11.4 Conic Sections

11.4.1 A parabola is the set of points in the plane which are equidistant from a given fixed point and a given fixed line.

11.4.3 A hyperbola is the set of points in the plane with the property that the difference of the distances from the point to two given fixed points is a given constant.

11.4.5

  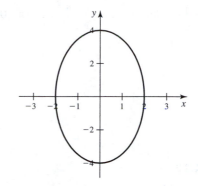

11.4.7 $\left(\dfrac{x}{a}\right)^2 + \dfrac{y^2}{a^2 - c^2} = 1.$

11.4.9 The foci for both are $(\pm ae, 0)$.

11.4.11 The asymptotes are $y = -\dfrac{b}{a} \cdot x$ and $y = \dfrac{b}{a} \cdot x$.

11.4.13   Directrix: $y = -3$. Focus: $(0, 3)$.

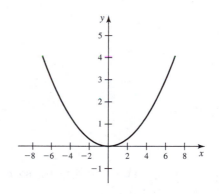

11.4.15   Directrix: $x = 4$. Focus: $(-4, 0)$.

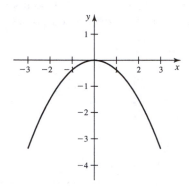

11.4.17 Directrix: $y = \frac{2}{3}$. Focus: $(0, -\frac{2}{3})$.

11.4.19

$y^2 = 16x$.

11.4.21

$y^2 = 12x$.

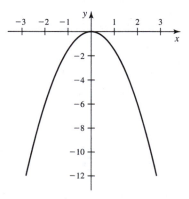

11.4.23 $x^2 = 4py$ and $4 = 4p(-6)$, so $p = -\frac{1}{6}$ and $x^2 = -\frac{2}{3} \cdot y$.

11.4.25 Because the vertex is $(-1, 0)$ and the parabola is symmetric about the x-axis, we have $y^2 = 4p(x+1)$ and because the directrix is one unit left of the vertex, we obtain $p = 1$ and $y^2 = 4(x + 1)$.

11.4.27 Vertices are $(\pm 2, 0)$, and the foci are $(\pm\sqrt{3}, 0)$. The major axis has length 4 and the minor axis has length 2.

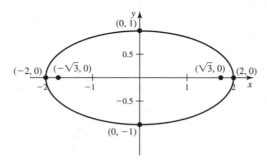

11.4.29 Vertices are $(0, \pm 4)$, and the foci are $(0, \pm 2\sqrt{3})$. The major axis has length 8 and the minor axis has length 4.

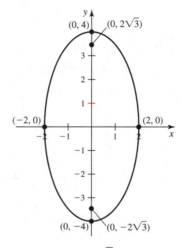

11.4.31 Vertices are $(0, \pm\sqrt{7})$, and the foci are $(0, \pm\sqrt{2})$. The major axis has length $2\sqrt{7}$ and the minor axis has length $2\sqrt{5}$.

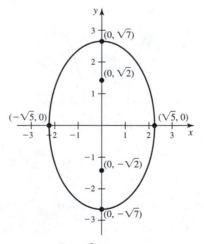

11.4.33 $a = 4$, and $b = 3$, so the equation is $\frac{x^2}{16} + \frac{y^2}{9} = 1$.

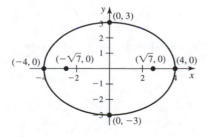

11.4.35 $a = 5$, and the equation is of the form $\frac{x^2}{25} + \frac{y^2}{b^2} = 1$. Because $(4, \frac{3}{5})$ is on the curve, we have $\frac{16}{25} + \frac{9}{25b^2} = 1$, so $b = 1$. The equation is $\frac{x^2}{25} + y^2 = 1$.

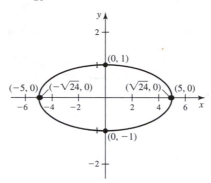

11.4.37 $a = 3$ and $b = 2$, so the equation is $\frac{x^2}{4} + \frac{y^2}{9} = 1$.

11.4.39 The vertices are $(\pm 2, 0)$, and the foci are $(\pm\sqrt{5}, 0)$. The asymptotes are $y = \frac{\pm 1}{2} \cdot x$.

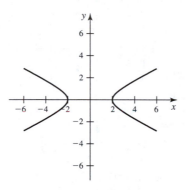

11.4.41 The vertices are $(\pm 2, 0)$, and the foci are $(\pm 2\sqrt{5}, 0)$. The asymptotes are $y = \pm 2x$.

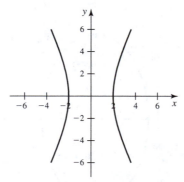

11.4.43 The vertices are $(\pm\sqrt{3}, 0)$, and the foci are $(\pm 2\sqrt{2}, 0)$. The asymptotes are $y = \pm\sqrt{\frac{5}{3}} \cdot x$.

11.4.45 We have $a = 4$ and $c = 6$, so $b^2 = c^2 - a^2 = 20$, so the equation is $\frac{x^2}{16} - \frac{y^2}{20} = 1$.

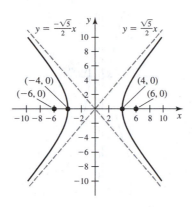

11.4.47 We have $a = 2$, and because the asymptoes are $y = \frac{\pm bx}{a}$, we have that $b = 3$, so the equation is $\frac{x^2}{4} - \frac{y^2}{9} = 1$.

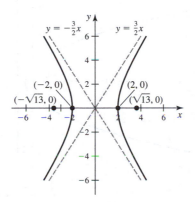

11.4.49 We have $a = 4$ and $c = 5$, so $b^2 = 25 - 16 = 9$, so $b = 3$ and the equation is $\frac{x^2}{16} - \frac{y^2}{9} = 1$.

11.4.51 We have $a = 9$ and $e = \frac{1}{3}$, so $c = ae = 3$, and $b^2 = a^2 - c^2 = 72$, so the equation is $\frac{x^2}{81} + \frac{y^2}{72} = 1$.

11.4.53 We have $a = 1$ and $e = 3$, so $c = ae = 3$ and $b^2 = c^2 - a^2 = 9 - 1 = 8$. Thus, the equation is $x^2 - \frac{y^2}{8} = 1$.

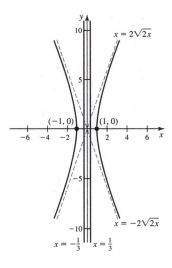

11.4.55 The vertex is $(2, 0)$. The focus is $(0, 0)$, and the directrix is the line $x = 4$.

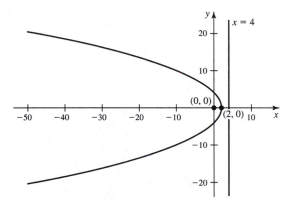

11.4.57 The vertices are $(1, 0)$ and $(-1/3, 0)$. The center is $(1/3, 0)$. The directrices are $x = -1$ and $x = 5/3$.

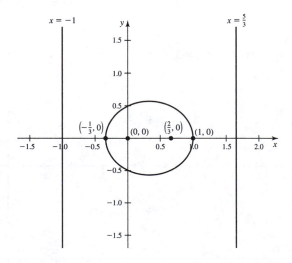

11.4.59 The vertex is $(0, -1/4)$, and the focus is $(0, 0)$. The directrix is the line $y = -\frac{1}{2}$.

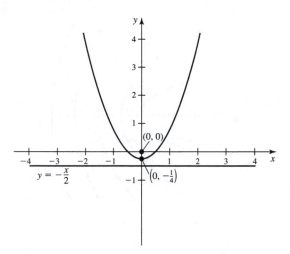

11.4.61 The parabola starts at $(1,0)$ and goes through quadrants I, II, and III for $\theta \in [0, 3\pi/2]$. It then approaches $(1,0)$ by traveling through quadrant IV for $\theta \in (3\pi/2, 2\pi)$.

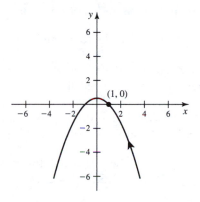

11.4.63 The parabola begins in the first quadrant and passes through the points $(0,3)$ and then $(-3/2, 0)$ and $(0-3)$ as θ ranges from 0 to 2π.

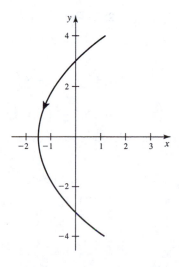

11.4.65 The parabolas open to the left due to the presence of a positive $\cos \theta$ term in the denominator. As d increases, the directrix $x = d$ moves to the right, resulting in wider parabolas.

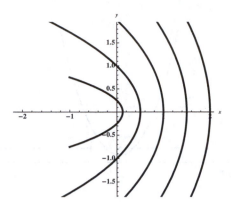

11.4.67

a. True. Note that if $x = 0$, the equation becomes $-y^2 = 9$, which has no solution.

b. True. The slopes of the tangent lines range continuously from $-\infty$ to 0 to ∞ and then back through 0 to $-\infty$ again.

c. True. Given c and d, one can compute a, b, and e. See the summary after Theorem 10.3.

d. True. The vertex is exactly halfway between the focus and the directrix.

11.4.69 Differentiating gives $2x = -6y'$, so at $(-6, -6)$ we obtain $-12 = -6y'$, so $y' = 2$. Thus $y - (-6) = 2(x - (-6))$, or $y = 2x + 6$ is the equation of the tangent line.

11.4.71 Differentiating implicitly, we have $2yy' - \frac{x}{32} = 0$, so at $(6, -5/4)$ we have $-\frac{5}{2}y' - \frac{3}{16} = 0$, so $y' = -\frac{3}{40}$. The equation of the tangent line is $y + \frac{5}{4} = -\frac{3}{40}(x - 6)$, or $y = -\frac{3x}{40} - \frac{4}{5}$.

11.4.73 We have a hyperbola with focal point at the origin and directrix $y = -2$. Furthermore $P = (0, -4/3)$ is a vertex. Thus, $e = \frac{|PF|}{|PL|} = \frac{4/3}{2/3} = 2$, and $r(\theta) = \frac{2(2)}{1-2\sin\theta} = \frac{4}{1-2\sin\theta}$.

11.4.75 The points on the intersection of the two circles are a distance of $2a + r$ from F_1 and a distance of r from F_2. So for P an intersection point, we have $|PF_1| - |PF_2| = 2a$ for all r, and the set of all such points form a hyperbola with foci F_1 and F_2.

11.4.77 Using implicit differentiation, we have $\frac{2x}{a^2} + \frac{2yy'}{b^2} = 0$, or $y' = -\frac{b^2 x}{a^2 y}$. If (x_0, y_0) is the point of tangency, then $-\frac{b^2 x_0}{a^2 y_0} = \frac{y - y_0}{x - x_0}$, so $\frac{x_0(x - x_0)}{a^2} = -\frac{y_0(y - y_0)}{b^2}$, so $\frac{x_0 x}{a^2} + \frac{y_0 y}{b^2} = \frac{x_0^2}{a^2} + \frac{y_0^2}{b^2} = 1$.

11.4.79 $V_x = \pi \int_{-a}^{a} \left(b^2 - \frac{b^2 x^2}{a^2}\right) dx = \pi b^2 \int_{-a}^{a} \left(1 - \frac{x^2}{a^2}\right) dx = \pi b^2 \left(x - \frac{x^3}{3a^2}\right)\Big|_{-a}^{a} = \frac{4\pi b^2 a}{3}$.

$V_y = \pi \int_{-b}^{b} \left(a^2 - \frac{a^2 y^2}{b^2}\right) dy = \pi a^2 \int_{-b}^{b} \left(1 - \frac{y^2}{b^2}\right) dy = \pi a^2 \left(y - \frac{y^3}{3b^2}\right)\Big|_{-b}^{b} = \frac{4\pi a^2 b}{3}$.

These are different if $a \neq b$. In the case $a = b$, both volumes give $\frac{4\pi a^3}{3}$, the volume of a sphere.

11.4.81

a. $V_x = \pi \int_a^c \left(\sqrt{\frac{b^2 x^2}{a^2} - b^2}\right)^2 dx = \pi \int_a^c \left(\frac{b^2 x^2}{a^2} - b^2\right) dx = \pi b^2 \left(\frac{x^3}{3a^2} - x\right)\Big|_a^c = \pi b^2 \left(\frac{c^3}{3a^2} - c - \frac{a}{3} + a\right) = \frac{\pi b^2}{3a^2}(c^3 - 3ca^2 + 2a^3) = \frac{\pi b^2}{3a^2}(a - c)^2(2a + c)$.

b. $V_y = 2 \cdot 2\pi \int_a^c a^2 b \sqrt{\frac{x^2}{a^2} - 1}\, dx = 2\pi \int_0^{b^2/a^2} a^2 b \sqrt{u}\, du = 2\pi a^2 b \left(\frac{2}{3} u^{3/2}\right)\Big|_0^{b^2/a^2} = 2\pi a^2 b \frac{2b^3}{3a^3} = \frac{4\pi b^4}{3a}$.

11.4.83

a. The slope of a line making an angle θ with the horizontal is $\tan\theta$. The slope of the tangent line at (x_0, y_0) is $y' = \frac{x}{2p}$, so $y' = \frac{x_0}{2p}$, so $\tan\theta = \frac{x_0}{2p}$.

b. The distance from $(0, y_0)$ to $(0, p)$ is $p - y_0$, and $\tan\phi = \frac{\text{opposite}}{\text{adjacent}} = \frac{p - y_0}{x_0}$.

c. Because l is perpendicular to $y = y_0$, we have $\alpha + \theta = \pi/2$, or $\alpha = \frac{\pi}{2} - \theta$, so $\tan\alpha = \cot\theta = \frac{2p}{x_0}$.

d. $\tan\beta = \tan(\theta + \phi) = \frac{\frac{x_0}{wp} + \frac{p - y_0}{x_0}}{1 - \frac{p - y_0}{2p}} = \frac{x_0^2 + 2p^2 - 2py_0}{x_0(p + y_0)}$. Now because $x_0^2 = 4py_0$, we obtain $\tan\beta = \frac{4py_0 + 2p^2 - 2py_0}{x_0(p + y_0)}$
$= \frac{2p(p + y_0)}{x_0(p + y_0)} = \frac{2p}{x_0}$.

e. Because α and β are acute, we have that $\tan\alpha = \tan\beta$, so $\alpha = \beta$.

11.4.85 Assume the two fixed points are at $(c, 0)$ and $(-c, 0)$. Let P be the point $(0, b)$, and note that P is equidistant from the two given points, so we must have $b^2 + c^2 = a^2$ by the Pythagorean theorem. Now let $Q = (u, 0)$ be on the ellipse for $u > c$. Then $u - c + (c + u) = 2a$, so $u = a$. Now let $R = (x, y)$ be an arbitrary point on the ellipse (assume $x > 0$ and $y > 0$ – the other cases are similar.) Using the triangles formed between the foci, R, and the projection of R onto the x-axis, we have $\sqrt{(x + c)^2 + y^2} = 2a - \sqrt{(c - x)^2 + y^2}$. Squaring both sides gives $(x + c)^2 + y^2 = 4a^2 - 4a\sqrt{(c - x)^2 + y^2} + (c - x)^2 + y^2$. Isolating the root gives $\sqrt{(c - x)^2 + y^2} = \frac{1}{4a}\left((c - x)^2 + y^2 - (c + x)^2 - y^2 + 4a^2\right)$, so $\sqrt{(c - x)^2 + y^2} = a - \frac{c}{a}x$. Squaring again yields $(c - x)^2 + y^2 = a^2 - 2xc + \frac{c^2}{a^2}x^2$, so $c^2 - 2cx + x^2 + y^2 = a^2 - 2cx + \frac{c^2}{a^2}x^2$, or $x^2\left(1 - \frac{c^2}{a^2}\right) + y^2 = a^2 - c^2$. Thus $\frac{x^2}{a^2} + \frac{y^2}{a^2 - c^2} = 1$, which can be written $\frac{x^2}{a^2} + \frac{y^2}{b^2} = 1$, because $b^2 = a^2 - c^2$.

11.4.87 Let the parabola be symmetric about the y-axis with vertex at the origin. Let the circle have radius r and be centered at $(r + a, 0)$, and let the line be $y = -a$. The distance form the point $P(x, y)$ to the line is $u = y + a$. The distance from the point P to the circle is $v = \sqrt{x^2 + (r + a - y)^2} - r$. Setting $u = v$ yields $y + a = \sqrt{x^2 + (r + a - y)^2} - r$, so $y + r + a = \sqrt{x^2 + (r + a - y)^2}$, and squaring gives $y^2 + 2(r + a)y + (r + a)^2 = x^2 + (r + a - y)^2$, so $y^2 + 2(r + a)y + (r + a)^2 = x^2 + (r + a)^2 - 2(r + a)y + y^2$, and thus $4(r + a)y = x^2$, so $y = \frac{1}{4(r + a)}x^2$, the equation of a parabola.

11.4.89 Let the hyperbolas be centered at the origin with equations $\frac{x^2}{a^2} - \frac{y^2}{b^2} = 1$ and $\frac{y^2}{B^2} - \frac{x^2}{A^2} = 1$ and eccentricities $e = \frac{c}{a}$ and $E = \frac{C}{B}$, respectively. Because the hyperbolas share a set of asymptotes $A = ra$ and $B = rb$ fro some $r > 0$, and

$$C^2 = A^2 + B^2 = (ra)^2 + (rb)^2$$
$$= r^2(a^2 + b^2) = r^2c^2.$$

Then we have

$$e^{-2} + E^{-2} = \left(\frac{c}{a}\right)^{-2} + \left(\frac{C}{B}\right)^{-2} = \frac{a^2}{c^2} + \frac{B^2}{C^2}$$
$$= \frac{a^2}{c^2} + \frac{r^2b^2}{r^2c^2} = \frac{a^2 + b^2}{c^2} = \frac{c^2}{c^2} = 1.$$

11.4.91 The latus rectum L intersects the parabola at $x = p$, $y = \pm 2p$. The distance between any point $P(x, y)$ on the parabola to the left of L and L is $p - x$. The distance from F to P is $\sqrt{(x - p)^2 + y^2} = \sqrt{x^2 - 2px + p^2 + 4px} = \sqrt{x^2 + 2px + p^2} = x + p$ (because both x and p are positive.) Thus $D + |FP| = p - x + x + p = 2p$.

11.4.93 Let P be a point on the intersection of the latus rectum and the ellipse. The length of the latus rectum is twice the distance from P to the focus. Let l be the length from P to the focus, and let L be the distance from P to the other focal point. Then $l + L = 2a$, so $L^2 = 4c^2 + l^2$, and thus $(2a - l)^2 = 4c^2 + l^2$, and solving for l yields $l = a - \frac{c^2}{a}$. Because $c^2 = a^2 - b^2$, this can be written as $l = a - \frac{a^2-b^2}{a} = a - (a - \frac{b^2}{a}) = \frac{b^2}{a}$. The length of the latus rectum is therefore $\frac{2b^2}{a}$. Now because $e = \frac{c}{a}$, we have $\sqrt{1 - e^2} = \sqrt{1 - \frac{a^2-b^2}{a^2}} = \sqrt{\frac{b^2}{a^2}} = \frac{b}{a}$. The length of the latus rectum can thus also be written as $2b \cdot \frac{b}{a} = 2b\sqrt{1 - e^2}$.

11.4.95 Let the equation of the ellipse be $\frac{x^2}{a^2} + \frac{y^2}{a^2-c^2} = 1$ and let the equation of the hyperbola be $\frac{x^2}{r^2} - \frac{y^2}{c^2-r^2} = 1$. Let (x_0, y_0) be a point of intersection. By evaluating both equations at the point of intersection and subtracting, we obtain the result

$$\frac{x_0^2}{a^2} - \frac{x_0^2}{r^2} + \frac{y_0^2}{a^2-c^2} + \frac{y_0^2}{c^2-r^2} = 0,$$

which can be written

$$\frac{r^2 x_0^2 - a^2 x_0^2}{a^2 r^2} + \frac{(c^2 - r^2)y_0^2 + (a^2 - c^2)y_0^2}{(a^2 - c^2)(c^2 - r^2)} = 0.$$

This equation can be rewritten in the form $\frac{x_0^2}{y_0^2} = \frac{a^2 r^2}{(a^2-c^2)(c^2-r^2)}$, which we will use later.

Now implicitly differentiating the equation for the ellipse yields $\frac{2x}{a^2} + \frac{2yy'}{a^2-c^2} = 0$, and thus the slope of the tangent line to the ellipse at (x_0, y_0) is $y_e' = -\frac{x_0}{y_0} \cdot \frac{a^2-c^2}{a^2}$. Differentiating the equation of the hyperbola gives $\frac{2x}{r^2} - \frac{2yy'}{c^2-r^2} = 0$, so the slope of the tangent line to the hyperbola at the point of intersection is $y_h' = \frac{x_0}{y_0} \cdot \frac{c^2-r^2}{r^2}$.

Now consider the product

$$-1 \cdot y_e' \cdot y_h' = \frac{x_0^2}{y_0^2} \cdot \frac{(a^2 - c^2)(c^2 - r^2)}{a^2 r^2}.$$

By the result of the first paragraph, this is equal to 1, and thus the two curves are perpendicular at the point of intersection.

11.4.97

a. The curve and the line intersect when $x^2 - m^2(x^2 - 4x + 4) - 1 = 0$, which occurs for $\frac{2m^2 \pm \sqrt{1+3m^2}}{m^2-1}$, assuming $m \neq \pm 1$. So there are two solutions in this case – but if $-1 < m < 1$, one of the solutions is negative (the intersection lies on the other branch of the hyperbola.) If $m^2 = 1$, then the equation becomes $4x - 5 = 0$, and there is only the solution $x = \frac{5}{4}$. So there are two intersection points on the right branch exactly for $|m| > 1$. We have $v(m) = \frac{2m^2+\sqrt{1+3m^2}}{m^2-1}$ and $u(m) = \frac{2m^2-\sqrt{1+3m^2}}{m^2-1}$.

b. $\displaystyle \lim_{m\to1^+} u(m) = \lim_{m\to1^+} u(m) \cdot \frac{2m^2 + \sqrt{1+3m^2}}{2m^2 + \sqrt{1+3m^2}} = \lim_{m\to1^+} \frac{4m^4 - 3m^2 - 1}{(m^2-1)(2m^2 + \sqrt{1+3m^2})} =$
$\displaystyle \lim_{m\to1^+} \frac{(m^2-1)(4m^2+1)}{(m^2-1)(2m^2 + \sqrt{1+3m^2})} = \frac{5}{4}$.

$\displaystyle \lim_{m\to1^+} v(m) = \lim_{m\to1^+} v(m) \cdot \frac{2m^2 - \sqrt{1+3m^2}}{2m^2 - \sqrt{1+3m^2}} = \lim_{m\to1^+} \frac{4m^4 - 3m^2 - 1}{(m^2-1)(2m^2 - \sqrt{1+3m^2})} =$
$\displaystyle \lim_{m\to1^+} \frac{(m^2-1)(4m^2+1)}{(m^2-1)(2m^2 - \sqrt{1+3m^2})} = \lim_{m\to1^+} \frac{(4m^2+1)}{(2m^2 - \sqrt{1+3m^2})} = \infty$.

c. $\displaystyle \lim_{m\to\infty} u(m) = \lim_{m\to\infty} \frac{2 - \sqrt{\frac{1}{m^4} + \frac{3}{m^2}}}{1 - \frac{1}{m^2}} = 2$.

$\displaystyle \lim_{m\to\infty} v(m) = \lim_{m\to\infty} \frac{2 + \sqrt{\frac{1}{m^4} + \frac{3}{m^2}}}{1 - \frac{1}{m^2}} = 2$.

d. The expression $\lim_{m\to\infty} A(m)$ represents the area of the region bounded by the hyperbola and the line $x = 2$. It is given by $2\int_1^2 \sqrt{x^2 - 1}\, dx = 2\left(\frac{x}{2}\sqrt{x^2-1} - \frac{1}{2}\ln(x + \sqrt{x^2-1}) \right)\Big|_1^2 = 2\sqrt{3} - \ln(2 + \sqrt{3})$.

11.4.99

a. With $x^2 = a^2\cos^2 t + 2ab\sin t\cos t + b^2\sin^2 t$, $y^2 = c^2\cos^2 t + 2cd\sin t\cos t + d^2\sin^2 t$, and $xy = ac\cos^2 t + (ad+bc)\sin t\cos t + bd\sin^2 t$, we have $Ax^2 + Bxy + Cy^2 = (Aa^2 + Bac + Cc^2)\cos^2 t + (2Aab + B(ad+bc) + 2Ccd)\sin t\cos t + (Ab^2 + Bbd + Cd^2)\sin^2 t = K$. Thus we have an equation of the desired form as long as there exist A, B, C, and K so that $A(a^2-b^2) + B(ac-bd) + C(c^2-d^2) = 0$ and $2Aab + B(ad+bc) + 2Ccd = 0$. This turns out to be the case when $ad - bc \neq 0$. Note that the value of K is $Aa^2 + Bac + Cc^2$.

b. Suppose that $ad - bc \neq 0$, but $ac + bd = 0$. Then $\frac{b}{a} = -\frac{c}{d}$, and $\tan^{-1}(b/a) = \tan^{-1}(-c/d)$.

Note that $x = \sqrt{a^2+b^2}\cos(t + \tan^{-1}(-b/a))$, $y = \sqrt{c^2+d^2}\sin(t + \tan^{-1}(c/d))$. This can be seen by applying the trigonometric identities for the sum of two angles. Then $\frac{x^2}{a^2+b^2} + \frac{y^2}{c^2+d^2} = \cos^2(t + \tan^1(b/a)) + \sin^2(\tan^{-1}(-c/d)) = 1$.

c. Using the work in part b), we see that the equation is $\frac{x^2}{a^2+b^2} + \frac{y^2}{c^2+d^2} = 1$, or $x^2 + y^2 = r^2$, where $r^2 = a^2 + b^2 = c^2 + d^2$.

Chapter Eleven Review

1

a. False. For example, $x = r\cos t$, $y = r\sin t$ for $0 \le t \le 2\pi$ and $x = r\sin t$, $y = r\cos t$ for $0 \le t \le 2\pi$ generate the same circle.

b. False. Because $e^t > 0$ for all t, this only describes the portion of that line where $x > 0$.

c. True. They both describe the point whose cartesian coordinates are $(3\cos(-3\pi/4), 3\sin(-3\pi/4)) = (-3\cos(\pi/4), -3\sin(\pi/4)) = (-3/\sqrt{2}, -3/\sqrt{2})$.

d. False. The given integral counts the inner loop twice.

e. True. This follows because the equation $0 - x^2/4 = 1$ has no real solutions.

f. True. Note that the given equation can be written as $(x-1)^2 + 4y^2 = 4$, or $\frac{(x-1)^2}{4} + y^2 = 1$.

3

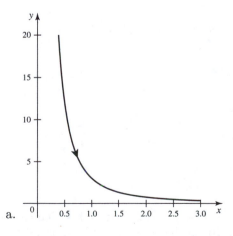

a.

b. $y = 3(e^t)^{-2} = \frac{3}{x^2}$.

c. The curve represents the portion of $\frac{3}{x^2}$ for $x > 0$.

d. $\frac{dy}{dx} = -\frac{6}{x^3}$, so at $(1,3)$ we have $\frac{dy}{dx} = -6$.

5

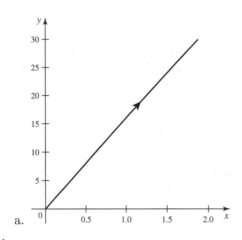

a.

b. Because $\ln t^2 = 2\ln t$ for $t > 0$, we have $y = 16x$ for $0 \le x \le 2$.

c. The curve represents a line segment from $(0,0)$ to $(2,32)$.

d. $\frac{dy}{dx} = 16$ for all value of x.

7 Note that $\left(\frac{x}{4}\right)^2 + \left(\frac{y}{3}\right)^2 = 1$. This represents an ellipse generated counterclockwise.

9 Note that $(x+3)^2 + (y-6)^2 = 1$. This is the right half of a circle of radius 1 centered at $(-3,6)$. It is generated clockwise.

11 $x = 3\sin t$, $y = 3\cos t$, $0 \le t \le 2\pi$.

13 $x = 3\cos t$, $y = 2\sin t$, $-\pi/2 \le t \le \pi/2$.

15 From P to Q, we use $(x(t), y(t)) = tQ + (1-t)P = (t,t) + (t-1,0) = (2t-1,t)$. So $x(t) = 2t-1$, $y(t) = t$, $0 \le t \le 1$.

From Q to P, we use $(x(t), y(t)) = tP + (1-t)Q = (-t,0) + (1-t,1-t) = (1-2t,1-t)$, for $0 \le t \le 1$. Thus $x(t) = 1-2t$, $y(t) = 1-t$, $0 \le t \le 1$.

17 $\frac{dy}{dx} = \frac{dy/dt}{dx/dt} = \frac{\sin t}{1-\cos t}$. At $t = \pi/6$, the slope of the tangent line is $\frac{1}{2-\sqrt{3}} = 2+\sqrt{3}$. So the equation of the tangent line is $y - (1 - \sqrt{3}/2) = (2+\sqrt{3})(x - (\pi/6 - 1/2))$, or $y = (2+\sqrt{3})x + (2 - \frac{\pi}{3} - \frac{\pi\sqrt{3}}{6})$.

At $t = 2\pi/3$, the slope of the tangent line is $\frac{\sqrt{3}}{3}$, so the equation of the tangent line is $y - \frac{3}{2} = \frac{\sqrt{3}}{3}(x - (\frac{2\pi}{3} - \frac{\sqrt{3}}{2}))$, or $y = \frac{x}{\sqrt{3}} + 2 - \frac{2\pi}{3\sqrt{3}}$.

19

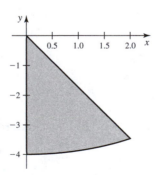

21 Liz should choose the cardioid, which is $r = 1 - \sin\theta$.

23 Letting $x = r\cos\theta$, $y = r\sin\theta$, and $r^2 = x^2 + y^2$, we have $x^2 + y^2 + 2y - 6x = 0$, which can be written as $x^2 - 6x + 9 + y^2 + 2y + 1 = 10$, or $(x-3)^2 + (y+1)^2 = 10$, so this is a circle of radius $\sqrt{10}$ centered at $(3,-1)$.

25 If $x = r\cos\theta$ and $y = r\sin\theta$, then $(r\cos\theta - 4)^2 + r^2\sin^2\theta = 16$, so $r^2\cos^2\theta - 8r\cos\theta + 16 + r^2\sin^2\theta = 16$, so $r^2 = 8r\cos\theta$, and thus $r = 8\cos\theta$. The complete circle can be described by $-\pi/2 \le \theta \le \pi/2$.

27

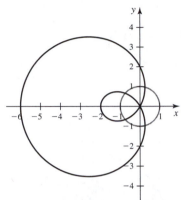

a.

There are 4 intersection points.

b. Note that $2 - 4\cos\theta = 1$ for $\theta = \cos^{-1}(1/4) \approx 1.32$, and $2 - 4\cos\theta = -1$ for $\theta = \cos^{-1}(3/4) \approx .73$. The points of intersection (in polar form) are approximately $(1, 1.32)$, $(1, 2\pi - 1.32) \approx (1, 4.96)$, $(-1, .73)$, and $(-1, 2\pi - .73) \approx (-1, 5.56)$.

29

a. $$\frac{dy}{dx} = \frac{dy/d\theta}{dx/d\theta} = \frac{2\cos\theta\sin\theta + (4 + 2\sin\theta)\cos\theta}{2\cos\theta\cos\theta - (4 + 2\sin\theta)\sin\theta} = \frac{4\cos\theta + 4\sin\theta\cos\theta}{2\cos^2\theta - 2\sin^2\theta - 4\sin\theta}.$$

This is 0 when $\cos\theta = 0$, and when $4\sin\theta = -4$, so the only solutions are $\theta = \pi/2, 3\pi/2$.

The denominator is 0 when $2 - 4\sin^2\theta - 4\sin\theta = 0$ which occurs (using the quadratic formula) for $\sin\theta = -\frac{1}{2} + \frac{\sqrt{3}}{2}$, so there are vertical tangent lines at $\theta = \sin^{-1}(-\frac{1}{2} + \frac{\sqrt{3}}{2})$ and $\theta = \pi - \sin^{-1}(-\frac{1}{2} + \frac{\sqrt{3}}{2})$.

b. The curve is never at the origin.

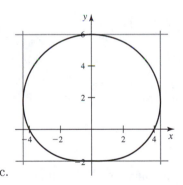

c.

31

a. Note that the whole curve is generated for $-\pi/4 \le \theta \le \pi/4$, so we restrict ourselves to that domain. Write the equations as $r = \sqrt{2\cos 2\theta}$. Then

$$\frac{dy}{d\theta} = \sqrt{2\cos 2\theta}\cos\theta - \sin\theta\frac{2\sin 2\theta}{\sqrt{2\cos 2\theta}} = \frac{\cos\theta}{\sqrt{2\cos 2\theta}}(2\cos 2\theta - 4\sin^2\theta) = \frac{\cos\theta}{\sqrt{2\cos 2\theta}}(2 - 8\sin^2\theta).$$

Also, $\frac{dx}{d\theta} = -\sqrt{2\cos 2\theta}\sin\theta + \cos\theta\frac{2\sin 2\theta}{\sqrt{2\cos 2\theta}} = \frac{\sin\theta}{\sqrt{2\cos 2\theta}}(-4\cos^2\theta - 2(\cos 2\theta)) = \frac{(2 - 8\cos^2\theta)\sin\theta}{\sqrt{2\cos(2\theta)}}$. Thus

$\frac{dy}{dx} = \frac{dy/d\theta}{dx/d\theta} = \cot\theta\left(\frac{1 - 4\sin^2\theta}{1 - 4\cos^2\theta}\right)$.

This expression is 0 on the given domain only for $\sin^2\theta = \frac{1}{4}$, so there are horizontal tangent lines at $\theta = \pm\frac{\pi}{6}$. There are vertical tangent lines on the given domain only for $\theta = 0$ In cartesian coordinates, the lines are $x = \pm\sqrt{2}$.

b. The curve is at the origin for $\theta = \pm\pi/4$, and because $\tan\pi/4 = 1$, the tangent lines have the equations $y = \pm x$.

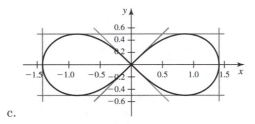

c.

33

The area is given by $A = \frac{1}{2}\int_0^{2\pi}(3 - \cos\theta)^2\,d\theta = \frac{1}{2}\int_0^{2\pi}(9 - 6\cos\theta + \cos^2\theta)\,d\theta = \frac{1}{2}\left(9\theta - 6\sin\theta + \frac{1}{2}(\cos\theta\sin\theta + \theta)\right)\Big|_0^{2\pi} = \frac{1}{2}(18\pi + \pi) = \frac{19\pi}{2}.$

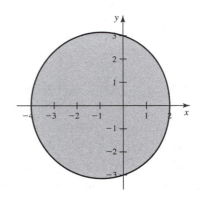

35

The curves intersect at $\theta = \pm\frac{1}{2}\cos^{-1}(1/16)$. By symmetry the total desired area is $A = 4 \cdot \frac{1}{2}\int_0^{\cos^{-1}(1/16)/2}(4\cos 2\theta - \frac{1}{4})\,d\theta = 2\left(2\sin 2\theta - \frac{\theta}{4}\right)\Big|_0^{\cos^{-1}(1/16)/2} = \frac{1}{4}\sqrt{255} - \frac{\cos^{-1}(1/16)}{4}.$

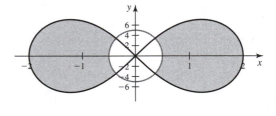

37 First we provide the solution for problem 36: By symmetry, we can compute the area within the curve $r = 1 - \cos\theta$ for $0 \le \theta \le \pi/2$ and then quadruple it. We have

$$4 \cdot \frac{1}{2}\int_0^{\pi/2}(1 - \cos\theta)^2\,d\theta = 2\int_0^{\pi/2}(1 - 2\cos\theta + (1/2) + (1/2)\cos 2\theta)\,d\theta$$

$$= (3\theta - 4\sin\theta + (1/2)\sin 2\theta)\Big|_0^{\pi/2} = \frac{3\pi}{2} - 4 + 0 - (0 - 0 + 0) = \frac{3\pi}{2} - 4.$$

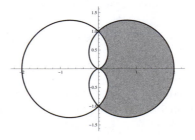

Note that we can compute the area within $1 + \cos\theta$ between 0 and $\pi/2$ and then subtract $1/4$ of the area from problem 36, and then double this difference. If we compute $\frac{1}{2}\int_0^{\pi/2}(1+\cos\theta)^2\,d\theta = \frac{1}{2}\int_0^{\pi/2}(1+2\cos\theta + \cos^2\theta)\,d\theta = \int_0^{\pi/2}(3/4 + \cos\theta + (1/4)\cos 2\theta)\,d\theta = (3\theta/4 + \sin\theta + (1/8)\sin 2\theta)\Big|_0^{\pi/2} = 3\pi/8 + 1 + 0 - (0+0+0) = 3\pi/8 + 1$.

If we subtract $1/4$ of the result from problem 36, we have $3\pi/8 + 1 - (3\pi/8 - 1) = 2$. Doubling this gives a final result of 4.

39

a. This represents a hyperbola with $a = 1$ and $b = \sqrt{2}$.

b. The vertices are $(\pm 1, 0)$, the foci are $(\pm c, 0)$ where $c^2 = a^2 + b^2 = 3$, so they are $(\pm\sqrt{3}, 0)$. The directrices are $x = \frac{\pm a^2}{c} = \frac{\pm 1}{\sqrt{3}}$.

c. The eccentricity is $e = \frac{c}{a} = \sqrt{3}$.

d.

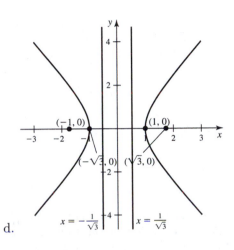

41

a. This can be written as $\frac{y^2}{16} - \frac{x^2}{4} = 1$. It is a hyperbola with $a = 4$ and $b = 2$.

b. The vertices are $(0, \pm 4)$. The foci are $(0, \pm c)$ where $c^2 = a^2 + b^2 = 16 + 4 = 20$, so they are $(0, \pm\sqrt{20})$. The directrices are $y = \frac{\pm a^2}{c} = \frac{\pm 16}{\sqrt{20}} = \frac{\pm 8}{\sqrt{5}}$.

c. The eccentricity is $e = \frac{c}{a} = \frac{\sqrt{20}}{4} = \frac{\sqrt{5}}{2}$.

d.

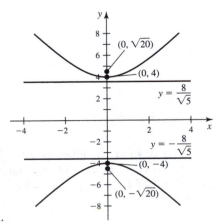

43

a. This can be written as $\frac{x^2}{4} + \frac{y^2}{2} = 1$, so it is an ellipse with $a = 2$ and $b = \sqrt{2}$.

b. The vertices are $(\pm 2, 0)$. The foci are $(\pm c, 0)$ where $c^2 = a^2 - b^2 = 4 - 2 = 2$, so they are $(\pm\sqrt{2}, 0)$. The directrices are $x = \frac{\pm a^2}{c} = \frac{\pm 4}{\sqrt{2}} = \pm 2\sqrt{2}$.

c. The eccentricity is $e = \frac{c}{a} = \frac{\sqrt{2}}{2}$.

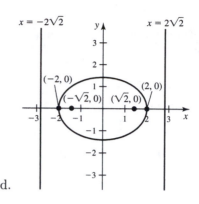

d.

45 $2y\frac{dy}{dx} = -12$, so at the point in question, $\frac{dy}{dx} = 3/2$. So the equation of the tangent line is $y + 4 = \frac{3}{2}\left(x + \frac{4}{3}\right)$, or $y = \frac{3}{2}x - 2$.

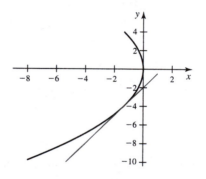

47 $\frac{x}{50} + \frac{y}{32} \cdot \frac{dy}{dx} = 0$, so at the given point, $\frac{dy}{dx} = -\frac{6}{10} = -\frac{3}{5}$. So the equation of the tangent line is $y + \frac{32}{5} = -\frac{3}{5}(x + 6)$, or $y = -\frac{3}{5}x - 10$.

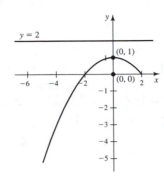

49   The eccentricity is 1, and the directrix is $y = 2$. The vertex is $(0,1)$ and the focus is $(0,0)$.

51 The eccentricity is $\frac{1}{2}$, and the directrices are $x = 4$ and $x = -\frac{20}{3}$. The vertices are $\left(\frac{4}{3}, 0\right)$ and $(-4,0)$ and the foci are $(0,0)$ and $\left(-\frac{8}{3}, 0\right)$.

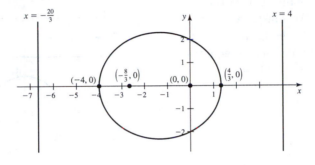

53

a. Recall that $\cos 2\theta = \cos^2\theta - \sin^2\theta$, so $r^2\cos(2\theta) = 1$ becomes $r^2(\cos^2\theta - \sin^2\theta) = x^2 - y^2 = 1$. The curve is a hyperbola.

b. With $a = b = 1$, we have $c^2 = 2$, so the vertices are $(\pm 1, 0)$ and the foci are $(\pm\sqrt{2}, 0)$. The directrices are $x = \pm\frac{a^2}{c} = \pm\frac{1}{\sqrt{2}}$. The eccentricity is $e = \frac{c}{a} = \sqrt{2}$.

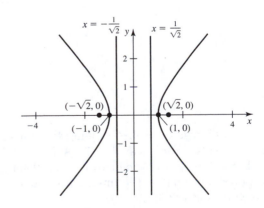

c. It does not have the form as in Theorem 11.4 because it does not have a focus at the origin.

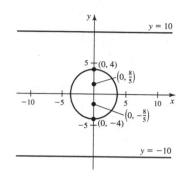

55 Because the center is halfway between the vertices, it is $(0,0)$. We must have $a = 4$ and because $\frac{a^2}{c} = d = 10$, we have $c = \frac{8}{5}$. So $b^2 = a^2 - c^2 = 16 - \frac{64}{25} = \frac{336}{25}$. The ellipse has equation $\frac{25x^2}{336} + \frac{y^2}{16} = 1$. The eccentricity is $\frac{c}{a} = \frac{8/5}{4} = \frac{2}{5}$.

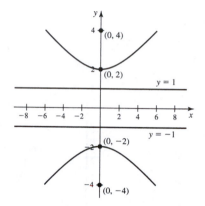

57 Because the center is halfway between the vertices, it is $(0,0)$. We must have $a = 2$ and because $\frac{a^2}{c} = d = 1$, we have $c = 4$. So $b^2 = 16 - 4 = 12$. The hyperbola has equation $\frac{y^2}{4} - \frac{x^2}{12} = 1$. The eccentricity is $\frac{c}{a} = \frac{4}{2} = 2$.

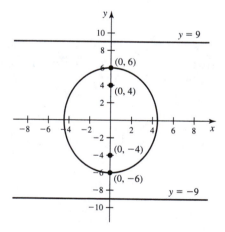

59 We have $a = 6$, $c = 4$ and $e = \frac{c}{a} = \frac{4}{6} = \frac{2}{3}$. Also, $b^2 = a^2 - c^2 = 36 - 16 = 20$, and the equation is $\frac{y^2}{36} + \frac{x^2}{20} = 1$. The vertices are $(\pm 2\sqrt{5}, 0)$. The directrices are $y = \pm \frac{a^2}{c} = \pm 9$.

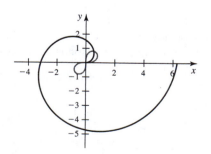

61 $\sin 2\theta = \theta^2$ when $\theta = 0$. Graphing the functions reveals a root near $\theta = 1$. A CAS reveals the intersection point to be $\theta \approx .9669$. In polar coordinates, the intersection points are $(0,0)$ and $(.9669, .9669)$.

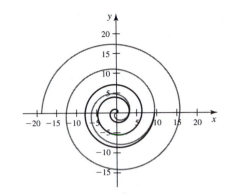

The curves intersect for $\theta = 0$. Note also that when $\theta = k\pi$ for k an odd integer, the curve $r = -\theta$ is at the polar point $(-k\pi, k\pi) = (k\pi, 0)$. And for $\theta = 2k\pi$, the curve $r = \frac{\theta}{2}$ is at the point $(k\pi, 0)$. So the curves intersect at these points.

65 By symmetry, we can focus on the region in the first quadrant. That area is given by $A = xy$ where $y = \sqrt{b^2 - \frac{b^2}{a^2}x^2}$. So

$$A(x) = x\sqrt{b^2 - \frac{b^2}{a^2}x^2},$$

so

$$A'(x) = \sqrt{b^2 - \frac{b^2}{a^2}x^2} - \frac{b^2 x^2}{a^2\sqrt{b^2 - \frac{b^2}{a^2}x^2}}.$$

Setting the derivative equal to 0 and clearing denominators yields $\left(b^2 - \frac{b^2}{a^2}x^2\right)a^2 - b^2x^2 = 0$, and solving for x gives the critical point $x = \frac{\sqrt{2}}{2}a$. Because this is the only critical point and it clearly does not give a minimum (because $A(0) = A(a) = 0$), it must yield a maximum. The whole rectangle has dimensions $\sqrt{2}a \times \sqrt{2}b$, and area $2ab$.

67 The area of the ellipse in the first quadrant is $\frac{\pi ab}{4}$, so we are seeking θ_0 so that

$$\frac{\pi ab}{8} = \frac{1}{2}\int_0^{\theta_0} \frac{a^2 b^2}{a^2\sin^2\theta + b^2\cos^2\theta}\,d\theta = \frac{a^2}{2}\int_0^{\theta_0}\frac{\sec^2\theta}{\frac{a^2}{b^2}\tan^2\theta + 1}\,d\theta.$$

Let $u = \frac{a}{b}\tan\theta$ so that $du = \frac{a}{b}\sec^2\theta\,d\theta$. Then we have $\frac{\pi ab}{8} = \frac{ab}{2}\int_0^{\frac{a}{b}\tan\theta_0}\frac{1}{1+u^2}\,du = \frac{ab}{2}\tan^{-1}(\frac{a}{b}\tan(\theta_0))$. Note that this equation is satisfied when $\tan(\theta_0) = \frac{b}{a}$, because then the expression on the right-hand side of that equation is $\frac{ab}{2}\cdot\frac{\pi}{4} = \frac{\pi ab}{8}$. So the desired value of m is $\tan(\theta_0) = \frac{b}{a}$.

69 Note that $Q = (a\cos\theta, a\sin\theta)$ and $R = (b\cos\theta, b\sin\theta)$, where θ is the angle formed by l and the x-axis. Then $P = (a\sin\theta, b\cos\theta)$ is a point on the ellipse $\frac{x^2}{a^2} + \frac{y^2}{b^2} = 1$, because it satisfies that equation.

71 The focal point is at the origin, the directrix is $y = -d$, so we have an equation of the form $r = \frac{ed}{1 - e\sin\theta}$. Because c is the distance from the center to the focal point, we have $c = 3/8$, and because a is the distance from the center to a vertex, we have $a = 9/8$. Then we have $e = \frac{c}{a} = \frac{3/8}{9/8} = \frac{1}{3}$, and $d = \frac{a^2}{c} - \frac{3}{8} = 3$. Thus $r = \frac{1}{1 - \frac{1}{3}\sin\theta} = \frac{3}{3 - \sin\theta}$.

Appendix A

A.1.1 This is the set of real numbers greater than -4 and less than or equal to 10; $(-4, 10]$.

A.1.3 The absolute value function is defined as a piecewise function in this manner: $|x| = \begin{cases} x & \text{if } x \geq 0 \\ -x & \text{if } x < 0. \end{cases}$

A.1.5 In order for $|2x - 4| \geq 3$ we must have either $2x - 4 \geq 3$ or $2x - 4 \leq -3$.

A.1.7 If (x_1, y_1) and (x_2, y_2) are two known points, then the distance between them is given by the expression $\sqrt{(x_2 - x_1)^2 + (y_2 - y_1)^2}$.

A.1.9 The set of points (x, y) whose distance from the origin is 6 is given by $x^2 + y^2 = 36$. Solving for y gives $|y| = \sqrt{36 - x^2}$. Note that $y = \sqrt{36 - x^2}$ represents the upper half of the circle while $y = -\sqrt{36 - x^2}$ represents the bottom half.

A.1.11 Using the point-slope form of the equation of a line, we write $y - (-2) = m(x-4)$ or $y = -2 + m(x-4)$.

A.1.13 If two lines are parallel, then either their slopes are equal (in the case of non vertical lines) or they both have slopes which are undefined (in the case of vertical lines).

A.1.15 $\left(\frac{1}{8}\right)^{-2/3} = 8^{2/3} = \sqrt[3]{8}^2 = 2^2 = 4$.

A.1.17 $(u + v)^2 - (u - v)^2 = u^2 + 2uv + v^2 - (u^2 - 2uv + v^2) = u^2 + 2uv + v^2 - u^2 + 2uv - v^2 = 4uv$.

A.1.19 $\frac{1}{x+h} - \frac{1}{x} = \frac{1}{x+h} \cdot \frac{x}{x} - \frac{1}{x} \cdot \frac{x+h}{x+h} = \frac{x-(x+h)}{x(x+h)} = -\frac{h}{x(x+h)}$.

A.1.21 $y^2 - y^{-2} = y^2 - (y^{-1})^2 = (y - y^{-1})(y + y^{-1})$.

A.1.23 Note that $u^4 - 11u^2 + 18 = (u^2 - 9)(u^2 - 2) = (u - 3)(u + 3)(u - \sqrt{2})(u + \sqrt{2})$, so this equation is equal to 0 when $u = 3$, $u = -3$, $u = \sqrt{2}$ or $u = -\sqrt{2}$.

A.1.25 $\frac{(x+h)^3 - x^3}{h} = \frac{x^3 + 3x^2h + 3xh^2 + h^3 - x^3}{h} = \frac{h(3x^2 + 3xh + h^2)}{h} = 3x^2 + 3xh + h^2$.

A.1.27 $x^2 - 6x + 5 < 0$ can be written as $(x-5)(x-1) < 0$. The product of two numbers is negative when one is positive and the other is negative, so we have either $x - 5 < 0$ and $x - 1 > 0$ or $x - 5 > 0$ and $x - 1 < 0$. The first case occurs when both $x < 5$ and $x > 1$, so for $1 < x < 5$. The second case occurs when $x > 5$ and $x < 1$ simultaneously, which isn't possible. So the solution set consists of the interval $(1, 5)$.

A.1.29 This can be written as $\frac{(x-5)(x-4)}{x-6} \le 0$. The expression on the left is zero for $x = 5$ and $x = 4$, while the expression is undefined for $x = 6$. These numbers partition the number line into the intervals $(-\infty, 4)$, $(4, 5)$, $(5, 6)$ and $(6, \infty)$. Using test values, we see that the expression is negative on the interval $(-\infty, 4)$ and on $(5, 6)$ while it is positive on the other intervals. Thus the given expression is less than or equal to 0 on $(-\infty, 4] \cup [5, 6)$.

A.1.31 In order for $|3x - 4| > 8$ we must have either $3x - 4 > 8$ or $3x - 4 < -8$. Thus we either have $3x > 12$ and thus $x > 4$, or we have $3x < -4$ or $x < -\frac{4}{3}$. The solution set is $\left(-\infty, -\frac{4}{3}\right) \cup (4, \infty)$.

A.1.33 If $2x - 1 > 0$ (which occurs for $x > \frac{1}{2}$), we have $3 < 2x - 1 < 5$ or $2 < x < 3$. If $2x - 1 < 0$ (which occurs for $x < \frac{1}{2}$), we have $3 < 1 - 2x < 5$ or $2 < -2x < 4$, or dividing through by -2 we have $-2 < x < -1$. The solution set is thus $(-2, -1) \cup (2, 3)$.

A.1.35 The circle centered at $(-1, 2)$ with radius 3 is given by $(x + 1)^2 + (y - 2)^2 = 9$. This can be written as $(y - 2)^2 = 9 - (x + 1)^2$, and taking square roots of both sides gives $|y - 2| = \sqrt{9 - (x + 1)^2}$, so $y = 2 \pm \sqrt{9 - (x + 1)^2}$. The lower half of the circle is described by $y = 2 - \sqrt{9 - (x + 1)^2}$.

A.1.37 Using the slope-intercept form of the equation of a line we have $y = \frac{5}{3}x + 4$.

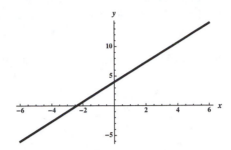

A.1.39 The slope of ℓ must be $\frac{-4-0}{0-5} = \frac{4}{5}$. Then using the slope-intercept form of the equation of a line, we have $y = \frac{4}{5}x - 4$.

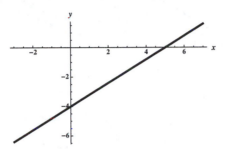

A.1.41 The line we are seeking must have the same slope as $x + 2y = 8$, which can be written as $y = -\frac{1}{2}x + 4$, so the slope must be $-\frac{1}{2}$. Using the slope-intercept form of the equation of a line, the line we are seeking must have equation $y = -\frac{1}{2}x + 12$.

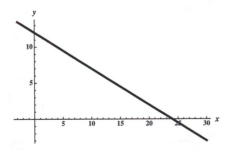

A.1.43 The line we are seeking is perpendicular to a line of slope -3, so it must have slope $\frac{1}{3}$. Using the point-slope form of the equation of a line, we have $y - (-6) = \frac{1}{3}(x - 3)$, or $y = -6 + \frac{1}{3}x - 1$, or $y = \frac{1}{3}x - 7$.

A.1.45

   a. False. $\sqrt{16} = 4$, not -4. Note that the equation $x^2 = 16$ does have two solutions, but this is solved properly by taking the square root of both sides to obtain $|x| = 4$, which then yields $x = \pm 4$.

   b. True. Both sides are equal to 4.

   c. False. No real numbers satisfy this equation as $|x| > 0$ for all x.

   d. False. $|x| \geq 0$ for all x.

   e. False. The distance from the point $(1, 1)$ to the origin is $\sqrt{(1 - 0)^2 + (1 - 0)^2} = \sqrt{2} > 1$, so the point $(1, 1)$ is outside the circle of radius 1 centered at the origin.

f. True. Note that $x^4 = (x^2)^2$, so $\sqrt{x^4} = |x^2| = x^2$ for all x.

g. False. For example, consider the case $a = 2$ and $b = -3$. Then $\sqrt{a^2} = 2$ and $\sqrt{b^2} = 3$, so $\sqrt{a^2} < \sqrt{b^2}$, but it isn't true that $a < b$.

A.1.47 The complement of this set is the interval $(-2, 4)$ which is the set of numbers within 3 units of the number 1. This can be written as $\{x : |x - 1| < 3\}$. The set we are seeking is the complement of this, so it can be written as $\{x : |x - 1| \geq 3\}$.

A.1.49 Because the absolute value function is only 0 at 0, we must have $x - y = 0$, or $y = x$, which is the line through the origin with slope 1.

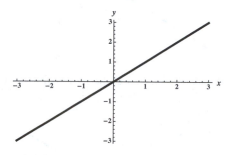